Developed by **Jane Heinze-Fry** with assistance from G. Tyler Mill
(For assistance in creating your own concept maps, see Appendix

W9-BEN-806

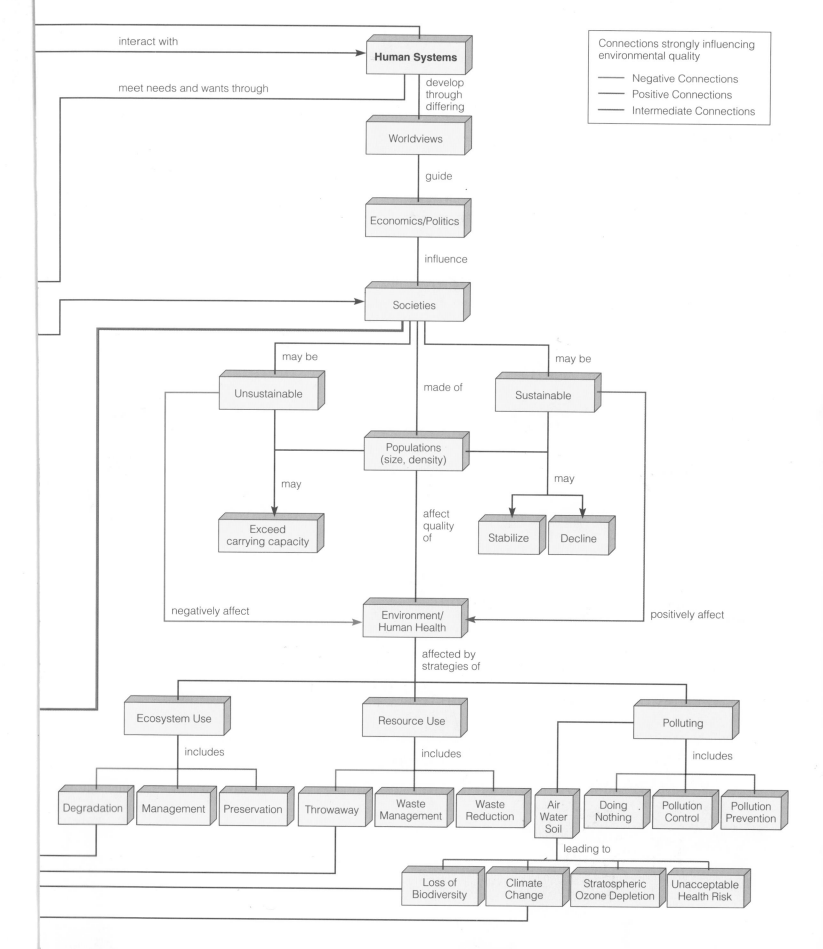

Environmental Science

Environmental Science

Working with the Earth

FIFTH EDITION

G. TYLER MILLER, JR.

Adjunct Professor of Human Ecology
St. Andrews Presbyterian College

Wadsworth Publishing Company
Belmont, California
A Division of Wadsworth, Inc.

Two trees have been planted in a tropical rain forest for every tree used to make this book, courtesy of G. Tyler Miller, Jr., and Wadsworth Publishing Company. The author also sees that 50 trees are planted to compensate for the paper he uses and that several hectares of tropical rain forest are protected.

Biology Publisher: Jack Carey
Editorial Assistant: Kristin Milotich
Production Editors: Karen Garrison, Vicki Friedberg
Managing Designer: Carolyn Deacy
Print Buyer: Barbara Britton
Art Editor: Nancy Spellman
Permissions Editor: Peggy Meehan
Copy Editor: George Dyke
Photo Researchers: Stephen Forsling, Roberta Broyer
Technical Illustrators: Darwin and Vally Hennings; Tasa Graphic Arts, Inc.; Alexander Teshin Associates; John and Judith Waller; Raychel Ciemma; and Victor Royer
Page Layout: Edie Williams, Vargas/Williams/Design
Composition: Brandon Carson, Wadsworth Digital Productions
Color Separator: H&S Graphics
Printer: Arcata Book Group/Hawkins
Cover Photograph: Bald Cypress, Trussum Pond, Delaware, © David Muench
Part Opening Photographs:
Part I: © 1990 Tom Van Sant/The GeoSphere Project
Part II: Brian Parker/Tom Stack & Associates
Part III: Silvestris Fotoservice/NHPA
Part IV: Gunter Ziesler/Bruce Coleman Ltd.
Part V: Hank Morgan/Photo Researchers, Inc.

 This book is printed on acid-free recycled paper.

 I(T)P ™

International Thomson Publishing
The trademark ITP is used under license.

Printed in the United States of America
2 3 4 5 6 7 8 9 10—99 98 97 96 95

Library of Congress Cataloging-in-Publication Data
Miller, G. Tyler (George Tyler)
    Environmental science : working with the Earth / G. Tyler Miller, Jr. — 5th ed.
      p.   cm. —
    Includes bibliographical references and index.
    ISBN 0-534-21588-2 (acid-free paper)
    1. Environmental sciences. 2. Human ecology.
    3. Environmental protection. I. Title.
GE105.M544 1995
363.7—dc20   94-1534

For Instructors and Students

How I Became Involved   In 1966 I heard a scientist give a lecture on the problems of overpopulation and environmental abuse. Afterward I went to him and said, "If even a fraction of what you have said is true, I will feel ethically obligated to give up my research on the corrosion of metals and devote the rest of my life to environmental issues. Frankly, I don't want to believe a word you have said, and I'm going into the literature to try to prove that your statements are either untrue or grossly distorted."

After six months of study I was convinced of the seriousness of these problems. Since then I have been studying, teaching, and writing about them. I have also attempted to live my life in an environmentally sound way—with varying degrees of success—by treading as lightly as possible on the earth (see pp. 511–512 for a summary of my own progress in attempting to work with nature). This book summarizes what I have learned in almost three decades of trying to understand environmental principles, connections, and solutions.

My Philosophy of Education   I believe that our lifelong pursuit of knowledge should be guided by the following principles:

- Recognize that the primary aim of education should be to help us develop a respect for life.

- Question everything and everybody, as any good scientist does.

- Develop a list of principles, concepts, and rules to be used as guidelines in making decisions, and continually evaluate and modify this list as a result of experience.

- Recognize that the primary goal of education should be to learn how to sift through mountains of facts and ideas and find the few that are useful and worth knowing. We need an *Earth-wisdom revolution*, not an information revolution. Facts and numbers are merely stepping-stones to ideas, laws, concepts, principles, and connections. And most statistics and facts are human beings with the tears wiped off and living things whose lives we are threatening.

- Interact with what you read—as a way to make learning more interesting and effective. I do this by marking key sentences and paragraphs with a highlighter or pen. I put an asterisk in the margin next to something I think is important, and double asterisks next to something that I think is especially important. I write comments in the margins, such as *Beautiful, Confusing, Bull, Wrong*, and so on. I fold down the top corner of pages with highlighted passages and the top and bottom corners of especially important pages. This way, I can flip through a book and quickly review the key passages.

Key Features   This book is designed to be used in introductory courses on environmental science. It treats environmental science as an *interdisciplinary* study, combining ideas and information from natural sciences (such as biology, chemistry, and geology) and social sciences (such as economics, politics, and ethics) to present a general idea of how nature works and how things are interconnected. This study of connections in nature examines how the environment is being used and abused—and what individuals can do to protect and improve it for themselves, for future generations, and for other living things.

This book is not just a description of the serious environmental problems we face. Instead, after each problem is discussed I list and discuss solutions to these problems—solutions proposed by a variety of scientists, environmental activists, and analysts—at global, national, local, and individual levels. Special boxes labeled *Solutions* give some of this information, along with entire chapter sections and subsections.

In this book I use scientific laws, principles, and concepts to help us understand environmental and resource problems and the possible solutions to these problems, and how these concepts, problems, and solutions are connected. I have introduced only the concepts and principles necessary to understanding material in this book, and I have tried to present them simply but accurately. The key principles and concepts used in this textbook are summarized inside the back cover.

My aim is to provide a readable and accurate introduction to environmental science without the use of mathematics or complex scientific information. To help make sure the material is accurate and up-to-date, I have consulted more than 10,000 research sources in the professional literature. In writing this book I have also profited from the more than 200 experts and teachers who have provided detailed reviews of the various editions of my other four books in this field (see list on pp. xiii–xv).

The book is divided into five major parts (see Brief Contents, p. xvii). After Parts I and II are covered, the rest of the book can be used in almost any order. For example, some instructors cover Part V, on Energy Resources, after Part II. Some find it useful to cover Part IV before Part III. In addition, most chapters and many sections within these chapters can be moved around or omitted to accommodate courses with different lengths and emphases. For example, some instructors move Chapter 7, which is on economics and politics, to Part I (as Chapter 3) or to the end of the book.

I have also written four other textbooks on environmental science—each with a different emphasis, organization, and length for use with various types of courses—*Living in the Environment*, 8th edition (701 pages, Wadsworth, 1994); *Resource Conservation and Management* (546 pages, Wadsworth, 1990); *Sustaining the Earth: An Integrated Approach* (325 pages, Wadsworth, 1994); and *Environment: Problems and Solutions* (150 pages, Wadsworth, 1994).

This book is an integrated study of environmental problems, connections, and solutions. The integrating themes in this book are *biodiversity and Earth capital, economics and environment, energy and energy efficiency, individual action and Earth citizens, politics and environmental laws, pollution prevention and waste reduction, population and exponential growth, science and technology, solutions and sustainability,* and *uncertainty and controversy* (see p. xvi).

After looking at the Brief Contents I urge you to look at the concepts and connections map inside the front cover. It is a summary of the key parts and concepts of environmental science and how they are connected to one another—in effect, a map of the book. I also suggest that you read the list of laws, concepts, and principles given inside the back cover. It is a two-page summary of the key ideas in this book.

I also relate the information in the book to the real world and to our individual lives, both in the main text and in various kinds of boxes sprinkled throughout the book. These include **(1)** *Spotlights* (21 in number) highlighting and giving further insights into environmental problems and concepts; **(2)** *Case Studies* (16) giving in-depth information about key issues; **(3)** *Connections* (19) showing how various environmental concepts, problems, and solutions are interrelated; **(4)** *Pro/Con* discussions (5) outlining both sides of controversial environmental issues; **(5)** *Solutions* (26) summarizing possible solutions to environmental problems or describing what individuals have done to help sustain the earth for us and all life; **(6)** *Guest Essays* (15) exposing readers to individual environmental researchers' or activists' points of view; and **(7)** *Individuals Matter* (15) giving examples of what we as individuals can do to help sustain the earth. More ways to evaluate individual lifestyles and campuses are given in *Green Lives, Green Campuses,* a supplement available with this book.

The book's 460 illustrations are designed to present complex ideas in understandable ways and to relate learning to the real world. They include 281 full-color diagrams (46 of them maps) and 179 carefully selected color photographs (17 of them satellite shots).

Major Changes in This Edition *This new edition is a major revision—the most extensive since the first edition was published.* Major changes include:

- Updating and revising material throughout the book. Because of rapid changes in data and information, a textbook in environmental science needs to be updated every two years.

- Improving readability by reducing sentence and paragraph length, omitting unnecessary details, and examining chapter organization and flow.

- Expanding the basic text from 470 pages to 540 pages to allow additional scientific content and many new topics. Instructors wanting shorter books covering this material can use *Sustaining the Earth: An Integrated Approach* (325 pages, Wadsworth, 1994) or *Environment: Problems and Solutions* (150 pages, Wadsworth, 1994).

- Adding brief *Earth Stories* at the beginning of each chapter to spark reader interest and provide useful information. Many describe what individuals have done to help us understand and work with the earth.

- Emphasizing solutions to environmental problems by adding 26 new *Solutions* boxes and adding sections and subsections on solutions within most chapters.

- Focusing on relationships among environmental concepts, problems, and solutions by adding 19 new *Connections* boxes and adding sections and subsections on connections within most chapters.

- Adding 19 new *Spotlights*, 6 new *Case Studies*, 8 new *Guest Essays*, 3 new *Individuals Matter* boxes, and 2 new *Pro/Con* boxes (see pp. x–xii).

- Introducing *concept mapping.* A concept map of the entire book (see inside front cover) has been developed and Appendix 4 shows students how to develop such maps. For instructors wishing to use this method for organizing and clarifying ideas, an end-of-chapter question asks students to develop one or more concept maps for each chapter. Concept maps for each chapter are provided in the instructor's manual and are available as overhead transparencies.

- Upgrading all diagrams to improve the color, add more detail, and give a more realistic 3-D effect.

- Adding 68 new color photos and 101 new diagrams and improving all previous diagrams.

- Adding 17 satellite photos of the earth.

- Adding 12 new maps and expanding a number of U.S. maps to include data on Canada, with the assistance of Daniel J. Boivin, Professor of Regional Planning, Université Laval, Quebec, Canada.

- Integrating the discussion of solid waste and hazardous waste into a new chapter (Chapter 13).

- Integrating the discussion of minerals and soil into one chapter (Chapter 12).

- Extensively rewriting and reorganizing Chapters 2, 7, 10, 16, 18, and the Epilogue (see Brief Contents, p. xvii).

- Providing new supplements, including an expanded *Instructor's Manual and Test Items*; a workbook supplement, *Green Lives, Green Campuses*; a *Laboratory Manual*; and *Environmental Articles*: more than 200 articles from which instructors can pick any number and have them bound as customized supplements for their courses.

- Greatly increasing the *scientific content* by adding or expanding material on more than 46 subjects, including: the nature of science and technology, limitations and misuse of science, half-life of radioisotopes, nature of life, five major kingdoms of organisms, importance of insects, amphibians, biodiversity, pyramids of numbers, pyramids of biomass, sulfur cycle, symbiotic relationships, interference competition, exploitation competition, energy flow in ecosystems, species interactions, feeding niches, coral reefs, thermal stratification and lake turnover, ecosystem diversity and stability, density-dependent and density-independent checks on population growth, r-strategists and K-strategists, survivorship curves, the origin of life and its diversity (chemical and biological evolution), adaptive radiation, inland wetlands, population dynamics in nature, mutations, adaptations, biological evolution and natural selection, ecological succession, structure and composition of the atmosphere, computer modeling and limits to growth, ecological land-use planning , geological hazards (earthquakes and volcanic eruptions), evaluation of models of possible global warming and ozone depletion, CFC substitutes, soil formation, humification and mineralization in soils, types of soil erosion (sheet, rill, and gully), riparian zones, toxicology and dose–response relationships, importance of spiders, lichens as air pollution indicators, toothed cetaceans and baleen whales, pollution and dissolved oxygen levels, waterborne diseases, mussel invaders in the Great Lakes, alien species in Australia, biological amplification of PCBs, bioremediation, perennial food crops, coyote problems, eucalyptus trees, the gray wolf, and "killer" bees.

- Adding more than 59 *new topics*, including: scientists' warning to humanity, Earth education, coral reef bleaching, population stabilization in Japan, population decline, computer models and limits to growth, ecocities, high-speed regional trains, land-use planning and control, the neem tree, population regulation in Thailand, the GATT trade agreement, getting political power, economic solutions to pollution and resource waste, types of economic systems, supply and demand, making sustaining the earth profitable, jobs and the environment, investing in the future, environmental investments by Germany, the anti-environmental movement, improving global environmental protection, the 1992 Rio Earth Summit, using regulation or market forces to control pollution, sustainable development, the Delaney clause, the asbestos controversy, floodplains and reducing the risks of flooding, cloud seeding and towing icebergs, mining with microbes, the materials revolution, resource exploitation, the Mining Law of 1872, Earthship houses, straw-bale houses, the diaper dilemma, recycled paper, paper made from kenaf, household hazardous materials and substitutes, the environmental justice movement, sustainable and optimum yields of fisheries, salmon ranching and farming, indigenous cultures and cultural extinction, the Everglades (problems and restoration), perennial crops, useful products from trees, rotation cycle of forest management, evaluation of monoculture forestry, metabolic reserve of rangeland grasses, wild game ranching, condition of the world's rangelands, riparian zones, the national trails system, reintroducing wolves to Yellowstone, effects of the ivory ban, exclusive economic zones, high seas in the oceans, and nuclear waste dumping in the former Soviet Union.

See pp. x–xii for a detailed summary of major changes for each chapter.

Welcome to Uncertainty, Controversy, and Challenge   We will never have scientific certainty or agreement over what we should do about the complex environmental problems and challenges we face for several reasons:

- Instead of proof or certainty, science provides us with varying degrees of uncertainty (high, medium, low) about the validity of scientific data, hypotheses (tentative explanations), and theories (well-tested and widely accepted hypotheses). *Scientists can disprove things, but they can never prove anything—only show that certain ideas have a high degree of certainty.*

- Science advances through controversy as scientists argue about the validity of data and hypotheses with the goal of converting hypotheses to widely accepted scientific theories. What is important is not what scientists disagree on (the frontiers of knowledge still being developed and argued about) but what they generally agree on— the scientific consensus on theories, concepts, problems, and possible solutions.

- Despite considerable research, we still know relatively little about how nature works at a time when we are altering nature at an accelerating pace.

This built-in uncertainty and the complexity and importance of environmental issues to present and future generations of humans and other species make them highly controversial. Intense controversy also arises because environmental science is a dynamic blend of natural and social sciences that sometimes questions the ways we view and act in the world around us. This interdisciplinary attempt to mirror reality asks us to evaluate our worldviews, values, and lifestyles and our economic and political systems. This can often be a threatening process.

Many environmental books and articles overwhelm us with the problems we face without suggesting ways we might deal with these problems. This book is loaded with solutions suggested by scientists and environmentalists with a wide range of viewpoints and expertise.

Don't take the numerous possible solutions given in this book as gospel. They are given to encourage you to think critically and to make up your own mind. In deciding how to walk more gently on the earth, don't feel guilty about all of the things you are not doing. No one can even come close to doing all of these things, and you may disagree with some of the actions suggested by a diverse array of environmentalists. Pick out the things you are willing to do or agree with and then try to expand your efforts.

People with widely different worldviews, political persuasions, and ideas can work together to help sustain the earth, and that is what counts. We are all in this together, and we need to respect our differences and work together to find a rainbow of solutions to the problems and challenges we face.

Rosy optimism and gloom-and-doom pessimism are traps that usually lead to denial, indifference, and inaction. I have tried to avoid these two extremes and give a realistic—and yet hopeful—view of the future. My reading of history reveals that hope converted into action has been the driving force of our species. This book is filled with stories of individuals who have acted to help sustain the earth for us and all life, and whose actions inspire us to do better. It's an exciting time to be alive as we struggle to enter into a new relationship with the planet that is our only home.

Study Aids   Each chapter begins with a few general questions to give you an idea of how the chapter is organized and what you will be learning. When a new term is introduced and defined, it is printed in **boldface type**. There is also a glossary of all key terms at the end of the book.

Factual recall questions (with answers) are listed at the bottom of most pages. You might cover the answer (on the right-hand page) with a piece of paper and then try to answer the question on the left-hand page. These questions are not necessarily related to the chapter in which they are found.

Each chapter ends with a set of questions designed to encourage you to think critically and apply what you have learned to your life. Some ask you to take sides on controversial issues and to back up your conclusions and beliefs. Individual and group projects also appear at the end of each chapter. These items are marked with an asterisk (*). Many additional projects are given in the *Instructor's Manual* and in the *Green Lives, Green Campuses* supplement available with this book.

Readers who become especially interested in a particular topic can consult the list of suggested readings for each chapter, given in the back of the book. Appendix 1 contains a list of publications to help keep up-to-date on the book's material, as well as a list of some key environmental organizations and government and international agencies.

Help Me Improve This Book   Let me know how you think this book can be improved; and if you find any errors, please let me know about them. Most errors can be corrected in subsequent printings of this edition, rather than waiting for a new edition. Send any errors you find and your suggestions for improvement to Jack Carey, Biology Publisher, Wadsworth Publishing Company, 10 Davis Drive, Belmont, CA 94002. He will send them on to me.

Supplements The following supplementary materials are available:

- *Instructor's Manual and Test Items,* written by Jane Heinze-Fry (Ph.D. in Science and Environmental Education). For each chapter, it has goals and objectives; one or more concept maps; key terms; teaching suggestions; multiple-choice test questions with answers; projects, field trips, and experiments; term-paper and report topics; and a list of audiovisual materials and computer software.

- *Green Lives, Green Campuses,* written by Jane Heinze-Fry. This workbook is designed to help students apply environmental concepts by investigating their lifestyles and by making an environmental audit of their campuses.

- *Laboratory Manual* written by C. Lee Rockett (Bowling Green State University) and Kenneth J. Van Dellen (Macomb Community College).

- A set of 50 color acetates and more than 400 black-and-white transparency masters for making overhead transparencies or slides of line art (including concept maps for each chapter), available to adopters.

- A special version of STELLA II software (a tool for developing critical thinking), together with an accompanying workbook, available to adopters.

- *Environmental Articles,* assembled by Jane Heinze-Fry. This is a collection of more than 200 articles that are indexed by topic and geographical location. Instructors may use the indexes to choose any combination of articles and have them bound as customized supplements for their courses. New articles will be added each year.

- *Watersheds: Classic Cases in Environmental Ethics* by Lisa H. Newton and Catherine K. Dillingham (Wadsworth, 1994). Nine readable case studies that amplify material in this textbook.

- *Environmental Ethics* by Joesph R. Des Jardins (Wadsworth, 1993). A very useful introduction.

- *Radical Environmentalism* by Peter C. List (Wadsworth, 1993). A series of readings on environmental politics and philosophy.

Annenberg/CPB Television Course This textbook is being offered as part of the Annenberg/CPB Project television series "Race to Save the Planet," a 10-part public television series and a college-level telecourse examining the major environmental questions facing the world today. The series takes into account the wide spectrum of opinion about what constitutes an environmental problem, as well as the controversies about appropriate remedial measures. It analyzes problems and emphasizes the successful search for solutions. The course develops a number of key themes that cut across a broad range of environmental issues, including sustainability, the interconnection of the economy and the ecosystem, short-term versus long-term gains, and the trade-offs involved in balancing problems and solutions.

A study guide and a faculty guide, both available from Wadsworth Publishing Company, integrate the telecourse and this text. The television program was developed as part of the Annenberg/CPB Collection.

For further information about available television course licenses and duplication licenses, contact PBS Adult Learning Service, 1320 Braddock Place, Alexandria, VA 22314-1698 (1-800-ALS-AL5-8).

For information about purchasing videocassettes and print material, contact the Annenberg/CPB Collection, P.O. Box 2284, South Burlington, VT 05407-2284 (1-800-LEARNER).

Acknowledgments I wish to thank the many students and teachers who responded so favorably to the eight editions of *Living in the Environment,* the four previous editions of *Environmental Science,* and the first edition of *Resource Conservation and Management* and who offered many helpful suggestions for improvement and corrected errors. I am also deeply indebted to the reviewers who pointed out errors and suggested many important improvements in this book. Any errors and deficiencies left are mine.

The members of Wadsworth's talented and dedicated production team, listed on the copyright page, have also made vital contributions. Their labors of love are also gifts to helping sustain the earth. I especially appreciate the competence and cheerfulness of Production Editors Karen Garrison and Vicki Friedberg and the superb inputs by two talented development editors, Mary Arbogast and Autumn Stanley. My thanks also go to Wadsworth's hard-working sales staff; to Kristin Milotich for her competence and cheerfulness in the midst of chaos; to George Dyke for superb copyediting; to Jane Heinze-Fry for her outstanding work on the *Instructor's Manual, Green Lives, Green Campuses,* and the *Environmental Articles;* and to C. Lee Rockett and Kenneth J. Van Dellen for developing the *Laboratory Manual* to accompany this book.

Special thanks go to Jack Carey, Biology Publisher at Wadsworth, for his encouragement, help, twenty-five years of friendship, and superb reviewing system. It helps immensely to work with the best and most experienced editor in college textbook publishing.

I also wish to thank Peggy Sue O'Neal, my spouse and best friend, for her love and support of me and the earth. I dedicate this book to her and to the earth that sustains us all.

G. Tyler Miller, Jr.

What's New in the Fifth Edition

PART I

HUMANS AND NATURE: AN OVERVIEW

1 Environmental Problems and Their Causes

1 color photo; 2 diagrams; opening Earth Story, "Living in an Exponential Age"; Section 1-1, "Living Sustainably"; 1 Spotlight; "World Scientists' Warning to Humanity"; Solutions, "Don't Kill the Goose"; 2 Guest Essays, "There Is No Crisis of Unsustainability," by Julian L. Simon and "Simple Simon Environmental Analysis," by Anne H. Ehrlich and Paul R. Ehrlich.

2 Cultural Changes, Worldviews, Ethics, and Sustainability

1 color photo; opening Earth Story, "2040 A.D.: Green Times on Planet Earth"; Sections 2-2 and 2-3 rewritten; Solutions, "Emotional Learning: Earth Wisdom"; expanded discussions of various types of worldviews; new discussion of Earth education; Guest Essay, "Launching the Environmental Revolution," by Lester R. Brown.

PART II

PRINCIPLES AND CONCEPTS

3 Matter and Energy Resources: Types and Concepts

1 color photo; opening Earth Story, "Saving Energy, Money, and Jobs in Osage, Iowa"; expanded discussion of nature of science and technology and limitations and misuse of science; new material on half-life of radioisotopes.

4 Ecosystems and How They Work

9 color photos; 7 diagrams; opening Earth Story, "Connections: Blowing in the Wind"; Connections, "Have You Thanked Insects Today?"; Connections, "The Mystery of the Vanishing Amphibians"; new discussions of nature of life, biodiversity, pyramids of numbers, pyramids of biomass, sulfur cycle, symbiotic relationships,

interference competition, exploitation competition; expanded treatment of five major kingdoms, energy flow in ecosystems, species interactions, feeding niches, and commensalism.

5 Ecosystems: What Are the Major Types, and What Can Happen to Them?

13 color photos; 9 diagrams; opening Earth Story, "Earth Healing: From Rice Back to Rushes"; Spotlight, "The Kangaroo Rat: Water Miser and Keystone Species"; Case Study, "The Importance of Coral Reefs"; Connections, "Earth: The Just-Right, Resilient Planet"; new discussions of coral reef bleaching, lake turnover, ecosystem diversity and stability, density-dependent and density-independent checks on population growth, r-strategists and K-strategists, survivorship curves, origin of life and its diversity (chemical and biological evolution), adaptive radiation; expanded discussions of coral reefs, inland wetlands, population dynamics in nature, biological evolution and natural selection, and ecological succession; Guest Essay, "The Ecological Design Arts," by David W. Orr.

6 The Human Population: Growth, Urbanization, and Regulation

3 color photos; 14 diagrams; opening Earth Story, "Cops and Rubbers Day in Thailand"; Case Study, "The Graying of Japan"; 2 Solutions, "High-Speed Regional Trains" and "Is Your City Green? The Ecocity Concept in Davis, California"; new discussions of high-speed regional trains, population decline, computer modeling and limits to growth, land-use planning and control, and ecological land-use planning; expanded discussion of transportation and urban development; updating of population data.

7 Environmental Economics and Politics

Entire chapter rewritten; 1 color photo; 6 diagrams; opening Earth Story, "To Grow or Not to

Guest Essayists and Reviewers

Guest Essayists   The following are the authors of Guest Essays:

Lester R. Brown, President, Worldwatch Institute; **Alberto Ruz Buenfil**, environmental activist, writer, and performer; **Robert D. Bullard**, Professor of Sociology, University of California, Riverside; **Vincent T. Covello**, Professor of Environmental Sciences, School of Public Health, and Director, Center for Risk Communication, Columbia University; **Anne H. Ehrlich**, Senior Research Associate, Department of Biological Sciences, Stanford University; **Paul R. Ehrlich**, Bing Professor of Population Studies, Stanford University; **Lois Marie Gibbs**, Director, Citizens' Clearinghouse for Hazardous Wastes; **Garrett Hardin**, Professor Emeritus of Human Ecology, University of California, Santa Barbara; **Jim Hightower**, populist, author, and radio commentator; **Amory B. Lovins**, Energy Policy Consultant and Director of Research, Rocky Mountain Institute; **Jessica Tuchman Mathews**, Vice President, World Resources Institute; **Peter Montague**, Senior Research Analyst, Greenpeace, and Director, Environmental Research Foundation; **Norman Myers**, consultant in environment and development; **David W. Orr**, Professor of Environmental Studies, Oberlin College; **David Pimentel**, Professor of Entomology, Cornell University; **Julian L. Simon**, Professor of Economics and Business Administration, University of Maryland.

Cumulative Reviewers   Barbara J. Abraham, Hampton College; Donald D. Adams, State University of New York at Plattsburgh; Larry G. Allen, California State University, Northridge; James R. Anderson, U.S. Geological Survey; Kenneth B. Armitage, University of Kansas; Gary J. Atchison, Iowa State University; Marvin W. Baker, Jr., University of Oklahoma; Virgil R. Baker, Arizona State University; Ian G. Barbour, Carleton College; Albert J. Beck, California State University, Chico; W. Behan, Northern Arizona University; Keith L. Bildstein, Winthrop College; Jeff Bland, University of Puget Sound; Roger G. Bland, Central Michigan University; Georg Borgstrom, Michigan State University; Arthur C. Borror, University of New Hampshire; John H. Bounds, Sam Houston State University; Leon F. Bouvier, Population Reference Bureau; Daniel J. Bovin, Université Laval; Michael F. Brewer, Resources for the Future, Inc.; Mark M. Brinson, East Carolina University; Patrick E. Brunelle, Contra Costa College; Terrence J. Burgess, Saddleback College North; David Byman, Pennsylvania State University, Worthington Scranton; Lynton K. Caldwell, Indiana University; Faith Thompson Campbell, Natural Resources Defense Council, Inc.; Ray Canterbery, Florida State University; Ted J. Case, University of San Diego; Ann Causey, Auburn University; Richard A. Cellarius, Evergreen State University; William U. Chandler, Worldwatch Institute; F. Christman, University of North Carolina, Chapel Hill; Preston Cloud, University of California, Santa Barbara; Bernard C. Cohen, University of Pittsburgh; Richard A. Cooley, University of California, Santa Cruz; Dennis J. Corrigan; George Cox, San Diego State University; John D. Cunningham, Keene State College; Herman E. Daly, The World Bank; Raymond F. Dasmann, University of California, Santa Cruz; Kingsley Davis, Hoover Institution; Edward E. DeMartini, University of California, Santa Barbara; Charles E. DePoe, Northeast Louisiana University; Thomas R. Detwyler, University of Wisconsin; Peter H. Diage, University of California, Riverside; Lon D. Drake, University of Iowa; T. Edmonson, University of Washington; Thomas Eisner, Cornell University; Michael Esler, Southern Illinois University; David E. Fairbrothers, Rutgers University; Paul P. Feeny, Cornell University; Nancy Field, Bellevue Community College; Allan Fitzsimmons, University of Kentucky; Andrew J. Friedland, Dartmouth College; Kenneth O. Fulgham, Humboldt State University; Lowell L. Getz, University of Illinois at Urbana-Champaign; Frederick F. Gilbert, Washington State University; Jay Glassman, Los Angeles Valley College; Harold Goetz, North Dakota State University; Jeffery J. Gordon, Bowling Green State University; Eville Gorham, University of Minnesota; Michael Gough, Resources for the Future; Ernest M. Gould, Jr., Harvard University; Katharine B. Gregg, West Virginia Wesleyan College; Peter Green, Golden West College; Paul K. Grogger, University of Colorado at Colorado Springs; L. Guernsey, Indiana State University; Ralph Guzman, University of California, Santa Cruz; Raymond Hames, University of Nebraska, Lincoln; Raymond E. Hampton, Central Michigan University; Ted L. Hanes, California State University, Fullerton; William S. Hardenbergh, Southern Illinois University at Carbondale; John P. Harley, Eastern Kentucky University; Neil A. Harriman, University of Wisconsin-Oshkosh; Grant A. Harris, Washington State University; Harry S. Hass, San Jose City College; Arthur N. Haupt, Population Reference Bureau; Denis A. Hayes, environmental consultant; Gene Heinze-Fry, Department of Utilities, State of Massachusetts; John G. Hewston, Humboldt State University; David L. Hicks, Whitworth College; Eric Hirst, Oak Ridge National Laboratory; S. Holling, University of British Columbia; Donald Holtgrieve, California State University, Hayward; Michael H. Horn, California State University, Fullerton; Mark A. Hornberger, Bloomsburg University; Marilyn Houck, Pennsylvania State University; Richard D. Houk, Winthrop College; Robert J. Huggett, College of William and Mary; Donald Huisingh, North Carolina State University; Marlene K. Hutt, IBM; David R. Inglis, University of Massachusetts; Robert Janiskee, University of South Carolina; Hugo H. John, University of Connecticut; Brian A. Johnson, University of Pennsylvania, Bloomsburg; David I. Johnson, Michigan State University; Agnes Kadar, Nassau Community College; Thomas L. Keefe, Eastern Kentucky University; Nathan Keyfitz, Harvard University; David Kidd, University of New Mexico; Edward J. Kormondy, University of Hawaii-Hilo/West Oahu College; John V. Krutilla, Resources for the Future, Inc.; Judith Kunofsky, Sierra Club; E. Kurtz; Theodore Kury, State University of New York, Buffalo; Steve Ladochy, University of Winnipeg; Mark B. Lapping, Kansas State University; Tom Leege, Idaho Department of Fish and Game; William S. Lindsay, Monterey Peninsula College; E. S. Lindstrom, Pennsylvania State University; M. Lippiman, New York University Medical Center; Valerie A. Liston, University of Minnesota; Dennis Livingston, Rensselaer Polytechnic Institute; James P. Lodge, air pollution consultant; Raymond C. Loehr, University of Texas at Austin; Ruth Logan, Santa Monica City College; Robert D. Loring, DePauw University; Paul F. Love, Angelo State University; Thomas Lovering, University of California, Santa Barbara; Amory B. Lovins, Rocky Mountain Institute; Hunter Lovins, Rocky Mountain Institute; Gene A. Lucas, Drake University; Claudia Luke; David Lynn; Timothy F. Lyon, Ball State University; Melvin G. Marcus, Arizona State University; Gordon E. Matzke, Oregon State University; Parker Mauldin, Rockefeller Foundation; Theodore R. McDowell, California State University; Vincent E. McKelvey, U.S. Geological Survey; John G. Merriam, Bowling Green State University; A. Steven Messenger, Northern Illinois University; John Meyers, Middlesex Community College; Raymond W. Miller, Utah State University; Arthur B. Millman, University of Massachusetts, Boston; Rolf Monteen, California Polytechnic State University; Ralph Morris, Brock University, St. Catharines, Ontario, Canada; William W. Murdoch, University of California, Santa Barbara; Norman Myers, environmental consultant; Brian C. Myres, Cypress College; A. Neale, Illinois State University; Duane Nellis, Kansas State University; Jan Newhouse, University of Hawaii, Manoa; John E. Oliver, Indiana State University; Eric Pallant, Allegheny College; Charles F. Park, Stanford University; Richard J. Pedersen, U.S. Department of Agriculture, Forest Service; David Pelliam, Bureau of Land Management, U.S. Department of Interior; Rodney Peterson, Colorado State University; William S. Pierce, Case Western Reserve University; David Pimentel, Cornell University; Peter Pizor, Northwest Community College; Mark D. Plunkett, Bellevue Community College; Grace L. Powell, University of Akron; James H. Price, Oklahoma College; Marian E. Reeve, Merritt College; Carl H. Reidel, University of Vermont; Roger Revelle, California State University, San Diego; L. Reynolds, University of Central Arkansas; Ronald R. Rhein, Kutztown University of Pennsylvania; Charles Rhyne, Jackson State University; Robert A. Richardson, University of Wisconsin; Benjamin F. Richason III, St. Cloud State University; Ronald Robberecht, University of Idaho; William Van B. Robertson, School of Medicine, Stanford University; C. Lee Rockett, Bowling Green State University; Terry D. Roelofs, Humboldt State University; Richard G. Rose, West Valley College; Stephen T. Ross, University of Southern Mississippi; Robert E. Roth, The Ohio State University; Floyd Sanford, Coe College; David Satterthwaite, I.E.E.D., London; Stephen W. Sawyer, University of Maryland; Arnold Schecter, State University of New York, Syracuse; Frank Schiavo, San Jose State University; William H. Schlesinger, Ecological Society of America; Stephen H. Schneider, National Center for Atmospheric Research; Clarence A. Schoenfeld, University of Wisconsin, Madison; Henry A. Schroeder, Dartmouth Medical School; Lauren A. Schroeder, Youngstown State University; Norman B. Schwartz, University of Delaware; George Sessions, Sierra College; David J. Severn, Clement Associates; Paul Shepard, Pitzer College and Claremont Graduate School; Michael P. Shields, Southern Illinois University at Carbondale; Kenneth Shiovitz; F. Siewert, Ball State University; E. K. Silbergold, Environmental Defense Fund; Joseph L. Simon, University of South Florida; William E. Sloey, University of Wisconsin-Oshkosh; Robert L. Smith, West Virginia University; Howard M. Smolkin, U.S. Environmental Protection Agency; Patricia M. Sparks, Glassboro State College; John E. Stanley, University of Virginia; Mel Stanley, California State Polytechnic University, Pomona; Norman R. Stewart, University of Wisconsin-Milwaukee; Frank E. Studnicka,

University of Wisconsin-Platteville; Chris Tarp, Contra Costa College; William L. Thomas, California State University, Hayward; John D. Usis, Youngstown State University; Tinco E. A. van Hylckama, Texas Tech University; Robert R. Van Kirk, Humboldt State University; Donald E. Van Meter, Ball State University; Gary Varner, Texas A&M University; John D. Vitek, Oklahoma State University; Lee B. Waian, Saddleback College; Warren C. Walker, Stephen F. Austin State University; Thomas D. Warner, South Dakota State University; Kenneth E. F. Watt, University of California, Davis; Alvin M. Weinberg, Institute of Energy Analysis, Oak Ridge Associated Universities; Brian Weiss; Anthony Weston, SUNY at Stony Brook; Raymond White, San Francisco City College; Douglas Wickum, University of Wisconsin-Stout; Charles G. Wilber, Colorado State University; Nancy Lee Wilkinson, San Francisco State University; John C. Williams, College of San Mateo; Ray Williams, Whittier College; Roberta Williams, University of Nevada at Las Vegas; Samuel J. Williamson, New York University; Ted L. Willrich, Oregon State University; James Winsor, Pennsylvania State University; Fred Witzig, University of Minnesota at Duluth; George M. Woodwell, Woods Hole Research Center; Robert Yoerg, Belmont Hills Hospital; Hideo Yonenaka, San Francisco State University; Malcolm J. Zwolinski, University of Arizona.

Integrating Themes

Ten themes are used to integrate the material in this book. You might think of them as connecting threads woven through the material. To follow each thread use the page numbers listed after each theme.

Biodiversity and Earth Capital

5, 6, 8, 12–13, 14–15, 21, 23, 61–63, 82, 91–128, 164, 166, 169, 188, 244, 279, 282, 288–89, 318, 319, 322, 324, 325–27, 372–406, 406–449, 450–478

Economics and Environment

8, 9, 10, 11, 16, 17, 32–33, 35, 40, 55, 98, 100–101, 102, 105–7, 111, 134, 138–39, 140, 142–49, 154–58, 163–79, 188, 189–90, 192, 193, 196, 200, 204, 224, 225, 227, 229, 246, 247, 252–53, 266–67, 270, 273, 274, 278–79, 281, 286, 290, 292, 301, 306, 313, 313–14, 315–17, 327, 333–34, 335, 337, 340, 345, 346–49, 350, 358, 360, 362, 363–64, 365, 366, 374, 375, 379, 380, 381, 386, 387, 390–91, 395, 403, 416, 417, 420–21, 423, 426, 427, 428, 429, 430, 431, 432, 436–39, 442, 443, 445, 451, 456, 459–61, 462, 463, 466, 470, 473, 476, 484, 485–86, 487, 488, 489, 490, 491, 492, 496, 500, 501–2, 504, 505, 507, 513–15, 518–20, 523, 524–25, 526, 529, 533, 534–35, 536–37, 538–39

Energy and Energy Efficiency

5, 8, 10, 12, 17, 25–29, 40, 45–55, 56, 58–60, 66–69, 71–77, 81–82, 94, 117, 126, 152, 176, 177, 179, 190, 229, 231, 233, 235, 237–38, 243, 246, 267–69, 272, 307, 309–10, 314, 340, 346, 349, 352, 375, 377, 390, 469, 480–515, 516–539

Individual Action and Earth Citizens

Garrett Hardin (16, 160–61); 21; Anne Ehrlich (22–23); Paul Ehrlich (22–23, 131, 157, 391); 25; Lester Brown (30–31, 259, 332, 373); David Orr (33, 128–29); David Ehrenfeld (33, 464); Peter Montague (37–38, 359, 360); Thomas Berry (33, 36); 35; Aldo Leopold (35, 36, 448, 451); Jesse Jackson (5); Henry David Thoreau (21, 36, 444); Mahatma Ghandi (35, 174, 179); Gary Snyder (36, 540); Michael J. Cohen (36, 213); Dave Foreman (36); Robert Cahn (36); E. O. Wilson (82, 124, 125, 127, 427, 456); John Todd (129, 259, 298–99); Mechai Viravidaiya (130); Gaylord Nelson (164); Robert Repetto (167); Herman E. Daly (168); John B. Cobb, Jr. (168); 180; 182–86; Al Gore (182); Jim Hightower (180, 189–90); 183; Kurt Vonnegut (184); Rachel Carson (193); Vincent T. Covello (209–10); Thomas Lovejoy (211, 235, 431); 232; Amory Lovins (246, 480, 484, 485, 513–15); Sherwood Rowland (248); Mario Molina (248); Ray Turner (255); Jessica Tuchman Mathews (256); Marion Stoddart (283); Jacques Costeau (287); 300; 332; David Pimentel (333–34); Lois Gibbs (338–39); Michael Reynolds (342–4); William Sanjour (359, 360–61); Robert Bullard (367–68); Joel Hirschorn (369);

Wes and Dana Jackson (372); 390; Alberto Ruz Buenfil (391–92); 403; Robert Horwich (408); Chico Mendes (426); 434; David Hopcraft (435); Jack Turrell (440); Daniel Janzen (444); John Muir (444); Wallace Stegner (444); Norman Myers (445, 446–47); Roderick Nash (447); Wangari Maathai (448); Theodore Roosevelt (467); 477; Hunter Lovins (480)

Politics and Environmental Laws

19, 21, 29, 35, 149, 155, 159, 167, 177, 179–90, 227–29, 232, 247, 254–61, 263, 270, 274, 275, 278–79, 283, 290, 293–94, 297, 301, 315, 330, 335, 338–39, 361–64, 364–68, 387–88, 399–402, 420–21, 422–23, 428, 436–39, 445–46, 447–48, 466–67, 469, 486, 509, 525–26, 530, Appendix 3

Pollution Prevention and Waste Reduction

16–17, 37–38, 49, 54, 152, 170–72, 187, 201, 204, 228, 229, 233, 284, 297–98, 337–43, 355, 356, 357–58, 365, 368–69,

Population and Exponential Growth

4, 5, 6–7, 8–9, 10, 11, 14–15, 17–19, 25–29, 48, 53, 55, 61, 71, 76, 85–88, 105, 112–15, 120–24, 126, 130–62, 165, 174, 176, 184, 188, 233, 234, 258, 260, 266, 271, 281, 293, 297, 313, 325, 377, 378, 384, 385, 391, 394, 427, 454–55, 456, 466, 471, 519

Science and Technology

16–19, 40–55, 56–129, 134, 146, 179, 193–207, 227, 230, 235–42, 244–45, 247–50, 251–53, 254, 267, 269, 273, 275–76, 279–82, 285, 291, 294–96, 298–99, 302–7, 310, 316–17, 318–23, 330–32, 342, 344–45, 350–53, 372, 373–74, 376, 378, 381–82, 385, 386–87, 393, 394–95, 396–97, 402–6, 408, 415–18, 425–27, 435, 436, 452–53, 454–55, 457–58, 459, 463, 470–71, 482–514, 517, 520–21, 521–22, 525–33

Solutions and Sustainability

5–8, 6–7, 8, 17–20, 24, 26, 29, 30–31, 33–38, 54–55, 69, 127–29, 145, 146, 148–61, 168–69, 169–79, 180, 182–84, 188–90, 194, 201–2, 204, 210, 227–33, 245–47, 248, 255, 256, 265–67, 268–79, 292–300, 315–18, 328–44, 337–52, 355, 357–61, 364–69, 380, 381–92, 402–6, 412–19, 421–23, 429–32, 433, 439–40, 441–43, 444, 445–47, 463–77, 482, 515, 519–20, 521–23, 535–39, 540

Uncertainty and Controversy

6, 16, 19, 20–24, 29–34, 36, 43, 112, 118, 154–55, 160–61, 163, 169–74, 181, 184–88, 190, 195–96, 206–10, 220–21, 222–24, 232, 238–54, 265, 267, 269, 270–71, 274, 275–76, 277–79, 282, 284–86, 301, 311, 312–14, 315, 341–42, 344–45, 347–48, 349–50, 351–53, 353–54, 355–56, 357–60, 361, 362–66, 387–89, 395–402, 405, 411–12, 413, 418, 420–21, 422–23, 428, 436–39, 444–45, 456, 467, 469, 473, 476–77, 481, 529, 532, 533–34, 534–35

Brief Contents

Detailed Contents

Working with the Earth: A Student's Visual Guide Through this Book

With the emergence of the environment as one of today's most pressing and controversial concerns, the interdisciplinary field of environmental science is becoming increasingly important and interesting. I assume that is one of the reasons why you are enrolled in this course.

In this fifth edition of **Environmental Science,** I try to present basic and easily understandable scientific principles to show you how various parts of Earth's life support systems, environmental problems, and possible solutions are connected and how all of us can do many things to help sustain the earth. The new subtitle for this edition, **Working with the Earth,** emphasizes this perspective.

The many changes in this edition of **Environmental Science** were made with you, the student, in mind. The art program is entirely new and includes many illustrations and photographs that are more three-dimensional, colorful, detailed, and useful. I've tried to make my writing clearer and more personal so you can enjoy the process of learning. And because I wanted to make sure that the information was as current, accurate, and scientifically sound as possible, I've added more scientific content. I've consulted well over 10,000 research sources in the professional literature, and I've listened to the more than 200 experts who have reviewed the material in this and earlier editions.

For the past 27 years, I've devoted my full time to studying, teaching, and writing about the environment and its connections. As you might imagine, environmental science is a passion that has changed my life. It is my hope that this book will change yours, or at the very least, challenge you to question your assumptions, think problems through, and begin to lead more environmentally responsible lives. The material that follows will enhance your ability to use and learn from this book.

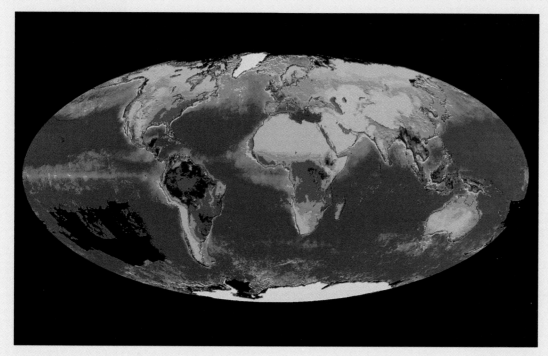

Three-year composite of satellite data on Earth's primary productivity. This is one of 17 satellite photographs in this edition.

3 Matter and Energy Resources: Types and Concepts

Saving Energy, Money, and Jobs in Osage, Iowa

Osage, Iowa (population about 4,000), has become the energy-efficiency capital of the United States. It began in 1974 when easygoing Wes Birdsall, general manager of Osage Municipal Gas and Electric Company, started going door-to-door preaching. He wanted the townspeople to save energy and reduce their natural gas and electric bills. The utility would save money, too, by not having to add a new power plant.

Wes started his crusade by telling homeowners about the importance of insulating walls and ceilings and of plugging of leaky windows and doors. He also advised people to replace their incandescent light bulbs with more efficient fluorescent bulbs and to turn down the temperature on water heaters and wrap them with insulation—an economic boon to the local hardware and lighting stores. Wes also suggested saving water and fuel by installing low-flow shower heads.

He stepped up his campaign by offering to give every building in town a free thermogram—an infrared scan that shows where heat escapes (Figure 3-1). When people could see the energy (and money) hemorrhaging out of their buildings, they took action to plug these leaks—again helping the local economy.

Since 1974 the town has cut its natural gas consumption by 45%; no mean feat in a place where winter temperatures can plummet to –103°C (–80°F). In addition, the utility company saved enough money to prepay all its debt, accumulate a cash surplus, and cut inflation-adjusted electricity rates by a third (which attracted two new factories). Furthermore, each household saves more than $1,000 per year. This money supports jobs, and most of it circulates in the local economy. Before the energy-efficiency revolution, about $1.2 million a year went out of town—usually out of state—to buy energy. What are your local utility and community doing to improve energy efficiency and stimulate the local economy?

Figure 3-1 An infrared photo showing heat loss around the windows, doors, roofs, and foundations (red, white, and yellow colors) of houses and stores in Plymouth, Michigan. Wes Birdsall provided similar thermograms made for houses in Osage, Iowa. The average U.S. house has heat leaks and air infiltration equivalent to leaving a window wide open during the heating season. Because of poor design, most U.S. office buildings and houses waste about half the energy used to heat and cool them. Americans pay about $300 billion a year for this wasted heat—more than the entire annual military budget. (VANSCAN® Continuous Mobile Thermogram by Daedalus Enterprises, Inc.)

Bats: A Bad Rap

CONNECTIONS

Despite their variety (950 species) and worldwide distribution, bats have several traits that expose them to extinction because of human activities. Many bats nest in huge cave colonies, which become vulnerable to destruction when people block the caves' entrances. And once the population of a bat species falls below a certain level, it may not recover because of its slow reproductive rate.

Bats play significant ecological roles and are also of great economic importance. Some bat species help control many crop-damaging insects and other pest species, such as mosquitoes and rodents (Figure 17-10). About 70% of all bat species feed on night-flying insects, making them the primary controls for such insects.

Other species of bats eat pollen; still others eat certain fruits. Bats with this kind of specialized feeding are the chief pollinators for certain trees, shrubs, and other plants; they also spread plants throughout tropical forests by excreting undigested seeds. If these keystone species are eliminated from an area, dependent plants will disappear. U.S. examples of bat-pollinated species are the giant saguaro cactus (Figure 4-11) and agaves (plants of the desert Southwest used in making fiber rope and tequila). In Southeast Asia, a cave-dwelling, nectar-eating bat species is the only known pollinator of durian trees, whose fruit is worth $120 million per year. If you enjoy bananas, cashews, dates, figs, avocados, or mangos, you can thank bats. Research on bats has contributed to the development of birth control and artificial insemination methods, and testing of drugs, studies of disease resistance and aging, vaccine production, and development of navigational aids for the blind.

People mistakenly fear bats as filthy, aggressive, rabies-carrying, bloodsuckers. But most bat species are harmless to people, livestock, and crops. In the United States only 10 people have died of bat-transmitted disease in four decades of record keeping. More Americans die each year from falling coconuts. Only three species of bats (none of them found in the United States) feed on blood, mostly that of cattle or wild animals. These bat species

Figure 17-10 Endangered ghost bat carrying a mouse in tropical northern Australia. This carnivorous nightfeeding bat is harmless to people. Bats are considered keystone species in many ecosystems because of their roles in pollinating plants, dispersing seeds, and controlling insect and rodent populations.

can be serious pests to domestic livestock but rarely affect humans.

Because of unwarranted fears of bats and misunderstanding of their vital ecological roles, several species have been driven to extinction, and others, such as the ghost bat (Figure 17-9), are endangered or threatened. We need to see bats as valuable allies, not as enemies.

are threatened with extinction, mostly because of habitat loss and fragmentation.

Commercial Hunting and Poaching Today subsistence hunting (for food) is rare because of the decline in hunting-and-gathering societies (Connections, p. 428). Sport hunting is closely regulated in most countries, and game species are endangered only where protective regulations do not exist or are not enforced.

However, a combination of habitat loss, legal commercial hunting, and illegal commercial hunting (poaching) has driven many species, such as the American bison (Case Study, p. 461) or over the brink of extinction (Figures 17-6 and 17-7). Bengal tigers are in trouble because a tiger fur sells for $100,000 in Tokyo. A mountain gorilla is worth $150,000 and a chimpanzee $50,000; an ocelot skin, $40,000; and an Imperial

Figure 17-11 Endangered ring-tailed lemur in Madagascar. Lemurs are the oldest distant relative of the human species. The 30 species of lemurs on this island are found nowhere else in the wild. Half of them are endangered.

A: 2.3–2.5 billion years ago

CHAPTER 17 **459**

Is Your City Green?—The Ecocity Concept in Davis, California

SOLUTIONS

In a sustainable and ecologically healthy city—called an *ecocity* or *green city*—matter and energy resources are used efficiently, and far less pollution and waste are produced than in conventional cities. Emphasis is on pollution prevention, reuse, recycling, and efficient use of energy and matter resources. Per capita solid waste is greatly reduced, and 60% of what is produced is recycled, composted, or reused. An ecocity takes advantage of locally available energy sources and requires that all buildings, vehicles, and appliances meet high energy-efficiency standards.

Trees and plants adapted to the local climate and soils are planted throughout the ecocity to provide shade and beauty, to reduce pollution and noise, and to supply habitats for wildlife. Abandoned lots and polluted creeks are cleaned up and restored. Nearby forests, grasslands, wetlands, and farms are preserved instead of being devoured by urban sprawl. Much of an ecocity's food comes from nearby organic farms, solar greenhouses, community gardens, and gardens on rooftops and in yards and window boxes.

An ecocity is a people-oriented city—not a car-oriented city. Its residents are able to walk or bike to most places, including work, and to take low-polluting mass transit. It is designed, retrofitted, and managed to provide a sense of community built around cooperative and vibrant neighborhoods.

The ecocity concept is not a futuristic dream. The citizens and elected officials of Davis, California—a city of about 40,000 people about 130 kilometers (80 miles) northeast of San Francisco—committed themselves in the early 1970s to making it an ecologically sustainable city.

City building codes encourage the use of solar energy for water and space heating. All new homes must meet high standards of energy efficiency, and when an existing home changes hands, the buyer must bring it up to the energy conservation standards for new homes. In Davis's Village Homes—America's first solar neighborhood—houses are heated by solar energy

(Figure 6-24). They face into a common open space reserved for people and bicycles; cars are restricted to streets, which are located only on the periphery of the development. The neighborhood also has orchards, vineyards, and a large community garden. Since 1975 the city has cut its use of energy for heating and cooling in half.

Davis has a solar power plant; some of the electricity it produces is sold to the regional utility company. Eventually the city plans to generate all of its own electricity.

The city discourages the use of automobiles and encourages the use of bicycles by closing some streets to automobiles, by building bike lanes on major streets, and by building bicycle paths. Any new housing tract must have a separate bike lane, and some city employees are given bikes. As a result, 28,000 bicycles account for 40% of all in-city transportation, and less land is needed for parking spaces. This heavy dependence on the bicycle is aided by the city's warm climate and flat terrain. Davis limits the type and rate of its growth, and it maintains a mix

---

problems and save money in the long run. Finally, it's difficult to get municipalities in the same general area to cooperate in planning efforts. Thus an ecologically sound development plan in one area may be disrupted by unsound development in nearby areas.

Making Urban Areas More Livable and Sustainable Since most people around the world now live, or will live, in urban areas, improving the quality of urban life is an urgent priority—and a few cities have started to do it (Solutions, above). Here are some ways various analysts have proposed to make urban areas more sustainable:

Economic Development and Population Regulation

■ Reduce population growth rates (Section 6-5).

■ Reduce the flow of people from rural to urban areas by increasing investments and social services in rural areas and by not giving higher food, energy, and other subsidies to urban dwellers than to rural dwellers.

■ Recognize that increased urbanization and urban density is better than spreading people out over the countryside, which would destroy more of the planet's biodiversity. The primary problem is not urbanization, but our failure to make cities more sustainable and livable.

■ Maintain employment and plug dollar drains from local economies by setting up "buy local" programs, greatly improving energy efficiency (p. 40), and instituting extensive recycling, reuse, and pollution prevention programs.

Q: What greenhouse gas is being emitted to the atmosphere in the largest quantity from human activities?

NEW SOLUTIONS BOXES

In addition to describing environmental problems, this new edition emphasizes more than ever before solutions to these problems. In each of 26 Solutions boxes, a new feature in this edition, I've highlighted a particular environmental problem and offered you possible ways to solve it. There are also sections and subsections of Solutions within most chapters.

Factual Recall Questions

Factual recall questions appear at the bottom of most left-hand pages so you can test your knowledge. Answers appear on the facing page.

The Kangaroo Rat: Water Miser and Keystone Species

SPOTLIGHT

The kangaroo rat (Figure 5-8) is a remarkable mammal that has mastered the art of water conservation in its desert environment. As the desert's chief seed eater, it is also a keystone species that helps support other desert species. By consuming seeds it helps keep desert shrubland from becoming grassland.

This rodent comes out of its burrow only at night, when the air is cool and water evaporation has slowed. Its main source of food is dry seeds, which it quickly stuffs into its cheek pouches. After a night of foraging, it returns to its burrow and empties its cache of seeds.

In the cool burrow the seeds soak up water exhaled in the rodent's breath. When the rodent eats these seeds it gets this water back. It does not drink water; its water comes from recycled moisture in the seeds it eats and from water produced when the sugars in the seeds undergo aerobic respiration. Some of the water vapor in the rat's breath condenses on a cool inside surface of the nose and diffuses

back to its body. Kangaroo rats have no sweat glands, so they don't lose water by perspiration. In addition, they save water by excreting hard, dry feces and thick, nearly solid urine from their super-efficient kidneys.

Figure 5-8 This nocturnal kangaroo rat of the California desert is an expert in water conservation. It is also a keystone species because it consumes such vast quantities of seeds, which prevents desert shrubland from becoming grassland.

Figure 5-9 Serengeti tropical savanna in Tanzania, Africa, an example of one type of tropical grassland. Most savannas consist of grasslands punctuated by stands of deciduous shrubs and trees, which shed their leaves during the dry season and thus avoid excessive water loss. More large, hoofed, plant-eating mammals (ungulates), such as the herd of wildebeest shown here, live in this biome than anywhere else.

Polar grasslands, or *arctic tundra*, occur just south of the Arctic polar ice cap (Figure 5-3). During most of the year these treeless plains are bitterly cold, swept by frigid winds, and covered with ice and snow. Winters are long and dark, and the low average annual precipitation falls mostly as snow. This biome is carpeted with a thick, spongy mat of low-growing plants (Figure 5-13). Most of the annual growth of these plants occurs during the summer, when sunlight shines almost around the clock.

Q: What percentage of the world's population own cars?

SPOTLIGHTS

I've included 21 Spotlights (19 new to this edition) to highlight and provide you with further insights into key environmental problems and concepts. This one, "The Kangaroo Rat: Water Miser and Keystone Species," discusses how this desert animal is an expert at conserving water. It also explains how this mammal keeps desert shrubland from becoming grassland.

In this edition, I've increased the use of scientific concepts and content significantly. The discussion shown here focuses on the two laws of energy and how they govern what we can and cannot do with energy.

Key Terms

Key terms are highlighted in boldface the first time they appear and are defined in the glossary at the end of the text. In addition, the index includes key terms and the page number where each term is first defined.

INTEGRATING THEMES

Ten major themes—*Biodiversity and Earth Capital, Economics and Environment, Energy and Energy Efficiency, Individual Action and Earth Citizens, Politics and Environmental Laws, Pollution Prevention and Waste Reduction, Population and Exponential Growth, Science and Technology, Solutions and Sustainability, Uncertainty and Controversy*—are used to integrate the book.

PRO/CON BOXES

5 Pro/Con boxes (2 new to this edition) present both sides of controversial environmental and resource issues to encourage you to think critically. Discussions of controversial issues are also found throughout much of the textual material. The Pro/Con box shown here presents arguments for and against drilling for oil in the Arctic National Wildlife Refuge. After reading it through, which side would you take?

Sample page (page 52):

Fuel ⟹ Reaction Conditions ⟹ Products

D-T Fusion

hydrogen-2 or deuterium nucleus
hydrogen-3 or tritium nucleus
100 million°C
neutron
helium-4 nucleus
energy

D-D Fusion

hydrogen-2 or deuterium nucleus
hydrogen-2 or deuterium nucleus
1 billion°C
helium-3 nucleus
neutron

proton   neutron

Figure 3-10 The deuterium-tritium (D-T) and deuterium-deuterium (D-D) nuclear fusion reactions, which take place at extremely high temperatures.

process is still at the laboratory stage. Even if it becomes technologically and economically feasible, it probably won't be a practical source of energy until 2050 or later.

3-6 THE TWO IRONCLAD LAWS OF ENERGY

First Law of Energy: You Can't Get Something for Nothing Scientists have observed energy being changed from one form to another in millions of physical and chemical changes, but they have never been able to detect any energy being created or destroyed. This summary of what happens in nature is called the **law of conservation of energy**, also known as the **first law of energy** or **first law of thermodynamics**. This law means that when one form of energy is converted to another form in any physical or chemical change *energy input always equals energy output: We can't get something for nothing in terms of energy quantity.*

Second Law of Energy: You Can't Break Even
Because the first law of energy states that energy can be neither created nor destroyed, you might think that there will always be enough energy; yet, if you fill a car's tank with gasoline and drive around, or if you use a flashlight battery until it is dead, you have lost something. If it isn't energy, what is it? The answer is **energy quality**, the amount of energy available that can perform useful work (Figure 3-6).

Countless experiments have shown that when energy is changed from one form to another, there is always a decrease in energy quality (the amount of useful energy). This summary of what we always find happening in nature is called the **second law of energy**, or the **second law of thermodynamics**: When energy is changed from one form to another, some of the useful energy is always degraded to lower-quality, more-dispersed, less-useful energy. This degraded energy is usually in the form of heat, which flows into the environment and is dispersed by the random motion of air or water molecules.

In other words *we can't break even in terms of energy quality because energy always goes from a more useful to a less useful form.* No one has ever found a violation of this fundamental scientific law.

Consider three examples of the second energy law in action. First, when a car is driven, only about 10% of the high-quality chemical energy available in its gasoline fuel is converted into mechanical energy (to propel the vehicle) and into electrical energy (to run its electrical systems). The remaining 90% is degraded to low-quality heat that is released into the environment and eventually lost into space. Second, when electrical energy flows through filament wires in an incandescent light bulb, it is changed into about 5% useful light and 95% low-quality heat that flows into the environment. What we call a light bulb is really a heat bulb. A third example is illustrated in Figure 3-11.

The second law of energy also means that *we can never recycle or reuse high-quality energy to perform useful work.* Once the concentrated energy in a serving of food, a liter of gasoline, a lump of coal, or a chunk of uranium is released, it is degraded to low-quality heat that becomes dispersed in the environment. We can heat air or water at a low temperature and upgrade it to high-quality energy, but the second law of energy tells us that it will take more high-quality energy to do this than we get in return.

Connections: Life and the Second Energy Law
To form and preserve the highly ordered arrangement of molecules and the organized network of chemical

---

Sample page (page 469):

Should We Develop Oil and Gas in the Arctic National Wildlife Refuge?

PRO/CON

The Arctic National Wildlife Refuge on Alaska's North Slope (Figure 17-22), which contains more than one-fifth of all the land in the U.S. wildlife refuge system, has been called the crown jewel of the system. During all or part of the year it is home for more than 160 animal species, including caribou, musk ox, snowy owls, threatened grizzly bears (Figure 17-7), arctic foxes (Figure 5-39), and migratory birds (including as many as 300,000 snow geese, Figure 5-1). It is also home for about 7,000 Inuit (Eskimos), who depend on the caribou for a large part of their diet.

The refuge's coastal plain, its most biologically productive part, is the only stretch of Alaska's arctic coastline not open to oil and gas development. U.S. oil companies hope to change this because they believe that the area *might* contain oil and natural gas deposits. Since 1985 they have been urging Congress to open to drilling some 607,000 hectares (1.5 million acres) along the coastal plain—roughly two-thirds the size of Yellowstone National Park. They argue that such exploration is needed to provide the United States with more oil and natural gas and reduce dependence on oil imports.

Environmentalists oppose this proposal and want Congress to designate the entire coastal plain as wilderness. They cite Interior Department estimates that there is only a 19% chance of finding as much oil in the coastal plain as the United States consumes every three years. Even if the oil does exist, environmentalists do not believe the potential degradation of any portion of this irreplaceable wilderness area would be worth it, especially considering that improvements in energy efficiency would save far more oil at a much lower cost (Section 18-2).

Oil company officials claim they have developed Alaska's Prudhoe Bay oil fields without significant harm to wildlife; they also contend that the area they want to open to oil and gas development is less than 1.5% of the entire coastal plain region—equivalent to an oil field the size of Dulles International Airport in Washington, D.C., within an area approximately the size of South Carolina.

However, the huge 1989 oil spill from the tanker *Exxon Valdez* in Alaska's Prince William Sound cast serious doubt on such claims (Case Study, p. 292). Moreover, a study leaked from the Fish and Wildlife Service in 1988 revealed that oil drilling at Prudhoe Bay has caused much more air and water pollution than was anticipated before drilling began in 1972. According to this study, oil development in the coastal plain could cause the loss of 20–40% of the area's 180,000-head caribou herd, 25–50% of the remaining musk oxen, 50% or more of the wolverines, and 50% of the snow geese that live there part of the year. A 1988 EPA study also found that "violations of state and federal environmental regulations and laws are occurring at an unacceptable rate" in the Prudhoe Bay area where oil fields and facilities have been developed.

Do you think this refuge should be explored and developed for oil and natural gas?

Figure 17-22 Proposed oil-drilling area in Alaska's Arctic National Wildlife Refuge. (Data from U.S. Fish and Wildlife Service)

ARCTIC OCEAN · Proposed oil-drilling area Arctic National Wildlife Refuge · BEAUFORT SEA · Prudhoe Bay · Arctic Circle · Alaska · Anchorage · CANADA · Valdez · GULF OF ALASKA

tivity with some returned to the wild include the California condor (Figure 17-2), the peregrine falcon (Figure 15-5), and the black-footed ferret (Figure 17-7). Endangered golden lion tamarins (Figure 17-7) bred at the National Zoo in Washington, D.C., have been released successfully in Brazilian rain forests.

Unfortunately, keeping populations of endangered animal species in zoos and research centers is limited by lack of space and money. The captive population of each species must number 100–500 to avoid extinction through accident, disease, or loss of genetic variability through inbreeding. Moreover, caring for

Why We Need Sharks

The world's 350 species of sharks range in size from the dwarf dog shark—about the size of a large goldfish—to the whale shark—the world's largest fish at 18 meters (60 feet) long. Various shark species are the key predators in the world's oceans, helping control the numbers of many other ocean predators. By feeding at the top of food webs, these shark species cull injured and sick animals from the ocean, thus keeping these species stronger and healthier.

Influenced by movies and popular novels, most people think of sharks as people-eating monsters. This is far from the truth. Every year a few species of shark—mostly great white, bull, bronze whaler, tiger, gray reef, blue (Figure 4-42), and oceanic whitetip—injure about 100 people worldwide and kill between 5 and 10. If you are a typical ocean-goer, you are 150 times more likely to be killed by lightning and thousands of times more likely to be killed when you drive a car than to be killed by a shark.

For every shark that injures a person, we kill 1 million sharks—a total of more than 100 million sharks each year. Sharks are killed mostly for their fins—widely used in Asia as a soup ingredient—at around $50 a bowl in some restaurants—and as a pharmaceutical cure-all. They are also killed for their livers, meat (especially mako and thresher), and jaws (especially great whites), or because we fear them. Some sharks (especially blue,

Figure 4-42 This blue shark and other types of sharks are key predators in the world's oceans. One of only a small number of shark species that occasionally attack swimmers, blue sharks prefer deep water and are a potential threat only to people swimming from boats in deep water.

mako, and oceanic whitetip) die when they are trapped in nets meant for swordfish, tuna, shrimp, and other commercially important species.

Why should we care how many sharks are killed? Because they perform valuable services for us and other species. Sharks save human lives by helping us learn how to fight cancer, bacteria, and viruses, because sharks seem to be free of almost all diseases (including cancer and eye cataracts) and are not affected by most toxic chemicals. Understanding why can help us improve human health.

Chemicals extracted from shark cartilage have killed cancerous tumors in laboratory animals

and may someday be the basis of cancer-treating drugs. Another chemical extracted from shark cartilage is being used as an artificial skin for burn victims. Sharks' highly effective immune system allows wounds to heal quickly without becoming infected, and it is being studied for protection against AIDS. And shark corneas have been transplanted into human eyes.

With more than 400 million years of evolution behind them, sharks have had a long time to get things right. We could undo most of this evolutionary wisdom in a few decades. Preventing this from happening begins with the recognition that sharks don't need us, but we and other species need them.

The essential features of the living and nonliving parts of individual terrestrial and aquatic ecosystems, and of the ecosphere, are interdependence and connectedness. Without the services performed by diverse communities of species, we would be starving, gasping for breath, and drowning in our own wastes. The next chapter shows how this interdependence is the key to understanding Earth's major life zones and ecosystems.

We sang the songs that carried in their melodies all the sounds of nature—the running waters, the sighing of winds, and the calls of the animals. Teach them to your children that they may come to love nature as we love it.

GRAND COUNCIL FIRE OF AMERICAN INDIANS

Q: How many of the world's people live in urban areas?

Rescuing a River

When Marion Stoddart (Figure 11-23) first moved to Groton, Massachusetts, in the early 1960s the nearby Nashua River was considered one of the nation's filthiest rivers. Dead fish bobbed on its waves, and at times the water was red, green, or blue from pigments discharged by paper mills.

Marion Stoddart was appalled. Instead of thinking nothing could be done, she committed herself to restoring the Nashua and establishing public parklands along its banks. She didn't start by filing lawsuits or organizing demonstrations. Rather, she created a careful cleanup plan and approached state officials with it in 1962. They laughed, but she was not deterred and began practicing

the most time-honored skill of politics—one-on-one persuasion. She identified the power brokers in the riverside communities and began to educate them, win them over, and get them to cooperate in cleaning up the river.

She got the state to ban open dumping in the river. When-promised federal matching funds for building the treatment plant failed to arrive, Stoddart got 13,000 signatures on a petition to President Nixon. The funds arrived in a hurry.

Stoddart's next success was getting a federal grant to beautify the river. She hired high school dropouts to clear away mounds of debris. When the river cleanup was completed, she persuaded communities along the river to create some 2,400 (6,000 acres) hectares of river-

side park and woodlands along both banks.

Now, over two decades later, the Nashua is still clean, and a citizens' group founded by Stoddart keeps watch on water quality. The river's waters support many kinds of fish and other wildlife, and they are used for canoeing (Figure 11-23) and other kinds of recreation. The project is considered a model for other states and is testimony to what a committed individual can do to change the world by getting people to work together.

For her efforts Stoddart has been named by the UN Environment Programme as an outstanding worldwide worker for the environment. She might say, however, that the naturally blue and canoeable Nashua is itself her best reward.

Figure 11-23 Earth citizen Marion Stoddart canoeing down the Nashua River near Groton, Massachusetts. She spent over two decades spearheading successful efforts to have this river cleaned up.

CASE STUDIES

I've included 16 Case Studies (6 new to this edition) to give you in-depth information about key issues while applying environmental concepts. This one from Chapter 4 explains the ecological importance of sharks to the earth and to us.

INDIVIDUALS MATTER BOXES

15 Individuals Matter boxes (3 new to this edition) demonstrate what individuals have already done or offer you tips on how you can work *with* the earth. A new workbook, *Green Lives, Green Campuses,* is being offered as a supplement to this edition. It contains projects to help you evaluate the environmental impact of your lifestyle and make an environmental audit of your campus.

To provide you with various perspectives on the environment, I've included 15 Guest Essays (8 new to this edition). The essays are written by both scholars and individuals actively engaged on the front lines of environmental change. Shown here is the first page of one of the new Guest Essays in this edition (see pages 367 and 368 of the text for the complete essay).

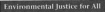

Environmental Justice for All

GUEST ESSAY

Robert D. Bullard

Robert D. Bullard is a professor of sociology at the University of California, Riverside. For more than a decade he has worked on and conducted research in the areas of urban land use, housing, community development, industrial facility siting, and environmental justice. His scholarship and activism have made him one of the leading experts on environmental racism—the systematic selection of communities of color for waste facilities and polluting industries. He is the author of four books and more than three dozen articles, monographs, and scholarly papers that address equity concerns. His book, Dumping in Dixie: Race, Class, and Environmental Quality *(Westview Press, 1990), has become a standard text in the field. His most recent book is* Confronting Environmental Racism *(South End Press, 1993).*

Despite widespread media coverage and volumes written on the U.S. environmental movement, environmentalism and social justice have seldom been linked. Nevertheless, an environmental revolution is now taking shape in the United States that combines the environmental and social justice movements into one framework.

People of color (African Americans, Latinos, Asians, Pacific Islanders, and Native Americans), working-class people, and poor people in the United States suffer disproportionately from industrial toxins, dirty air and drinking water, unsafe work conditions, and the location of noxious facilities such as municipal landfills, incinerators, and toxic waste dumps. Despite the government's attempts to level the playing field, all communities are not created equal.

The environmental justice movement attempts to dismantle exclusionary zoning ordinances, discriminatory land-use practices, differential enforcement of environmental regulations, disparate siting of risky technologies, and the dumping of toxic waste on the poor and people of color in the United States and in LDCs.

All communities are not treated as equals when it comes to resolving environmental and public health concerns, either. Over 300,000 farm workers (over 90% of whom are people of color) and their children are poisoned by pesticides sprayed on crops in the United States. Some 3–4 million children (many of them African Americans or Latinos living in the inner city) are poisoned by lead-based paint in old buildings, lead-soldered pipes and water mains, lead-tainted soil contaminated by industry, and air pollutants from smelters. Lead poisoning is considered the number one environmental health problem facing children in the United States. Yet little has been done over the past 20 years to rid the nation of this preventable childhood hazard.

All communities do not bear the same burden or reap the same benefits from industrial expansion. This is true in the case of the mostly African American Emelle, Alabama (home of the nation's largest hazardous-waste landfill); Navajo lands in Arizona where uranium is mined; and the 2,000 factories known as *maquiladores,* located just across the U.S. border in Mexico.

Communities, states, and regions that contain hazardous-waste disposal facilities (importers) receive far fewer economic benefits (jobs) than the geographic locations that generate the wastes (exporters). Nationally 60% of African Americans and 50% of Latinos live in communities with at least one uncontrolled toxic-waste site. Three of the five largest hazardous-waste landfills are located in communities that are predominantly African American or Latino.

The marginal status of many people of color in the United States makes them prime actors in the movement for environmental and social justice. For example, the organizing theme of the 1991 First National People of Color Environmental Summit, held in Washington, D.C., was justice, fairness, and equity. More than 650 delegates from all 50 states, as well as Puerto Rico, Mexico, Chile, Colombia, and the Marshall Islands, participated in this historic four-day gathering.

Environmental justice does not stop at the U.S. borders. Environmental injustices exist from the *favelas* of Rio de Janeiro [Figure 7-9] to the shantytowns of Johannesburg. Members of the environmental justice movement are also questioning the wasteful and nonsustainable development models being exported to the developing world.

It is no mystery why grass-roots environmental justice groups in Louisiana's "Cancer Alley," [Figure 13-16], Chicago's southside, and Los Angeles's East and South Central neighborhoods are attacking the institutions they blame for their underdevelopment, disenfranchisement, and poisoning. Some people see these threats to their communities as a form of genocide.

Grass-roots leaders are demanding justice. Residents of communities such as West Dallas and Texarkana (Texas), West Harlem (New York), Rosebud (South Dakota), Kettleman City (California), and Sunrise, Lions, and Wallace (Louisiana) see their struggle for environmental justice as a life-and-death matter. Unfortunately their stories of environmental racism are not piped into the nation's living rooms during the nightly news, nor are they blasted across the front pages of national newspapers and magazines. To a large extent the communities that are the victims of environmental injustice remain "invisible" to the larger society.

The environmental justice movement is led, planned, and to a large extent funded by individuals who are not part of the established environmental community or the "Big 10" environmental organizations. Most environ-

(continued)

A: Natural gas — CHAPTER 13 **367**

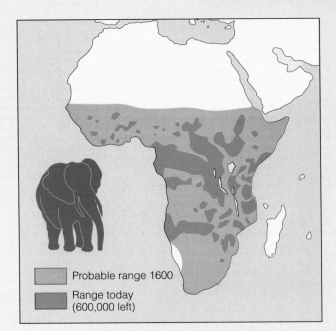

African Elephant

Probable range 1600
Range today (600,000 left)

This is one of several maps showing how the range of various endangered species has been drastically reduced, mostly from habitat loss and over hunting.

I've used location maps throughout the book to enhance your knowledge of geography.

Shown below are three of 179 photographs found in the text. Although they are visually appealing, I selected them first and foremost because of their ability to illustrate important concepts and enhance learning. I strongly believe that photographs should be used not as decorative elements, but as useful tools to augment the text.

This clownfish living amongst deadly sea anemones in the Coral Sea off Australia demonstrates the concept of *commensalism*, where two species interact and "live together" with one species benefiting while the other is neither helped nor harmed to any great degree.

Controversial plans call for reintroducing the gray wolf—an endangered species in 48 states—to its former habitat in the Yellowstone National Park area.

The world's largest solar power facility, in California's Mojave Desert. These mirrored, curved collectors focus solar energy on an oily substance flowing through a pipe running along the center of the collectors. The hot oil is sent to a nearby facility where it produces steam, which spins a turbine and produces electricity.

Completely New Art Program

For this new fifth edition of **Environmental Science** each piece of existing art was redesigned to provide more realistic details, more vibrant color, and when possible, more three-dimensional effects. The result is 281 full-color diagrams (101 of them new) and 179 carefully selected photographs (68 of them new) that simplify and demonstrate key concepts and ideas. This is an example of the realistic and useful new art rendered for this edition.

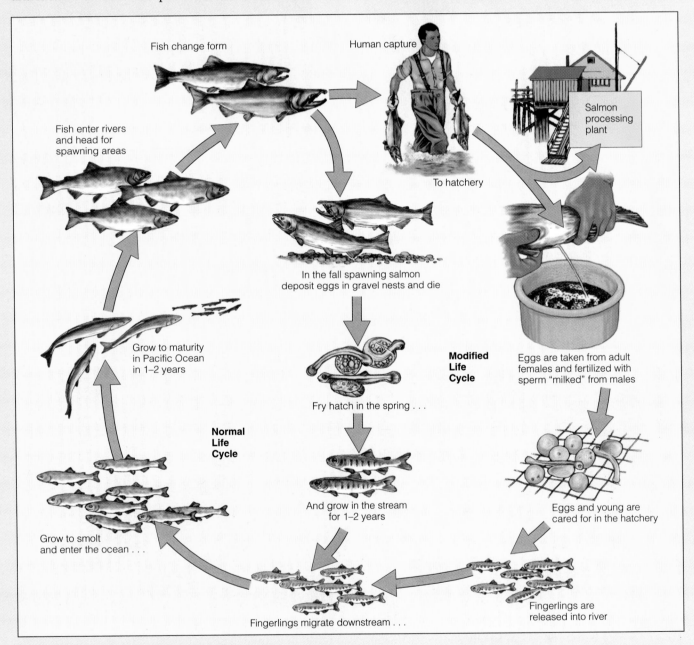

Normal (left) and modified (right) life cycle of an anadromous salmon species.

COMPARE THESE EXAMPLES FROM THE FOURTH AND FIFTH EDITIONS

From the Fifth Edition

An illustration of secondary ecological succession of plant communities on an abandoned farm field in North Carolina over about 150 years.

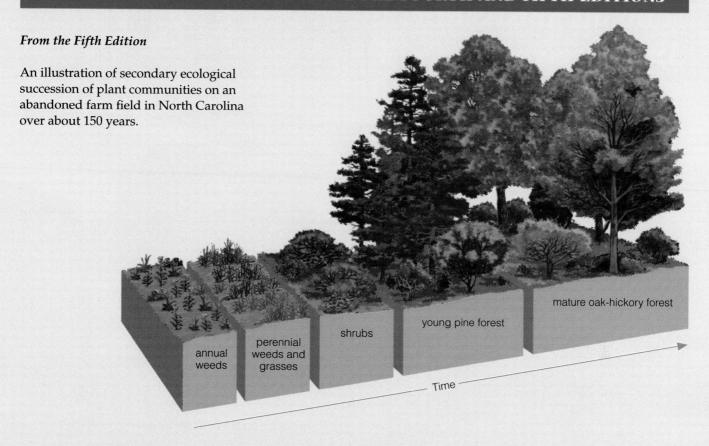

From the Fourth Edition

The environmental crisis is an outward manifestation of a crisis of mind and spirit. There could be no greater misconception of its meaning than to believe it is concerned only with endangered wildlife, human-made ugliness, and pollution. These are part of it, but more importantly, the crisis is concerned with the kind of creatures we are and what we must become in order to survive.

LYNTON K. CALDWELL

1 Environmental Problems and Their Causes

Living in an Exponential Age

Once there were two kings who enjoyed playing chess, with the winner claiming a prize from the loser. After their match was over the winning king asked the loser to place one grain of wheat on the first square of the chessboard, two on the second, four on the third, and so on. The number of grains was to double each time until all 64 squares were filled.

The losing king, thinking he was getting off easy, agreed with delight. It was the biggest mistake he ever made. He bankrupted his kingdom and still could not produce the 2^{64} grains of wheat he had promised. In fact, it's probably more than all the wheat that has ever been harvested! This is an example of **exponential growth**, in which a quantity increases by a fixed percentage of the whole in a given time. As the loser learned, exponential growth is deceptive. It starts off slowly, but after only a few doublings it rises to enormous numbers, because each doubling is more than the total of all earlier growth.

Here is another example. Fold a piece of paper in half to double its thickness. If you could manage to do this 42 times, the stack would tower from Earth to the moon, 386,400 kilometers (240,000 miles) away. If you could double it 50 times, the folded paper would almost reach the sun, 149 million kilometers (93 million miles) away!

The danger and challenge that we face is that *the major environmental problems—population growth* (Figure 1-1), *excessive and wasteful resource use, wildlife extinction, and pollution—are interconnected and are growing exponentially.*

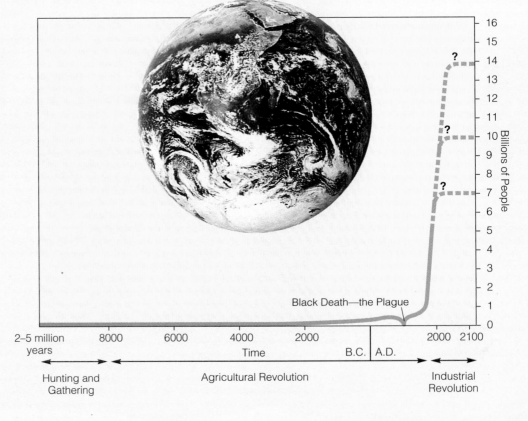

Figure 1-1 The J-shaped curve of past exponential world population growth with projections to 2100. Notice that exponential growth starts off slowly, but as time passes the curve becomes increasingly steep. World population has more than doubled in only 43 years, from 2.5 billion in 1950 to 5.5 billion in 1993. Unless death rates rise sharply, it may reach 11 billion by 2045 and 14 billion by 2100. (Data from World Bank and United Nations)

Black Death—the Plague

Billions of People

2–5 million years

8000 6000 4000 2000

Time

B.C. | A.D.

2000 2100

Hunting and Gathering

Agricultural Revolution

Industrial Revolution

We need to come together and choose a new direction. We need to transform our society into one in which people live in true harmony—harmony among nations, harmony among the races of humankind, and harmony with nature.... We will either reduce, reuse, recycle, and restore—or we will perish.

REV. JESSE JACKSON

This chapter is an overview of environmental problems and their root causes. It will discuss these questions:

- What is Earth capital? How are we depleting it?

- How fast is the human population increasing?

- What are Earth's main types of resources? How can they be depleted or degraded?

- What are the principal types of pollution? How can pollution be reduced?

- What are the root causes of the environmental problems we face?

1-1 LIVING SUSTAINABLY

Solar Capital and Earth Capital Our existence, lifestyles, and economies depend totally on the sun and the earth. We can think of energy from the sun as **solar capital**, and we can think of the planet's air, water, fertile soil, forests, grasslands, wildlife, minerals, and natural purification and recycling processes as **Earth capital**.

The basic problem we face is that we are depleting and degrading Earth's natural capital at an accelerating rate (Figure 1-2). Each year more forests, grasslands, and wetlands disappear, and some deserts grow larger. Vital topsoil, washed or blown away from farmland and cleared forests (Figure 1-3), clogs streams, lakes, and reservoirs with sediment. Underground water is pumped from wells faster than it can be replenished. Oceans are poisoned with our wastes. According to some estimates, every hour, we drive as many as four wildlife species to extinction.

Burning fossil fuels and cutting down and burning forests raise the concentrations of carbon dioxide and other heat-trapping gases in the lower atmosphere. Within the next 40 to 50 years, Earth's climate may become warm enough to disrupt agricultural productivity, alter water distribution, and drive countless species to extinction.

Extracting and burning fossil fuels pollute the air and water and disrupt land. Other chemicals we add to the air drift into the upper atmosphere and deplete ozone gas, which now filters out much of the sun's harmful ultraviolet radiation. Toxic wastes from facto-

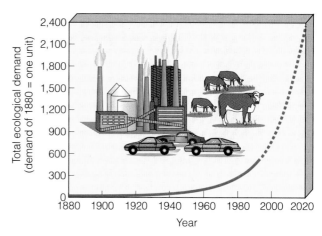

Figure 1-2 J-shaped curve of exponential growth in the total ecological demand on Earth's resources from agriculture, mining, and industry between 1880 and 1993. Projections to 2020 assume that resource use will continue to grow at the current rate of 5.5% per year. At that rate our total ecological demand on Earth's resources doubles every 13 years. If global economic output grew by only 3% a year, resource consumption would still double every 23 years. (Data from United Nations, World Resources Institute, and Carrol Wilson, *Man's Impact on the Global Environment*, Cambridge, Mass.: MIT Press, 1970)

Figure 1-3 Severe soil erosion on a hillside in Spain. Removal of forest cover often leads to this type of erosion. Wildlife habitat is lost, and streams can be polluted with eroded sediment.

ries and homes poison the air, water, and soil. Agricultural pesticides contaminate some of our drinking water and food.

Sooner or later there is a limit to exponentially growing resource use and the capacity of Earth's life-support systems to absorb, dilute, and degrade the

The following urgent warning to humanity was issued on November 18, 1992, and sent to government leaders of all nations. It was signed by over 1,600 scientists from 70 countries, including 102 out of the 196 living scientists who are Nobel laureates.

Introduction

The environment is suffering critical stress:

The Atmosphere Stratospheric ozone depletion threatens us with enhanced ultraviolet radiation at the earth's surface, which can be damaging or lethal to many forms of life. Air pollution near ground level, and acid deposition, are already causing widespread injury to humans, forests, and crops.

Water Resources Heedless exploitation of depletable groundwater supplies endangers food production and other essential human systems. Heavy demands on the world's surface waters have resulted in serious shortages in some 80 countries, containing 40% of the world's population. Pollution of rivers, lakes, and groundwater further limits the supply.

Oceans Destructive pressure on the oceans is severe, particularly on coastal regions which produce most of the world's food fish. The total marine catch is now at or above the estimated maximum sustainable

*Used by permission from Union of Concerned Scientists, Cambridge, MA 02238.

yield. Some fisheries have already shown signs of collapse. Rivers carrying heavy burdens of eroded soil into the seas carry industrial, municipal, agricultural, and livestock waste—some of it toxic.

Soil Loss of soil productivity, which is causing extensive land abandonment, is a widespread byproduct of current practices in agriculture and animal husbandry. Since 1945, 11% of the earth's vegetated surface has been degraded—an area larger than India and China combined—and per capita food production in many parts of the world is decreasing.

Forests Tropical rain forests, as well as tropical and temperate dry forests, are being destroyed rapidly. At present rates, some critical forest types will be gone in a few years, and most of the tropical rain forests will be gone before the end of the next century. With them will go large numbers of plant and animal species.

Living Species The irreversible loss of species, which by 2100 may reach one-third of all species now living, is especially serious. We are losing the potential they hold for providing medicinal and other benefits, and the contribution that genetic diversity of life forms gives to the robustness of the world's biological systems and to the astonishing beauty of the earth itself.

Much of this damage is irreversible on a scale of centuries, or it is permanent. Other processes appear to pose additional threats. Increasing levels of gases in the atmosphere from human activities, including carbon dioxide released

from fossil fuel burning and from deforestation, may alter climate on a global scale. Predictions of global warming are still uncertain—with projected effects ranging from tolerable to very severe—but the potential risks are very great.

Our massive tampering with the world's interdependent web of life—coupled with the environmental damage inflicted by deforestation, species loss, and climate change—could trigger widespread adverse effects, including unpredictable collapses of critical biological systems whose interactions and dynamics we only imperfectly understand. Uncertainty over the extent of these effects cannot excuse complacency or delay in facing the threats.

Population The earth is finite. Its ability to absorb wastes and destructive effluent is finite. Its ability to provide food and energy is finite. Its ability to provide for growing numbers of people is finite. And we are fast approaching many of the earth's limits. Current economic practices which damage the environment, in both developed and undeveloped nations, cannot be continued without the risk that vital global systems will be damaged beyond repair.

Pressures resulting from unrestrained population growth put demands on the natural world that can overwhelm any efforts to achieve a sustainable future. If we are to halt destruction of our environment, we must accept limits to that growth. A World Bank estimate indicates that world population will stabilize at 12.4 billion, while the United Nations concluded that the eventual total could reach 14 billion, a near tripling of today's 5.4 billion. But, even at this

resulting waste and pollution. Whether we have exceeded such limits and how close we are to them are hotly debated issues. Environmentalists and a number of the world's leading scientists, however, warn us that such limits exist—and that we are depleting and degrading Earth capital for us and other species (Spotlight, above).

Working with the Earth A sustainable society is one that manages its economy and population size

Q: How many people are added to the world's population each day?

moment, one person in five lives in absolute poverty without enough to eat, and one in ten suffers serious malnutrition.

No more than one or a few decades remain before the chance to avert the threats we now confront will be lost and the prospects for humanity immeasurably diminished.

Warning We the undersigned, senior members of the world's scientific community, hereby warn all humanity of what lies ahead. A great change in the stewardship of the earth and the life on it, is required, if vast human misery is to be avoided and our global home on this planet is not to be irretrievably mutilated.

What We Must Do Five inextricably linked areas must be addressed simultaneously:

1. **We must bring environmentally damaging activities under control to restore and protect the integrity of the earth's systems we depend on.** We must, for example, move away from fossil fuels to more benign, inexhaustible energy sources to cut greenhouse emissions and the pollution of our air and water. Priority must be given to the development of energy sources matched to third world needs—small scale and relatively easy to implement. We must halt deforestation, injury and loss of agricultural land, and the loss of terrestrial and marine plant and animal species.

2. **We must manage resources crucial to human welfare more effectively.** We must give high priority to efficient use of energy, water, and other materials, including expansion of conservation and recycling.

3. **We must stabilize population. This will be possible only if all nations recognize that it requires improved social and economic conditions, and the adoption of effective, voluntary family planning.**

4. **We must reduce and eventually eliminate poverty.**

5. **We must ensure sexual equality, and guarantee women control over their own reproductive decisions.**

The developed nations are the largest polluters in the world today. They must greatly reduce their overconsumption, if we are to reduce pressures on resources and the global environment. The developed nations have the obligation to provide aid and support to developing nations, because only the developed nations have the financial resources and technical skills for these tasks.

Acting on this recognition is not altruism, but enlightened self-interest: Whether industrialized or not, we all have but one lifeboat. No nation can escape injury when global biological systems are damaged. No nations can escape from conflicts over increasingly scarce resources. In addition, environmental and economic instabilities will cause mass migrations with incalculable consequences for developed and undeveloped nations alike.

Developing nations must realize that environmental damage is one of the gravest threats they face, and that attempts to blunt it will be overwhelmed if their populations go unchecked. The greatest peril is to become trapped in spirals of environmental decline, poverty, and unrest, leading to social, economic, and environmental collapse.

Success in this global endeavor will require a great reduction in violence and war. Resources now devoted to the preparation and conduct of war—amounting to over $1 trillion annually—will be badly needed in the new tasks and should be diverted to the new challenge.

A new ethic is required—a new responsibility for caring for ourselves and for the earth. We must recognize the earth's limited capacity to provide for us. We must recognize its fragility. We must no longer allow it to be ravaged. This ethic must motivate a great movement, convincing reluctant leaders and reluctant governments and reluctant peoples themselves to effect the needed changes.

The scientists issuing this warning hope that our message will reach and affect people everywhere. We need the help of many.

We require the help of the world community of scientists—natural, social, economic, political;

We require the help of the world's business and industrial leaders;

We require the help of the world's religious leaders; and

We require the help of the world's people.

We call on all to join us in this task.

without doing irreparable environmental harm. It satisfies the needs of its people without depleting Earth capital and thus jeopardizing the prospects of future generations of humans or other species. This is done by **(1)** regulating population growth; **(2)** taking no more potentially renewable resources from the natural world than can be replenished naturally; **(3)** not overloading the capacity of the environment to cleanse and renew itself by natural processes; **(4)** encouraging Earth-sustaining rather than Earth-degrading forms of

A: 247,000

Don't Kill the Goose

SOLUTIONS

Imagine you inherit $1 million. If you invest this capital at 10% interest, you will have a sustainable annual income of $100,000. That is, you can spend up to $100,000 a year without touching your capital.

But suppose you develop a taste for diamonds or a yacht—or all your relatives move in with you. If you spend $200,000 a year, your million will be gone early in the seventh year. Even if you spend just $110,000 a year, you will be bankrupt early in the eighteenth year.

The lesson here is a very old one: Don't kill the goose that lays the golden eggs. Deplete your capital, and you move from a sustainable to an unsustainable lifestyle. Get too greedy and you'll soon be needy.

The same lesson applies to Earth capital. With the help of solar energy, natural processes developed over billions of years can renew the topsoil, water, air, forests, grasslands, and wildlife that we and other species depend on—as long as we don't use these resources faster than they are renewed. Some of our wastes can also be diluted, decomposed, and recycled by natural processes, as long as these processes are not overloaded.

Likewise, most of Earth's nonrenewable minerals, such as aluminum and glass, can be recycled or reused if we don't contaminate them or mix them in landfills so that recycling and reuse are too costly. History also shows that we can often find substitutes when certain nonrenewable minerals become scarce.

Conserving energy and relying on virtually inexhaustible solar energy in the form of heat, wind, flowing water, and renewable wood constitute a sustainable lifestyle. The brief "fossil-fuel age" we now live in is unsustainable. Fixed supplies of coal, oil, and natural gas produced underground over millions of years from decayed matter are being depleted within a few hundred years—hundreds of thousands of times faster than they were formed. By wasting less energy we could make these fuels last longer, reduce their massive environmental impact, and make an easier transition to a new, renewable-energy age.

economic development; and **(5)** reducing poverty (which can cause people to use land and other resources unsustainably for short-term survival).

The good news is that we can help sustain Earth for human beings and other species indefinitely by learning how to live off Earth income instead of Earth

capital (Solutions, at left). It's not too late to replace our Earth-degrading actions with Earth-sustaining ones, as discussed throughout this book. The key is *Earth wisdom*—learning how Earth sustains itself—and integrating such lessons from nature into the ways we think and act.

1-2 GROWTH AND THE WEALTH GAP

Linear and Exponential Growth　Suppose you hop on a commuter train that accelerates by 1 kilometer (0.6 mile) an hour every second. After 5 seconds you would be traveling at 5 kilometers (3 miles) per hour; after 30 seconds your speed would be 30 kilometers (19 miles) per hour. This is an example of **linear growth**, in which a quantity increases by a constant amount per unit of time, as 1, 2, 3, 4, 5—or 1, 3, 5, 7, 9—and so on.

But suppose the train had a magic motor strong enough to increase its speed by 100% every second. After 5 seconds you would be moving at 32 kilometers per hour, and 1 second later you would be doing 64. After only 30 seconds you would be traveling a billion kilometers (620 million miles) per hour! This is an example, like the wheat on the chessboard (p. 4), of the astounding power of **exponential growth**, in which a quantity increases by a fixed percentage of the whole in a given time.

In dealing with a quantity such as resource use, population size, or money in a savings account that is growing exponentially, we may want to know how soon it will double. A quick way to calculate this **doubling time** in years is to use the **rule of 70**: 70/percentage growth rate = doubling time in years. For example, in 1993 the world's population grew by 1.63%. If that rate continues, Earth's population will double in 42.5 years (70/1.63 = 42.5 years)—more growth than has occurred in all of human history.

The J-Shaped Curve of Human Population Growth　The increasing size of the human population provides an example of exponential growth. As the population base grows, the number of people on Earth soars. The population growth curve rounds a bend and heads almost straight up, creating a J-shaped curve (Figure 1-1).

It took 2 million years to reach a billion people; 130 years to add the second billion; 30 years for the third; 15 years for the fourth; and only 12 years for the fifth billion. At present growth rates the sixth billion will be added during the 11-year period between 1987 and 1998, and the seventh only 10 years later, in 2008. The relentless ticking of this population clock means that in 1993 the world's population of 5.5 billion peo-

Q: Where does everything that supports your life come from?

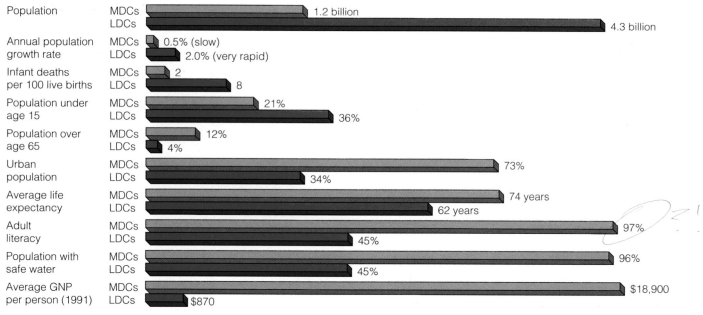

Population	MDCs	1.2 billion
	LDCs	4.3 billion
Annual population growth rate	MDCs	0.5% (slow)
	LDCs	2.0% (very rapid)
Infant deaths per 100 live births	MDCs	2
	LDCs	8
Population under age 15	MDCs	21%
	LDCs	36%
Population over age 65	MDCs	12%
	LDCs	4%
Urban population	MDCs	73%
	LDCs	34%
Average life expectancy	MDCs	74 years
	LDCs	62 years
Adult literacy	MDCs	97%
	LDCs	45%
Population with safe water	MDCs	96%
	LDCs	45%
Average GNP per person (1991)	MDCs	$18,900
	LDCs	$870

Figure 1-4 Some characteristics of more developed countries (MDCs) and less developed countries (LDCs) in 1993. (Data from United Nations and Population Reference Bureau)

ple grew by about 90 million—an average increase of 247,000 people a day, or 10,300 an hour. At this 1.63% annual rate of exponential growth, it takes about

- 5 days to add the number of Americans killed in all U.S. wars

- 10 months to add 75 million people—the number killed by the bubonic plague (the Black Death) in the fourteenth century

- 1.8 years to add 165 million people—the number killed in all wars fought during the past 200 years

- 2.9 years to add 260 million people—more than the population of the United States in 1993

This should give you some idea of what it means to go around the bend of the J curve of exponential growth.

Economic Growth and the Wealth Gap Virtually all countries seek **economic growth**: increasing their capacity to provide goods and services for final use. Such growth is accomplished by maximizing the flow of matter and energy resources through an economy (throughput)—by means of population growth (more consumers), more consumption per person, or both.

Economic growth is usually measured by an increase in a country's **gross national product (GNP)**: the market value in current dollars of all goods and services produced by an economy for final use during

a year. To show one person's slice of the economic pie, economists often calculate the **GNP per capita** (per person): the GNP divided by the total population.

The United Nations broadly classifies the world's countries as more developed or less developed. The **more developed countries (MDCs)**, or *developed countries*, are highly industrialized, and most have high average GNPs per capita. They include the United States, Canada, Japan, the former USSR, Australia, New Zealand, and the Western European countries. These countries, with 1.2 billion people (22% of the world's population), command about 85% of the world's wealth and income, use 88% of its natural resources and 73% of its energy, and generate most of its pollution and wastes.

All other nations are classified as **less developed countries (LDCs)**, or *developing countries*, with low to moderate industrialization and GNPs per capita. Most are in Africa, Asia, and Latin America. Their 4.3 billion people, or 79% of the world's population, have only about 15% of the wealth and income, and they use only about 12% of the natural resources and 27% of the energy. The "less developed" label, invented by MDC political scientists and economists, is in some ways a misnomer because LDCs are highly developed culturally and in other ways. The label also assumes that Western-style industrialization is the best model for all countries.

Figure 1-4 shows some general differences between MDCs and LDCs. Most of the projected increase in world population will take place in LDCs, where 1

A: The sun and the earth (Earth capital)

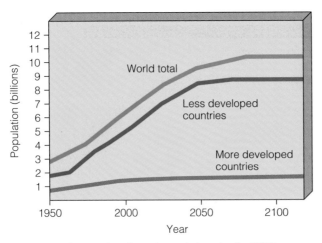

Figure 1-5 Past and projected population size for MDCs, LDCs, and the world, 1950–2120. (Data from United Nations)

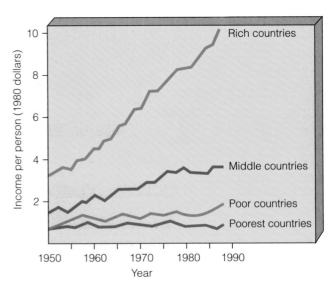

Figure 1-6 The wealth gap: changes in the distribution of global income per person in four types of countries, 1950–1990. Instead of trickling down, most of the income from economic growth has flowed up, with the situation worsening since 1980. (Data from United Nations)

million people are added every four days (Figure 1-5). By 2010 the combined populations of Asia and Africa are projected to be 5.3 billion—almost as many as now live on the entire planet.

The growing gap since 1960 between rich and poor in GNP per capita has widened further since 1980 (Figure 1-6). The rich have grown much richer, while the poor have stayed poor or grown even poorer. Today, one in five people lives in luxury; the next three "get by;" but the fifth struggles to survive on less than $1 a day (Spotlight, p. 11).

1-3 RESOURCES

Types of Resources A **resource** is anything we get from the living or nonliving environment to meet our needs and wants. We usually define resources in terms of humans, but resources are needed by all forms of life for their survival and good health. Some resources, such as solar energy, fresh air, fresh surface water, fertile soil, and wild edible plants, are directly available for use by us and other organisms. Most human resources, such as petroleum (oil), iron, groundwater (water occurring underground), and modern crops, aren't directly available, and their supplies are limited. They become useful to us only with some effort and technological ingenuity. Petroleum, for example, was a mysterious fluid until we learned how to find it, extract it, and refine it into gasoline, heating oil, and other products at affordable prices. On our short human time scale we classify material resources as renewable, potentially renewable, or nonrenewable (Figure 1-9).

Nonrenewable Resources **Nonrenewable**, or **exhaustible**, **resources** exist in fixed quantities in Earth's crust. On a time scale of millions to billions of years, such resources can be renewed by geological processes. However, on a much shorter human time scale of hundreds to thousands of years, these resources can be depleted much faster than they are formed.

These exhaustible resources include *energy resources* (coal, oil, natural gas, uranium), *metallic mineral resources* (iron, copper, aluminum), and *nonmetallic mineral resources* (salt, gypsum, clay, sand, phosphates, water, and soil) (Figure 1-9). We know how to find and extract more than 100 nonrenewable minerals from the earth's crust. We convert these raw materials into many everyday items we use, and then we discard, reuse, or recycle them.

We never completely run out of any nonrenewable mineral. But a mineral becomes *economically depleted* when finding, extracting, transporting, and processing the remaining deposits cost more than the results are worth. At that point we have five choices: recycle or reuse existing supplies, waste less, use less, try to find a substitute, or do without—and wait millions of years for more to be produced.

Some nonrenewable material resources, such as copper and aluminum, can be recycled or reused to extend supplies. **Recycling** involves collecting and reprocessing a resource into new products. For example,

Q: How many years does it take to add 1 billion people at current growth rates?

The Desperately Poor

Because of population growth and the wealth gap,

- One out of five persons is hungry or malnourished (Figure 1-7) and lacks clean drinking water, decent housing (Figure 1-8), and adequate health care.

- One out of three persons lacks enough fuel to keep warm and to cook food.

- One out of four adults cannot read or write.

- More than half of humanity lacks sanitary toilets.

- One of every five people on Earth is desperately poor—un-

able to grow or buy enough food to stay healthy or work.

Life for the desperately poor is a harsh daily struggle for survival. Parents—some with seven to nine children—are lucky to have an annual income of $300 (82¢ a day). Having many children makes good sense to most poor parents. Their children are a form of economic security, helping them grow food, tend livestock, work, or beg in the streets. The two or three who live to adulthood will help their parents survive in old age (forties or fifties).

Each year an estimated 40 million of the desperately poor die from malnutrition (lack of protein and other nutrients needed for

good health) or related diseases, and from contaminated drinking water. This death toll of 110,000 people *per day* is equivalent to 275 planes, each carrying 400 passengers, crashing every day! Half of those who die are children under 5, with most dying from diarrhea and measles because they are weakened by malnutrition. While some put this death toll at about 20 million a year, even this lower estimate is appalling. *Those dying are unique human beings, not mere numbers or things.*

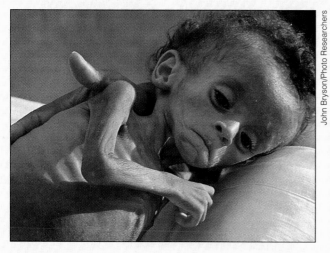

Figure 1-7 One out of every three children under age five, like this Brazilian child, suffers from malnutrition. Each day an estimated 40,000 children die—an average of 28 preventable deaths each minute.

Figure 1-8 One-fifth of the people in the world have inadequate housing, and at least 100 million have no housing. These homeless people in Calcutta, India, must sleep on the street.

aluminum cans can be collected, melted, and made into new beverage cans or other aluminum products. And glass bottles can be crushed and melted to make new bottles or other glass items. **Reuse** involves using a resource over and over in the same form. For example, glass bottles can be collected, washed, and refilled many times.

Other nonrenewable fuel resources—such as coal, oil, and natural gas—can't be recycled or reused. Once

burned, the useful energy in these fossil fuels is gone, leaving behind only waste heat and polluting exhaust gases. Most of the economic growth per person shown in Figure 1-6 has been fueled by nonrenewable oil, which is expected to be depleted within 40 to 80 years.

Most published supply estimates for a given nonrenewable resource refer to **reserves**: known deposits from which a usable mineral can be extracted profitably at current prices. Reserves can be increased

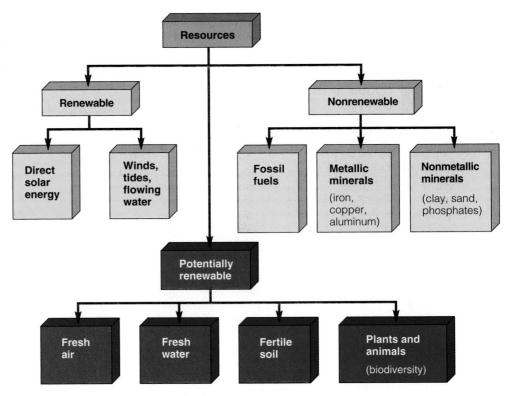

Figure 1-9 Major types of material resources. This scheme isn't fixed; potentially renewable resources can become nonrenewable resources if used for a prolonged time at a faster rate than they are renewed by natural processes.

when new deposits are found—or when price raises make it profitable to extract identified deposits that were previously too expensive to exploit.

Renewable Resources   Solar energy is called a **renewable resource** because on a human time scale it is essentially inexhaustible. It is expected to last at least 4 billion years while the sun completes its life cycle.

A **potentially renewable resource*** can be renewed fairly rapidly through natural processes. Examples of such resources include forest trees, grassland grasses, wild animals, fresh lake and stream water, groundwater, fresh air, and fertile soil. One important potentially renewable resource for us and other species is **biological diversity**, or **biodiversity**. It consists of the forms of life that can best survive the variety of conditions currently found on Earth, and it includes (**1**) *genetic diversity* (variability in the genetic makeup among individuals within a single species), (**2**) *species diversity* (the variety of species on Earth and in different habitats of the planet, Figure 1-10), and (**3**) *ecological diversity* (the variety of forests, deserts, grasslands, streams, lakes, oceans, and other biological

communities that interact with one another and with their nonliving environments).

But potentially renewable resources can be depleted. The highest rate at which a potentially renewable resource can be used without reducing its available supply is called its **sustainable yield**. If this natural replacement rate is exceeded, the available supply begins to shrink—a process known as **environmental degradation**.

Several types of environmental degradation can change potentially renewable resources into nonrenewable or unusable resources (Figure 1-11 and Connections, p. 16):

- *Covering productive land with water, concrete, asphalt, or buildings so that plant growth declines and wildlife habitats are lost.*

- *Cultivating land without proper soil management, causing soil erosion and depletion of plant nutrients.* Topsoil is now eroding faster than it forms on about 33% of the world's cropland—a loss of about 23 billion metric tons (25 billion tons) of topsoil a year (Figure 1-3).

- *Irrigating cropland without good drainage, causing salinization or waterlogging.* Salt buildup has cut yields on one-fourth of all irrigated cropland, and waterlogging has reduced productivity on at least one-tenth.

---

*Most sources use the term *renewable resource*. I have added the word *potentially* to emphasize that these resources can be depleted if we use them faster than natural processes renew them.

Q: How long does it take to add people equal to all those killed in all wars fought during the past 200 years?

Figure 1-10 Two species found in tropical forests are part of Earth's precious biodiversity. On the left is the world's largest flower, the flesh flower (*Rafflesia arnoldi*), growing in a tropical rain forest in Sumatra. The flower of this leafless plant can be as large as 1 meter (3.3 feet) in diameter and weigh 7 kilograms (15 pounds). The plant gives off a smell like rotting meat, presumably to attract flies and beetles that pollinate its flower. After blossoming for a few weeks the flower dissolves into a slimy black mass. On the right is a cotton top tamarin.

- *Taking fresh water from underground sources (aquifers) and from streams and lakes faster than it is replaced by natural processes.* In the United States one-fourth of the groundwater withdrawn each year is not replenished.

- *Destroying wetlands and coral reefs.* Between 25% and 50% of the world's wetlands have been drained, built upon, or seriously polluted. The United States has lost 55% of its wetlands and loses another 150,000 hectares (371,000 acres) each year. Coral reefs are being destroyed or damaged in 93 of the 109 significant locations.

- *Cutting trees from large areas (deforestation) without adequate replanting* (Figure 1-3). Almost half of the world's tropical forests have been cleared. If current deforestation rates continue, within 30–50 years there may be little of these forests left. In MDCs many of the remaining diverse, old-growth forests are being cleared and replaced with single-species tree farms, or with often much less diverse second-growth forests. This reduces wildlife habitats and forms of wildlife that make up the planet's biodiversity.

- *Overgrazing of grassland by livestock, which converts productive grasslands into unproductive land or deserts (desertification).* Each year almost 60,000 square kilometers (23,000 square miles) of new desert are formed.

- *Eliminating or decimating wild species (biodiversity) through destruction of habitats, commercial hunting, pest control, and pollution.* Each year thousands of wildlife species become extinct, mostly because of human activities. If habitat destruction continues at present rates, as many as 1.5 million species could disappear over the next 25 years—a drastic loss in vital Earth capital.

- *Polluting renewable air, water, and soil so that they are unusable.*

Figure 1-11 A sampling of how some of Earth's natural systems that support all life and all economies are being assaulted at an accelerating rate as a result of the exponential growth of population and resource use. (Data from The World Conservation Union, World Wildlife Fund, Conservation International, United Nations, Population Reference Bureau, U.S. Fish and Wildlife Service, and Daniel Boivin)

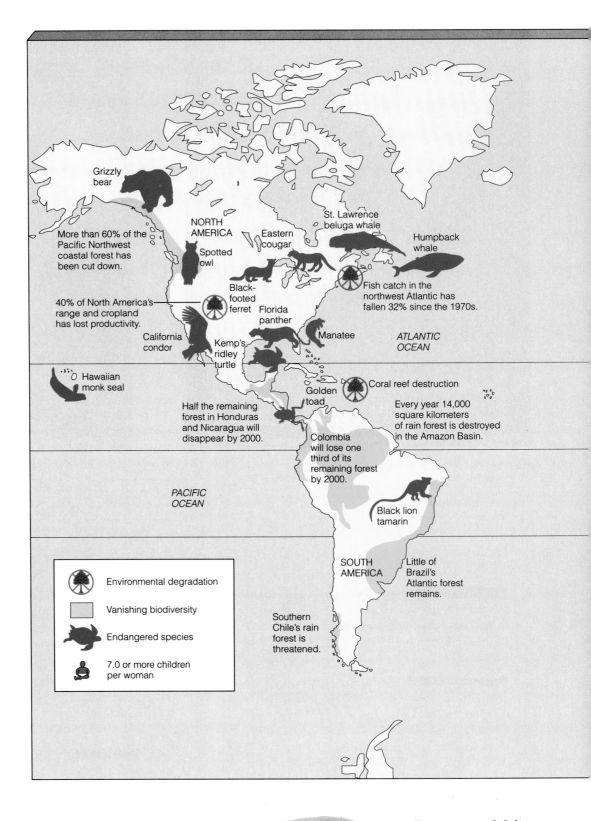

1-4 POLLUTION

What Is Pollution? Any addition to air, water, soil, or food that threatens the health, survival capability, or activities of humans or other living organisms is called **pollution**. Most pollutants are solid, liquid, or gaseous by-products or wastes produced when a resource is extracted, processed, made into products, or used. Pollution can also take the form of unwanted energy emissions, such as excessive heat, noise, or radiation.

Q: What percentage of the world's resources are used by people in more developed countries?

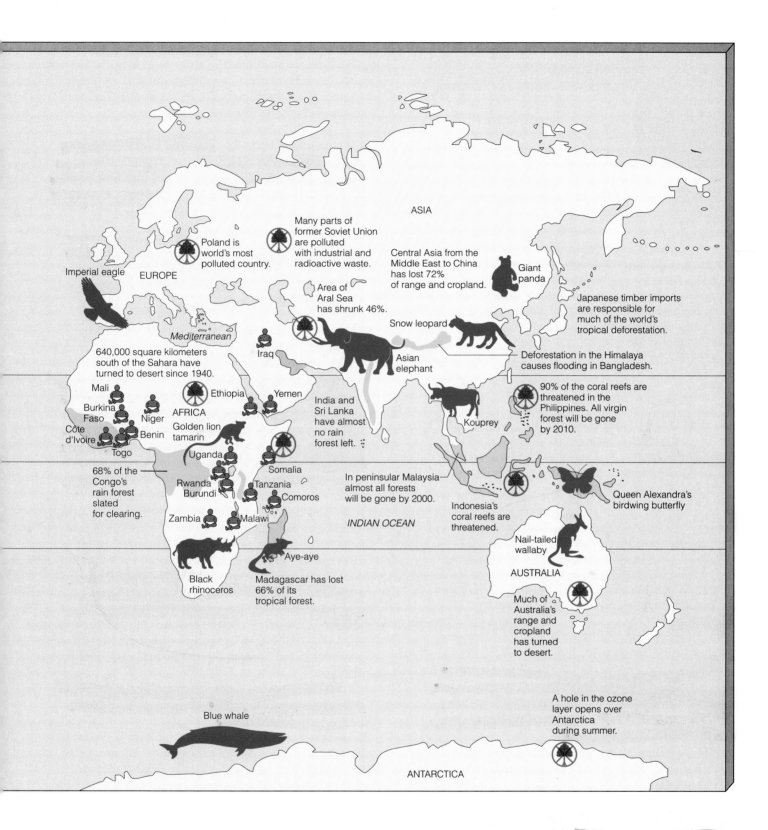

ASIA

Poland is world's most polluted country.

Many parts of former Soviet Union are polluted with industrial and radioactive waste.

Central Asia from the Middle East to China has lost 72% of range and cropland.

Giant panda

Imperial eagle

EUROPE

Mediterranean

Area of Aral Sea has shrunk 46%.

Japanese timber imports are responsible for much of the world's tropical deforestation.

Snow leopard

640,000 square kilometers south of the Sahara have turned to desert since 1940.

Iraq

Asian elephant

Deforestation in the Himalaya causes flooding in Bangladesh.

Mali

Ethiopia

Yemen

Burkina Faso

AFRICA

Niger

Golden lion tamarin

India and Sri Lanka have almost no rain forest left.

90% of the coral reefs are threatened in the Philippines. All virgin forest will be gone by 2010.

Côte d'Ivoire

Benin

Kouprey

Togo

Uganda

68% of the Congo's rain forest slated for clearing.

Somalia

Rwanda
Burundi

Tanzania

In peninsular Malaysia almost all forests will be gone by 2000.

Queen Alexandra's birdwing butterfly

Zambia

Malawi

Comoros

INDIAN OCEAN

Indonesia's coral reefs are threatened.

Nail-tailed wallaby

AUSTRALIA

Black rhinoceros

Aye-aye

Madagascar has lost 66% of its tropical forest.

Much of Australia's range and cropland has turned to desert.

A hole in the ozone layer opens over Antarctica during summer.

Blue whale

ANTARCTICA

A major problem is that people differ on the definition of a pollutant and on acceptable levels of pollution—especially if they have to choose between pollution control and their jobs. As the philosopher Hegel pointed out, tragedy is not the conflict between right and wrong, but the conflict between right and right.

Sources of Pollution Pollutants can enter the environment naturally (for example, from volcanic eruptions) or through human activities (for example, from burning coal). Most natural pollution is dispersed over a large area and either diluted or broken down to harmless levels by natural processes. By

The Tragedy of the Commons

CONNECTIONS

One cause of environmental degradation is the overuse of **common-property resources**, which are owned by none and available to all. Most are potentially renewable. Examples include clean air, fish in the open ocean, migratory birds, Antarctica, gases of the lower atmosphere, and the ozone content of the upper atmosphere.

In 1968 biologist Garrett Hardin (Guest Essay, p. 160) called this the **tragedy of the commons**. It happens because each user reasons, "If I don't use this resource, someone else will. The little bit I use or pollute is not enough to matter." With few users, this logic works. However, the cumulative effect of many people trying to exploit a common-property resource eventually exhausts or ruins it. Then no one can benefit from it. Therein lies the tragedy.

The obvious solution is to use common-property resources at rates below their sustainable yields or overload limits, by reducing population, regulating access, or both. Unfortunately, it is difficult to determine the sustainable yield of a forest, grassland, or an animal population, partly because yields vary with weather, climate, and unpredictable biological factors. These uncertainties mean that *it is best to use a potentially renewable resource at a rate well below its estimated sustainable yield.* This is a *prevention or precautionary approach,* designed to reduce the risk of environmental degradation. However, this approach is rarely taken because it conflicts with the drive for short-term profit, which has no or little regard for future consequences.

contrast, most serious pollution from human activities occurs in or near urban and industrial areas, where pollutants are concentrated in small volumes of air, water, and soil. Industrialized agriculture is also a major source of pollution.

Some pollutants contaminate the areas where they are produced. Others are carried by winds or flowing water to other areas. Pollution does not respect state or national boundaries.

Some pollutants come from single, identifiable sources, such as the smokestack of a power plant, the drainpipe of a meat-packing plant, the chimney of a house, or the exhaust pipe of an automobile. These are called **point sources**. Other pollutants enter the air, water, or soil from dispersed, and often hard-to-identify, **nonpoint sources**. Examples are the runoff of fertilizers and pesticides (from farmlands and sub-

urban lawns and gardens) into streams and lakes, and pesticides sprayed into the air or blown by the wind into the atmosphere. It is much easier and cheaper to identify and control pollution from point sources than from widely dispersed nonpoint sources.

Effects of Pollution Unwanted effects of pollutants are **(1)** disruption of life-support systems for us and other species; **(2)** damage to wildlife; **(3)** damage to human health; **(4)** damage to property; and **(5)** nuisance effects such as noise and unpleasant smells, tastes, and sights.

Three factors determine how severe the effects of a pollutant will be. One is its *chemical nature*—how active and harmful it is to living organisms. Another is its **concentration**—the amount per volume unit of air, water, soil, or body weight. One way to lower the concentration of a pollutant is to dilute it in a large volume of air or water. Until we started overwhelming the air and waterways with pollutants, dilution was the solution to pollution. Now it is only a partial solution.

A third factor is a pollutant's *persistence*—how long it stays in the air, water, soil, or body. **Degradable**, or **nonpersistent**, **pollutants** are broken down completely or reduced to acceptable levels by natural physical, chemical, and biological processes. Those broken down by living organisms (usually by specialized bacteria) are called **biodegradable pollutants**. Human sewage in a river, for example, is biodegraded fairly quickly by bacteria if it is not added faster than it can be broken down.

Unfortunately, many of the substances we introduce into the environment take decades or longer to degrade. Examples of these **slowly degradable** or **persistent pollutants** include the insecticide DDT, most plastics, aluminum cans, and chlorofluorocarbons (CFCs)—these latter used as coolants in refrigerators and air conditioners, as spray propellants (in some countries), and as foaming agents for making some plastics.

Nondegradable pollutants cannot be broken down by natural processes. Examples include the toxic elements lead and mercury. The best ways to deal with nondegradable pollutants (and slowly degradable hazardous pollutants) are to not release them into the environment, to recycle them, or to remove them from contaminated air, water, or soil (an expensive process).

We know little about the possible harmful effects of 80% of the 70,000 synthetic chemicals now in commercial use. Our knowledge about the effects of the other 20% of these chemicals is limited, mostly because it is quite difficult, time-consuming, and expensive to get this knowledge.

Dealing with Pollution: Prevention and Cleanup Pollution prevention, or input pollution

Q: How many children under age 5 die each day in poor countries of causes that could be prevented?

control, reduces or eliminates the input of pollutants and wastes into the environment. Pollution can be prevented, or at least reduced, by the three R's of resource use: *Reduce, Reuse, Recycle*.

Pollution cleanup, or **output pollution control**, deals with pollutants after they have been produced. Relying primarily on pollution cleanup causes several problems. One is that, as long as population and consumption levels continue to grow, cleanup is only a temporary bandage. For example, adding catalytic converters to cars has reduced air pollution. However, the increase in the number of cars has made this cleanup approach less effective.

Another problem is that cleanup often removes a pollutant from one part of the environment only to cause pollution in another part. We can collect garbage, but garbage is then burned (perhaps causing air pollution and leaving a toxic ash that must be put somewhere); dumped into streams, lakes, and oceans (perhaps causing water pollution); buried (perhaps causing soil and groundwater pollution); recycled; or reused.

So far, most efforts to improve environmental quality in the United States and other MDCs have used pollution cleanup. Countries making up the former USSR, many Eastern European countries (especially Poland, the former East Germany, and Czechoslovakia), and most LDCs near the bottom of the economic ladder lag far behind in any kind of pollution control.

Both pollution prevention and pollution cleanup are needed, but environmentalists urge us to emphasize prevention because it works better and is cheaper than cleanup or repair. As Benjamin Franklin reminded us long ago, "An ounce of prevention is worth a pound of cure."

An increasing number of businesses have found that *pollution prevention pays*. It saves them money and at the same time helps the earth—a win-win strategy. So far, however, about 99% of environmental spending in the United States is devoted to pollution cleanup and only 1% to pollution prevention—a situation that environmentalists believe must be reversed as soon as possible.

1-5 UNSUSTAINABILITY: PROBLEMS AND CAUSES

Key Problems A number of interconnected environmental and resource problems make up the overall *crisis of unsustainability* many of the world's leading scientists believe we face (Spotlight, p. 6). Four of these problems—possible climate change from global warming, acid rain, depletion of stratospheric ozone, and urban air pollution—result from the chemicals we have put into the atmosphere (mostly from burning fossil fuels). A fifth problem, the continued poisoning of the soil and water by pesticides and numerous other toxic wastes, is the result of not relying on pollution prevention. Another six problems—depletion of nonrenewable minerals (especially oil), depletion and contamination of groundwater, deforestation, soil erosion, conversion of productive cropland and grazing to desert (desertification), and species loss (biodiversity depletion)—result from exponentially growing depletion and degradation of Earth capital.

Population growth (more consumers) and environmentally harmful forms of economic growth (more resource depletion, pollution, and environmental degradation) can intensify these problems. Poverty can also increase environmental degradation by forcing poor people to destroy potentially renewable resources such as soils, forests, and wildlife for survival.

Root Causes The first step in dealing with the crisis of unsustainability is to identify its underlying causes. These include:

- *Overpopulation.*
- *Overconsumption of resources, especially by the affluent.*
- *Poverty.*
- *Resource waste.*
- *Widespread use of environmentally damaging fossil fuels* (especially oil and coal).
- *Loss of biodiversity through oversimplification of Earth's life-support systems.*
- *Failure to encourage Earth-sustaining forms of economic development and discourage Earth-degrading forms of economic growth.*
- *Failure to have market prices represent the overall environmental cost of an economic good or service.* This promotes inefficiency and depletion of Earth capital for short-term profit by concealing the harmful effects of the products we buy.
- *Our urge to dominate and control nature for our use.*

Connections Between Root Causes and Problems Once we have identified the causes of our problems, the next step is to understand how they are connected to one another. The three-factor model in Figure 1-12 is a good starting point.

According to this model, total environmental degradation and pollution—that is, the environmental impact of population—in a given area depends on three factors: **(1)** the number of people (population size, P), **(2)** the average number of units of resources each person uses (consumption per capita or affluence,

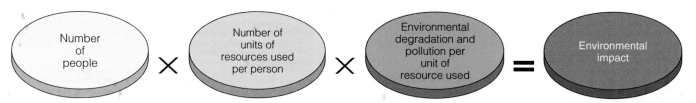

Figure 1-12 Simplified model of how three factors—population (P), affluence (A), and technology (T)—affect the environmental impact (environmental degradation and pollution) of population.

People Overpopulation

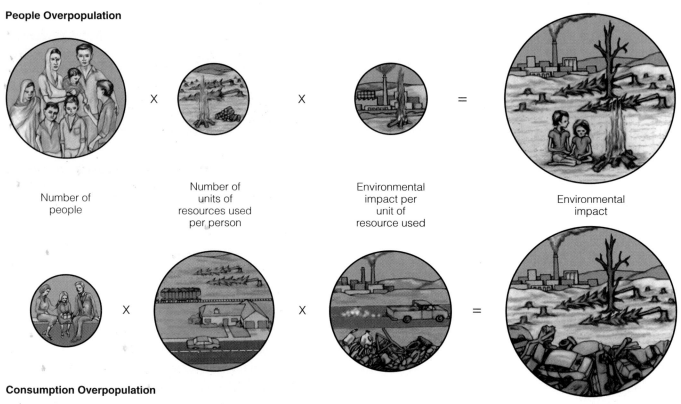

Consumption Overpopulation

Figure 1-13 Two types of overpopulation based on the relative importance of the factors in the model shown in Figure 1-12. Circle size shows relative importance of each factor.

A), and **(3)** the amount of environmental degradation and pollution produced for each unit of resource used (the environmental destructiveness of the technologies used to provide and use resources, T). This model, developed in the early 1970s by biologist Paul Ehrlich (Guest Essay, p. 22) and physicist John Holdren, can be summarized in simplified form as

Impact = Population × Affluence × Technology

or

$$I = P \times A \times T$$

Overpopulation occurs when too many people deplete the resources that support life and economies, and when they introduce more wastes than the environment can handle. This happens when people exceed the **carrying capacity** of an area: the number of people an area can support given its resource base and the way those resources are used. The three factors in Figure 1-12 can interact to produce two types of overpopulation (Figure 1-13).

People overpopulation exists where there are more people than the available supplies of food, water, and other important resources can support at a minimal level. Here, population size and the resulting degradation of potentially renewable resources—as the poor struggle to stay alive—tend to be the key factors determining total environmental impact. In the world's poorest LDCs, people overpopulation causes premature death for 40 million people each year and absolute poverty for 1.2 billion (Spotlight, p. 11).

Consumption overpopulation, or **overconsumption**, exists in industrialized countries, where only one-fifth of the world's people use resources at such a

Q: What percentage of environmental spending in the United States is devoted to preventing pollution?

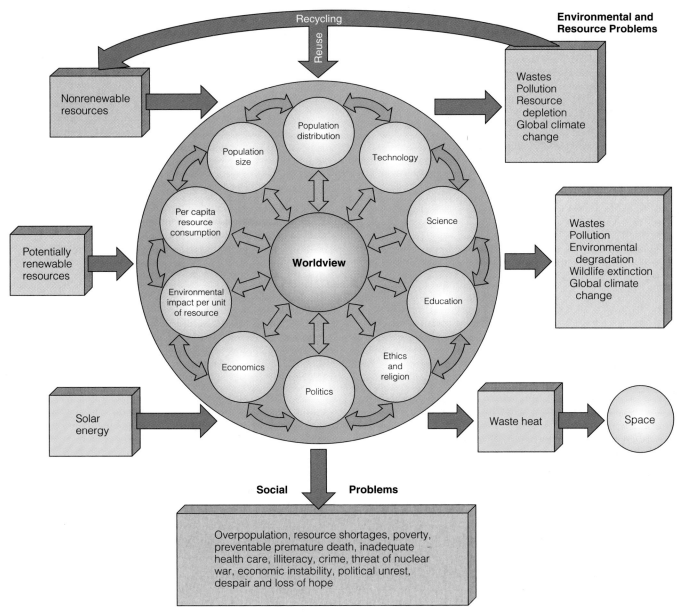

Figure 1-14 Environmental, resource, and social problems are caused by a complex, poorly understood mix of interacting factors, as illustrated by this simplified model.

high rate that significant pollution, environmental degradation, and resource depletion occur. With this type of overpopulation, high rates of resource use per person (and the resulting high levels of pollution and environmental degradation per person) are the key factors determining overall environmental impact.

For example, the average U.S. citizen consumes 50 times as much as the average citizen of India (and 100 times as much as the average person in some LDCs). *This means that poor parents in an LDC would need 100–200 children to have the same lifetime environmental impact as 2 children in a typical family in the United States.*

We know from studying other species that when a population exceeds or *overshoots* the carrying capac-

ity of its environment, it suffers a *dieback* that reduces its population to a sustainable size. How long will we be able to continue our exponential growth in population (Figure 1-1) and resource use (Figure 1-2) without suffering overshoot and dieback? No one knows, but warning signals from the earth are forcing us to consider this question (Spotlight, p. 6 and Figure 1-11).

The three-factor model in Figure 1-12 is useful in understanding how key environmental problems and some of their causes are connected. However, the interconnected environmental, resource, and social problems we face are much more complex, and they involve a number of poorly understood interactions among a number of factors (Figure 1-14).

There Is No Crisis of Unsustainability

Julian L. Simon

GUEST ESSAY

Julian L. Simon is professor of economics and business administration at the University of Maryland. He has effectively presented and defended the position that continued economic growth and technological advances will produce a less crowded, less polluted, and more resource-rich world. His many articles and books on this subject include The Ultimate Resource, The Resourceful Earth, *and* Population Matters *(see Further Readings).*

This textbook, like most others discussing environmental and resource problems, begins with the proposition that there is an environmental and resource crisis. If this means that the situation of humanity is worse now than in the past, then the idea of a crisis—and all that follows from it—is dead wrong. In almost every respect important to humanity, the trends have been improving, not deteriorating.

Our world now supports 5.5 billion people. In the nineteenth century, the earth could sustain only 1 billion. And 10,000 years ago, only 1 million people could keep themselves alive. People are living more healthily than ever before, too.

One would expect lovers of humanity—people who hate war and worry about famine in Africa—to jump with joy at this extraordinary triumph of the human mind and human organization over the raw forces of nature. Instead, they lament that there are so many human beings and wring their hands about the problems that more people inevitably bring.

The recent extraordinary decrease in the death rate—to my mind, the greatest miracle in history—accounts for the bumper crop of humanity. It took thousands of years to increase life expectancy at birth from the 20s to the 30s. Then, in just the last 200 years, life expectancy in the advanced countries jumped from the mid-30s to the 70s. And, starting well after World War II, life expectancy at birth in the poor countries, even the very poorest, has leaped upward (averaging 63 in 1993) because of progress in agriculture, sanitation, and medicine. Average life expectancy at birth in China, the world's most populous country, was 70 in 1993, an increase of 26 years since the 1950s. Is this not an astounding triumph?

In the short run another baby reduces income per person by causing output to be divided among more people. And as the British economist Thomas Malthus argued in 1798, more workers laboring with existing capital results in less output per worker. However, if resources are not fixed, then the Malthusian doctrine of diminishing resources, resurrected by today's doom-and-gloom analysts, does not apply. Given some time to adjust to shortages with known methods and new inventions, free people create additional resources.

It is amazing but true that a resource shortage resulting from population or income growth usually leaves us better off than if the shortage had never arisen. If firewood had not become scarce in seventeenth-century England, coal would not have been developed. If coal and whale oil shortages hadn't loomed, oil wells would not have been dug.

The prices of food, metals, and other raw materials have been declining by every measure since the beginning of the nineteenth century, and as far back as we know. That is, raw materials have been getting less scarce instead of more scarce throughout history, defying the commonsense notion that if one begins with an inventory of a resource and uses some up, there will be less left. This is despite, and indirectly because of, increasing population.

All statistical studies show that population growth doesn't lead to slower economic growth, though this defies common sense. Nor is high population density a drag on economic development. Statistical comparison across nations reveals that higher population density is associated with faster instead of slower growth. Drive around on Hong Kong's smooth-flowing highways for an hour or two. You will then realize that a large concentration of human beings in a small area does not make comfortable existence impossible. It also allows for exciting economic expansion, if the system gives individuals the freedom to exercise their talents and pursue economic opportunities. The experience of densely populated Singapore makes it clear that Hong Kong is not unique, either.

Because so much is at stake there are conflicting views about how serious our environmental problems really are and what should be done about them (Guest Essays on pp. 20 and 22). Some analysts—mostly economists—contend that the world is not overpopulated and that people are our most important resource—as consumers, to fuel continued economic growth, and as sources of technological ingenuity. They believe that technological advances will allow us to clean up pollution to acceptable levels and find substitutes for any resources that become scarce. To these analysts, there is no actual or potential crisis of sustainability, and they accuse environmentalists of overstating the seriousness of the problems we face.

Working with the Earth This chapter has presented you with an overview of the serious problems environmentalists believe we face—and their root

Q: How much of the world's population is in the United States?

In 1984 a blue-ribbon panel of scientists summarized their wisdom in *The Resourceful Earth*. Among the findings, besides those I have noted previously, were these:

- Many people are still hungry, but the food supply has been improving since at least World War II, as measured by grain prices, production per consumer, and the death rate from famines.

- Land availability won't increasingly constrain world agriculture in coming decades.

- In the United States the trend is toward higher-quality cropland, suffering less from erosion than in the past.

- The widely published report of increasingly rapid urbanization of U.S. farmland was based on faulty data.

- Trends in world forests are not worrisome, though in some places deforestation is troubling.

- There is no statistical evidence for rapid loss of plant and animal wildlife species in the next two decades. An increased rate of extinction cannot be ruled out if tropical deforestation continues unabated, but the linkage has not yet been demonstrated.

- Water does not pose a problem of physical scarcity or disappearance, although the world and U.S. situations do call for better institutional management through more rational systems of property rights.

- There is no compelling reason to believe that world oil prices will rise in coming decades. In fact, prices may fall well below current levels.

- Compared with coal, nuclear power is no more expensive and is probably much cheaper under most circumstances. It is also much cheaper than oil.

- Nuclear power gives every evidence of costing fewer lives per unit of energy produced than does coal or oil.

- Solar energy sources (including wind and wave power) are too dilute to compete economically for much of humankind's energy needs, though for specialized uses and certain climates they can make a valuable contribution.

- Threats of air and water pollution have been vastly overblown. Air and water in the United States have been getting cleaner, rather than dirtier.

We don't say that all is well everywhere, and we don't predict that all will be rosy in the future. Children are hungry and sick; people live out lives of physical or intellectual poverty and lack of opportunity; war or some other pollution may do us in. *The Resourceful Earth* does show that for most relevant matters we've examined, total global and U.S. trends are improving instead of deteriorating.

Also, we do not say that a better future happens automatically or without effort. It will happen because men and women—sometimes as individuals, sometimes as enterprises working for profit, sometimes as voluntary nonprofit groups, and sometimes as governmental agencies—will address problems with muscle and mind, and will probably overcome, as has been usual through history.

We are confident that the nature of the physical world permits continued improvement in humankind's economic lot in the long run, indefinitely. Of course, there are always newly arising local problems, shortages, and pollution, resulting from climate or increased population and income and new technologies. Sometimes temporary large-scale problems arise. But the world's physical conditions and the resilience of a well-functioning economic and social system enable us to overcome such problems, and the solutions usually leave us better off than if the problem had never arisen. That is the great lesson to be learned from human history.

Critical Thinking

1. Do you agree with the author's contention that there is no environmental, population, or resource crisis? Explain. How is it compatible with the data presented in Figure 1-11? After you've finished this course and this text, come back and answer this question again to see if your views have changed.

2. Do you feel you will be better off than your parents? What do you mean by *better off*? Do you think any children you might have will be better off than you?

causes. The rest of this book presents a more detailed analysis of these problems, along with solutions proposed by various scientists and environmentalists. There are also encouraging examples of things individuals and groups are doing to help sustain the earth, as well as a number of "Individuals Matter" boxes that suggest what you can do to work with the earth.

The key to dealing with the problems we face is recognizing that *individuals matter*. Anthropologist Margaret Mead has summarized our potential for change: "Never doubt that a small group of thoughtful, committed citizens can change the world. Indeed it is the only thing that ever has."

What's the use of a house if you don't have a decent planet to put it on?

HENRY DAVID THOREAU

Simple Simon Environmental Analysis

Paul R. Ehrlich and Anne H. Ehrlich

GUEST ESSAY

Paul R. Ehrlich is Bing Professor of population studies at Stanford University. His many honors include membership in the U.S. National Academy of Sciences, the Fellowship of the American Academy of Arts and Sciences, and the American Philosophical Society, the Crafoord Prize of the Swedish Academy of Sciences (given as the equivalent of the Nobel Prize), the American Association for the Advancement of Science/Scientific American Award for Science in the Service of Mankind, and a MacArthur Prize Fellowship. Anne H. Ehrlich is a senior research associate in Stanford's Department of Biological Sciences. Her many honors have included an honorary degree from Bethany College, the Global 500 Environmental Roll of Honor, and the Humanists' Award. The Ehrlichs have authored or coauthored more than 30 books and 600 articles dealing with population, the environment, ecology, and evolution.

Today there are over 5.5 billion people in the human population, and population experts now project that it will reach a maximum size of 10–14 billion before it stops growing. It is not the number of people per se that causes concern, but the ways in which numbers, patterns of consumption, and choices of technology are now combining to destroy civilization's life support systems [Figure 1-11]. In 1992 a joint statement by the U.S. National Academy of Sciences and the British Royal Society on Population Growth, Resource Consumption, and a Sustainable World stated, among other warnings: "It is not prudent to rely on science and technology alone to solve problems created by rapid population growth, wasteful resource consumption, and harmful human practices."

Nevertheless, a few uninformed people claim that population growth is beneficial, the ozone hole is a hoax, global warming is too uncertain to justify action, the extinction of other organisms is no problem, and the degree of crowding possible in Hong Kong, Singapore, and the Netherlands can be accommodated over the entire planet. Julian Simon is the leading spokesperson for this view. He believes that a finite earth can hold an almost infinite number of people.

He also has maintained that resources are getting cheaper because they are infinite in supply. He believes, incorrectly, that resources are infinitely subdivisible (for example, petroleum, once it is divided into atoms, is no longer petroleum). But even if they were infinitely subdivisible, that would not make them infinite in quantity. Simon has simply resurrected a mathematical error known to the ancient Greeks (Zeno's paradox, which concluded that there is an infinite distance between any two points). He has even asserted that humanity could convert the entire universe (including itself and Simon, presumably) into copper!

Moving on from his clever "analysis," what are the facts?

- The connections between economic growth, population growth, and quality of life are much more subtle and complicated than Simon imagines. In many poor nations with rapidly growing populations, GNP *per person* has recently been shrinking, and quality of life is clearly declining.

- In *absolute numbers* more people are hungry today than ever—over a billion, according to the World Bank—although the *proportion* of hungry people has probably been reduced somewhat in recent decades. But can food production continue to increase at a faster rate than population growth? Agriculture is already running into problems such as a "cap" on rice yields and diminishing returns from green revolution technology. Moreover, per person food production has been falling in Africa for more than two decades.

- Land availability is very likely to constrain world agriculture, especially given that widespread degradation is occurring on a major portion of the world's farmland.

- In the United States production is now concentrated on the better-quality farmland for conservation rea-

Critical Thinking

1. Is the world overpopulated? Explain. Is the United States suffering from consumption overpopulation? Explain.

2. Do you favor instituting policies designed to reduce population growth and stabilize:
   a. the size of the world's population as soon as possible and
   b. the size of the U.S. population as soon as possible?
   Explain. What policies do you believe should be implemented?

3. Explain why you agree or disagree with the following propositions:
   a. High levels of resource use by the United States and other MDCs are beneficial.
   b. MDCs stimulate the economic growth of LDCs by buying their raw materials. ~~Does not trickle down~~
   c. High levels of resource use also stimulate economic growth in MDCs.
   d. Economic growth provides money for more financial aid to LDCs and for reducing pollution, environmental degradation, and poverty.

4. Explain why you agree or disagree with the following proposition: The world will never run out of re-

sons; but the amount of good land is not increasing, and even it is subject to degradation.

■ The world (and the United States) is steadily losing prime farmland to urbanization. It is estimated that an area equivalent in size to Indiana, much of it farmland, will be built on in poor nations between 1980 and 2000.

■ Trends in world forests are very worrisome—ask a biologist. Simon apparently doesn't know the difference between an old-growth virgin forest (with its critical biodiversity intact) and a tree farm. Sadly, in most temperate regions, only a tiny fraction of old-growth forests remains even though the total area of "forest" is approximately as large as a century ago. And the diversity-rich tropical forests are being cut down at unprecedented rates.

■ Biologists have uncovered convincing evidence that major losses of biodiversity have already occurred through the widespread destruction of habitats on which other organisms are totally dependent.

■ Yes, there is a lot of water on this planet, but it is often not available where or when needed, and much of it is polluted. Groundwater in many areas is being removed faster than it can be recharged. Since water is essential for high-yield agriculture, its scarcity in important agricultural regions is a serious concern. Some Earth scientists feel that water shortages will be the main factor limiting human population growth.

■ The main problem now with oil is neither its supply nor its price, but the environmental costs of burning it and other fossil fuels, especially the injection of greenhouse gases into the atmosphere.

■ Nuclear power is not cheap if the costs of development, waste disposal, decommissioning of worn-out plants, and insurance are properly factored in. The spread of nuclear weapons from use of nuclear power technology also adds enormous military defense costs.

■ Nuclear power may cost fewer lives than coal if there are no major nuclear accidents; but the Chernobyl (which may eventually cost many more lives than the coal industry) and Three Mile Island accidents convinced the public that the risk was too high. Still, the possibility that a safe fission power technology could be developed should remain part of our thinking about future energy supplies.

■ A solar energy economy in which electricity is generated (using some of it to make hydrogen as a versatile, portable fuel) is considered by most experts to be a feasible major energy option.

■ Air pollution is not a minor matter. Besides causing serious direct threats to human health, agriculture, and forestry, it threatens to degrade the ozone shield, which is essential to the persistence of human and many other forms of life, and may lead to climate change severe enough to cause billions of premature deaths in the next century. Air and water pollution problems in the United States were abated by pollution control efforts in the 1970s, but in the 1980s they worsened again because the population and the economy continued to grow (producing more pollutants) with little improvement in pollution abatement.

You don't have to take our word for all this. The facts are readily available in numerous sources, and you can make your own assessment as to whether population growth, resource consumption, and environmental deterioration are real problems. But remember, while putting huge effort into solving nonexistent or trivial problems would be wasteful, failing to address or postponing actions on the truly serious problems we have outlined in this essay would be very costly indeed.

Critical Thinking

1. How can Simon and the Ehrlichs take mostly the same data and come to such different conclusions?

2. Whom do you believe? Why?

sources because technological innovations will produce substitutes or allow use of lower grades of scarce resources.

*5. What are the major resource and environmental problems in:
   a. the city, town, or rural area where you live and
   b. the state where you live?

Which of these problems affect you directly?

*6. Make a list of the resources you truly need. Then make another list of the resources that you use each day only because you want them. Finally, make a third list of resources you want and hope to use in the future.

*7. Make a concept map of the key ideas in this chapter using the section heads and subheads and the key terms (shown in boldface type in the chapter). See the inside front cover and Appendix 4 for information on concept maps.

---

*Critical Thinking topics preceded by an asterisk are either laboratory exercises or individual or class projects. Additional exercises and projects are found in the laboratory manual and in the *Green Lives, Green Campuses* workbook, which can be used with this textbook.

2 Cultural Changes, Worldviews, Ethics, and Sustainability

2040 A.D.: Green Times on Planet Earth*

Mary Wilkins sat in the living room of the passive solar earth-sheltered house she shared with her daughter Jane and her family. It was July 4, 2040—Independence Day.

She walked into the greenhouse and gazed out at the large blue and green Earth flag (Figure 2-1) they proudly flew beside the American flag. Her eyes shifted to her grandchildren, Jessica and Jeffrey, running and hollering as they played tag on the neighborhood commons. She heard the hum of solar-powered pumps trickling water to rows of organically grown vegetables, and she glanced at the fish in the aquaculture and waste treatment tanks.

Things began changing rapidly in 2000 with the formation of the Earth Coalition. Environmentalists with a diversity of worldviews and agendas decided to stop bickering with one another and work together to help sustain the Earth. They finally decided to live by two of nature's most fundamental lessons: Sustainability depends on cooperation and pursuing a diversity of strategies.

She returned to the cavelike coolness of her earth-sheltered house and began putting the finishing touches on the children's costumes for this afternoon's pageant in Aldo Leopold Park. It would honor Earth heroes who began the Age of Ecology in the twentieth century and those who continued this tradition in the twenty-first century. Mary smiled when she thought of Jessica's delight at being chosen to play Rachel Carson, heroine of the 1960s skit.

Even in her most idealistic dreams she had never guessed she would see a halt in the hemorrhaging of global biodiversity, no more wars over oil, the end of nuclear power plants and nuclear weapons—not to mention the disappearance of most air pollution when energy from the sun, wind, and hydrogen replaced fossil and nuclear fuels, and the replacement of most cars with walking, bicycles, tricycles, and efficient mass transportation. World population had stabilized at 8 billion in 2030 and then begun a slow decline. And the rate of global warming had slowed significantly.

Two hours later she, her daughter Jane, and her son-in-law Gene watched with pride as 40 beautiful children honored the leaders of the Age of Ecology. At the end Jessica stepped forward and said, "Today we have honored many Earth heroes, but the real heroes are the ordinary people in this audience and throughout the world who worked to help sustain the earth. Thank you Grandma, Mom, Dad, and everyone here for giving us such a wonderful gift. We promise to leave the earth even better for our children and grandchildren and all living creatures." There wasn't a dry eye in the audience.

---

*Compare this hopeful scenario with the worst-case scenario that opens Chapter 10.

Figure 2-1 Growing numbers of people are pledging allegiance to the planet that keeps them alive. They see themselves as common citizens of One Earth who represent every culture, every race, every species, and every living creature in this and future generations. Some display the Earth Flag as a symbol of their commitment to sustaining the earth. (Courtesy of Earth Flag Co., 33 Roberts Road, Cambridge, MA 02138)

A continent ages quickly once we come.

ERNEST HEMINGWAY

This chapter will answer the following questions:

- What major impacts have hunter-gatherer societies, early agricultural societies, and industrialized societies had on the environment?

- What conflicting worldviews underlie today's societies?

- What ethical guidelines can be used to help sustain the earth, and how can we live sustainably?

2-1 CULTURAL CHANGES

Cultural Change: An Overview Fossil evidence suggests that the current form of our species, *Homo sapiens sapiens*, has walked the earth for only about 40,000 years, an instant in the planet's estimated 4.6-billion-year existence. During the first 30,000 years we were mostly hunter-gatherers. Since then there have been two major cultural shifts, the *Agricultural Revolution*, which began 10,000–12,000 years ago, and the *Industrial Revolution*, which began about 275 years ago.

These cultural revolutions have given us much more energy (Figure 2-2) and new technologies with which to alter and control more of the planet to meet our basic needs and increasing wants. By expanding food supplies, lengthening average life spans, and raising average living standards, each shift increased

the human population (Figure 2-3). The result has been skyrocketing resource use (Figure 1-2) and pollution, and accelerating environmental degradation (Figure 1-11).

Most environmentalists see an urgent need for a new cultural shift before we are overwhelmed by the

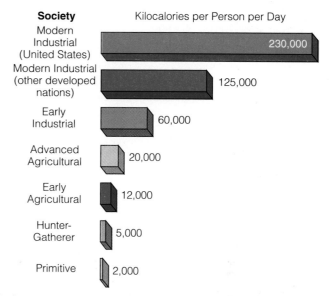

Society	Kilocalories per Person per Day
Modern Industrial (United States)	230,000
Modern Industrial (other developed nations)	125,000
Early Industrial	60,000
Advanced Agricultural	20,000
Early Agricultural	12,000
Hunter-Gatherer	5,000
Primitive	2,000

Figure 2-2 Average direct and indirect daily energy use per person at various stages of human cultural development. A *calorie* is the amount of energy needed to raise the temperature of 1 gram of water 1°C (1.8°F). A *kilocalorie* is 1,000 calories. Food calories or calories expended during exercise are kilocalories (sometimes represented by spelling Calories with a capital C).

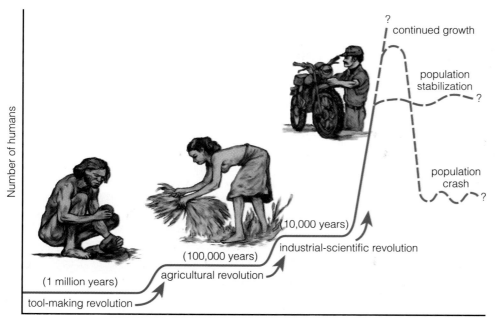

Figure 2-3 Expansion of Earth's carrying capacity for our species. Technological innovation has led to major cultural changes, and we have displaced and depleted numerous species that compete with us for—and provide us with—resources. Dashed lines represent three alternative futures: **(1)** uninhibited human population growth; **(2)** population stabilization; and **(3)** growth followed by a crash and stabilization at a much lower level. Time spans shown represent periods over which each revolution took place.

? continued growth

population stabilization ?

population crash ?

Number of humans

?

(10,000 years) industrial-scientific revolution

(100,000 years) agricultural revolution

(1 million years)

tool-making revolution

Time

Figure 2-4 Most people who have lived on Earth have survived by hunting wild game and gathering wild plants. These !Kung san (left) in Africa are going hunting. The woman and the child (right) are gathering wild plants in a tropical forest in the Amazon Basin of Brazil.

exponential growth of people, pollution, and environmental degradation (Figure 1-12). This **Earth-Wisdom** or **Environmental Revolution** (Guest Essay, p. 30) calls for us to halt population growth and to change our lifestyles, our political and economic systems, and how we treat the earth—so that we can heal and help sustain it for ourselves and other species. This revolution involves learning how to work with the rest of nature by learning how nature sustains itself.

Hunting-and-Gathering Societies   During about three-fourths of our 40,000-year existence, we were **hunter-gatherers** who survived by gathering edible wild plants and by hunting and fishing (including shellfish) (Figure 2-4). Archaeological and anthropological evidence indicates that our hunter-gatherer ancestors normally lived in small bands of rarely more than 50 people who worked together to get enough food to survive. If food became scarce, they picked up their few possessions and moved on.

The earliest hunter-gatherers (and those still living this way today) survived through *Earth wisdom*—having expert knowledge about their natural surroundings. They discovered that a variety of plants and animals could be eaten and used as medicines. They used stone-sharpened and -shaped sticks, other stones, and animal bones as primitive weapons for hunting—and as tools for harvesting plants and for scraping hides for clothing and shelter. These dwellers in nature had only three energy sources: **(1)** sunlight captured by plants (which also served as food for the animals they hunted), **(2)** fire, and **(3)** their own muscle power.

Population-control practices varied with different cultures, but they included abstention from sexual intercourse, herbal contraceptives, abortion, late marriage, and prolonged breast-feeding of infants (which provides some degree of birth control by inhibiting ovulation). Because of these measures, high infant mortality, and an average life expectancy of about 30 years, hunter-gatherer populations grew very slowly.

Gradually the hunter-gatherers refined their tools and hunting practices. Some worked together to hunt herds of reindeer, woolly mammoths, European bison, and other big game. They used fire to flush game from thickets toward waiting hunters and to stampede herds into traps or over cliffs. Some learned that burning vegetation promoted the growth of food and forage plants.

Advanced hunter-gatherers had a greater impact on their environment than did early hunter-gatherers, especially in using fire to convert forests into grasslands. There is also evidence that they contributed to, and perhaps even caused, the extinction of some large game animals.

Because of their small numbers, nomadism, and dependence on their own muscle power to modify the environment, however, their environmental impact was fairly limited and localized. Both early and advanced hunter-gatherers were *dwellers in nature*, who trod lightly on the earth because they were not capable of doing more. They survived by learning to work with nature and with one another.

The Agricultural Revolution   Some 10,000–12,000 years ago a cultural shift known as the **Agricul-**

Q: What percentage of the world's pollution is produced by the United States?

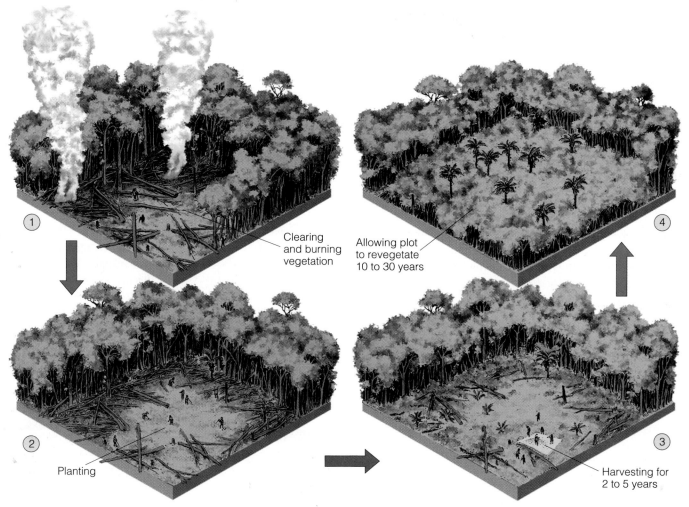

Figure 2-5 The first crop-growing technique may have been a combination of slash-and-burn and shifting cultivation in tropical forests. This method is sustainable if only small plots of the forest are cleared, cultivated for no more than 5 years, and then allowed to lie fallow for 10–30 years to renew soil fertility.

1 — Clearing and burning vegetation

2 — Planting

3 — Harvesting for 2 to 5 years

4 — Allowing plot to revegetate 10 to 30 years

tural Revolution began in several regions of the world. This food-producing revolution involved a gradual move from a lifestyle based on nomadic hunting-and-gathering bands to one of settled agricultural communities, where people learned how to domesticate wild animals and cultivate wild plants. This shift may have resulted from climate changes that forced humans to adapt or perish.

Plant cultivation may have begun in tropical forests. People discovered that they could grow various wild food plants from roots or tubers (fleshy underground stems). To prepare for planting they cleared small patches of forests by cutting down trees and other vegetation, and then burning the underbrush (Figure 2-5). The ashes fertilized the nutrient-poor soils. This was **slash-and-burn cultivation**.

These early growers also used **shifting cultivation** (Figure 2-5). After a garden had been used for several years, the soil would be depleted of nutrients or

reinvaded by the forest. Then growers moved and cleared a new plot. Each abandoned patch had to be left fallow (unplanted) for 10–30 years before the soil became fertile enough to grow crops again. By doing this, early growers practiced sustainable cultivation.

These growers practiced **subsistence farming**, growing only enough food to feed their families. Their dependence on human muscle power and crude stone or stick tools meant that they could cultivate only small plots; thus, they had relatively little impact on their environment.

About 7,000 years ago the invention of the metal plow, pulled by domesticated animals, allowed farmers to cultivate larger plots of land and to break up fertile grassland soils, which previously couldn't be cultivated because of their dense root systems. In some arid (dry) regions early farmers further increased crop output by diverting water from nearby streams into hand-dug ditches and canals to irrigate crops.

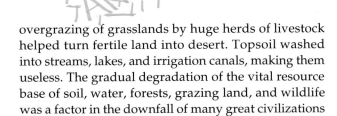

The gradual shift from hunting and gathering to farming had several significant effects:

- *Using domesticated animals to haul loads and perform other tasks increased the average energy use per person* (Figure 2-2).

- *Birth rates rose faster than death rates and population increased, mostly because of a larger, more reliable food supply* (Figure 2-3).

- *People cleared increasingly larger fields and built irrigation systems to transfer water from one place to another.*

- *People began accumulating material goods.* Nomadic hunter-gatherers could not carry many possessions in their travels, but farmers living in one place could acquire as much as they could afford.

- *Farmers could grow more than enough food for their families.* They could use the surplus to barter with craftspeople who specialized in weaving, tool making, and pottery.

- *Urbanization—the formation of villages, towns, and cities—became practical.* Craftspeople and others with specialized occupations moved into permanent villages. Some villages grew into towns and cities, which served as centers for trade, government, and religion.

- *Conflict between societies became more common as ownership of land and water rights became a valuable economic resource.* Armies and their leaders rose to power and conquered large areas of land. These rulers forced powerless people—slaves and landless peasants—to do the hard, disagreeable work of producing food and constructing things such as irrigation systems, temples, and walled fortresses.

- *The survival of wild plants and animals, once vital to humanity, no longer seemed to matter.* Wild animals competing with livestock for grass and feeding on crops became enemies to be killed or driven from their habitats. Wild plants invading cropfields became weeds to be eliminated.

The growing populations of these emerging civilizations needed more food and more wood for fuel and buildings, so people cut down vast forests and plowed up large expanses of grassland. Such extensive land clearing degraded or destroyed the habitats of many wild plants and animals, causing or hastening their extinction.

Many of these cleared lands were poorly managed. Soil erosion, salt buildup in irrigated soils, and overgrazing of grasslands by huge herds of livestock helped turn fertile land into desert. Topsoil washed into streams, lakes, and irrigation canals, making them useless. The gradual degradation of the vital resource base of soil, water, forests, grazing land, and wildlife was a factor in the downfall of many great civilizations (Guest Essay, p. 30).

The gradual spread of agriculture meant that most of the world's human population shifted from hunter-gatherers working with nature to shepherds, farmers, and urban dwellers trying to tame and control wild nature.

The Industrial Revolution The next great cultural shift, the **Industrial Revolution**, began in England in the mid-1700s and spread to the United States in the 1800s. It multiplied energy consumption per capita and thus the power of humans to shape the earth to their will and to fuel economic growth (Figure 2-2). Production, trade, and distribution of goods all expanded.

The Industrial Revolution represented a shift from dependence on renewable wood (with supplies dwindling in some areas because of unsustainable cutting) and flowing water as energy sources to dependence on nonrenewable fossil fuels. The availability of coal led to the invention of steam engines to pump water and perform other tasks, and eventually to an array of new machines powered by coal and later by oil and natural gas. The new machines in turn led to a switch from small-scale, localized production of handmade goods to large-scale production of machine-made goods in centralized factories in rapidly growing industrial cities.

Factory towns grew into cities as rural people came to the factories for work. There they worked long hours in noisy, dirty, and hazardous conditions. Other workers toiled in coal mines, which were also dangerous. In these industrial cities coal smoke belching from a profusion of chimneys was so heavy that many died—mainly from lung ailments. Ash and soot covered everything, and some days the smoke was so thick that it blotted out the sun.

Fossil-fuel-powered farm machinery, commercial fertilizers, and new plant-breeding techniques increased the crop yields per acre, meaning that fewer farmers were needed. With a more reliable food supply, the size of the human population began the sharp upturn still going on today (Figures 1-1 and 2-3).

After World War I (1914–18), more efficient machines and mass-production techniques were developed, forming the basis of today's advanced industrial societies in the United States, Canada, Japan, and western Europe. These societies are characterized by

Q: About how long has the latest version of the human species been on the earth?

- Higher production and consumption of goods
- Greater dependence on nonrenewable resources such as fossil fuels, iron, and aluminum
- Increased use of synthetic chemicals and materials, many of which break down slowly and are sometimes toxic to humans and wildlife
- A jump in the amount of energy used per person for transportation, manufacturing, agriculture, lighting, and heating and cooling (Figure 2-2)

Advanced industrial societies benefit most people living in them. These benefits include

- Creation and mass production of many useful and economically affordable products
- A sharp increase in average agricultural productivity per person because of industrialized agriculture, in which a small number of farmers produce large amounts of food
- Lower infant mortality and higher average life expectancy because of better sanitation, hygiene, nutrition, and medical care
- A decrease in the rate of population growth in MDCs (Figure 1-5)
- Better health, birth control methods, education, average income, and old-age security

People in MDCs experience far more of these benefits than do people in LDCs. These benefits to industrialized societies have been accompanied by the resource and environmental problems we face today. Industrialization also reinforces the idea that our role is to dominate and manage nature for our species.

Making a New Cultural Change. Most environmentalists believe that there is an urgent need to begin shifting from our present array of industrialized and partially industrialized societies to a variety of *sustainable-Earth* or *Earth-wisdom* *societies* throughout the world, by launching an Earth-Wisdom or Environmental Revolution (Guest Essay, p. 30).

Dealing with global environmental and resource problems will involve much controversy and require us to make some trade-offs and significant changes in our worldviews, economic and political systems, and lifestyles. Specific actions environmentalists believe are needed to help sustain the earth are presented throughout this book.

Environmentalists call for us to shift our efforts from pollution cleanup to pollution prevention (Guest Essay on p. 37); waste disposal (mostly burial and burning of wastes) to waste prevention and reduction;

species protection to habitat protection; and increased resource use to increased resource conservation. They urge us to use our political and economic systems to reward Earth-sustaining economic activities and discourage those that harm the earth. They also believe that we should allow many parts of the world we have damaged to heal, help restore severely damaged areas, and protect remaining wild areas from destructive forms of economic development.

The Agricultural Revolution took place over some 10,000 years, and the Industrial Revolution over more than 200 years. Because of mushrooming population, resource consumption, pollution, and environmental degradation, environmentalists believe that we have only a few decades to bring about this new Environmental Revolution.

2-2 WORLDVIEWS: A CLASH OF VALUES AND CULTURES

How Shall We Live? There are conflicting views about how serious our environmental problems really are and what should be done about them (Guest Essays on pp. 20 and 22). These conflicts arise mostly out of differing **worldviews**: how individuals think the world works, what they think their role in the world should be, and what they believe is right and wrong behavior (ethics). People with widely differing worldviews can take the same data and arrive at quite different conclusions because they start with entirely different assumptions (Guest Essays, pp. 20 and 22). Worldviews come in many flavors. But they can be divided into two groups according to whether they put humans at the center of things or not.

Clashing Worldviews in North America
When European colonists began settling North America in 1607, they found a vast continent with seemingly unlimited resources. Not surprisingly, these settlers had a human-centered **frontier worldview**. They saw a hostile wilderness to be conquered and cleared as fast as possible. This frontier attitude led to enormous resource waste because the colonists believed there would always be more.

This worldview was in sharp contrast to the life-centered worldview of many of the Native Americans whose land was taken over and whose cultures were fragmented or destroyed as European settlers spread across the continent. For at least 10,000 years these indigenous peoples had practiced mostly hunting and gathering, and they had survived by being immersed in and learning to work with nature. Although there were exceptions, most Native American cultures were

Launching the Environmental Revolution*

Lester R. Brown

Lester R. Brown is President of the World-watch Institute, a private nonprofit research institute he founded in 1974 that is devoted to analysis of global environmental issues. Under his leadership the institute publishes the annual State of the World *report, considered by environmentalists and world leaders as the best way to become informed about key environmental issues. It also publishes monographs on specific topics,* World Watch *magazine, and a series of Environmental Alert books. He is author of a dozen books, recipient of the McArthur Foundation "Genius Award," and winner of the United Nations' 1989 environment prize. He has been described by the* Washington Post *as "one of the world's most influential thinkers."*

Our world of the mid-1990s faces potentially convulsive change. The question is, In what direction will it take us? Will the change come from strong worldwide initiatives that reverse the degradation of the planet and restore hope for the future, or will it come from continuing environmental deterioration that leads to economic decline and social instability?

Muddling through will not work. Either we will turn things around quickly or the self-reinforcing internal dynamic of the deterioration-and-decline scenario will take over. The policy decisions we make in the years immediately ahead will determine whether our children live in a world of development or decline.

There is no precedent for the rapid and substantial change we need to make. Building an environmentally sustainable future depends on restructuring the global economy, major shifts in human reproductive behavior, and dramatic changes in values and lifestyles. Doing all this quickly adds up to a revolution that is driven and defined by the need to restore and preserve the earth's environmental systems. If this *Environmental Revolution* succeeds, it will rank with the Agricultural and Industrial Revolutions as one of the great economic and social transformations in human history.

Like the Agricultural Revolution, it will dramatically alter population trends. While the former set the stage for enormous increases in human numbers, this revolu-

*Excerpt from Brown's expanded version of these ideas in "Launching the Environmental Revolution," *State of the World 1992* (New York: W. W. Norton, 1992).

tion will succeed only if it stabilizes human population size, reestablishing a balance between people and natural systems on which they depend [Figure 2-3]. In contrast to the Industrial Revolution, which was based on a shift to fossil fuels, this new transformation will be based on a shift away from fossil fuels.

The two earlier revolutions were driven by technological advances—the first by the discovery of farming and the second by the invention of the steam engine, which converted the energy in coal into mechanical power. The Environmental Revolution, while it will obviously need new technologies, will be driven primarily by the restructuring of the global economy so that it does not destroy its natural support systems.

The pace of the Environmental Revolution needs to be far faster than that of its predecessors. The Agricultural Revolution began some 10,000 years ago, and the Industrial Revolution has been under way for about two centuries. But if the Environmental Revolution is to succeed, it must be compressed into a few decades.

Progress in the Agricultural Revolution was measured almost exclusively in the growth in food output that eventually enabled farmers to produce a surplus that could feed city dwellers. Similarly, industrial progress was gained by success in expanding the output of raw materials and manufactured goods. The Environmental Revolution will be judged by whether it can shift the world economy into an environmentally sustainable development path, one that leads to greater economic security, healthier lifestyles, and a worldwide improvement in the human condition.

Many still do not see the need for such an economic and social transformation. They see the earth's deteriorating physical condition as a peripheral matter that can be dealt with by minor policy adjustments. But 20 years of effort have failed to stem the tide of environmental degradation. There is now too much evidence on too many fronts [Figure 1-11] to take these issues lightly.

Already the planet's degradation is damaging human health, slowing the growth in world food production, and reversing economic progress in dozens of countries. By the age of 10, thousands of children living in southern California's Los Angeles basin have respiratory systems that are permanently impaired by polluted air. Some 300,000 people in the former Soviet Union are being treated for radiation sickness caused by the Chernobyl nuclear power plant accident. The accelerated depletion of ozone in the stratosphere in the northern hemisphere

based on a deep respect for the land and its animals. The conflict between their life-centered worldviews and the human-centered worldview of the settlers has been summarized by a medicine woman of California's Wintu tribe:

The white people never cared for the land or deer or bear. When the Indians kill meat, we eat it all up. When we dig roots, we make little holes. When we build houses we make little holes.... We don't chop down trees. We only use dead wood. But the white peo-

Q: Do scientists establish absolute proof or truth?

will lead to an estimated additional 20,000 skin cancer fatalities over the next half century in the United States alone. Worldwide, millions of lives are at stake. These examples, and countless others, show that our health is closely linked to that of the planet.

A scarcity of new cropland and fresh water plus the negative effects of soil erosion, air pollution, and hotter summers on crop yields is slowing the growth of the world grain harvest. Combined with continuing rapid population growth, this has reversed the steady rise in grain output per person that the world had become accustomed to. Between 1950 and 1984, the historical peak year, world grain production per person climbed by nearly 40%. Since then, it has fallen roughly 1% a year, with the drop concentrated in poor countries. With food imports in these nations restricted by rising external debt, there are far more hungry people today than ever before.

On the economic front, the signs are equally ominous: soil erosion, deforestation, and overgrazing are adversely affecting productivity in the farming, forestry, and livestock sectors, slowing overall economic growth in agriculturally based economies. The World Bank reports that after three decades of broad-based economic gains, incomes fell during the 1980s in 40 developing countries. Collectively, these nations contain more than 800 million people—almost three times the population of North America and nearly one-sixth that of the world. In Nigeria, the most populous country in the ill-fated group, the incomes of 123 million people fell a painful 29%, exceeding the fall in U.S. incomes during the depression decade of the 1930s.

Anyone who thinks these environmental, agricultural, and economic trends can easily be reversed need only look at population projection. Those of us born before the middle of this century have seen the world population more than double to 5.5 billion. We have witnessed the environmental effects of adding 3 billion people, especially in developing countries. We can see the loss of tree cover, the devastation of grasslands, the soil erosion, the crowding and poverty, the land hunger, and the air and water pollution associated with this addition of people. But what if 4.2 billion more people are added by 2050, over 90% of them in developing countries, as now projected by U.N. population experts.

The decline in living standards that was once predicted by some ecologists from the combination of continuing rapid population growth, spreading environmental

degradation, and rising external debt has become a reality for one-sixth of humanity. Moreover, if a more comprehensive system of national economic accounting were used—one that incorporated losses of natural capital, such as topsoil and forests, the destruction of productive grasslands, the extinction of plant and animal species, and the health costs of air and water pollution, nuclear radiation, and increased ultraviolet radiation—it might well show that most of humanity suffered a decline in living conditions in the 1980s.

Today we study archaeological sites of civilizations that were undermined by environmental deterioration. The wheatlands that made North Africa the granary of the Roman Empire are now largely desert. The early civilizations of the Tigris–Euphrates Basin declined as the waterlogging and salting of irrigation systems slowly shrank their food supply. And the collapse of the Mayan civilization that flourished in the Guatemalan lowlands from the third century B.C. to the ninth century A.D. may have been triggered by deforestation and soil erosion.

No one knows for certain why centers of Mayan culture and art fell into neglect, nor whether the population of 1 million to 3 million moved or died off, but recent progress in deciphering hieroglyphs in the area adds credence to the environmental decline hypothesis. One of those involved with the project, Linda Schele of the University of Texas, observes: "They were worried about war at the end. Ecological disasters, too. Deforestation. Starvation. I think the population rose to the limits their technology could bear. They were so close to the edge, if anything went wrong, it was all over."

Whether the Mayan economy had become environmentally unsustainable before it actually began to decline, we do not know. We do know that ours is.

Critical Thinking

1. Do you agree with the author that we need to bring about an Environmental Revolution within a few decades? Explain.

2. Do you believe that this can be done by making minor adjustments in the global economy or that the global economy needs to be restructured to put less stress on Earth's natural systems?

ple plow up the ground, pull down the trees, kill everything. The tree says: "Don't. I am sore. Don't hurt me." But they chop it down and cut it up. The spirit of the land hates them.... The white people destroy all. They blast rocks and scatter them on the ground. The

rock says: "Don't. You are hurting me." But the white people pay no attention.... How can the spirit of the Earth like the white man?... Everywhere the white man has touched the Earth it is sore.

A: No. Scientific laws and theories are based on statistical probabilities, not on certainties.

Human-Centered Worldviews Today   Most people in today's industrial-consumer societies have a **planetary management worldview**, which has gained wide acceptance during the past 50 years. According to this human-centered worldview, human beings, as the planet's most important and dominant species, can and should manage the planet mostly for their own benefit. Other species have only *instrumental value*; that is, their value depends on whether or not they are useful to us.

These are the basic beliefs of this worldview:

- *We are the planet's most important species, and we are apart from and in charge of the rest of nature.*

- *There is always more, and it's all for us.* Earth has an essentially unlimited supply of resources, which we gain access to through use of science and technology. If we deplete a resource, we will find substitutes. If resources become scarce or a substitute can't be found, we can mine the moon, asteroids, or other planets. If we pollute an area, we can invent a technology to clean it up, dump it into space, or move into space ourselves. If we extinguish other species, we can use genetic engineering to create new and better ones.

- *All economic growth is good, more economic growth is better, and the potential for economic growth is limitless.*

- *A healthy environment depends on a healthy economy.*

- *Our success depends on how well we can understand, control, and manage the planet for our benefit.*

There are several variations of this worldview:

- *No-problem.* We don't face a crisis of unsustainability (Guest Essay, p. 20). There are no environmental, population, or resource problems that can't be solved by more economic growth, better management, and better technology.

- *Free-market school.* The best way to manage the planet is through a truly free-market global economy with minimal government interference. Free-market advocates would convert essentially all public property resources to private property resources and let the marketplace, governed by true free-market competition (pure capitalism), decide essentially everything.

- *Responsible planetary management.* We have serious environmental, resource, and population problems that we must deal with by becoming better and more responsible planetary managers. People believing this follow the pragmatic principle of *enlightened self-interest*. Better Earth care is better self-care. They believe we can sustain our species by using a mixture of market-based competition, better technology, and some government intervention to promote sustainable forms of economic growth, prevent abuse of power in the marketplace, and protect and manage public and common-property resources.

- *Spaceship-Earth.* A variation of responsible planetary management in which Earth is seen as a spaceship—a complex machine that we can understand, dominate, change, and manage to prevent environmental overload and provide a good life for everyone.

- *Stewardship.* We have an ethical responsibility to be caring and responsible managers or stewards who tend the earth as if it were a garden. We can and should make the world a better place for ourselves and other species through love, care, and knowledge.

Can We Manage the Planet?   Some people believe that ever-increasing population, production, and consumption will severely stress the natural processes that not only renew the air, water, and soil but also support all life and economies. They compare our pursuit of unlimited economic growth and growth in the human population to being on a treadmill that moves faster and faster. They believe that sooner or later we will fall off the treadmill or cause it to break down because of our limited knowledge and managerial skills compared to the incredible complexity of Earth's life-support systems.

These individuals believe that human-centered worldviews won't lead to sustainability. They argue that the unhindered free-market approach won't work because it is based on mushrooming losses of Earth capital (Spotlight, p. 8) and focused on short-term economic benefits regardless of the long-term harmful consequences. They agree with economist Alfred Kahn, head of the Council on Wage and Price Stability under President Carter: "No one in his or her right mind would argue that the competitive market system takes care of protecting the environment—it does not."

These critics also contend that the spaceship-Earth and stewardship versions of planetary management won't work. To such people, thinking of Earth as a spaceship or a garden—simplified human constructs—is an oversimplified and misleading way to view an incredibly complex and ever-changing planet that is the result of billions of years of evolution.

For example, they point out that we don't even know how many species live on Earth, much less what their roles are and how they interact with one another

Q: How much of the commercial energy used in the world comes from nonrenewable resources?

and their nonliving environment. We have only an inkling of what goes on in a handful of soil, a meadow, a patch of forest, a pond, or any other part of the earth. They liken us to technicians who think they can build and repair car motors after a couple minutes of superficial training.

As biologist David Ehrenfeld puts it, "In no important instance have we been able to demonstrate comprehensive successful management of the world, nor do we understand it well enough to manage it even in theory." Environmental educator David Orr (Guest Essay, p.128) says we are losing rather than gaining the knowledge and wisdom needed to adapt creatively to continually changing environmental conditions:

> On balance, I think, we are becoming more ignorant because we are losing cultural knowledge about how to inhabit our places on the planet sustainably, while impoverishing the genetic knowledge accumulated through millions of years of evolution.... Most research is aimed to further domination of the planet. Considerably less of it is directed at understanding the effects of domination. Less still is aimed to develop ecologically sound alternatives that enable us to live within natural limits.

These are sobering thoughts for those who see planetary management as the solution to our problems.

Even if we had enough knowledge and wisdom to manage spaceship Earth, some critics see this as a threat to individual freedom in order to survive. We can compare the life of astronauts on spaceships with what life might be like on spaceship Earth under a comprehensive system of planetary management. The astronauts have virtually no individual freedom, with virtually all of their actions dictated by a central command (ground control). Otherwise, they cannot survive.

Others such as theologian Thomas Berry call the industrial-consumer society built upon the human-centered, planetary management worldview the "supreme pathology of all history":

> We can break the mountains apart; we can drain the rivers and flood the valleys. We can turn the most luxuriant forests into throwaway paper products. We can tear apart the great grass cover of the western plains, and pour toxic chemicals into the soil and pesticides onto the fields, until the soil is dead and blows away in the wind. We can pollute the air with acids, the rivers with sewage, the seas with oil—all this in a kind of intoxication with our power for devastation.... We can invent computers capable of processing ten million calculations per second. And why? To increase the volume and speed with which we move natural resources through the consumer economy to the junk pile or the waste heap.... If, in these activities, the topography of the planet is damaged, if the environment is made inhospitable for a multitude of living species, then so be it. We are, supposedly, creating a technological wonderworld.... But our supposed progress ... is bringing us to a wasteworld instead of a wonderworld.

Life-Centered Worldviews: Working with the Earth Critics of human-centered worldviews believe that such worldviews should be expanded to recognize *inherent* or *intrinsic value* to all forms of life (that is, value that exists regardless of these life-forms' potential or actual use to us). Proponents of such *life-centered* or *biocentric* worldviews believe that all species have an inherent right to live and flourish, or at least to struggle to exist—to play their roles in evolution.

There are many biocentric or life-centered worldviews, several of them overlapping in some of their beliefs. One such *biocentric* worldview, known as the **Earth-wisdom worldview**, has the following beliefs, which are in sharp contrast to the major beliefs of human-centered worldviews:

- *Nature exists for all of Earth's species, not just for us, and we are not apart from or in charge of the rest of nature.* We need the earth, but the earth does not need us.

- *There is not always more, and it's not all for us.* Earth's limited resources should not be wasted and should be used sustainably for us and all species.

- *Some forms of economic growth are beneficial, and some are harmful.* Our goals should be to design economic and political systems that encourage Earth-sustaining forms of growth and discourage or prohibit Earth-degrading forms.

- *A healthy economy depends on a healthy environment.* Our survival, life quality, and economies are totally dependent on the rest of nature (Earth capital, Solutions, p. 8).

- *Our success depends on learning to cooperate with one another and with the rest of nature instead of trying to dominate and manage Earth for our own use.* Because nature is so incredibly complex and always changing, we will never have enough information and understanding to effectively manage the planet.

People with this or other life-centered worldviews reject claims that they are antipeople or against celebrating humanity's special qualities and achievements. Rather, they call for us to expand our sense of compassion and caring to individuals, species, and ecosystems—not just humans—including future generations of all forms of life. To them, recognizing the intrinsic value of all life is the best way to serve and love people.

Others say we don't need to be biocentrists to value life. Human-centered stewardship also calls for us to value individuals, species, and ecosystems as part of our responsibility as Earth's caretakers.

2-3 SOLUTIONS: LIVING SUSTAINABLY

Mindquake   Questioning and changing one's worldview is difficult and threatening. It is a cultural *mindquake* that involves examining many of our most basic beliefs. However, once individuals change their worldview, it no longer makes sense for them to do things in the old ways. If enough people do this, then tremendous cultural change, once considered impossible, can take place rapidly.

Earth Ethics: Respecting Life   Here are some ethical guidelines various biocentrists have proposed for helping us work with the earth:

Ecosphere and Ecosystems

- We should not deplete or degrade Earth's physical, chemical, and biological capital, which supports all life and human economic activities.

- We should try to understand and cooperate with the rest of nature. The earth does not belong to us; we belong to the earth.

- We should work with the rest of nature to sustain the ecological integrity, biodiversity, and adaptability of Earth's life-support systems for us and other species.

- We have to alter nature to meet our needs or wants, but we should choose methods that do the least possible harm to us and other living things.

- Before altering nature, we should carry out an Environmental Impact Analysis (EIA) and a Grandchild Impact Analysis (GIA) to help us decide whether to intervene and to discover how to inflict the minimum short- and long-term harm.

Species and Cultures

- Every species has a right to live, or at least to struggle to live, simply because it exists.

- We have the right to defend ourselves against individuals of species that do us harm and to use

individuals of species to meet vital needs, but we should strive not to cause the premature extinction of any wild species.

- The best way to protect species and individuals of species is to protect the ecosystems in which they live.

- No human culture should become extinct because of our actions.

Individual Responsibility

- We should not inflict unnecessary suffering or pain on any animal we raise or hunt for food or use for scientific or other purposes.

- We should leave wild things in the wild unless their survival depends on human protection.

- To prevent excessive deaths of people and other species, people should prevent excessive births.

- We should leave the earth in as good (or better) shape than we found it.

- We should strive to live more lightly on the earth, not because of guilt or fear, but because of a desire to make the world a better place.

- We should get to know, care about, and defend a piece of the earth.

Earth Education: A Personal View   Learning how to work with the earth requires a foundation of Earth education. I believe that the main goals of such an education should include the following:

- *Developing respect for all life.*

- *Understanding what we know about how the earth works and sustains itself.*

- *Understanding connections.* These include connections within nature, between people and the rest of nature, between people with different cultures and beliefs, between generations, among the problems we face, and among the solutions to these problems.

- *Understanding and evaluating one's worldview.*

- *Becoming a wisdom seeker instead of an information vessel.* We need a *wisdom revolution* that enriches our minds and lives, not an *information revolution* that clogs our brains and dims our capacity to care and share and that too often is used to dominate, deplete, and degrade the earth.

- *Learning how to evaluate and resist advertising.* Most of the $250 billion spent worldwide on advertising each year (an average of $48 per person or

Q: How much of the commercial energy used in the United States comes from nonrenewable resources?

$448 per American) is designed to get us to consume more by making us unhappy with what we have. As humorist Will Rogers put it, "Too many people spend money they haven't earned to buy things they don't want, to impress people they don't like." Learning how to detect psychological manipulation by analyzing TV and print ads, beginning in elementary school, is a superb way to teach critical thinking.

- *Learning to live sustainably in a place.* This would be a piece of Earth to which we are rooted or emotionally attached and whose sustainability and adaptability we feel driven to nurture and defend.

- *Fostering a desire to make the world a better place and to act on this desire.* As David Orr puts it, education should help students "make the leap from 'I know' to 'I care' to 'I'll do something.'"

Making Earth education the center of the learning process will not be an easy task because most teachers and members of the educational establishment are trained to think primarily in terms of disciplines, and they vigorously guard their turfs (and budgets). Shifting the focus of our education system involves the following:

- *Exposing teachers, media people, and corporate and government leaders to Earth education.* This would involve the use of summer workshops, visiting teachers, team teaching, week-long retreats, and other devices.

- *Inserting examples of Earth thinking into teaching materials at all levels, beginning in kindergarten.*

- *Requiring every student graduating from high school, college, and any professional school to take one or more courses in Earth education.*

- *Developing and widely using measures of ecological literacy.*

- *Listening to the earth.* In 1948 Aldo Leopold said, "We can be ethical only in relation to something we can see, feel, understand, love, or otherwise have faith in." This requires experiencing the earth not only with our minds but also with our senses and our hearts (Solutions, p. 36).

- *Listening to children: bottom-up education.* We need to take the time to listen—truly listen—to children. Many children that I have talked with believe that much of what we are doing to the earth (and thus to them) is stupid and wrong, and they don't buy the excuses we give for not changing the way we do things.

- *Living more simply.* Although attaining happiness through material acquisition is denied by virtually every major religion and philosophy, it is preached incessantly by modern advertising. Some affluent people in MDCs, however, are adopting a lifestyle of *voluntary simplicity*, based on doing and enjoying more with less by learning to live more simply but richly. This is based on Mahatma Gandhi's *principle of enoughness.* "The earth provides enough to satisfy every person's need but not every person's greed." It means asking oneself, "How much is enough?" This is not an easy question to answer because affluent people are conditioned to want more and more, and they often think of such wants as vital needs. Voluntary simplicity by those who have more than they really need should not be confused with the *forced simplicity* of the poor, who do not have enough to meet their basic needs for food, clothing, shelter, clean water and air, and good health.

- *Avoiding common traps that lead to denial, indifference and inaction.* These include: **(1)** *gloom-and-doom pessimism* (it's hopeless), **(2)** *blind technological optimism* (science and technofixes will always save us), **(3)** *fatalism* (we have no control over our actions and the future), **(4)** *extrapolation to infinity* ("If I can't change the entire world quickly, I won't try to change any of it."), **(5)** *paralysis-by-analysis* (searching for the perfect worldview, philosophy, solutions, and scientific information before doing anything), and **(6)** *faith in simple, easy answers.* We should be guided by philosopher and mathematician Alfred North Whitehead, who advised, "Seek simplicity and distrust it," and by writer and social critic H. L. Mencken, who warned, "For every problem there is a solution—simple, neat, and wrong."

Working with the earth requires each of us to make a personal commitment to strive to live an environmentally ethical life—not because it is mandated by law but because it is the right thing to do. It is our responsibility to ourselves, our children and grandchildren, our neighbors, and the earth.

Many people now see themselves as members of a global community with ultimate loyalty to the planet, not merely a particular country (Figure 2-1). These Earth citizens urge individual citizens to *think globally and act locally* to work with the earth. They are guided by historian Arnold Toynbee's observation: "If you make the world ever so little better, you will have done splendidly, and your life will have been worthwhile," and by George Bernard Shaw's reminder that "indifference is the essence of inhumanity."

A: 91% (84% from fossil fuels and 7% from nuclear power)

Emotional Learning: Earth Wisdom

SOLUTIONS

Formal education is important, but Aldo Leopold, Henry David Thoreau, Gary Snyder, and many others believe it's not enough. Such Earth thinkers and many religious thinkers believe that much of the essence, rhythms, and pulse of the earth within and around us must also be experienced by our intuitive senses and emotions.

They urge us to take the time to escape the cultural and technological "body armor" we use to insulate ourselves from wild nature and experience nature directly. They suggest that we reenchant our senses and kindle a sense of awe, wonder, and humility by standing under the stars, sitting in a forest, taking in the majesty and power of an ocean, or experiencing a stream, lake, or other part of untamed nature. We might pick up a handful of soil and try to sense the teeming microscopic life forms in it that keep us alive. We might look at a tree, a mountain, a rock, a bee and try to sense how they are a part of us, and we a part of them.

Earth thinker Michael J. Cohen suggests that each of us recognize who we really are by saying:

I am a desire for water, air, food, love, warmth, beauty, freedom, sensations, life, community, place, and spirit in the natural world. These pulsating feelings are the Planet Earth, alive and well within me. I have two mothers: my human mother and my planet mother, Earth. The planet is my womb of life.

Earth philosophers call for us to recognize that the technological cocoon we enclose ourselves in and the feeling of self-importance we hold as a species have given us a severely distorted picture of what is really important. As theologian Thomas Berry put it:

So long as we are under the illusion that we know best what is good for the earth and for ourselves, then we will continue our present course, with its devastating consequences on the entire earth community.... We need only listen to what the Earth is telling us ... the time has come when we will listen, or we will die.

Experiencing nature emotionally allows us to get in touch with our deepest self, which has sensed from birth that when we destroy and degrade the natural systems that support us, we are attacking ourselves. As wilderness protector Dave Foreman puts it:

When a chain saw slices into the heartwood of a two-thousand-year-old Coast Redwood, it's slicing into my guts. When a bulldozer rips through the Amazon rain forests, it's ripping into my side. When a ... whaler fires an exploding harpoon into a great whale, my heart is blown to smithereens. I am the land, the land is me. Why shouldn't I be emotional, angry, passionate? Madmen and madwomen are wrecking this beautiful, blue-green, living Earth.... We must love Earth and rage against her destroyers.... The Earth is crying. Do we hear?

Many psychologists believe that consciously or unconsciously we spend much of our lives in a search for roots—something to anchor us in a bewildering and frightening sea of change. As philosopher Simone Weil observed, "To be rooted is perhaps the most important and least recognized need of the human soul."

Earth philosophers say that to be rooted, each of us needs to find a *sense of place*—a stream, a mountain, a yard, a neighborhood lot, or any piece of the Earth we feel truly at one with. It can be a place where we live or a place we occasionally visit and experience in our inner being. When we become part of a place, it becomes a part of us. Then we are driven to defend it from harm and to help heal its wounds.

Emotionally experiencing our connectedness with the Earth leads us to recognize that the healing of the earth and the healing of the human spirit are one and the same. We need to discover and tap into the green fire that burns in our hearts and use this as a force for working with the Earth.

Some analysts, however, consider such emotional learning to be unscientific, mystical poppycock. They believe that better scientific understanding of how the earth works is the only way to achieve sustainability. What do you think?

The main ingredients of an environmental ethic are caring about the planet and all of its inhabitants, allowing unselfishness to control the immediate self-interest that harms others, and living each day so as to leave the lightest possible footprints on the planet.

ROBERT CAHN

Critical Thinking

1. What obligations, if any, concerning the environment do you have to future generations? How many generations ahead do we have responsibilities? List the most important environmental benefits and harmful conditions passed on to you by the previous two generations.

Q: Upon what three factors does life on Earth depend?

GUEST ESSAY

Peter Montague

Peter Montague is director of the Environmental Research Foundation in Washington, D.C. The Foundation studies environmental problems and informs the public about these problems and the technologies and policies that might help solve them. He has served as project administrator of a hazardous waste research program at Princeton University and has taught courses in environmental impact analysis at the University of New Mexico. He is the coauthor of two books on toxic heavy metals in the natural environment and is editor of Rachel, *an informative and readable newsletter on environmental problems, focusing on hazardous waste.*

Environmentalism as we have known it for over 25 years is dead. The environmentalism of the 1970s advocated strict numerical controls on releases of *dangerous wastes* (any unwanted or uncontrolled materials that can harm living things or disrupt ecosystems) into the environment. Industry's ability to create new hazards, however, quickly outstripped government's ability to establish adequate controls and enforcement programs.

After so many years of effort by government and by concerned citizens (the environmental movement), the overwhelming majority of dangerous chemicals is still not regulated in any way. Even those few that are covered by regulations have not been adequately controlled.

In short, the *pollution management* approach to environmental protection has failed and stands discredited; *pollution prevention* is our only hope. An ounce of prevention really is worth a pound of cure.

Here, in list form, is the new environmentalism that is emerging:

- *All waste disposal—landfilling, incineration, deep-well injection—is polluting because disposal means dispersal into the environment.* Once wastes are created they cannot be contained or controlled because of the scientific laws of matter and energy [Chapter 3]. The old environmentalism failed to recognize this important truth, and thus squandered enormous resources trying to achieve the impossible. We in the United States presently spend about $90 billion per year on pollution control. Yet, the global environment is increasingly threatened by a buildup of heat because of heat-trapping gases we emit into the atmosphere; at least half the surface of the planet is being subjected to damaging ultraviolet radiation from the sun as a result of ozone-depleting chemicals we have discharged into the atmosphere; and vast regions of the United States, Canada, and Europe are suffering from loss of forests, crop productivity, and fish as a result of acid rain (caused by releases of sulfur and nitrogen compounds, chiefly by power plants and automobiles) and other air pollutants. Soil and water are danger-

ously polluted at thousands of locales where municipal garbage and industrial wastes have been (and continue to be) dumped or incinerated; thousands of such sites remain to be discovered, according to U.S. government estimates.

- *The inevitable result of our reliance upon waste treatment and disposal systems has been an unrelenting buildup of exotic synthetic toxic materials in humans and other forms of life worldwide.* For example, breast milk of women in industrialized countries like the United States is so contaminated with pesticides and industrial hydrocarbons that, if human milk were bottled and sold commercially, it could be banned by the Food and Drug Administration (FDA) as unsafe for human consumption. If a whale today beaches itself on the shores of the United States and dies, its body must be treated as a "hazardous waste" because whales contain legally hazardous concentrations of PCBs (polychlorinated biphenyls—a class of industrial toxins).

- *The ability of humans and other life forms to adapt to changes in their chemical environment is strictly limited by the genetic code each form of life inherits.* Continued contamination occurring hundreds of times faster than we can adapt will drive humans to increasingly widespread sickness, to degradation of the species, and could ultimately lead to extinction.

- *Damage to humans (and to other life forms) is abundantly documented.* Birds, fish, and humans in industrialized countries like the United States are enduring steadily rising levels of cancer and other serious disorders attributable to pollution. An astonishing 88% of children under 6 years old in the United States have enough toxic lead in their blood that they perform below par on standardized tests of physical, mental, and emotional development.

If we will but look, the handwriting is on the wall everywhere.

To deal with these problems, industrial societies must abandon their reliance upon waste treatment and disposal and upon the regulatory system of numerical standards created by government to manage the damage that results from relying on waste disposal instead of waste prevention. We must—relatively quickly—move the industrialized and industrializing countries to new technical approaches accompanied by new industrial goals—namely, clean production or zero discharge systems.

The concept of "clean production" involves industrial systems that avoid or eliminate dangerous wastes and dangerous products and minimize the use of raw materials, water, and energy. Goods manufactured in a clean production process must not damage natural ecosystems throughout their entire life cycle, including **(1)** raw mate-

(continued)

rials selection, extraction, and processing; **(2)** product conceptualization, design, manufacture, and assembly; **(3)** materials transport during all phases; **(4)** industrial and household usage; and **(5)** reintroduction of the product into industrial systems or into the environment when it no longer serves a useful function.

Clean production does not include "end-of-pipe" pollution controls such as filters or scrubbers or chemical, physical, or biological treatment. Measures that pretend to reduce the volume of waste by incineration or concentration, that mask the hazard by dilution, or that transfer pollutants from one environmental medium to another are also excluded from the concept of "clean production."

A new industrial pattern, and a new environmentalism, is thus emerging. It insists that the long-term well-being of humans and other species must be factored into our production and consumption plans. These new requirements are not optional; human survival depends upon our willingness to make, and pay for, the necessary changes.

Critical Thinking

1. Do you agree with the author that the *pollution management* approach to environmental protection practiced during the past 25 years has failed and must be replaced with a *pollution prevention* approach? Explain.

2. List key economic, health, consumption, and lifestyle changes you might experience as a consequence of switching from pollution control to pollution prevention. What changes might the next generation face?

---

2. Do you believe that all chemicals we release or propose to release into the environment should be assumed to be potentially harmful until shown to be safe? Explain. What major effects might adopting this principle have on your life and lifestyle? On the national economy?

3. Which of the various worldviews discussed in this chapter comes closest to your own worldview? Does the way you live reflect your worldview?

4. Which (if any) of the ethical guidelines on pp. 34–35 do you disagree with? Explain. Can you suggest any additional ethical guidelines for working with the earth?

*5. Make a concept map of the key ideas in this chapter using the section heads and subheads and the key terms (shown in boldface type in the chapter). See the inside front cover and Appendix 4 for information on concept maps.

P A R T I I
Principles and Concepts

Animal and vegetable life is too complicated a problem for human intelligence to solve, and we can never know how wide a circle of disturbance we produce in the harmonies of nature when we throw the smallest pebble into the ocean of organic life.

GEORGE PERKINS MARSH

3 Matter and Energy Resources: Types and Concepts

Saving Energy, Money, and Jobs in Osage, Iowa

Osage, Iowa (population about 4,000), has become the energy-efficiency capital of the United States. It began in 1974 when easygoing Wes Birdsall, general manager of Osage Municipal Gas and Electric Company, started going door-to-door preaching. He wanted the townspeople to save energy and reduce their natural gas and electric bills. The utility would save money, too, by not having to add a new power plant.

Wes started his crusade by telling homeowners about the importance of insulating walls and ceilings and of plugging of leaky windows and doors. He also advised people to replace their incandescent light bulbs with more efficient fluorescent bulbs and to turn down the temperature on water heaters and wrap them with insulation—an economic boon to the local hardware and lighting stores. Wes also suggested saving water and fuel by installing low-flow shower heads.

He stepped up his campaign by offering to give every building in town a free thermogram—an infrared scan that shows where heat escapes (Figure 3-1). When people could see the energy (and money) hemorrhaging out of their buildings, they took action to plug these leaks—again helping the local economy.

Since 1974 the town has cut its natural gas consumption by 45%; no mean feat in a place where winter temperatures can plummet to –103°C (–80°F). In addition, the utility company saved enough money to prepay all its debt, accumulate a cash surplus, and cut inflation-adjusted electricity rates by a third (which attracted two new factories). Furthermore, each household saves more than $1,000 per year. This money supports jobs, and most of it circulates in the local economy. Before the energy-efficiency revolution, about $1.2 million a year went out of town— usually out of state—to buy energy. What are your local utility and community doing to improve energy efficiency and stimulate the local economy?

Figure 3-1 An infrared photo showing heat loss around the windows, doors, roofs, and foundations (red, white, and yellow colors) of houses and stores in Plymouth, Michigan. Wes Birdsall provided similar thermograms made for houses in Osage, Iowa. The average U.S. house has heat leaks and air infiltration equivalent to leaving a window wide open during the heating season. Because of poor design, most U.S. office buildings and houses waste about half the energy used to heat and cool them. Americans pay about $300 billion a year for this wasted heat—more than the entire annual military budget. (VANSCAN® Continuous Mobile Thermogram by Daedalus Enterprises, Inc.)

The laws of thermodynamics control the rise and fall of political systems, the freedom or bondage of nations, the movements of commerce and industry, the origins of wealth and poverty, and the general physical welfare of the human race.

FREDERICK SODDY (Nobel Laureate, Chemistry)

This chapter will answer the following questions:

- What is science? What is technology? What is environmental science?

- What are the basic forms of matter? What is matter made of? What makes matter useful to us as a resource?

- What are the major forms of energy? What energy resources do we rely on? What makes energy useful to us as a resource?

- What are physical and chemical changes? What scientific law governs changes of matter from one physical or chemical form to another?

- What are the three main types of nuclear changes that matter can undergo?

- What two scientific laws govern changes of energy from one form to another?

- How are the scientific laws governing changes of matter and energy from one form to another related to resource use and environmental disruption?

3-1 SCIENCE, TECHNOLOGY, AND ENVIRONMENTAL SCIENCE

What Is Science?   Which of the following statements are true?

- Science emphasizes facts or data.

- Science establishes absolute truth about nature.

- Science has a method—a how-to scheme—for learning about nature.

- Science emphasizes logic over creativity, imagination, and intuition.

The answer is that they are all false or mostly false. Let's see why.

Science is an attempt to discover order in nature and then use that knowledge to make predictions or projections about what will happen in nature. In this search for order scientists try to answer two basic questions: **(1)** *What events happen in nature over and over with the same results?* and **(2)** *How or why do things happen this way?*

What Do Scientists Do?   Scientists collect **scientific data**, or facts, by making observations and taking measurements, but this is not the main purpose of science. As French scientist Henri Poincaré put it, "Science is built up of facts, but a collection of facts is no more science than a heap of stones is a house."

Scientists try to describe what is happening in nature by organizing data into a generalization or scientific law. Thus scientific data are stepping stones to a **scientific law**, a description of the orderly behavior observed in nature—a summary of what we find happening in nature over and over in the same way. For example, after making thousands of measurements involving changes in matter, chemists concluded that in any physical change (such as converting liquid water to water vapor) or any chemical change (such as burning coal) no matter is created or destroyed. This summary of what we always observe in nature is called the *law of conservation of matter*, as discussed in more detail later in this chapter.

Scientists then try to explain how or why things happen the way a scientific law describes them. For example, why does the law of conservation of matter work? To answer such questions, investigators develop a **scientific hypothesis**, an educated guess that explains a scientific law or certain scientific facts. More than 2,400 years ago Greek philosophers proposed that all matter is composed of tiny particles called atoms, but they had no experimental evidence to back up their *atomic hypothesis*. Scientists also develop and use various types of **scientific models** to simulate complex processes and systems. Many are mathematical models that are run and tested using computers.

If many experiments by different scientists support the hypothesis or model, it becomes a **scientific theory**—a well-tested and widely accepted scientific hypothesis or model. During the last two centuries scientists have done experiments that elevated the atomic hypothesis to the *atomic theory of matter*. This theory, in turn, explains the law of conservation of matter with the idea that in any physical or chemical change no atoms can be created or destroyed. Establishing a scientific law or converting a scientific hypothesis or model to a widely accepted theory is an arduous and rigorous undertaking, and such laws and theories should not be taken lightly.

Scientific Methods   The ways scientists gather data and formulate and test scientific hypotheses, laws, and theories are called **scientific methods**. A scientific method is a set of questions with no particular rules for answering them. The major questions a scientist attempts to answer are these:

- What questions about nature should I try to answer?

- What relevant facts are already known, and what new data should I collect?

- How should I collect these data?

- How can I organize and analyze the data I have collected to develop a pattern of order or scientific law?

- How can I come up with a hypothesis to explain the law and use it to predict some new facts?

- Is this the simplest and only reasonable hypothesis?

- What new experiments should I run to test the hypothesis (and modify it if necessary) so it can become a scientific theory?

New discoveries happen in many ways. Some follow a data → law → hypothesis → theory sequence. At other times scientists simply follow a hunch or a bias and then do experiments to test it. Some discoveries occur when an experiment gives totally unexpected results and the scientist insists on finding out what happened. So, in reality, there are many methods of science rather than one scientific method.

Trying to discover order in nature requires logical reasoning, but it also requires imagination and intuition. As Albert Einstein once said, "Imagination is more important than knowledge, and there is no completely logical way to a new scientific idea." Intuition and creativity are as important in science as they are in poetry, art, music, and other great adventures of the human spirit that awaken us to the wonder, mystery, and beauty of the universe, the earth, and life.

What Is Technology?   **Technology** is the creation of new products and processes that are supposed to improve our chances for survival, our comfort level, and our quality of life. In many cases technology develops from known scientific laws and theories. Scientists invented the laser, for example, by applying knowledge about the internal structure of atoms. Applied scientific knowledge about chemistry has given us nylon, pesticides, laundry detergents, pollution control devices, and countless other products.

Some technologies arose long before anyone understood the underlying scientific principles. Aspirin, originally extracted from the bark of a willow tree, relieved pain and fever long before anyone found out how it did so. Similarly, photography was invented by people who had no inkling of its chemistry. Farmers crossbred new strains of livestock and crops long before biologists understood the principles of genetics.

Science and technology differ in the way the information and ideas they produce are shared. Many of the results of scientific research are published and passed around freely to be tested, challenged, verified, or modified, a process that strengthens the validity of scientific knowledge and helps expose cheaters. In contrast, technological discoveries are often kept secret until the new process or product is patented.

Environmental Science   **Environmental science** is the study of how we and other species interact with one another and with the nonliving environment of matter and energy. It is a holistic *physical and social science* that uses and integrates knowledge from physics, chemistry, biology (especially ecology), geology, geography, resource technology and engineering, resource conservation and management, demography (the study of population dynamics), economics, politics, and ethics. In other words it is a study of how everything works and interacts—a study of *connections* in the common home of all living things. This is an incredibly complex, uncertain, and controversial task (Spotlight, p. 43).

3-2   MATTER: FORMS, STRUCTURE, AND QUALITY

Nature's Building Blocks   **Matter** is anything that has mass (the amount of material in an object) and takes up space. It includes the solids, liquids, and gases around you and within your body. Matter is found in three *chemical forms*: **elements** (the distinctive

Q: How do most producer organisms get the nutrients they need?

Limitations and Misuse of Science

In supporting or opposing various environmental policies people have sometimes distorted facts and misused science. One problem involves arguments over the validity of data. There is no way to accurately measure how many metric tons of soil are eroded worldwide, how many hectares of tropical forest are cut, how many species become extinct, or how many metric tons of certain pollutants are emitted into the atmosphere or aquatic systems per year. Sometimes the estimates scientists make of such quantities may be off (either way) by as much as a factor of two or more. Even estimates of the number of people in the world or in a particular country or city are typically off by 5–10% either way.

We may legitimately argue over the numbers, but the point environmentalists want to make is that the trends in these areas are significant enough to be evaluated and addressed. They should not be dismissed because they are estimates (which is all we can ever have). This, however, does not relieve us from the responsibility of getting the best estimates possible.

Another limitation is that most environmental problems involve such complex mixtures of data, lack of data, hypotheses, and theories in the physical and social sciences that we don't have enough information to understand them very well.

This allows advocates of any proposed action or inaction on an environmental problem to support their beliefs, and it can lead to *scientific overkill*. It can also lead to the *paralysis-by-analysis* trap that insists that we fully understand a problem before taking any action—an impossible dream because of the inherent limitations of science and the complexities of environmental problems.

Since environmental problems won't go away, at some point we have to evaluate available information and make a political or economic decision about what to do (or not do), often based primarily on gut feelings, intuition, common sense (which each side claims it is using), and values. This is why differing worldviews and values are at the heart of most environmental controversies (Section 2-2). People with different worldviews and values can take the same information, come to completely different conclusions, and still be logically consistent (Guest Essays, p. 20 and 22).

A third problem is that some people misunderstand the nature and limitations of science, and others may deliberately misuse it. Those who say that something has or has not been "scientifically proven" imply falsely that science yields absolute proof or certainty. *Scientists can disprove things, but they can never establish absolute proof or truth.*

Instead of certainty, science gives us information in the following form: If we do so-and-so (say, add certain chemicals to the atmosphere at particular rates), there is a certain chance (high, moderate, or low) that we will cause various effects (such as change the climate or deplete ozone in the stratosphere). The more complex the system or problem being studied, the less certain the hypotheses, models, and theories used to describe and explain it.

Science advances by debate, argument, speculation, and controversy. Disputes among scientists over the validity of data and hypotheses and models (tentative ideas still being rigorously evaluated and tested) are what the media usually report. Such disagreements make juicier stories, but what's really important is the *consensus* among scientists about various scientific laws, theories, and issues (Spotlight, p. 6). This substantial agreement—the real knowledge of science—rarely gets reported, giving the public a false idea of the nature of science and of scientific knowledge. The best course is to seek out reports by bodies such as the U.S. National Academy of Sciences and the U.S. Office of Technology Assessment that attempt to summarize scientific consensus in key areas of science and technology.

---

building blocks of matter that make up every material substance), **compounds** (two or more different elements held together in fixed proportions by attractive forces called *chemical bonds*), and **mixtures** (combinations of elements, compounds, or both).

All matter is built from the 109 known chemical elements. Ninety-two of them occur naturally, and the other 17 have been synthesized in laboratories. To simplify things, chemists represent each element by a one- or two-letter symbol—for example: hydrogen (H), carbon (C), oxygen (O), nitrogen (N), phospho-

rus (P), sulfur (S), chlorine (Cl), fluorine (F), bromine (Br), sodium (Na), calcium (Ca), and uranium (U).

If you had a supermicroscope to look at elements and compounds, you would discover that they are made up of three types of building blocks: **atoms** (the smallest unit of matter that is unique to a particular element), **ions** (electrically charged atoms), and **molecules** (combinations of atoms of the same or different elements, held together by chemical bonds). Since ions and molecules are formed from atoms, atoms are the ultimate building blocks for all matter.

A: They produce them through photosynthesis.

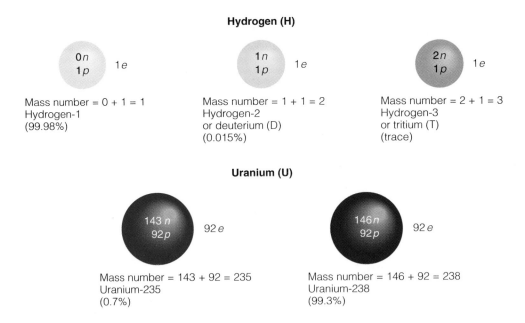

Figure 3-2 Isotopes of hydrogen and uranium. All isotopes of hydrogen have an atomic number of 1 because each has one proton in its nucleus; similarly, all uranium isotopes have an atomic number of 92. However, each isotope of these elements has a different mass number because its nucleus contains a different number of neutrons. Figures in parentheses show the percentage abundance by weight of each isotope in a natural sample of the element.

Hydrogen (H)

$0n$
$1p$ $1e$

Mass number = 0 + 1 = 1
Hydrogen-1
(99.98%)

$1n$
$1p$ $1e$

Mass number = 1 + 1 = 2
Hydrogen-2
or deuterium (D)
(0.015%)

$2n$
$1p$ $1e$

Mass number = 2 + 1 = 3
Hydrogen-3
or tritium (T)
(trace)

Uranium (U)

$143n$
$92p$ $92e$

Mass number = 143 + 92 = 235
Uranium-235
(0.7%)

$146n$
$92p$ $92e$

Mass number = 146 + 92 = 238
Uranium-238
(99.3%)

Some elements are found in nature as molecules. Examples are nitrogen and oxygen, which make up about 99% of the volume of the air we breathe. Two atoms of nitrogen (N) combine to form a nitrogen gas molecule with the shorthand formula N_2 (read as "N-two"). The subscript after the symbol of the element gives the number of atoms of that element in a molecule. Similarly, most of the oxygen in the atmosphere exists as O_2 (read as "O-two") molecules. A small amount of oxygen, found mostly in the second layer of the atmosphere (stratosphere), exists as O_3 (read as "O-three") molecules; this type of oxygen is called *ozone.*

Elements can combine to form an almost limitless number of compounds, just as the letters of our alphabet have been combined to form almost a million English words. So far, chemists have identified more than 10 million compounds.

Matter is also found in three *physical states*: solid, liquid, and gas. Water, for example, exists as ice, liquid water, and water vapor, depending on its temperature and pressure. The differences among the three physical states of a sample of matter are in the relative orderliness of its atoms, ions, or molecules, with solids having the most orderly arrangement and gases the least orderly.

Atoms and Ions   If you increased the magnification of your supermicroscope, you would find that each different type of atom is composed of a certain number of *subatomic particles*. The main building blocks of an atom are positively charged **protons** (represented by the symbol *p*), uncharged **neutrons** (*n*), and negatively charged **electrons** (*e*). Many other sub-

atomic particles have been identified in recent years, but they need not concern us here.

Each atom consists of a relatively small center, or **nucleus**, containing protons and neutrons, and one or more electrons in rapid motion somewhere around the nucleus. We can describe electrons only in terms of the probability that they might be at various locations outside the nucleus.

The distinguishing feature of an atom of any given element is the number of protons in its nucleus, called its **atomic number**. The simplest element, hydrogen (H), has only 1 proton in its nucleus; its atomic number is 1. Carbon (C), with 6 protons, has an atomic number of 6, while uranium (U), a much larger atom, has 92 protons and an atomic number of 92.

The mass of an electron is almost negligible compared with the mass of a proton or a neutron. As a result most of an atom's mass is concentrated in its nucleus. We describe the mass of an atom in terms of its **mass number**: the number of neutrons plus the number of protons in its nucleus. An atom of hydrogen with 1 proton and no neutrons has a mass number of 1, and an atom of uranium with 92 protons and 143 neutrons has a mass number of 235.

Although all atoms of an element have the same number of protons in their nuclei, they may have different numbers of uncharged neutrons in their nuclei and thus different mass numbers. These different forms of an element with the same atomic number but a different mass number are called **isotopes** of that element. Isotopes are identified by attaching their mass numbers to the name or symbol of the element. Hydrogen, for example, has three isotopes: hydrogen-1, or H-1; hydrogen-2, or H-2 (common name, deuterium); and hydrogen-3, or H-3 (common name, tritium). A natural sample of an element contains a mixture of its isotopes in a fixed proportion or percent abundance by weight (Figure 3-2).

Atoms of some elements can lose or gain one or more electrons to form **ions**: atoms or groups of atoms with one or more net positive (+) or negative (–) electrical charges. For example, an atom of sodium (Na) can lose one of its electrons and become a sodium ion with a positive charge of one (Na^+). An atom of chlorine (Cl) can gain an electron and become a chlorine ion with a negative charge of one (Cl^-). The number of positive or negative charges on an ion is shown as a superscript after the symbol for an atom or a group of atoms. Examples of other positive ions are calcium ions (Ca^{2+}) and ammonium ions (NH_4^+). Other common negative ions are nitrate ions (NO_3^-), sulfate ions (SO_4^{2-}), and phosphate ions (PO_4^{3-}).

Compounds Most matter exists as compounds. Chemists use a shorthand **chemical formula** to show the number of atoms (or ions) of each type found in the basic structural unit of a compound. The formula contains the symbols for each of the elements present and uses subscripts to show the number of atoms (or ions) of each element in the compound's basic structural unit. Each molecule of water, for example, consists of two hydrogen atoms chemically bonded to an oxygen atom, giving H_2O (read as "H-two-O") molecules. The subscript after the symbol of the element gives the number of atoms of that element in a molecule. Other examples you will encounter in this book are oxygen (O_2), ozone (O_3), nitrogen (N_2), nitrous oxide (N_2O), nitric oxide (NO), carbon monoxide (CO), carbon dioxide (CO_2), nitrogen dioxide (NO_2), sulfur dioxide (SO_2), ammonia (NH_3), hydrogen sulfide (H_2S), sulfuric acid (H_2SO_4), nitric acid (HNO_3), methane (CH_4), and glucose ($C_6H_{12}O_6$).

Matter Quality **Matter quality** is a measure of how useful a matter resource is, based on its availability and concentration. **High-quality matter** is organized and concentrated, and is usually found near the earth's surface. It has great potential for use as a matter resource. **Low-quality matter** is disorganized, dilute, or dispersed, and it is often found deep underground or dispersed in the ocean or in the atmosphere. It usually has little potential for use as a matter resource (Figure 3-3).

An aluminum can is a more concentrated, higher-quality form of aluminum than aluminum ore with the same amount of aluminum. That's why it takes less energy, water, and money to recycle an aluminum can than to make a new can from aluminum ore.

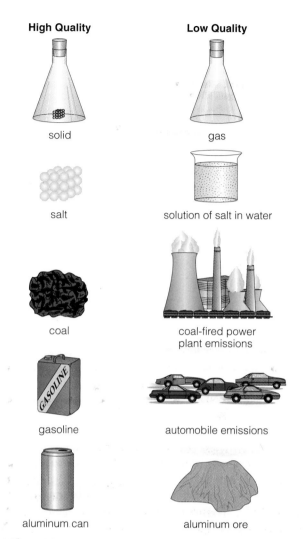

Figure 3-3 Examples of differences in matter quality. High-quality matter is fairly easy to extract and is concentrated. Low-quality matter is more difficult to extract and is more dispersed than high-quality matter.

3-3 ENERGY: TYPES, FORMS, AND QUALITY

Types **Energy** is the capacity to do work. You cannot pick up or touch energy, but you can use it to do work. You do work when you move matter, such as your arm or this book. Work, or matter movement, also is needed to boil liquid water and change it into steam—or to burn natural gas to heat a house or cook food. Energy is also the heat that flows automatically from a hot object to a cold object. Touch a hot stove and you experience this energy flow in a painful way.

Energy comes in many forms: light; heat; electricity; chemical energy stored in the chemical bonds in coal, sugar, and other materials; moving matter such as water, wind (air masses), and joggers; and nuclear energy emitted from the nuclei of certain isotopes.

A: 5–20% (that is, a loss of 80–95%)

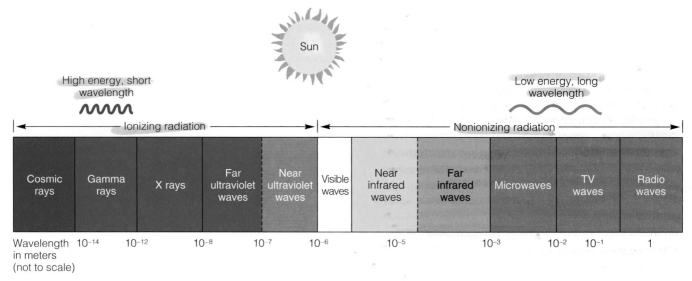

Figure 3-4 The electromagnetic spectrum: the range of electromagnetic waves, which differ in wavelength (distance between successive peaks or troughs) and energy content.

Scientists classify energy as either kinetic or potential. **Kinetic energy** is the energy that matter has because of its motion and mass. Wind (a moving mass of air), flowing streams, falling rocks, heat, electricity (flowing charged particles), moving cars—all have kinetic energy.

Radio waves, TV waves, microwaves, infrared radiation, visible light, ultraviolet radiation, X rays, gamma rays, and cosmic rays are all forms of kinetic energy traveling as waves and known as **electromagnetic radiation**. These forms of energy make up a wide band or spectrum of electromagnetic waves that differ in their wavelength (distance between two consecutive peaks or troughs) and energy content (Figure 3-4).

Cosmic rays, gamma rays, X rays, and ultraviolet radiation have enough energy to knock electrons from atoms and change them to positively charged ions. The resulting highly reactive electrons and ions can disrupt living cells, interfere with body processes, and cause many types of sickness, including various cancers. These potentially harmful forms of electromagnetic radiation are called **ionizing radiation**.

The other forms of electromagnetic radiation do not contain enough energy to form ions and are called **nonionizing radiation**. Some controversial evidence now suggests that long-term exposure to nonionizing radiation emitted by radios, TV sets, the video display terminals of computers, overhead electric power lines, electrically heated water beds, electric blankets, motors, and other electrical devices may also damage living cells. Some scientists dispute such claims.

Potential energy is stored energy that is potentially available for use. A rock held in your hand, an unlit stick of dynamite, still water behind a dam, and nuclear energy stored in the nuclei of atoms all have potential energy because of their position or the position of their parts. When you drop a rock held in your hand its potential energy changes into kinetic energy. When you burn gasoline in a car engine, the potential energy stored in the chemical bonds of its molecules changes into heat, light, and mechanical (kinetic) energy that propels the car.

An electric power plant burns some kind of fuel to make heat, which is then used to boil water into steam. The steam expands through turbines, where its thermal energy (the energy of heat) is converted into kinetic energy. The turbines turn generators, and electromagnetic energy—electricity—comes out and is transmitted by wire to factories and buildings. When you flip a light switch you are at the end of the following energy chain: fuel → heat → steam → kinetic energy → electricity.

Energy Resources Used by People *Some 99% of the energy used to heat Earth, and all our buildings, comes directly from the sun.* Without this direct input of solar energy, Earth's average temperature would be –240°C (–400°F), and life as we know it would not have arisen. Solar energy also helps recycle the carbon, oxygen, water, and other chemicals we and other organisms need to stay alive and healthy.

Broadly defined, **solar energy** includes both renewable direct energy from the sun and several forms of renewable energy produced indirectly by the sun's energy: wind, falling and flowing water (hydropower), and biomass (solar energy converted to chemical energy stored in the chemical bonds of organic compounds in trees and other plants).

The 99% of energy that comes directly from the sun is not sold in the marketplace. The remaining 1%,

Q: How much of the world's net primary productivity on land is used by the world's 5.6 billion people?

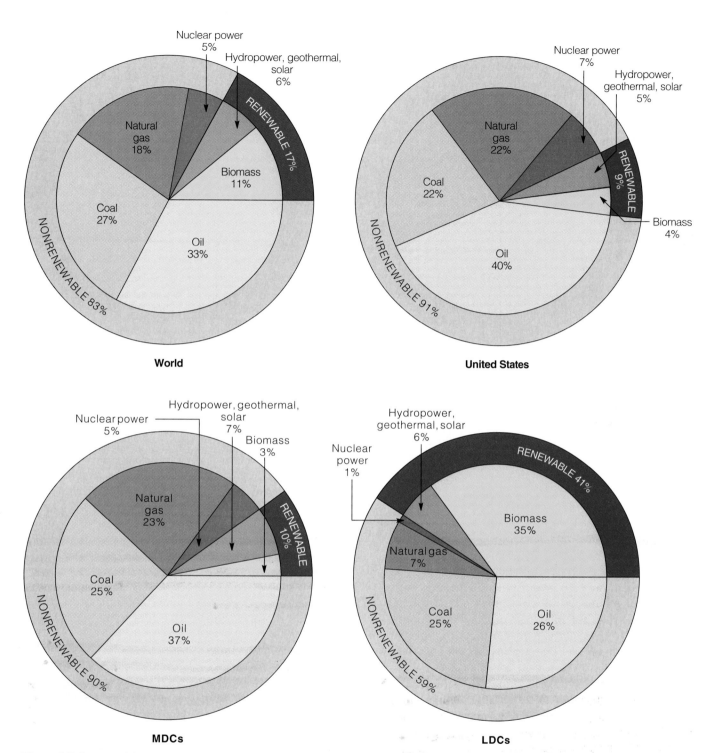

Figure 3-5 Commercial energy use by source in 1991 for the world, the United States, MDCs, and LDCs. This amounts to only 1% of the energy used in the world. The other 99% of the energy used to heat Earth comes from the sun and is not sold in the marketplace. (Data from U.S. Department of Energy, British Petroleum, and Worldwatch Institute)

the portion we generate to supplement the solar input, is either *commercial energy* sold in the marketplace or *noncommercial energy* used by people who gather fuelwood, dung, and crop wastes for their own use. Most commercial energy comes from extracting and burn-

ing mineral resources in the earth's crust, primarily nonrenewable fossil fuels.

MDCs and LDCs differ greatly in their sources of energy (Figure 3-5), the total amount used, and average energy use per person. The most important source

A: 40% (27% including terrestrial and aquatic productivity)

Figure 3-6 Generalized categories of the quality or usefulness (for performing various energy tasks) of different sources of energy. *High-quality energy* is concentrated and has great ability to perform useful work. *Low-quality energy* is dispersed and has little ability to do useful work. To avoid unnecessary energy waste, it's best to match the quality of an energy source with the quality of energy needed to perform a task.

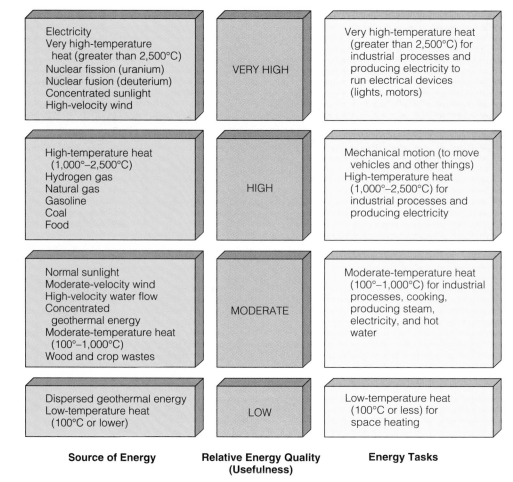

Source of Energy	Relative Energy Quality (Usefulness)	Energy Tasks
Electricity Very high-temperature heat (greater than 2,500°C) Nuclear fission (uranium) Nuclear fusion (deuterium) Concentrated sunlight High-velocity wind	VERY HIGH	Very high-temperature heat (greater than 2,500°C) for industrial processes and producing electricity to run electrical devices (lights, motors)
High-temperature heat (1,000°–2,500°C) Hydrogen gas Natural gas Gasoline Coal Food	HIGH	Mechanical motion (to move vehicles and other things) High-temperature heat (1,000°–2,500°C) for industrial processes and producing electricity
Normal sunlight Moderate-velocity wind High-velocity water flow Concentrated geothermal energy Moderate-temperature heat (100°–1,000°C) Wood and crop wastes	MODERATE	Moderate-temperature heat (100°–1,000°C) for industrial processes, cooking, producing steam, electricity, and hot water
Dispersed geothermal energy Low-temperature heat (100°C or lower)	LOW	Low-temperature heat (100°C or less) for space heating

of energy for LDCs is biomass—especially fuelwood—the main source of energy for heating and cooking for roughly half the world's population. Much of this energy, which accounts for about 15% of the energy consumed by humans, is not sold in the marketplace. Within a few decades one-fourth of the world's population in MDCs may face an oil shortage, but half the world's population in LDCs already faces a fuelwood shortage.

The United States is the world's largest user (and waster) of energy. With only 4.7% of the population, it uses 25% of the commercial energy, mostly by getting 91% of its energy from nonrenewable sources of energy—84% from fossil fuels (coal, oil, natural gas), and 7% from nuclear power (Figure 3-5). In contrast, India, with almost 16% of the world's people, uses only about 3% of the world's commercial energy. In 1993, 258 million Americans used more electricity for air conditioning than 1.2 billion Chinese used for all purposes. The average American uses as much energy as 6 Mexicans, 153 Bangladeshis, or 500 Ethiopians.

Energy Quality Energy quality is a measure of energy's ability to do useful work (Figure 3-6). **High-quality energy** is organized or concentrated and can perform a great deal of useful work. Examples of high-quality energy are electricity, coal, gasoline, concentrated sunlight, nuclei of uranium-235 used as fuel in nuclear power plants, and heat concentrated in fairly small amounts of matter so that its temperature is high.

By contrast, **low-quality energy** is disorganized or dispersed and has little ability to do useful work. An example is heat dispersed in the moving molecules of a large amount of matter, such as the atmosphere or a large body of water, so that its temperature is relatively low. For instance, the total amount of heat stored in the Atlantic Ocean is greater than the amount of high-quality chemical energy stored in all the oil deposits of Saudi Arabia. However, the ocean's heat is so widely dispersed that it can't be used to move things or to heat things to high temperatures.

We use energy to accomplish certain tasks, each requiring a certain minimum energy quality (Figure 3-6). Electrical energy, which is very high-quality energy, is needed to run lights, electric motors, and electronic devices. We need high-quality mechanical energy to move a car, but we need only low-temperature

Q: What two gases make up 99% of the volume of air in the troposphere?

air (less than 100°C) to heat homes and other buildings. It makes sense to match the quality of an energy source to the quality of energy needed to perform a particular task (Figure 3-6)—this saves energy and usually money.

3-4 PHYSICAL AND CHEMICAL CHANGES AND THE LAW OF CONSERVATION OF MATTER

Physical and Chemical Changes   A **physical change** is one that involves no change in chemical composition. For example, cutting a piece of aluminum foil into small pieces is a physical change—each cut piece is still aluminum. Changing a substance from one physical state to another is also a physical change. For example, when solid water, or ice, is melted or liquid water is boiled, none of the H_2O molecules involved are altered; instead, the molecules are organized in different spatial patterns.

In a **chemical change**, or **chemical reaction**, there is a change in the chemical composition of the elements or compounds involved. Chemists use shorthand chemical equations to represent what happens in a chemical reaction. A chemical equation shows the chemical formulas for the *reactants* (starting chemicals) and the *products* (chemicals produced) with an arrow placed between them. For example, when coal burns completely, the solid carbon (C) it contains combines with oxygen gas (O_2) from the atmosphere to form the gaseous compound carbon dioxide (CO_2):

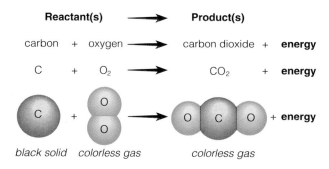

Reactant(s)	⟶	Product(s)
carbon + oxygen ⟶		carbon dioxide + **energy**
C + O_2 ⟶		CO_2 + **energy**

black solid   colorless gas   colorless gas

Energy is given off in this reaction, making coal a useful fuel. The reaction also shows how the burning of coal or any carbon-containing compounds, such as those in wood, natural gas, oil, and gasoline, adds carbon dioxide gas to the atmosphere.

The Law of Conservation of Matter: There Is No "Away"   You, like most people, probably talk about consuming or using up material resources, but the truth is that we don't consume matter. We only use some of Earth's resources for a while. We take materi-als from the earth, carry them to another part of the globe, and process them into products. These products are used and then discarded, burned, buried, reused, or recycled.

We may change various elements and compounds from one physical or chemical form to another, but in all physical and chemical changes we can't create or destroy any of the atoms involved. All we can do is rearrange them into different spatial patterns (physical changes) or different combinations (chemical changes). This fact, based on many thousands of measurements of matter undergoing physical and chemical changes, is known as the **law of conservation of matter**.

The law of conservation of matter means that there is no "away" to throw things to. *Everything we think we have thrown away is still here with us in one form or another.* We can collect dust and soot from the smokestacks of industrial plants, but these solid wastes must then be put somewhere. We can remove substances from polluted water at a sewage treatment plant, but the gooey sludge must either be burned (producing some air pollution), buried (possibly contaminating underground water supplies), or cleaned up and applied to the land as fertilizer (dangerous if the sludge contains nondegradable toxic metals, such as lead and mercury). High smokestacks can reduce some types of local air pollution but can increase air pollution in downwind areas. Banning the pesticide DDT in the United States and still selling it abroad means that it comes back to us as DDT residues in imported coffee, fruit, and other foods or by deposition from air masses moved long distances by winds—something environmentalists call the *circle of poison.*

The law of conservation of matter means that we will always be faced with the problem of what to do with some quantity of wastes. By placing much greater emphasis on pollution prevention and waste reduction, however, we can greatly reduce the amount of wastes we add to the environment (Guest Essay, p. 37).

3-5 NUCLEAR CHANGES

Natural Radioactivity   In addition to physical and chemical changes, matter can undergo a third type of change, known as a **nuclear change**. This occurs when nuclei of certain isotopes spontaneously change or are forced to change into one or more different isotopes. Three types of nuclear change are natural radioactive decay, nuclear fission, and nuclear fusion.

Natural radioactive decay is a nuclear change in which unstable isotopes spontaneously shoot out fast-moving particles, high-energy radiation, or both at a fixed rate. The unstable isotopes are called **radioactive isotopes**, or **radioisotopes**. This spontaneous process

Figure 3-7 The three principal types of ionizing radiation emitted by radioactive isotopes vary considerably in their penetrating power.

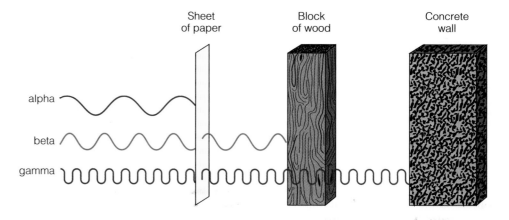

is called *radioactive decay*. It continues until the original isotope is changed into a new stable isotope, one that is not radioactive.

Radiation emitted by radioisotopes is damaging ionizing radiation. The most common form of ionizing energy released from radioisotopes is **gamma rays**, a form of high-energy electromagnetic radiation (Figure 3-4). High-speed particles emitted from the nuclei are a different form of ionizing radiation. The two most common types of ionizing particles emitted by radioactive isotopes are **alpha particles** (fast-moving, positively charged chunks of matter that consist of two protons and two neutrons) and **beta particles** (high-speed electrons). Figure 3-7 shows the relative penetrating power of alpha, beta, and gamma ionizing radiation. We are all exposed to small amounts of harmful ionizing radiation from both natural and human sources.

Each type of radioisotope decays spontaneously at a characteristic rate into a different isotope. This rate of decay can be expressed in terms of **half-life**—the time needed for *one-half* of the nuclei in a radioisotope to emit its radiation. Each radioisotope has a characteristic half-life, which may range from a few millionths of a second to several billion years (Table 3-1).

Half-life can also be used to estimate how long a sample of a radioisotope must be stored in a safe enclosure before it decays to what is considered a safe level. A general rule of thumb is that this takes about ten half-lives. Thus people would have to be protected from radioactive waste containing iodine-131 (which concentrates in the thyroid gland) for 80 days (10 × 8 days). Plutonium-239, which is produced in nuclear reactors and is used as the explosive in some atomic weapons, can cause lung cancer when its particles are inhaled in minute amounts. It must be stored safely for 240,000 years (10 × 24,000 years)—six times longer than the latest version of our species has existed.

Nuclear Fission: Splitting Nuclei
Nuclear fission is a nuclear change in which nuclei of certain iso-

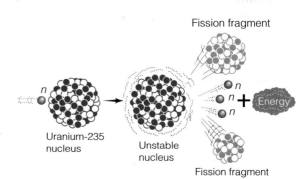

Figure 3-8 Fission of a uranium-235 nucleus by a neutron (*n*).

Table 3-1 Half-Lives of Selected Radioisotopes

Isotope	Half-Life	Radiation Emitted
Potassium-42	12.4 hours	Alpha, beta
Iodine-131	8 days	Beta, gamma
Cobalt-60	5.27 years	Beta, gamma
Hydrogen-3 (tritium)	12.5 years	Beta
Strontium-90	28 years	Beta
Carbon-14	5,370 years	Beta
Plutonium-239	24,000 years	Alpha, gamma
Uranium-235	710 million years	Alpha, gamma
Uranium-238	4.5 billion years	Alpha, gamma

topes with large mass numbers (such as uranium-235; Figure 3-2) are split apart into lighter nuclei when struck by neutrons; each fission releases two or three more neutrons and energy (Figure 3-8). Each of these neutrons, in turn, can cause an additional fission. For these multiple fissions to take place, there must be

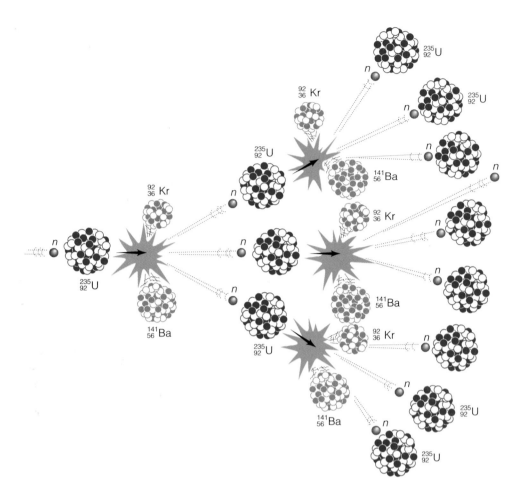

Figure 3-9 A nuclear chain reaction initiated by one neutron triggering fission in a single uranium-235 nucleus. This shows only a few of the trillions of fissions caused when a single uranium-235 nucleus is split within a critical mass of uranium-235 nuclei. The elements krypton (Kr) and barium (Ba) shown here as fission fragments are only two of many possibilities.

enough fissionable nuclei present to provide the **critical mass** needed for efficient capture of these neutrons.

Multiple fissions within a critical mass form a **chain reaction**, which releases an enormous amount of energy (Figure 3-9). Living cells can be damaged by the ionizing radiation released by the radioactive lighter nuclei and by high-speed neutrons produced by nuclear fission.

In an atomic or nuclear fission bomb an enormous amount of energy is released in a fraction of a second in an uncontrolled nuclear fission chain reaction. This reaction is initiated by an explosive charge, which suddenly pushes two masses of fissionable fuel together from all sides, causing the fuel to reach the critical mass needed for a chain reaction.

In the reactor of a nuclear electric power plant the rate at which the nuclear fission chain reaction takes place is controlled, so that under normal operation only one of each two or three neutrons released is used to split another nucleus. In conventional nuclear-fission reactors, nuclei of uranium-235 are split apart and release heat. The heat is used to produce high-pressure steam, which spins turbines, which in turn generate electricity.

Nuclear Fusion: Forcing Nuclei to Combine

Nuclear fusion is a nuclear change in which two isotopes of light elements, such as hydrogen (Figure 3-2), are forced together at extremely high temperatures until they fuse to form a heavier nucleus, releasing energy in the process. Temperatures of at least 100 million °C are needed to force the positively charged nuclei (which strongly repel one another) to fuse.

Nuclear fusion is much harder to initiate than nuclear fission, but once started, it releases far more energy per unit of fuel than does fission. Fusion of hydrogen nuclei to form helium nuclei is the source of energy in the sun and other stars.

After World War II the principle of *uncontrolled nuclear fusion* was used to develop extremely powerful hydrogen, or thermonuclear, weapons. These weapons use the D-T fusion reaction, in which a hydrogen-2, or deuterium (D), nucleus and a hydrogen-3, or tritium (T), nucleus are fused to form a larger, helium-4 nucleus, a neutron, and energy (Figure 3-10).

Scientists have also tried to develop *controlled nuclear fusion*, in which the D-T reaction is used to produce heat that can be converted into electricity. Despite more than 40 years of research, however, this

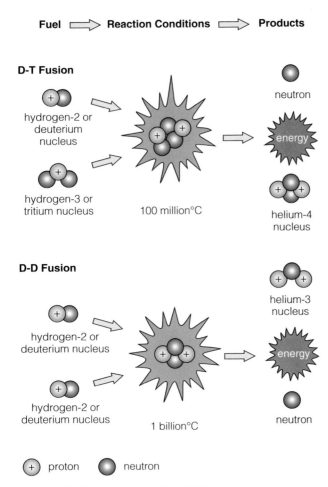

Fuel ⟹ Reaction Conditions ⟹ Products

D-T Fusion

hydrogen-2 or
deuterium
nucleus

hydrogen-3 or
tritium nucleus

100 million°C

neutron

energy

helium-4
nucleus

D-D Fusion

hydrogen-2 or
deuterium nucleus

hydrogen-2 or
deuterium nucleus

1 billion°C

helium-3
nucleus

energy

neutron

⊕ proton   ● neutron

Figure 3-10 The deuterium-tritium (D-T) and deuterium-deuterium (D-D) nuclear fusion reactions, which take place at extremely high temperatures.

process is still at the laboratory stage. Even if it becomes technologically and economically feasible, it probably won't be a practical source of energy until 2050 or later.

3-6 THE TWO IRONCLAD LAWS OF ENERGY

First Law of Energy: You Can't Get Something for Nothing   Scientists have observed energy being changed from one form to another in millions of physical and chemical changes, but they have never been able to detect any energy being created or destroyed. This summary of what happens in nature is called the **law of conservation of energy**, also known as the **first law of energy** or **first law of thermodynamics**. This law means that when one form of energy is converted to another form in any physical or chemical change *en-*

ergy input always equals energy output: We can't get something for nothing in terms of energy quantity.

Second Law of Energy: You Can't Break Even
Because the first law of energy states that energy can be neither created nor destroyed, you might think that there will always be enough energy; yet, if you fill a car's tank with gasoline and drive around, or if you use a flashlight battery until it is dead, you have lost something. If it isn't energy, what is it? The answer is **energy quality**, the amount of energy available that can perform useful work (Figure 3-6).

Countless experiments have shown that when energy is changed from one form to another, there is always a decrease in energy quality (the amount of useful energy). This summary of what we always find happening in nature is called the **second law of energy**, or the **second law of thermodynamics**: When energy is changed from one form to another, some of the useful energy is always degraded to lower-quality, more-dispersed, less-useful energy. This degraded energy is usually in the form of heat, which flows into the environment and is dispersed by the random motion of air or water molecules.

In other words we *can't break even in terms of energy quality because energy always goes from a more useful to a less useful form.* No one has ever found a violation of this fundamental scientific law.

Consider three examples of the second energy law in action. First, when a car is driven, only about 10% of the high-quality chemical energy available in its gasoline fuel is converted into mechanical energy (to propel the vehicle) and into electrical energy (to run its electrical systems). The remaining 90% is degraded to low-quality heat that is released into the environment and eventually lost into space. Second, when electrical energy flows through filament wires in an incandescent light bulb, it is changed into about 5% useful light and 95% low-quality heat that flows into the environment. What we call a light bulb is really a heat bulb. A third example is illustrated in Figure 3-11.

The second law of energy also means that *we can never recycle or reuse high-quality energy to perform useful work.* Once the concentrated energy in a serving of food, a liter of gasoline, a lump of coal, or a chunk of uranium is released, it is degraded to low-quality heat that becomes dispersed in the environment. We can heat air or water at a low temperature and upgrade it to high-quality energy, but the second law of energy tells us that it will take more high-quality energy to do this than we get in return.

Connections: Life and the Second Energy Law
To form and preserve the highly ordered arrangement of molecules and the organized network of chemical

52

Q: What are the three major types of biomes?

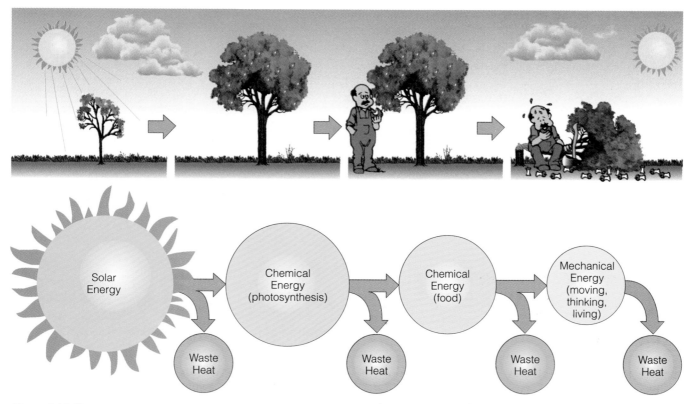

Figure 3-11 The second energy law in action in living systems. When energy is changed from one form to another, some of the initial input of high-quality energy is degraded, usually to low-quality heat, which disperses in the environment.

changes in your body, you must continually get and use high-quality matter and energy resources from your surroundings. As you use these resources, you add low-quality heat and waste matter to your surroundings. For example, your body continuously gives off heat equal to that of a 100-watt light bulb; this is the reason a closed room full of people gets warm. You also continuously give off molecules of carbon dioxide gas and water vapor, which become dispersed in the atmosphere.

Planting, growing, processing, and cooking food all require high-quality energy and matter resources that add low-quality heat and various chemicals to the environment. In addition, enormous amounts of low-quality heat and waste matter are added to the environment when concentrated deposits of minerals and fuels are extracted from the earth's crust, processed, and used to make roads, clothes, shelter, and other items, or when they are burned to heat or cool buildings or to transport you.

Because of the second energy law, the more energy we use (and waste), the more heat and waste we add to the environment. This is the reason that reducing energy waste (p. 40) and switching from harmful nonrenewable energy resources to less harmful renewable energy resources are the keys to a sustainable future for us and many other species.

3-7 CONNECTIONS: MATTER AND ENERGY LAWS AND ENVIRONMENTAL PROBLEMS

Throwaway Societies Because of the law of conservation of matter and the second law of energy, resource use by each of us automatically adds some waste heat and waste matter to the environment. Your individual use of matter and energy resources and your additions of waste heat and matter to the environment may seem small and insignificant. But you are only one of the 1.2 billion individuals in the MDCs using large quantities of matter and energy resources at a rapid rate. Meanwhile, the 4.3 billion people in the LDCs hope to be able to use more of these resources. And each year there are 90 million more consumers of Earth's energy and matter resources.

Most of today's advanced industrialized countries are largely *high-waste societies*, or **throwaway societies**, sustaining ever-increasing economic growth by increasing the flow or *throughput* of planetary *sources* of materials and energy. These resources flow through the economy to planetary *sinks* (air, water, soil, organisms) where pollutants and wastes end up (Figure 3-12). There is an important lesson to be learned from the scientific laws of matter and energy discussed in this chapter. They tell us that if more and more people continue

A: Deserts, grasslands, forests

CHAPTER 3   53

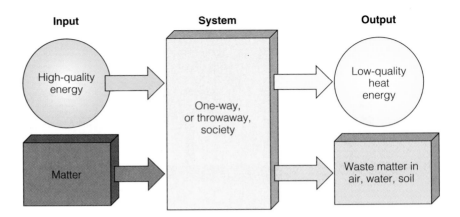

Figure 3-12 The high-waste or throwaway society of most MDCs is based on maximizing the rates of energy and matter flow, thereby rapidly converting the world's high-quality matter and energy resources into waste, pollution, and low-quality heat.

to use and waste more and more energy and matter resources at an increasing rate, sooner or later the capacity of the local, regional, and global environments to dilute and degrade waste matter and absorb waste heat will be exceeded.

Matter-Recycling Societies A stopgap solution to this problem is to convert from a throwaway society to a **matter-recycling society**. The goal of such a shift would be to allow economic growth to continue without depleting matter resources and without producing excessive pollution and environmental degradation. As we have learned, however, there is no free lunch when it comes to energy.

Recycling matter saves energy. However, the two laws of energy tell us that *recycling matter resources always requires high-quality energy, which cannot be recycled*. In the long run a matter-recycling society based on continuing population growth and per capita resource consumption must have an inexhaustible supply of affordable high-quality energy. The environment must also have an infinite capacity to absorb and disperse waste heat and to dilute and degrade waste matter. Also, there is a physical limit to the number of times some materials, such as paper fiber, can be recycled before they become unusable.

The second energy law tells us that the faster we use energy to transform matter into products and to recycle those products, the faster low-quality heat and waste matter are dumped into the environment. Thus, *the more we use energy to "conquer" Earth, the more stress we put on the environment*. Experts argue over how close we are to environmental overload, but the scientific laws of matter and energy indicate that limits do exist.

Thus, *shifting from a throwaway society to a matter-recycling society is only a temporary solution to our problems*. Nevertheless, making such a shift is necessary to give us more time to convert to a sustainable-Earth or Earth-wisdom society.

Learning from Nature: Sustainable-Earth Societies The three scientific laws governing matter and energy changes indicate that the best long-term solution to our environmental and resource problems is to shift from a society based on maximizing matter and energy flow (throughput) to a **sustainable-Earth** or **low-waste society** (Figure 3-13).

Using these lessons from nature as guidelines, a sustainable-Earth society would:

- Reduce the throughput of matter and energy resources to prevent excessive depletion and degradation of planetary sources and overload of planetary sinks

- Use energy more efficiently and not use high-quality energy for tasks that require only moderate-quality energy (Figure 3-6)

- Shift from exhaustible and potentially polluting fossil and nuclear fuels to less harmful renewable energy obtained directly or indirectly from the sun

- Not waste potentially renewable resources, and use them no faster than the rate at which they are regenerated

- Not waste nonrenewable resources, and use them no faster than the rate at which a renewable resource, used sustainably, can be substituted for it

- Recycle and reuse at least 60% of the matter we now discard as trash

- Reduce use and waste of matter resources by making things that last longer and are easier to recycle, reuse, and repair

- Add wastes and pollutants to environmental sinks no faster than the rate at which they can be recycled, reused, absorbed, or rendered harmless to us and other species by natural processes

Q: How much of Earth's tropical forests have been cleared or damaged?

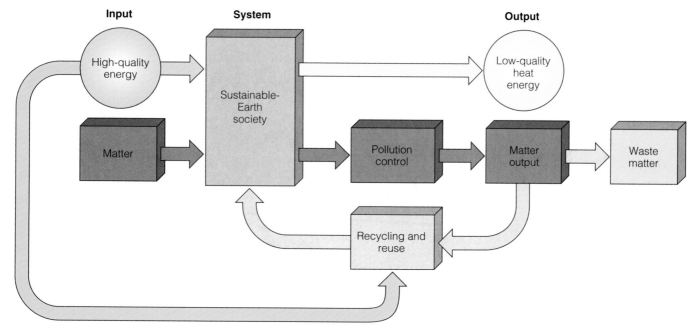

Figure 3-13 A sustainable-Earth or low-waste society—based on energy flow and matter recycling—works with nature by reusing and recycling nonrenewable matter resources; by using potentially renewable resources no faster than they are replenished; by using matter and energy resources efficiently; by reducing unnecessary consumption; by emphasizing pollution prevention and waste reduction; and by controlling population growth.

- Emphasize pollution prevention and waste reduction instead of pollution cleanup and waste management

- Slow (and eventually halt) human population growth to help reduce stress on global life-support systems

- Greatly reduce poverty, which degrades humans and the environment by forcing people to use resources unsustainably to stay alive

Because of the three basic scientific laws of matter and energy, we are all dependent on one another and on the rest of nature for our survival. We are all in it together. In the next chapter we will apply these laws to living systems and look at some biological principles that can teach us how to work with nature.

The second law of thermodynamics holds, I think, the supreme position among laws of nature.... If your theory is found to be against the second law of thermodynamics, I can give you no hope.

ARTHUR S. EDDINGTON

Critical Thinking

1. Do you think fraud is more or less likely in science than in other areas of knowledge? Explain.

2. To what extent are scientists responsible for the applications of knowledge they discover? Should scientists abandon research because of its possible harmful uses? Explain.

3. Explain why we don't really consume anything and why we can never really throw matter away.

4. A tree grows and increases its mass. Explain why this isn't a violation of the law of conservation of matter.

5. If there is no "away," why isn't the world filled with waste matter?

6. Use the second energy law to explain why a barrel of oil can be used only once as a fuel.

7. a. Use the law of conservation of matter to explain why a matter-recycling society will sooner or later be necessary.
   b. Use the first and second laws of energy to explain why, in the long run, a sustainable-Earth or low-waste society—not just a matter-recycling society—will be necessary.

*8. Make a concept map of the key ideas in this chapter using the section heads and subheads and the key terms (shown in boldface type in the chapter). See the inside front cover and Appendix 4 for information on concept maps.

Ecosystems and How They Work

Connections: Blowing in the Wind

Environmental science is a study of connections. One of the things that connects all life on Earth is wind. Without wind most of Earth would be uninhabitable. The tropics would be unbearably hot and the rest of the planet would freeze. Winds also transport nutrients from one place to another. Dust rich in phosphates blows across the Atlantic from the Sahara Desert in Africa (Figure 4-1), replenishing rain forest soils in Brazil. That's the good news.

The bad news is that wind also transports harmful substances. Sulfur compounds and soot from oil-well fires in Kuwait have been detected over Wyoming, and deposits of toxic DDT and PCBs have been turning up in Antarctica for decades. Furthermore, cesium-137 blowing from the 1986 Chernobyl nuclear power plant disaster in the Ukraine has made the lichen food of Lapland's reindeer radioactive.

Reindeer meat, milk, and cheese, has, in turn, become unfit to eat for the herders who depend on it.

There's mixed news as well. Clouds of particles from volcanic eruptions ride the winds, encircle the globe, and change Earth's climate for a while. After Indonesia's Tambora blew up in 1815, for example, distant Europe had a "year without a summer" in 1816 because the ash in the atmosphere reduced the amount of sunlight reaching the earth for several years. And emissions from the 1991 eruption of Mount Pinatubo in the Philippines—the largest eruption this century and the third largest in two centuries—cooled the earth for three years. On the other hand, volcanic ash, like the blowing desert dust, adds valuable trace minerals to the soil where it settles.

The lesson in this wind news is the same: *There is no "away,"* and wind—acting as part of the planet's circulatory system for heat, moisture, and plant nutrients—is one reason. As the Roman poet Virgil wrote over 2,000 years ago: "Before we plow an unfamiliar patch/It is well to be informed about the winds."

Figure 4-1 Some of the dust shown here blowing from Africa's Sahara Desert can end up as soil nutrients in Amazonian rain forests. (NASA)

When we try to pick out anything by itself, we find it hitched to everything else in the universe.

JOHN MUIR

This chapter focuses on answering the following questions about *ecosystems* (communities of species interacting with one another and with their nonliving environment of matter and energy):

- What basic processes keep us and other organisms alive?
- What are the major living and nonliving parts of an ecosystem?
- What happens to energy in an ecosystem?
- What happens to matter in an ecosystem?
- What roles do different types of organisms play, and how do they interact in an ecosystem?

4-1 LIFE AND EARTH'S LIFE-SUPPORT SYSTEMS

What Is Life? The **cell** is the basic unit of life. Each cell is encased in an outer membrane or wall and contains genetic material (DNA) and other parts to perform its life functions. Organisms like bacteria consist of only one cell, but most of the organisms we are familiar with contain many cells. All forms of life

- *have a highly organized internal structure and organization.*
- *have characteristic types of deoxyribonucleic acid, or DNA, molecules in each cell.* DNA is the stuff genes—the basic units of heredity—are made of. These self-replicating molecules contain the instructions both for making new cells from "lifeless" molecules and also for assembling proteins and other molecules each cell needs in order to survive and reproduce.
- *can capture and transform matter and energy from their environment to supply their needs for survival, growth, and reproduction.*
- *can maintain favorable internal conditions despite changes in their external environment, if not overwhelmed.*
- *arise through reproduction—the production of offspring by one or more parents—and in turn are capable of reproduction.*
- *can adapt to external change by* **mutations**—*random changes in their DNA molecules—and through combinations of existing genes during reproduction.* Most mutations are harmful, but some give rise to new

genetic traits that allow the organisms to adapt to changing environmental conditions and have more offspring than those without such traits—a process called **natural selection**. Through successive generations these adaptive traits become more common. When such genetic changes in existing organisms form distinctly different new organisms, what biologists call **evolution** has occurred. Life-forms that cannot adapt become fewer and may become extinct. These ongoing processes of extinction and evolution in response to environmental changes over billions of years have led to the diversity of life-forms found on Earth today. This *biodiversity*—a vital part of Earth capital—sustains life and provides the genetic raw material for adaptation to future changes in environmental conditions.

Earth: A Dynamic Planet We can think of earth as being made up of several layers or spheres (Figure 4-2):

- The **atmosphere**—a thin, gaseous envelope of air around the planet. Its inner layer, the **troposphere**, extends only about 17 kilometers (11 miles) above sea level but contains most of the planet's air—mostly nitrogen (78%) and oxygen (21%). The next layer, stretching 17–48 kilometers (11–30 miles) above Earth's surface, is called the **stratosphere**. Its lower portion contains enough ozone (O_3) to filter out most of the sun's harmful ultraviolet radiation, thus allowing life on land to exist.
- The **hydrosphere**—liquid water (both surface and underground), frozen water (polar ice, icebergs, permafrost in soil), and water vapor in the atmosphere.
- The **lithosphere**—Earth's crust and upper mantle. It contains the fossil fuels and minerals we use and the soil chemicals (nutrients) needed to support plant life.
- The **ecosphere** or **biosphere**—the portion of the earth where living (biotic) organisms are found and interact with one another and with their nonliving (abiotic) environment. This zone of life includes most of the hydrosphere and parts of the lower atmosphere and upper lithosphere. It reaches from the deepest ocean floor 20 kilometers (12 miles) upward to the tops of the highest mountains. If Earth were an apple, the ecosphere would be no thicker than the apple's skin, a haven for life between Earth's molten interior and the lifeless cold of space. *The goal of ecology is to*

Figure 4-2 Our life-support system: the general structure of the earth.

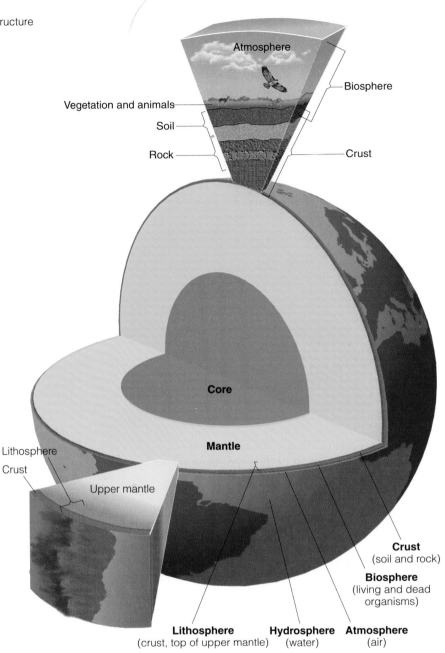

learn how this thin global skin of air, water, soil, and organisms works.

Connections: Energy Flow, Matter Cycling, and Gravity Life on Earth depends on three connected factors (Figure 4-3):

- The *one-way flow of high-quality (usable) energy* from the sun, through materials and living things on or near the earth's surface, then into the environment as low-quality energy (mostly heat dispersed into air or water molecules at a low tem-

perature), and eventually back into space as infrared radiation (Figure 3-4)

- The *cycling of matter* required by living organisms through parts of the ecosphere

- *Gravity*, caused mostly by the attraction between the sun and the earth, allows the planet to hold onto its atmosphere and causes the downward movement of chemicals in the matter cycles

The Sun: Source of Energy for Life The sun is a middle-aged star that lights and warms the planet. It also supplies the energy for *photosynthesis*, the process

Q: How much of North America's original of temperate deciduous forests has been cleared?

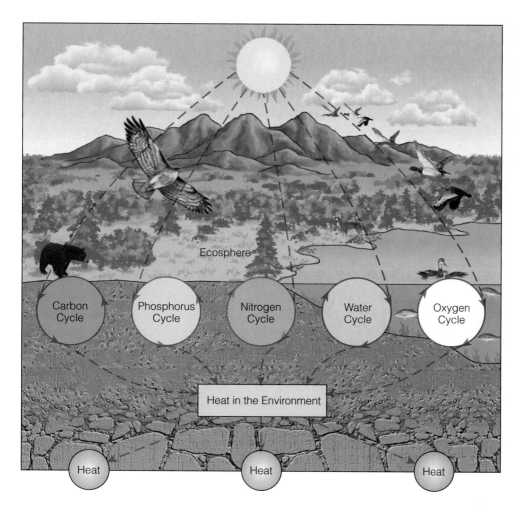

Figure 4-3 Life on Earth depends on the *one-way flow of energy* (dashed lines) from the sun through the ecosphere, the *cycling of critical elements* (solid lines around circles), and *gravity*, which keeps atmospheric gases from escaping into space and draws chemicals downward in the matter cycles. This simplified overview shows only a few of the many cycling elements.

used by green plants and some bacteria to synthesize the compounds that keep them alive and feed most other organisms. Solar energy also powers the cycling of matter and drives the climate and weather systems that distribute heat and fresh water over Earth's surface. It is expected to provide Earth with energy for at least another 4 billion years.

The sun is a gigantic fireball of hydrogen (72%) and helium (28%) gases. Temperatures and pressures in its inner core are so high that hydrogen nuclei fuse to form helium nuclei (Figure 3-10), releasing enormous amounts of energy.

This nuclear fusion reactor radiates energy in all directions as electromagnetic radiation (Figure 3-4). At the speed of light, this radiation makes the 150-million-kilometer (93-million-mile) trip between Sun and Earth in slightly more than 8 minutes. Earth, a tiny target in the vastness of space, receives only about one-billionth of this output, and much of that is either reflected away or absorbed by chemicals in the atmosphere. Most of the harmful, high-energy cosmic rays, gamma rays, X rays, and ultraviolet ionizing radiation never reach the planet's surface. Most of what reaches the troposphere

is visible light and infrared radiation (heat) plus a small amount of ultraviolet radiation not absorbed by ozone in the stratosphere.

About 34% of the solar energy reaching the troposphere is reflected right back to space by clouds, chemicals, and dust and by the earth's surface of land and water (Figure 4-4).

Most of the remaining 66% of solar energy warms the troposphere and land, evaporates water and cycles it through the ecosphere, and generates winds. A tiny fraction (0.023%), captured by green plants and bacteria, fuels photosynthesis to make the organic compounds that most life-forms need to survive.

Most of this unreflected solar radiation is degraded into infrared radiation (which we experience as heat) as it interacts with the earth. How fast this heat flows through the atmosphere and back into space is affected by heat-trapping (greenhouse) gases, such as water vapor, carbon dioxide, methane, nitrous oxide, and ozone, in the troposphere. Without this atmospheric thermal "blanket," known as the *natural greenhouse effect*, Earth would be nearly as cold as Mars, and life as we know it could not exist.

Figure 4-4 The flow of energy to and from Earth. The ultimate source of energy in most ecosystems is sunlight.

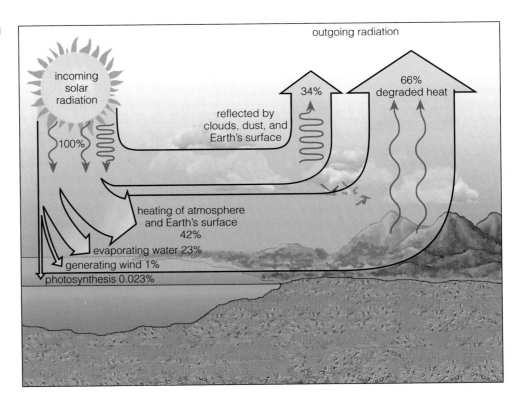

Connections: Nutrient Cycles Any chemical element or compound an organism must take in to live, grow, or reproduce is called a **nutrient**. Some elements such as carbon, oxygen, hydrogen, nitrogen, and phosphorus are needed in fairly large amounts. Others such as iron, copper, chlorine, and iodine are needed in small, or trace, amounts.

These nutrient elements and their compounds are continuously cycled from the nonliving environment (air, water, soil) to living organisms and then back to the nonliving environment in what are called **nutrient cycles**, or **biogeochemical cycles** (literally, "life-earth-chemical" cycles). These cycles, driven directly or indirectly by incoming solar energy and gravity, include the carbon, oxygen, nitrogen, phosphorus, sulfur, and hydrologic (water) cycles (Figure 4-3).

Some nutrients, such as oxygen and carbon, cycle quickly and so are readily available for use by organisms. Others like phosphorus cycle slowly, explaining why availability of phosphorus in soil often limits plant growth.

Earth's chemical cycles connect past, present, and future forms of life. Thus, some of the carbon atoms in the skin of your right hand may once have been part of a leaf, a dinosaur's skin, or a layer of limestone rock. And some of the oxygen molecules you just inhaled may have been inhaled by your grandmother, by Plato, or by a hunter-gatherer who lived 25,000 years ago.

4-2 ECOSYSTEM COMPONENTS

Ecology: Connections in Nature What organisms live in a coral reef, a field, or a pond? How do they get enough food and energy to stay alive? How do these organisms interact with one another? What changes might this coral reef, field, or pond undergo through time?

Ecology—created as a discipline about 100 years ago by M.I.T. chemist Ellen Swallow—is the science that attempts to answer such questions. **Ecology**, from the Greek *oikos* ("house" or "place to live") and *logos* ("study of"), is the study of how organisms interact with one another and with their physical and chemical environments. Ecology deals mainly with interactions among organisms, populations, communities, ecosystems, and the ecosphere (Figure 4-5). It is a study of connections in nature.

An **organism** is any form of life. Organisms can be classified into **species**, groups of organisms that resemble one another in appearance, behavior, chemistry, and genetic structure. Most organisms reproduce sexually and are classified in the same species if they can breed with one another and produce fertile offspring under natural conditions.

We don't have the foggiest notion how many species exist on Earth. Estimates range from 5 million to 100 million, most of them insects, microscopic or-

Q: How much of Earth's surface is covered by oceans?

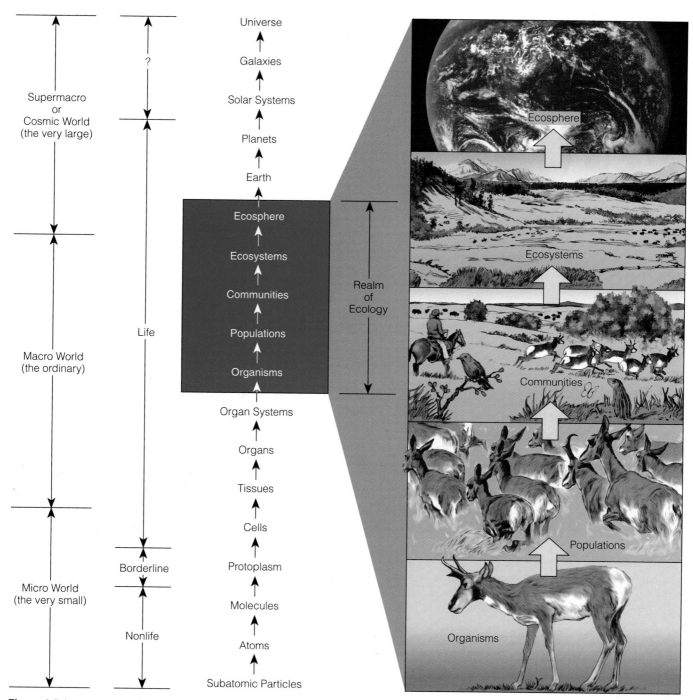

Figure 4-5 Levels of organization of matter according to size and function. This is one way scientists classify patterns of matter found in nature.

ganisms, and tiny sea creatures. So far biologists have identified and named only about 1.4 million species. They know a fair amount about roughly one-third of these species and the detailed roles and interactions of only a few.

A **population** is a group of individuals of the same species occupying a given area at the same time (Figure 4-6). Examples are all sunfish in a pond, white oak trees in a forest, and people in a country. All members of the same population have certain structural, functional, and behavioral traits in common. In most natural populations individuals vary slightly in their genetic makeup so that they don't all look or behave alike—something called **genetic diversity** (Figure 4-7).

A: 70%

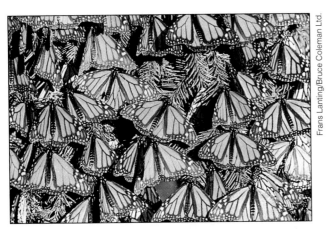

Figure 4-6 A population of monarch butterflies wintering in Michoacán, Mexico. Monarchs winter in parts of Mexico and Central America, as the map shows. They head north during the spring and breed along the way. In the fall their offspring migrate south.

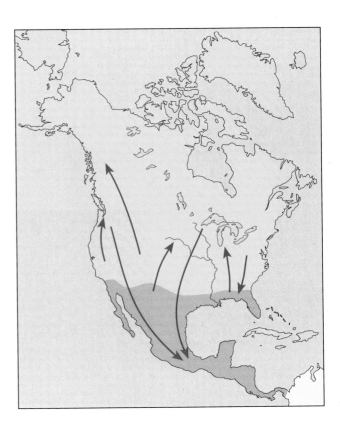

The place where a population (or an individual organism) normally lives is known as its **habitat**. Populations of all the different species occupying a particular place make up a **community** or **biological community**.

An **ecosystem** is a community of different species interacting with one another and with their nonliving environment of matter and energy. The size of an ecosystem—and thus a community—is arbitrary and is defined by the system we wish to study. All of Earth's ecosystems together make up the *biosphere* or *ecosphere.*

Climate—long-term weather—is the main factor determining what type of life, especially what plants, will thrive in a given land area. Biologists have divided the terrestrial (land) portion of the ecosphere into **biomes**, large regions such as forests, deserts, and grasslands characterized by certain climatic conditions and inhabited by certain types of life, especially vegetation (Figure 4-8). Each biome consists of many ecosystems whose communities have adapted to small differences in climate, soil, and other environmental factors. Marine and freshwater portions of the ecosphere can be divided into aquatic life zones, containing numerous ecosystems.

Biodiversity   As environmental conditions have changed over billions of years, many species have become extinct and new ones have formed. The result of these changes is **biological diversity**, or **biodiversity**. It consists of the forms of life that can best survive the variety of conditions currently found on Earth and includes **(1) genetic diversity** (variability in the genetic makeup among individuals within a single species,

Figure 4-7), **(2) species diversity** (the variety of species on Earth and in different habitats of the planet), and **(3) ecological diversity** (the variety of forests, deserts, grasslands, streams, lakes, wetlands, oceans, and other biological communities that interact with one another and with their nonliving environments).

We are utterly dependent on this mostly unknown "bio-capital." This rich variety of genes, species, and ecosystems gives us food, wood, fibers, energy, raw materials, industrial chemicals, and medicines, and it pours hundreds of billions of dollars yearly into the world economy.

Earth's vast library of life-forms and ecosystems also provides free recycling and purification services and natural pest control. Every species here today contains stored genetic information representing thousands to millions of years of adaptation to Earth's changing environmental conditions, and that is the raw material for future adaptations. Biodiversity is nature's "insurance policy" against disasters.

Some people also include *human cultural diversity* as part of Earth's biodiversity. The variety of human cultures on the planet represents our social and technological "solutions" to survival and can help us adapt to changing conditions.

Types of Organisms   Earth's diverse organisms are classified as eukaryotic or prokaryotic on the basis of their cell structure. All organisms except bacteria

Q: How much of Earth's wetlands have been destroyed or polluted?

Alan Solem

Figure 4-7 Genetic diversity. Variation in shell color and banding patterns among populations of one species of snail found on Caribbean islands.

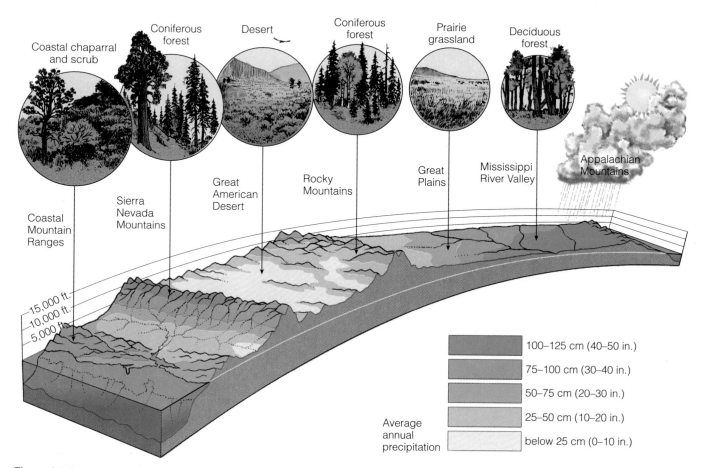

Coastal chaparral and scrub

Coniferous forest

Desert

Coniferous forest

Prairie grassland

Deciduous forest

Coastal Mountain Ranges

Sierra Nevada Mountains

Great American Desert

Rocky Mountains

Great Plains

Mississippi River Valley

Appalachian Mountains

15,000 ft.
10,000 ft.
5,000 ft.

	100–125 cm (40–50 in.)
	75–100 cm (30–40 in.)
	50–75 cm (20–30 in.)
	25–50 cm (10–20 in.)
Average annual precipitation	below 25 cm (0–10 in.)

Figure 4-8 Major biomes found along the 39th parallel crossing the United States. The differences mostly reflect changes in climate, mainly differences in average annual temperature and precipitation.

A: 25–50% (55% in the United States, 91% in California)

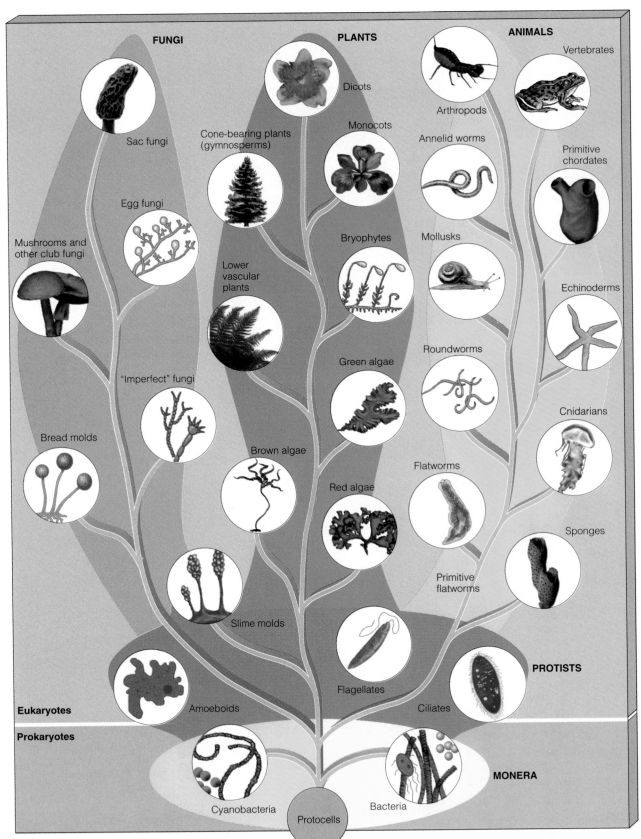

FUNGI

Sac fungi

Egg fungi

Mushrooms and other club fungi

"Imperfect" fungi

Bread molds

Slime molds

Amoeboids

PLANTS

Dicots

Monocots

Cone-bearing plants (gymnosperms)

Bryophytes

Lower vascular plants

Green algae

Brown algae

Red algae

Flagellates

Ciliates

ANIMALS

Vertebrates

Arthropods

Primitive chordates

Annelid worms

Mollusks

Echinoderms

Roundworms

Cnidarians

Flatworms

Sponges

Primitive flatworms

PROTISTS

MONERA

Eukaryotes

Prokaryotes

Cyanobacteria

Protocells

Bacteria

Figure 4-9 Kingdoms of the living world. This is one of several ways that biologists classify Earth's diverse species into major groups. Most biologists believe that protocells gave rise to single-celled prokaryotes and that these in turn evolved into more complex protists, fungi, plants, and animals that make up the planet's stunning biodiversity.

Q: How old is the earth?

Figure 4-10 Giant sequoia trees are evergreen conifers found in just 75 groves along the western slopes of California's Sierra Nevada. Although coastal redwoods can grow taller, giant sequoias are Earth's most massive species. The 2,500–3,000-year-old General Sherman Tree, shown here, is about 84 meters (275 feet) high and 11 meters (36 feet) in diameter. Notice how the people standing at its base are dwarfed. Its bark, which is more than 61 centimeters (2 feet) thick, helps protect it from fire, predators, and disease.

Figure 4-11 These saguaro (pronounced sa-<u>WA</u>-ro) cacti in Arizona store water and produce food in their spongy stems and branches. They reduce water loss in the hot desert climate by having no leaves and by opening their pores only at night. Their thorns discourage predators. These plants can reach heights of 15 meters (50 feet) and can swell to twice their normal size during the rainy season by absorbing water from a vast network of shallow roots. Their night-opening flowers are pollinated by nectar-eating bats.

Figure 4-12 This white orchid, an epiphyte from the tropical forests of Latin America, roots in the fork of a tree rather than the soil. It gets water from rain and humid air, and nutrients from bits of organic matter falling from the thick leaf canopy above. Epiphytes provide hiding and breeding sites for entire communities of frogs, birds, rodents, snakes, and insects—some of which live their entire lives in these aerial gardens. In this interaction between the epiphytes and their host tree, the epiphytes gain access to water, other nutrients, and sunlight; the tree is apparently unharmed.

are **eukaryotic**: Their cells have a *nucleus* (genetic material surrounded by a membrane) and several other internal parts surrounded by membranes. Bacterial cells are **prokaryotic**: They have no distinct nucleus or other internal parts enclosed by membranes. Although most of the organisms we see are eukaryotic, they could not exist without hordes of microscopic prokaryotic organisms (bacteria) toiling away unseen.

In this book Earth's organisms are classified into five kingdoms (Figure 4-9):

- **Monera (bacteria** and **cyanobacteria)** are single-celled, microscopic, prokaryotic organisms.

- **Protista (protists)** are mostly single-celled eukaryotic organisms such as diatoms, amoebas, golden brown and yellow-green algae, protozoans, and slime molds.

- **Fungi** are mostly many-celled (some microscopic) and eukaryotic organisms such as mushrooms, molds, and yeasts.

- **Plantae (plants)** are mostly many-celled, eukaryotic organisms such as red, brown, and green algae; mosses, ferns, and conifers (cone-bearing trees such as pine, juniper, redwood, and yew); and flowering plants (cacti, grasses, beans, roses, bromeliads, and maple trees). **Evergreens** like ferns, magnolias, pines, spruces, and sequoias (Figure 4-10) keep some leaves or needles all year. This allows them to carry out photosynthesis year-round in tropical climates or take maximum advantage of a short growing season in colder climates. **Deciduous plants**, such as oak and maple trees, survive drought and cold by shedding their leaves during such periods. **Succulent plants**, such as desert cacti (Figure 4-11), survive in dry climates by having no leaves, thus conserving scarce water. **Epiphytes**, or "air plants," grow on tree branches rather than in soil. Examples are some species of mosses, ferns, lichens, and orchids in tropical forests (Figure 4-12).

- **Animalia (animals)** are many-celled, eukaryotic organisms. Most, called **invertebrates**, have no backbones. They include sponges, jellyfish, worms, arthropods (insects, shrimp, spiders), mollusks (snails, clams, octopuses), and echinoderms (sea urchins, sea stars; Figure 4-13). The **vertebrates**, animals with backbones, include fish

Figure 4-13 This cobalt blue sea star found on a barrier coral reef in Indonesia is an invertebrate.

Figure 4-14 The peacock is a vertebrate. This courtship display of plumage can also scare off predators.

(sharks, tuna), amphibians (frogs, salamanders), reptiles (turtles, crocodiles, snakes), birds (eagles, penguins, robins, peacocks; Figure 4-14), and mammals (kangaroos, moles, bats, cats, rabbits, elephants, whales, seals, rhinos, humans). Insects and other arthropods such as spiders are vital to our existence (Connections, above).

Connections: What Do We Find in Ecosystems?
The ecosphere and its ecosystems can be studied by being broken down into two parts: the living or **biotic** components (Figure 4-9) and the nonliving or **abiotic** components like water, air, nutrients, and solar energy needed to sustain life. Figures 4-15 and 4-16 are greatly simplified diagrams showing a few of the components of ecosystems in a freshwater pond and in a field. Living organisms in ecosystems are usually classified as either *producers* or *consumers,* based on how they get food.

Have You Thanked Insects Today?

CONNECTIONS

Insects play important and largely unrecognized roles. A large fraction of Earth's plant species depend on insects for pollination and reproduction. Plants also owe their lives to the insects that help turn the soil around their roots and decompose dead tissue into nutrients the plants need. In turn, we and other land-dwelling animals depend on plants for food, either by eating them or by consuming plant-eating animals.

Indeed, if all insects were to disappear, we and most of Earth's amphibians, reptiles, birds, and other mammals would become extinct within a year because of the disappearance of most plant life. Earth would be covered with dead and rotting vegetation and animal carcasses being decomposed by unimaginable hordes of bacteria and fungi. The land would be largely devoid of animal life and covered by mats of wind-pollinated vegetation and clumps of small trees and bushes here and there.

This, however, is not a realistic scenario. Insects, which originated on land nearly 400 million years ago, are so diverse, abundant, and adaptable that they are virtually invincible. Some have asked whether insects will take over if the human race extinguishes itself. This is the wrong question. The roughly billion billion (10^{18}) insects alive at any given time around the world have been in charge for millions of years. Insects can thrive without newcomers such as us, but we and most other land organisms would quickly perish without them.

Producers—sometimes called **autotrophs** (self-feeders)—can make the nutrients they need to survive from compounds in their environment. In most land ecosystems green plants are the producers. In aquatic ecosystems most of the producers are *phytoplankton,* floating and drifting bacteria and protists, mostly microscopic.

Most producers use sunlight to make complex nutrient compounds (such as the carbohydrate glucose) by **photosynthesis**. Although hundreds of chemical changes take place in sequence during photosynthesis, the overall net change can be summarized as follows:

carbon dioxide + water + **solar energy** $\longrightarrow$ glucose + oxygen

Although plants do most of the planet's photosynthesis today, bacteria "invented" this process. Photosynthetic bacteria still exist, including the highly successful cyanobacteria (once called blue-green algae; Figure 4-9).

Q: When did the first forms of life arise on the earth?

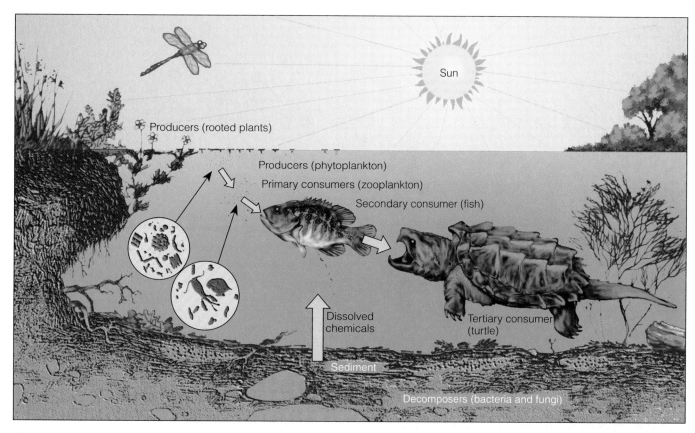

Figure 4-15 Major components of a freshwater pond ecosystem.

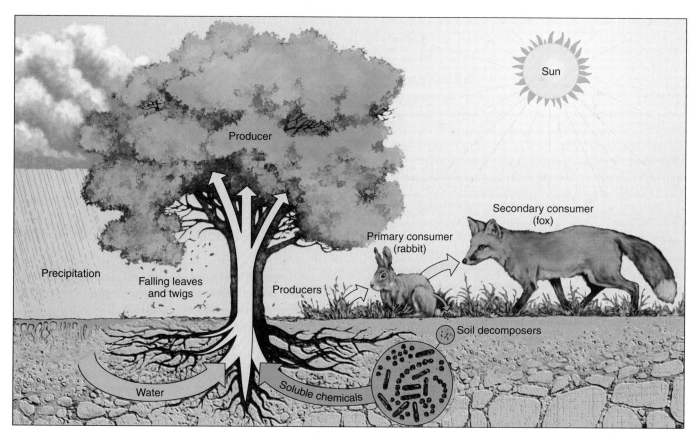

Figure 4-16 Major components of an ecosystem in a field.

A: 3.6–3.8 billion years ago (probably bacterial cells)

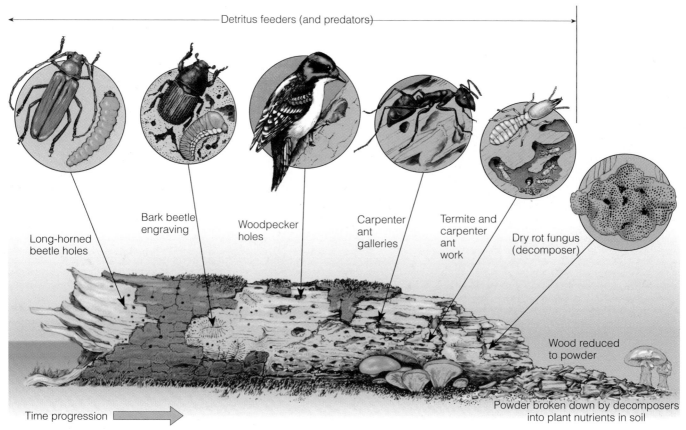

Detritus feeders (and predators)

Long-horned beetle holes

Bark beetle engraving

Woodpecker holes

Carpenter ant galleries

Termite and carpenter ant work

Dry rot fungus (decomposer)

Wood reduced to powder

Powder broken down by decomposers into plant nutrients in soil

Time progression

Figure 4-17 Some detritivores, called *detritus feeders*, directly consume fragments of this log. Other detritivores, called *decomposers* (mostly fungi and bacteria), digest complex organic chemicals in fragments of the log into simpler inorganic nutrients. If these nutrients are not washed away or otherwise removed from the system, they can be used again by producers. The woodpecker shown in this diagram is not a detritivore. In its search for insects, it pecks out fragments of organic matter that are consumed by detritivores.

A few producers, mostly specialized bacteria, can convert simple compounds from their environment into more complex nutrient compounds without sunlight—a process called **chemosynthesis**. In one such case, the source of energy is heat generated by the decay of radioactive elements deep in the earth's core and released at hot-water vents in the ocean's depths. In the pitch-darkness around such vents specialized producer bacteria use the heat to convert dissolved hydrogen sulfide (H_2S) and carbon dioxide into organic nutrient molecules.

All other organisms in ecosystems are **consumers**, or **heterotrophs** (other-feeders), which get their organic nutrients by feeding on the tissues of producers or other consumers. There are several classes of consumers, depending on their food sources.

- **Herbivores** (plant-eaters) are called **primary consumers** because they feed directly on other producers.

- **Carnivores** (meat eaters) feed on other consumers. Those called **secondary consumers** feed only on

primary consumers. Most secondary consumers are animals, but a few plants, such as the Venus flytrap, trap and digest insects. **Tertiary (higher-level) consumers** feed only on other carnivores.

- **Omnivores** eat both plants and animals. Examples are pigs, rats, foxes, cockroaches, and humans.

- **Detritivores** (decomposers and detritus feeders) live off **detritus**, parts of dead organisms and cast-off fragments and wastes of living organisms (Figure 4-17).

Decomposers digest the complex organic molecules in detritus into simpler inorganic compounds and absorb the soluble nutrients. These decomposers—mostly bacteria and fungi (Figure 4-18)—are an important source of food for worms and insects in the soil and water. **Detritus feeders**, such as crabs, carpenter ants, termites, and earthworms, extract nutrients from partly decomposed organic matter.

Both producers and consumers use the chemical energy stored in glucose and other nutrients to drive

Q: How many species are there on the earth?

Figure 4-18 Two types of decomposers are shelf fungi (left) and *Boletus luridus* mushrooms (right).

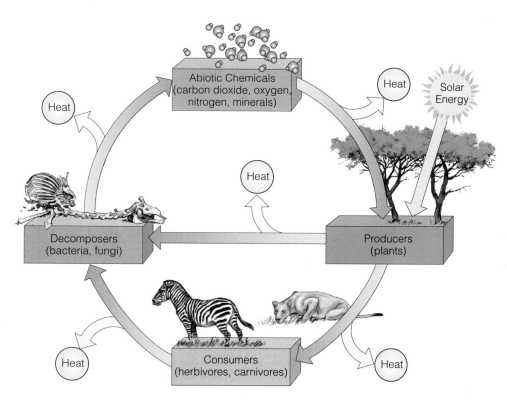

Figure 4-19 The main structural components (energy, chemicals, and organisms) of an ecosystem are linked by energy flow and matter recycling.

their life processes. In most cells, this energy is released by the process of **aerobic respiration** (not the same as the breathing process called respiration), which uses oxygen to convert organic nutrients back into carbon dioxide and water. The net effect of the hundreds of changes in this complex process is

glucose + oxygen ⟶ carbon dioxide + water + **energy**

Thus, although the detailed steps differ, the net chemical change for aerobic respiration is the opposite of that for photosynthesis.

The survival of any individual organism depends on the *flow of matter* and *energy* through its body. However, the community of organisms in an ecosystem survives primarily by a combination of *matter recycling* and *one-way energy flow* (Figure 4-19). Decomposers

A: We don't know. Estimates range from 5–100 million.

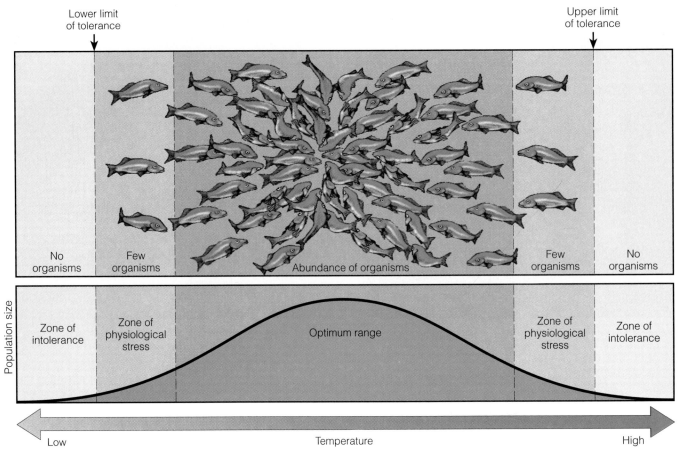

Figure 4-20 Range of tolerance for a population of organisms of the same species to an abiotic environmental factor—in this case, temperature.

complete the cycle of matter by breaking detritus into inorganic nutrients usable by producers. Without decomposers the entire world would soon be knee-deep in plant litter, dead animal bodies, animal wastes, and garbage.

Tolerance Ranges of Species Each population in an ecosystem has a **range of tolerance** to variations in its physical and chemical environment (Figure 4-20). Individuals within a population may have slightly different tolerance ranges for temperature, for example, because of small differences in their genetic makeup, health, and age. Thus, although a trout population may do best at one narrow band of temperatures (optimum level), a few individuals can survive both above and below that band. But, as Figure 4-20 shows, tolerance has its limits. Beyond them, none of the trout will survive.

These observations are summarized in the **law of tolerance**: *The existence, abundance, and distribution of a species in an ecosystem are determined by whether the levels of one or more physical or chemical factors fall within the range tolerated by the species.* A species may have a wide range of tolerance to some factors and a narrow range of tolerance to others. Most organisms are least tolerant during juvenile or reproductive stages of the life cycle. Highly tolerant species can live in a range of habitats with different conditions.

Some species can adjust their tolerance to physical factors such as temperature if change is gradual, just as you can tolerate a hotter bath by slowly adding hotter and hotter water. This adjustment to slowly changing new conditions, or **acclimation**, is a useful protective device. However, acclimation has limits. At each step the species comes closer to its absolute limit. Suddenly, without warning, the next small change triggers a **threshold effect**, a harmful or even fatal reaction as the tolerance limit is exceeded—like adding that proverbial single straw that breaks the already heavily loaded camel's back.

The threshold effect explains why many environmental problems seem to arise suddenly. For example,

Q: How many of Earth's species have been identified?

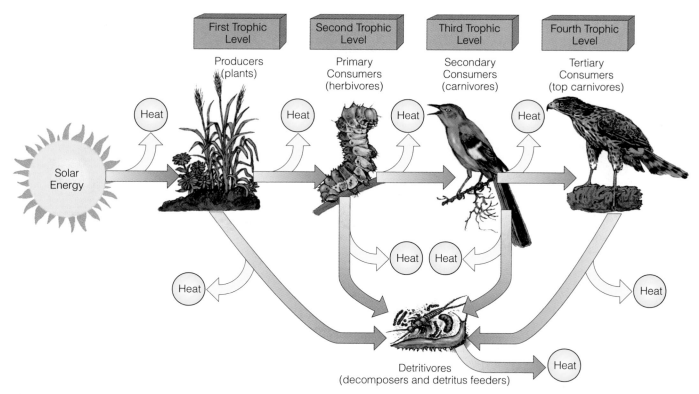

First Trophic Level	Second Trophic Level	Third Trophic Level	Fourth Trophic Level
Producers (plants)	Primary Consumers (herbivores)	Secondary Consumers (carnivores)	Tertiary Consumers (top carnivores)

Figure 4-21 A food chain. The arrows show how chemical energy in food flows through various trophic levels, with most of it degraded to heat in accordance with the second law of energy.

spruce trees suddenly begin dying in droves, but the cause may be exposure to numerous air pollutants for decades. The threshold effect also explains why we must prevent pollution to keep thresholds from being exceeded.

Limiting Factors in Ecosystems An ecological principle related to the law of tolerance is the **limiting factor principle**: *Too much or too little of any abiotic factor can limit or prevent growth of a population, even if all other factors are at or near the optimum range of tolerance.* Such a factor is called a **limiting factor**.

Limiting factors in land ecosystems include temperature, water, light, and soil nutrients. For example, suppose a farmer plants corn in phosphorus-poor soil. Even if water, nitrogen, potassium, and other nutrients are at optimum levels, the corn will stop growing when it uses up the available phosphorus. Here, phosphorus determines how much corn will grow in the field. Growth can also be halted by too much of an abiotic factor. For example, plants can be killed by too much water or too much fertilizer.

In aquatic ecosystems **salinity** (the amounts of various salts dissolved in a given volume of water) is a limiting factor. It determines the species found in both marine (oceans) and freshwater ecosystems.

Aquatic ecosystems can also be divided into surface, middle, and bottom layers or life zones. Important limiting factors for these layers are temperature, sunlight, **dissolved oxygen (DO) content** (the amount of oxygen gas dissolved in a given volume of water at a particular temperature and pressure), and availability of nutrients.

4-3 CONNECTIONS: ENERGY FLOW IN ECOSYSTEMS

Connections: Food Chains and Food Webs

All organisms, dead or alive, are potential sources of food for other organisms. A caterpillar eats a leaf; a robin eats the caterpillar; a hawk eats the robin. When plant, caterpillar, robin, and hawk all die, they in turn are consumed by decomposers.

The sequence of who eats or decomposes whom in an ecosystem is called a **food chain** (Figure 4-21). It determines how energy moves from one organism to another through the ecosystem. Ecologists assign every organism in an ecosystem to a *feeding level*, or **trophic level** (from the Greek *trophos*, "nourishment"), depending on whether it is a producer or a consumer and on what it eats or decomposes. Producers belong

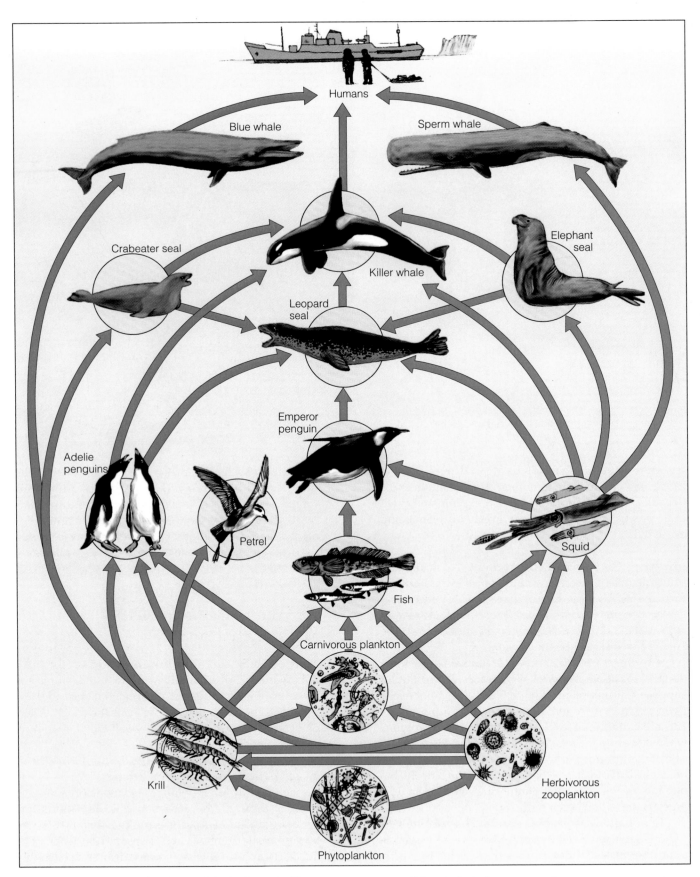

Figure 4-22 Greatly simplified food web in the Antarctic. There are many more participants, including an array of decomposer organisms.

Q: What percentage of all species that have ever lived have become extinct?

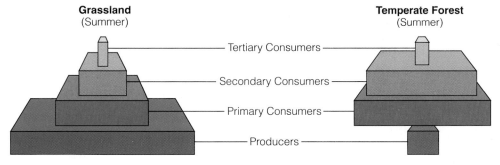

Figure 4-23 Generalized graphs of numbers for two ecosystems. Numbers for grasslands and many other ecosystems taper off from the producer level to the higher trophic levels, forming a pyramid (left). For some ecosystems, however, the numbers take different shapes (right). For example, a redwood forest has a few large producers (the trees) that support a much larger number of small primary consumers (insects) that feed on the trees.

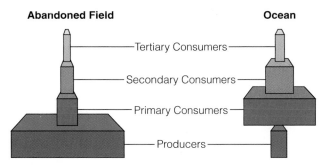

Figure 4-24 Generalized graphs of biomass for two ecosystems. The size of each tier represents the dry weight per square meter of all organisms at that trophic level. For most land ecosystems the total biomass at each successive trophic level decreases. This yields a pyramid of biomass with a large base of producers, topped by a series of increasingly smaller trophic levels of consumers (left). In aquatic ecosystems the biomass of consumers can exceed that of producers (right). Here the producers are microscopic phytoplankton that grow and reproduce rapidly, not large plants that grow and reproduce slowly.

to the first trophic level, primary consumers to the second trophic level, secondary consumers to the third trophic level, and so on. Detritivores process detritus from all trophic levels.

Real ecosystems are more complex than this. Most consumers eat—and are eaten by—two or more types of organisms. Some animals feed at several trophic levels. Thus, the organisms in most ecosystems form a complex network of feeding relationships called a **food web**. Figure 4-22 shows a simplified Antarctic food web. Trophic levels can be assigned in food webs just as in food chains.

Ecological Pyramids By counting the organisms at each trophic level, ecologists can graph this information to yield a **pyramid of numbers** for an ecosystem (Figure 4-23). Since the typical relationship is many producers to not so many primary consumers to just a few secondary consumers, the graph usually looks like a pyramid. For example, a million phyto-

plankton in a small pond may support 10,000 zooplankton, which in turn may support 100 perch, which might feed 1 person for a month or so. However, Figure 4-23 also shows there are exceptions to the pyramid shape.

Each trophic level in a food chain or web contains a certain amount of **biomass**, the weight of all organic matter contained in its organisms. Ecologists estimate biomass by harvesting random patches or narrow strips in an ecosystem. The sample organisms are then sorted according to trophic levels, dried, and weighed. These data are used to plot a **pyramid of biomass** for the ecosystem (Figure 4-24). Here too, the shape is not always pyramidal.

In a food chain or web, biomass (and thus chemical energy) is transferred from one trophic level to another, with some usable energy lost in each transfer. At each successive trophic level some of the available biomass is not eaten or is not digested or absorbed. Usable energy is also lost at each level, mostly as a result of the inevitable loss in energy quality imposed by the second law of energy.

The percentage of usable energy transferred from one trophic level to the next varies from 5% to 20% (that is, a loss of 80–95%), depending on the types of species and the ecosystem involved. The pyramid in Figure 4-25 illustrates this energy loss for a simple food chain, assuming a 90% energy loss with each transfer. This **pyramid of energy flow** shows that the more trophic levels or steps in a food chain or web, the greater the cumulative loss of usable energy. Figure 4-26 shows the actual pyramid and details of energy flow during one year for an aquatic ecosystem in Silver Springs, Florida.

The energy flow pyramid explains why Earth can support more people if they eat at lower trophic levels by consuming grains directly (for example, rice → human) rather than eating grain-eaters (grain → steer → human). The large energy loss in going to each consecutive trophic level explains why food chains and webs rarely have more than four consecutive links. This explains why top carnivores like eagles,

Figure 4-25 Generalized pyramid of energy flow, showing the decrease in usable energy available at each succeeding trophic level in a food chain or web. This diagram shows a 90% loss in usable energy with each transfer from one trophic level to another. In nature such losses vary from 80% to 95%.

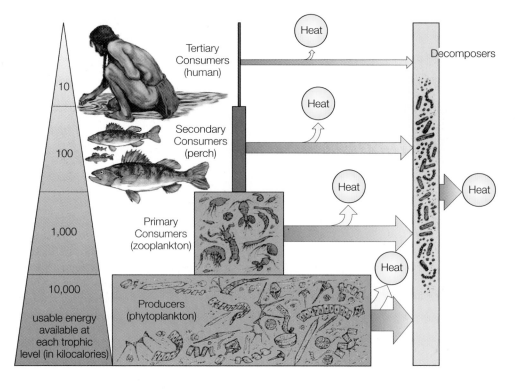

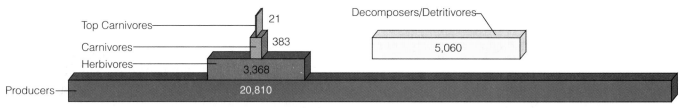

Figure 4-26 Annual pyramid of energy flow (in kilocalories per square meter per year) for an aquatic ecosystem in Silver Springs, Florida. (Used by permission from Cecie Starr and Ralph Taggart, *Biology: The Unity and Diversity of Life*, 6th ed., Belmont, Calif.: Wadsworth, 1992)

tigers, and white sharks are sparse in numbers and are usually the first to suffer when the ecosystems that support them are disrupted.

Productivity of Producers   The *rate* at which an ecosystem's producers capture and store chemical energy as biomass is the ecosystem's **gross primary productivity**. Figure 4-27 shows how this productivity varies in different parts of Earth. However, since the producers must use some of this biomass to stay alive, what we need to know is an ecosystem's **net primary productivity**.

net primary productivity = rate at which producers produce chemical energy stored in biomass through photosynthesis − rate at which producers use chemical energy stored in their biomass through aerobic respiration

Net primary productivity, usually reported as the energy output of a specified area of producers over a given time (typically as kilocalories per square meter per year), is the basic food source or "income" for an ecosystem's consumers. It is the rate at which energy is stored in new biomass—cells, leaves, roots, stems—available for use by consumers.

Estuaries, swamps and marshes, and tropical rain forests are highly productive; open ocean, tundra (arctic grasslands), and desert are the least productive (Figure 4-28). You might conclude that we should harvest plants in estuaries, swamps, and marshes to feed our hungry millions—or clear tropical forests and plant crops. Wrong. The grasses in estuaries, swamps, and marshes cannot be eaten by people, but they are vital food sources (and spawning areas) for fish, shrimp, and other aquatic life that provide us and other consumers with protein. Thus we should protect, not harvest or destroy, these plants.

Q: What two countries have the world's largest populations?

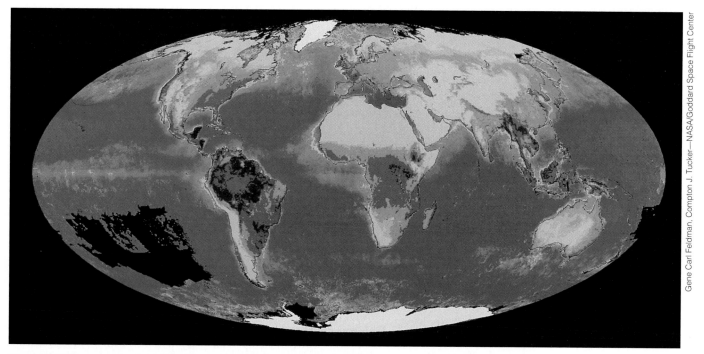

Figure 4-27 Three years of satellite data on Earth's primary productivity. Rain forests and other highly productive areas appear as dark green, deserts as yellow. The concentration of phytoplankton, a primary indicator of ocean productivity, ranges from red (highest) to orange, yellow, green, and blue (lowest).

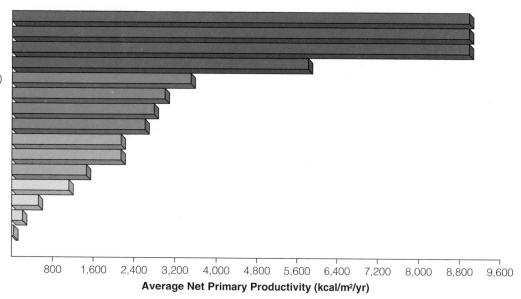

Type of Ecosystem

Estuaries
Swamps and marshes
Tropical rain forest
Temperate forest
Northern coniferous forest (taiga)
Savanna
Agricultural land
Woodland and shrubland
Temperate grassland
Lakes and streams
Continental shelf
Open ocean
Tundra (arctic and alpine)
Desert scrub
Extreme desert

800 1,600 2,400 3,200 4,000 4,800 5,600 6,400 7,200 8,000 8,800 9,600

Average Net Primary Productivity (kcal/m²/yr)

Figure 4-28 Estimated average net primary productivity in major life zones and ecosystems. Values are given in kilocalories of energy produced per square meter per year.

In tropical forests most nutrients are stored in vegetation rather than in the soil. When the trees are cleared, the nutrient-poor soils are rapidly depleted of their nutrients by frequent rains and by growing crops. Thus food crops can be grown only for a short time without massive and expensive applications of commercial fertilizers. Again, as with estuaries, swamps, and marshes, we should protect, not clear, these forests.

It is estimated that humans now use or waste about 27% of the world's potential net primary productivity (40% for land systems). To do this our

A: China and India

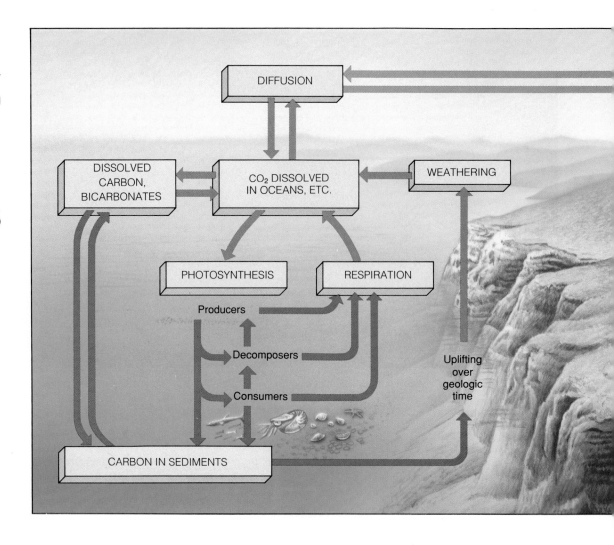

Figure 4-29 Simplified diagram of the global carbon cycle. The left portion shows the movement of carbon through marine ecosystems, and the right portion shows its movement through terrestrial ecosystems. (Used by permission from Cecie Starr and Ralph Taggart, *Biology: The Unity and Diversity of Life*, 6th ed., Belmont, Calif.: Wadsworth, 1992)

species has crowded out or eliminated other species. Many biologists contend that the resulting biodiversity impoverishment is reducing Earth's carrying capacity for *all* species, including ourselves.

When earlier civilizations failed because they exceeded the carrying capacity of the land, people could expand into new areas. If we take over (or try to take over) 80% of the planet's primary land productivity as our population doubles over the next 40 years, most biologists believe we will exceed the carrying capacity in vast areas. There will be no place to expand to and no place to hide from the consequences of such biological impoverishment.

4-4 CONNECTIONS: MATTER CYCLING IN ECOSYSTEMS

Carbon Cycle The carbon cycle is based on carbon dioxide gas, which makes up almost 0.036% by volume of the troposphere and is also dissolved in water. Producers remove CO_2 from the atmosphere

(terrestrial producers) or water (aquatic producers) and use photosynthesis to convert it into complex carbohydrates such as glucose ($C_6H_{12}O_6$). Then the cells in oxygen-consuming producers and consumers carry out aerobic respiration, which breaks down glucose and other complex organic compounds and converts the carbon back to CO_2 in the atmosphere or water for reuse by producers. This linkage between photosynthesis in producers and aerobic respiration in producers and consumers circulates carbon in the ecosphere and is a major part of the global carbon cycle (Figure 4-29). Oxygen and hydrogen, the other elements in carbohydrates, cycle almost in step with carbon.

Carbon dioxide is the key component of nature's thermostat. If the carbon cycle removes too much CO_2 from the atmosphere, Earth will cool; if the cycle generates too much, Earth will get hotter. Thus even slight changes in the carbon cycle can affect climate and ultimately the types of life that can exist on the planet.

As Figure 4-29 shows, some carbon lies deep in the earth in fossil fuels—coal, petroleum, and natural gas—and is released to the atmosphere as carbon diox-

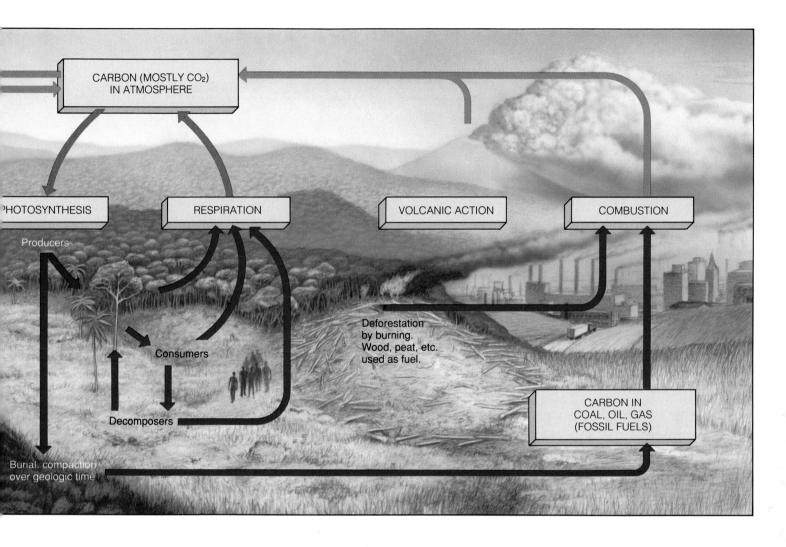

CARBON (MOSTLY CO₂) IN ATMOSPHERE

PHOTOSYNTHESIS — Producers

RESPIRATION — Consumers — Decomposers

VOLCANIC ACTION

COMBUSTION

Deforestation by burning. Wood, peat, etc. used as fuel.

Burial, compaction over geologic time

CARBON IN COAL, OIL, GAS (FOSSIL FUELS)

ide only when these fuels are extracted and burned. Some CO_2 also enters the atmosphere from aerobic respiration and from volcanic eruptions, which free carbon from rocks deep in the earth's crust.

The oceans also play a major role in regulating the level of carbon dioxide in the atmosphere. Carbon dioxide gas is readily soluble in water. Some of this dissolved CO_2 stays in the sea, and some is removed by photosynthesizing producers. As water warms, more dissolved CO_2 returns to the atmosphere.

In marine ecosystems some organisms take up dissolved CO_2 molecules, carbonate ions (CO_3^{2-}), or bicarbonate ions (HCO_3^{-}) from ocean water. These ions can then react with calcium ions (Ca^{2+}) in seawater to form slightly soluble carbonate compounds such as calcium carbonate ($CaCO_3$) to build shells and the skeletons of marine organisms. When these organisms die, tiny particles of their shells and bone drift slowly to the ocean depths and are buried for eons (as long as 400 million years) in bottom sediments (Figure 4-29).

In fact, most of the earth's carbon—10,000 times that in the total mass of all life on Earth—is stored in the ocean floor sediments and on the continents. This carbon reenters the cycle very slowly when some of the sediments dissolve and form dissolved CO_2 gas that can enter the atmosphere. Geologic processes can also bring bottom sediments to the surface, exposing the carbonate rock to chemical attack by oxygen and conversion to carbon dioxide gas.

Especially since 1950, as world population and resource use have soared, we have disturbed the carbon cycle in two ways that add more carbon dioxide to the atmosphere than oceans and plants have been able to remove:

- Forest and brush clearing, leaving less vegetation to absorb CO_2

- Burning fossil fuels and wood, which produces CO_2 that flows into the atmosphere

Computer models of Earth's climate systems suggest that this CO_2, along with other heat-trapping gases we're adding to the atmosphere, could enhance the planet's natural greenhouse effect. This could alter

Figure 4-30 Simplified diagram of the nitrogen cycle in a terrestrial ecosystem. (Adapted by permission from Carolina Biological Supply)

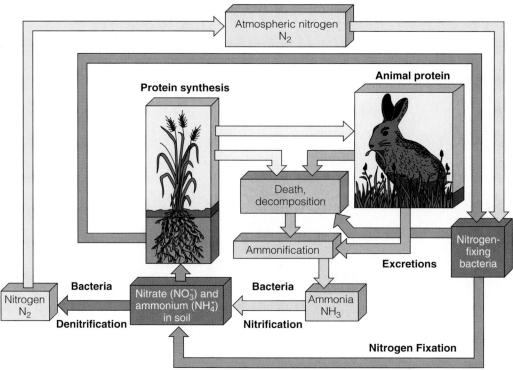

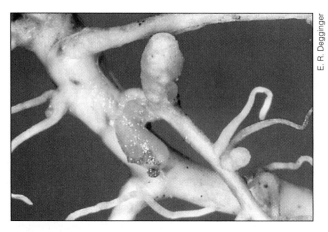

Figure 4-31 Plants in the legume family have root nodules where *Rhizobium* bacteria "fix" nitrogen. That is, they change nitrogen gas (N_2) into ammonia (NH_3), which in soil water forms ammonium ions (NH_4^+) that are taken up by the roots of plants. This process benefits both species. The bacteria convert atmospheric nitrogen into a form usable by the plants, and the plants provide the bacteria with sugar.

climate patterns, disrupt global food production and wildlife habitats, and possibly raise the average sea level.

Nitrogen Cycle: Bacteria in Action Nitrogen gas (N_2) makes up 78% of the volume of the troposphere. Yet, this largest reservoir of nitrogen cannot be used directly as a nutrient by multicellular plants or animals. Fortunately, lightning and certain bacteria convert nitrogen gas into compounds that can enter food webs as part of the **nitrogen cycle** (Figure 4-30).

The conversion of atmospheric nitrogen gas into other chemical forms useful to plants is called **nitrogen fixation**. It is done mostly by cyanobacteria in soil and water and by several other specific microorganisms, and by *Rhizobium* bacteria living in small nodules (swellings) on the root systems of a wide variety of plant species (Figure 4-31).

Plants convert inorganic nitrate ions and ammonium ions from soil water into DNA, proteins, and other, nitrogen-containing nutrients. Animals get their nitrogen by eating plants or plant-eating animals.

After nitrogen has served its purpose in living organisms, armies of specialized decomposer bacteria convert the nitrogen-rich organic compounds, wastes, cast-off particles, and dead bodies of organisms into simpler inorganic compounds—such as water-soluble salts containing ammonium ions (NH_4^+). Other specialized bacteria then convert these inorganic forms of nitrogen back into nitrite (NO_2^-) and nitrate (NO_3^-) ions in the soil and then into nitrogen gas, which is released to the atmosphere to begin the cycle again.

We intervene in the nitrogen cycle in several ways:

■ Emitting large quantities of nitric oxide (NO) into the atmosphere when any fuel is burned. (Most of this NO is produced when nitrogen and oxygen

Q: Worldwide, what is the average number of children per woman?

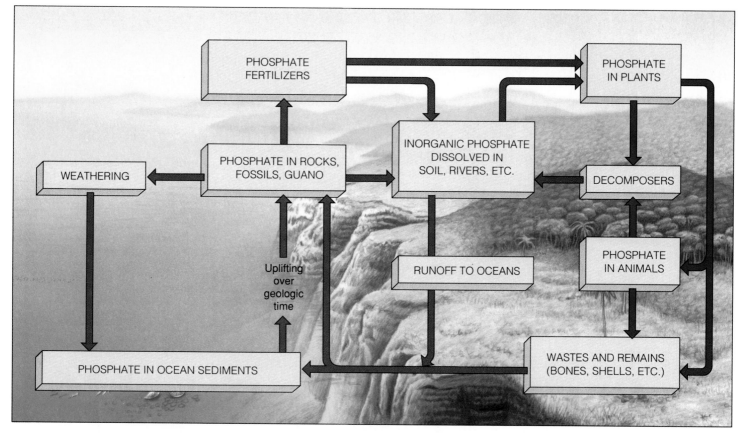

Figure 4-32 Simplified diagram of the phosphorus cycle. (Used by permission from Cecie Starr and Ralph Taggart, *Biology: The Unity and Diversity of Life*, 6th ed., Belmont, Calif.: Wadsworth, 1992)

molecules in the air combine at high temperatures.) This nitric oxide combines with oxygen to form nitrogen dioxide (NO_2) gas, which can react with water vapor to form nitric acid (HNO_3). This acid is a component of acid deposition (commonly called acid rain), which can damage or weaken trees and upset aquatic ecosystems.

- Emitting heat-trapping nitrous oxide (N_2O) gas into the atmosphere by the bacteria in livestock wastes and by commercial inorganic fertilizers applied to the soil.

- Mining nitrogen-containing mineral deposits for fertilizers.

- Depleting nitrogen from soil by harvesting nitrogen-rich crops.

- Leaching water-soluble nitrate ions from soil through irrigation.

- Losing soil nitrogen and adding nitrogen oxides to the atmosphere when grasslands and forests are burned before planting crops.

- Adding excess nitrogen compounds to aquatic ecosystems in agricultural runoff and discharge

of municipal sewage. This excess of plant nutrients stimulates rapid growth of algae and other aquatic plants. The subsequent breakdown of dead algae by aerobic decomposers depletes the water of dissolved oxygen gas and can disrupt aquatic ecosystems.

Phosphorus Cycle Phosphorus moves through water, Earth's crust, and living organisms in the **phosphorus cycle** (Figure 4-32). In this cycle phosphorus moves slowly from phosphate deposits on land and shallow ocean sediments to living organisms, and then back to the land and ocean. Bacteria are less important here than in the nitrogen cycle. Unlike carbon and nitrogen, phosphorus does not circulate in the atmosphere.

Phosphorus released by the slow breakdown, or weathering, of phosphate rock deposits is dissolved in soil water and taken up by plant roots. Wind can also carry phosphate particles long distances (Figure 4-1). Most soils contain little phosphorus because phosphate compounds are only slightly soluble in water and are found in few kinds of rocks. Thus phosphorus is the limiting factor for plant growth in many (but not all) soils and aquatic ecosystems.

A: 3.3 (1.8 in MDCs, 2.0 in the U.S., and 3.7 in LDCs) in 1993

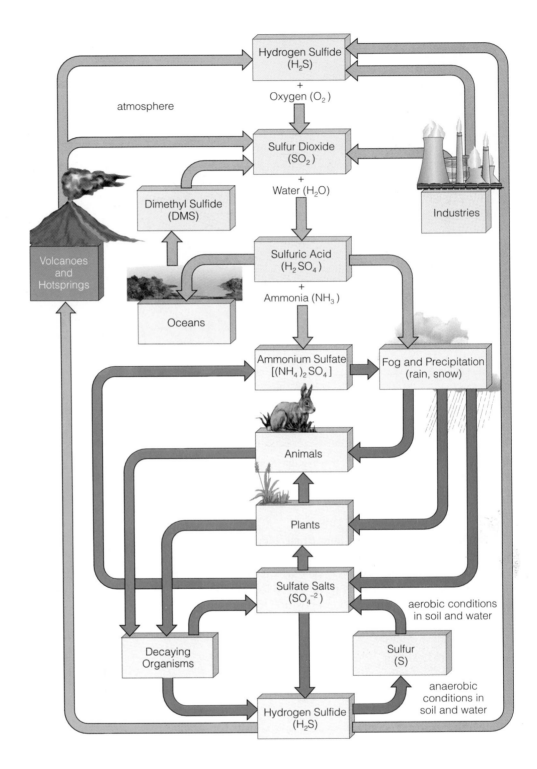

Figure 4-33 Simplified diagram of the sulfur cycle.

Animals get their phosphorus by eating producers or animals that have eaten producers. Animal wastes and the decay products of dead animals and producers return much of this phosphorus to the soil, to streams, and eventually to the ocean bottom as deposits of phosphate rock.

Some phosphate returns to the land as guano—the phosphate-rich manure of fish-eating birds such as pelicans and cormorants. This return is small, though, compared with the phosphate transferred from the land to the oceans each year by natural processes and human activities.

Over millions of years geologic processes may push up and expose the seafloor. Weathering then slowly releases phosphorus from the exposed rocks and allows the cycle to begin again.

Q: How many teenage women (ages 15 to 19) become pregnant each year in the United States?

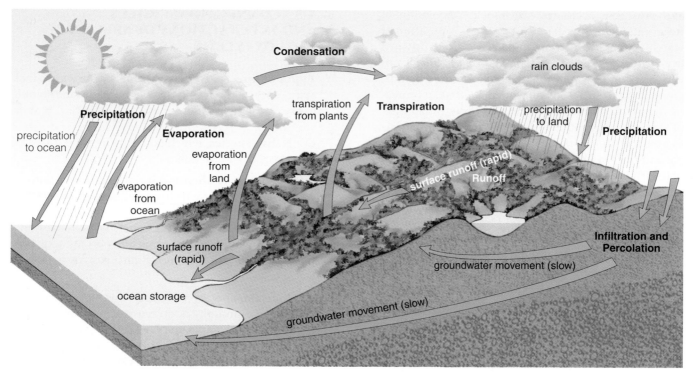

Figure 4-34 Simplified diagram of the hydrologic cycle.

We intervene in the phosphorus cycle chiefly in two ways:

- Mining large quantities of phosphate rock to produce commercial inorganic fertilizers and detergents.

- Adding excess phosphorus to aquatic ecosystems in runoff of animal wastes from livestock feedlots, runoff of commercial phosphate fertilizers from cropland, and discharge of municipal sewage. Too much of this nutrient causes explosive growth of cyanobacteria, algae, and aquatic plants, disrupting life in aquatic ecosystems. This depletes dissolved oxygen.

Sulfur Cycle   Sulfur circulates through the ecosphere in the **sulfur cycle** (Figure 4-33). Much of Earth's sulfur is tied up underground in sulfide rocks (such as iron disulfide or pyrite) and minerals, including sulfate salts (such as hydrous calcium sulfate or gypsum) buried deep under ocean sediments.

However, sulfur also enters the atmosphere from several natural sources. Hydrogen sulfide (H_2S), a colorless, highly poisonous gas with a "rotten-egg" smell, comes from active volcanoes and from the breakdown of organic matter in swamps, bogs, and tidal flats caused by decomposers that don't use oxygen (anaerobic decomposers). Sulfur dioxide (SO_2), a colorless, suffocating gas, also comes from volcanoes. Particles of sulfate (SO_4^{2-}) salts, such as ammonium sulfate, enter the atmosphere from sea spray.

In the atmosphere, sulfur dioxide reacts with oxygen to produce sulfur trioxide gas (SO_3), which in turn reacts with water vapor to produce tiny droplets of sulfuric acid (H_2SO_4). It also reacts with other chemicals in the atmosphere to produce tiny particles of sulfate salts. These droplets of sulfuric acid and particles of sulfate salts fall to Earth as components of acid deposition, which can harm trees and aquatic life. Droplets of sulfuric acid are also produced from dimethylsulfide (DMS) emitted into the atmosphere by many species of plankton.

About a third of all sulfur, including 99% of the sulfur dioxide, that reaches the atmosphere comes from human activities. We intervene in the atmospheric phase of the sulfur cycle in two ways:

- Burning sulfur-containing coal and oil to produce electric power, which produces about two-thirds of the human inputs of sulfur dioxide

- Refining petroleum; smelting sulfur compounds of metallic minerals into free metals such as copper, lead, and zinc; and other industrial processes

Hydrologic Cycle   The **hydrologic cycle** or **water cycle**, which collects, purifies, and distributes Earth's fixed supply of water, is shown in simplified form in Figure 4-34. The main processes in this water recycling and purifying cycle are *evaporation* (conversion of water into water vapor), *transpiration* (evaporation of water extracted by roots and transported upward from leaves or other parts of plants), *condensation* (conversion of water vapor into droplets of liquid water), *precipitation* (rain, sleet, hail, snow), *infiltration* (movement of water into soil), *percolation* (downward flow of water through soil and permeable rock formations to groundwater storage areas), and *runoff* downslope back to the sea to begin the cycle again. At different phases of the cycle water is stored for varying amounts of time on the planet's surface (oceans, streams, reservoirs, glaciers) or in the ground.

The water cycle is powered by energy from the sun and by gravity. Incoming solar energy evaporates water from oceans, streams, lakes, soil, and vegetation into the atmosphere. About 84% of the moisture comes from the oceans, and the rest from land. Winds and air masses transport this water vapor over various parts of Earth's surface, often over long distances. Falling temperatures cause the water vapor to condense into tiny droplets that form clouds or fog and precipitation.

Some of the fresh water returning to Earth's surface as precipitation becomes locked in glaciers. Much of it collects in puddles and ditches and runs off into nearby lakes and into streams, which carry water back to the oceans, completing the cycle.

Besides replenishing streams and lakes, surface water running off the land also causes soil erosion, which moves various nutrients through portions of other biogeochemical cycles. Water is the primary sculptor of Earth's landscape.

Some of the water returning to the land seeps into or infiltrates surface soil layers. Some percolates downward into the ground, dissolving minerals from porous rocks on the way. There this mineral-laden water is stored as groundwater in the pores and cracks of rocks. This underground water, like surface water, flows downhill and seeps out into streams and lakes or comes out in springs. Eventually this water evaporates or reaches the sea to begin the cycle again.

We intervene in the water cycle in two main ways:

- Withdrawing large quantities of fresh water from streams, lakes, and underground sources. In heavily populated or heavily irrigated areas, withdrawals have led to groundwater depletion or intrusion of ocean salt water into underground water supplies.

- Clearing vegetation from land for agriculture, mining, roads, construction, and other activities. This reduces seepage, which recharges groundwater supplies. This reduction also increases the risk of flooding and speeds surface runoff, producing more soil erosion and landslides.

4-5 CONNECTIONS: ROLES AND INTERACTIONS OF SPECIES IN ECOSYSTEMS

Types of Species in Ecosystems   One way to look at an ecosystem's species from a human standpoint is to divide them into four types:

- **Native species**, which normally live and thrive in a particular ecosystem.

- **Immigrant** or **alien species**, which migrate into an ecosystem or which are deliberately or accidentally introduced into an ecosystem by humans. Some of these species are beneficial to humans in the form of crops and game for sport hunting, but some also take over and eliminate many native species.

- **Indicator species**, which serve as early warnings that a community or an ecosystem is being damaged. For example, the present decline of migratory, insect-eating songbirds in North America indicates that their summer habitats there and their winter habitats in the tropical forests of Latin America and the Caribbean are rapidly disappearing. Some frogs, toads, salamanders, and other amphibians that live part of their lives in water and part on land are indicator species (Connections, p. 83).

- **Keystone species**, which affect many other organisms in an ecosystem.* For example, in tropical forests, various species of bees, bats, ants, and hummingbirds play keystone roles in pollinating flowering plants, dispersing seed, or both. Some keystone species, such as the alligator (Connections, p. 84), the wolf, the leopard, the lion, the giant anteater, and the giant armadillo, are top predators that exert a stabilizing effect on their ecosystems by feeding on and regulating the populations of certain species. The loss of a keystone species can lead to population crashes and extinctions of other species that depend on it for certain services—a ripple or domino effect that spreads throughout an ecosystem. According to biologist E. O. Wilson, "The loss of a keystone species is like a drill accidentally striking a power line. It causes lights to go out all over."

A given species may be more than one of these types in a particular ecosystem.

---

*All species play some role in their ecosystem and thus are important. Some scientists consider all species equally important, but others consider certain species to be keystone species or more important than others, at least in helping maintain the ecosystems they are a part of.

Q: What percentage of the world's population are under age 15?

CONNECTIONS

Amphibians first appeared about 350 million years ago. These cold-blooded creatures range in size from a frog that can sit on your thumb to a Japanese salamander at about 1.5 meters (5 feet) long.

Fossil records suggest that frogs and toads, the oldest of today's amphibians, were living as long as 150 million years ago. Such endurance testifies to the adaptability of these organisms. Within the last decade or two, however, hundreds of the world's estimated 5,100 amphibian species (including 2,700 species of frogs and toads) have been vanishing or dying back, even in protected wildlife reserves and parks (Figure 4-35).

Scientists have not identified any single reason for this decline. In some cases, such as Costa Rican golden toads, diebacks may be caused by prolonged drought. This causes breeding pools to dry up so that few if any tadpoles survive. Dehydration can also weaken amphibians, making them more susceptible to fatal viruses, bacteria, fungi, or protozoans.

In other cases the culprit may be pollution. Because amphibians live part of their lives in water and part on land, they are exposed to water, soil, and air pollutants. The soft, permeable skin that allows them to absorb oxygen from water also makes them extremely sensitive to water-borne pollutants. Their skin may also make them highly sensi-

Figure 4-35 A pair of golden toads mating. Populations of this species have dropped sharply in recent years, even in Costa Rica's protected Monteverde Cloud Forest Reserve. Prolonged drought may explain this population decline, but other factors may also be involved. The male's dazzling reddish-orange color helps it attract its mate.

tive to increases in ultraviolet radiation caused by depletion of ozone in the stratosphere. Their insect diet guarantees them abundant food, but in farming areas it also exposes them to pesticides.

In Asia, where frog legs are a delicacy, overhunting may play a part. In other areas migration or introduction of predators and competitors can threaten amphibian populations. Another possibility in some areas is loss of habitat or fragmentation of habitat into pieces too small to support populations of some amphibians. Once isolated small populations decline to a certain level they may not be able to

recover or can easily be wiped out by a chance event.

Scientists are concerned about the amphibians' decline for two reasons. First, it suggests that the world's environmental health is deteriorating rapidly because amphibians are tough survivors. Second, amphibians, which outnumber birds and eat more insects than birds, play an important role in the world's ecosystems. In some habitats their extinction could also spell death for other species such as the reptiles, birds, and other amphibians that feed on them. As indicator species, amphibians are sending us an important message.

Niche The **ecological niche**, or simply **niche** (pronounced *nitch*), of a species is its total way of life or its role in an ecosystem. It includes all the physical, chemical, and biological conditions a species needs to live and reproduce in an ecosystem.

Species can be broadly classified as specialists or generalists, according to their niches. **Specialist species** have narrow niches. They may be able to live in only one type of habitat, tolerate only a narrow range of climatic and other environmental conditions, or use only one or a few types of food. Examples of specialists are tiger salamanders, which can breed only in fishless ponds so their larvae won't be eaten; red-cockaded woodpeckers, which carve nest-holes almost exclusively in longleaf pines at least 75 years old; and the highly endangered giant panda, which gets 99% of its food from bamboo (Figure 4-37). Within your lifetime, the giant panda may disappear from the wild.

The American Alligator: A Keystone Species

The American alligator (Figure 4-36), North America's largest reptile, has no natural predators except people. Hunters once killed large numbers of these animals for their exotic meat and for the supple belly skin used to make items such as shoes, belts, and pocketbooks. People also considered alligators to be useless, dangerous vermin and hunted them for sport or out of hatred. Between 1950 and 1960 hunters wiped out 90% of the alligators in Louisiana, and by the 1960s the alligator population in the Florida Everglades was also near extinction.

People who say, "So what?" are overlooking the alligator's keystone role in subtropical wetland ecosystems such as Florida's Everglades. Alligators dig deep depressions, or "gator holes," which collect fresh water during dry spells. These holes are refuges for aquatic life and supply fresh water and food for birds and other animals. Large alligator nesting mounds also serve as nest sites for herons and egrets.

As alligators move from gator holes to nesting mounds, they help keep areas of open water free of invading vegetation. Alligators also eat large numbers of predatory gar fish and thus help maintain populations of game fish such as bass and bream.

In 1967 the U.S. government placed the American alligator on the endangered species list. Protected from hunters, the alligator population had made a strong comeback in many areas by 1975—too strong, according to people who find alligators in their backyards and swimming pools.

The problem is that people are invading the alligator's natural habitats. And while the gator's diet consists mainly of snails, apples, sick fish, ducks, raccoons, and turtles, a pet or a person who falls into or swims in a canal, a pond, or some other area where a gator lives is subject to attack.

In 1977 the U.S. Fish and Wildlife Service reclassified the American alligator from *endangered* to *threatened* in Florida, Louisiana, and Texas, where 90% of the animals live. In 1987 this reclassifica-

Figure 4-36 The American alligator is a keystone species in its marsh and swamp habitats in the southeastern United States. In 1967 it was classified as an endangered species. This protection allowed the species to recover enough for its status to be changed from endangered to threatened. Because of its thick skin, speed in the water, and powerful jaws, this species has no natural predators except humans.

tion was extended to seven other states. It is illegal to kill alligators as a threatened species, but limited hunting by licensed professional game wardens is allowed in some areas to control the population. The comeback of the American alligator is an important success story in wildlife conservation.

Figure 4-37 The giant panda, which is found in China, is a specialist species. It has a narrow food niche, surviving primarily on bamboo stalks and leaves. Today only about 800 giant pandas survive in the wild, in about 20 isolated "islands" of bamboo forest in western China. These isolated populations of 10–50 animals are vulnerable to being wiped out by illegal hunting, habitat loss, and inbreeding. They are also biologically vulnerable to extinction because only about one cub per female survives every other year. They are also quite finicky about picking mates, which becomes critical with their low numbers and isolated habitats. Within your lifetime this species may disappear from the wild. About 220 giant pandas are found in zoos and research centers in China and elsewhere, but more captive pandas die than are born.

Q: How many legal immigrants are admitted to the United States each year?

Figure 4-38 Stratification of specialized plant and animal niches in various layers of a tropical rain forest. These specialized niches allow species to avoid or minimize competition for resources with other species and results in coexistence of a great variety of species (biodiversity). Niche specialization is promoted by the adaptation of plants to the different levels of light available in the forest's layers and by hundreds of thousands of years of evolution in a fairly constant climate.

In a tropical rain forest an incredibly diverse array of species survives by occupying specialized ecological niches in distinct layers of the forest's vegetation (Figure 4-38). The widespread clearing and degradation of such forests is dooming large numbers of such specialized species to extinction.

Generalist species have broad niches. They can live in many different places, eat a variety of foods, and tolerate a wide range of environmental conditions. Flies, cockroaches, mice, rats, white-tail deer, raccoons, and human beings are all generalist species.

Connections: An Overview of Ways Species Interact When any two species in an ecosystem have some activities or requirements in common, they may interact to some degree. The principal types of species

interactions are *interspecific competition, predation, parasitism, mutualism,* and *commensalism.* Three of these interactions—parasitism, mutualism, and commensalism—are **symbiotic relationships** in which two or more kinds of organisms live together in an intimate association, with members of one or both species benefiting from the association. In mutualism and commensalism, neither species is harmed by the interaction.

Competition Between Species for Limited Resources As long as commonly used resources are abundant, different species can share them. This allows each species to come closer to occupying its **fundamental niche**: the full potential range of physical, chemical, and biological factors it could use if there were no competition from other species.

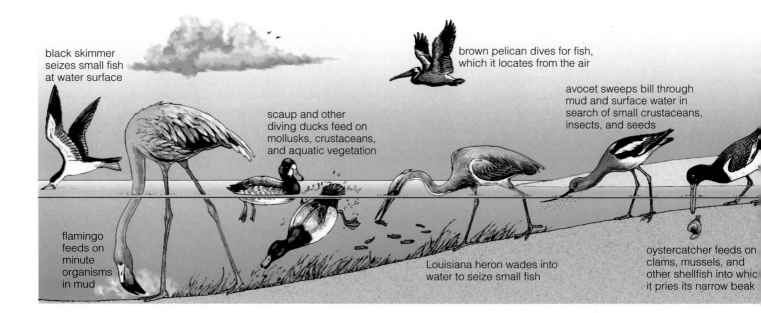

black skimmer seizes small fish at water surface

brown pelican dives for fish, which it locates from the air

avocet sweeps bill through mud and surface water in search of small crustaceans, insects, and seeds

scaup and other diving ducks feed on mollusks, crustaceans, and aquatic vegetation

flamingo feeds on minute organisms in mud

Louisiana heron wades into water to seize small fish

oystercatcher feeds on clams, mussels, and other shellfish into whic it pries its narrow beak

In most ecosystems each species faces competition from one or more other species for one or more of the same limited resources (such as food, sunlight, water, soil nutrients, or space) it needs. Because of such **interspecific competition**, parts of the fundamental niches of different species overlap significantly. When the fundamental niches of two competing species do overlap, one species may occupy more of its fundamental niche than the other species through competitive interactions.

In **interference competition**, one species may limit another's access to some resource, regardless of whether the resource is abundant or scarce. For example, one species of hummingbird may defend patches of spring wildflowers from which it gets nectar by chasing away individuals of other hummingbird species. Coral animals kill other nearby species of coral by poisoning and growing over them.

In pure **exploitation competition**, two competing species have equal access to a specific resource but differ in how fast or efficiently they exploit it. In this way one species gets more of the resource and thus hampers the growth, reproduction, or survival of the other species. When shared resources are abundant this type of interspecific competition does not occur.

Another process that reduces the degree of fundamental niche overlap is **resource partitioning**, the dividing up of scarce resources so that species with similar requirements use them at different times, in different ways, or in different places (Figures 4-38 and 4-39). In effect, they "share the wealth," with each competing species occupying a **realized niche**, the portion of the fundamental niche that a species actually occupies. For example, hawks and owls feed on

similar prey, but hawks hunt during the day and owls hunt at night. Where lions and leopards live in the same area, lions take mostly larger animals as prey and leopards take smaller ones. Some species of birds, such as warblers and tanagers, avoid competition by hunting for insects in different parts of the same coniferous trees.

Experiments have shown that no two species can occupy exactly the same fundamental niche indefinitely in a habitat where there is not enough of a particular resource to meet the needs of both species. This is called the **competitive exclusion principle**. As a result, one of the competing species must migrate to another area (if possible), shift its feeding habits or behavior, suffer a sharp population decline, or become extinct in that area.

Predation   The most obvious form of species interaction is **predation**. Members of a **predator** species feed on parts or all of an organism of a **prey** species, but they do not live on or in the prey. Together, the two kinds of organisms, such as lions and zebras, are said to have a **predator–prey relationship**. Examples of predators and their prey are shown in Figures 4-15, 4-16, 4-21, and 4-22. Many shark species are key predators in the world's oceans (Case Study, p. 88).

Parasitism   Another type of predator–prey interaction is parasitism. A **parasite** is a predator that preys on another organism—its **host**—by living on or in the host for all or most of the host's life. The parasite is smaller than its host and draws nourishment from and gradually weakens the host, sometimes killing it. Tapeworms, disease-causing organisms (pathogens),

Q: How many illegal immigrants enter the United States each year?

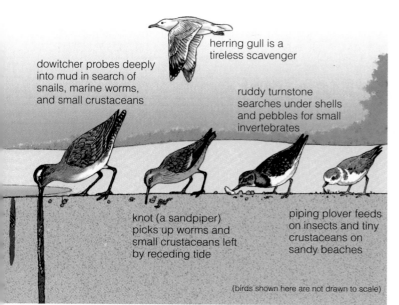

dowitcher probes deeply into mud in search of snails, marine worms, and small crustaceans

herring gull is a tireless scavenger

ruddy turnstone searches under shells and pebbles for small invertebrates

knot (a sandpiper) picks up worms and small crustaceans left by receding tide

piping plover feeds on insects and tiny crustaceans on sandy beaches

(birds shown here are not drawn to scale)

Figure 4-39 Specialized feeding niches of various species of birds in a wetland. This resource partitioning allows them to reduce competition and share limited resources.

Figure 4-40 Parasitism. Sea lampreys are parasites that use their suckerlike mouths to attach themselves to the sides of fish on which they prey. They then bore a hole in the fish with their teeth and feed on its blood.

Figure 4-41 Mutualism. Oxpeckers feed on the ticks that infest the endangered black rhinoceros in Kenya. The rhino benefits by having parasites removed from its body, and oxpeckers benefit by having a dependable source of food. This and other species of rhinoceros face extinction because they are illegally killed for their horns and because their habitat is shrinking.

and other parasites live inside their hosts. Lice, ticks, mosquitoes, mistletoe plants, and lampreys (Figure 4-40) attach themselves to the outside of their hosts. Some parasites move from one host to another, as fleas and ticks do. Others spend their adult lives attached to a single host. Examples are mistletoe, which feeds on oak tree branches, and tapeworms, which feed in the intestines of humans and other animals.

Mutualism **Mutualism** is a type of species interaction in which both participating species generally benefit. The honeybee and certain flowers have a mutualistic relationship. The honeybee feeds on a flower's nectar and in the process picks up pollen, pollinating female flowers when it feeds on them. Other examples are the mutualistic relationships between rhinos and oxpeckers (Figure 4-41) and between legume plants and *Rhizobium* bacteria that live in nodules on the roots of these plants (Figure 4-31).

Other important mutualistic relationships exist between animals and the vast armies of bacteria—in their stomachs or intestines—that break down (digest) their food. The bacteria gain a safe home with a steady food supply; the animal gains access to a large source of energy.

Commensalism In another type of species interaction, called **commensalism**, one species benefits while the other is neither helped nor harmed to any great degree. An example is the relationship between various species of clownfish and sea anemones, marine animals with stinging tentacles that paralyze most fish that touch them (Figure 4-43). The clownfishes gain protection by living unharmed among the deadly ten-

tacles and feed on the detritus left from the meals of the host anemone. The sea anemones seem to neither benefit nor suffer harm from this relationship.

On land there are commensalistic relationships between various trees and epiphytes or "air plants" that attach themselves to tree branches (Figure 4-12). The epiphytes benefit by obtaining water and nutrients from air or bark surfaces without penetrating or harming their hosts.

Why We Need Sharks

The world's 350 species of sharks range in size from the dwarf dog shark—about the size of a large goldfish—to the whale shark—the world's largest fish at 18 meters (60 feet) long. Various shark species are the key predators in the world's oceans, helping control the numbers of many other ocean predators. By feeding at the top of food webs, these shark species cull injured and sick animals from the ocean, thus keeping these species stronger and healthier.

Influenced by movies and popular novels, most people think of sharks as people-eating monsters. This is far from the truth. Every year a few species of shark—mostly great white, bull, bronze whaler, tiger, gray reef, blue (Figure 4-42), and oceanic whitetip—injure about 100 people worldwide and kill between 5 and 10. If you are a typical ocean-goer, you are 150 times more likely to be killed by lightning and thousands of times more likely to be killed when you drive a car than to be killed by a shark.

For every shark that injures a person, we kill 1 million sharks—a total of more than 100 million sharks each year. Sharks are killed mostly for their fins—widely used in Asia as a soup ingredient—at around $50 a bowl in some restaurants—and as a pharmaceutical cure-all. They are also killed for their livers, meat (especially mako and thresher), and jaws (especially great whites), or because we fear them. Some sharks (especially blue,

Figure 4-42 This blue shark and other types of sharks are key predators in the world's oceans. One of only a small number of shark species that occasionally attack swimmers, blue sharks prefer deep water and are a potential threat only to people swimming from boats in deep water.

mako, and oceanic whitetip) die when they are trapped in nets meant for swordfish, tuna, shrimp, and other commercially important species.

Why should we care how many sharks are killed? Because they perform valuable services for us and other species. Sharks save human lives by helping us learn how to fight cancer, bacteria, and viruses, because sharks seem to be free of almost all diseases (including cancer and eye cataracts) and are not affected by most toxic chemicals. Understanding why can help us improve human health.

Chemicals extracted from shark cartilage have killed cancerous tumors in laboratory animals and may someday be the basis of cancer-treating drugs. Another chemical extracted from shark cartilage is being used as an artificial skin for burn victims. Sharks' highly effective immune system allows wounds to heal quickly without becoming infected, and it is being studied for protection against AIDS. And shark corneas have been transplanted into human eyes.

With more than 400 million years of evolution behind them, sharks have had a long time to get things right. We could undo most of this evolutionary wisdom in a few decades. Preventing this from happening begins with the recognition that sharks don't need us, but we and other species need them.

The essential features of the living and nonliving parts of individual terrestrial and aquatic ecosystems, and of the ecosphere, are interdependence and connectedness. Without the services performed by diverse communities of species, we would be starving, gasping for breath, and drowning in our own wastes. The next chapter shows how this interdependence is the key to understanding Earth's major life zones and ecosystems.

We sang the songs that carried in their melodies all the sounds of nature—the running waters, the sighing of winds, and the calls of the animals. Teach these to your children that they may come to love nature as we love it.

GRAND COUNCIL FIRE OF AMERICAN INDIANS

Q: How many of the world's people live in urban areas?

Figure 4-43 Commensalism. This clownfish in the Coral Sea, Australia, has a commensalistic relationship with deadly sea anemones, whose tentacles quickly paralyze most other fishes that touch them. The clownfish gains protection and food by feeding on scraps of food left over from fish killed by the sea anemones, which seem to be neither helped nor harmed by this relationship. All 26 species of clownfish are found only in association with various species of sea anemones.

Carl Roessler/FPG International

Critical Thinking

1. a. A bumper sticker asks, "Have you thanked a green plant today?" Give two reasons for appreciating a green plant.
   b. Trace the sources of the materials that make up the bumper sticker and see whether the sticker itself is a sound application of the slogan.
   c. Explain how decomposers help keep you alive.

2. a. How would you set up a self-sustaining aquarium for tropical fish?
   b. Suppose you have a balanced aquarium sealed with a clear glass top. Can life continue in the aquarium indefinitely as long as the sun shines regularly on it?
   c. A friend cleans out your aquarium and removes all the soil and plants, leaving only the fish and water. What will happen?

3. Using the second law of energy, explain why there is such a sharp decrease in usable energy as energy flows through a food chain or web. Doesn't an energy loss at each step violate the first law of energy? Explain.

4. Using the second law of energy, explain why many poor people in LDCs exist mostly on a vegetarian diet.

5. Using the second law of energy, explain why, on a per weight basis, steak costs more than corn.

*6. Make a concept map of the key ideas in this chapter using the section heads and subheads and the key terms (shown in boldface type in the chapter). See the inside front cover and Appendix 4 for information on concept maps.

Earth Healing: From Rice Back to Rushes

Something there is that doesn't love a marsh. Our deepest fears are linked somehow with swamps, quicksand, and the "things" that lurch out of them. Driven by such fears and ignorance about the ecological importance of wetlands, as well as by desire for land and hunger for profit, we have drained swamps and marshes relentlessly for centuries. Now, belatedly, we begin to question that campaign against wild nature. Can we turn back the clock to restore or rehabilitate vanished marshes?

California rancher Jim Callender decided to try. In 1982 he bought 20 hectares (50 acres) of Sacramento Valley ricefield that had been a marsh until the early 1970s. The previous owner had been thorough in his destruction—bulldozing, draining, leveling, uprooting the native plants, and spraying with chemicals to kill the snails and other food of the waterfowl.

Jim and his friends set out to work in reverse. They hollowed out low areas, built up islands, re-planted tules and bulrushes, brought back smartweed and other plants needed by birds, and planted fast-growing Peking willows.

After six years of care, hand-planting, and annual seeding with a mixture of watergrass, smartweed, and rice, the marsh is once again a part of the Pacific Flyway used by migratory waterfowl (Figure 5-1). Many birds pass through on their way south in autumn and north in the spring, and some even stay. Mallards and widgeons nest there, for example, as do wood ducks. All kinds of hawks, shorebirds, and songbirds live or forage in the marsh.

Jim Callender and a few others have shown that at least part of the continent's wetlands heritage can be reclaimed, with canny planning and hard work. Such Earth healing is vital, but the real challenge is to protect remaining wetlands from having to be healed.

Figure 5-1 Migrating snow geese in eastern Oregon along the Pacific flyway. (Courtesy of Pat and Tom Lesson/Photo Researchers, Inc.)

We cannot command nature except by obeying her.

SIR FRANCIS BACON

This chapter focuses on the following questions:

- What are the principal types of biomes?

- What are the basic types of aquatic life zones, and what influences the kinds of life they contain?

- How are living systems affected by stress?

- How can populations of species change and adapt to environmental stress?

- How can communities and ecosystems change and adapt to small- and large-scale environmental stress?

- What impacts do human activities have on populations, communities, and ecosystems?

5-1 BIOMES: LIFE ON LAND

Connections: Climate and Vegetation   Why is one area of Earth's land surface a desert, another a grassland, and another a forest? Why are there different types of deserts, grasslands, and forests? What determines the types of life in these biomes when they are undisturbed by human activities?

The general answer to these questions is differences in **climate**: the average long-term weather of an area; it is a region's general pattern of atmospheric or weather conditions, seasonal variations, and weather extremes (such as hurricanes and prolonged drought or rain) averaged over a long period (at least 30 years). The two most important factors determining the climate of an area are temperature and precipitation (Figure 5-2).

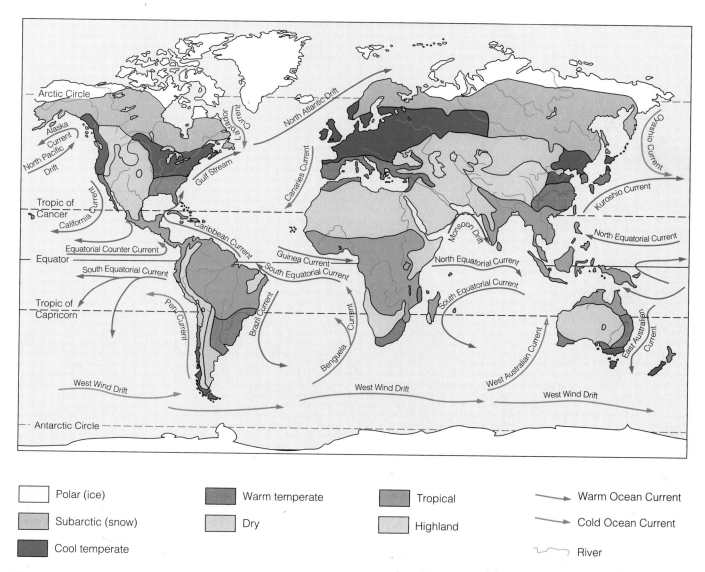

Polar (ice)	Warm temperate	Tropical	→ Warm Ocean Current
Subarctic (snow)	Dry	Highland	→ Cold Ocean Current
Cool temperate			River

Figure 5-2 Generalized map of global climate zones and major ocean currents and drifts. Major variations in climate are dictated mainly by two variables: **(1)** the temperature, with its seasonal variations, and **(2)** the quantity and distribution of precipitation.

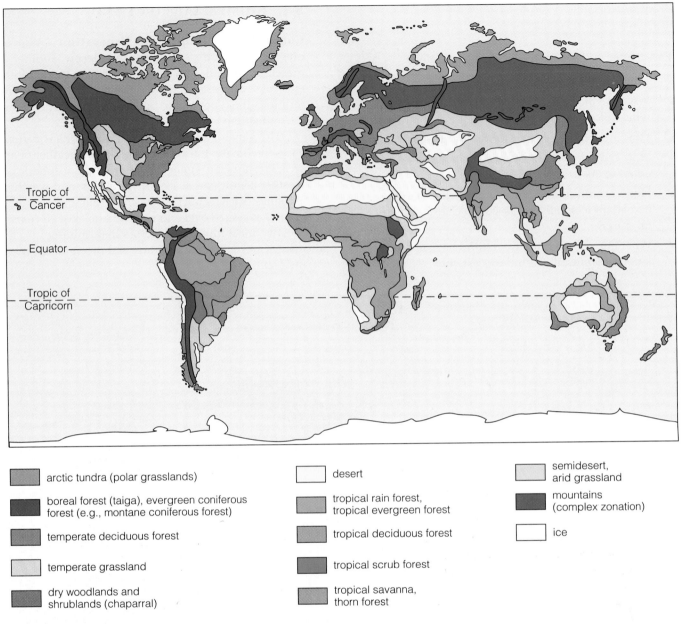

arctic tundra (polar grasslands)	desert	semidesert, arid grassland
boreal forest (taiga), evergreen coniferous forest (e.g., montane coniferous forest)	tropical rain forest, tropical evergreen forest	mountains (complex zonation)
temperate deciduous forest	tropical deciduous forest	ice
temperate grassland	tropical scrub forest	
dry woodlands and shrublands (chaparral)	tropical savanna, thorn forest	

Figure 5-3 Earth's major biomes. This map indicates the main types of natural vegetation we would expect to find in different undisturbed land areas, mostly because of differences in climate. Each biome contains many ecosystems whose communities have adapted to smaller differences in climate, soil, and other environmental factors within the biome. In reality, people have removed or altered much of this natural vegetation for farming, livestock grazing, lumber and fuelwood, mining, water transfer, villages, and cities, thereby altering the biomes.

Figure 5-3 shows the distribution of eleven **bi-omes** (excluding mountains and ice)—regions with characteristic types of natural, undisturbed plant communities. By comparing this figure with Figure 5-2, you can see how the world's major biomes vary with climate.

For plants, *precipitation is generally the limiting factor that determines whether a land area is desert, grassland, or forest* (Figures 5-3 and 5-4). A **desert** is an area where evaporation exceeds precipitation and the average amount of precipitation is less than 25 centimeters (10 inches) a year. Such areas have little vegetation or have widely spaced, mostly low vegetation.

Grasslands are regions where the average annual precipitation is great enough to allow grass, and in some areas a few trees, to prosper, yet the precipitation is so erratic that drought and fire prevent large stands of trees from growing. Grasses in these biomes are renewable resources if not overgrazed. They grow out from the bottom instead of at the top, and their stems can grow again after being nibbled off by grazing animals. Undisturbed areas with moderate to high

Q: What percentage of the world's cars are found in the United States?

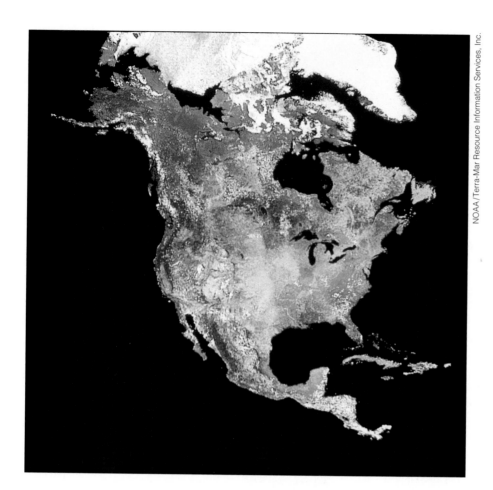

NOAA/Terra-Mar Resource Information Services, Inc.

Figure 5-4 View from space showing major biomes and surface features of most of North America.

average annual precipitation tend to be covered with **forest**, containing various species of trees and smaller forms of vegetation.

Average annual precipitation and temperature, along with soil type, are the most important factors determining the type of desert, grassland, or forest in a particular area. Acting together, these factors lead to tropical, temperate, and polar deserts, grasslands, and forests (Figures 5-5 and 4-8).

Climate and vegetation both vary with **latitude** (distance from the equator) and **altitude** (height above sea level). If you travel from the equator toward either pole, you will generally encounter ever colder and wetter climates (Figure 5-2) and zones of vegetation adapted to each (Figure 5-6). Similarly, as elevation or height above sea level increases, the climate becomes colder and is often wetter. Thus, if you climb a tall mountain from its base to its summit, you will find changes in plant life similar to those you would find in traveling from equator to poles (Figure 5-6).

Deserts and Semideserts   *Tropical deserts*, such as the southern Sahara and the Namib in Africa (Figure 5-7), make up about one-fifth of the world's desert area. They are the driest places on Earth and typically have few plants and a hard, windblown surface

strewn with rocks and some sand. In *temperate deserts* (Figure 4-11), such as the Mojave in southern California, daytime temperatures are hot in summer and cool in winter. In *cold deserts*, such as the Gobi in China, winters are cold and summers are warm or hot. In the typically semiarid zones between deserts and grasslands we find *semidesert*, dominated by thorn trees and shrubs adapted to a long dry season followed by brief, sometimes heavy, rains.

Desert plants and animals have evolved strategies to capture and conserve scarce water. Some of the plants are evergreens. Their wax-coated leaves (creosote bush) cut down on evaporation. Some desert plants (mesquite) send down deep roots to tap into groundwater, while fleshy-stemmed, short (prickly pear) and tall (saguaro, Figure 4-11) cacti spread their shallow roots wide to collect scarce water for storage in their spongy tissues.

Most desert animals escape the heat by hiding in burrows or rocky crevices by day and being active during night or early morning hours. They also have physical adaptations for conserving water (Spotlight, p. 96). Insects and reptiles have thick outer coverings to minimize water loss through evaporation. Some desert animals become dormant during periods of extreme heat or drought.

A: 35% (with only 4.7% of the world's population)

Figure 5-5 Average precipitation and average temperature, acting together as limiting factors over a period of 30 or more years, determine the type of desert, grassland, or forest biome in a particular area. Although the actual situation is much more complex, this simplified diagram explains in a general way how climate determines the types and amounts of natural vegetation found in an area that has not been disturbed by human activities. (Used by permission of Macmillan Publishing Company from Derek Elsom, *Earth*, New York: Macmillan, 1992. Copyright © 1992 by Marshall Editions Developments Limited.)

You might expect desert soils to be poor in plant nutrients, but some are nutrient-rich. We have converted many areas of such deserts into productive farmland by bringing water to them.

Because of the slow growth rate of plants, low species diversity, and shortages of water, deserts take a long time to recover from disruptions. For example, vegetation destroyed by livestock grazing and off-road vehicles may take decades to grow back. Vehicles can also cause the collapse of some underground burrows that are habitats for many desert animals.

Grasslands *Tropical grasslands*—such as most savannas with scattered shrubs and isolated trees (Figure 5-9)—are found in areas with high average temperatures and low-to-moderate average precipitation. They occur in a wide belt on either side of the equator beyond the borders of tropical rain forests (Figure 5-3). Tropical savannas have warm temperatures year-round, with two prolonged dry seasons and abundant rain the rest of the year. African tropical savannas contain enormous herds of *grazing* (grass- and herb-eating) and *browsing* (twig- and leaf-nibbling) hoofed animals, including wildebeest (Figure 5-9), gazelle, zebra, giraffe, and antelope. These and other large herbivores have specialized eating habits that minimize their competition for resources. During the dry season fires often sweep the savannas, and great herds of grazing animals migrate to find food. The remains of these herbivores and their predators are picked over by hyenas, jackals, vultures, and other scavengers.

Temperate grasslands cover vast expanses of flat and rolling hills in the interiors of North and South America, Europe, and Asia (Figure 5-3). Winters are bitterly cold, but the hot, dry summers, as well as drought, occasional fires, and intense grazing, help prevent the growth of trees and bushes, except near rivers.

Types of temperate grasslands are the *tall-grass prairies* (Figure 5-10) and *short-grass prairies* (Figure 5-11) of the midwestern and western United States and Canada, the South American *pampas*, the African *veld*, and the *steppes* of central Europe and Asia. Here winds blow almost continuously, and evaporation is rapid, leading to recurring fires in the summer and fall. As long as the soil is not plowed, it is held in place by a thick network of grass roots; because of their fertile soils, however, many of the world's temperate grasslands have been cleared of their native grasses and used for growing crops (Figure 5-12). Overgrazing, mismanagement, and occasional prolonged droughts sometimes lead to severe wind erosion and loss of topsoil, which can convert these grasslands into desert or semidesert.

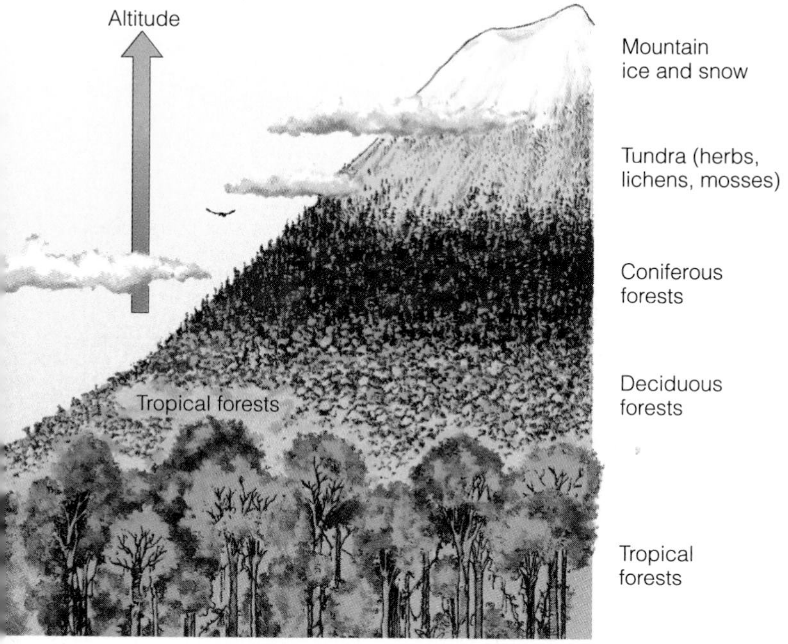

Figure 5-7 Tropical desert. Only a few of the world's deserts are such vast expanses of sand and dunes, as shown in this photograph of the Namib of southwest Africa. Most are covered with rock or small stones, as well as some forms of scattered vegetation.

Gert Behrens/Ardea London

Figure 5-6 Generalized effects of latitude and altitude on climate and biomes. Parallel changes in types of vegetation occur when we travel from equator to poles or from plain to mountaintop.

Altitude

Mountain ice and snow

Tundra (herbs, lichens, mosses)

Coniferous forests

Deciduous forests

Tropical forests

Tropical forests

The Kangaroo Rat: Water Miser and Keystone Species

The kangaroo rat (Figure 5-8) is a remarkable mammal that has mastered the art of water conservation in its desert environment. As the desert's chief seed eater, it is also a keystone species that helps support other desert species. By consuming seeds it helps keep desert shrubland from becoming grassland.

This rodent comes out of its burrow only at night, when the air is cool and water evaporation has slowed. Its main source of food is dry seeds, which it quickly stuffs into its cheek pouches. After a night of foraging, it returns to its burrow and empties its cache of seeds.

In the cool burrow the seeds soak up water exhaled in the rodent's breath. When the rodent eats these seeds it gets this water back. It does not drink water; its water comes from recycled moisture in the seeds it eats and from water produced when the sugars in the seeds undergo aerobic respiration. Some of the water vapor in the rat's breath condenses on a cool inside surface of the nose and diffuses

back to its body. Kangaroo rats have no sweat glands, so they don't lose water by perspiration. In addition, they save water by excreting hard, dry feces and thick, nearly solid urine from their super-efficient kidneys.

Figure 5-8 This nocturnal kangaroo rat of the California desert is an expert in water conservation. It is also a keystone species because it consumes such vast quantities of seeds, which prevents desert shrubland from becoming grassland.

Figure 5-9 Serengeti tropical savanna in Tanzania, Africa, an example of one type of tropical grassland. Most savannas consist of grasslands punctuated by stands of deciduous shrubs and trees, which shed their leaves during the dry season and thus avoid excessive water loss. More large, hoofed, plant-eating mammals (ungulates), such as the herd of wildebeest shown here, live in this biome than anywhere else.

Polar grasslands, or *arctic tundra,* occur just south of the Arctic polar ice cap (Figure 5-3). During most of the year these treeless plains are bitterly cold, swept by frigid winds, and covered with ice and snow. Winters are long and dark, and the low average annual precipitation falls mostly as snow. This biome is carpeted with a thick, spongy mat of low-growing plants (Figure 5-13). Most of the annual growth of these plants occurs during the summer, when sunlight shines almost around the clock.

Q: What percentage of the world's population own cars?

Figure 5-10 A patch of tall-grass prairie in Mason County, Illinois, in early September. Grasses in this biome may be more than 2 meters (6.5 feet) high. Only about 1% of the original tall-grass prairie that once thrived in the midwestern United States and Canada remains. Because of their fertile soils, most have been cleared for crops such as corn, wheat, and soybeans, and for hog farming.

Figure 5-11 American bison grazing on a short-grass prairie to the east of the Rocky Mountains. Grasses in this biome are less than 0.6 meter (2 feet) high. Precipitation is too light and soils are too low in some plant nutrients to support taller grasses. These grasslands are widely used to graze unfenced cattle and sheep and, in some areas, to grow wheat and irrigated crops.

Figure 5-12 Replacement of a temperate grassland with a monoculture cropland near Blythe, California. When the tangled network of natural grasses is removed, the fertile topsoil is subject to severe wind erosion unless it is kept covered with some type of vegetation. If global warming accelerates as projected over the next 50 years, many of these grasslands may become too hot and dry for farming, thus threatening the world's food supply.

Figure 5-13 Polar grassland (arctic tundra) in Alaska in summer. During the long, dark, cold winter this land is covered with snow and ice. Its low-growing plants are adapted to the lack of sunlight and water, to freezing temperatures, and to constant high winds. Below the surface layer of soil is a thick layer of permafrost, which stays frozen year-round.

A: 8%

Figure 5-14 Tropical rain forest in Monteverde Cloud Forest Reserve in Costa Rica. Although tropical rain forests cover only about 2% of Earth's land surface, they contain almost half of the world's growing wood and a third of its plant matter. They are also habitats for 50–80% of Earth's species. These biomes are being cut down and degraded at a rapid rate.

One effect of the extreme cold is **permafrost**—a thick layer of ice beneath the soil surface that remains frozen year-round. The permafrost and the icy winter weather prevent the establishment of trees. In summer, water near the surface thaws, but the permafrost layer below stays frozen and keeps water melted at the surface from seeping into the ground. During this period the tundra is dotted with shallow lakes, marshes, bogs, and ponds. Hordes of mosquitoes, deerflies, blackflies, and other insects thrive in the shallow surface pools. They feed large colonies of migratory birds, especially waterfowl, which migrate from the south to nest and breed in the bogs and ponds. Caribou herds also arrive to feed on the summer vegetation, bringing along their predators, wolves.

The low rate of decomposition, the shallow soil, and the slow growth rate of plants make the arctic tundra especially vulnerable to disruption. For example, vegetation destroyed by human activities can take decades to grow back.

Forests _Tropical rain forests_ are a type of evergreen broadleaf forest (Figure 5-14) found near the equator (Figure 5-3), where hot, moisture-laden air rises and dumps its moisture. They cover about 2% of Earth's land surface. They have a warm annual mean temperature (that varies little daily or seasonally), high humidity, and heavy rainfall almost daily. The almost unchanging climate means that water and temperature are not limiting factors as in other biomes. Here soil nutrients are the main limiting factors.

A mature rain forest has more plant and animal species per unit of area than any other biome. These diverse life-forms occupy a variety of specialized niches in distinct layers, based mostly on their need for sunlight (Figure 4-38). Because of the dense vegetation, little wind blows in tropical rain forests, eliminating the possibility of wind pollination. Instead, many of the plants have evolved elaborate flowers (Figure 1-10, left) that attract particular insects, birds, or bats as pollinators.

Undisturbed tropical rain forests can sustain themselves indefinitely. Unfortunately, as we will see in Chapter 16, these storehouses of biodiversity are being cleared or degraded at an alarming rate. Within 50 years only scattered fragments might remain—causing a massive loss of Earth's vital biodiversity, which some biologists have called a biocrisis or extinction crisis.

Moving a little farther from the equator, we find _tropical deciduous forests_ (sometimes called tropical monsoon forests or tropical seasonal forests), usually located between tropical rain forests and tropical savannas (Figure 5-3). These forests are warm year-round. Most of their plentiful rainfall occurs during a wet (monsoon) season that is followed by a long dry season. They contain a mixture of drought-tolerant evergreen trees and deciduous trees, which lose their leaves to help survive the dry season. Many tropical seasonal forests are being cleared for timber, grazing land, and agriculture, subjecting them to erosion which can lead to desertification. Where the dry season is even longer, we find _tropical scrub forests_ (Figure 5-3) containing mostly small deciduous trees and shrubs.

Temperate deciduous forests grow in areas with moderate average temperatures that change significantly during four distinct seasons (Figure 5-15). These

Q: What percentage of the cars carrying people to and from work in the United States have only one passenger?

Figure 5-15 Temperate deciduous forest in Rhode Island during (clockwise from top left) winter, spring, summer, and fall.

A: 75%

Figure 5-16 Tree farm, or plantation, in North Carolina. Converting a diverse temperate deciduous forest to an even-aged stand of a single species (monoculture) increases the production of wood for timber or pulpwood but results in a loss of biodiversity. Such monocultures are more vulnerable to attacks by pests, disease, and air pollution than are the more diverse forests they replaced.

areas have long summers, cold but not too severe winters, and abundant precipitation, often spread fairly evenly throughout the year. This biome is dominated by a few species of broadleaf deciduous trees, such as oak, hickory, maple, poplar, sycamore, and beech. They survive the winter by dropping their leaves in the fall and becoming dormant. Each spring they sprout new leaves that change in the fall into a blazing array of reds and golds before dropping.

All but about 0.1% of the original stands of temperate deciduous forests in North America have been cleared for farms, orchards, timber, and urban development. Some have been converted to managed *tree farms* or *plantations*, where a single species is grown for timber, pulpwood, or Christmas trees (Figure 5-16).

Evergreen coniferous forests, also called *boreal forests* (meaning "northern forests") and *taigas* (meaning "swamp forests"), are found just south of the arctic tundra in northern regions across North America, Asia, and Europe with a subarctic climate (Figure 5-3). Winters are long, cold, and dry, and sunlight is available only 6–8 hours a day. Summers are short, with mild-to-warm temperatures, and the sun typically shines 19 hours a day.

These forests are dominated by a few species of coniferous evergreens such as spruce, fir, cedar, hemlock, and pine (Figure 5-17). The tiny, needle-shaped, waxy-coated leaves of these trees can withstand the intense cold and drought of winter. Plant diversity is fairly low in these forests.

Beneath the dense stands of trees, a carpet of fallen needles and leaf litter covers the nutrient-poor

Figure 5-17 Evergreen coniferous forest (taiga or boreal forest) in Washington State. Many of the ancient forests along the coast from Canada to northern California have been clear-cut and replaced with tree plantations. There is intense pressure to clear-cut many of the remaining stands of these forests located in the national forests.

Q: What percentage of Americans use public transportation to get to and from work?

soil, making the soil acidic and preventing most other plants from growing on the dim forest floor. During the brief summer the soil becomes waterlogged, forming bogs, or muskegs, in low-lying areas of these forests. Warblers and other insect-eating birds arrive to feed on hordes of flies, mosquitoes, and caterpillars.

Loggers have cut the trees from large areas of taiga in North America, and many of these remaining ancient forests may soon be cut. Most of the vast boreal forests that once covered Finland and Sweden have been cut, and some have been replaced with even-aged tree plantations. Within a decade the vast boreal forests of Siberia may also disappear as the Russian republics log them to earn hard currency. Boreal forests in Canada are also being cleared rapidly, mostly for export to Japan. Because trees grow slowly in the cold northern climate, these forests take a long time to recover from disruption.

5-2   LIFE IN WATER ENVIRONMENTS

Why Are the Oceans Important?   A more accurate name for Earth would be Ocean, because oceans cover more than 70% of its surface (Figure 5-18). The oceans play key roles in the survival of virtually all life on Earth. Because solar heat is distributed through ocean currents (Figure 5-2) and because ocean water evaporates as part of the global hydrologic cycle (Figure 4-34), oceans play a major role in regulating Earth's climate. They also participate in other important nutrient cycles.

By serving as a gigantic reservoir for carbon dioxide (Figure 4-29), oceans help regulate the temperature of the troposphere. Oceans provide habitats for about 250,000 species of marine plants and animals, which are food for many other organisms, including human beings. They also serve as a source of iron, sand, gravel,

phosphates, magnesium, oil, natural gas, and many other valuable resources. Because of their size and currents, oceans absorb and dilute many human-produced wastes that flow or are dumped into them to less harmful or even harmless levels, so long as the oceans are not overloaded.

Ocean Zones   Oceans have two major life zones: the coastal zone and the open sea (Figure 5-19). The **coastal zone** is the relatively warm, nutrient-rich, shallow water that extends from the high-tide mark on land to the gently sloping, relatively shallow edge of the *continental shelf*, the submerged part of the continents. Although it makes up less than 10% of the ocean's area, the coastal zone contains 90% of all marine species and is the site of most of the large commercial marine fisheries. Because of ample sunlight and nutrients deposited from land and stirred up by wind and ocean currents, this coastal zone has a very high net primary productivity per unit of area (Figures 4-27 and 4-28). Examples of highly productive ecosystems in the coastal zone include the following:

- **Coral reefs**—often found in warm tropical and subtropical oceans (Figure 5-20). They are formed by massive colonies containing billions of tiny coral animals called *polyps*. Algae and other producers, which give corals their bright colors, grow on the outside surfaces of coral animals and provide plentiful food for a variety of marine animals. These reefs provide habitats for a diversity of aquatic life and help protect 15% of the world's coastlines from erosion by reducing the energy of incoming waves These ecosystems grow slowly and are easily disrupted. Human activities are causing widespread destruction and damage to these vital ecosystems, making them the most threatened ecosystems in the coastal zone (Case Study, p. 104).

- **Estuaries**—coastal areas at the mouths of rivers, whose fresh water, carrying fertile silt and runoff from the land, mixes with salty seawater (Figure 5-21).

- **Coastal wetlands**—land in a coastal area covered all or part of the year with salt water. They are breeding grounds and habitats for waterfowl and other wildlife, and they dilute and filter out large amounts of nutrients and waterborne pollutants, helping protect the quality of waters used for swimming, fishing, and wildlife habitats. In temperate areas these wetlands usually consist of a mix of bays, lagoons, salt flats, mud flats, and salt marshes (Figure 5-23, p. 104) where grasses are the dominant vegetation. In warm tropical

Figure 5-18 The ocean planet. The oceans cover about 71% of Earth's surface. About 97% of Earth's water is in the interconnected oceans, which cover 90% of the planet's mostly ocean hemisphere (left) and 50% of its land-ocean hemisphere (right). The average depth of the world's oceans is 3.8 kilometers (2.4 miles).

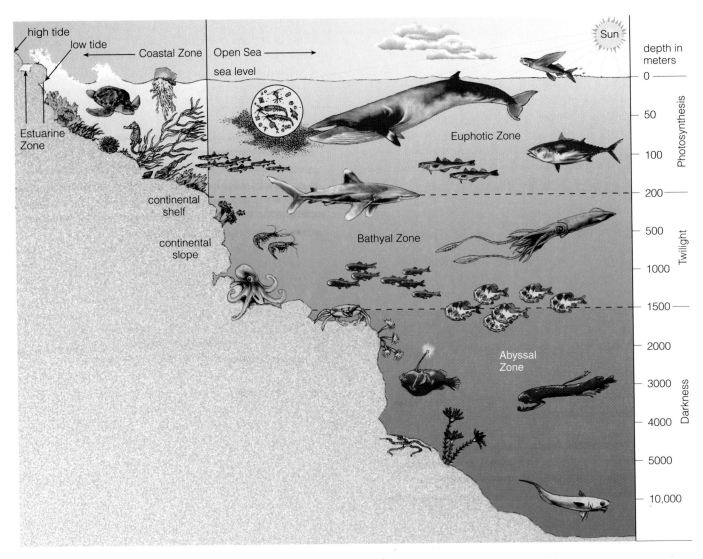

Figure 5-19 Major life zones in an ocean. (Actual depths of zones may vary.)

climates we find highly productive mangrove swamps dominated by mangrove trees, any of about 55 species of trees and shrubs that can live partly submerged in the relatively salty environment of coastal swamps (Figure 5-24, p. 105). Since 1900 the world may have lost half of its coastal wetlands (55% in the United States), primarily by coastal development. Currently, 54% of the U.S. population lives in coastal counties.

■ **Beaches**—Some coasts have steep *rocky shores* pounded by waves (Figure 5-25, p. 106). Many organisms live in the numerous intertidal pools in the rocks. Other coasts have gently sloping *barrier beaches* at the water's edge. If not destroyed by human activities, one or more rows of natural sand dunes on such beaches (with the sand held

in place by the roots of grasses) serve as the first line of defense against the ravages of the sea (Figure 5-26, p. 106). However, such beaches are prime sites for development, which usually destroys their protective dunes. When coastal developers remove the dunes or build behind the first set of dunes, minor hurricanes and sea storms can flood and even sweep away houses and other buildings.

■ **Barrier islands**—long, thin, low, offshore islands of sediment that generally run parallel to the shore along some coasts (such as most of North America's Atlantic and Gulf coasts, Figure 5-27, p. 107). These islands help protect the mainland, estuaries, lagoons, and coastal wetlands by dispersing the energy of approaching storm waves.

Q: What percentage of Americans walk or use a bicycle to get to and from work?

Karl & Jill Wallin/FPG International

Figure 5-20 A healthy coral reef in the Philippines covered by colorful algae. These diverse and productive ecosystems are being destroyed and damaged at an alarming rate.

NASA

Figure 5-21 Space shuttle view of sediment plume at the mouth of Madagascar's Betsiboka River flowing into the estuary of the Mozambique Channel, which separates this huge island from the African coast. The massive amount of sediment results from topography, heavy rain, and clearing of forests for agriculture, making Madagascar the world's most eroded country.

A: 2%

The Importance of Coral Reefs

Coral reefs are among the world's oldest and most diverse and productive ecosystems—the marine equivalent of tropical rain forests (Figure 5-20). Coral reefs are a joint venture. Tiny, single-celled, photosynthesizing protists (dinoflagellates), living in or between the cells of coral animals, synthesize organic food compounds for the polyps. In this mutualistic partnership the protists gain a protected habitat and some minerals from the polyps' body fluids and wastes. Algae (especially zooxanthellae) and other producers, which give corals their bright colors, grow on the outside surfaces of coral animals and provide plentiful food for fish, starfish (Figure 4-13), and other marine animals.

Coral polyps build reefs by cementing themselves to the ocean floor and converting chemicals in seawater into white crystals of calcium carbonate (limestone), which they build up layer by layer to create the reef structure. Covering only 0.17% of the ocean floor—a total area the size of Texas—coral reefs support at least one-fourth of all marine species and account for 20–25% of the fish catch of LDCs (Figure 5-20). For their beauty alone, coral reefs rank as one of the planet's greatest treasures.

Coral reefs reduce the energy of incoming waves and help protect 15% of the world's coastlines from erosion. By forming limestone shells, coral polyps take up carbon dioxide as part of the carbon cycle (Figure 4-29).

Reef organisms produce chemicals useful for cancer and AIDS research. Corals produce a natural sunscreen chemical, which chemists are developing for sale in Australia. And the porous limestone skeletons of corals have been used for bone grafts in humans.

Coral reefs are one of Earth's most endangered ecosystems. The reefs grow slowly and are easily disrupted. They thrive only in clear, clean, warm, and fairly shallow water of constant high salinity. They also need ample sunlight and enough wave action to provide dissolved oxygen and nutrients.

Natural threats to coral reefs include hurricanes, predation by crown-of-thorn starfish, and ocean warming. Normally, reefs gradually recover from the ravages of nature, but human activities are upsetting this healing process.

The greatest human threats to these delicate ecosystems come from eroded soil produced by deforestation, construction, agriculture, mining, dredging, and poor land management along increasingly populated coastlines. The suspended soil sediment that washes downriver to the sea or erodes from coastal areas smothers coral polyps or blocks their sunlight. The sediments also reduce the ability of coral larvae to form new colonies on the reef. Such silting is one cause of coral reef bleaching. It occurs when a reef loses its colorful algae and other producers, exposing the colorless coral animals and the underlying white skeleton of calcium carbonate (Figure 5-22).

Other threats to coral reefs include viruses, global warming, chemical pollution, nuclear weapons testing, anchor damage, dredging, overfishing, the dynamiting of fish, oil spills (and some detergents used to disperse spills), mining coral (limestone) for use as building material, collecting coral for sale to local tourists and for ex-

Figure 5-23 Salt marsh on Cape Cod coast of Massachusetts. These and other temperate coastal wetlands trap nutrients and sediment flowing in from rivers and nearby land and thus have a high net primary productivity. They also filter out and degrade some of the pollutants deposited by rivers and land runoff.

E. R. Degginger

Q: How many people have been killed by motor vehicles since the first automobile was built in 1885?

port (especially for use in home saltwater aquariums), and damage from tourists who walk on reefs and from recreational divers.

Reefs can gradually recover, but if the stress is prolonged the corals die. Without corals the reef erodes, and populations of most of the organisms it supports decline sharply. It takes centuries for large corals and missing reef sections to be rebuilt.

There are 109 countries whose shores are lined with more than 100,000 kilometers (62,000 miles) of reefs. Marine biologists and reef experts estimate that humans have directly or indirectly caused the death of 5–10% of the world's living coral reefs. Already nearly half of the world's population (70% in parts of Southeast Asia) live in coastal regions. And world population is expected to double over the next 40 years. If the current rates of destruction continue, another 60% of the reefs could be gone in the next 20–40 years.

Some 300 coral reefs in 65 countries are protected as reserves or parks, and another 600 have been recommended for protection. Also, some marine biologists are trying to

Figure 5-22 A bleached coral reef in the Bahamas that has lost most of its colorful protists because of changes in the environment, such as cloudy water or extreme temperatures. With the protists gone, the white limestone of the coral skeleton becomes visible. If the stress is not removed, the corals die.

replenish dead or dying reefs by scattering or gluing pieces of live coral onto the reef surface. These are important steps, but protecting and restoring reefs is difficult and expensive, and only half the countries with coral reefs have set aside reserves. Unless we act now to pro-

tect the world's diminishing coral reefs, another important part of Earth's vital biodiversity will be greatly depleted.

Figure 5-24 Mangrove swamp in Colombia. These swamps help protect the coastline from erosion, reduce damage from storms, trap sediment washed off the land, and provide breeding, nursery, and feeding grounds for some 2,000 species of plants and animals. Since the mid-1960s some tropical coastal countries have lost half or more of their mangrove forests because of industrial logging for timber and fuelwood, conversion to ponds for raising fish and shellfish (aquaculture), conversion to rice fields and other agricultural land, and urban development. Mangroves are disappearing fastest in Asia, especially in the Philippines, Indonesia, and Java.

Despite their ecological importance, these coastal ecosystems are under severe stress from human activities. Coastal zones are among our most densely populated and most intensely used—and polluted—ecosystems. Since 1900 the world may have lost half of its coastal wetlands. Asia has lost as much as 60% of its original coastal wetlands and Africa almost 30%. During the past 200 years nearly 55% of the area of estuaries and coastal wetlands in the United States has been destroyed or damaged, primarily by dredging

and filling and waste contamination. California alone has lost 91% of its original coastal and inland wetlands, but Florida has lost the largest area of wetlands in the United States.

These ecosystems are particularly vulnerable to toxic contamination because they trap pesticides, heavy metals, and other pollutants, concentrating them to very high levels. On any given day one-third of U.S. shellfish beds are closed to commercial or sport fishing because of contamination.

Figure 5-25 Rocky shore beach in Acadia National Park, Maine. Organisms of most seashores must be able to withstand the tremendous force of incoming waves and the pull of the outgoing tide. Coasts like this one provide rocks to which organisms can attach themselves.

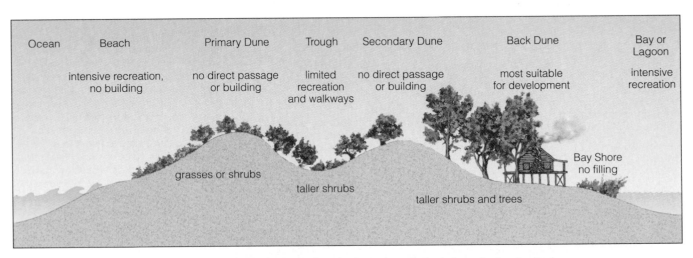

Figure 5-26 Primary and secondary dunes on a gently sloping beach play an important role in protecting the land from erosion by the sea. The roots of various grasses that colonize the dunes help hold the sand in place. Ideally, construction and development should be allowed only behind the second strip of dunes, with walkways to the beach built over the dunes to keep them intact. Not only does this help preserve barrier beaches, but it helps protect structures from being damaged and washed away by wind, high tides, beach erosion, and flooding from storm surges. This type of protection is rare, however, because the short-term economic value of limited oceanfront land is considered to be much higher than its long-term ecological and economic values.

Q: Worldwide, how many people are killed each year by motor vehicles?

Fortunately, about 45% of the area of estuaries and coastal wetlands remains undeveloped. Each year, however, more of this land is developed or severely degraded, especially in the southeastern United States, where 83% of the remaining wetlands in the lower 48 states are found.

The **open sea** is divided into three vertical zones—euphotic, bathyl, and abyssal—based primarily on the penetration of sunlight (Figure 5-19). This vast volume of ocean contains only about 10% of all marine species. Except at occasional areas where currents bring up nutrients from the ocean bottom, average net primary productivity per unit of area in the open sea is quite low (Figure 4-28). This is because sunlight cannot penetrate the lower layers and because the surface layer normally has fairly low levels of nutrients for phytoplankton, which are the main photosynthetic producers of the open ocean.

Freshwater Lakes   **Lakes** are large natural bodies of standing fresh water formed when precipitation, land runoff, or groundwater flowing from underground springs fills depressions in the earth. Causes of such depressions include glaciation (the Great Lakes of North America), crustal displacement accompanied by or resulting in earthquakes (Lake Nyasa in East Africa), and volcanic activity (Crater Lake in Oregon). Lakes normally consist of distinct zones (Figure 5-28), providing habitats and niches for various species.

A lake with a large or excessive supply of nutrients (mostly nitrates and phosphates) needed by producers is called a **eutrophic** (well-nourished) **lake** (Figure 5-29). A lake with a small supply of nutrients is called an **oligotrophic** (poorly nourished) **lake** (Figure 5-29). Many lakes fall somewhere between the two extremes of nutrient enrichment and are called **mesotrophic lakes**.

Most substances become denser as they go from gaseous to liquid to solid physical states. Water, however, doesn't follow this normal behavior. It is densest as a liquid at 4°C (39°F). In other words, solid ice at 0°C (32°F) is less dense than liquid water at 4°C (39°F), which is why ice floats on water. This is fortunate for us and most freshwater organisms. Otherwise, lakes and other bodies of standing fresh water in cold climates would freeze from the bottom up instead of from the surface down, pushing fish and other organisms to the top and killing them.

This unusual property of water causes *thermal stratification* of deep lakes in northern temperate areas with cold winters and warm summers (Figure 5-30, p. 110). In summer such lakes become stratified into layers with different temperatures that hinder mixing. These lakes have an *epilimnion,* an upper layer of warm water with high levels of dissolved oxygen, and a *hypolimnion,* a lower layer of colder, denser water, usually with a lower concentration of dissolved oxygen. These layers are separated by a middle layer called a *thermocline,* where the water temperature changes rapidly with increased depth. The thermocline acts as a barrier to the transfer of nutrients and dissolved oxygen from the epilimnion to the hypolimnion.

In fall, when temperatures begin to drop, the surface layer sinks to the bottom when it cools to 4°C (39°F), and the thermocline disappears. This mixing, or *fall turnover*, brings nutrients from bottom sediments to the top and dissolved oxygen from the top to

Figure 5-27 Developed barrier island, Ocean City, Maryland—host to 8 million visitors a year. To keep up with shifting sands, officials spend millions of dollars to pump sand onto the beaches and to rebuild natural sand dunes; they may end up spending millions more to keep buildings from sinking. There is no effective protection against flooding and damage from severe storms, as residents on barrier islands in South Carolina learned when a devastating hurricane hit in 1989. Within a few hours a barrier island may be cut in two or destroyed by a hurricane. If global warming raises average sea levels as projected sometime in the next century, most of these valuable pieces of real estate will be under water.

G. H. Demetrakas/O. C. Camera

A: About 250,000 (10 million more are injured or permanently disabled)

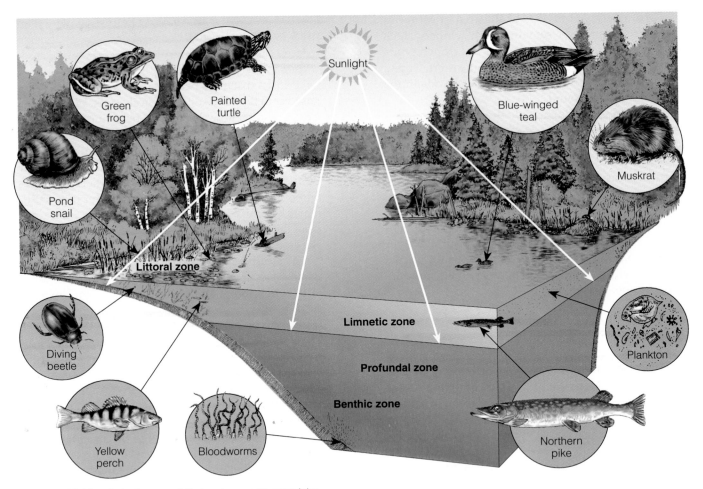

Figure 5-28 The distinct zones of life in a temperate-zone lake.

the bottom. In spring, when the atmosphere warms, the lake's surface water warms to 4°C (39°F), reaches maximum density, and sinks through and below the cooler, less dense water. This brings the bottom water to the surface. During this *spring turnover* dissolved oxygen in the surface layer is moved downward, and nutrients released by decomposition on the lake bottom are moved toward the surface.

Freshwater Streams Precipitation that doesn't sink into the ground or evaporate is **surface water**. This water becomes **runoff**, which flows into streams and eventually to the oceans to continue circulating in the hydrologic cycle (Figure 4-34). The entire land area that delivers water, sediment, and dissolved substances via small streams to a larger stream (or river), and ultimately to the sea, is called a **watershed**, or a **drainage basin**.

The downward flow of water from mountain highlands to the sea takes place in three zones in a *river system* (Figure 5-31). Because of different environmen-

tal conditions in each zone, a river system is a series of different ecosystems. First, narrow headwater or mountain highland streams with cold, clear water rush over waterfalls and rapids. As this turbulent water flows and tumbles downward, it dissolves large amounts of oxygen from the air. Here the plants are attached to rocks, and the fish are cold-water fish such as trout, which need lots of dissolved oxygen.

In the second zone the headwater streams merge to form wider, deeper streams that flow down gentler slopes with fewer obstacles. The warmer water and other conditions in this zone support a variety of cold-water and warm-water fish species with slightly lower oxygen requirements.

In the third zone, these streams join into wider and deeper rivers that meander across broad, flat valleys. The main channels of these rivers support distinctive varieties of fish, whereas their backwaters support species similar to those in lakes. Meandering streams are sometimes straightened, deepened, and widened to improve navigation and to help reduce

Q: Worldwide, how much of urban land is devoted to roads and parking?

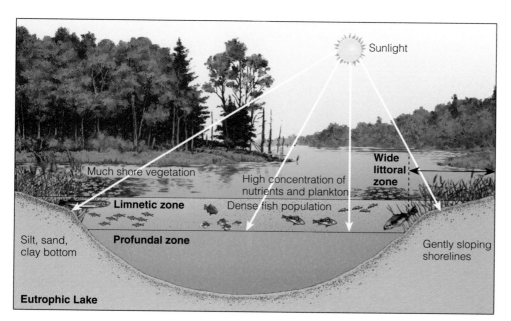

Figure 5-29 Eutrophic, or nutrient-rich, lake (top) and oligotrophic, or nutrient-poor, lake (bottom). Mesotrophic lakes fall between these two extremes of nutrient enrichment.

Sunlight

Much shore vegetation

High concentration of nutrients and plankton

Wide littoral zone

Limnetic zone

Dense fish population

Silt, sand, clay bottom

Profundal zone

Gently sloping shorelines

Eutrophic Lake

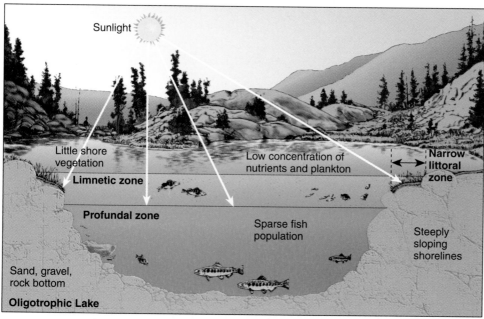

Sunlight

Little shore vegetation

Low concentration of nutrients and plankton

Narrow littoral zone

Limnetic zone

Profundal zone

Sparse fish population

Sand, gravel, rock bottom

Steeply sloping shorelines

Oligotrophic Lake

flooding and bank erosion, but such *stream channelization* is controversial. At its mouth a river may divide into many channels as it flows across a delta and coastal wetlands and estuaries, where the river water mixes with ocean water and deposits (Figures 5-21 and 5-31).

As streams flow downhill, they become powerful shapers of land. Over millions of years the friction of moving water levels mountains and cuts deep canyons. The rock and soil the water removes are deposited as sediment in low-lying areas.

Inland Wetlands Lands covered with fresh water at least part of the year (excluding lakes, reservoirs, and streams) and located away from coastal areas are called **inland wetlands**. They include bogs, marshes, prairie potholes, swamps, mud flats, floodplains, bogs, fens, wet meadows, and the wet arctic tundra (Figure 5-13) in summer. Shallow marshes and swamps are among the world's most productive ecosystems per unit of area (Figure 4-28).

Some wetlands are always covered with water. Others, such as prairie potholes, floodplain wetlands,

A: At least 33% (50% in the United States)

Figure 5-30 (a) Thermal stratification and restriction on water mixing during summer for a lake in a temperate climate region. **(b)** There is little thermal stratification during the winter because once or twice a year such stratified lakes undergo a mixing or *turnover*, in which seasonal temperature changes cause changes in the densities of water in the top and bottom layers. (Used by permission from Cecie Starr and Ralph Taggart, *Biology: The Unity and Diversity of Life*, 6th ed., Belmont, Calif.: Wadsworth, 1992)

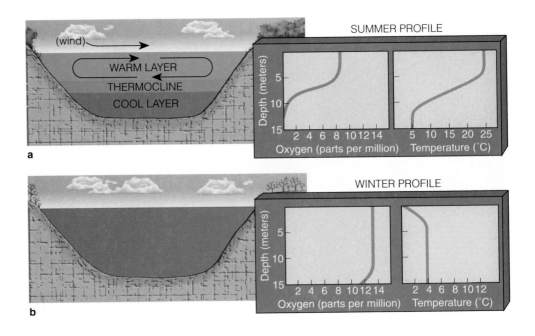

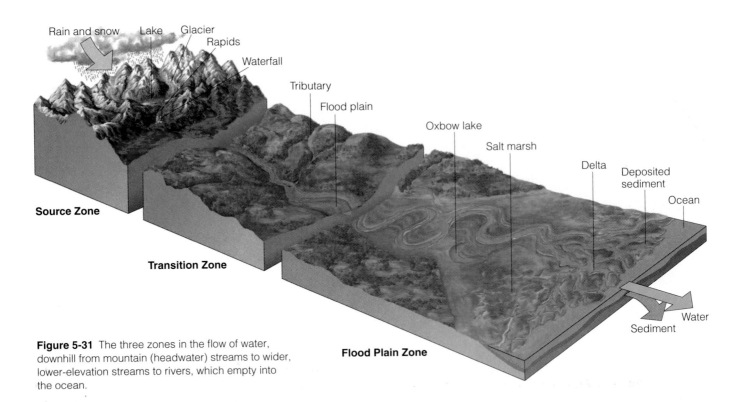

Figure 5-31 The three zones in the flow of water, downhill from mountain (headwater) streams to wider, lower-elevation streams to rivers, which empty into the ocean.

and bottomland hardwood swamps are *seasonal wetlands* that are underwater or soggy for only a short time each year.

Inland wetlands provide habitats for fish, waterfowl, and other wildlife, and they improve water quality by filtering, diluting, and degrading sediments and pollutants as water flows through. Floodplain wetlands near rivers help regulate stream flow by storing water during periods of heavy rainfall and releasing it slowly, which reduces riverbank erosion and flood damage. By storing water, many seasonal and year-round wetlands allow increased infiltration, thus helping recharge groundwater supplies.

Despite the ecological importance of year-round and seasonal inland wetlands, many are drained, dredged, filled in, or covered. Each year some 1,200

Q: What is the average government subsidy per car in the United States?

square kilometers (470 square miles) of inland wetland in the United States are lost, about 80% to agriculture and the rest to mining, forestry, oil and gas extraction, highways, and urban development. Other countries have suffered similar losses.

The stated goal of current federal policy is no net loss of the function and value of wetlands. Unfortunately, this allows destruction of existing wetlands as long as an equal area of the same type of wetland is created or restored. Exceptions are also allowed. As a result, at least 14 hectares (34 acres) of U.S. wetlands are being destroyed every hour.

Some inland wetlands can be partially restored (p. 90), and there have been attempts to create new wetlands to replace some of those allowed to be developed. Experience has shown, however, that many of these attempts fail or that the resulting wetlands bear little resemblance to natural wetlands, resulting in a net loss of wetland ecological functions and values. Restoring and creating wetlands is also expensive. The best and cheapest thing to do is to keep these valuable natural ecosystems from being degraded or destroyed by human activities.

5-3 RESPONSES OF LIVING SYSTEMS TO ENVIRONMENTAL STRESS

Change: The Driving Force of Nature There is no such thing as "natural balance" or "balance of nature." The key thing going on in nature is constant change, caused by natural and human-related forces, and adjustments to such environmental stresses. Populations, communities, and ecosystems constantly change, and have always done so.

Sustainability does not imply a static situation. Instead, it includes the capacity of populations, communities, ecosystems, and human economic, political, and social systems to adapt to new conditions.

Homeostasis and Stability of Living Systems To survive you must keep your body temperature within certain ranges (Figure 4-20), whether the temperature outside is steamy or freezing. This phenomenon is called **homeostasis**: the maintenance of favorable internal conditions despite fluctuations in external conditions.

Organisms, populations, communities, and ecosystems have some ability to withstand or recover from externally imposed changes or stresses (Table 5-1)—provided those stresses are not too severe. In other words, they have some degree of *stability.*

This stability, however, is maintained only by constant dynamic change. Although an organism's structure is fairly stable over its life span, it is continually

Table 5-1 Unfavorable Changes Affecting Ecosystems

Natural Changes

Catastrophic	Drought
	Flood
	Fire
	Volcanic eruption
	Earthquake
	Hurricane
	Landslide
	Change in stream course
	Disease
Gradual	Changes in climate
	Immigration
	Adaptation and evolution
	Changes in plant and animal life (ecological succession)

Human-Caused Changes

Catastrophic	Deforestation
	Overgrazing
	Plowing
	Erosion
	Pesticides
	Fires
	Mining
	Toxic releases (can also be gradual)
	Urbanization (can also be gradual)
Gradual	Salt buildup in soil from irrigation (salinization)
	Waterlogging of soil from irrigation
	Compaction of soil from agricultural equipment
	Depletion of groundwater (aquifers)
	Water pollution (can also be catastrophic)
	Air pollution (can also be catastrophic)
	Loss and degradation of wildlife habitat (can also be catastrophic)
	Killing of predator and "pest" species
	Introduction of alien species
	Overhunting
	Overfishing
	Excessive tourism

A: $1,600 per year

gaining and losing matter and energy. Similarly, in a mature tropical rain forest, some trees will die and others will take their place. Some species may disappear, and the number of individual species in the forest may change. Unless it is cut, burned, or blown down, however, you will recognize it as a tropical rain forest 50 years from now.

It's useful to distinguish between three aspects of stability in living systems. **Inertia, or persistence,** is the ability of a living system to resist being disturbed or altered. **Constancy** is the ability of a living system, such as a population, to maintain a certain size or keep its numbers within certain limits. **Resilience** is the ability of a living system to bounce back after an outside disturbance that is not too drastic.

Communities and ecosystems are so complex and variable that ecologists have little understanding of how they maintain some degree of inertia and resilience while undergoing continual change in response to changes in environmental conditions.

Ecologists also find it difficult to predict which one or combination of environmental factors can stress ecosystems beyond their levels of tolerance. However, here are some signs of ill health in stressed ecosystems:

- Primary productivity often drops.

- Nutrient leakage increases.

- Sensitive (indicator) species decline in numbers or become extinct.

- Diseases and populations of insect pests increase.

- Species diversity drops, although this may be offset by invasions of insect pests, disease organisms, and other species.

- Contaminants are present.

Studying Ecosystems Is Difficult Greatly simplified ecosystems can be set up and observed under laboratory conditions, but extrapolating the results of such experiments to much more complex natural communities and ecosystems is difficult, if not impossible.

At one time ecologists believed that the higher the species diversity in an ecosystem, the greater its stability. According to this hypothesis, an ecosystem with a diversity of species has more ways to respond to most environmental stresses because it does not have "all its eggs in one basket." Research indicates, however, that there are numerous exceptions to this intuitively appealing idea.

There is, of course, some minimum threshold of species diversity below which ecosystems cannot function. For example, no ecosystem can function without some plants and decomposers (Figure 4-19). Beyond this it is hard to know whether simple ecosystems are less stable than complex ones, or what the threshold is below which complex ecosystems fail.

Part of the problem is that ecologists disagree on how to define stability and diversity. Does an ecosystem need both high inertia and high resilience to be considered stable? Evidence indicates that some ecosystems have one of these properties but not the other. For example, tropical rain forests (Figures 4-38 and 5-14) have high species diversity and high inertia. This means they are hard to alter significantly or to destroy. However, once a large tract of redwood forest or tropical forest is severely degraded, the ecosystem resilience is so low that the forest may never be restored. Soil nutrients and other conditions needed for recovery may no longer be present.

Grasslands, by contrast, are much less diverse than most forests, burn easily, and thus have low inertia. However, because most of their plant matter is underground, in the roots, these ecosystems have high resilience, which allows them to recover quickly. A grassland can be destroyed only if its roots are plowed up and something else is planted in its soil (Figure 5-12), or if it is severely overgrazed by livestock or other herbivores.

Another difficulty is that populations, communities, and ecosystems are rarely, if ever, at equilibrium. Instead, nature is in a continuing state of disturbance and fluctuation. Change and turmoil, more than constancy and balance, are the rule. When disturbed, natural systems may change and operate within new limits, rather than returning to some "perfect" equilibrium state when the source of change is removed. Indeed, ecologists now recognize that disturbances of ecosystems are not abnormal but are integral parts of the way nature works. Clearly, we have a long way to go in understanding how the factors in natural communities and ecosystems interact and change in response to changes in environmental conditions.

5-4 POPULATION RESPONSES TO STRESS: POPULATION DYNAMICS

Changes in Population Size Populations change in size, density, dispersion, and age distribution (proportion of individuals of each age in a population) in response to changes in environmental conditions, such as availability of food. These changes are called **population dynamics**.

Four variables—births, deaths, immigration, and emigration—govern changes in population size. A population gains individuals by birth and immigration and loses them by death and emigration:

Q: What is the best way to prevent or reduce pollution?

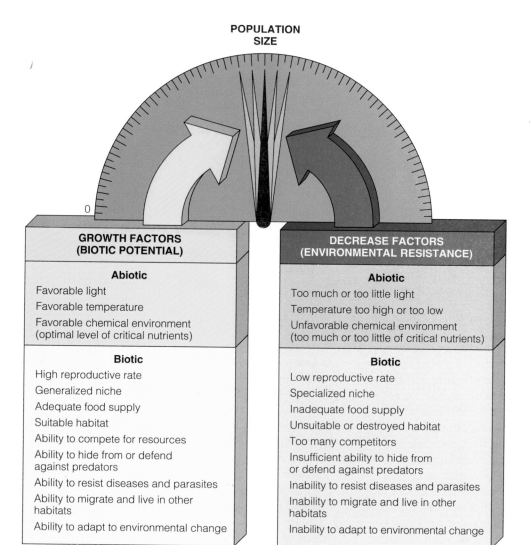

POPULATION SIZE

GROWTH FACTORS (BIOTIC POTENTIAL)	DECREASE FACTORS (ENVIRONMENTAL RESISTANCE)
Abiotic	**Abiotic**
Favorable light	Too much or too little light
Favorable temperature	Temperature too high or too low
Favorable chemical environment (optimal level of critical nutrients)	Unfavorable chemical environment (too much or too little of critical nutrients)
Biotic	**Biotic**
High reproductive rate	Low reproductive rate
Generalized niche	Specialized niche
Adequate food supply	Inadequate food supply
Suitable habitat	Unsuitable or destroyed habitat
Ability to compete for resources	Too many competitors
Ability to hide from or defend against predators	Insufficient ability to hide from or defend against predators
Ability to resist diseases and parasites	Inability to resist diseases and parasites
Ability to migrate and live in other habitats	Inability to migrate and live in other habitats
Ability to adapt to environmental change	Inability to adapt to environmental change

$$\text{population change rate} = \begin{pmatrix} \text{births} \\ + \\ \text{immigration} \end{pmatrix} - \begin{pmatrix} \text{deaths} \\ + \\ \text{emigration} \end{pmatrix}$$

These variables in turn depend on changes in resource availability or other environmental changes (Figure 5-32).

Populations vary in their capacity for growth. The **biotic potential** of a population is the *maximum* rate (r_{max}) at which it can increase when there are no limits on its growth. Animal species have different biotic potentials because of variations in **(1)** when reproduction starts and stops (reproductive span), **(2)** how often reproduction occurs, **(3)** how many live offspring are born each time (litter size), and **(4)** how many offspring survive to reproductive age.

No population can grow exponentially indefinitely. In the real world an exploding population reaches some size limit imposed by a shortage of one or more limiting factors, such as light, water, space, and nutrients. There are always limits to growth in nature.

Environmental resistance consists of all the factors acting jointly to limit the growth of a population. They determine the **carrying capacity (K)**, the number of individuals of a given species that can be sustained indefinitely in a given area. Carrying capacity is determined by many factors, including predation, competition between species, migration, and climate. The carrying capacity for a population can vary with seasonal or abnormal changes in the weather or supplies of food, water, nesting sites, or other crucial environmental resources.

Because of environmental resistance, any population growing exponentially starts out slowly, goes through a rapid growth phase, and then levels off once the carrying capacity of the area is reached. In most

A: Full-cost pricing, with the market price of any item including its internal and external costs

Figure 5-33 (a) Idealized S-shaped curve of population growth. **(b)** Population curve crash or dieback occurs. These idealized curves only approximate what goes on in nature.

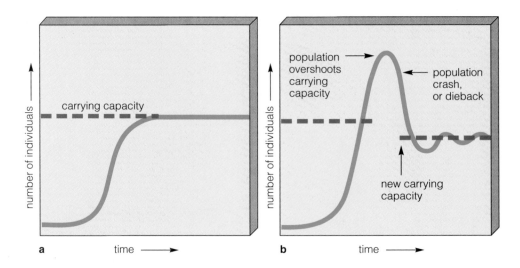

cases the size of such a population fluctuates slightly above and below the carrying capacity. A plot of this type of growth yields an *S-shaped curve* (Figure 5-33a). Sometimes, however, a population temporarily overshoots the carrying capacity (Figure 5-33b). This happens because of a *reproductive time lag*, the time required for the birth rate to fall and the death rate to rise in response to resource limits. Unless large numbers of individuals can move to an area with more favorable conditions, the population will suffer a *dieback* or *crash*, falling back to a lower level that typically fluctuates around the area's carrying capacity. An area's carrying capacity can also be lowered because of resource destruction and degradation during the overshoot period.

Humans are not exempt from this phenomenon. Ireland, for example, experienced a population crash after a fungus infection destroyed the potato crop in 1845. About 1 million people died, and 3 million people emigrated to other countries.

Technological, social, and other cultural changes have extended Earth's carrying capacity for the human species (Figure 2-3). We have increased food production, controlled many diseases, and used energy and matter resources at a rapid rate to make normally uninhabitable areas of Earth habitable. A crucial question is how long we will be able to keep doing this on a planet with finite size and resources.

Density-Dependent and Density-Independent Checks on Population Growth   Some limiting factors become more influential as a population's density increases. Examples of such *density-dependent population controls* are competition for resources, predation, parasitism, and disease. When prey populations become increasingly dense, for example, their members

face greater competition for resources. They also have a greater risk of being killed by predators or being invaded by parasites or disease-causing organisms.

Infectious disease is a classic example of this type of population control. In dense populations disease spreads faster, increasing the death rate and leaving fewer individuals who can reproduce; in sparse populations disease spreads more slowly and has less effect on population growth.

In some species physiological and sociological control mechanisms limit reproduction as population density rises. For example, overcrowding in populations of mice and rats causes hormonal changes that can inhibit sexual maturity, lower sexual activity, and cut milk production in nursing females. Stress caused by crowding in these and several other species also reduces the number of offspring produced per litter through mechanisms such as spontaneous abortion. Crowding may also lead to population control in some species through cannibalism and killing of the young.

Density-independent population controls exert their effects on populations more or less regardless of population density. Examples include floods, hurricanes, severe drought, unseasonal temperature changes, fire, human destruction of habitat (such as clearing a forest of its trees), and changes in the chemical environment that increase the death rate and lower the reproduction rate.

Connections: Population Dynamics in Nature
The idealized S-shaped and crash or dieback curves of population growth shown in Figure 5-33 can be observed in laboratory experiments. In nature, however, we find three general types of population change curves: relatively stable, irruptive, and cyclic (Figure 5-34).

Q: How many of the world's people live in absolute poverty?

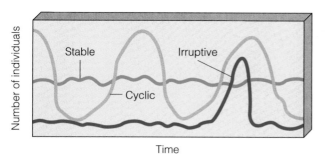

Figure 5-34 General types of idealized population change curves found in nature.

A species whose population size fluctuates slightly above and below its carrying capacity (as long as that capacity doesn't change significantly) has a relatively stable population size. Such stability is characteristic of many species found in habitats of undisturbed tropical rain forests (Figure 4-38), where there is little variation in average temperature and rainfall.

Some species, such as the raccoon, normally have a fairly stable population. However, the population may occasionally explode, or *irrupt*, to a high peak and then crash to a relatively stable lower level. The population explosion is due to some factor that temporarily increases carrying capacity for the population. Examples are better weather, more food, or fewer predators, including humans.

Some species undergo sharp increases in their numbers followed by seemingly periodic crashes. Predators are sometimes blamed, but the actual causes of such "boom-bust" cycles are poorly understood.

Reproductive Strategies and Survival   Each species has a characteristic mode of reproduction. At one extreme are species that produce hordes of offspring early in their life cycle. The offspring, usually small and short-lived, mature rapidly with little or no parental care. Typically, many of them die before they can reproduce.

Species with this capacity for a high rate of population growth (r) (many offspring in a short time) are called **r-strategists**. Algae, bacteria, rodents, annual plants, many fish, and most insects are examples. Such species tend to be *opportunists*, reproducing rapidly when conditions are favorable or when a new habitat or niche becomes available—a cleared forest or a newly plowed field, for example. Unfavorable environmental conditions, however, can cause such populations to crash. Hence, most r-strategists go through "boom-bust" cycles.

At the other extreme are **K-strategists**, species that produce a few, often fairly large offspring and often look after them for a long time to ensure that most reach reproductive age. Living in fairly stable environments and tending to maintain their population size near their habitat's carrying capacity (K), they are called K-strategists. Their populations typically follow an S-shaped growth curve (Figure 5-33). Examples are sharks, most large mammals—including humans, elephants, rhinoceroses, and whales—birds of prey, and large, long-lived plants, such as the saguaro cactus (Figure 4-11). The reproductive strategies of most species fall somewhere between these two extremes.

Species with different reproductive strategies tend to have different life expectancies, the length of time an individual of a certain age in a given species can expect to survive. This information is often presented as a **survivorship curve**, which shows the number of survivors in each age group for a particular species.

Three types of survivorship curves are common in nature (Figure 5-35). *Late-loss (type I)* curves are typical for K-strategists, and *early-loss (type III)* curves for r-strategist species. Species with *constant-loss (type II)* survivorship curves typically have intermediate reproductive strategies.

5-5 RESPONSES TO CHANGING CONDITIONS: THE RISE OF LIFE ON EARTH

Origins of Life and Its Diversity   How did life on Earth evolve to its present system of diverse species living in an interlocking network of matter cycles, energy flow, and species interactions? We don't know the full answers to these questions, but a widely accepted body of evidence suggests what might have happened. Some of this evidence comes from chemical experiments simulating possible atmospheric compositions and energy sources on the primitive Earth to see whether this could lead to the formation of the molecules necessary for life. Most of what we know of Earth's life history, however, comes from **fossils**—mineralized or petrified skeletons, bones, shells, body parts, leaves, and seeds (or impressions of such items)—that provide recognizable evidence of ancient organisms. The fossil record, however, is uneven. Some life-forms left no fossils, some fossils have decomposed, and others are yet to be found.

Other sources of information include chemical analysis of ancient rocks and of cores drilled out of buried ice, and analysis of the genetic makeup of fossil specimens. When this diverse evidence is pieced together, an important fact emerges: *The evolution of life has been linked, from the beginning, to the physical and*

A: At least 1.2 billion—about one in every five people on Earth

Figure 5-35 Three generalized types of survivorship curves for populations of different species. For a type I population (elephant in East Africa) there is high survivorship to a certain age, then high mortality. A type II population (song thrush) shows a fairly constant death rate at all ages. For a type III population (purple sea star), survivorship is low early in life. Many animal and protist populations have survivorship curves that lie somewhere between those characteristic of type II and type III. Many plant species have curves closer to type III. (Used by permission from Cecie Starr and Ralph Taggart, *Biology: The Unity and Diversity of Life*. 6th ed. Belmont, Calif.: Wadsworth, 1992)

chemical evolution of the earth. This shows that Earth has the right physical and chemical conditions for life as we know it to exist (Connections, p. 117).

A second conclusion from this body of evidence is that life on Earth developed in two phases over 4.6–4.8 billion years:

Phase 1: Chemical Evolution (about 1 billion years)

Part 1: Formation of Earth and its early crust and atmosphere

Part 2: Formation of the nonliving biological molecules necessary for life—primarily

Q: Will improving environmental quality in the United States cause a net gain or a net loss of jobs?

Earth: The Just-Right, Resilient Planet

CONNECTIONS

Like Goldilocks tasting porridge at the house of the three bears, life as we know it is picky about temperature. Venus is much too hot, Mars is much too cold, and Earth is just right. Otherwise, you wouldn't be reading these words.

Life as we know it also depends on water. Again, temperature is crucial. Life on Earth needs average temperatures between the freezing and boiling points of water—between 0°C and 100°C (32°F and 212°F)—at Earth's atmospheric pressures.

Earth orbit is the right distance from the sun to have these conditions. If it were much closer, it would be too hot—like Venus—for water vapor to condense to form rain. If it were much farther, its surface would be so cold—like Mars—that its water would exist only as ice. Earth also spins. Otherwise, the side facing the sun would be too hot (and the other side too cold) for water-based life to exist. So far, the temperature has been, like Baby Bear's porridge, just right.

Earth is also the right size. That is, it has enough gravitational mass to keep its iron-nickel core molten and to keep its atmosphere from disappearing into space. The slow transfer of its internal heat (geothermal energy) to the surface also helps keep the planet at the right temperature for life. A much smaller Earth would not have enough gravitational mass to hold onto an atmosphere consisting of light molecules such as N_2, O_2, CO_2, and H_2O. And thanks to the development of photosynthesizing bacteria over 2 billion years ago, an ozone sunscreen protects us from an overdose of ultraviolet radiation.

On a time scale of millions of years, Earth is also enormously resilient and adaptive. Earth's average temperatures have remained between the freezing and boiling points of water, even though the sun's energy output has increased by about 30% over the 3.6 billion years since life arose. In short, the earth is just right for life as we know it.

---

the polymers DNA, RNA, proteins, and carbohydrates

Part 3: Development of systems of chemical reactions needed for linking these biopolymers in ways to produce the first living cells

Phase 2: Biological Evolution (about 3.6–3.8 billion years)

Development of diverse species through genetic modification and natural selection of Earth's primitive cells and of the organisms that followed

Chemical Evolution: Setting the Stage for Life
Billions of years ago, explosions of dying stars ripped through our galaxy and left behind a dense, swirling cloud of dust particles and hot gas extending trillions of kilometers across space in one of the spiral arms of our galaxy. As the cloud cooled, countless bits of matter gravitated toward one another. By 4.6 billion years ago, the cloud had flattened into a slowly rotating disk. Our sun was born at the extremely dense, hot center of the disk. Extremely high temperatures in this core, which consisted of isotopes of hydrogen, forced these nuclei together in nuclear fusion reactions (Figure 3-10) that would give off immense quantities of energy for the next 10 billion years.

Farther out from the center, the earth and other planets were forming as innumerable bits of matter were drawn together and coalesced. At some stage in its early life, Earth was a completely molten mass that did not cool for several hundred million years. As cooling took place, the outermost portion of the molten sphere solidified to form a thin, hardened crust of rocks, devoid of atmosphere and oceans.

The planet's first atmosphere formed as molten rock frequently erupted through the thin crust, releasing gases from the molten interior or from below its crust. At first, water vapor released from the breakdown of rocks during volcanic eruptions evaporated in the intense heat blanketing the crust. Eventually, however, the crust cooled enough for this water vapor to condense and fall to the surface as rain, dissolving minerals from rocks and collecting in depressions to form the early oceans that covered most of the earth's surface.

Studying the light given off by the sun and other stars indicates that hydrogen (H), helium (He), carbon (C), nitrogen (N), and oxygen (O) are the most abundant elements in the universe. Thus, C, H, N, and O—the four elements that are the major components of living organisms—were probably abundant on the prebiotic Earth. Basic chemistry predicts that three gases—methane (CH_4), ammonia (NH_3), and water vapor (H_2O)—are likely to have formed from these elements and may have comprised most of Earth's primitive atmosphere, although other gases such as carbon dioxide (CO_2) and hydrogen sulfide (H_2S) may also

A: A net gain

CHAPTER 5 **117**

have been present. Whatever the composition of this primitive atmosphere, chemists agree that it had almost no free oxygen gas (O_2). This element is so chemically reactive that it would have combined into compounds. The only reason today's atmosphere has so much O_2 is that plants and some bacteria produce it in vast quantities through photosynthesis. But this is getting ahead of the story.

Energy was readily available to drive the synthesis of biological molecules from the chemicals found in Earth's primitive atmosphere—an idea first proposed in 1923 by Russian biochemist Alexander Oparin. There was no ozone layer, so the planet was exposed to intense ultraviolet (UV) light, cosmic rays, and other forms of solar radiation (Figure 3-4). Electrical discharges (lightning), radioactivity, heat from volcanoes, and even the shock waves from meteorite impact may also have been important.

In a number of experiments carried out since 1953 various mixtures of gases believed to have been in Earth's early atmosphere have been put in sterilized glass containers and subjected to spark discharge to simulate lightning and heat. In these experiments various amino acids (the building-block molecules of proteins), simple carbohydrates, and small organic compounds necessary for life formed from the inorganic gaseous molecules. Many details are missing or hotly debated, but the basic hypothesis—that the molecules necessary for life can be produced by exposing various simple inorganic molecules believed to have existed on the prebiotic Earth to energy—remains viable. Another possibility is that simple organic molecules necessary for life formed elsewhere in the universe, perhaps on dust particles in space, and reached Earth on meteorites.

A more recent hypothesis is that bubbles floating on the ancient ocean trapped carbon-containing molecules and other chemicals necessary for life. When the bubbles popped they injected droplets rich in the precursors of life into the atmosphere, allowing them to be converted by lightning and other forms of energy into the building-block molecules of life. Scientists will never be sure how life originated, but they have developed and to varying degrees tested a number of possibilities.

During the 300 million years after the first rains began, organic compounds formed in the early atmosphere (or arriving from space) dissolved and accumulated in the shallow waters of the earth. Indirect evidence suggests that the first living cells, probably bacteria, emerged between 3.6 and 3.8 billion years ago. No one knows how these first bacterial cells arose from the slow-cooking organic soup. But after several hundred million years of different chemical combina-

tions, conglomerates of proteins and other biopolymers may have combined to form membrane-bound *protocells*, small globules that could take up materials from their environment, grow, and divide, much like living cells.

Over time, it is believed, these protocells developed into simple prokaryotic cells having the properties we describe as life (Section 4-1). These cells probably developed at least 10 meters (30 feet) below the ocean's surface, protected by water from the intense UV radiation that bathed the earth. With these first cells the stage was set for the drama of life to begin 3.6–3.8 billion years ago (Figures 4-9 and 5-36).

Life and the Atmosphere: The Oxygen Revolution   These single-celled bacteria multiplied and underwent genetic changes (mutated) for about a billion years in Earth's warm, shallow seas. This led to a variety of new types of prokaryotic cells (bacteria) and the emergence of the first *eukaryotic cells* between 2.6 billion and 700 million years ago. Numerous genetic changes in these eukaryotic cells spawned an amazing variety of protists and fungi and (beginning about 600 million years ago) plants and animals (Figures 4-9 and 5-36).

During this long, early period, life could not have developed on land. There was no ozone layer to shield the DNA and other molecules of early life from bombardment by intense ultraviolet radiation.

Then, about 2.3–2.5 billion years ago, something happened in the ocean that would change Earth forever—the development of photosynthetic bacteria. These cells could remove carbon dioxide from the atmosphere and, powered by sunlight, combine it with water to make the carbohydrates they needed. In the process they gave off oxygen (O_2) into the ocean, with some of it released into the atmosphere.

This *oxygen revolution* taking place over almost 2 billion years led to our current atmosphere. As oxygen proliferated, some was converted by incoming solar energy into ozone (O_3), which began forming a shield in the lower stratosphere that protected lifeforms from the sun's deadly UV radiation, allowing green plants to live closer to the surface of the ocean. As the production of oxygen snowballed, more UV light was filtered out, more life-forms developed, and these produced still more oxygen. About 400–500 million years ago it was safe to go to the beach, so to speak, and the first plants began existing on land. Over the next several hundred million years this led to a variety of land plants and animals, mammals, and eventually the first humans (Figure 5-36).

Q: What is the world's largest environmental group?

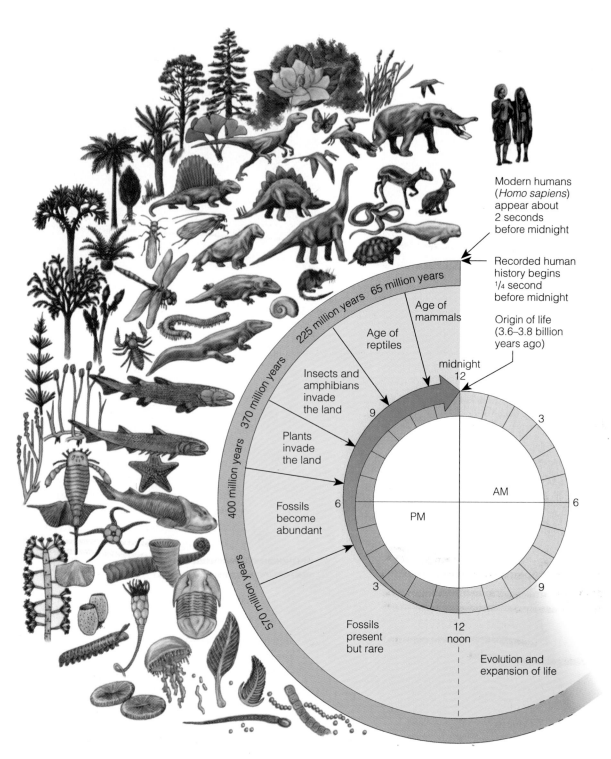

Modern humans (*Homo sapiens*) appear about 2 seconds before midnight

Recorded human history begins ¼ second before midnight

Origin of life (3.6–3.8 billion years ago)

65 million years

225 million years

370 million years

400 million years

570 million years

Age of mammals

Age of reptiles

Insects and amphibians invade the land

Plants invade the land

Fossils become abundant

Fossils present but rare

midnight 12

Evolution and expansion of life

AM

PM

Figure 5-36 Greatly simplified history of biological evolution of life on Earth, which was preceded by about 1 billion years of chemical evolution. The early span of biological evolution on Earth, between 3.8 billion and about 570 million years ago, was dominated by microorganisms (mostly bacteria and later protists). Plants and animals evolved only about 570 million years ago, and we arrived on the scene only a short time ago. If we compress Earth's roughly 3.8-billion-year history of life to a 24-hour time scale, our closest human ancestors (*Homo sapiens*) appeared about 2 seconds before midnight, and our species (*Homo sapiens sapiens*) appeared less than 1 second before midnight. Agriculture began only ¼ second before midnight, and the Industrial Revolution has been around for only seven-thousandths of a second. Despite our brief time on Earth, humans may be hastening the extinction of more species in a shorter time than ever before in Earth's long history. (Adapted from George Gaylord Simpson and William S. Beck, *Life: An Introduction to Biology*, 2d ed., New York: Harcourt Brace Jovanovich, 1965.)

Michael Tweedie/NHPA

Kim Taylor/Bruce Coleman Ltd.

Figure 5-37 Two color varieties of peppered moths found in England. In the mid-1800s, before the Industrial Revolution, the speckled light-gray form of this moth was prevalent. These night-flying moths rested on light-gray speckled lichens on tree trunks during the day. Their color camouflaged them from their bird predators. A dark-gray form also existed but was quite rare. However, during the Industrial Revolution, this dark form became the common one, especially near industrial cities, where soot and other pollutants from factory smokestacks began killing lichens and darkening tree trunks. In this new environment the dark form blended in with the blackened trees, while the light form was highly visible to its bird predators. Through natural selection the dark form began to survive and reproduce at a greater rate than its light-colored kin. (Both varieties appear in each photo. Can you spot them?)

5-6 POPULATION RESPONSES TO STRESS: EVOLUTION, ADAPTATION, AND NATURAL SELECTION

Adapting to Changes in Environmental Conditions Populations can, within limits, adapt to changes in environmental conditions. The major driving force of adaptation to environmental change is **biological evolution**, or **evolution**, the change in the genetic makeup of a population of a species in successive generations. Note that populations—not individuals—evolve. According to the **theory of evolution**, all life-forms developed from earlier life-forms. Although this theory conflicts with the creation stories of most religions, it is the way biologists explain how life has changed over the past 3.6–3.8 billion years (Figure 5-36) and why it is so diverse today.

Evolutionary change occurs as a result of the interplay of genetic variation and changes in environmental conditions. The first step in evolution is the development of *genetic variability* in a population (Figure 4-7). Genetic information in *chromosomes* is contained in various sequences of chemical units (called *nucleotides*) in DNA molecules. Distinctive DNA segments that are found in chromosomes and that impart certain inheritable traits are called **genes**.

A population's **gene pool** is the sum total of all genes possessed by the individuals of the population of a species. Individuals of a population of a particular species, however, don't have exactly the same genes. The source of this genetic variability is **mutations**—random changes in DNA molecules (which make up genes) that can yield changes in anatomy, physiology, or behavior in offspring. Such mutations, or random changes in genetic makeup, can occur and be transmitted to offspring in several ways:

- When DNA molecules are copied, every time a cell divides.

- During *sexual reproduction*, when offspring are produced by the union of a male sperm and a female ovum (egg). For example, when a human sperm and egg unite there are over 70 billion potentially different chromosomal combinations.

- During the *production of ova (eggs) or sperm cells*, known as *gametes*. When gametes are formed, the genetic deck is "reshuffled," and the number of chromosomes in each cell is reduced by half.

- *By crossing over*. This occurs when pieces of chromosomes break off and become attached to other chromosomes during the production of germ cells (which make sperm in males and ova in females).

- *By exposure to an external agent.* Examples are chemicals (called mutagens) and high-energy

Q: What environmental and lifestyle factor causes the most death and suffering?

Figure 5-38 Predator avoidance by behavior. When touched, this snake caterpillar alters its body shape to look like the head of a snake. The puffed-up head "strikes" at whatever touches it.

electromagnetic radiation (Figure 3-4) such as ultraviolet light, cosmic rays, X rays, and radio-activity (Figure 3-7).

The net result of millions of such random and unpredictable changes in the DNA molecules of individuals in a population is genetic variability.

Most mutations are harmful, but some result in new genetic traits that give their bearer and most of its offspring better chances for survival and reproduction. Any genetically controlled structural, physiological, or behavioral characteristic that enhances the chances for members of a population to survive and reproduce in their environment is called an **adaptation**.

Structural adaptations include *coloration* (allowing more individuals to sneak up on prey or to hide from predators, Figure 5-37), *mimicry* (looking like a poisonous or dangerous species), *protective cover* (shell, thick skin, bark, thorns), and *gripping mechanisms* (hands with opposable thumbs). *Physiological adaptations* include the ability to hibernate during cold weather, to poison predators (the Monarch butterfly, Figure 4-6), and to give off chemicals that repel prey. The ability to fly to a warmer climate during winter or to change shape (Figure 5-38); resource partitioning (Figures 4-38 and 4-39); and species interactions such as parasitism (Figure 4-40), mutualism (Figure 4-41), and commensalism (Figure 4-43)—all are examples of *behavioral adaptations*.

Individuals with one or more adaptations that allow them to survive under changed environmental conditions are more likely to reproduce, and they will leave behind more offspring with the same favorable adaptations than will individuals without such adaptations. This phenomenon is known as **differential reproduction**.

The process by which a particular beneficial gene or set of beneficial genes is reproduced more than others in a population through adaptation and differential reproduction is called **natural selection**. Natural selection is the major (but not necessarily the only) mechanism that leads to evolution through a combination of mutation-caused genetic variability and changes in environmental conditions.

Limits to Adaptation Shouldn't adaptations to new environmental conditions allow our skin to be able to become more resistant to the harmful effects of ultraviolet radiation, our lungs better able to cope with air pollutants, and our liver more capable of detoxifying pollutants we are exposed to? The answer is *no*, because there are limits to adaptations in nature:

- A change in environmental conditions can lead to adaptation only for traits already present in the gene pool of a population.

- Even if a beneficial heritable trait is present in a population, its ability to adapt can be limited by its reproductive capacity. If members of a population can't reproduce fast enough to adapt to a particular environmental change (or group of environmental changes), all of its members can die. For example, populations of genetically diverse r-strategists (weeds, mosquitoes, rats, or bacteria) that can produce hordes of offspring early in their life cycle can adapt to a change in environmental conditions through natural selection in a short time. By contrast, populations of K-strategist species (elephants, tigers, sharks, and humans) cannot produce large numbers of offspring rapidly and take long times (typically thousands or even millions of years) to adapt through natural selection.

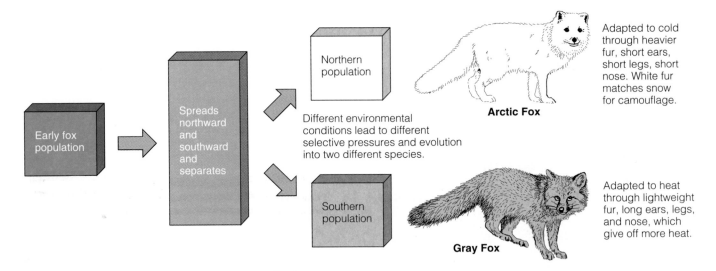

Figure 5-39 Geographic isolation leading to reproductive isolation and speciation.

Even if a favorable genetic trait is present, most members of the population would have to die or become sterile so that individuals with the trait could dominate and pass the trait on—hardly a desirable solution to the environmental problems humans face.

Connections: Coevolution　Physical and chemical changes in environmental conditions are the most familiar agents of evolutionary change. But some biologists have proposed that interactions between species (Section 4-5) can also result in natural selection and evolution. According to this hypothesis, when two species interact over a long period, changes in one species can lead to changes in the other. This process is called **coevolution**. For example, a carnivore may become increasingly efficient at hunting. However, if certain individuals of its prey have traits that allow them to get away, they pass these adaptive traits on to their offspring. Then the predator may evolve ways to overcome this new trait, leading the prey to new adaptations, and so on.

Similarly, plants may evolve defenses, such as camouflage, thorns, or poisons against efficient herbivores. This, in turn, can lead their herbivore consumers to evolve ways to counteract those defenses. Animals such as golden toads (Figure 4-35) may also evolve poisons that help protect them from their predators. Through coevolution animals may also develop camouflage to hide them or to make them more effective predators.

Hummingbirds, for example, have coevolved with plants whose flowers attract them and whose flower shapes are adapted to their beak lengths. The birds get nectar that only they can reach, and the plant gets pollinated.

Speciation, Extinction, and Biodiversity

Earth's 5–100 million species are believed to be the result of two processes over the past 3.6–3.8 billion years. One is **speciation**: the formation of two or more species from one as a result of divergent natural selection in response to changes in environmental conditions.

A common speciation mechanism is *geographic isolation,* the situation in which populations of a species become separated in areas with different environmental conditions for fairly long periods, as when part of the group migrates in search of food and doesn't return (Figure 5-39). Populations may also become separated by a physical barrier or change, such as a highway or volcanic eruption or when a few individuals are carried to a new area by wind, water, or humans.

After long geographic separation during which the two groups don't interbreed, they may begin to diverge in their genetic makeup because of different selection pressures. If this *reproductive isolation* continues long enough, members of the separated populations may become so different that they can't interbreed and produce fertile offspring. Then one species has become two (Figure 5-39).

In a few rapidly reproducing organisms (mostly r-strategists) speciation may take place in thousands or even hundreds of years. With most species (especially K-strategists), however, it takes from tens of thousands to millions of years. Given this time scale, it is difficult to observe and document the appearance of a new species.

The second process affecting Earth's species is **extinction**: A species ceases to exist because it cannot genetically adapt and successfully reproduce under new environmental conditions. When the rules of the game change, a species may vanish entirely, or it may evolve into a new and better-adapted species.

　　　Q: How many Americans die because of exposure to other people's smoke (passive smoke)?

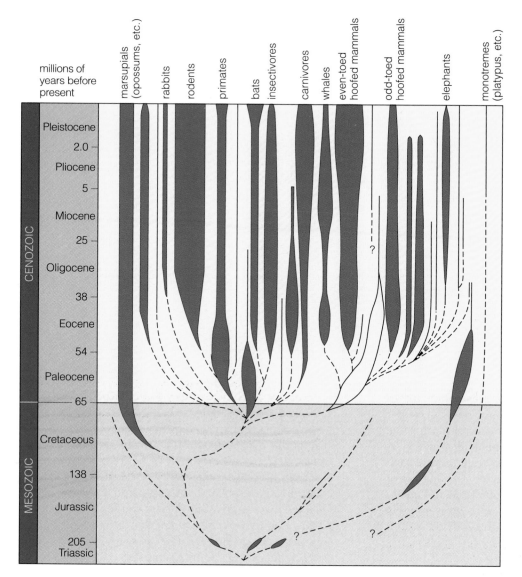

Figure 5-40 Adaptive radiation of mammals in the first 10–12 million years of the Cenozoic Era. This radiation began about 65 million years ago and continues today. This development of a large number of new species is thought to have resulted when huge numbers of vacant and new environmental niches became available after the mass extinction of dinosaurs near the end of the Mesozoic Era. (Used by permission from Cecie Starr. *Biology: Concepts and Applications*, Belmont, Calif.: Wadsworth, 1991)

Extinction is nothing new on Earth. Biologists estimate that 99% of all the species that have ever lived are now extinct. Some species inevitably disappear as local conditions change; this is a process called *background extinction*.

In contrast, **mass extinction** is an abrupt rise in extinction rates above the background level. It is a catastrophic, widespread—often global—event in which not just one species but large groups of species are wiped out simultaneously. Fossil and geological evidence indicates that Earth's species have experienced five great mass extinctions—at roughly 26-million-year intervals—with smaller ones in between.

A crisis for one species, however, is an opportunity for another. The fact that 5–100 million species exist today means that speciation, on average, has kept ahead of extinction. Evidence shows that Earth's mass extinctions have been followed by periods of recovery and **adaptive radiations**, in which numerous new species evolved to fill new or vacant ecological niches in changed environments. The disappearance of dinosaurs at the end of the Mesozoic Era about 65 million years ago, for example, was followed by an evolution explosion for mammals (Figure 5-40). This adaptive radiation marked the beginning of the Cenozoic Era (the past 65 million years).

Speciation minus extinction equals *biodiversity*, one of the planet's most important resources. Extinction is a natural process, but we have become a major force in the premature extinction of species—causing the greatest mass extinction since dinosaurs vanished 65 million years ago. As population and resource consumption increase over the next 30 years, we may cause the extinction of up to a quarter of Earth's species, each the product of millions to billions of years of evolution.

A: Up to 40,000 per year

Characteristic	Immature Ecosystem	Mature Ecosystem
Ecosystem Structure		
Plant size	Small	Large
Species diversity	Low	High
Trophic structure	Mostly producers, few decomposers	Mixture of producers, consumers, and decomposers
Ecological niches	Few, mostly generalized	Many, mostly specialized
Community organization (number of interconnecting links)	Low	High
Ecosystem Function		
Food chains and webs	Simple, mostly plant → herbivore with few decomposers	Complex, dominated by decomposers
Efficiency of nutrient recycling	Low	High
Efficiency of energy use	Low	High

Table 5-2 Ecosystem Characteristics at Immature and Mature Stages of Ecological Succession

On our short time scale such catastrophic losses cannot be recouped by formation of new species; it took tens of millions of years after each of Earth's five great mass extinctions for evolution to allow life to recover to at least the level of biodiversity before each extinction. Genetic engineering cannot stop this loss of biodiversity because genetic engineers do not create new genes. They transfer genes or gene fragments from one organism to another and thus rely on natural biodiversity for their raw material.

5-7 COMMUNITY–ECOSYSTEM RESPONSES TO STRESS

Connections: Ecological Succession One characteristic of most communities and ecosystems is that the types of species in a given area are usually changing. This gradual process of change in the composition and function of communities is called **ecological succession**, or **community development.**

Succession is a normal process in nature. It reflects the results of the continuing struggle among species with different adaptations for food, light, space, nutrients, and other survival resources. Ecological succession can result in a progression from immature, rapidly changing, unstable communities to more mature, self-sustaining communities when this process is not disrupted by large-scale natural events or human actions. Immature communities or ecosystems at an early stage of succession have strikingly different characteristics from mature ones at a later stage (Table 5-2).

Primary and Secondary Succession Ecologists recognize two types of ecological succession, primary and secondary, depending on the conditions at a particular site at the beginning of the process. **Primary succession** involves the development of biotic communities in a barren habitat with very little or no topsoil. Examples of such areas include the rock or mud exposed by a retreating glacier (Figure 5-41) or a mudslide, newly cooled lava, a new sandbar deposited by a shift in ocean currents, and surface-mined areas from which all topsoil has been removed.

After such a large-scale disturbance, life usually returns first with a few hardy **pioneer species**—microbes, mosses, and lichens. They are usually r-strategists with the ability to establish large populations quickly in a new area—species that Edward O. Wilson calls "nature's sprinters."

Sometimes the species of this pioneer community make the area suitable for species with different niche requirements, a process called *facilitation*. More common is *inhibition*, in which the early species create conditions that hinder invasions and growth by other species. Then succession can proceed only when a fire, heavy grazing, bulldozing, or some other disturbance removes most of the pioneer species. In other cases later species are largely unaffected by earlier species, a phenomenon known as *tolerance*.

The more common type of succession is **secondary succession**, which begins in an area where the natural vegetation has been removed or destroyed but where the soil or bottom sediment has not been covered or removed. Examples of candidates for sec-

Q: How many U.S. workers die prematurely from exposure to toxic substances?

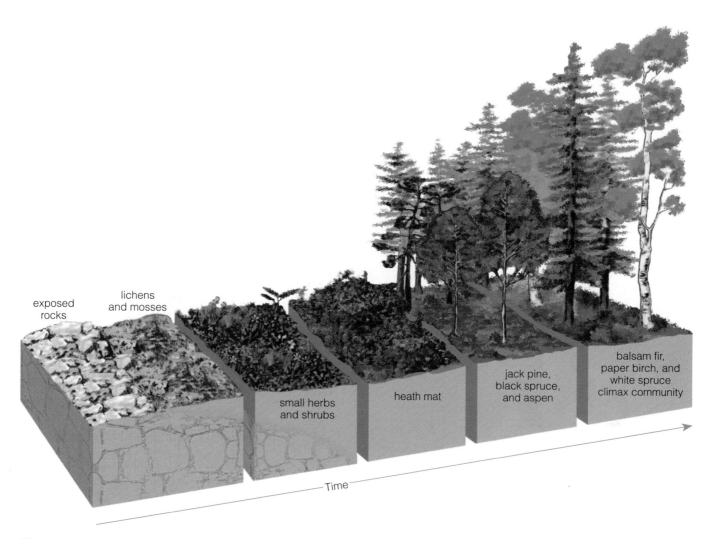

labels in figure:
exposed rocks — lichens and mosses — small herbs and shrubs — heath mat — jack pine, black spruce, and aspen — balsam fir, paper birch, and white spruce climax community

Time

Figure 5-41 Primary succession of plant communities over several hundred years on bare rock exposed by a retreating glacier, on Isle Royal in northern Lake Superior.

ondary succession include abandoned farmlands, burned or cut forests, heavily polluted streams, and land that has been dammed or flooded to produce a reservoir or pond. Because some soil or sediment is present, new vegetation can usually sprout within only a few weeks. In the central (Piedmont) region of North Carolina, for example, European settlers cleared the mature forests of native oak and hickory, and replanted the land with crops. Later some of the land was abandoned. Figure 5-42 shows how such abandoned farmland, covered with a thick layer of soil, has undergone secondary succession. After about 150 years, the area again supports a mature oak and hickory forest.

It is tempting to conclude that ecological succession is an orderly sequence, with each successional stage leading predictably to the next, more stable stage until an area is occupied by a *mature or climax community*, dominated by plant species Edward O. Wilson calls "nature's long-distance runners." Research has

shown, however, that this does not necessarily happen. The exact sequence of species and community types that appears during primary or secondary succession can be highly variable. We cannot predict the course of a given succession or view it as some preordained progress toward an ideally adapted climax community.

5-8 HUMAN IMPACTS ON ECOSYSTEMS: WORKING WITH NATURE

The Condition of Earth's Ecosystems   From a human standpoint, we can classify the general condition of Earth's ecosystems into five main categories:

- *Natural ecosystems* which have not been seriously damaged or degraded by human activities.

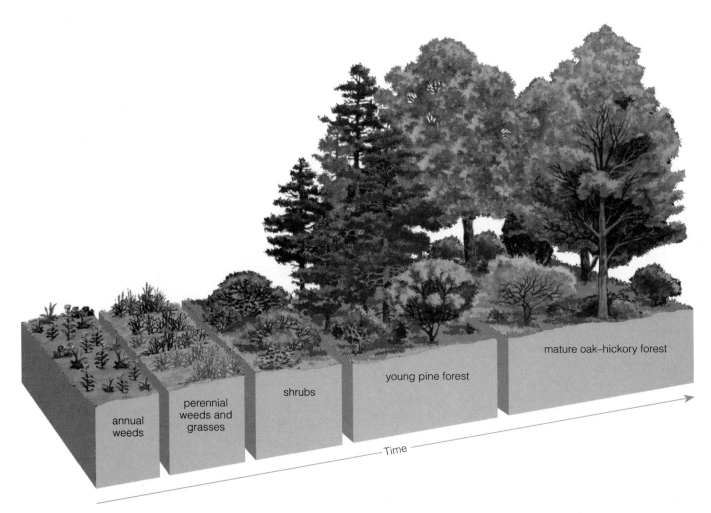

Figure 5-42 Secondary ecological succession of plant communities on an abandoned farm field in North Carolina. After the farmland was abandoned it took about 150 years for the area to be covered again with a mature oak and hickory forest.

- *Modified systems* where human impact is greater than that of native species but where cultivation has not taken place. An example is a forest regenerating by secondary ecological succession after being cut (Figure 5-42).

- *Cultivated systems* where certain species of plants (crops or trees) have been planted and are regularly harvested.

- *Built systems* where humans have constructed buildings, roads, parking lots, or other structures.

- *Degraded systems* where habitats, productivity, and biodiversity have been substantially reduced.

To survive and support growing numbers of people and rising resource consumption we have greatly increased the amount of Earth's natural systems that we have modified, cultivated, built on, or degraded.

Simplifying Ecosystems   In modifying natural ecosystems for our use, we usually simplify them: We plow grasslands, clear forests, and fill in wetlands. Then we replace their thousands of interrelated plant and animal species with one crop (Figure 5-12) or one kind of tree (Figure 5-16), called **monocultures**—or with buildings, highways, and parking lots.

We spend a lot of time, energy, and money trying to protect such monocultures from continual invasion by pioneer species. We call such species *weeds* if they are plants (because weeds and annual crop plants typically occupy the same niches); *pests* if they are insects or other animals; and *pathogens* if they are fungi, viruses, or disease-causing bacteria. Weeds, pests, or pathogens can wipe out an entire monoculture crop unless it is protected by pesticides or by some form of biological control.

When fast-breeding insect species begin to undergo natural selection and develop genetic resistance

Q: What percentage of the 70,000 chemicals in commercial use have been thoroughly screened for toxicity?

The Day the Roof Fell In

Malaria once infected 9 out of 10 people in North Borneo, now known as Brunei. In 1955 the World Health Organization (WHO) began spraying the island with dieldrin (a DDT relative) to kill malaria-carrying mosquitoes. The program was so successful that the dread disease was virtually eliminated.

Other, unexpected things began to happen, however. The dieldrin killed other insects, including flies and cockroaches living in houses.

At first the islanders applauded this turn of events, but then small lizards that also lived in the houses died after gorging themselves on dead insects. Next, cats began dying after feeding on the dead lizards. Then, without cats, rats flourished and overran the villages. Now people were threatened by sylvatic plague carried by the rat fleas. WHO then parachuted healthy cats onto the island to help control the rats.

Then the villagers' roofs began to fall in. The dieldrin had killed wasps and other insects that fed on

a type of caterpillar that either avoided or was not affected by the insecticide. With most of its predators eliminated, the caterpillar population exploded. The larvae munched their way through one of their favorite foods, the leaves used in thatched roofs.

Ultimately, the Borneo episode ended happily: Both malaria and the unexpected effects of the spraying program were brought under control. However, the chain of unforeseen events shows the unpredictability of interfering in an ecosystem.

---

to pesticides, the salespeople urge farmers to use stronger doses or switch to a new pesticide. As a result natural selection in the pests increases to the point that these chemicals eventually become ineffective.

This process illustrates two important ideas. One is what biologist Garrett Hardin calls the **first law of human ecology**: *We can never do merely one thing.* The other is the **principle of connectedness**: *Everything is connected to and intermingled with everything else; we are all in it together.* Because of these principles, any human intrusion into nature has multiple effects. Many (perhaps most) of these effects are unpredictable because of our limited understanding of how nature works and of which connections are the strongest and most important for the sustainability and adaptability of ecosystems (Connections, above).

Cultivation is not the only way people simplify ecosystems. Ranchers, who don't want bison or prairie dogs competing with sheep for grass, eradicate those species, as well as wolves, coyotes, eagles, and other predators that occasionally kill sheep. In addition, far too often, ranchers allow livestock to overgraze grasslands until erosion converts these ecosystems to simpler and less productive deserts. The cutting of vast areas of diverse tropical rain forests destroys part of Earth's biodiversity forever. And people tend to overfish and overhunt some species to extinction or near extinction, another way of simplifying ecosystems. Finally, the burning of fossil fuels in industrial plants, homes, and vehicles creates air pollutants that simplify both forest ecosystems (by killing or weakening trees) and aquatic ecosystems (by killing fish).

The challenge is to maintain a balance between simplified human ecosystems and the neighboring, more complex natural ecosystems on which our systems and other forms of life depend—and to slow down the rates at which we are altering nature for our purposes. If we alter and simplify too much of the planet to meet our needs and wants, what's at risk is not the earth—which will eventually evolve new forms of life—but our current social systems and our own species. According to biodiversity expert E. O. Wilson, "If this planet were under surveillance by biologists from another world, I think they would look at us and say, 'Here is a species in the mid-stages of self-destruction.'" The evolutionary lesson to be learned from nature is that no species can get too big for its britches, at least not for long.

Solutions: Working with Nature   The brief discussion of scientific principles in this chapter reveals that living systems have six key features: *interdependence, diversity, resilience, adaptability, unpredictability, and limits.* This suggests that the best way for us to live sustainably is to understand and mimic how nature is perpetuated—based on basic scientific laws, concepts, and principles (see inside back cover) and by instituting ecological design (Guest Essay, p. 128). This begins by recognizing three things: **(1)** We are part of—not apart from—Earth's dynamic web of life; **(2)** our survival, lifestyles, and economies are totally dependent on the sun and Earth; and **(3)** everything is connected to everything.

The Ecological Design Arts

David W. Orr

GUEST ESSAY

Since 1990 David W. Orr has been professor of environmental studies at Oberlin College. In 1979 he cofounded Meadowcreek, a nonprofit 600-hectare (1,500-acre) laboratory in north central Arkansas for the study of environmentally sound means of agriculture, forestry, renewable energy systems, architectural design, and livelihood. He served as its director until 1990. He has written numerous environmental articles and three books, including Ecological Literacy *and* The Campus and the Biosphere. *He is education editor for* Conservation Biology, *coeditor of the environmental policy series for SUNY press, and a member of the editorial advisory board of* Orion Nature Quarterly.

If *Homo sapiens* entered industrial civilization in an intergalactic design competition, it would be tossed out at the qualifying round. It doesn't fit. It won't last. The scale is wrong. And even its defenders admit that it's not very pretty. The most glaring design failures of industrial/technologically driven societies are the loss of diversity of all kinds, impending climate change, pollution, and soil erosion.

Industrial civilization, of course, wasn't designed at all. It was mostly imposed by single-minded individuals, armed with one doctrine of human progress or another, each requiring a homogenization of nature and society. These individuals for the most part had no knowledge of "ecological design arts." By this I mean the set of perceptual and analytic abilities, ecological wisdom, and practical wherewithal needed to make things that fit into a world of microbes, plants, animals, and energy laws.

Good ecological design incorporates understanding about how nature works into the ways we design, build, and live. It is required in our designs of farms, houses, neighborhoods, cities, transportation systems, technologies, economies, energy policies, and just about anything that directly or indirectly requires energy or materials or governs their use.

When human artifacts and systems are well designed, they are in harmony with the ecological patterns in which they are embedded. When poorly designed, they undermine those larger patterns, creating pollution, higher costs, and social stress. Bad design is not simply an engineering problem, although better engineering would often help. Its roots go deeper.

Good ecological design has certain common characteristics including right scale, simplicity, efficient use of resources, a close fit between means and ends, durability, redundancy, and resilience. They are often place specific, or in John Todd's words: "elegant solutions predicated on the uniqueness of place." Good design also solves more than one problem at a time and promotes human competence (instead of addiction), efficient and frugal use of resources, and sound regional economies. Where good design becomes part of the social fabric at all levels, unanticipated positive side effects multiply. When people fail to design with ecological competence, unwanted side effects and disasters multiply.

The pollution, violence, social decay, and waste all around us indicate that we have designed things badly. Why? There are, I think, three primary reasons. First, as long as land and energy were cheap and the world was relatively empty, we did not have to master the discipline of good design. The result: sprawling cities, wasteful economies, waste dumped into the environment, bigger and less efficient automobiles and buildings, and conversion of entire forests into junk mail and Kleenex—all in the name of economic growth and convenience.

Second, design intelligence fails when greed, narrow self-interest, and individualism take over. Good design is

---

What has gone wrong, probably, is that we have failed to see ourselves as part of a large and indivisible whole. For too long we have based our lives on a primitive feeling that our "God-given" role was to have "dominion over the fish of the sea and over the fowl of the air and over every living thing that moveth upon the earth." We have failed to understand that the earth does not belong to us, but we to the earth.

ROLF EDBERG

Critical Thinking

1. What factors in your lifestyle contribute to the destruction and degradation of wetlands?

2. Someone tries to sell you several brightly colored pieces of coral. Explain in biological terms why this is a rip-off.

3. Someone tells you not to worry about air pollution because through natural selection the human species will develop lungs that can detoxify pollutants. How would you reply?

4. Explain why a simplified ecosystem such as a cornfield is much more vulnerable to harm from insects, plant diseases, and fungi than a more complex, natural ecosystem such as a grassland. Why are natural ecosystems normally less vulnerable?

Q: Worldwide, how many people die from sexually transmitted diseases.

a cooperative community process requiring people who share common values and goals that bring them together and hold them together. American cities with their extremes of poverty and opulence are products of people who believe they have little in common with one another. Greed, suspicion, and fear undermine good community and good design alike.

Third, poor design results from poorly equipped minds. Good design can only be done by people who understand harmony, patterns, and systems. Industrial cleverness, on the contrary, is mostly evident in the minutiae of things, not in their totality or in their overall harmony. Good design requires a breadth of view that causes people to ask how human artifacts and purposes fit within a particular culture and place. It also requires ecological intelligence, by which I mean an intimate familiarity with how nature works.

An example of good ecological design is found in John Todd's "living machines," which are carefully orchestrated ensembles of plants, aquatic animals, technology, solar energy, and high-tech materials to purify wastewater, but without the expense, energy use, and chemical hazards of conventional sewage treatment technology. Todd's living machines resemble greenhouses filled with plants and aquatic animals [Figure 11-34]. Wastewater enters at one end and purified water leaves at the other. In between, an ensemble of organisms driven by sunlight remove and use nutrients, break down toxins, and incorporate heavy metals in plant tissues.

Ecological design standards also apply to the making of public policy. For example, the Clean Air Act of 1970 required car manufacturers to install catalytic converters to remove air pollutants. Two decades later emissions per vehicle are down substantially, but since more cars are on the road, air quality is about the same—an example of inadequate ecological design. A sounder design approach to transportation would create better access among housing, schools, jobs, stores, and recreation areas; build better public transit systems; restore and improve railroads; and create bike trails and walkways.

An education in the ecological design arts would foster the ability to see things in their ecological context, integrating firsthand experience and practical competence with theoretical knowledge about how nature works. It would aim to equip learners to build households, institutions, farms, communities, corporations, and economies that (1) do not emit carbon dioxide or other heat-trapping gases; (2) operate on renewable energy; (3) preserve biological diversity; (4) recycle material and organic wastes; and (5) promote sustainable local and regional economies.

The outline of a curriculum in ecological design arts can be found in recent work in ecological restoration, ecological engineering, solar design, landscape architecture, sustainable agriculture, sustainable forestry, energy efficiency, ecological economics, and least-cost, end-use analysis. A program in ecological design would weave these and similar elements together around actual design objectives that aim to make students smarter about systems and about how specific things and processes fit in their ecological context. With such an education we can develop the habits of mind, analytical skills, and practical competence needed to help sustain the earth for us and other species.

Critical Thinking

1. Does your school offer courses or a curriculum in ecological design? If not, why?

2. Use the principles of good ecological design to evaluate how well your campus is designed and suggest ways to improve its design.

*5. What type of biome do you live in or near? What effects have human activities had on the characteristic vegetation and animal life normally found in this biome? How is your own lifestyle affecting this biome?

*6. Visit a nearby land area or pond and look for signs of ecological succession. If possible, compare the types of species found on an abandoned farm field or lot with another nearby area that is at a more mature stage of ecological succession.

*7. Make a concept map of the key ideas in this chapter using the section heads and subheads and the key terms (shown in boldface type in the chapter). See the inside front cover and Appendix 4 for information on concept maps.

A: An estimated 750,000 per year

6 The Human Population: Growth, Urbanization, and Regulation

Cops and Rubbers Day in Thailand

In 1960 the population of Thailand (Figure 6-13, p. 139) was growing rapidly at a rate of 3.2% per year, and the average Thai family had 6.4 children. Today its population is growing at a rate of 1.4%, and the average number of children per family is 2.4. And since 1960 the country's average per capita income has doubled.

There are several reasons for this impressive feat: the creativity of the government-supported family-planning program; the openness of the Thai people to new ideas; the willingness of the government to work with the private, nonprofit Population and Community Development Association (PCDA); and support of family planning by the country's Buddhist religious leaders (95% of Thais are Buddhist). Buddhist scripture teaches that "many children make you poor."

This remarkable transition was catalyzed by the charismatic leadership of Mechai Viravidaiya, a former government economist and public relations genius, who launched the PCDA. Anywhere there was a crowd, PCDA workers handed out condoms—at festivals, movie theaters, even traffic jams. Schoolchildren held condom-blowing championships (Figure 6-1). Mechai showed that condoms—now commonly called "mechais"—could be used as a tourniquet for deep cuts and snake bites and as a coin or beverage container. Humorous songs were written about condom use and the reasons to have no more than two children.

Mechai also persuaded traffic police to hand out condoms on New Year's Eve, now known as "Cops and Rubbers Day." On the Thai king's birthday the PCDA offers free vasectomies; sterilization is now the most widely used form of birth control in the country.

All is not completely rosy. While Thailand has done well in slowing population growth, it has been less successful in improving public health, especially maternal health and control of AIDS and other sexually transmitted diseases. And the capital, Bangkok, remains one of the world's most polluted and congested cities.

Figure 6-1 Thai children blowing up condoms in Bangkok, Thailand. (Courtesy of Population and Community Development Association, Bangkok, Thailand)

We shouldn't delude ourselves: The population explosion will come to an end before very long. The only remaining question is whether it will be halted through the humane method of birth control, or by nature wiping out our surplus.

PAUL EHRLICH

This chapter is devoted to answering the following questions:

- How is population size affected by rates of birth, death, fertility, and migration?

- How is population size affected by the percentage of males and females at each age level?

- How is the world's population distributed between rural and urban areas?

- How do transportation systems shape urban areas and growth?

- How can cities be made more livable and sustainable?

- How can we control population growth?

- What success have India and China had in controlling population growth?

6-1 FACTORS AFFECTING HUMAN POPULATION SIZE

Birth Rates and Death Rates Populations grow or decline through the interplay of three factors: births, deaths, and migration. The **birth rate**, or **crude birth rate**, is the number of live births per 1,000 people in a population in a given year. The **death rate**, or **crude death rate**, is the number of deaths per 1,000 people in a population in a given year. Figure 6-2 shows the crude birth and death rates for various groups of countries in 1993. There are more births than deaths; every time your heart beats three more babies are added to the world's population—amounting to roughly a quarter million more people each day. When the death rate equals the crude birth rate and migration is not a factor, population size remains stable, a condition known as **zero population growth (ZPG)**.

The annual rate at which the size of a population changes (excluding migration) is called the **annual rate of natural population change**. It is usually expressed as a percentage.

$$\text{annual rate of population change (\%)} = \frac{\text{birth rate} - \text{death rate}}{1{,}000 \text{ persons}} \times 100$$

$$= \frac{\text{birth rate} - \text{death rate}}{10}$$

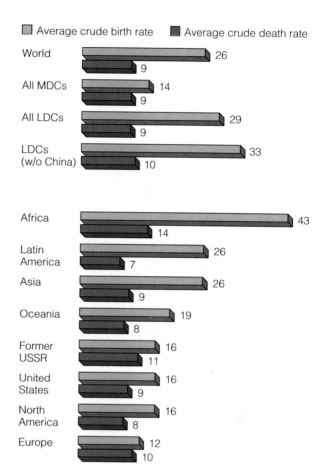

Figure 6-2 Average crude birth and death rates for various groups of countries in 1993. (Data from Population Reference Bureau)

The world's annual population growth rate dropped 18% between 1965 and 1993, from 2% to 1.63% (0.37 drop/2 × 100 = 18% drop). This is good news, but during the same period the population base rose by 72%, from 3.2 billion to 5.5 billion. This 18% drop in the growth rate of population is akin to learning that the truck heading straight at you has slowed from 100 kilometers per hour to 82 kilometers per hour while its weight has increased by almost three-fourths.

Figure 6-3 gives the annual population change rates for major parts of the world in 1993. An annual population growth rate of 1–3% may seem small, but such exponential rates lead to enormous increases in population size over a 100-year period (p. 4). The current annual growth rate of 1.63% adds 90 million people per year—the equivalent of adding the population of Mexico each year and that of the United States about every three years.

Africa has the highest rate of population growth, but Asia has the largest increase in numbers of people.

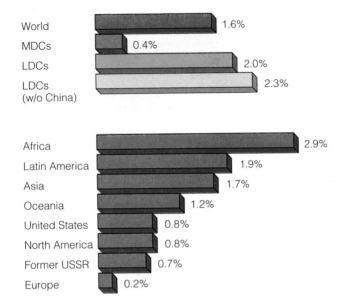

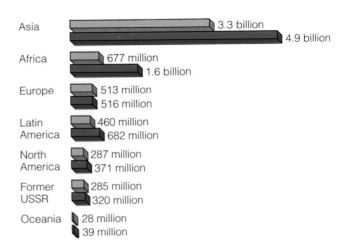

Figure 6-5 Population projections by region, 1993–2025. (Data from United Nations and Population Reference Bureau)

Figure 6-3 Average annual rate of population change for various groups of countries in 1993. (Data from Population Reference Bureau)

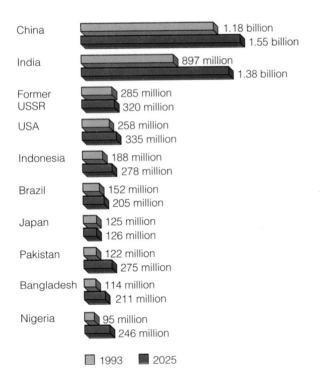

■ 1993   ■ 2025

Figure 6-4 The world's 10 most populous countries in 1993, with projections of their population size in 2025. (Data from Population Reference Bureau and World Bank)

In terms of sheer numbers of people, China and India dwarf all other countries, together making up 38% of the world's population (Figure 6-4). One person in five is Chinese, and 60% of the world's population is Asian. Even though the United States has the world's fourth largest population, it has only 4.7% of the world's people. Figure 6-5 shows how the population in various regions is projected to grow between 1993 and 2025.

Fertility Rates Two types of fertility rates affect a country's population size and growth rate. The first type, **replacement-level fertility**, is the number of children a couple must bear to replace themselves. The actual average replacement-level fertility rate is slightly higher than two children per couple (2.1 in MDCs and as high as 2.5 in some LDCs), mostly because some female children die before reaching their reproductive years.

The second type of fertility rate, and the most useful measure for projecting future population change, is the **total fertility rate (TFR)**, an estimate of the average number of children a woman will have during her childbearing years.

In 1993 the worldwide average TFR was 3.3 children per woman. It was 1.8 in MDCs (down from 2.5 in 1950) and 3.7 in LDCs (down from 6.5 in 1950). If the world's TFR remains at 3.3, its population will reach 694 billion by the year 2150—391 times the current population! Clearly this is not possible, but it does illustrate the enormous power of exponential population growth. Population experts expect TFRs in MDCs to remain around 1.8 and those in LDCs to drop to around 2.3 by 2025—the basis of the population projections in Figure 6-5. That is good news, but it will still lead to a projected world population of

Q: Worldwide, how many people have malaria?

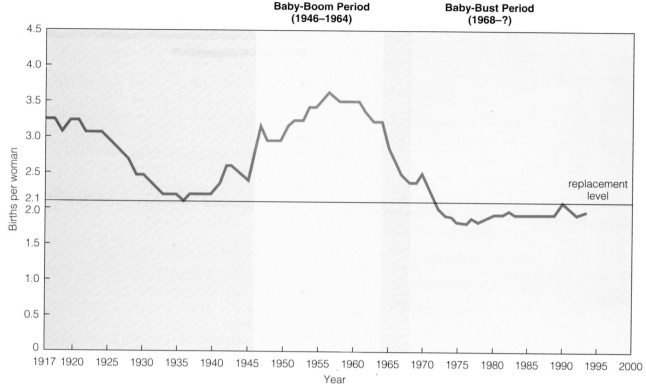

Figure 6-6 Total fertility rate for the United States between 1917 and 1993. (Data from Population Reference Bureau and U.S. Census Bureau)

around 9 billion by 2025, with most of this growth taking place in LDCs (Figure 1-5).

The population of the United States has grown from 4 million in 1790 to 258 million in 1993—a 65-fold increase—even though the country's total fertility rate has oscillated wildly (Figure 6-6). At the peak of the post–World War II baby boom (1946–64) in 1957, the TFR reached 3.7 children per woman. Since then it has generally declined, remaining at or below replacement level since 1972. Various factors contributed to this decline in TFR:

- Widespread use of effective birth control methods (Figure 6-7).
- Availability of legal abortions.
- Social attitudes favoring smaller families.
- Greater social acceptance of childless couples.
- Increasing cost of raising a family. It will cost between $151,000 and $293,000 to raise to age 18 a child born in 1992.
- The rise in the average age at marriage between 1958 and 1991 from 20 to 24 for women and from 23 to 26 for men.
- More women working outside the home. In 1960, only 37% of American women ages 25 to 54 were in the work force. In 1993, the figure was 75%;

their childbearing rate was one-third that of women not in the paid labor force.

- Delayed reproduction. Many women marrying at a later age and working outside the home have children at a later age and thus tend to have fewer children.

The drop in the total fertility rate has led to a decline in the annual rate of population growth in the United States, but the country has not reached zero population growth (ZPG), nor is it even close. In 1993 the U.S. population of 258 million grew by 1.2%—faster than that of any other MDC. This added 3.1 million people: 2.0 million more births than deaths, 900,000 legal immigrants, and 200,000–500,000 illegal immigrants. This is equivalent to adding another California every ten years.

The three main reasons for this continued growth are: **(1)** the large number of women (58 million) born during the baby-boom period who are still moving through their childbearing years (even though the total fertility rate has remained at or below replacement level for 21 years, there has been a large increase in the number of potential mothers); **(2)** high levels of legal and illegal immigration; and **(3)** an increase in the number of unmarried women (including teenagers) having children.

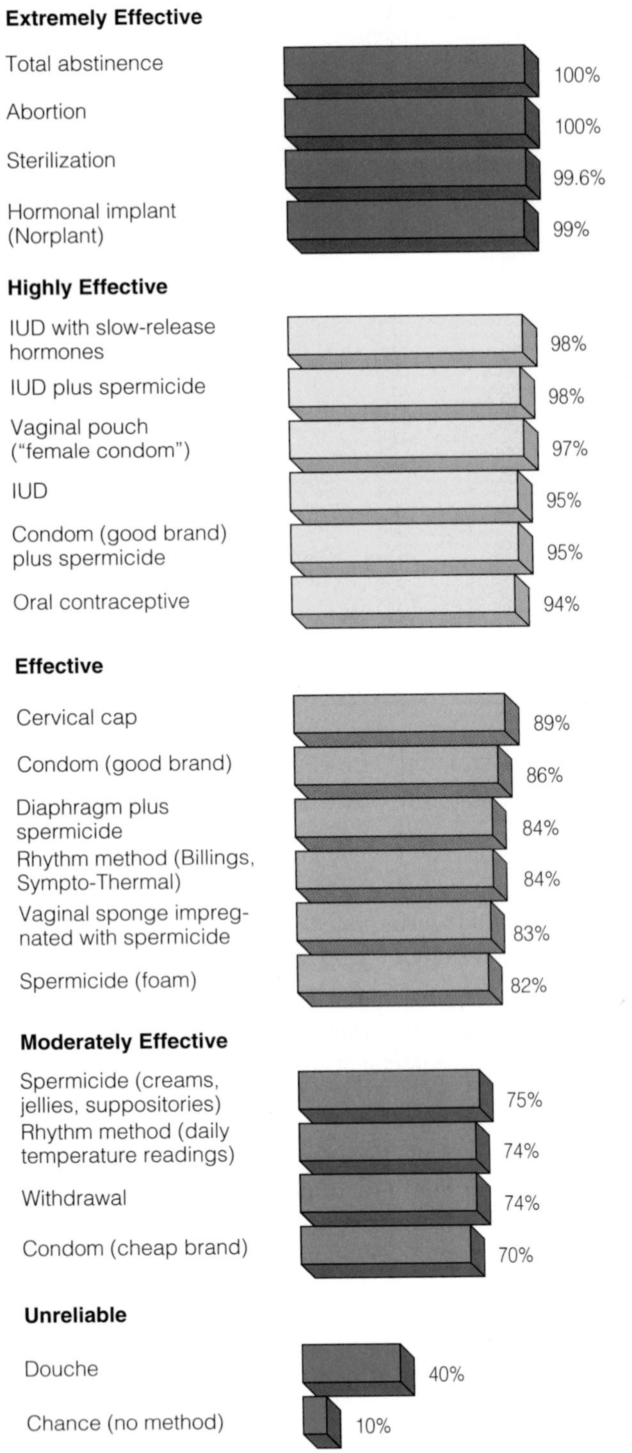

Extremely Effective

Total abstinence — 100%

Abortion — 100%

Sterilization — 99.6%

Hormonal implant (Norplant) — 99%

Highly Effective

IUD with slow-release hormones — 98%

IUD plus spermicide — 98%

Vaginal pouch ("female condom") — 97%

IUD — 95%

Condom (good brand) plus spermicide — 95%

Oral contraceptive — 94%

Effective

Cervical cap — 89%

Condom (good brand) — 86%

Diaphragm plus spermicide — 84%

Rhythm method (Billings, Sympto-Thermal) — 84%

Vaginal sponge impregnated with spermicide — 83%

Spermicide (foam) — 82%

Moderately Effective

Spermicide (creams, jellies, suppositories) — 75%

Rhythm method (daily temperature readings) — 74%

Withdrawal — 74%

Condom (cheap brand) — 70%

Unreliable

Douche — 40%

Chance (no method) — 10%

Figure 6-7 Typical effectiveness of birth control methods in the United States. Percentages are based on the number of undesired pregnancies per 100 couples using a specific method as their sole form of birth control for a year. For example, a 94% effectiveness rating for oral contraceptives means that for every 100 women using the pill regularly for one year, 6 will get pregnant. Effectiveness rates tend to be lower in LDCs because of human error and lack of education. (Data from Alan Guttmacher Institute)

Connections: Factors Affecting Birth Rates and Fertility Rates The most significant and interrelated factors affecting a country's average birth rate and total fertility rate are:

- *Average level of education and affluence.* Rates are usually lower in MDCs, where levels of both education and affluence are higher than in LDCs.

- *Importance of children as a part of the family labor force.* Rates tend to be lower in MDCs and higher in LDCs (especially in rural areas).

- *Urbanization.* People living in urban areas usually have better access to family planning services and tend to have fewer children than those living in rural areas, where children are needed to help grow food, collect firewood and water, and perform other essential tasks.

- *Cost of raising and educating children.* Rates tend to be lower in MDCs, where raising children is much more costly because children don't enter the labor force until their late teens or early twenties.

- *Educational and employment opportunities for women.* Rates tend to be low when women have access to both education and paid employment outside the home.

- *Infant mortality rate.* In areas with low infant mortality rates, people tend to have fewer children because fewer of their children die at an early age.

- *Average age at marriage* (or more precisely, the average age at which women give birth to their first child). People have fewer children when the women's average age at marriage is 25 or older.

- *Availability of private and public pension systems.* Pensions eliminate the need for parents to have many children to support them in old age.

- *Availability of reliable methods of birth control.* Widespread availability tends to reduce birth and fertility rates (Figure 6-7).

- *Religious beliefs, traditions, and cultural norms that influence the number of children couples want to have.* In many LDCs these factors favor large families.

Connections: Factors Affecting Death Rates
The rapid growth of the world's population over the past 100 years was not caused by a rise in crude birth rates. Rather, it was due largely to a decline in crude death rates, especially in the LDCs (Figure 6-8). The principal interrelated reasons for this general drop in death rates are:

- *Better nutrition* because of greater food production and better distribution

Q: How much of the money spent on health care in the United States is used to prevent disease?

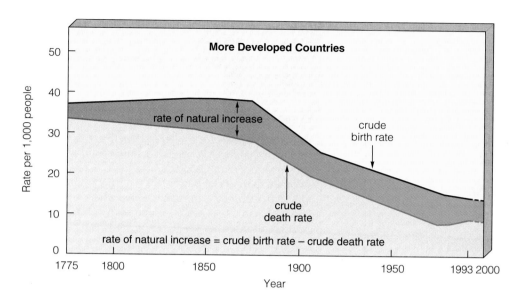

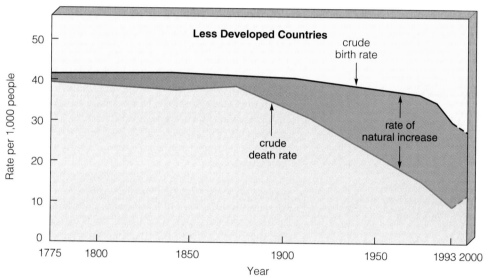

Figure 6-8 Changes in crude birth and death rates for MDCs and LDCs between 1775 and 1993, and projected rates (dashed lines) to 2000. (Data from Population Reference Bureau and United Nations)

■ *Fewer infant deaths and longer average life expectancy* because of improved personal hygiene, sanitation, and water supplies, which have curtailed the spread of many infectious diseases

■ *Improvements in medical and public health technology*, including antibiotics, immunizations, and insecticides

Two useful indicators of overall health in a country or region are **life expectancy**—the average number of years a newborn infant can be expected to live—and the **infant mortality rate**—the number of babies out of every 1,000 born each year that die before their first birthday (Figure 6-9). In most cases a low life expectancy in an area is the result of high infant mortality.

It is encouraging that life expectancy has increased since 1965, and averaged 75 in MDCs and 62 in LDCs

in 1993. But in the world's 41 poorest countries, mainly in Asia and Africa, average life expectancy is only 47 years.

Between 1900 and 1993 average life expectancy at birth rose sharply in the United States from 47 to 76 (79 for females and 72 for males); yet, in 1993, the average life expectancy at birth for people in 18 countries was higher than that for people in the United States. Japan has the highest life expectancy (79), followed by Sweden (78).

Because it reflects the general level of nutrition and health care, infant mortality is probably the single most important measure of a society's quality of life. A high infant mortality rate usually indicates insufficient food (undernutrition), poor nutrition (malnutrition), and a high incidence of infectious disease (usually from contaminated drinking water). Between 1965 and 1993,

A: 5%

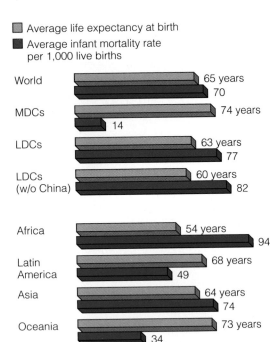

■ Average life expectancy at birth
■ Average infant mortality rate
   per 1,000 live births

World	65 years / 70
MDCs	74 years / 14
LDCs	63 years / 77
LDCs (w/o China)	60 years / 82
Africa	54 years / 94
Latin America	68 years / 49
Asia	64 years / 74
Oceania	73 years / 34
Former USSR	72 years / 28
United States	75 years / 9
North America	75 years / 8
Europe	75 years / 10

Figure 6-9 Average life expectancy at birth and average infant mortality rate for various groups of countries in 1993. (Data from Population Reference Bureau)

the world's infant mortality rate dropped 31% in MDCs and 35% in LDCs. This is an impressive achievement, but it still means that at least 12 million infants die each year of preventable causes.

Although the U.S. infant mortality rate of 9 per 1,000 in 1993 was low by world standards, 23 other countries had lower rates. Several factors keep the U.S. infant mortality rate higher than it could be, including:

🔹 Inadequate health care for poor women during pregnancy and for their babies after birth

🔹 Drug addiction among pregnant women

🔹 The high birth rate among teenage women (Case Study, at right)

Migration The population change for a specific geographic area is also affected by movement of people into (immigration) and out of (emigration) that area, according to the following equation:

Teenage Pregnancy in the United States

CASE STUDY

The United States has the highest teenage pregnancy rate of any industrialized country, about six times higher than the rates in Japan and in most European countries and two times higher than Canada's. Among U.S. teenage women there are about 406,000 abortions and 490,000 births (68% to unmarried mothers—up from 48% in 1980), and 3 million sexually transmitted diseases per year.

Babies born to teenagers are more likely to have a low birth weight—the most important factor in infant deaths—thus increasing the country's infant mortality rate. Half of these teenage mothers go on welfare within a year; 77% in five years.

UN studies show that American teenagers aren't more sexually active than those in other MDCs, but they are less likely to have learned what precautions to take to prevent pregnancy and in some cases less willing to use them. In Sweden, by contrast, which has a teenage pregnancy rate about one-fifth that of the United States, every child receives a thorough grounding in basic reproductive biology by age 7. By age 12 each child has been told about the various types of contraceptives.

Many analysts urge that effective sex education programs be developed in which American children are made aware of the values of abstinence and the various types of contraceptives by age 12 and that school-based health clinics be opened in all junior and senior high schools. Studies in the United States and several other MDCs show that early reproductive and sex education (beginning in elementary school) and availability of contraceptives in high schools reduce teenage pregnancies and abortions. Such proposals, however, are vigorously opposed on religious grounds or by groups who fear that early sex education will lead to increased sexual activity. What do you think should be done about sex education and teenage pregnancy?

Most countries control their rates of population growth to some extent by restricting immigration. Only a few countries accept large numbers of immigrants or refugees. Thus population change for most countries is determined mainly by the difference between their birth rates and death rates.

Q: What does the EPA consider to be the three most dangerous indoor air pollutants in MDCs?

Figure 6-10 Population age structure diagrams for countries with rapid, slow, zero, and negative population growth rates. Bottom portions represent prereproductive years (ages 0–14), middle portions represent reproductive years (ages 15–44), and top portions represent postreproductive years (ages 45–85+). (Data from Population Reference Bureau)

Migration within countries, especially from rural to urban areas, plays an important role in the population dynamics of cities, towns, and rural areas, as discussed in Section 6-3.

6-2 POPULATION AGE STRUCTURE

Age Structure Diagrams Even if the replacement-level fertility rate of 2.1 were magically achieved globally by tomorrow, the world's population would keep on growing for at least another 60 years! Why? The answer lies in an understanding of the **age structure, or age distribution, of a population**—that is, the percentage of the population (or the number of people of each sex) at each age level.

Demographers typically construct a population age structure diagram by plotting the percentages or numbers of males and females in the total population in each of three age categories: *prereproductive* (ages 0–14), *reproductive* (ages 15–44), and *postreproductive* (ages 45–85+). Figure 6-10 shows the age structure diagrams for countries with rapid, slow, zero, and negative population growth rates.

Connections: Age Structure and Population Growth Momentum Any country with many people below age 15 (represented by a wide base in Figure 6-10) has powerful built-in momentum toward an increase in population size unless death rates rise sharply. The number of births rises even if women

have only one or two children because of the large number of women moving into their reproductive years.

Today half of the world's 2.6 billion women are in the reproductive age category of 15–44, and one-third of the people on this planet are under 15 years old and are poised to move into their prime reproductive years. In LDCs the number is even higher—36% compared with 21% in MDCs. Asia alone has more children under age 15 than Africa has people. Figure 6-11 shows the powerful momentum for population growth in LDCs caused by these large numbers of young people. Even if each female in this group has only two children, world population will still grow for 60 years unless the death rate rises sharply. And women in LDCs now average 3.7 children, well above replacement level. This powerful force for continued population growth, mostly in LDCs, will be slowed only by an effective program to reduce birth rates—or by a catastrophic rise in death rates.

Connections: Making Projections from Age Structure Diagrams A large increase occurred in the U.S. population between 1946 and 1964 (Figure 6-6). This 80-million-person bulge, known as the *baby-boom generation*, will move upward through the country's age structure between 1946 and 2040 as baby boomers leave young adulthood and enter their middle and then old age (Figure 6-12).

Today baby boomers make up nearly half of all adult Americans. They dominate the population's

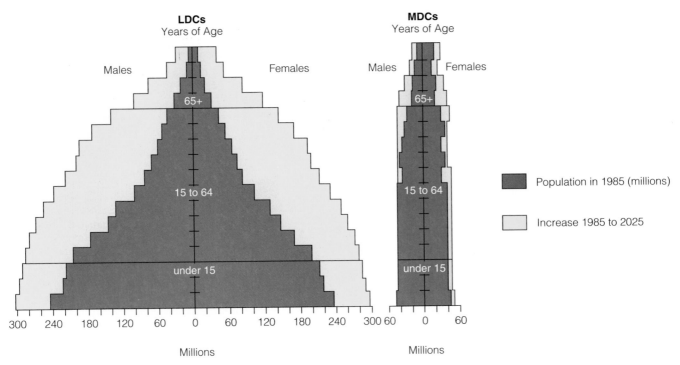

Figure 6-11 Age structure and projected population growth in LDCs and MDCs, 1985–2025. (Data from Population Reference Bureau)

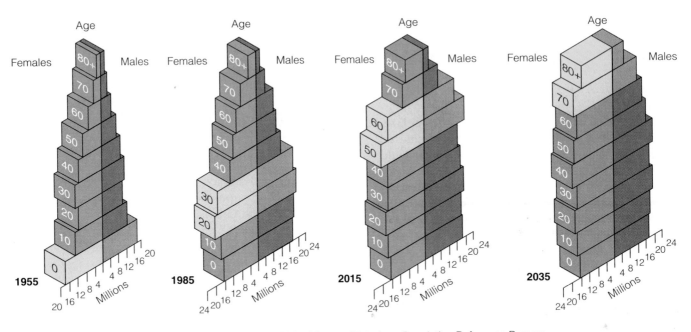

Figure 6-12 Tracking the baby-boom generation in the United States. (Data from Population Reference Bureau and U.S. Census Bureau)

demand for goods and services and play an increasingly important role in deciding who gets elected and what laws are passed.

During their working years baby boomers will create a large surplus of money in the Social Security trust fund. However, unless changes in funding procedures are made, the retired baby boomers will use up this surplus by 2048. Elderly baby boomers will also put severe strains on health care services.

The economic burden of helping support so many retired baby boomers will fall on the *baby-bust generation*, the much smaller group of 47 million people born in the period between 1970 and 1985, during which total fertility rates fell sharply (Figure 6-6). Retired

138

Q: What is the most dangerous indoor air pollutant in LDCs?

The Graying of Japan

CASE STUDY

In only seven years, between 1949 and 1956, Japan (Figure 6-13) cut its birth rate, total fertility rate, and population growth rate in half, mostly because of a liberal abortion law and access to family planning implemented by the post–World War II U.S. occupation forces and the Japanese government. Since 1956 these rates have declined further. In 1949 Japan's total fertility rate was 4.5. In 1993, it was 1.5—one of the world's lowest—and is projected to fall to 1.35 in 1996.

Average life expectancy at birth is 79 years—the highest in the world. Japan also has one of the world's lowest death rates for infants under one year of age. Japan's population of 125 million in 1993 is growing very slowly and is projected to be only 126 million in 2025. Its population could shrink to 65–96 million by 2090, depending on its fertility rate and immigration rate (which is currently negligible).

As Japan approaches zero population growth, it is beginning to face some of the problems of an aging population. Between 1993 and 2010 Japan's population age 65 or older is expected to increase from 13% to 21%, and by 2045 reach 27%.

Japan's universal health insurance and pension systems used

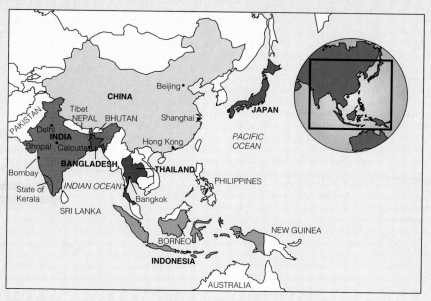

Figure 6-13 Where are Japan, Thailand, Indonesia, India, China, and Bangladesh?

about 41% of the national income in 1993. This economic burden is projected to rise to 60% or higher in 2020, with at least two-thirds of the expenditures for pensions and the virtually free health care provided to the elderly. Economists worry that the steep taxes needed to fund these services could discourage economic growth.

Since 1980 Japan has been feeling the effects of a declining work force. This is one reason it has invested heavily in automation and has encouraged women to work outside the home.

The population of Japan is 99% Japanese. Fearing a breakdown in its social cohesiveness, the government has been unwilling to increase immigration to provide more workers. Despite this official policy, the country is becoming increasingly dependent on illegal immigrants to keep its economic engines running. How Japan deals with these problems will provide ideas for the United States and other countries as they make the transition to zero population growth and, eventually, to population decline.

baby boomers may use their political clout to force members of the baby-bust generation to pay higher income, health care, and Social Security taxes.

In other respects the baby-bust generation should have an easier time than the baby-boom generation. Fewer people will be competing for educational opportunities, jobs, and services, and labor shortages may drive up their wages, at least for jobs requiring education or technical training beyond high school. On the other hand, members of the baby-bust group may find it hard to get job promotions as they reach middle age because most upper-level positions will be occupied by the much larger baby-boom group. And many baby boomers may delay retirement because of

improved health and the need to accumulate adequate retirement funds.

From these few projections we can see that any booms or busts in the age structure of a population create social and economic changes that ripple through a society for decades.

Population Decline Natural population decrease (more deaths than births) is already a reality in Hungary and Germany. Most other European countries are at or near zero population growth and could undergo population declines in the next century. Japan also faces this prospect sometime after 2025 (Case Study, above).

A: Smoke from unvented or poorly vented stoves for cooking and heating

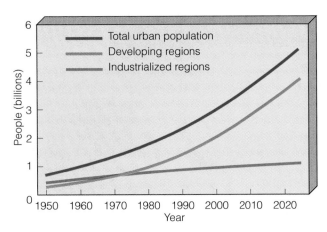

Figure 6-14 Urban population growth in MDCs and LDCs, 1950–2025. (Data from United Nations and Population Reference Bureau)

If population decline is gradual, its negative effects can usually be managed. But rapid population decline, like rapid population growth, can lead to severe economic and social problems. For example, once a country reaches zero population growth and experiences a rapid population decline, the proportion of the population made up of older people rises sharply. Older people consume a large share of medical, social security, and other costly public services. A country with a declining population can also face labor shortages unless it relies on greatly increased automation, immigration of foreign workers, or both.

6-3 POPULATION DISTRIBUTION: URBANIZATION AND URBAN PROBLEMS

The Future Is Urban The quality of urban life affects every person and every species on the planet. Since 1950 the number of people living in urban areas has more than tripled and by 2025 is projected to reach 5.5 billion—equal to the world's current population (Figure 6-14). Today about 42% of the world's population lives in urban areas, and by 2025 this figure is expected to increase to 61%.

An **urban area** is often defined as a town or city with a population of more than 2,500 people, although some countries set the minimum at 10,000–50,000. A country's **degree of urbanization** is the percentage of its population living in an urban area. **Urban growth** is the rate of growth of urban populations, which grow in two ways: **(1)** by natural increase (more births than deaths) and **(2)** by immigration (mostly from rural areas). During the 1990s about 83% of the world's population increase is expected to occur in urban areas.

People are drawn to urban areas in search of jobs and a better life. They may also be pushed into urban areas by modern mechanized agriculture, which uses fewer farm laborers and allows large landowners to buy out subsistence farmers who cannot afford to modernize. Without jobs or land, these people are forced to move to cities. The poor people fortunate enough to get a job usually must work long hours for low wages. These jobs may also expose them to dust, hazardous chemicals, excessive noise, and dangerous machinery.

At current rates the world's population will double in 41 years, the urban population in 22 years, and the urban population of LDCs in only 15 years. Several trends are important in understanding the problems and challenges of urbanization and urban growth on our rapidly urbanizing planet:

- The percentage of the population living in urban areas increased from 14% to 42% (72% in MDCs and 34% in LDCs) between 1900 and 1993. By 2025 it is projected that 61% of the world's people will be living in urban areas. During the 1990s about 83% of the world's population increase is expected to take place in urban areas—adding about 78 million people a year to these already overburdened areas.

- The number of large cities is mushrooming. Today 1 person out of every 10 lives in a city with a million or more inhabitants, and many live in *megacities* with 10 million or more people. The United Nations projects that by 2000 there will be 26 megacities, more than two-thirds of them in LDCs.

- LDCs, with 34% urbanization, contain 1.5 billion people—more than the total population of Europe, North America, Latin America, and Japan combined. LDCs are projected to reach at least 57% urbanization by 2025.

- In MDCs, with 72% urbanization, urban growth is slower than in LDCs, but MDCs should reach 84% urbanization by 2025. As they grow outward, some urban areas merge with other urban areas to form *megalopolises*. For example, the remaining open space between Boston and Washington, D.C., is rapidly urbanizing and merging. This sprawling, 800-kilometer-long (500-mile) urban area, sometimes called *Bowash*, has almost 60 million people—more than twice Canada's entire population (Figure 6-15).

- Poverty is becoming urbanized as more poor people migrate from rural to urban areas (Case Study, p. 142). At least 1 billion people—18% of the world's population—live in the crowded

Q: How much air pollution is emitted into the atmosphere each year by a typical motor vehicle in the United States?

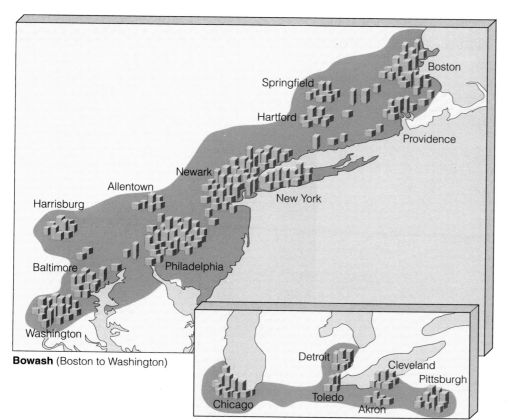

Figure 6-15 Bowash—an example of urban sprawl and coalescence leading to a megalopolis between Boston and Washington, D.C.—and Chipitts, extending from Chicago to Pittsburgh.

Bowash (Boston to Washington)

Chipitts (Chicago to Pittsburgh)

slums of inner cities and in the vast, mostly illegal squatter settlements or shantytowns that ring the outskirts of most cities in LDCs (Figure 7-9).

■ Despite being centers of commerce and industry, many cities suffer from extreme poverty and social and environmental decay.

Population Distribution in the United States
In 1800 only 5% of Americans lived in cities. Since then three major internal population shifts have taken place in the United States:

■ *Migration to large central cities.* About 75% of Americans live in 350 *metropolitan areas*—cities and towns with at least 50,000 people (Figure 6-18, p. 144). Nearly half of the country's population lives in large metropolitan areas containing 1 million or more residents (Figure 6-19, p. 145).

■ *Migration from large central cities to suburbs and smaller cities.* Since 1970 this type of migration has followed new jobs to such areas. Today about 41% of the country's urban dwellers live in central cities, and 59% in suburbs.

■ *Migration from the North and East to the South and West.* Since 1980 about 80% of the U.S. population

increase has occurred in the South and West, particularly near the coasts. This shift is expected to continue.

Since 1920, many of the worst urban environmental problems in the United States and other MDCs have been significantly reduced. Most people have better working and housing conditions; air and water quality have improved. Better sanitation, public water supplies, and medical care have slashed death rates and the prevalence of sickness from malnutrition and transmittable diseases such as measles, diphtheria, typhoid fever, pneumonia, and tuberculosis.

The biggest problems facing numerous cities in the United States and other MDCs are: deteriorating services; aging infrastructures (streets, schools, bridges, housing, sewers); budget crunches from lost tax revenues and rising costs as businesses and more affluent people move out; environmental degradation; and neighborhood collapse.

Connections: Transportation and Urban Development If a city cannot spread outward, it grows upward and downward (below ground), occupying a relatively small area with a high population density. These are the only avenues of expansion for cities such as Manhattan, Tokyo, and Hong Kong.

A: 0.9 metric ton (1 ton)

Mexico City, Mexico

In 1993 the population of Mexico City (Figure 6-16) was 16.2 million—the world's fourth most populous city. Every day an additional 1,000 poverty-stricken rural peasants pour into the city, hoping to find a better life.

The city suffers from severe air pollution, high unemployment (close to 50%), deafening noise, congestion, and a soaring crime rate. One-third of the city's people live in crowded slums (called *barrios*) or squatter settlements, without running water or electricity. And at least 8 million people—as many as live in New York City—have no sewer facilities, which means huge amounts of human waste are left in gutters and vacant lots every day. When the winds pick up dried excrement, a "fecal snow" often falls on parts of the city—leading to widespread salmonella and hepatitis infections, especially among children. About half of the city's garbage is left in the open to rot, attracting armies of rats and swarms of flies. People living in the slums have to buy their water by the bucket from vendors. Many of those who have access to city water won't drink it, complaining that it is yellow and full of worms.

Some 3.5 million motor vehicles and 30,000 factories spew pollutants into the atmosphere. Air pollution is intensified because the city lies in a basin surrounded by mountains, and frequent thermal inversions trap pollutants at ground level. Since 1982 the amount of contamination in the city's smog-choked air has more

Figure 6-16 Where are Brazil, Mexico, Belize, and Costa Rica?

Most people living in such compact cities walk, ride bicycles, or use energy-efficient mass transit. Residents often live in multistory apartment buildings; with few outside walls in many apartments, heating and cooling costs are reduced. Many European cities are compact and tend to be more energy-efficient than the dispersed cities in the United States, Canada, and Australia, where there is often ample land for outward expansion.

A combination of cheap gasoline, plentiful land, and a network of highways leads to dispersed, car-culture cities with a low population density—often called *urban sprawl*. Most people living in such urban areas live in single-family houses, with unshared walls that lose and gain heat rapidly unless they are well insulated and airtight. Urban sprawl also gobbles up unspoiled natural habitats, paves over fertile farmland, and promotes considerable dependence on the automobile.

The decision to build highways between cities—and to build freeways and beltways within urban areas—is probably the most far-reaching land-use, energy policy, and environmental decision a country, state, or city makes. These methods of moving people and goods by car and truck profoundly shape where we live and work, how we get from place to place, how urban areas grow, how much energy we waste, and how much pollution we produce.

The Automobile Despite having only 4.7% of the world's people, the United States has 35% of the world's 550 million cars and trucks. In LDCs most people cannot afford a car; they travel mostly by foot, bicycle, or motor scooter. In the United States the car is used for 86% of all trips (compared to about 45% in most western European countries), for 98% of all urban transportation, and for 86% of travel to work. Public transit is used for only 2.5% of all trips. Almost

Q: How many people in the United States die prematurely each year because of air pollution?

than tripled (Figure 6-17). Breathing the air is like smoking two packs of cigarettes a day, and birds have dropped dead flying through the smog.

The city's air and water pollution cause an estimated 100,000 premature deaths a year. Some doctors have advised parents to take their children and leave the city—permanently. These problems, already at crisis levels, will become even worse if the city grows as projected to 25.6 million people by the end of this century.

The Mexican government is industrializing other parts of the country in an attempt to slow migration to Mexico City. In 1991, the government closed the city's huge state-run oil refinery and ordered many of the industrial plants in the basin to go elsewhere by 1994. Cars have also been banned from a 50-block central zone. Taxis built before 1985 have been taken off the streets, and trucks can run only on liquefied petroleum gas (LPG). The government began phasing in unleaded gasoline in 1991, but it will be years before millions of older lead-burning vehicles are eliminated. If you were in charge of Mexico City, what would you do?

United Nations

Figure 6-17 The air in Mexico City ranks with the dirtiest in the world. This is due to a combination of topography, a large population, industrialization, large numbers of motor vehicles, and too little emphasis on reducing rural-to-urban migration and preventing and controlling pollution. Breathing the city's air has been compared to smoking two packs of cigarettes a day. This photo was taken during the morning of a bright, sunny day.

75% of commuting cars carry only one person, and only 13% of commuters use carpools. Only about 5% of Americans use public transportation, and only 2% walk or use a bicycle to get to and from work. Americans drive 3 billion kilometers (2 billion miles) each year—equal to more than 10 round trips to the sun and as far as the rest of the world combined. No wonder the British author J. B. Priestley remarked, "In America, the cars have become the people."

The automobile provides convenience and undreamed-of mobility. To many people cars are also symbols of power, sex, excitement, and success.

Moreover, much of the world's economy is built on producing motor vehicles and supplying roads, services, and repairs for them. In the United States, one of every six dollars spent and one of every six non-farm jobs are connected to the automobile.

In spite of their important economic and personal benefits, motor vehicles have many destructive effects on both human lives and on air, water, land, and wildlife resources. Since 1885, when Karl Benz built the first automobile, almost 18 million people have been killed by motor vehicles. This death toll increases by about 250,000 people per year, and each year about 10 million people are injured or permanently disabled in motor vehicle accidents.

In the United States almost 15 million motor vehicle accidents kill around 39,000 people each year and injure almost 5 million people, at least 300,000 of them severely. More Americans have been killed by cars than were killed in all the country's wars. Each year these accidents cost the United States more than $350 billion, 8% of the GNP.

Motor vehicles are also the largest source of air pollution, laying a haze of smog over the world's car-clogged cities (Figure 6-17). In the United States they produce at least 50% of the air pollution, even though emission standards are as strict as any in the world.

A: 150,000–350,000—an average of 410–960 per day

Figure 6-18 Computer composite of current major urban regions in the lower 48 states as revealed by night illumination detected by satellite sensors.

Gains in fuel efficiency and emission reductions have been largely offset by the increase in cars and a doubling of the distance Americans traveled by car between 1970 and 1990.

By making long commutes and distant shopping possible, automobiles and highways have helped create urban sprawl and reduced use of mass transit, bicycling, and walking. Worldwide, at least a third of urban land is devoted to roads, parking lots, gasoline stations, and other automobile-related uses. In the United States more land is now devoted to cars than to housing. Half the land in an average American city is used for cars, prompting urban expert Lewis Mumford to suggest that the U.S. national flower should be the concrete cloverleaf.

In 1907, the average speed of horse-drawn vehicles through the borough of Manhattan was 18.5 kilometers (11.5 miles) per hour. Today cars and trucks with the potential power of 100–300 horses creep along Manhattan streets at 8 kilometers (5 miles) per hour. In Paris and Tokyo, average auto speeds are even lower. If current trends continue, U.S. motorists will spend an average of two years of their lifetimes in traffic jams. The U.S. economy loses at least $100 billion a year because of time lost in traffic delays. Building more roads is not the answer because, as economist Robert Samuelson put it, "Cars expand to fill available concrete."

One way to break this cycle is to make drivers pay directly for most of the true costs of auto use. Federal, state, and local government auto subsidies in the United States amount to at least $300 billion a year (almost 5% of the country's GNP)—an average of $1,600 per vehicle. Taxpayers (drivers and nondrivers) foot this bill mostly without knowing this is part of their automobile-use bill. If drivers had to pay these hidden

Q: Worldwide, how many people live in cities where outdoor air is unhealthy to breathe?

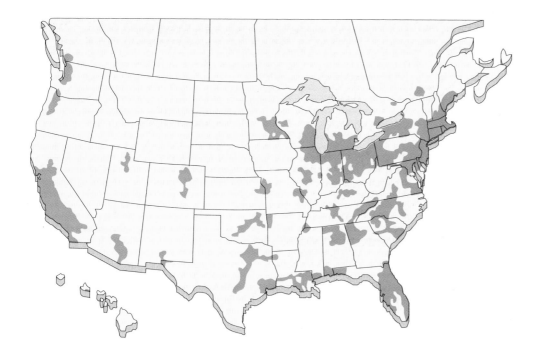

Figure 6-19 Major urban regions in the United States by 2000. Nearly half (48%) of Americans live in *consolidated metropolitan areas* with 1 million or more people. (Data from U.S. Census Bureau)

costs directly as a gasoline tax, the tax on each gallon of gas would be about $5.

Although including these hidden costs up front in the market price of cars and gasoline makes economic and environmental sense, this approach faces massive political opposition from the general public and the powerful automobile-oil-industrial complex.

Alternatives to the Car Cars are an important form of transportation, but some countries are encouraging alternatives. One alternative is the *bicycle*. Worldwide there are twice as many bicycles as cars, and three times as many bicycles as cars are sold each year. It is an inexpensive form of transportation, burns no fossil fuels, produces no pollution, takes few resources to make, and is the most energy-efficient form of transportation (including walking).

In urban traffic, cars and bicycles move at about the same average speed. Using separate bike paths or lanes running along roads, bicycle riders can make most trips shorter than 8 kilometers (5 miles) faster than a car.

Many Chinese cities have exclusive pedestrian/ bicycle lanes and bridges for the country's 300 million cyclists. In the Netherlands (with more bicycle paths than any other country) bicycle travel makes up 30% of all urban trips, and in Japan 15% of all commuters ride bicycles to work or to commuter-rail stations.

Another alternative is *mass transit,* in which large numbers of people are moved by subways, trains, trolleys, and buses. Although the U.S. population increased by 127 million between 1946 and 1994, the number of

riders on all forms of mass transit dropped from 23.4 million to about 9 million. In the United States mass transit accounts for only 7% of all passenger travel, compared with 15% in Germany and 47% in Japan. Only 20% of the federal gasoline tax goes to mass transit, the rest to highways. This encourages states and cities to invest in highways instead of in mass transit.

Rapid-rail, suburban train, and trolley systems can transport large numbers of people at high speed, but they are efficient only where many people live along a narrow corridor and can easily reach properly spaced stations. However, high-speed trains between cities can greatly reduce the need for travel by car and plane (Solutions, p. 146).

Bus systems are more flexible than rail systems. They can be routed throughout sprawling cities and rerouted overnight if transportation patterns change. Bus systems also require less capital and have lower operating costs than rail systems. However, because they must offer low fares to attract riders, bus systems often cost more to operate than they bring in. Because buses are cost-effective only when full, they are sometimes supplemented by car pools, van pools, and jitneys (small vans or minibuses traveling along regular routes).

Connections: Urban Resource and Environmental Problems Most of today's cities aren't sustainable, but rather are heavily dependent on distant sources for their food, water, energy, and materials. Their massive use of resources damages nearby and distant air, water, soil, and wildlife (Figure 6-21).

High-Speed Regional Trains

In western Europe and Japan a new generation of streamlined, comfortable, and low-polluting high-speed-rail (HSR) lines are being developed for medium-distance travel between cities within a region. These "bullet" or supertrains travel on new or upgraded existing tracks at speeds up to 320 kilometers (200 miles) per hour. They are ideal for trips of intermediate length—240–480 kilometers (150–300 miles). Journeys between major cities within a region on these trains are smoother and cheaper than by airplane. However, such systems are expensive to run and maintain, and they must operate along heavily used transportation routes to be profitable.

A future alternative to the automobile and the airplane for medium-distance travel of 970 kilometers (600 miles) or less between cities is the magnetic-levitation (MAGLEV) train, which uses powerful superconducting electromagnets to suspend the train on a cushion of air a few centimeters above a guiding rail. Such trains zoom along without rail friction at speeds up to 500 kilometers (310 miles) per hour. Germany and Japan have prototypes in operation.

Because these trains never touch the track, they make little noise, require little maintenance; and can be elevated over existing median strips or other rights-of-way along highways, avoiding costly and disruptive land acquisitions. MAGLEV and bullet trains reduce energy consumption as well as greenhouse and other air emissions.

A U.S. MAGLEV network could replace airplanes, buses, and private cars for most medium-distance travel between major American cities (Figure 6-20). Such systems could use the airport of a major metropolitan area as a hub to provide rapid transportation to cities.

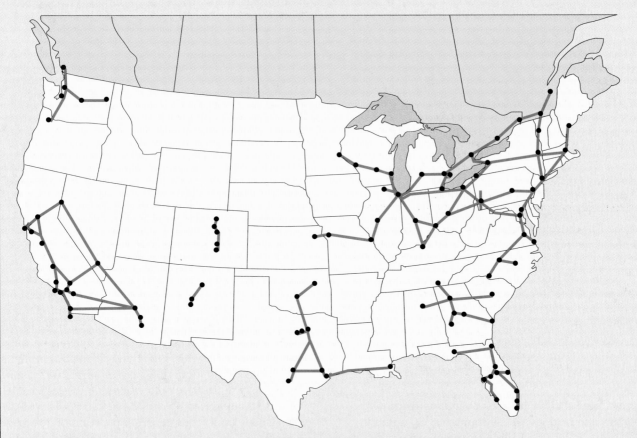

Figure 6-20 Potential routes for high-speed MAGLEV and bullet trains in the United States and parts of Canada. Such a system would allow rapid, comfortable, safe, and affordable travel between major cities in a region, and it would slash dependence on cars, buses, and airplanes. (Data from High Speed Rail Association)

Q: Is there doubt about the validity of the greenhouse effect?

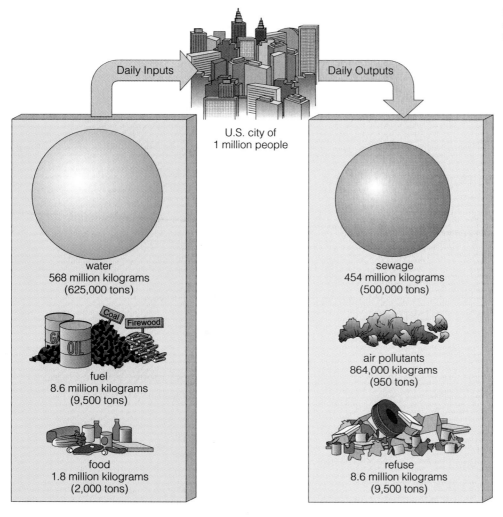

Figure 6-21 Typical daily input and output of matter and energy for a U.S. city of 1 million people.

Daily Inputs

Daily Outputs

U.S. city of
1 million people

water
568 million kilograms
(625,000 tons)

fuel
8.6 million kilograms
(9,500 tons)

food
1.8 million kilograms
(2,000 tons)

sewage
454 million kilograms
(500,000 tons)

air pollutants
864,000 kilograms
(950 tons)

refuse
8.6 million kilograms
(9,500 tons)

Major urban resource and environmental problems are:

- *Scarcity of trees, shrubs, and other natural vegetation.* Plants absorb air pollutants, give off oxygen, help cool the air as water evaporates from their leaves, muffle noise, provide wildlife habitats, and give aesthetic pleasure. As one observer remarked, "Most cities are places where they cut down the trees and then name the streets after them." According to the American Forestry Association (AFA) one city tree provides over $57,000 worth of air conditioning, erosion and storm-water control, wildlife shelter, and air pollution control over a 50-year lifetime.

- *Alteration of local and, sometimes, regional climate.* Generally cities are warmer, rainier, foggier, and cloudier than suburbs and nearby rural areas. The enormous amounts of heat generated by cars, factories, furnaces, lights, air conditioners, and people in cities create an **urban heat island** surrounded by cooler suburban and rural areas. The

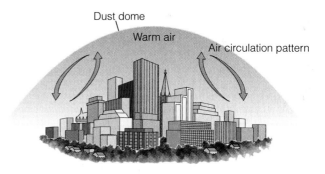

Figure 6-22 An urban heat island causes patterns of air circulation that create a dust dome over the city. Winds elongate the dome toward downwind areas. A strong cold front can blow the dome away and lower urban pollution levels.

dome of heat also traps pollutants, especially tiny solid particles (suspended particulate matter), creating a **dust dome** above urban areas (Figure 6-22). If wind speeds increase, the dust dome elongates downwind to form a **dust plume**, which can spread the city's pollutants for hundreds of kilometers. As cities grow and merge

A: No. Without it, Earth would be too cold for life as we know it to exist.

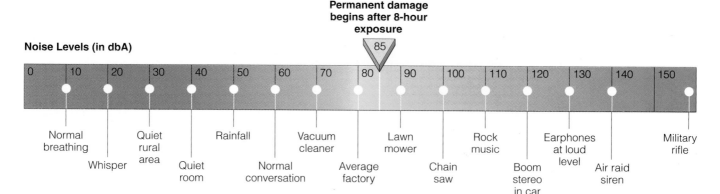

Permanent damage begins after 8-hour exposure

85

Noise Levels (in dbA)

| 0 | 10 | 20 | 30 | 40 | 50 | 60 | 70 | 80 | 90 | 100 | 110 | 120 | 130 | 140 | 150 |

Normal breathing — Whisper — Quiet rural area — Quiet room — Rainfall — Normal conversation — Vacuum cleaner — Average factory — Lawn mower — Chain saw — Rock music — Boom stereo in car — Earphones at loud level — Air raid siren — Military rifle

Figure 6-23 Noise levels (in decibel-A or dbA sound pressure units) of common sounds. You are being exposed to a sound level high enough to cause permanent hearing damage if you need to raise your voice to be heard above the racket, if a noise causes your ears to ring, or if nearby speech seems muffled. Prolonged exposure to lower noise levels and occasional loud sounds may not damage your hearing but can greatly increase internal stress. Noise pollution can be reduced by **(1)** modifying noisy activities and devices, **(2)** shielding noisy devices or processes, **(3)** shielding workers from noise, and **(4)** using anti-noise—a new technology that cancels out one noise with another. The control of noise pollution in the United States has lagged behind that in the former Soviet Union and many western European and Scandinavian countries.

(Figure 6-15), individual heat islands also merge. This can affect the climate of a large area and keep polluted air from being adequately diluted and cleansed.

■ *Lack of water.* This requires expensive reservoirs, canals, and deep wells (Chapter 11).

■ *Rapid runoff of water from asphalt and concrete.* This can overload sewers and storm drains, contributing to water pollution and flooding in cities and downstream areas.

■ *Production of large quantities of air pollution (Chapters 9 and 10), water pollution (Chapter 11), hazardous waste (Chapter 13), and garbage and other solid waste (Chapter 13).*

■ *Excessive noise* (Figure 6-23). Every day, one of every nine Americans (28 million people) lives, works, or plays around noise of sufficient duration and intensity to cause some permanent hearing loss, and that number is rising rapidly. The lives of 40 million more Americans, mostly urban dwellers, are significantly disrupted by noise pollution.

■ *Loss of rural land, fertile soil, and wildlife habitats as cities expand.* As urban areas expand, they swallow up rural land, especially flat or gently rolling cropland with well-drained, fertile soil. Once prime agricultural land is paved over or built upon, it is lost for food production. As land values near urban areas rise, taxes on nearby farmland rise to the point where many farmers are forced to sell their land. They realize that they

can make much more money growing houses and shopping centers than corn and cows.

6-4 SOME SOLUTIONS TO URBAN PROBLEMS

Urban Land-Use Planning and Control Most urban areas and some rural areas use some form of **land-use planning** to decide the best present and future use of each parcel of land in the area. The most widely used approach is **zoning**, in which various parcels of land are designated for certain uses. In addition to zoning, local governments can control the rate of development by limiting the number of building permits, sewer hookups, roads, and other services.

Local, state, and federal governments can also take any of the following measures to protect cropland, forestland, wetlands, and other nonurban lands near expanding urban areas from degradation and ecologically unsound development:

■ *Require an environmental impact analysis for proposed roads and development projects.*

■ *Require developers to pay a fee that includes the full cost of additional services such as roads and water and sewer lines for all development projects.*

■ *Tax land on the basis of its actual use as agricultural land or forestland rather than on the basis of its economically most profitable potential use. This keeps farmers and other landowners from being forced to sell land to pay their tax bills.*

Q: Why is there concern over a potentially enhanced greenhouse effect?

- *Give tax breaks, or conservation easements, to landowners who agree to use land only for specified purposes, such as agriculture, wilderness, wildlife habitat, or nondestructive forms of recreation.*

- *Give subsidies to farmers who take highly erodible cropland out of production, or eliminate subsidies for farmers who farm such land or who convert wetlands to cropland.*

- *Purchase land development rights that restrict the way land can be used.*

- *Use land trusts to buy and protect ecologically valuable land.* Such purchases can be made by private groups such as the Nature Conservancy, the Audubon Society, and by local nonprofit, tax-exempt, charitable organizations, as well as by public agencies.

Most land-use planning is based on the assumption that substantial future growth in population and economic development should be encouraged, regardless of the environmental and other consequences. Typically this leads to uncontrolled or poorly controlled urban growth and sprawl.

A major reason for allowing this usually destructive process is that 90% of the revenue that local governments use to provide schools, roads, police and fire protection, public water and sewer systems, welfare services, and other obligations comes from *property taxes*, taxes on all buildings and property proportional to their economic value.

However, the costs of providing more services to accommodate population and economic growth often exceed the revenue from increased property values. Because local governments can rarely raise property taxes enough to meet expanding needs, they often try to raise more money by promoting further economic growth. Typically the long-term result is a treadmill of economic growth, eventually leading to environmental decay. Then businesses and residents move away, decreasing the tax base, reducing tax income, and causing further environmental and social decay as governments are forced to cut the quantity and quality of services or to raise the tax rate (which drives more people away and worsens the situation).

Environmentalists challenge the "all-growth-is-good" dogma that is the basis of most land-use planning. They urge communities to use comprehensive, regional **ecological land-use planning**, in which all major variables are considered and integrated into a model designed to anticipate present and future needs and problems, and to propose solutions for an entire region. It is a complex process taking into account geological, ecological, economic, health, and social factors. Six basic steps are involved:

- *Make an environmental and social inventory.* Experts survey geological factors (soil type, fault lines, floodplains, water availability), ecological factors (forest types and quality, wildlife habitats, stream quality, pollution), economic factors (housing, transportation, utilities, industrial development), and health and social factors (disease and crime rates, ethnic distribution, illiteracy). A top priority is to identify and protect areas that are critical for preserving water quality, supplying drinking water, and reducing erosion.

- *Determine goals and their relative importance.* Experts, public officials, and the general public decide on goals and rank them in order of importance. For example, is the primary goal to encourage or to discourage further economic development and population growth? To preserve prime cropland from development? To reduce soil erosion?

- *Develop individual and composite maps.* Data for each factor surveyed in the environmental and social inventory are plotted on separate transparent plastic maps. The transparencies are then superimposed or combined by computer into three composite maps—one each for geological, ecological, and socioeconomic factors.

- *Develop a master composite.* The three composite maps are combined to form a master composite, which shows how the variables interact and indicates the suitability of various areas for different types of land use.

- *Develop a master plan.* The master composite (or a series of alternative master composites) is evaluated by experts, public officials, and the general public, and a final master plan is drawn up and approved.

- *Implement the master plan.* The plan is set in motion and monitored by the appropriate governmental, legal, environmental, and social agencies.

Ecological land-use planning sounds good on paper, but it is not widely used for several reasons. There is intense pressure by economically and politically powerful people to develop urban land for short-term economic gain, with little regard for long-term ecological and economic losses. In addition, local officials seeking election every few years usually focus on short-term rather than long-term problems and can often be influenced by economically powerful developers. Officials—and often the majority of citizens—are frequently unwilling to pay for costly ecological land-use planning and implementation, even though a well-designed plan can prevent or ease many urban

Is Your City Green?—The Ecocity Concept in Davis, California

SOLUTIONS

In a sustainable and ecologically healthy city— called an *ecocity* or *green city*—matter and energy resources are used efficiently, and far less pollution and waste are produced than in conventional cities. Emphasis is on pollution prevention, reuse, recycling, and efficient use of energy and matter resources. Per capita solid waste is greatly reduced, and 60% of what is produced is recycled, composted, or reused. An ecocity takes advantage of locally available energy sources and requires that all buildings, vehicles, and appliances meet high energy-efficiency standards.

Trees and plants adapted to the local climate and soils are planted throughout the ecocity to provide shade and beauty, to reduce pollution and noise, and to supply habitats for wildlife. Abandoned lots and polluted creeks are cleaned up and restored. Nearby forests, grasslands, wetlands, and farms are preserved instead of being devoured by urban sprawl. Much of an ecocity's food comes from nearby organic farms, solar greenhouses, community gardens, and gardens on rooftops and in yards and window boxes.

An ecocity is a people-oriented city—not a car-oriented city. Its residents are able to walk or bike to most places, including work, and to take low-polluting mass transit. It is designed, retrofitted, and managed to provide a sense of community built around cooperative and vibrant neighborhoods.

The ecocity concept is not a futuristic dream. The citizens and elected officials of Davis, California—a city of about 40,000 people about 130 kilometers (80 miles) northeast of San Francisco—committed themselves in the early 1970s to making it an ecologically sustainable city.

City building codes encourage the use of solar energy for water and space heating. All new homes must meet high standards of energy efficiency, and when an existing home changes hands, the buyer must bring it up to the energy conservation standards for new homes. In Davis's Village Homes—America's first solar neighborhood— houses are heated by solar energy (Figure 6-24). They face into a common open space reserved for people and bicycles; cars are restricted to streets, which are located only on the periphery of the development. The neighborhood also has orchards, vineyards, and a large community garden. Since 1975 the city has cut its use of energy for heating and cooling in half.

Davis has a solar power plant; some of the electricity it produces is sold to the regional utility company. Eventually the city plans to generate all of its own electricity.

The city discourages the use of automobiles and encourages the use of bicycles by closing some streets to automobiles, by building bike lanes on major streets, and by building bicycle paths. Any new housing tract must have a separate bike lane, and some city employees are given bikes. As a result, 28,000 bicycles account for 40% of all in-city transportation, and less land is needed for parking spaces. This heavy dependence on the bicycle is aided by the city's warm climate and flat terrain.

Davis limits the type and rate of its growth, and it maintains a mix

---

problems and save money in the long run. Finally, it's difficult to get municipalities in the same general area to cooperate in planning efforts. Thus an ecologically sound development plan in one area may be disrupted by unsound development in nearby areas.

Making Urban Areas More Livable and Sustainable Since most people around the world now live, or will live, in urban areas, improving the quality of urban life is an urgent priority—and a few cities have started to do it (Solutions, above). Here are some ways various analysts have proposed to make urban areas more sustainable:

Economic Development and Population Regulation

- Reduce population growth rates (Section 6-5).

- Reduce the flow of people from rural to urban areas by increasing investments and social services in rural areas and by not giving higher food, energy, and other subsidies to urban dwellers than to rural dwellers.

- Recognize that increased urbanization and urban density is better than spreading people out over the countryside, which would destroy more of the planet's biodiversity. The primary problem is not urbanization, but our failure to make cities more sustainable and livable.

- Maintain employment and plug dollar drains from local economies by setting up "buy local" programs, greatly improving energy efficiency (p. 40), and instituting extensive recycling, reuse, and pollution prevention programs.

Q: What greenhouse gas is being emitted to the atmosphere in the largest quantity from human activities?

Figure 6-24 Solar home in Village Homes in Davis, California.

of homes for people with low, medium, and high incomes. Development of the fertile farmland surrounding the city for residential or commercial use is restricted.

Davis and other cities—San Jose, California; Arcosanti, Arizona; Cerro Gordo, Oregon; Osage, Iowa (p. 40); Horsen, Denmark; Tuggelite, Sweden; and Tapiola, Finland, for

example—are blazing the trail by trying to make cities more sustainable and enjoyable places to live. What is being done to make the area where you live more sustainable?

Land Use and Maintenance

- Encourage well-planned growth to conserve energy, water, land, and wildlife resources; to arrest destructive sprawl; to create more diverse and socially integrated communities; to promote smaller, more affordable housing; and to take back urban areas from cars.

- Repair, modernize, and improve the resource-use efficiency of existing urban infrastructures (streets, bridges, housing, water and sewer systems, and energy delivery systems).

- Stimulate the development of ecologically and economically sustainable new towns.

- Rely on comprehensive, regional ecological land-use planning and control.

- Adopt ecocity rezoning strategies to reshape cities, shrink urban sprawl, and establish ecologically healthy town and city centers.

- Give squatters legal title to land they have lived on, and provide them with support and low-cost loans to improve their communities.

- Plant lawns with low-maintenance ground cover vegetation and public areas with native plants to lessen the use of water, fertilizer, and pesticides.

- Plant large numbers of trees on unused lots and along streets to reduce air pollution and noise and to provide recreational areas and wildlife habitats.

- Establish greenbelts of undeveloped forestland and open space within and around urban areas

A: Carbon dioxide (CO_2)

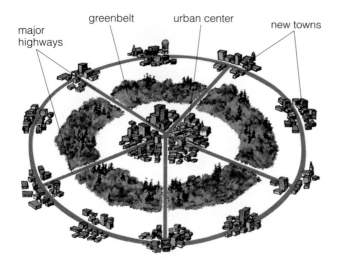

major highways greenbelt urban center new towns

Figure 6-25 A greenbelt around a large city. This arrangement can control urban growth and provide open space for recreation and other nondestructive uses. Satellite towns are sometimes built outside the belt. Rail systems transport people around the periphery or into the central city.

(Figure 6-25), and preserve nearby wetlands and agricultural land.

- Restore degraded natural areas and farm lands everywhere feasible. At the same time, protect and expand existing natural reserves and build wildlife corridors between such reserves.

Transportation

- Discourage excessive dependence on motor vehicles within urban areas. Do this by: providing efficient bus (traveling in express lanes) and trolley service; building bike lanes; charging single-occupant vehicles higher fees for tolls and parking; abolishing tax-free status for company cars; giving businesses tax write-offs for providing employees with transit passes instead of free parking spaces; establishing express lanes solely for vehicles with three or more people; providing tax incentives for individuals and businesses using car and van pooling or walking or cycling to work; and establishing car-free zones in downtown areas, with mass transit systems carrying people to and from the edges of these zones.

- End automobile subsidies. Make automobile owners pay for their environmental and harmful social costs directly (up front) instead of indirectly (in higher income, sales, and other taxes and higher insurance and health costs).

- Raise gasoline taxes sharply (with tax relief for the poor and lower middle class) and use at least half of the revenue for mass transit, bike paths,

pedestrian paths, and improving the energy efficiency of motor vehicles.

Improving Energy Efficiency

- Get more energy from locally available renewable resources (Chapter 18).

- Require all new cars to have fuel efficiencies of at least 17 kilometers per liter (40 miles per gallon) by 2000 and 21 kilometers per liter (50 miles per gallon) by 2010.

- Give tax rebates to businesses and individuals who use gas-sipping vehicles, and impose heavy taxes on those using gas guzzlers.

- Enact building codes that require new and existing buildings to be energy-efficient and responsive to climate.

- Retrofit public buildings to obtain all or most of their energy from renewable sources.

Water and Food

- Encourage water conservation by installing water meters in all buildings and raising the price of water to reflect its true cost.

- Establish small neighborhood water-recycling plants.

- Enact building codes that require water conservation in new and existing buildings and businesses.

- Grow food (with emphasis on sustainable organic methods) in abandoned lots, community garden plots, small fruit-tree orchards, rooftop gardens and greenhouses, apartment window boxes, school yards, and solar-heated fish ponds and tanks, and on some of the land in greenbelts.

Pollution and Wastes

- Withdraw government subsidies from—and impose taxes on—industries that produce large quantities of pollution and that use large amounts of water or energy.

- Give tax breaks and other economic incentives to businesses that recycle and reuse resources and that emphasize pollution prevention.

- Enact and enforce strict noise control laws to reduce stress from rising levels of urban noise.

- Establish urban composting centers to convert yard and food wastes into soil conditioner and to convert the effluents and sludge from sewage treatment plants into fertilizer for use on parks, roadsides, flower gardens, and forests.

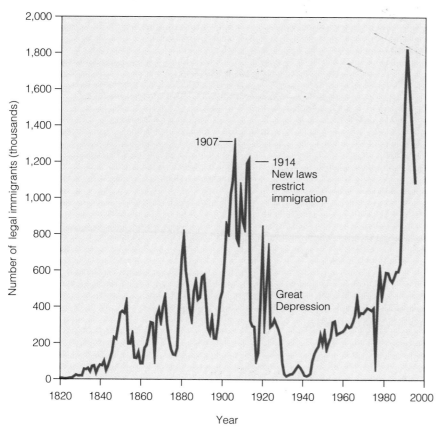

Figure 6-26 Legal immigration to the United States, 1820–1993. The much higher rates since 1989 represent a combination of normal immigration levels of around 700,000 per year, plus illegal immigrants who had been living in the country for years and were granted legal status under the Immigration Reform and Control Act of 1986. This explains the large increase in immigration in 1991, when this new policy was implemented. (Data from U.S. Immigration and Naturalization Service)

- Recycle or reuse at least 60% (and ideally 80%) of urban solid waste and some types of hazardous waste instead of burying it or burning it. Because solid wastes are concentrated in cities, collection costs are lower—and recycling and reuse are more efficient—than in less densely populated areas.

6-5 SOLUTIONS: INFLUENCING POPULATION SIZE

A society can influence the size and rate of growth or decline of its population by encouraging a change in any of the three basic demographic variables: immigration, births, and deaths. In the following sections, we will consider the first two of these variables.

Controlling Migration   Only a few countries—chiefly Canada, Australia, and the United States (Figure 6-26)—allow large annual increases in population from immigration, and some countries encourage emigration to reduce population pressures.

Between 1960 and 1993, the annual number of legal immigrants into the United States rose from 265,000 to about 1 million—twice as many legal immigrants as into all other countries combined. Each year

200,000–500,000 more people enter illegally, most from Mexico and other Latin American countries. This means that in 1993 legal and illegal immigration accounted for 37–43% of the country's population growth. These figures do not include refugees, who are admitted under other regulations, and the estimated 3.2 million illegal immigrants living in the United States. Soon, immigration is expected to be the primary factor increasing the population of the United States (Figure 6-27).

Between 1820 and 1960 most legal immigrants came from Europe. Since then most have come from Asia and Latin America. Almost two-thirds of new U.S. immigrants live in just five states: California, New York, Texas, Florida, and Illinois. By 2050 almost half of the U.S. population will be Spanish-speaking (up from 10% from 1993). At current immigration rates and projected total fertility rates, the U.S. population will increase from 258 million in 1993 to more than 400 million by 2050. Some demographers and environmentalists call for an annual ceiling of no more than 450,000 people for all categories of legal immigration, including refugees. They argue that this will allow the United States to reach ZPG sooner and reduce the country's enormous environmental impact from consumption overpopulation (Figure 1-13). Others oppose such limits, arguing that it would diminish the

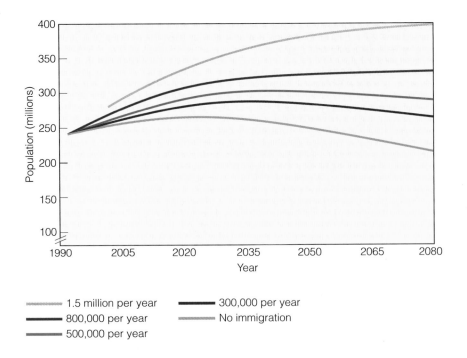

Figure 6-27 Immigration and projected population growth of the United States, 1990–2080. These projections assume that fertility levels will remain around 1.9 children per woman during this period. These projections may be too low, because traditionally most immigrants have tended to have large families. (Data from U.S. Bureau of the Census and Population Reference Bureau)

historical role of the United States as a place of opportunity for the world's poor and oppressed.

Reducing Births: A Controversial Issue

Because raising the death rate is not ethically acceptable, lowering the birth rate is the focus of most efforts to slow population growth. Today about 93% of the world's population and 91% of the people in LDCs live in countries with fertility reduction programs. The funding for and effectiveness of these programs vary widely from country to country. Few governments spend more than 1% of the national budget on them.

There is intense controversy over whether the earth is overpopulated and over measures, if any, used to reduce population growth. To some the planet is already overpopulated (Figures 1-13 and 6-28). But others project that if everyone existed at a minimum survival level, the earth could support 20–48 billion people. Other analysts believe the planet could support 7–12 billion people at a decent standard of living if land and food were distributed more equally. However, even if such optimistic estimates are technologically and environmentally feasible, many analysts doubt that our social and political structures can adapt to such a crowded and stressful world.

Others believe that asking how many people the world can support is the wrong question—like asking how many cigarettes one can smoke before getting lung cancer. Instead, they say, we should be asking what the *optimum sustainable population* of Earth might be. Such an optimum level would allow most, if not all, people to live with reasonable comfort and freedom without impairing the ability of the planet to sustain future generations. Some consider it a meaningless question; some put it at 20 billion, others at 8 billion, and others at a level below today's population size.

Those who disagree with the view that the earth is overpopulated point out that the average life span of the world's 5.5 billion people is longer than at any time in the past. Things are getting better, not worse, for many of the world's people. Those holding this view say that talk of a population crash is alarmist, that the world can support billions more people, and that people are the world's most valuable resource for solving the problems we face (Guest Essay, p. 20). Having more people increases economic productivity by creating and applying new knowledge, and people stimulate economic growth by becoming consumers.

Some people believe that if population declines in the MDCs, their economies will suffer and their power will decrease. To these analysts the primary cause of poverty is not population growth but the lack of free and productive economic systems in LDCs. And they argue that without more babies, MDCs with declining populations will face a shortage of workers, taxpayers, scientists and engineers, consumers, and soldiers needed to maintain healthy economic growth, national security, and global power and influence (Case Study, p. 139). They also contend that the aging societies in MDCs will be less innovative and dynamic. These analysts urge the governments of the United States and other MDCs to prevent this by giving tax breaks and other economic incentives to couples who have more than two children.

Q: How many trees per person would have to be planted each year to absorb the CO_2 we put into the atmosphere?

State of the World　　　　　　　　　　**Material Standard of Living**

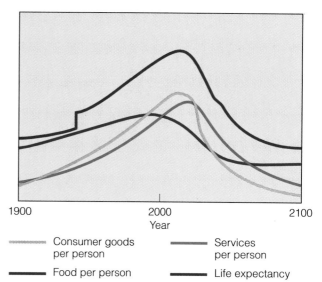

Industrial output	Population
Food	Pollution
Resources	

Consumer goods per person	Services per person
Food per person	Life expectancy

Figure 6-28 Computer-model scenario projecting what might happen if the world's population and economy continue growing exponentially at 1990 levels, assuming no major policy changes or technological innovations. This scenario projects that the world has already overshot some of its limits and that if current trends continue unchanged, we face global economic and environmental collapse sometime in the next century. (Used by permission from Donella Meadows et al., *Beyond the Limits: Confronting Global Collapse, Envisioning a Sustainable Future.* Post Mills, Vt.: Chelsea Green Publishing Co., 1992)

Some people view population regulation as a violation of their deep religious beliefs, whereas others see it as an intrusion into their privacy and personal freedom. They believe that all people should be free to have as many children as they want. And some members of minorities regard population control as a form of genocide to keep their numbers and power from rising.

Proponents of slowing and eventually stopping population growth point out that currently we are not providing adequate basic necessities for one out of five people on Earth. They also see people overpopulation in LDCs and consumption overpopulation in MDCs (Figure 1-13) as threats to Earth's life-support systems. Although they concede that population growth is not the only cause of our environmental and resource problems (Figure 6-28), they contend that adding several hundred million more people in MDCs and several billion more in LDCs will intensify many environmental and social problems by increasing resource use and waste, increasing environmental degradation and pollution, increasing the threat of climate change, and reducing biodiversity, all of which could lead to economic and political chaos. They call for drastic changes to prevent accelerating environmental decline (Figure 6-29).

Those favoring population regulation point out that technological innovation, not sheer numbers of people, is the key to military and economic power. Otherwise, England, Germany, Japan, and Taiwan, with fairly small populations, should have little global economic and military power, and China and India should rule the world.

Proponents of population regulation believe that the United States and other MDCs, rather than encouraging births, should establish an official goal of stabilizing their populations by 2025 and then begin a gradual population decline. Proponents also believe that MDCs have a better chance of influencing LDCs to reduce their population growth more rapidly if the more affluent countries officially recognize the need to stabilize their populations.

These analysts believe that people should have the freedom to produce as many children as they want only if this does not reduce the quality of other people's lives now and in the future—by impairing the ability of the earth to sustain life. They point out that limiting an individual's freedom in order to protect the freedom of other individuals is the basis of most laws in modern societies.

Reducing Births Through Economic Development Demographers have examined the birth and death rates of western European countries that industrialized during the nineteenth century, and from these data they developed a hypothesis of population

A: 1,000 (4,500 per American because of their larger input)

State of the World   **Material Standard of Living**

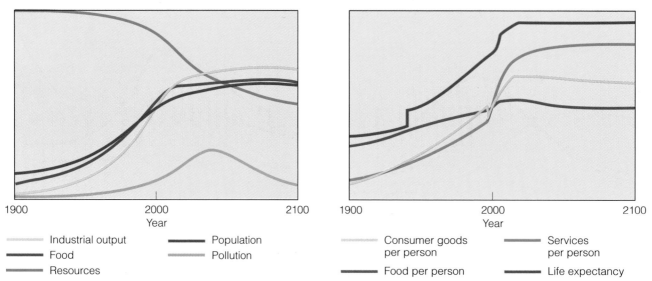

Industrial output   Population
Food   Pollution
Resources

Consumer goods per person   Services per person
Food per person   Life expectancy

Figure 6-29 Computer model scenario projecting how we can avoid overshoot and collapse and make a fairly smooth transition to a sustainable future. This scenario assumes that **(1)** technology allows us to double supplies of nonrenewable resources, double crop and timber yields, cut soil erosion in half, and double the efficiency of resource use within 20 years; **(2)** 100% effective birth control is made available to everyone by 1995; **(3)** no couple has more than two children beginning in 1995; and **(4)** industrial output per capita is stabilized at 1990 levels. Another computer run projects that waiting until 2015 to implement these changes would lead to collapse and overshoot sometime around 2075, followed by a transition to sustainability by 2100. (Used by permission from Donella Meadows et al., *Beyond the Limits: Confronting Global Collapse, Envisioning a Sustainable Future.* Post Mills, Vt.: Chelsea Green Publishing Co., 1992)

change known as the **demographic transition:** As countries become industrialized, first their death rates and then their birth rates decline.

According to this hypothesis, the transition takes place in four distinct phases (Figure 6-30). In the *preindustrial stage* harsh living conditions lead to a high birth rate (to compensate for high infant mortality) and a high death rate. Thus there is little population growth.

In the *transitional stage* industrialization begins, food production rises, and health care improves. Death rates drop, but birth rates remain high, so the population grows rapidly (typically 2.5–3% a year).

In the *industrial stage* industrialization is widespread. The birth rate drops and eventually approaches the death rate. Reasons for this include better access to birth control, drops in the infant mortality rate, more job opportunities for women, and the high costs of raising children who don't enter the workforce until after high school or college. Population growth continues, but at a slower and perhaps fluctuating rate, depending on economic conditions. Most MDCs are now in this third phase.

In the *postindustrial stage* the birth rate declines even further to equal the death rate, thus reaching zero population growth. Then the birth rate falls below the

death rate, and total population size slowly decreases. Emphasis shifts from unsustainable to sustainable forms of economic development.

In most LDCs today death rates have fallen much more than birth rates (Figure 6-8). In other words, these LDCs, mostly in Southeast Asia, Africa, and Latin America, are still in the transitional stage (Figure 6-30), halfway up the economic ladder, with high population growth rates. Some economists believe that LDCs will make the demographic transition over the next few decades without increased family-planning efforts. Many population analysts fear that the rapid population growth in many LDCs will outstrip economic growth and will overwhelm local life-support systems—causing many of these countries to be caught in a *demographic trap*—something that is currently happening in a number of LDCs, especially in Africa.

Furthermore, some of the conditions that allowed the MDCs to develop are not available to today's LDCs. Even with large and growing populations, many LDCs do not have enough skilled workers to produce the high-technology products needed to compete in today's economic environment. Most low- and middle-income LDCs also lack the capital and resources needed for rapid economic development. Also, the amount of money being given or loaned to LDCs—

Q: How long do CFCs stay in the atmosphere?

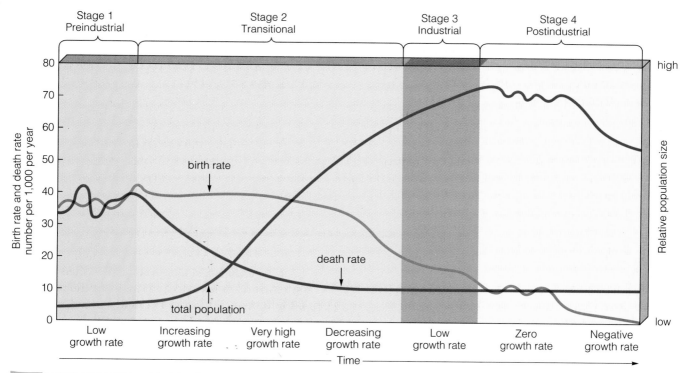

Figure 6-30 Generalized model of the demographic transition.

struggling under tremendous debt burdens—has been decreasing since 1980. Indeed, since the mid-1980s the LDCs have paid MDCs about $50 billion a year more (mostly in debt interest) than they have received from these countries.

Reducing Births Through Family Planning

Family planning programs provide educational and clinical services that help couples choose how many children to have and when to have them. Such programs vary from culture to culture, but most provide information on birth spacing, birth control (Figure 6-7), and prenatal care—and distribute contraceptives as well. In some cases, they also perform abortions and sterilizations, often without charge or at low rates.

Family planning has been an important factor both in increasing the percentage of married women in LDCs using contraception (from 10% in the 1960s to 54% in 1993) and also in lowering TFRs (total fertility rates) in LDCs (from 6 in 1960 to 3.7 in 1993). Family planning saves a society money by reducing the need for various social services. It also has health benefits. In LDCs, for example, about 1 million women die each year from pregnancy-related causes. Half of these deaths could be prevented by effective family-planning and health care programs.

The effectiveness of family planning has varied depending on program design and funding. It has been a significant factor in reducing birth and fertility rates in populous countries such as China, Indonesia, and Brazil. Family planning has also played a major role in reducing population growth in Japan (Case Study, p. 139), Thailand (p. 130), Mexico, South Korea, Taiwan, and several other countries with moderate-to-small populations. These successful programs based on committed leadership, local implementation, and wide availability of contraceptives show that population rates can be decreased significantly within two to three decades.

Family planning, however, has had moderate-to-poor results in more populous LDCs such as India, Egypt, Pakistan, and Nigeria. Results have also been poor in 79 less populous LDCs—especially in Africa and Latin America—where population growth rates are usually very high (Figure 6-3).

Family planning could be provided in LDCs to all couples who want it for about $10 billion a year—the equivalent of less than four days' worth of worldwide military expenditures. If MDCs provided two-thirds of the $10 billion, each person in the MDCs would spend only $7.70 a year to help reduce world population by 2.7 billion, by shrinking average family size from 3.7 to 2.1 children. According to biologist Paul Erhlich (Guest Essay, p. 22), "Anyone who is fighting providing people with contraception and getting family size down is fighting to get millions or hundreds of million or billions to die early, and in very nasty ways, in the not-too-distant future."

A: 65–110 years depending on the type

Using Economic Rewards and Penalties to Reduce Births Some population experts argue that family planning, even coupled with economic development, cannot lower birth and fertility rates fast enough to avoid a sharp rise in death rates in many LDCs. The main reason for this is that most couples in LDCs want three or four children—well above the 2.1 fertility rate needed to bring about eventual population stabilization.

These analysts believe that we must go beyond family planning and offer economic rewards and penalties to help slow population growth. About 20 countries offer small payments to individuals who agree to use contraceptives or to be sterilized. They also pay doctors and family-planning workers for each sterilization they perform and each IUD they insert. In India, for example, a person receives about $15 for being sterilized, the equivalent of about two weeks' pay for an agricultural worker. Such payments are most likely to attract people who already have all the children they want, however.

Some countries, such as China (Section 6-6), penalize couples who have more than a certain number of children—usually one or two—by raising their taxes, charging other fees, or not allowing income tax deductions for a couple's third child (as in Singapore, Hong Kong, Ghana, and Malaysia). Families who have more children than the prescribed limit may also lose health care benefits, food allotments, and job choice. Such economic penalties can be psychologically coercive for the poor. Programs that withhold food or increase the cost of raising children punish innocent children for the actions of their parents.

Experience has shown that economic rewards and penalties designed to reduce fertility work best if they nudge rather than push people to have fewer children; reinforce existing customs and trends toward smaller families; do not penalize people who produced large families before the programs were established; and increase a poor family's income or land.

Once population growth is out of control, a country may be forced to use coercive methods to prevent mass starvation and hardship. This is what China has had to do (Section 6-6).

Reducing Births by Empowering Women

Research has shown that women tend to have fewer children when they have access to education and to jobs outside the home, and when they live in societies where their individual rights are not suppressed. Today women do almost all of the world's domestic work and child care. They also do more than half the work associated with growing food, gathering fuelwood, and hauling water. Women also provide more health care with little or no pay than all the world's organized health services put together. As one Brazilian woman put it, "For poor women the only holiday is when you are asleep."

The worldwide economic value of women's domestic work is estimated at $4 trillion annually. This unpaid work is not included in the GNP of countries, so the central role of women in an economy is both unrecognized and unrewarded.

Despite their vital economic and social contributions, most women in LDCs don't have a legal right to own land or to borrow money to increase agricultural productivity. Although women work two-thirds of all hours worked in the world, they get only one-tenth of the world's income and own a mere 1% of the world's land. In some LDCs in which male children are strongly favored, young girls are sometimes sold as work slaves or sex slaves or are even killed. Furthermore, women make up about 60% of the more than 900 million adults who can neither read nor write. And women suffer the most malnutrition, because men and children are usually fed first where food supplies are limited.

Giving women everywhere the opportunity to become educated and to work at paid jobs outside the home should be done not only as a way to slow population growth but more importantly as a way to promote human rights and freedom. However, this will require some major social changes that will be difficult to bring about in male-dominated societies.

6-6 CASE STUDIES: POPULATION CONTROL IN INDIA AND CHINA

India India (Figure 6-13) started the world's first national family-planning program in 1952, when its population was nearly 400 million. In 1993, after 41 years of population control efforts, India was the world's second most populous country, with a population of 897 million.

In 1952 India was adding 5 million people to its population each year. In 1993 it added 18 million—49,300 more mouths to feed each day. Future population growth is fueled by the 36% of India's population under age 15. India's population is projected to reach 1.4 billion by 2025, and possibly 1.9 billion before leveling off early in the twenty-second century.

India's people are among the poorest in the world. The average income per person is about $330 a year, and for at least one-third of the population it is less than $100 a year—27¢ a day. To add to the problem, nearly half of India's labor force is unemployed or can find only occasional work. Almost one-third of the present population goes hungry. Life expectancy is

Q: What is the largest source of CFC emission in the United States?

only 59 years, and the infant mortality rate is 91 deaths per 1,000 live births.

Some analysts fear that India's already serious hunger and malnutrition problems will increase as its population continues to grow rapidly. About 40% of India's cropland is degraded as a result of soil erosion, waterlogging, salinization, overgrazing, and deforestation; and roughly 80% of the country's land is subject to repeated droughts—often lasting two to five years.

Without its long-standing family-planning program, India's numbers would be growing even faster. However, the results of the program have been disappointing. Factors contributing to this failure have been poor planning, bureaucratic inefficiency, the low status of women (despite constitutional guarantees of equality), extreme poverty, and a lack of administrative and financial support.

For years the government has provided information about the advantages of small families; yet, Indian women still have an average of 3.9 children because most couples believe they need many children as a source of cheap labor and old-age survival insurance. Almost one-third of Indian children die before age 5, reinforcing that belief. And although 90% of Indian couples know of at least one method of birth control, only 45% actually use one. Finally, many social and cultural norms favor large families—as does the strong preference for male children, which means that some couples will keep having children until they produce one or more boys.

China   Since 1970, China (Figure 6-13) has made impressive efforts to feed its people and bring its population growth under control. Between 1972 and 1993 China achieved a remarkable drop in its crude birth rate, from 32 to 18 per 1,000 people; and its total fertility rate dropped from 5.7 to 1.9 children per woman. Since 1985 China's infant mortality rate has been almost one-half the rate in India. Life expectancy in China is 71 years—12 years higher than in India. China's average per capita income of $370 is similar to that in India.

To achieve a sharp drop in fertility, China has established the most extensive, intrusive, and strict population control program in the world. Couples are strongly urged to postpone the age at which they marry and to have no more than one child. Married couples have ready access to free sterilization, contraceptives, and abortion. Paramedics and mobile units ensure access even in rural areas.

Couples who pledge not to have more than one child are given extra food, larger pensions, better housing, free medical care, and salary bonuses; their child will be provided with free school tuition and with pref-

erential treatment in employment when he or she enters the job market. Those who break the pledge lose all the benefits. All leaders are expected to set an example by limiting their own family size. The result is that 83% of married women in China are using contraception compared to only 42% in other LDCs.

These are drastic measures, but government officials realized in the 1960s that the only alternative to strict population control was mass starvation. China is a dictatorship and thus, unlike India, has been able to impose a unified population policy from the top down. Moreover, Chinese society is fairly homogeneous and has a widespread common written language. India, by contrast, has over 1,600 languages and dialects and numerous religions, which makes it more difficult to educate people about family planning and to institute population regulation policies.

Most countries do not want to use the coercive elements of China's program. Coercion is not only incompatible with democratic values and notions of basic human rights but ineffective in the long run because sooner or later people resist being coerced. Other parts of this program, however, could be used in many LDCs. Especially useful is the practice of localizing the program, rather than asking the people to go to distant centers. Perhaps the best lesson that other countries can learn from China's experience is not to wait to curb population growth until the choice is between mass starvation and coercive measures that severely restrict human freedom.

Cutting Global Population Growth   Birth rates and death rates are coming down, but death rates have fallen more sharply than birth rates, especially in LDCs (Figure 6-8). If this trend continues, one of two things will probably happen during your lifetime: **(1)** the number of people on Earth will at least double and perhaps almost triple (Figure 1-1), or **(2)** the world will experience an unprecedented population crash, with hundreds of millions of people—perhaps billions— dying prematurely (Figures 5-33b and 6-29).

Lester Brown, president of the Worldwatch Institute, urges world leaders to adopt a goal of cutting world population growth in half during the 1990s by reducing the average global birth rate from 26 to 18 per 1,000 people. The experience of countries such as Japan (Case Study, p. 139), Thailand (p. 130) and China indicate that this could be achieved.

We need the size of population in which human beings can fulfill their potentialities; in my opinion we are already over-populated from that point of view; not just in places like India and China and Puerto Rico, but also in the United States and Western Europe.

Geoorge Wald (Nobel Laureate, Biology)

A: Leaky air conditioners in motor vehicles (76%)

Moral Implications of Cultural Carrying Capacity

Garrett Hardin

GUEST ESSAY

As longtime professor of human ecology at the University of California at Santa Barbara, Garrett Hardin made important contributions to the joining of ethics and biology. He has raised hard ethical questions, sometimes taken unpopular stands, and forced people to think deeply about environmental problems and their possible solutions. He is best known for his 1968 essay "The Tragedy of the Commons," which has had a significant impact on the disciplines of economics and political science, and on the management of potentially renewable resources. His many books include Promethean Ethics; *and* Filters Against Folly: How to Survive Despite Economists, Ecologists, and the Merely Eloquent; *and* Living Within Limits *(see Further Readings).*

For many years Angel Island in San Francisco Bay was plagued with too many deer. A few animals transplanted there early in this century lacked predators and rapidly increased to nearly 300 deer—far beyond the carrying capacity of the island. Scrawny, underfed animals tugged at the heartstrings of Californians, who carried extra food for them from the mainland to the island.

Such charity worsened the plight of the deer. Excess animals trampled the soil, stripped the bark from small trees, and destroyed seedlings of all kinds. The net effect was to lower the carrying capacity, year by year, as the deer continued to multiply in a deteriorating habitat.

State game managers proposed that the excess deer be shot by skilled hunters. "How cruel!" some people protested. Then the managers proposed that coyotes be imported to the island. Though not big enough to kill adult deer, coyotes can kill defenseless young fawns, thus reducing the size of the herd. However, the Society for the Prevention of Cruelty to Animals was adamantly opposed to such human introduction of predators.

In the end, it was agreed that some deer would be exported to some other area suitable for deer life. A total of 203 animals were caught and trucked many miles away. From the fate of a sample of animals fitted with radio collars, it was estimated that 85% of the transported deer died within a year (most of them within two months) from various causes: predation by coyotes, bobcats, and domestic dogs; shooting by poachers and legal hunters; and being run over by automobiles.

The net cost (in 1982 dollars) for relocating each animal surviving for a year was $2,876. The state refused to continue financing the program, and no volunteers stepped forward to pay future bills.

Angel Island is a microcosm of the planet as a whole. Organisms reproduce exponentially, but the environment doesn't increase at all. The moral is a simple ecological commandment: *Thou shalt not transgress the carrying capacity.*

Now let's look at the human situation. A competent physicist has placed the human carrying capacity of the globe at 50 billion—about 10 times the present world population. Before you are tempted to urge women to have more babies, however, consider what Robert Malthus said nearly 200 years ago: "There should be no more people in a country than could enjoy daily a glass of wine and piece of beef for dinner."

A diet of grain or bread and water is symbolic of minimum living standards; wine and beef are symbolic of higher living standards that make greater demands on the environment. When land that could produce plants for direct human consumption is used to grow grapes for wine or corn for cattle, fewer calories get to the human population. Since carrying capacity is defined as the *maximum* number of animals (humans) an area can support, using part of the area to support such cultural luxuries as wine and beef reduces the carrying capacity. This reduced capacity is called the *cultural carrying capacity.* Cultural carrying capacity is always less than simple carrying capacity.

Energy is the common coin in which all competing demands on the environment can be measured. Energy saved by giving up a luxury can be used to produce more bread and support more people. We could increase the simple carrying capacity of the earth by giving up any (or all) of the following "luxuries": street lighting, vacations, private cars, air conditioning, and artistic performances of all sorts—drama, dance, music, and lectures. Since the heating of buildings is not as efficient as multiple layers of clothing, space heating could be forbidden as a luxury.

Is that all? By no means: To come closer to home, look at this book. Its production and distribution consumes a great deal of energy. In fact, the energy bill for higher education as a whole is very high (which is one reason tuition costs so much). By giving up all education beyond the eighth grade, we could free enough energy to sustain millions of additional human lives.

At this point a skeptic might well ask: "Does God give a prize for the maximum population?" From this brief

Q: If all ozone-depleting substances were banned now, when might the ozone layer to return to 1985 levels?

analysis we can see that there are two choices. We can maximize the number of human beings living at the lowest possible level of comfort, or we can try to optimize the quality of life for a much smaller population.

What is the carrying capacity of the earth? is a scientific question. Scientifically it may be possible to support 50 billion people at a "bread" level. Is that what we want? The question What is the cultural carrying capacity? requires that we debate questions of value, about which opinions differ.

An even greater difficulty must be faced. So far, we have been treating the capacity question as a *global* question, as if there were a global sovereignty to enforce a solution on all people. However, there is no global sovereignty ("one world"), nor is there any prospect of one in the foreseeable future. We must make do with nearly 200 national sovereignties. That means, as concerns the capacity problem, that we must ask how nations are to coexist in a finite global environment if different sovereignties adopt different standards of living.

Consider a redwood forest which produces no human food. Protected in a park, the trees do not even produce lumber for houses. Because people have to travel many kilometers to visit it, the forest is a net loss in the national energy budget. However, those who are fortunate enough to wander quietly through the cathedral-like aisles of soaring trees report that the forest does something precious for the human spirit.

Now comes an appeal from a distant land where millions are starving because their population has overshot the carrying capacity. We are asked to save lives by sending food. So long as we have surpluses, we may safely indulge in the pleasures of philanthropy. But the typical population in such poor countries increases by 2.3% a year—*or more*; that is, the country's population doubles every 30 years—*or less*. After we have run out of our surpluses, then what?

A spokesperson for the needy makes a proposal: "If you would only cut down your redwood forests, you could use the lumber to build houses and then grow potatoes on the land, shipping the food to us. Since we are all passengers together on Spaceship Earth, are you not duty bound to do so? Which is more precious, trees or human beings?"

This last question may sound ethically compelling, but let's look at the consequences of assigning a preemptive and supreme value to human lives. At least 2 billion people in the world are poorer than the 34 million legally "poor" in America, and their numbers are increasing by about 40 million per year. Unless this increase is brought to a halt, sharing food and energy on the basis of need would require the sacrifice of one amenity after another in rich countries. The final result of sharing would be complete poverty everywhere on the face of the earth to maintain the earth's simple carrying capacity. Is that the best humanity can do?

To date there has been an overwhelmingly negative reaction to all proposals to make international philanthropy conditional upon the stopping of population growth by the poor, overpopulated recipient nations. Foreign aid is governed by two apparently inflexible assumptions:

- The right to produce children is a universal, irrevocable right of every nation, no matter how hard it presses against the carrying capacity of its territory.

- When lives are in danger, the moral obligation of rich countries to save human lives is absolute and undeniable.

Considered separately, each of these two well-meaning doctrines might be defended; together, they constitute a fatal recipe. If humanity gives maximum carrying capacity precedence over problems of cultural carrying capacity, the result will be universal poverty and environmental ruin.

Or do you see an escape from this harsh dilemma?

Critical Thinking

1. What population size do you believe would allow the world's people to have good quality of life? What do you believe is the cultural carrying capacity of the United States? Should the United States have a national policy to establish this population size as soon as possible? Explain.

2. Do you support the two principles this essay lists as the basis of foreign aid to needy countries? If not, what changes would you make in the requirements for receiving such aid?

Critical Thinking

1. Why is it rational for a poor couple in India to have six or seven children? What changes might induce such a couple to think of their behavior as irrational?

2. Project what your own life may be like at ages 25, 45, and 65 on the basis of the present population age structure of the country in which you live.

3. What conditions, if any, would encourage you to rely less on the automobile? Would you regularly travel to school or work by bicycle or motor scooter, on foot, by mass transit, or by a car (or van) pool? Explain.

4. Are there physical limits to growth on Earth (Figure 6-28)? Are there social limits to growth? Explain your answers.

5. Do you believe that all U.S. high schools and colleges should have health clinics that make contraceptives available to students and provide prenatal care for pregnant teenage women? Explain. Should such services be available at the junior-high school level, when many teenagers first become sexually active? Explain.

6. a. Should the number of legal immigrants and refugees allowed into the United States each year be sharply reduced? Explain your answer.

   b. Should illegal immigration into the United States be sharply decreased? Explain. If so, how would you go about achieving this?

7. Should families in the United States be given financial incentives and be persuaded to have more children to prevent population decline? Explain.

8. Should the United States adopt an official policy to stabilize its population and reduce unnecessary resource waste and consumption as rapidly as possible? Explain.

9. Why has China been more successful than India in reducing its rate of population growth? Do you agree with China's present population control policies? Explain. What alternatives, if any, would you suggest?

*10. As a class project, evaluate land use and land-use planning by your school, draw up an improved plan based on ecological principles, and submit the plan to school officials.

*11. Make a concept map of the key ideas in this chapter using the section heads and subheads and the key terms (shown in boldface type in the chapter). See the inside front cover and Appendix 4 for information on concept maps.

7 Environmental Economics and Politics

To Grow or Not to Grow: Is That the Question?

In the American Northwest, the timber-based economy is in trouble, and recent efforts to save old-growth forests and the northern spotted owl have run into an economic buzz saw. Embattled loggers and environmentalists alike have too often framed their positions in either-or terms: "Trees or jobs?" "Owls or people?" What we really need to ask is, "How can we have trees *and* jobs?" "How can we save the owls and the forests without putting people out of work?" "What happens to jobs after all the trees are cut?"

The same types of polarized positions apply to the question of economic growth in general. On one side, some economists and investors argue that we must have unlimited economic growth to create jobs, satisfy people's economic needs and wants, clean up the environment, and help eliminate poverty. To peo-ple with this view, environmentalists put endangered species above endangered people, threaten jobs, and are against growth.

On the other hand, environmentalists and some economists argue that economic systems depend on resources and services provided by the sun and by Earth's natural processes (Figure 7-1). A healthy economy is ultimately dependent on a healthy planet. If these beliefs are correct, we must replace the economics of unlimited growth with the economics of sustainability over the next few decades.

The question then is not so much, "To grow or not to grow?" but, "How can we grow without plundering the planet?" or, "How can we grow as if Earth Matters?"

Figure 7-1 Earth, air, fire, water, and life. These basic components of Earth capital for us and other species are undervalued in the marketplace. (Greg Vaughn/Tom Stack & Associates)

When it is asked how much it will cost to protect the environment, one more question should be asked: How much will it cost our civilization if we do not?

GAYLORD NELSON

In this chapter we will seek answers to the following questions:

- What types of economic systems are found in the world, and what supports them?

- How should we measure economic growth?

- How can economics be used to improve environmental quality?

- How can we sharply reduce poverty?

- How can we shift to an Earth-sustaining economy?

- How do political decisions affect resource use and environmental quality?

- How can people bring about change?

7-1 ECONOMIC SYSTEMS AND ENVIRONMENTAL PROBLEMS

What Supports and Drives Economies?

An **economy** is a system of production, distribution, and consumption of economic goods—any material item or service that gives people satisfaction. In such a system individuals, businesses, and societies make **economic decisions** about what goods and services to produce, how to produce them, how much to produce, how to distribute them, and what to buy and sell.

The kinds of capital used in an economy to produce material goods and services, and thus to sustain economic growth, are called **economic resources**. They fall into three groups:

1. **Earth capital** or **natural resources**: goods and services produced by Earth's natural processes. They include the planet's air, water, and land; nutrients and minerals in the soil and deeper in the earth's crust; wild and domesticated plants and animals (biodiversity); and nature's dilution, waste disposal, pest control, and recycling services.

2. **Manufactured capital**: items made from Earth capital. These include tools, machinery, equipment, factory buildings, and transportation and distribution facilities.

3. **Human capital**: people's physical and mental talents. Workers sell their time and talents for wages. Managers take responsibility for combining Earth capital, manufactured capital, and workers to produce an economic good. In market-based systems entrepreneurs and investors put up the monetary capital needed to produce an economic good in the hope of making a profit on their investment.

Economic Systems and Economic Growth

There are two major types of economic systems: centrally planned and market-based.

In a **pure command economic system**, or **centrally planned economy**, all economic decisions are made by the government. This command-and-control system is based on the belief that government control and ownership of the means of production is the most efficient and equitable way to produce, use, and distribute goods and services.

In a **pure market economic system**, also known as **pure capitalism**, all economic decisions are made in *markets*, where buyers (demanders) and sellers (suppliers) of economic goods freely interact without government or other interference. All economic resources are owned by private individuals and private institutions, rather than by the government. All buying and selling is based on *pure competition*, in which no seller or buyer is powerful enough to control the supply, demand, or price of a good; and all sellers and buyers have full information about and access to the market.

Economic decisions in a pure market system are governed by interactions of demand, supply, and price. Buyers want to pay as little as possible for an economic good, and sellers want to set as high a price as possible so as to maximize profit. **Market equilibrium** occurs when the quantity supplied equals the quantity demanded, and the price is no higher than buyers are willing to pay and no lower than sellers are willing to accept (Figure 7-2). Various factors can change the demand and supply of an economic good and establish a new market equilibrium.

Economists often represent pure capitalism as a circular flow of economic goods and money between households and businesses operating essentially independently of the ecosphere (Figure 7-3). By contrast, environmentalists emphasize the dependence of this or any economic system on the ecosphere (Figure 7-4).

In a pure capitalist system a company has no legal allegiance to a particular nation, no obligation to supply any particular good or service, and no obligation to provide jobs, safe workplaces, or environmental protection. The company's only obligation is to produce the highest possible short-term economic return (profit) for the owners or stockholders, whose capital the company is using to do business.

In reality, all countries have **mixed economic systems** that fall somewhere between the pure market and pure command systems. The economic systems of countries such as China and North Korea fall toward

Q: What percentage of your body weight consists of water?

the command-and-control end of the economic spectrum, while those of countries such as the United States and Canada fall toward the market-based end of the spectrum. Most other countries fall somewhere in between.

Pure market economies don't exist because they have flaws that require government intervention in the marketplace to:

- Prevent one seller or buyer (*monopoly*) or one group of sellers or buyers (*oligopoly* or *cartel*) from dominating the market and thus controlling supply or demand and price. (The goal of almost every business lobbyist in Washington and in other capitals of market economy countries is to subvert free-market competition by getting a subsidy or tax break.)

- Provide national security, education, and other public goods

- Redistribute some income and wealth, especially to people unable to attain their basic needs

- Protect people from fraud, trespass, theft, and bodily harm

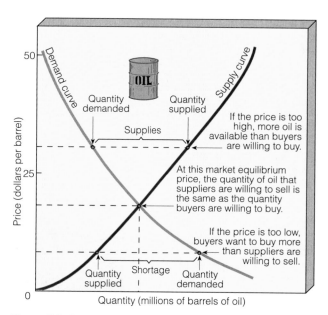

Figure 7-2 Supply, demand, and market equilibrium for gasoline in a pure market system. If price, supply, and demand are the only factors involved, the market equilibrium point occurs where the demand and supply curves intersect.

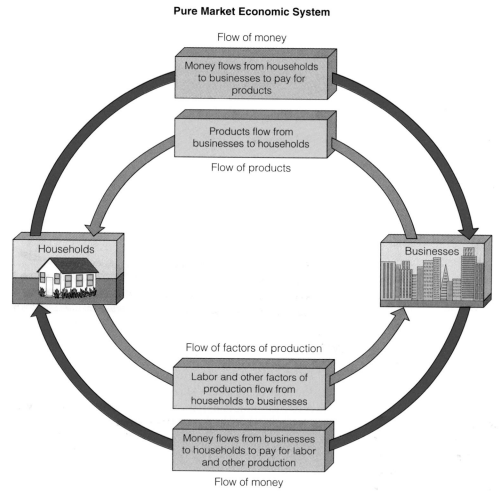

Pure Market Economic System

Figure 7-3 In a pure market economic system, economic goods and money would flow in a closed loop of households and businesses. People in households spend money to buy goods that firms produce, and firms spend money to buy factors of production (natural capital, manufactured capital, and human capital). In many economics textbooks this and other economic systems are depicted, as here, as if they were self-contained and thus not dependent on the ecosphere—a model that reinforces the idea that unlimited growth of any kind is sustainable.

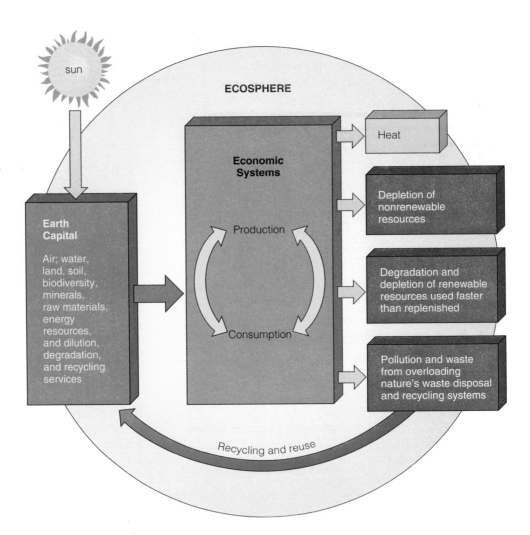

Figure 7-4 Environmentalists see all economies as artificial subsystems dependent on resources and services provided by the sun and the ecosphere. A consumer society devoted to unlimited economic growth to satisfy ever-expanding wants assumes that ecosphere resources and services are essentially infinite or that our technological cleverness will allow us to overcome any ecosphere limits.

- Protect the health and safety of workers and consumers

- Ensure economic stability by trying to control cycles of boom and bust that afflict a pure market system

- Help compensate owners for large-scale destruction of assets by floods, earthquakes, hurricanes, and other natural disasters (Section 8-3)

- Prevent or reduce pollution

- Prevent or reduce depletion of natural resources, which are undervalued in market economies

- Manage public land resources (Chapter 16)

Pure command economies don't exist either. Experience has shown that governments cannot efficiently control all economic activity from the top down.

Virtually all economies seek **economic growth**: an increase in the capacity of the economy to provide goods and services for people's final use. Such growth is accomplished by maximizing the flow of matter and energy resources (throughput) by means of population growth (more consumers), more consumption per per-

son, or both (Figures 3-12 and 7-3). Nature is seen as a "superstore" stocked with an infinite supply of goods to be used to meet the needs and ever-expanding wants of a consumer society. All economic growth is seen as good, an idea that many environmentalists and some economists challenge (p. 163 and Section 2-2).

Gross National Product: Faulty Radar

Economic growth is usually measured by the increase in a country's **gross national product (GNP)**: the market value in current dollars of all goods and services produced by an economy for final use during a year (an indicator introduced in the late 1940s). To get a clearer picture economists use the **real GNP**: the GNP adjusted for *inflation* (any increase in the average price level of final goods and services). And to show how the average person's slice of the economic pie is changing, economists use the **real GNP per capita**: the real GNP divided by the total population. If population expands faster than economic growth, the real GNP per capita falls.

We are urged to buy and consume more and more so that the GNP will rise, making the country and the world a better place for everyone (see cartoon). The

Q: How much of Earth's enormous supply of water is available to us as usable fresh water?

truth is that GNP indicators were never intended to be measures of social well-being, environmental health, or even economic health, because:

- *They hide the negative impact on humans and the rest of the ecosphere of producing many goods and services.* For example, each year in the United States the estimated $150 billion in health care expenses and other damages caused by air pollution raise the GNP and GNP per capita. So do the funeral expenses for the 150,000–350,000 Americans killed prematurely each year from air pollution in the United States. The $2.2 billion that Exxon spent partially cleaning up the oil spill from the *Exxon Valdez* tanker pushed up the GNP, as did the $1 billion spent because of the accident at the Three Mile Island nuclear power plant.

- *They don't include depletion and degradation of natural resources or Earth capital upon which all economies ultimately depend* (Solutions, p. 8 and Figure 7-4). A country can be headed toward ecological bankruptcy—exhausting its mineral resources, eroding its soils, polluting its water, cutting down its forests, destroying its wetlands and estuaries, and depleting its wildlife and fisheries—yet have a rapidly rising GNP—at least for a while until its ecological debts come due.

- *They hide or underestimate some positive effects on society.* For example, more energy-efficient light bulbs, appliances, and cars reduce electric and gasoline bills and pollution, but these register as a drop in GNP. GNP indicators also do not include the labor we put into volunteer work, the health care we give loved ones, the food we grow for ourselves, or the cooking, cleaning, and repairs we do for ourselves.

- *They tell us nothing about economic justice.* They don't tell us how resources, income, or the harmful effects of economic growth (pollution, waste dumps, land degradation) are distributed among the people in a country. UNICEF suggests that countries should be ranked not by average GNP per capita but by average income of the lowest 40% of their people. The environmental justice movement calls not only for a fairer distribution of income but also for a fairer distribution of the harmful effects of economic growth instead of concentrating these effects on the poor and on minority groups (Guest Essay, p. 367). The experience of Kerala in India shows that life quality can be improved by cooperation, education, and a more equitable distribution of land and resource wealth instead of relying on unrestricted economic growth to boost per capita income (Case Study, p. 168).

Russell Petersen, former head of the White House Council on Environmental Quality, summed up the problem with using GNP as a guideline for progress:

If we produce a million dollars worth of carcinogens, this weighs as much on the GNP scale as a million dollars worth of antibiotics. If we hire a housekeeper, this counts in the GNP, but when one's spouse manages the household, this doesn't count. Teaching counts, but learning doesn't. GNP gives no measure of the hungry, the unemployed, the sick, the ill-housed,

A: 0.003%

Kerala: Improving Life Quality Without Conventional Economic Growth

The state of Kerala in southwest India (Figure 6-13) has shown how quality of life can be improved without emphasizing economic growth. By conventional measures it is one of the world's poorest areas, with a per capita income of less than $200 (compared to $330 for India and $870 for all LDCs) and 18% unemployment. Yet in terms of life quality it has some of the highest scores among LDCs.

Life expectancy is 68 years compared with 63 years in LDCs and 59 years in India. Infant mortality is 27 per 1,000 births compared with 77 for LDCs and 91 for India. And the total fertility rate is 2.3 in Kerala compared with 3.7 for MDCs and 3.9 in India. Almost 94% of Kerala's citizens can read or write compared with less than 50% (29% for women) for the rest of India. It is the sole Indian state where women are as literate as men, and in Kerala women have far more rights than in the rest of India.

Kerala's high educational level has attracted industrial jobs, but so far not at the expense of its natural environment. It is one of the world's most densely populated places, with a population larger than California's squeezed into one-tenth the area. Yet, forest still covers 29% of its area—the third highest rate in India and well ahead of other states with half its population. Since 1960 a land reform program has meant that 90% of the people own the land on which their house stands.

Kerala also leads India in the quality of its roads, schools, hospitals, public housing, drinking water, sanitation, immunization programs, and nutrition programs for infants and pregnant and lactating women. Its 29 million people have access to free or inexpensive medical care, and all households can buy rice and certain basic commodities at subsidized prices.

These gains in life quality and social justice are the results of decades of political struggles. Many observers see culturally diverse Kerala as a shining example of how people can help themselves by choosing self-reliance and economic redistribution of local wealth (mostly by land reforms) over conventional economic growth and dependence on outside loans.

the illiterate, the oppressed, the frightened, the unhappily employed, or those who have reached the highest level of fulfillment. Furthermore, it does not measure the waste of resources, the spending of our natural

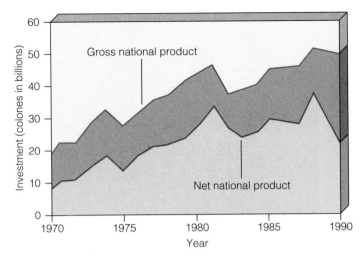

Figure 7-5 GNP and NNP for Costa Rica, 1970–90. The NNP (Net National Product) was calculated by adjusting the GNP to include depletion of the country's forests, soils, and fisheries. If depletion of coal, mineral ores, and other nonrenewable resources had been included, Costa Rica's NNP would have grown even more slowly. (Data from World Resources Institute)

capital such as oil, or the befoulment of our life support systems.

Solutions: Better Indicators Environmentalists and some economists believe that GNP indicators should be replaced or supplemented with indicators that measure the quality of life and our harmful impacts on the ecosphere. Such indicators exist, though they are not widely used. Here are three:

■ *Net economic welfare (NEW).* This approach, developed in 1972 by economists William Nordhaus and James Tobin (and used to some extent in Japan), estimates the costs of pollution and other "negative" goods and services included in the GNP, and it subtracts them from the GNP to give the NEW. Economist David Pearce estimates that pollution and natural resource degradation subtract 1–5% from the GNP of MDCs and 5–15% for LDCs, and that in general these percentages are rising. Dividing a country's NEW by its population gives the *per capita NEW.* Since 1940 the real NEW per capita in the United States has risen at about half the rate of the real GNP per person, and since 1968 the gap between these two indicators has been widening.

■ *Net national product (NNP).* This approach, developed by economist Robert Repetto and other researchers at the World Resources Institute, includes the depletion or destruction of natural resources as a factor in GNP (Figure 7-5).

Q: What is the largest global use of water withdrawn from surface or groundwater sources?

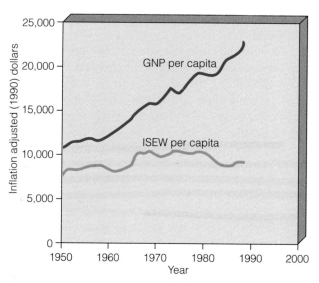

Figure 7-6 Comparison of GNP and ISEW (index of sustainable economic welfare) per person in the United States, 1950–88. After rising by 42% between 1950 and 1976, the ISEW fell 12% between 1977 and 1988. (Data from Herman E. Daly, John B. Cobb, Jr., and Clifford W. Cobb)

 Index of sustainable economic welfare (ISEW). This comprehensive indicator of well-being was developed in the 1980s by economists Herman E. Daly and John B. Cobb, Jr. It measures per capita GNP adjusted for inequalities in income distribution, depletion of nonrenewable resources, loss of wetlands, loss of farmland from soil erosion and urbanization, the cost of air and water pollution, and estimates of long-term environmental damage from ozone depletion and possible global warming. It has been falling in the United States since 1977 (Figure 7-6).

None of these new indicators is perfect, but they are much better than relying on GNP. Without such indicators we know too little about what is happening to people, the environment, and the planet's natural resource base; what needs to be done; and what types of policies work. In effect, we are trying to guide national and global economies through treacherous economic and environmental waters at ever-increasing speed using faulty radar. Because changing the indicators used by countries and the United Nations is so difficult, economists and the media should work together to widely publish several state-of-the-Earth indicators.

Connections: Internal and External Costs

All economic goods and services have both internal and external costs. For example, the price we pay for a car reflects the costs of the factory, raw materials, labor, marketing, shipping, and company and dealer profits. After we buy the car, we must pay for gasoline, maintenance, and repair. All these direct costs, paid for by the seller and the buyer of an economic good, are called **internal costs**.

Making, distributing, and using any economic good or service also involve **externalities**. These are social costs or benefits not included in the market price of an economic good or service. For example, if a car dealer builds an aesthetically pleasing showroom and grounds, that is an **external benefit** to people who enjoy the sight at no cost.

On the other hand, extracting and processing raw materials to make and propel cars disturbs land, pollutes the environment, reduces biodiversity, and harms people. These harmful effects are **external costs** passed on to workers, to the general public, and in some cases to future generations. Air pollution from cars also kills or weakens some types of trees, raising the price of lumber and paper. In addition, taxes may go up because the public demands that the government regulate the pollution and degradation associated with producing and operating motor vehicles.

You add to the external costs when you throw trash out of a car, drive a gas-guzzler (which adds more air pollution than a more efficient car), disconnect or don't maintain a car's air pollution control devices, drive with faulty brakes, or don't keep your motor tuned. Because these harmful costs aren't included in the market price, you don't connect them with the car or type of car you are driving. As a consumer and taxpayer, however, you pay these hidden costs sooner or later in the form of higher taxes, higher health costs, higher health insurance, and higher cleaning and maintenance bills.

These external costs also show up as environmental problems—loss of biodiversity (Chapters 14–17), soil erosion (Chapter 12), depletion of nonrenewable resources (Chapters 1 and 12), loss of stratospheric ozone and possible climate change (Chapter 10), air pollution (Chapter 9), water pollution (Chapter 11), hazardous waste (Chapter 13), and solid waste (Chapter 13). The health (Chapter 8) and economic costs of these problems are passed on to society at large or to future generations.

To pro-growth economists, external costs are minor defects in the flow of production and consumption in a self-contained economy (Figure 7-3); they assume that these defects can be cured from the profits made from more economic growth in a free-market economy (Guest Essay, p. 20). To environmentalists the rising number of harmful externalities is a warning sign that our economic systems are stressing the ecosphere and depleting Earth capital (Solutions, p. 8 and Figure 1-11), and that they need to be restructured.

A: Irrigation (69% of global withdrawal, 41% in the United States)

7-2 SOLUTIONS: USING ECONOMICS TO IMPROVE ENVIRONMENTAL QUALITY

Full-Cost Pricing As long as businesses get subsidies and tax breaks for extracting and using virgin resources and are not taxed for the pollutants they produce, few will volunteer to commit economic suicide by changing. Suppose you own a company and believe it's wrong to subject your workers to hazardous conditions and to pollute the environment any more than can be handled by Earth's natural processes. If you voluntarily improve safety conditions for your workers and install pollution controls but your competitors don't, your product will cost more and you will be at a competitive disadvantage. Your profits will decline, sooner or later you may go bankrupt, and your employees will lose their jobs.

One way to deal with the problem of harmful external costs is for the government to add taxes, pass laws, provide subsidies, or use other strategies that force or entice producers to include all or most of this expense in the market price of economic goods and services. Then that price would be the **full cost** of these goods and services: internal costs plus its short- and long-term external costs. Full-cost pricing involves *internalizing the external costs*. This requires government action because few companies will increase their cost of doing business unless their competitors must do so as well.

What would happen if such a policy were phased in over the next 10 to 20 years? Economic growth would be redirected. We would increase the beneficial parts of the GNP, decrease the harmful parts, increase production of beneficial goods, and raise the net economic welfare. Preventing pollution would become more profitable than cleaning it up; and waste reduction, recycling, and reuse would be more profitable than burying or burning most of the waste we produce.

We would pay more for most things because their market prices would be closer to their true costs, but everything would be "up front." External costs would no longer be hidden. We would also have the information we need to make informed economic decisions (as called for by the theory behind a true free-market economy) about the effects of our lifestyles on the planet's life-support systems.

Moreover, real market prices wouldn't always be higher, and some things might even cost less. Internalizing external costs encourages producers to find ways both to cut costs (by inventing more resource-efficient and less environmentally harmful methods of production), and to offer less harmful, Earth-sustaining (or *green*) products. Jobs would be lost in Earth-degrading businesses, but at least as many jobs—some analysts say more jobs—could be created in Earth-sustaining businesses.

As external costs are internalized, governments must reduce income and other taxes and must withdraw subsidies formerly used to hide and pay for these external costs. Otherwise, consumers will face higher market prices without tax relief—a policy guaranteed to fail.

Full-cost pricing makes so much sense you might be wondering why it's not more widely done. One reason is that many producers of harmful and wasteful goods fear they would have to charge so much that they couldn't stay in business or would have to give up government subsidies that have helped hide the external costs. And why should they want to change a system that's been so good to them? Another reason is that it's difficult to internalize external costs because it's not easy to put a price tag on every harmful effect of providing and using an economic good or service. People also disagree on the values they attach to various costs and benefits. However, making difficult choices about how resources should be used and distributed is what economics and politics are all about.

Full-cost pricing will be imperfect and controversial. Also, it cannot be applied to such things as biodiversity, wilderness, and the release of highly toxic substances into the environment—things that can't be effectively evaluated in terms of money. Such things require protective regulations. However, proponents of full-cost pricing contend that for most things it is far better than continuing the pricing system that gives too little information about the harmful and beneficial effects of the goods and services people buy. Initiating full-cost pricing won't be easy, but such a change in the way we do business can occur if enough citizens become politically active and insist that it be done.

Cost-Benefit Analysis: Panacea or Smokescreen? One of the chief tools corporations and governments use in making economic decisions is **cost-benefit analysis**. This controversial approach involves comparing the estimated short-term and long-term costs (losses) and benefits (gains) resulting from various alternatives in making economic decisions. Used properly, cost-benefit analyses can be reasonable guides for action and can indicate the cheapest way to go.

However, cost-benefit analyses can be misused and even do great harm. Here are some controversial aspects of this economic tool:

- *What discount rate should we assign to resources?* The size of the **discount rate**—an estimate of how much economic value a resource will have in the future compared with its present value—is a primary factor affecting the outcome of any cost-

Q: How much of the world's cropland is irrigated?

benefit analysis. At a zero discount rate, a stand of redwood trees worth $1 million today will still be worth $1 million 50 years from now; so there is no need to cut them down for short-term economic gain. However, at a 10% per year discount rate (normally used by most businesses and the U.S. Office of Management and Budget), the stand will be worth only $10,000 so it makes economic sense to cut them down now. Proponents of high (5–10%) discount rates argue that inflation will make the value of their earnings less in the future than now. Why then should the current generation pay higher prices and taxes (based on a low discount rate) to benefit future generations who will be better off anyway because of increased economic growth (Guest Essay, p. 20)? Environmentalists question this belief, pointing out that current economic systems are based upon depleting the natural capital that supports them (Figure 7-4). High discount rates worsen this situation by encouraging such rapid exploitation of resources for immediate payoffs that sustainable use of most natural resources is virtually impossible. They believe that unique and scarce resources should be protected by having a 0% or a negative discount rate, and that discount rates of 1–3% would make it profitable to use other resources sustainably or slowly. Thus, the choice of a discount rate is an *ethical decision* about our responsibility to future generations.

■ *Who benefits and who is harmed?* An estimated 100,000 Americans die each year from exposure to hazardous chemicals and other safety hazards at work, and an additional 400,000 are seriously injured from such exposure. In many other countries (especially LDCs) the situation is much worse. Is that a necessary or an unnecessary (and unethical) cost of doing business?

■ *Can everything be reduced to dollars and cents?* How do we put meaningful price tags on human life, good health, clean air and water, pollution accidents that are prevented, beautiful scenery, wilderness, the northern spotted owl, and the ability of natural systems to dilute and degrade some of our wastes and to replenish timber, fertile soil, and other vital potentially renewable resources? The dollar values we assign to such things will vary widely because of different assumptions, discount rates, and value judgments, leading to a wide range of projected costs and benefits. For example, monetary values assigned to a human life vary from nothing to about $7 million, but typically range from $200,000–$500,000. Faced with regulation in 1971, the oil

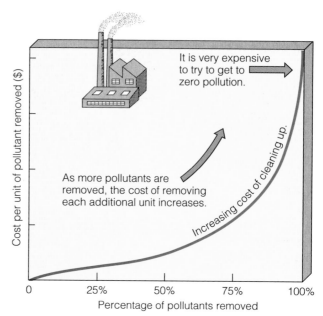

Figure 7-7 The cost of removing each additional unit of pollution rises exponentially, which explains why it is usually cheaper to prevent pollution than to clean it up.

industry estimated that phasing out leaded gasoline in the United States would cost $7 billion per year until the task was completed, although actual costs were only $150–$500 million per year.

In short, cost-benefit analyses are useful only if decision makers realize their limitations and use them as only one of several guidelines for making economic decisions.

Control or Prevention? Shouldn't our goal be zero pollution? Ideally, yes; in the real world, not necessarily. First, nature can handle some of our wastes, so long as we don't destroy, degrade, or overload these natural processes. But ideally toxic products that cannot be degraded by natural processes or that break down very slowly in the environment should not be produced and released into the environment—or should be used only in small amounts regulated by special permit.

Second, so long as we continue to rely on pollution control, we can't afford zero pollution. We can remove a certain percentage of the pollutants in air, water, or soil, but when we remove more, the cleanup cost per unit of pollutant rises sharply (Figure 7-7). Beyond a certain point, the cleanup costs will exceed the harmful effects of pollution. As a result, some businesses could go bankrupt, and some people could lose jobs, homes, and savings. If we don't go far enough, dealing with the harmful external effects of pollution will cost more than pollution reduction.

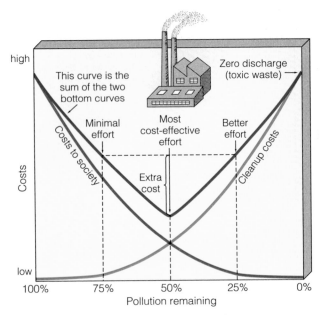

Figure 7-8 Finding the optimal level of pollution.

How do we achieve this balance? Theoretically we plot a curve of the estimated economic costs of cleaning up pollution and a curve of the estimated social (external) costs of pollution. Adding the two curves together, we get a third curve showing the total costs. At the lowest point on this third curve is the balance point or *optimal level of pollution* (Figure 7-8).

On a graph this looks neat and simple, but environmentalists and business leaders often disagree in their estimates of the harmful costs of pollution. This approach assumes that we know which substances are harmful and how much each part of the environment can handle without serious environmental harm—things we most likely will never know.

Some environmentalists believe that relying mostly on this regulatory, end-of-pipe approach is doomed to failure. They argue that as long as population and per capita resource use continue to rise, any gains based on pollution control will eventually be overwhelmed (Figure 1-12). Instead of spending a lot of money and time setting and enforcing standards, arguing over "optimal" levels, and cleaning up chemicals we release into the environment, they argue, we should start rewarding people for preventing pollution and penalizing those who don't.

These environmentalists also call for reversing the present assumption that a chemical or new technology is safe until it is shown to be harmful. By contrast, if a chemical or new technology were considered potentially harmful until shown to be safe—and if this were coupled with full-cost pricing—emphasis would shift from costly pollution cleanup to less costly and more effective pollution prevention (Guest Essay, p. 37).

Critics argue that this approach would wreck the economy, putting large numbers of people out of work. Proponents counter that our present course will further deplete the Earth capital upon which our economies and survival depend—wrecking the environment and thus the economy, putting even more people out of work, and killing large numbers of people.

In 1992 the ministers of 13 western European nations made a binding commitment to try to achieve zero discharge of persistent toxic substances. In doing this they replaced the *safe-until-shown-harmful* approach to pollution control with the *harmful-until-shown-safe* approach.

Regulation or Market Forces? Controlling or preventing pollution and reducing resource waste require government intervention in the marketplace. This can be done either by regulation or by using market forces.

Regulation is a *command-and-control* approach that involves passing and enforcing laws that set pollution standards, establish deadlines and penalties, regulate harmful activities, ban the release of toxic chemicals into the environment, and require that certain resources be protected from use or unsustainable use.

Studies estimate that the current U.S. bill for environmental protection—over $120 billion per year—based mostly on regulation, could be cut by one-third to one-half with a mix of market-based policies. Ways to use market forces to improve environmental quality and reduce resource waste are:

- *Providing subsidies that encourage desirable behavior.*

- *Withdrawing subsidies that encourage harmful behavior.*

- *Granting tradable pollution and resource-use rights.* This would be done by setting a total limit for a pollutant or resource use and allocating that total among manufacturers or users by permit. Permit holders not using their entire allocation could use it as a credit against future expansion; use it in another part of their operation; or sell it to other companies. Tradable rights could also be established between countries to preserve biodiversity and to reduce emissions of greenhouse gases, ozone-destroying chemicals, lead, and other chemicals with regional or global effects.

- *Enacting green taxes.* This could include taxes on each unit of pollution discharged into the air or water, each unit of hazardous or nuclear waste produced, each unit of specified virgin resources used, each unit of pesticide used, each unit of fossil fuel used, and each unit of solid waste produced. To allow economic adjustment, such

Q: Worldwide, how much of the water withdrawn for irrigation is wasted?

Table 7-1 Economic Solutions to Pollution and Resource Waste					
Solution	Internalizes External Costs	Innovation	International Competitiveness	Administrative Costs	Increases Government Revenue
Regulation	Partially	Can encourage	Decreased*	High	No
Subsidies	No	Can encourage	Increased	Low	No
Withdrawing harmful subsidies	Yes	Can encourage	Decreased*	Low	Yes
Tradable rights	Yes	Encourages	Decreased*	Low	Yes
Green taxes	Yes	Encourages	Decreased*	Low	Yes
User fees	Yes	Can encourage	Decreased*	Low	Yes
Pollution-prevention bonds	Yes	Encourages	Decreased*	Low	No

*Unless more cost-effective and productive technologies are developed.

ecotaxes should be phased in over 5–10 years. At the same time, income and other taxes would be reduced so that low- and middle-income people and businesses (especially small ones) would not be penalized.

- *Charging user fees.* Users would pay fees to cover all or most costs for grazing livestock, extracting lumber and minerals from public lands, and using water provided by government-financed projects.

- *Requiring manufacturers to post a pollution prevention bond when they open a plant, incinerator, or landfill.* After a reasonable time the deposit (with interest) would be returned *minus* estimated environmental costs.

Each of these approaches has advantages and disadvantages (Table 7-1).

Encouraging Global Free Trade: Solution or Problem?   Since 1986 the United States and 117 other nations have been meeting to extend the 1948 General Agreement on Tariffs and Trade (GATT) by abolishing or reducing barriers to trade among them. In 1993 a new GATT agreement was reached and will go into effect when approved by the legislatures or other governing bodies of the countries involved. The 1989 Free Trade Agreement (FTA) between Canada and the United States and the North American Free Trade Agreement (NAFTA) between Canada, the United States, and Mexico have similar goals. Proponents argue that such agreements will

- Benefit LDCs whose products are often at a competitive disadvantage in the global marketplace

because of trade barriers erected by MDCs. Mostly because of such barriers the LDCs' share of world trade shrank from 37% to 28% between 1980 and 1990.

- Stimulate economic growth in all countries.

- Encourage competition and reduce costs and resource use by increasing efficiency of production.

- Raise the overall global levels of environmental protection and worker health and safety.

Because of such potential benefits, some environmental groups generally support NAFTA and GATT.

However, other environmentalists see the following serious drawbacks to existing and proposed trade agreements:

- Globalizing trade does not necessarily reduce resource waste and pollution. For example, more than half of all international trade involves the simultaneous import and export of essentially the same goods, with large quantities of energy used and pollution produced to transport similar goods between countries.

- Increased competition and lower costs may be achieved through lower standards for things such as wages, pollution control, worker safety, and health care.

- Attempts by MDCs to strengthen their intellectual property rights could make the transfer of environmentally advanced technologies (such as pollution control devices and solar cells) too costly for LDCs and also not adequately pay LDCs for biotechnology, pharmaceuticals, and agricultural products developed from biological resources extracted from LDCs.

A: 70–80% evaporates or seeps into the ground before reaching crops

- Tariff reductions might increase exports of unsustainably produced commodities such as timber and mineral products.

- Global environmental and worker safety and health standards are set and enforced by a secretive, undemocratic process dominated by industrial interests.

Existing trade agreements have already lowered some environmental standards. For example,

- British Columbia ended a government-funded tree-planting program because the United States (under pressure from U.S. lumber companies) argued that it was an unfair subsidy to Canada's timber industry and violated the 1989 agreement.

- Canada had to relax its regulations on pesticides and food irradiation to bring them in line with the weaker ones in the United States.

- In 1993, Austria was forced to drop plans to introduce a 70% tax on tropical timber.

- The Canadian timber industry is urging its government to challenge a U.S. law requiring the use of recycled fiber in newsprint as a trade barrier.

- The European community has complained to a GATT panel that U.S. gas-efficiency standards discriminate against European luxury cars.

- In 1991 a GATT panel of judges ruled that a U.S. requirement—that tuna imported to the United States had to be caught in ways that didn't ensnare and kill dolphins—was an unfair and illegal trade barrier. To some people in LDCs the ban also represented an attempt to impose certain values on other countries caring more about dolphins than about people as a cover for keeping goods from LDCs out of markets in MDCs. They ask what U.S. citizens would say if India said that no beef or leather could be imported because it involved killing cows, which many Indians consider to be sacred?

Here are some other possibilities:

- Government bans on the export or import of raw logs to slow the destruction of rain forests or ancient forests could be overturned.

- Germany could be forced to repeal its law requiring recycling of all beverage containers.

- An LDC banning dirty industries could be accused of violating free trade. Indeed, Lawrence Summers, Vice President and Chief Economist of the World Bank, asked in a 1991 memorandum, "Shouldn't the World Bank be encouraging more migration of dirty industries to the less developed countries?" He also noted that in LDCs fewer people live long enough to get cancer from exposure to toxic chemicals.

- Countries might not be able to restrict imports of hazardous wastes, banned medicines, and dangerous pesticides.

- International moratoriums and quotas on harvesting whales and bans on trading ivory to protect elephants could be overturned.

- Small and fledgling businesses would be pitted against large multinational and transnational companies in the name of free trade.

Environmentalists critical of current free trade agreements believe that such problems could be largely overcome by

- *Allowing any country to set higher environmental, worker health and safety, or resource depletion standards than other countries.*

- *Requiring bodies setting and enforcing GATT environmental and worker health and safety standards to have environmental and health representatives, and that all discussions and findings be open to global public scrutiny.*

- *Ensuring that any trade agreement provide adequate funding for monitoring and rapid enforcement of violations of environmental, worker safety, or resource depletion standards.*

- *Allowing countries to impose economic sanctions on countries with consistent lax enforcement of environmental laws.*

- *Requiring that international environmental agreements prevail when they conflict with trade agreements between countries.*

- *Giving preferential tariff treatment for imports of environmentally sound goods.*

- *Urging governments to require any businesses opening operations in other countries with lower environmental standards to meet the higher standards of their own country.*

- *Judging any trade agreement primarily on how it benefits the environment, workers, and the poorest 40% of humanity.* This means that such concerns should be included as primary components of such agreements, not as side agreements (the current approach) that are much more difficult and time consuming to enforce.

- *Making full-cost pricing and zero discharge of hazardous materials key components of economic systems and all trade agreements.*

Q: How much of the world's population live in areas that have prolonged droughts?

7-3 SOLUTIONS: REDUCING POVERTY

The Trickle-Down Approach   Poverty is usually defined as not being able to meet one's basic economic needs for clean air and water, food, shelter, and health care (Spotlight, p. 11). Currently, 1.2 billion people—one of every five persons on the planet—in LDCs have an annual income of less than $370 per year—roughly $1 per day—classified by the World Bank as poverty (Figure 7-9). This number is projected to rise to 1.3 billion by 2000 and 1.5 billion by 2025. Poverty causes premature deaths and preventable health problems, tends to increase birth rates, and often pushes people to make unsustainable use of potentially renewable resources in order to survive.

Most economists believe that a growing economy is the best way to help the poor. This is the so-called *trickle-down hypothesis*: Economic growth creates more jobs, enables more of the increased wealth to reach workers, and provides more tax revenues for helping the poor help themselves.

The facts, suggest either that the hypothesis is wrong or that it has not been applied. Instead of trickling down, most of the benefits of economic growth have flowed up since 1950 to make the top fifth of the world's people much richer and the bottom fifth poorer; most of those in between have lost or gained only slightly in real income per capita (Figure 1-6).

Encouraging Sustainable Development

Some critics of the current approach to economic development believe that dealing with the interrelated problems of poverty, environmental degradation, and population growth requires new forms of development based on Mahatma Gandhi's concept of *antyodaya*: putting the poor and their environment first, not last. They believe that more beneficial and sustainable economic development can be achieved by:

- *Protecting what works.* This involves learning where and how people are living sustainably, and not disrupting such cultures.

- *Involving local residents—including women—and private nongovernmental organizations (NGOs) in the planning and execution of all projects.* This means asking the poor what they need, giving it to them, putting them in leadership positions, and telling others about what works.

- *Making use of local wisdom, skills, and resources.* The poor generally know far more about poverty, survival, environmental sustainability, local needs, and what will work locally than do outside bureaucrats or experts.

- *Learning from other cultures about sustainable living and sharing this knowledge with people in other LDCs and in MDCs.*

Figure 7-9 Extreme poverty forces hundreds of millions of people to live in slums such as this one in Rio de Janeiro, Brazil, where adequate water supplies, sewage disposal, and other services don't exist. These people, and the much larger number of the desperately poor trying to survive in rural areas, are pushed deeper into poverty by local, national, and global economic and political forces beyond their control. Yet, these people that the world's economic systems are throwing away are the world's experts on recycling, reuse, and living sustainably on the land. If they weren't, they'd die.

Ways to Reduce Poverty   Reducing poverty requires governments of most LDCs to make drastic, difficult, and controversial policy changes. They include:

- *Shifting more of the national budget to aid the rural and urban poor*

- *Giving villages, villagers, and the urban poor title to common lands and to crops and trees they plant on common lands*

- *Redistributing some of the land owned by the wealthy or by governments to the poor*

- *Extending full human rights to women, with special emphasis on poor women*

Analysts urge MDCs and the rich in LDCs to help reduce poverty. Controversial ways to do this include:

- *Forgiving at least 60% of the $1.35 trillion that LDCs owe to MDCs and international lending agencies.* Some of this debt can be forgiven in exchange for carefully monitored agreements by the governments of LDCs to increase expenditures for rural development, family planning, health care, education, land redistribution, protection of biodiversity, ecological restoration, and sustainable use of renewable resources.

- *Increasing the nonmilitary aid to LDCs from MDCs.* This aid should go directly to the poor to help them become more self-reliant.

- *Shifting most international aid from large-scale to small-scale projects targeted to benefit local communities of the poor.*

- *Requiring international lending agencies to use an environmental and social impact statement (developed by standardized guidelines) to evaluate any proposed development project.* No project should be supported unless its net environmental impact is favorable, most of its benefits go to the poorest 40% of the people affected, and the local people it affects are involved in planning and executing the project. All projects should be carefully monitored, and further funding should be halted immediately when environmental safeguards are not followed.

- *Lifting trade barriers that hinder the export of commodities from LDCs to MDCs.* Trade barriers in rich countries cost LDCs about $100 billion annually in lost sales and depressed prices. But many environmentalists believe that all trade policies should be judged primarily on how well they benefit the environment, workers, and the poorest 40% of humanity.

- *Establishing policies that will encourage MDCs and LDCs to slow population growth and stabilize their populations as soon as possible (Section 6-5).*

7-4 SOLUTIONS: MAKING THE TRANSITION TO A SUSTAINABLE-EARTH ECONOMY

Sustainable-Earth Economies: A New Vision

What's wrong with today's economies? Most economists, business leaders, and government officials would answer, Not much. They believe any problems caused by current market-based mixed economic systems can be cured by further unlimited growth in an expanded global economy built around less—not more—government interference into economic matters. To them the best way to sustain the earth and its

people is to free the market and let it work on a global basis.

Most environmentalists and some economists and business leaders disagree. They believe that today's economies, based on depleting Earth capital (Figure 1-11) and producing huge amounts of pollution and waste (Figure 3-12), are unsustainable and must be converted to *sustainable-Earth economies* over the next few decades. This new vision of economics would:

- *Change the system of economic rewards (subsidies) and penalties (taxes and regulations) in today's mixed-market economies so that the highest profits and largest source of jobs lie in Earth-sustaining economic activities.*

- *Be guided by economic, social, and environmental indicators that distinguish between harmful and helpful forms of growth (Figures 7-5 and 7-6).*

- *Use full-cost pricing, in which externalities are included in the market prices of all goods and services.* This would be accomplished by withdrawing Earth-degrading subsidies, adding Earth-sustaining subsidies, and using a mix of regulation, tradable pollution and resource use rights, green taxes, and user fees (Table 7-1).

- *Slow human population growth and then gradually reduce population in all countries.*

- *Greatly reduce poverty by meeting the basic needs of all.*

- *Require all agencies of federal, state, and local governments to purchase products with the highest feasible percentage of post-consumer recycled materials and energy efficiency—and to minimize use of disposable products.*

- *Require all products to be audited (using standardized guidelines) for environmental impact from cradle to grave, and to carry green labels summarizing this information in easily understandable form.*

- *Not allow the concepts of free trade and the global marketplace to restrict the freedom of any country or region to impose higher environmental, consumer, worker safety, or resource-depletion standards than found in other countries.*

- *Repair past damage and create jobs by using local people to replant forests and grasslands, restore soil fertility, and rehabilitate streams and wetlands.*

Even if one believes that sustainable-Earth economics is desirable, is it possible to make such a drastic change in the way people think and act? Some environmentalists, economists, and business leaders say it's not only possible but imperative and that it can be done over the next 40–50 years (the time span of the radical transformation of the current economies in

Q: How much drinking water in the United States is withdrawn from groundwater?

most MDCs since 1950). They call for *all* government subsidies encouraging resource depletion, waste, pollution, and environmental degradation to be phased out over the next 10–20 years and replaced with taxes on such activities. During that same period, new government subsidies would be phased in for businesses built around recycling and reuse, reducing waste, preventing pollution, improving energy efficiency, and using renewable energy. Income and other taxes would be reduced to compensate for the increase in taxes leading to full-cost pricing.

Economic models indicate that after this first phase is completed the entire economy would be transformed within another 30–40 years. The system for change just described is economically feasible because it represents a shift in determining which economic actions are rewarded (profitable) and which ones are discouraged. It doesn't go against the profit motive, but uses it to redirect the economy. Because these plans would be well publicized and would take effect over decades, businesses would have time to adjust. Because businesses go where the profits are, most of today's Earth-degrading businesses would be tomorrow's Earth-sustaining businesses—a win-win solution for business and the earth.

The problem in making this shift is not economics but politics. It involves the difficult task of convincing business leaders and elected officials to begin changing the current system of rewards and penalties that is profitable and that has given them economic and political power.

Another important element in shifting to a sustainable-Earth economy is to improve environmental management in businesses. To do this colleges and universities training economists, business leaders, and lawyers need to educate their students and those already in these professions about how the earth works, what we are doing to it, how things are connected, how we need to change the way we think and do business, and what the principles of good environmental management are (Spotlight, at right).

Connections: Jobs and the Environment

Evidence indicates that improvements in environmental quality will not lead to a net loss of jobs. Telling workers that they must choose between their jobs and a cleaner environment is a false choice used by some business and political leaders to weaken unions, undermine improvements in worker safety and health, create fear, pit workers against environmentalists, and hide the real causes of most job losses.

Studies have shown that the major reasons for job loss in the United States are unsustainable use of potentially sustainable resources (clear-cutting old-growth forests); rapid depletion of nonrenewable

What Is Good Environmental Management?

SPOTLIGHT Good environmental management requires:

- Providing leadership, beginning with the corporate board of directors

- Giving a clear statement of corporate environmental principles and objectives with the full backing of the board

- Making a commitment to quality; correcting defects before products are sold; providing a safe and healthy workplace; preventing pollution and waste; recycling and reusing materials within the production process; disposing of hazardous waste safely; using energy-efficient production methods; using renewable resources at sustainable rates; protecting biodiversity; marketing products that are safe, durable, energy-efficient, and easy to recycle; and asking all employees to find innovative and better ways of accomplishing these objectives

- Involving employees, environmental groups, customers, and members of local communities in developing and evaluating corporate environmental policies and strategies for improvement, and evaluating progress

- Making improving environmental quality and worker safety and health a major priority for every employee

- Conducting an annual cradle-to-grave environmental audit of all operations and products, with a detailed strategy for making improvements, and publicizing the results to employees and stockholders

- Applying the same environmental and safety and health standards to company facilities and workers everywhere in the world

- Helping customers safely distribute, store, use, and dispose of or recycle company products (including picking them up and transporting them to remanufacturing plants)

- Recognizing that carrying out these policies is the best way to encourage innovation, expand markets, improve profit margins, develop happy and loyal customers, attract and keep the best-qualified employees, and help sustain the earth—a win-win strategy resources (oil and minerals); automation; declining sales because of more efficient and innovative competitors; higher energy costs without improvements in energy efficiency (Section 18-2); cheaper labor in other countries; decline of unsustainable "sunset" industries; failure to modernize or to invest in emerging "sunrise" industries; decreased research and development by government and business; and a reduction in defense contracts. A Bureau of Labor study found that only 0.1% of the job loss in the United States in 1988 was linked to environmental causes.

Jobs can be lost in any economy as businesses decline or disappear because of changes in technologies, markets, and resource substitution. This can have tragic impacts on individuals and on communities dependent on sunset businesses.

The real issues are whether a country and investors are investing in new technologies and sunrise businesses (so more jobs are created than are lost), and whether people losing their jobs are retrained and helped financially until they can find new jobs.

Promoting investments in Earth-sustaining businesses will create a variety of planet-friendly jobs requiring low-, moderate-, and high-level skills. Here are some examples:

- One million jobs are expected to be added to the U.S. environmental protection industry between 1995 and 2005.

- Increasing the aluminum recycling rate to 75% would create 350,000 more jobs.

- Collecting and refilling reusable containers creates many more jobs per dollar of investment than using throwaway containers, with most of the jobs created in local communities.

- According to the Council for an Energy-Efficient Economy, improving the fuel economy of new cars in the United States to 17 kilometers per liter (40 miles per gallon) by 2000 would lead to a net gain of 70,000 jobs by spurring development of new technology and by putting more money in the hands of consumers.

- A congressional study concluded that investing $115 billion per year in solar energy and improving energy efficiency would eliminate about 1 million jobs in oil, gas, coal, and electricity production but would create 2 million new jobs. And investment of the money saved by reducing energy waste could create another 2 million jobs.

- Large numbers of jobs could also be created by establishing a national and global network of Civilian Conservation Corps supported by government funds, private enterprise, or both. They could reforest degraded areas, help restore degraded wetlands and streams, and weatherize low-income housing units.

Although a host of new jobs would be created in a sustainable-Earth economy, jobs would be lost in some industries, regions, and communities. Ways to ease the transition include (1) providing tax breaks to make it more profitable for companies to keep or hire more workers instead of replacing them with machines; (2) using incentives to encourage location of sunrise industries in hard-hit communities and helping such areas diversify their economic base; and (3) providing income and retraining assistance for workers displaced from environmentally destructive businesses (a *Superfund for Workers*).

Can We Change Economic Gears?   Critics claim that a shift toward an Earth-sustaining economy won't happen because it would be opposed by people whose subsidies were being eliminated and whose activities were being taxed. However, investors and business people have just as much interest in sustaining the earth as anyone else. Forward-looking investors, corporate executives, and the governments of countries such as Japan and Germany (Solutions, p. 179) recognize that Earth-sustaining businesses with good environmental management (Spotlight, p. 177), will prosper as the environmental revolution proceeds. They see environmental responsibility to their workers, customers, and society as part of a broadening concept of total quality.

The United States invests only 4% of its GNP in future growth, compared to 8% in Germany and 16% in Japan. Companies and countries that fail to invest in a green future may find that they do not have one. The environmental revolution is also an economic revolution.

Consuming Less and Living Better   Buying environmentally friendly or green products and boycotting harmful or wasteful products are ways to reduce our environmental impact and stimulate companies to make less harmful products. However, green consuming buys some time, but it is not a cure-all. And if population and per capita consumption (even if it's all green) continue rising, the environmental benefits of such consumption will eventually be overwhelmed. (Figure 1-12).

Some environmentalists urge us to move beyond green consuming to *Earth-sustaining living* by:

- *Asking ourselves whether we really need a particular product.* Recognize that even green consuming is still consuming, much of it devoted to filling harmful and unsatisfying wants.

Q: How much of the world's water use is provided by desalination?

Germany: Investing in the Future and the Earth

German (and Japanese) political and business leaders see sales of environmental goods and services—already a more than $250 billion per year business (and expected to rise to $600 billion by 2000)—as a major source of new markets and income in the next century, as environmental standards and concerns will rise everywhere.

Stricter environmental standards in Germany have paid off in a cleaner environment and the development of cutting-edge technologies that can be sold at home and abroad. In 1977 the German government started the Blue-Angel product-labeling program to inform consumers about products that cause the least environmental harm. Since 1985 a more complete cradle-to-grave environmental audit of products has been available in *Okotest* (Ecotest)—the environmental equivalent of *Consumer Reports* magazine. The result has been a torrent of new environmentally friendly products that gives German producers a competitive edge in the rapidly growing global market for such products.

Mostly because of stricter air pollution regulations, German companies sell some of the world's cleanest and most efficient gas turbines, and they have developed the world's first steel mill that uses no coal to make steel. Germany has also cut energy waste and air pollution by requiring large and medium-sized industries and utilities to use cogeneration. Germany sells this improved technology globally.

Germany has also revolutionized the recycling business. German car companies are required to pick up and recycle all domestic cars they make. Bar-coded parts and predesigned disassembly plants can dismantle an auto for recycling in 20 minutes. Such "take-back" requirements are being extended to almost all products to reduce use of energy and virgin raw materials. Germany plans to sell its newly developed recycling technologies to other countries. According to Carl Hahn, Chairman of Volkswagen, "We must adopt the cyclical processes on which the whole of nature is based."

The German government has also supported research and development aimed at making Germany the world's leader in solar-cell technology and hydrogen fuel, which it expects will provide a rapidly increasing share of the world's energy (Chapter 18). Finally, Germany provides about $1 billion per year in green foreign aid to LDCs. Much of the aid is designed to stimulate demand for German technologies such as solar-powered lights, solar cells, and wind-powered water pumps.

- *Seeing improved energy efficiency, pollution prevention, waste reduction, and environmental protection as needs, not luxuries.*

- *Seeking fulfillment in love, friendship, cooperation, sharing, and caring for others and the earth rather than in things bought in the marketplace.*

Is this unrealistic? Too idealistic? Risky? Perhaps. However, it may be more unrealistic and dangerous to maintain our present course. As Mohandas Gandhi reminds us, "When we take more than we need, we are simply taking from each other, borrowing from the future, or destroying the environment and other species."

7-5 POLITICS AND ENVIRONMENTAL POLICY: PROBLEMS AND SOLUTIONS

How Government Works in a Democracy

Politics is the process by which individuals and groups try to influence or control the policies and actions of governments, whether local, state, national, or international. Politics is concerned with who has power over the distribution of resources and benefits—who gets what, when, and how. Thus it plays a significant role in regulating the world's economic systems, influencing economic decisions, and persuading people to work together toward common goals.

Democracy is literally government "by the people." In a *representative democracy* people govern through elected officials and representatives. In a *constitutional democracy* a constitution provides the basis of governmental authority and limits governmental power through free elections and freely expressed public opinion.

Political systems in constitutional democracies are designed for gradual change to promote economic and political stability. Rapid change in the United States, for example, is curbed by the system of checks and balances that distributes power among the three branches of government—executive, legislative, and judicial—and among federal, state, and local governments.

Most political decisions in democracies are made by bargaining, accommodation, and compromise

A: 0.1%

Getting and Keeping Power: The Dark Side

People in power use several classic tactics to stay on top:

- *Divide and conquer.* Keeping people and interest groups fighting with one another so they can't get together on vital issues that threaten the status quo.

- *Use a real or trumped-up enemy to divert people's attention and energy from the real issues.*

- *Set up smokescreens or decoys to divert attention away from real issues.*

- *Mount smear and fear campaigns.* What counts is public perception, not the facts or truth.

- *Intimidate and weaken.* This is accomplished by infiltrating opposing groups, slapping them with nuisance lawsuits, and having opponents arrested and given large fines or jail sentences.

- *Threaten or use violence.* If charismatic leaders become too effective, intimidate, injure, or even kill them.

- *Deny that a problem exists.*

- *Delay and decay.* This means setting up legal and bureaucratic roadblocks to forestall change and tire reformers out.

- *Practice paralysis-by-analysis.* This involves delaying action by appointing blue ribbon panels to study problems and make recommendations, paying little attention to their advice, and then calling for more research (sometimes needed, but also used as an excuse for inaction).

- *Don't collect data, or keep it secret, and claim nothing can be done about a problem because of a lack of information.*

- *Support unenforceable legislation.* This undermines confidence in governments, splits up opposition groups, and helps convince the public that real change is impossible.

- *Convince the public that it is hopeless to work for significant change.*

These often effective tactics can be countered only by the efforts of individuals and leaders at all levels of society who bring out the good in people and work together to focus on key issues and bring about real changes.

---

among leaders of competing *elites*, or power groups. The overarching goal of government by competing elites is maintaining the overall stability of the system by making only gradual change and not questioning or changing the rules of the game or the fundamental societal beliefs that gave them political or economic power.

One disadvantage of this deliberate design is that such governments mostly react to crises instead of acting to prevent them. Ralph Waldo Emerson once said, "Democracy is a raft which will never sink, but then your feet are always in the water." In other words, there is what amounts to a built-in bias against policies for protecting the environment because they often call for prevention instead of reacting to crises, require integrated planning into the future, and sometimes call for fundamental changes in societal beliefs that can threaten the power of existing elites in government and business. Democracy is designed to limit the abuse of power, but this is a constant struggle against the dark side of human nature (Spotlight, above).

Some say that we can keep on doing business and politics as usual and that no fundamental changes need to be made. Others argue the old ways of doing business and politics are so threatening to our life-support systems that we must have the wisdom and courage to reshape them into Earth-sustaining economic and political systems based on how nature works. Most people believe that a democracy is the best or only feasible system for dealing with environmental problems, but that doesn't mean that current democratic systems can't be improved. This improvement includes making them more democratic and more responsive to people at large instead of becoming increasingly dominated by powerful economic interests.

Because of the built-in bias against significant change from the top down, such change must occur from the bottom up, forced by people working together to insist that those in power either change or lose power. As grass-roots populist Jim Hightower (Guest Essay, p. 189) puts it:

> To move America from greed to greatness we must once again tap into the genius and gumption of the ... workaday majority of Americans who sweat, plow, invent, repair, teach, construct, nurse, and do the myriad of other productive tasks that sustain society from the bottom up. You can't keep a tree alive by fertilizing it at the top.... We want a government that quits doing things to us. But neither do we want a government that does things for us. We want a government that is

Q: Worldwide, how much of the water withdrawn is unnecessarily wasted?

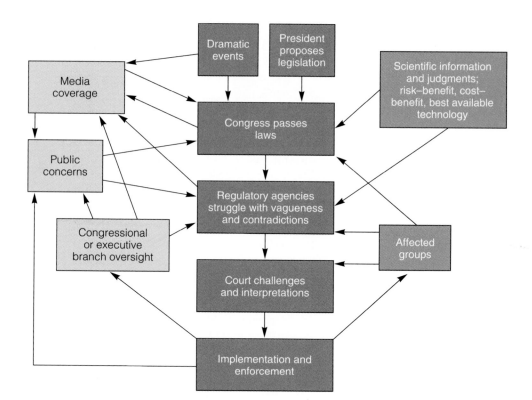

us, that involves us and empowers us so we can do for ourselves and do for the country.

Environmental Policy Making in the United States Environmental protection is highly controversial because it involves the transfer of considerable control over many very valuable resources (Earth capital) from one group of people (traditional industry) to another (the general public). It is a fierce political and economic struggle over who will benefit the most from Earth's resources. Figure 7-10 summarizes the primary forces in environmental policy making at the federal level in the United States. Similar factors are found at the state level. The first step in establishing environmental policy (or any other policy) is to persuade lawmakers that a problem exists and that the government has a responsibility to find solutions to the problem. Once over that hurdle, lawmakers try to pass laws to deal with the problem. Most environmental bills are evaluated by as many as 10 committees in both the House and the Senate. Effective proposals are often weakened by this fragmentation and by lobbying from groups opposing the law.

Even if a tough environmental law is passed, Congress must appropriate enough funds to implement and enforce it. Indeed, developing and adopting a budget for spending limited tax revenue is the most important and controversial thing members of the executive and legislative branches do. Developing a bud-

get involves answering two key questions: **(1)** What resource use and distribution problems will be addressed? and **(2)** How much limited tax revenue will be used to address each problem?

Next, regulations designed to implement the law are drawn up by the appropriate government department or agency. Groups favoring or opposing the law try to influence how the regulations are written and enforced. Some of the affected parties may challenge the final regulations in court. Then the agency implements and enforces the approved regulations. Proponents or affected groups may take the agency to court for failing to implement and enforce the regulations or for enforcing them too rigidly.

Environmentalists, with backing from many other citizens and members of Congress, have pressured the U.S. Congress to enact a number of important federal environmental and resource protection laws, as discussed throughout this text and listed in Appendix 3. These laws seek to protect the environment by the following approaches:

- *Setting standards for pollution levels or limiting emissions or effluents for various classes of pollutants* (Federal Water Pollution Control Act and Clean Air Act)

- *Screening new substances for safety before they are widely used* (Toxic Substances Control Act)

- *Requiring comprehensive evaluation of the environmental impact of an activity before it is undertaken by a federal agency* (National Environmental Policy Act)

- *Setting aside or protecting various ecosystems, resources, and species from harm* (Wilderness Act and Endangered Species Act)

- *Encouraging resource conservation* (Resource Conservation and Recovery Act and National Energy Act)

Some environmental laws contain glowing rhetoric about goals but little guidance about how to meet the goals, leaving this up to regulatory agencies and the courts. In other cases the laws specify general principles for setting regulations, such as the following:

- *No unreasonable risk*—for example, food regulations in the Food, Drug, and Cosmetic Act

- *No-risk*—for example, the Delaney clause, which prohibits the deliberate use of any food additive shown to cause cancer in test animals or people (Spotlight, p. 400), and the zero-discharge goals of the Safe Drinking Water Act and the Clean Water Act

- *Risk-benefit balancing* (Section 8-5)—for example, pesticide regulations

- *Standards based on best available technology*—for example, the Clean Air, Clean Water, and Safe Drinking Water acts

- *Cost-benefit balancing*—for example the Toxic Substances Control Act and Executive Order 12291, which gives the Office of Management and Budget the power to delay indefinitely, or even veto, any federal regulation not proven to have the least cost to society

The net result of this fragmentation process among competing interests is *incremental decision making*, in which only small changes are made in existing policies and programs.

Perhaps the greatest challenge we face is to use or modify political systems to anticipate and prevent serious long-term problems, many of them global, instead of reacting to crises. We cannot predict the future, but we can look at trends and project an array of possible results. The challenge is to see that the best possible futuristic thinking and analysis are incorporated into all government economic and political decision making (Solutions, at right).

Influencing Public Policy   A major theme of this book is that individuals matter. History shows that significant change comes from the bottom up, not the top down. Without grass-roots political action by mil-

Thinking for Tomorrow

SOLUTIONS

In 1985 then-Senator Al Gore proposed that the federal government establish an Office of Critical Trends Analysis (OCTA) in the executive branch to help guide public- and private-sector decision making. OCTA would identify and analyze major economic, environmental, social, and political trends, use them in models to project the long-term effects of existing and proposed government policies, and outline alternative futures for the next 20 years. An OCTA report in easily understandable language would be published every four years and widely distributed to government officials, business leaders, universities, the media, and citizens to help "futurize" decision making at all levels of society. OCTA would also hold press conferences and workshops throughout the country to present its findings and get feedback for improving projections. It could also evaluate proposed legislation for positive and negative effects on economic and environmental sustainability. State governments could set up similar offices.

lions of individual citizens and organized groups, the air you breathe and the water you drink today would be much more polluted, and much more of Earth's biodiversity would have disappeared.

Individuals can influence and change government policies in constitutional democracies by:

- *Voting for candidates and ballot measures*

- *Contributing money and time to candidates running for office*

- *Lobbying, writing, or calling elected representatives, asking them to pass or oppose certain laws, establish certain policies, and fund various programs (Solutions, next page)*

- *Using education and persuasion*

- *Exposing fraud, waste, and illegal activities in government (whistle-blowing)*

- *Filing lawsuits*

- *Participating in grass-roots activities*

- *Using consumer buying power to help convert brown (Earth-degrading) corporations into green (Earth-sustaining) corporations*

Q: How many people don't have a safe supply of drinking water?

Communicating with Elected Officials

SOLUTIONS

Here are some guidelines for communicating effectively with elected officials:

- Find out their names and addresses and write or call them.*

- When you write a letter use your own words, be brief and courteous, cover only one subject, and ask the elected official to do something specific (such as co-sponsoring, supporting, or opposing certain bills†). Give reasons for your position, explain its impact on you and your district, try to offer alternatives, share any expert knowledge you have, and ask for a response. Be sure to include your name and return address.

- After a vote supporting your position, write your representative (and others) a short note of thanks.‡

- Call and ask to speak to a staff member who works on the issue you are concerned about: the White House, 202-456-1414; the U.S. Senate, 202-224-3121; the House of Representatives 202-456-1414.

- Once a desirable bill is passed, call or write to the president about not vetoing it and to the members of the appropriations committee asking that enough money be appropriated to implement the law.

- Monitor and influence action at the state and local levels, where all federal and state laws are either ignored or enforced. As Thomas Jefferson said, "The execution of laws is more important than the making of them."

- Get others who agree with your position to write or call.

*Each year, the League of Women Voters (1730 M St. N.W., Washington, DC 20036) publishes *When You Write to Washington*, which lists all elected federal officials and includes a list of all committee members and chairpersons.

†If possible, identify the bill by number (for example, "H.R. 123" or "S. 313" for federal legislation). You can get a free copy of any federal bill or committee report by writing to the House Document Room, U.S. House of Representatives, Washington, DC 20515, or the Senate Document Room, U.S. Senate, Washington, DC 20510. Call or write your state representative to find out how to get copies of state bills and committee reports.

‡Each year the League of Conservation Voters (P.O. Box 500, Washington, DC 20077; 202-785-VOTE) publishes an *Environmental Scorecard* rating all members of Congress on how they voted on environmental issues.

Environmental Careers

INDIVIDUALS MATTER

Besides committed Earth citizens, the environmental movement needs dedicated professionals working to help sustain the earth. You will find environmental career opportunities in a large number of fields: sustainable forestry and range management; parks and recreation; air and water quality control; solid-waste and hazardous-waste management; urban and rural land-use planning; ecological restoration; and soil, water, fishery, and wildlife conservation and management.

Environmental careers can also be found in education, planning, health and toxicology, geology, ecology, conservation biology, chemistry, climatology, population dynamics and regulation (demography), law, accounting, journalism and communication, engineering, design and architecture, energy conservation and analysis, renewable-energy technologies, hydrology, consulting, activism and lobbying, economics, diplomacy, development and marketing, and law enforcement (pollution detection and enforcement teams). You can also run for an elected office on an environmental platform.

Many employers are now scrambling for environmentally educated graduates. They are especially interested in people with scientific and engineering backgrounds and in people with double majors (business and ecology, for example) or double minors.

For details on these careers, consult the Environmental Careers Organization (formerly the CEIP Fund), *The Complete Guide to Environmental Careers* (Covelo, Calif.: Island Press, 1990); Nicholas Basta, *The Environmental Career Guide* (New York: Wiley, 1991).

Environmental Leadership There are three types of environmental leadership:

- *Leading by example.* Individuals can use their own lives and lifestyles to show others that change is possible and beneficial.

- *Working within existing economic and political systems to bring about environmental improvement, often in new, creative ways.* Individuals can influence political elites by campaigning and voting for pro-Earth candidates and by communicating with elected officials (Solutions, at left). They can also choose an environmental career (Individuals Matter, above).

A: 1.5 billion—about one in every four persons on the planet

Challenging the system and basic societal values and proposing and working for better solutions to environmental problems. Leadership is more than being against something. It also involves showing people a better way to accomplish various goals.

All three types of leadership are needed to sustain the earth. Writer Kurt Vonnegut has suggested a prescription for choosing Earth-sustaining leaders:

> I hope you have stopped choosing abysmally ignorant optimists for positions of leadership.... The sorts of leaders we need now are not those who promise ultimate victory over Nature, but those with the courage and intelligence to present what appear to be Nature's stern, but reasonable surrender terms:
>
> 1. *Reduce and stabilize your population.*
>
> 2. *Stop poisoning the air, the water, and the topsoil.*
>
> 3. *Stop preparing for war and start dealing with your real problems.*
>
> 4. *Teach your kids, and yourselves too, while you're at it, how to inhabit a small planet without killing it.*
>
> 5. *Stop thinking science can fix anything, if you give it a trillion dollars.*
>
> 6. *Stop thinking your grandchildren will be OK no matter how wasteful or destructive you may be, since they can go to a nice new planet on a spaceship. That is really mean and stupid.*

7-6 ENVIRONMENTAL AND ANTI-ENVIRONMENTAL GROUPS

Mainstream and Grass-Roots Environmental Groups   There are many types of environmental groups working at the local, state, national, and international levels (Appendix 1). These groups generally fall into two categories: mainstream and grass-roots. Mainstream environmental groups are active mostly at the national level and to a lesser extent at the state level. Often they form coalitions to work together on issues. Mainstream groups do important work within the system and have been major forces in persuading Congress to pass environmental laws. However, if they become too dependent on high salaries, large budgets, and donations from Earth-degrading businesses, they can end up spending too much time and money on fundraising (in competition with other mainstream environmental groups). They can also have their goals corrupted by corporate lobbyists and lose touch with ordinary people and nature.

The base of the environmental movement in the United States and in other countries consists of thousands of grass-roots groups of citizens who have organized to protect themselves from pollution and environmental damage at the local level. The motto of such groups is *think globally and act locally*. They take to the streets, forests, oceans, and other front-line sites to stop environmental abuse, make harmful activities economically unattractive, and raise public awareness of environmental abuse and the need for change.

The activist group Greenpeace is the world's largest environmental group, with 2.3 million members (up from 30,000 in 1980) in 26 countries. Greenpeace members have risked their lives by putting small boats between whales and the harpoon guns of whaling ships, and placing themselves between seals and the clubs of hunters. Others have dangled from a New York bridge to stop traffic and protest a garbage barge heading out to sea (Figure 13-12), hung banners to protest the dumping of toxic wastes into rivers by industries and sewage treatment plants (Figure 7-11), skydived from the smokestacks of coal-burning power plants to protest acid rain, and sneaked into plants to document illegal pollution and dumping.

A small number of environmental activists engage in *nonviolent civil disobedience* by disobeying laws they believe to be unjust. They block bulldozers with their bodies (Figure 7-12); obstruct illegal whaling ships; and dye the fur of harp seals to keep them from being killed for their furs. They photograph or videotape illegal activities involving pollution, mistreatment of animals, harvesting of whales (Sea Shepherd Society), and harvesting of trees and wildlife.

Other citizens have worked to restore wetlands (p. 90), save forests, and restore degraded rivers (Individuals Matter p. 283). In the United States people of color, working-class people, and poor people, who often bear the brunt of pollution and environmental degradation, have formed a growing coalition known as the *grass-roots movement for environmental justice* (Guest Essay, p. 367) to change the system. Some unions and nonunionized workers are also forming coalitions with environmental groups to improve worker safety and health.

Many local grass-roots organizations are unwilling to compromise or negotiate. Instead of dealing with environmental goals and abstractions, they are fighting perceived threats to their lives, the lives of their children and grandchildren, and the value of their property. They want pollution and environmental degradation stopped and prevented rather than merely controlled. They are inspired by the words of ecoactivist Edward Abbey: "At some point we must draw a line across the ground of our home and our being, drive a spear into the land, and say to the bull-

Q: How many people die every year from preventable waterborne diseases?

Figure 7-11 Greenpeace activists protesting discharge of toxic waste from the "Pig's Eye" sewage treatment plant in St. Paul, Minnesota. This plant is the largest contributor of toxic chemicals into the Mississippi River north of St. Louis.

Figure 7-12 Earth First! activists blocking a logging road in the Siskiyou National Forest in Oregon.

dozers, earthmovers, and corporations, 'this far and no further.'"

The Anti-Environmental Movement   Increasingly the *East-West* polarization is being replaced by a *North-South* or *Rich-Poor* clash over how Earth's limited resources and wealth should be shared and the related *Green-Brown* clash between proponents of the planetary-management and Earth-wisdom worldviews (Section 2-2). The clash of such opposing worldviews will dominate our political and economic lives in coming decades.

A small but growing number of political and business leaders in the United States and in other MDCs see improving environmental quality as a way to stimulate innovation, increase profits, and create jobs while helping sustain the earth (Solutions, p. 179). However, leaders of some corporations attempt to ensure that environmental laws and regulations do little to damage corporate profit margins (a responsibility they have to shareholders) by:

- *Making donations to the election campaigns of politicians favoring their positions. A run for federal office in the United States costs $10–150 million, with most of the money coming from corporate PACs (political action committees).*

- *Putting lobbyists in Washington and in state capitals.*

- *Getting industry-friendly people appointed to regulatory agencies.*

- *Making donations or giving research grants to environmental organizations to influence how far they go.*

- *Influencing the media by threatening to withdraw vital advertising income if they probe too deeply.*

- *Establishing industry-funded grass-roots groups* (Spotlight, next page).

- *Using the "green menace" scare tactic.* Brand environmentalists as radical, communist, anti-American terrorists who threaten jobs, the economy, national security, and traditional values—instead of people who are trying to make the planet a safer, better, and more just place to live.

- *Intimidating environmental activists.* Fire or harass whistle-blowers in industries and government service; sue individuals and environmental groups; persuade government officials to put environmental activists under surveillance and to arrest or harass them. A study by the Center for Investigative Reporting uncovered more than 100 death threats, firebombings, shootings, and assaults on environmental activists in the United States between 1988 and 1992. This increase in violence against environmentalists parallels the growth of the anti-environmental Wise-Use movement during this period (Spotlight, next page).

Have Environmentalists Gone Too Far?

In some cases environmental dangers have been exaggerated. Here are a few examples of what can be considered overkill on the part of some environmentalists and government officials:

- *Stating that acid rain is causing widespread destruction of forests and aquatic life in lakes and streams.* While tree diebacks and acidic lakes are serious problems (especially in parts of Europe), they involve a mix of poorly understood natural and human-related factors and interactions, including deposition of acidic compounds. Decreasing human inputs of air into the atmosphere can be justified on the basis of improving human health and reducing costly corrosion and degradation of materials, regardless of their possible roles in degrading forests and streams (Chapter 9).

- *Requiring that all forms of asbestos be removed from buildings.* Experience has shown that removing asbestos that is not crumbling or releasing fibers can increase levels of asbestos fibers inside buildings and cost enormous amounts of money. In 1992 the EPA reversed its policy of requiring asbestos removal in most buildings (largely because of court rulings against this approach), but only after billions of dollars had been spent.

- *Stating that global warming is occurring now.* Any changes in average global temperature so far are too small to break out of normal climate fluctuations. The real questions are how might global climate change in the future, what roles do our inputs into the atmosphere have, and how might climate change in specific areas—all controversial and poorly understood issues (Section 10-2)? To most environmentalists the basic remedies for slowing *possible* global warming such as improving energy efficiency and relying more on renewable energy resources (Chapter 18), and protecting biodiversity (especially forests, Chapter 16) are things we should be doing, regardless of what one believes about the possibility of global warming (or cooling) from natural processes or human activities.

- *Excessive and too costly government regulations.* In a world with many people with conflicting goals, laws and regulations are necessary to protect health, security, and individual freedom. However, the regulatory approach can go too far and waste enormous amounts of limited funds with too little results. An example is the Superfund cleanup of hazardous waste dumps where most of the billions of dollars spent so far have gone to lawyers and consultants (Section 13-7), with few dumps being cleaned up (basically an impossible and extremely expensive goal that illustrates the need for hazardous waste prevention). Superfund cleanup liability has forced some small landowners and owners of small businesses that are easy targets into bankruptcy and loss of property unjustly while some large businesses involved in creating toxic waste sites have the money and lawyers to escape or negotiate fines and liability. People should be protected from toxic chemicals in dumps but this should be done more efficiently and fairly by improving the original superfund law (Section 13-7).

Such exaggerations and excessive regulation have triggered an environmental backlash lead by popular writers (such as the late Dixy Ray Lee and Rush Limbaugh) and by anti-environmental groups (Spotlight, next page). The problem is that many of these critics of environmental excesses go too far in the other direction and mislead the public by dismissing legitimate concerns about most major environmental problems (such as possible global warning, ozone depletion, toxic waste, and loss of biodiversity) as hoaxes.

Q: What is the largest source of water pollution in the United States?

Browns vs. Greens: The Wise-Use Movement

SPOTLIGHT

Since 1988 several hundred local and regional grass-roots groups in the United States have formed a national anti-environmental coalition called the *wise-use movement*. Much of their money comes from developers and from timber, mining, oil, coal, and ranching interests.

According to one of its leaders, Ron Arnold, the goal of this movement is to "eradicate the environmental movement in the United States within a decade." According to Arnold, its specific goals are to:

- *Cut all old-growth forests in the national forests and replace them with tree plantations*

- *Modify the Endangered Species Act so that economic factors override preservation of endangered and threatened species*

- *Eliminate government restrictions on wetlands development*

- *Open all public lands—including national parks and wilderness areas—to mineral and energy production*

- *Open three-fourths of the land in the National Wilderness Preservation System for mineral and energy production, off-road vehicles (ORVs), developed campsites, and commercial development*

- *Do away with the National Park Service and launch a 20-year construction program of new concessions in national parks to be run by private firms*

- *Continue mining on public lands under the provisions of the 1872 Mining Law, which allows mining interests not only to pay no royal-*

ties to taxpayers for hardrock minerals they remove, but also to buy public lands for a pittance

- *Recognize private property rights to mining claims, water, grazing permits, and timber contracts on public lands, and not raise fees for these activities*

- *Provide civil penalties against anyone who legally challenges economic action or development on federal lands*

- *Allow pro-industry (wise-use) groups or individuals to sue as "harmed parties" on behalf of industries threatened by environmentalists*

Leaders of this anti-environmental movement raise funds and whip up support by branding all environmentalists as antipeople, anti-Christian, antijob, and antigrowth radical extremists who are crippling the U.S. economy, robbing landowners of the right to do what they want with their land, and trying to lock up all natural resources and remove people from all public lands in the United States.

When accused of using wild exaggeration and smear tactics against environmentalists to raise money and spread distrust, hate, and fear, Ron Arnold said, "Facts don't really matter. In politics, perception is reality."

Examples of these groups* include the Center for the Defense of Free Enterprise (strategic planning, and implementation center for the wise-use movement, run by Allan Gottlieb and Ron Arnold and partially funded by South Korean Sun Myung Moon's International Unification Church); U.S. Council for Energy Awareness (nuclear power

industry); America the Beautiful (packaging industry); Partnership for Plastics Progress (plastics industry); American Council on Science and Health (food and pesticide industries); and People for the West (mining industry). Also included are the Sahara Club, which advocates violence against environmentalists in defense of its members' "right" to dirt-bike in wilderness areas (one of its leaders tells audiences to "Throw environmentalists off the bridge. Water optional."), and Citizens for the Environment, an industry-backed education group that counters environmental ideas with such slogans as "Recycling doesn't save forests," "Packaging prevents waste," and "Global warming and ozone depletion are hoaxes."

To environmentalists the well-funded, mostly industry-backed wise-use movement should be called the *Earth- and people-abuse movement*. However, it could have two beneficial effects on the environmental movement: **(1)** expand its membership and financial support, and **(2)** force mainstream environmental organizations to work more at the grass-roots level by getting in touch with the concerns of ordinary people. These people are struggling to keep their jobs or to find work, and they fear excessive government regulation and threats to their property values.

*For lists of these organizations see *The Greenpeace Guide to Anti-Environmental Organizations* (1993. Odonian Press, Box 7776, Berkeley, CA 94707) and *Fronting for Big Business in America* by Andy Friedman and Mark Megalli (available for $20 from Essential Information, P.O. Box 19405, Washington, DC 20036).

Those supporting and opposing environmental policies and laws can and should root out overkill from all sides and oppose excessive, unfair, and wasteful forms of government regulation. This is one reason why an increasing number of environmentalists sup-

port emphasis on pollution and waste prevention (which can save individuals and businesses money and eliminate or reduce certain risks and the need for certain types of regulations) instead of relying mostly on complex and costly pollution control and waste

A: Agriculture (responsible for about two-thirds)

management based on command-and-control regulations that if too prescriptive can hinder innovation and market-based solutions (Sections 7-2 and 7-3 and Table 7-1).

A major dilemma is that most environmental problems are so complex that we will never come close to having enough scientific information to understand them very well. Another problem is not recognizing that science yields only probability or various levels of confidence instead of absolute proof or certainty. When proponents of either side of an issue say that "it has not been scientifically proven" or "that it has been scientifically proven" that so-and-so is true, they are misusing or misunderstanding the nature and limitations of science.

A third problem involves arguments over the validity of data. When environmentalists say that each year so many metric tons of soil are eroded worldwide, so many hectares of tropical forest are cut, so many species are lost, and so many metric tons of certain pollutants are emitted into the atmosphere or into aquatic systems, these are of course only estimates. We may legitimately argue over the numbers, but the point environmentalists are trying to make is that the trends in these areas are significant and need to be evaluated and addressed not dismissed merely because they are estimates (which is all we can ever have for such complex problems). This, however, does not relieve environmentalists from trying to get the best estimates possible and emphasizing that they are only estimates.

7-7 CONNECTIONS AND SOLUTIONS: GLOBAL ENVIRONMENTAL POLICY

Progress at the International Level Since the United Nations Conference on the Human Environment was held in Stockholm, Sweden, in 1972, some progress has been made at addressing environmental issues at the global level. Today 115 nations have environmental protection agencies, and more than 170 international environmental treaties have been signed. These treaties cover a range of subjects including endangered species, ozone depletion, ocean pollution, global warming, biodiversity, acid deposition, preservation of Antarctica, and export of hazardous waste.

The 1972 conference also created the United Nations Environment Programme (UNEP) to negotiate environmental treaties and help implement them.

The 1992 Rio Earth Summit In June 1992 the second United Nations Conference on the Human Environment was held in Rio de Janeiro, Brazil. More than 100 heads of state, thousands of officials, and more than 1,400 accredited nongovernmental organizations (NGOs) from 178 nations met to develop plans for addressing environmental issues. The official results included

- An Earth Charter, a nonbinding statement of broad principles for guiding environmental policy
- Agenda 21, an action plan for developing the planet sustainably during the twenty-first century
- A broad statement of principles on how to protect forests
- A convention on climate change (signed by 154 nations)
- A convention on protecting biodiversity (signed by 153 nations, but not by the United States)
- Establishment of the UN Commission on Sustainable Development, composed of high-level government representatives, to carry out and oversee implementation of these agreements

There were also failures:

- The convention on climate change lacks the targets and timetables for stabilizing carbon dioxide emissions favored by all major industrial countries except the United States (which only signed the treaty when such items were eliminated).
- The forest-protection statement was so watered down that most environmentalists consider it almost useless.
- Officials largely avoided addressing the issues of population and its relation to poverty and economic development; fairer distribution of Earth's wealth and income; forgiving much of the massive debt of LDCs; giving LDCs preferential access to modern environmental and energy-saving technology; creating an international tax on carbon emissions; and the environmental and economic-justice impacts of free trade.
- Countries did not commit even the minimum amount of money that conference organizers said was needed to begin implementing Agenda 21.
- A treaty on reversing desertification (Figure 12-25) was not developed, mainly because of opposition by the United States.
- The new UN Commission on Sustainable Development lacks the power to enforce any of its recommendations (although world leaders can take steps to make sure the commission is effective).
- The United States was sharply criticized for its views and lack of leadership at a time when

Q: What is the largest source of water pollution from oil?

If I Were President*

Jim Hightower

GUEST ESSAY

Jim Hightower is a populist, author, radio commentator, and political spark plug who has spent more than two decades battling Washington and Wall Street on behalf of consumers, children, working families, environmentalists, and just-plain-folks who sustain society from the ground up. As Texas commissioner of agriculture from 1983 to 1991, he saw that Texas enacted the country's most stringent certification standards for organically grown food as well as a law that provided migrant workers with a right to know about the pesticides they were exposed to. He regularly travels the country working with local citizens and offering inspiration as a "Johnny Appleseed of grass-roots democracy."

Let's just do it.

No more Presidential Commissions on Environmental Evasion. We know what needs to be done. Indeed, the "Blueprint for the Environment," a federal action plan laid out by the environmental community in 1988, still serves nicely as a point-by-point program in a gamut of environmental issues. Having never been used, it remains perfectly fresh.

What's been missing is any real presidential commitment to the environment. To get America moving on a pro-Earth agenda requires some heavy-duty pushing, pulling, and maybe a bit of shoving by a president. I'll take five approaches to moving our agenda.

First, I'll go to the people. You can't have a mass movement without the masses, so the place to start is out in the countryside, using the presidential campaign as a

mechanism for rallying a mandate for environmental action. The people are way ahead of politicians and pundits in wanting strong action and leadership on these issues. We know this from the polls, the breadth and intensity of local environmental activism, and simply from chatting with Americans at Chat and Chew cafes throughout the land.

In my campaign, I'd ask them for the political power needed to implement the three central issues of my campaign: a pro-Earth agenda, economic regeneration, and social justice. A presidential campaign can invigorate, organize, and focus a national pro-Earth constituency, bringing it to bear on Washington and Wall Street, starting at the opening bell when I assumed office.

Second, I'll use the presidential "Bullypulpit" (to borrow Theodore Roosevelt's adroitly coined phrase) to propound a new ethic of environmental responsibility. As surely as Reagan preached greed from the White House, I'll preach conservation, regeneration, and accountability.

I'll challenge status quo assumptions. For example, we must challenge the assumption that pollution and environmental degradation by industry are "externalities." This allows corporations to take no responsibility for environmental costs, passing them on to society rather than putting them on their own balance sheets. Where's the disincentive to pollute if you don't have to pay? Indeed, this gives a competitive advantage to those who don't give a damn.

In presidential speeches and actions, I'll try to instill in the public mind and in public policy the assumption that polluters and degraders must put the messes on

(continued)

Used with permission from Jim Hightower and Environmental Action.

other MDCs and LDCs desperately needed such leadership.

Does this mean that the conference was a failure? No, for two reasons. First, it gave the world a forum for talking about and seeking solutions to environmental problems. Second, paralleling the official meeting was a Global Forum that brought together more than 7,000 other nongovernmental organizations from 175 countries. These NGOs worked behind the scenes to influence official policy, formulated their own agendas for sustaining the earth and reducing poverty, learned from one another, and developed a series of new global networks, alliances, and projects. In the long run these newly formed networks and alliances may play the greatest role in monitoring and supporting the commitments and plans developed by

the formal conference—and determining whether they are carried out and strengthened.

Improving Global Environmental Protection

Suggestions various analysts have made for improving environmental protection at the global level include:

- *Expanding the concepts of national security and economic security to recognize that both ultimately depend on global environmental security based on sustainable use of the ecosphere (Figure 7-4).*

- *Expanding the role and budget of the United Nations Environmental Programme in negotiating environmental treaties and in monitoring and overseeing their implementation.*

- *Requiring regular compliance reports from countries signing environmental treaties.*

A: Waste oil dumped onto land by cities, individuals, and industries their own tabs. I will also make clear our society will have to just say no to production of things we cannot recycle, reuse, or manage safely, from foam coffee cups to nuclear waste.

Third, I'll invest in the green conversion of our economy. In defiance of conventional wisdom, this approach properly assumes that environmentalism is not anti-business, it is business. Converting our economy from Earth pollution to Earth solution—including conservation, alternative energy, sustainable agriculture, pollution prevention, toxic cleanup, and ecological restoration—will spawn thousands of new inventions and hundreds of new industries and huge numbers of new jobs. And it will open to American "bioneers" a long-term world-wide growth market in Earth-sustaining processes and products. This approach to economic regeneration will call forth not the speculators and spoilers, but the genius and gumption of the workaday people of our country.

Fourth, I'll offer a plan and process for global ecological security. If we can muster the good will and resources to restore Kuwait to its royal family and protect access of the United States and its allies to Middle East oil, then we can surely do as much to help protect and restore the health of this old Earth for future generations.

This will not be an effort to batter down the economic aspirations of any nation or region, but rather I will promote new ways for all people to participate fully in a global economy that is in harmony with global environmental security. In addition to such daunting problems as CO_2 emissions and population growth, it is also time to confront multinational conglomerates and the world's greedy elites with their globe-hopping exploitation both of low-income people and of the environment, all under the false rubric of "free trade." I believe it is presidential cowardice to avoid engaging in this Herculean struggle any longer.

Fifth, I will act locally while thinking globally. We need a president who will set a personal example. Why not an organic garden on the Ellipse, biological controls for White House roaches and Rose Garden aphids, energy-saving fluorescent lights and windows throughout the people's house, solar cells for producing some of the electricity, a presidential bicycle to compensate however slightly for the excess of Air Force One, a comprehensive White House recycling program, and other efforts to make the presidency as Earth-friendly as possible. Being a leader requires putting yourself first on the line, then asking others to join you. Let's do it.

Critical Thinking

1. Would you campaign for, vote for, and actively support a presidential candidate running on the platform proposed in this Guest Essay? Explain. Do you think a candidate running on such a platform could be elected? Explain.

2. List any items you would delete or add to those proposed by Jim Hightower, and explain why.

- *Establishing an International Environmental Court to settle disputes between nations over violations of international agreements.*

- *Having heads of all countries (not just the rich countries) meet together in an Earth Summit at least every two years.*

- *Creating an Environmental Security Council as part of the United Nations.* Its mission would be to discuss environmental issues, seek solutions to environmental problems, coordinate UN responses to environmental issues, and respond to environmental emergencies.

- *Greening international lending agencies such as the World Bank and the International Monetary Fund.* This might involve having them shift most of their investment to small-scale grass-roots projects; having local people and NGOs participate in a meaningful way in the planning and implementation of all projects; opening agency documents to public scrutiny; and requiring environmental impact assessments and monitoring for all projects.

A central paradox of international environmental agreements is that the world is better off if all nations cooperate, but individual nations have a strong incentive not to cooperate. Environmental treaties cannot be strictly enforced without intervening in the national affairs of individual countries. Thus, the primary mechanism for compliance is widely publicizing cheaters or free riders. Most countries and leaders dislike being considered environmental villains and, under the glare of public opinion, can often be persuaded to change their ways.

As the wagon driver said when they came to a long, hard hill, "Them that's going on with us, get out and push. Them that ain't, get out of the way."

ROBERT FULGHUM

Q: In the United States, how many underground tanks storing gasoline and other hazardous chemicals are leaking?

Critical Thinking

1. The primary goal of all current economic systems is to maximize growth by maximizing the production and consumption of economic goods. Do you agree with that goal? Explain. What are the alternatives?

2. Do you favor internalizing the external costs of pollution and unnecessary resource waste? Explain. How might it affect your lifestyle? Wildlife? Any children you might have?

3. a. Do you believe that we should establish optimal levels or zero-discharge levels for most of the chemicals we release into the environment? Explain. What effects would adopting zero-discharge levels have on your life and lifestyle?

   b. Do you believe that all chemicals we release or propose to release into the environment should be assumed to be potentially harmful until proven otherwise? Explain. What effects would adopting this principle have on your life and lifestyle?

4. Do you favor making a shift to a sustainable-Earth economy? Explain. How might this affect your lifestyle? The lifestyle of any children you might have?

5. Do you agree or disagree with the proposals some analysts have made for sharply reducing poverty, as listed on pages 174–176? Explain.

6. Suppose a presidential candidate ran on a platform calling for the federal government to phase in a tax on gasoline over 5–10 years to the point where gasoline would cost about 3–5 dollars a gallon (as is the case in Japan and most western European nations). The candidate argues that this tax increase is necessary to encourage conservation of oil and gasoline, to reduce air pollution, and to enhance future economic, environmental, and military security. Some of the tax revenue would be used to provide tax relief or other aid to all people with incomes below a certain level (the poor and lower middle class), who would be hardest hit by such a consumption tax. Would you vote for this candidate who promises to triple the price of gasoline? Explain.

7. What do you believe are the greatest strengths and weaknesses of the system of government in your country related to protecting the environment and working with the earth? What substantial changes, if any, would you make in this system?

*8. Make a concept map of the key ideas in this chapter using the section heads and subheads and the key terms (shown in boldface type in the chapter). See the inside front cover and Appendix 4 for information on concept maps.

A: At least 1 million

8 Risk, Toxicology, and Human Health

Smoking: The Big Killer

What is roughly the size of a 30-caliber bullet, can be bought almost anywhere, and kills about 8,200 people every day? It's a cigarette. *Cigarette smoking is the single most preventable major cause of death and suffering among adults.* The World Health Organization estimates that tobacco kills at least 3 million people each year from heart disease, lung cancer, other cancers, bronchitis, emphysema, and stroke. By 2050 the death toll from smoking-related diseases is projected to be 12 million annually—an average of almost 33,000 preventable deaths per day.

In 1992, smoking killed about 435,000 Americans—an average of 1,190 per day (Figure 8-1). This death toll is equivalent to three fully loaded jumbo jets crashing every day with no survivors. Smoking kills more people each year in the United States than do all illegal drugs, alcohol, automobile accidents, suicide, and homicide combined (Figure 8-1).

Studies show that passive smoke (inhaled by nonsmokers) is also a killer, causing up to 40,000 premature deaths in the United States a year (3,000 of them from lung cancer), although tobacco companies dispute these findings. According to the EPA, each year secondhand smoke contributes to 150,000–300,000 respiratory infections like bronchitis and pneumonia.

Smoking is also highly addictive. A British government study showed that adolescents who smoke more than one cigarette have an 85% chance of becoming smokers. About 75% of smokers who quit start smoking again within six months, about the same relapse rate as for recovering alcoholics and heroin addicts.

Smoking costs the United States at least $68 billion (some estimate $100 billion) per year in expenses related to premature death, disability, medical treatment, increased insurance costs, and lost earnings and productivity because of illness (accounting for 19% of all absenteeism in industry). These harmful social costs amount to at least $3.00 per pack of cigarettes sold.

All of us face certain hazards to our health and well-being; some of them—like smoking—are avoidable, but others are not. When we evaluate the risks we face, the key questions we need to ask are whether the risks of damage from each hazard outweigh the short- and long-term benefits, and how we can reduce the hazards and minimize the risks.

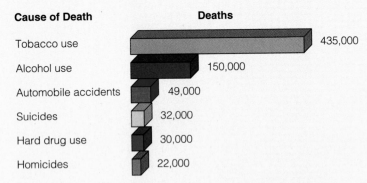

Cause of Death	Deaths
Tobacco use	435,000
Alcohol use	150,000
Automobile accidents	49,000
Suicides	32,000
Hard drug use	30,000
Homicides	22,000

Figure 8-1 Deaths in the United States from tobacco use and other causes in 1992. Smoking is by far the nation's leading cause of preventable death, causing more premature deaths each year than all the other categories in this figure combined. (Data from National Center for Health Statistics)

For the first time in the history of the world, every human being is now subjected to dangerous chemicals from the moment of conception until death.

RACHEL CARSON

The discussion in this chapter answers several general questions:

- What types of hazards do people face?
- What chemical hazards do people face, and how can they be measured?
- What physical hazards do we face from earthquakes and volcanic eruptions?
- What types of disease threaten people living in LDCs and in MDCs?
- How can risks be estimated, managed, and reduced?

8-1 TYPES OF HAZARDS

Risk   Risk is the possibility of suffering harm from a *hazard* that can cause injury, disease, economic loss, or environmental damage. It is expressed in terms of **probability**—a mathematical statement about how likely it is that something will happen. **Risk assessment** involves using data, assumptions, and models to estimate the probability of harm to human health or to the environment that may result from exposures to specific hazards.

Common Hazards   Here are some hazards that people face:

- *Cultural hazards.* These include unsafe living and working conditions (Connections, at right), smoking, poor diet, drugs, drinking, driving, criminal assault, unsafe sex, and poverty (Spotlight, p. 11).
- *Chemical hazards.* These result from harmful chemicals in the air (Chapter 9), water (Chapter 11), soil (Chapter 12), and food (Chapter 14).
- *Physical hazards.* These include ionizing radiation (Figures 3-4 and 3-7), noise (Figure 6-23), fires, floods and drought (Chapter 11), tornadoes, hurricanes, landslides, and earthquakes and volcanic eruptions (Section 8-3).
- *Biological hazards.* These are disease-causing bacteria and viruses, pollen, parasites, and animals such as pit bulls and poisonous snakes.

Working Can Be Hazardous to Your Health

CONNECTIONS

Roughly one-fourth of U.S. workers risk some illness from routine exposure to one or more toxic compounds. The National Institute for Occupational Safety and Health estimates that as many as 100,000 deaths per year in the United States—at least half from cancer—are linked to workers' exposure to toxic agents. The most dangerous occupation is farming, followed by construction, mining, and factory work. Every day an average of 28 Americans are killed in on-the-job accidents.

Most work-related illnesses and premature deaths could be prevented by stricter laws and by enforcement of existing laws governing exposure of workers to ionizing radiation and dangerous chemicals. However, political pressure by industry officials has hindered effective enforcement of these laws.

Industry representatives argue that U.S. industries have some of the world's safest workplaces and that stricter health and safety standards would cut profits and reduce competitiveness in the global marketplace. If such standards are enacted, they claim, they would have to shut down U.S. plants and move to other countries where health and safety standards—and wages—are lower.

Environmentalists call this form of job blackmail *greenmail.* They believe that pollution taxes—and more of a company's profits—could and should be used for improving worker health and safety standards, as is done in many Japanese, Scandinavian, and German industries.

8-2 CHEMICAL HAZARDS AND TOXICOLOGY

Dose and Response   The amount of a potentially harmful substance that an individual has ingested, inhaled, or absorbed through the skin is called the **dose**, and the amount of resulting health damage is called the **response.** Whether a chemical is harmful depends on **(1)** how big the dose is during a certain period of time, **(2)** how often an exposure occurs, **(3)** who is exposed (adult or child, for example), and **(4)** how well the body's detoxification system (liver, lungs, and kidneys) works. If the body gets a large dose in a short time, its repair mechanisms can be overwhelmed. The

The Prudent Diet

SOLUTIONS

Improper diet plays a key role in an estimated 20–30% of all cancer deaths. The National Academy of Sciences, the Surgeon General, and the American Heart Association advise that the risk of certain types of cancer (lung, stomach, colon, breast, and esophageal), heart disease, and diabetes can be significantly reduced by a daily diet that cuts down on certain foods and includes others. The guidelines are:

- Limit total fat intake to 25% or less of total calories, with no more than 10% from saturated fats and the remaining 15% divided about equally between polyunsaturated (safflower oil and corn oil) and monounsaturated (olive oil) fats.

- Limit protein (particularly meat) to 15% of total calories, or about 171 grams (6 ounces) per day (about the amount in one hamburger).

- Limit alcohol consumption to 15% of total caloric intake—no more than two drinks, glasses of wine, or beers per day. Pregnant women should not consume any alcohol.

- Limit cholesterol consumption to no more than 300 milligrams per day, the goal being to keep blood cholesterol levels below 200 milligrams per deciliter.

- Limit sodium intake to no more than 6 grams (about 1 teaspoon of salt) per day to help lower blood pressure, which should not exceed 140 over 90.

- Eat more poultry, fish, beans, whole grains, cereals, fruits, and vegetables—but much less processed foods and red meat (which recently was linked to a higher risk of colon cancer).

- Achieve and maintain the ideal body weight for your frame size and age by a combination of diet and 20 minutes of exercise at least three days per week.

same total dose spread out over a much longer time might cause little harm. On the other hand, some substances—such as lead—accumulate in the body, so that the total dose over a long period of time can be harmful or even fatal.

An *acute effect* is an immediate or rapid reaction to an exposure. It can range from dizziness or a rash to death. A *chronic effect* is a long-lasting effect from exposure to a harmful substance. It can result from a single large or small dose or many doses over a long period of time. Examples include kidney and liver damage.

Types of Chemical Hazards The main types of chemical hazards are the following:

- **Toxic chemicals** are generally defined as substances fatal to over 50% of test animals at stated concentrations. Many are *neurotoxins*, which attack nerve cells. Nerve gases, potassium cyanide, heroin, chlorinated hydrocarbons (DDT, PCBs, dioxins), organophosphate pesticides (malathion, parathion), carbamate pesticides (Sevin, zineb), and various compounds of arsenic, mercury, lead, and cadmium are all neurotoxins. Most toxic chemicals are discharged into the environment by industrial and agricultural activities, but others—such as hemlock, botulinus toxin, and cobra venom—occur naturally.

- **Hazardous chemicals** are literally dangerous chemicals. They cause harm because they: are flammable or explosive; irritate or damage the skin or lungs (such as strong acidic or alkaline substances); interfere with or prevent oxygen uptake and distribution (asphyxiants such as carbon monoxide and hydrogen sulfide); or induce allergic reactions of the immune system (allergens).

- **Carcinogens** are chemicals, radiation, or viruses that cause or promote the growth of a malignant (cancerous) tumor, in which certain cells multiply uncontrollably. Many cancerous tumors **metastasize**; that is, they release malignant cells that travel in body fluids to various parts of the body and start new tumors there—making treatment much more difficult. According to the World Health Organization, environmental and lifestyle factors play a key role in causing or promoting up to 80% of all cancers. Major sources of carcinogens are cigarette smoke (35–40% of cancers, p. 192), diet (20–30%, Solutions, at left), occupational exposure (5–15%, Connections, p. 193), and environmental pollutants (1–10%). About 10–20% of cancers are believed to be caused by inherited genetic factors or by certain viruses. Typically 10–40 years may elapse between the initial exposure to a carcinogen and the appearance of detectable symptoms. Healthy teenagers and young adults have trouble believing that their smoking, drinking, eating, and other lifestyle habits today could kill them before they reach age 50.

- **Mutagens** are agents, such as a chemical or radiation, that cause *mutations*—changes in the DNA molecules of the genes that can be transmitted from parent to offspring. A few mutations are beneficial; some are neutral; but many are harmful.

- **Teratogens** are chemicals, radiation, or viruses that cause birth defects during the growth and

Q: What can be done when groundwater becomes contaminated?

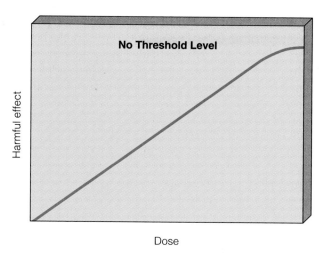

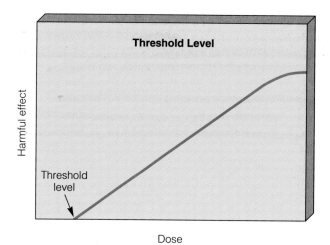

Figure 8-2 Different hypothetical dose–response curves. The curve on the left represents harmful effects that occur with increasing doses of a chemical or ionizing radiation. No dose is considered to be safe. The curve on the right shows the response from exposure to a chemical or ionizing radiation in which harmful effects appear only when the dose is above a certain threshold level. There is considerable uncertainty and controversy over which of these models applies to various harmful agents because of the difficulty in estimating the response to very low doses.

development of the human embryo during pregnancy. Chemicals known to cause birth defects in laboratory animals include PCBs, thalidomide, and metals such as arsenic, cadmium, lead, and mercury.

A growing body of research on wildlife and laboratory animals indicates that long-term (often low-level) exposure to various chemicals can cause damage by disrupting the *endocrine system* (which regulates hormones), the *immune system* (which defends the body against infectious disease and cancer), and the *nervous system*. Observed effects include lowered fertility, abnormal sexual development, lowered resistance to disease, nerve disorders, and abnormal behavior.

Some people have the mistaken idea that all natural chemicals are safe and all synthetic chemicals are harmful. In fact, many synthetic chemicals are quite safe if used as intended, and a great many natural chemicals are deadly.

The Great Unknown Avoiding or minimizing exposure to harmful or potentially harmful chemicals is not easy. According to the National Academy of Sciences, only about 10% of the 70,000 chemicals in commercial use have been thoroughly screened for toxicity, and only 2% have been adequately tested to determine whether they are carcinogens, teratogens, or mutagens. Only 2% of cosmetic ingredients, 5% of food additives, 10% of pesticides, and 18% of drugs used in medicines have been thoroughly tested. Fur-

thermore, each year, about 1,000 new chemicals are introduced into the marketplace with little knowledge about their potentially harmful effects.

Even less research has been done to determine the potential of widely used chemicals for damage to the endocrine, immune, and nervous systems. The primary reason for this is that the EPA (under laws enacted by Congress) does not require makers of industrial chemicals to run specific tests (except for pesticides, food additives, cosmetics, and drugs) to determine any adverse effects of their products before putting them on the market. In other words, such industrial chemicals are considered "innocent until proven guilty" and no one is required to investigate if they might be guilty. Even if we determine the biggest risks associated with a particular technology or chemical, we know little about its possible interactions with other technologies and chemicals, or about the effects of such interactions on human health and ecosystems.

What Is Toxic? Determining the level at which a substance poses a health threat is done by laboratory investigations (*toxicology*) and by studies of human populations (*epidemiology*). Both approaches are difficult and costly, and both have certain limitations.

Toxicity is usually determined by tests on live laboratory animals (especially mice and rats); on bacteria; and on cell and tissue cultures. Tests are run to develop a **dose–response curve**, which shows the effects of various doses of a toxic agent on a group of test organisms (Figure 8-2). The results are extrapolated for low doses on the test organisms and then extrapolated to humans.

A: Usually nothing

An experiment to assess acute effects often involves determining the **lethal dose**: the amount of material per unit of body weight of the test organism that kills all of the test population in a certain time. Then the dose is reduced until an exposure level is found that kills half the test population in a certain time. This is the **median lethal dose** or **LD$_{50}$**, a standard measurement in toxicity research. Using high dose levels reduces the number of test animals needed, cuts the time needed to obtain results, and lowers costs. Because it is difficult to get data on responses to low doses, the results of high dose exposures are usually extrapolated to low dose levels.

There are problems with animal tests. Extrapolating test-animal data from high to low dose levels is uncertain and controversial. According to the *linear dose–response model*, any dose of ionizing radiation or of a toxic chemical is harmful; and the harm rises as the dose increases, until it reaches a saturation level (Figure 8-2, left). With the *threshold dose–response model* there is a threshold dose, below which no detectable harmful effects occur, presumably because the body can repair the damage caused by low doses of some substances (Figure 8-2, right). It's very difficult, however, to establish which of these models applies at low doses.

Some scientists challenge the validity of extrapolating data from test animals to humans because human physiology and metabolism are different from those of the test animals. Others counter that such tests work fairly well, especially for revealing cancer risks when the correct experimental animal is chosen. However, animal tests take two to five years and cost $200,000–$2 million per substance. Furthermore, they are coming under increasing fire from animal rights groups. As a result, scientists are looking for substitute methods.

Also controversial are bacteria tests, as well as cell and tissue culture tests, for toxic agents. One of the most widely used bacterial tests, the Ames test, is considered an accurate predictor of whether a substance is mutagenic; it is also quick (two weeks) and cheap ($1,000 to $1,500 per substance). Some believe this test will also detect carcinogens, but the evidence for this is controversial. Cell and tissue culture tests have similar uncertainties, take several weeks to months, and cost about $18,000 per substance.

Chronic toxicity is much harder to assess than acute toxicity. The problem lies in establishing that a particular substance is responsible for chronic effects when people are exposed to hundreds, even thousands, of potentially harmful substances over a long period of time.

Another approach to testing for toxicity and determining the agents causing diseases such as cancer is **epidemiology**—an attempt to find out why some people get sick and others do not. Typically the health of people exposed to a particular toxic agent from an industrial accident—or of people working under high exposure levels, or of people in certain geographic areas—is compared with the health of people not exposed to these conditions, to see if there are statistically significant differences.

This approach also has limitations. For many toxic agents, not enough people have been exposed to high enough levels to detect statistically significant differences. Because people are exposed to many different toxic agents and disease-causing factors throughout their lives, it is often impossible to link an observed epidemiological effect with exposure to a particular toxic agent. And because epidemiology can be used only to evaluate hazards to which people have already been exposed, it is rarely useful for predicting the effects of new technologies or substances.

Thus all methods for estimating toxicity levels have serious limitations, but they are also all we have. To take this uncertainty into account and minimize harm, standards for allowed exposure to toxic substances and radiation are typically set at levels 10, 100, or even 1,000 times lower than the estimated harmful level.

This still may not be enough to protect some people, particularly those vulnerable to a particular hazard because of an allergic reaction or acute sensitivity. Ideally, all products likely to be inhaled, ingested, or absorbed would list their ingredients so that people allergic to or sensitive to certain substances could avoid exposure.

8-3 PHYSICAL HAZARDS: EARTHQUAKES AND VOLCANIC ERUPTIONS*

Earthquakes Stress can cause solid rock to deform elastically until it suddenly fractures and is displaced along the fracture, producing a *fault* (Figure 8-3). The faulting or a later abrupt movement on an existing fault causes an **earthquake**.

An earthquake has certain features and impacts (Figure 8-4). Earthquakes set up shock waves that radiate out from the center of movement like ripples in a pool of water. Reports of an earthquake commonly mention the focus and the epicenter. The *focus* is the point of initial movement, and the *epicenter* is the point on the surface directly above the focus. The epicenter

---

*Kenneth J. Van Dellen, professor of geology and environmental science, Macomb Community College, is the primary author of this section, with assistance from G. Tyler Miller, Jr.

may not be on the fault, either because the fault does not reach the surface or because the fault is not vertical (and, therefore, it surfaces some distance away from the epicenter). When the stressed parts of the earth suddenly fracture or shift, energy is released as shock waves that move outward from the earthquake's focus.

One way of measuring the severity of an earthquake is by its *magnitude* on the Richter scale. The magnitude is a measure of the amount of energy released in the earthquake, as indicated by the amplitude (size) of the vibrations when they reach the recording instrument. Using this scale, seismologists rate earthquakes as *insignificant* (less than 4 on the Richter scale), *minor* (4–4.9), *damaging* (5–5.9), *destructive* (6–6.9), *major* (7–7.9), and *great* (over 8). The northern California earthquake of 1989 had a Richter magnitude of 7.1 and caused damage within a radius of 97 kilometers (60 miles) from its epicenter (Figure 8-5). Earthquakes often have *aftershocks* that gradually decrease in frequency over a period of up to several months, and some have *foreshocks* from seconds to weeks before the main shock.

The primary effects of earthquakes include shaking and sometimes permanent vertical or horizontal displacement of the ground. These may have serious

Figure 8-3 San Andreas Fault as it crosses part of the Carrizo Plain between San Francisco and Los Angeles, California. This fault extends almost the full length of California. When part of the fault locks, strain builds until the fault ruptures and movement occurs along it. The sudden release of strain produces an earthquake.

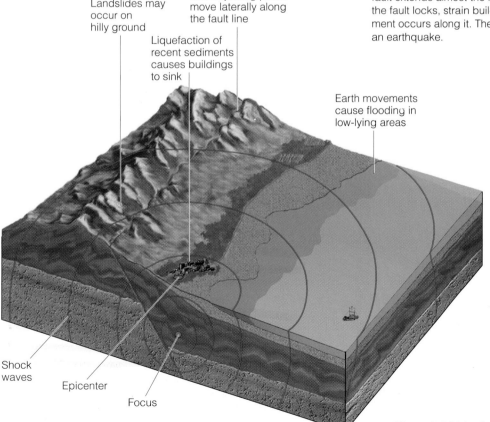

Landslides may occur on hilly ground

Two adjoining plates move laterally along the fault line

Liquefaction of recent sediments causes buildings to sink

Earth movements cause flooding in low-lying areas

Shock waves

Epicenter

Focus

Figure 8-4 Major features of an earthquake and its effects.

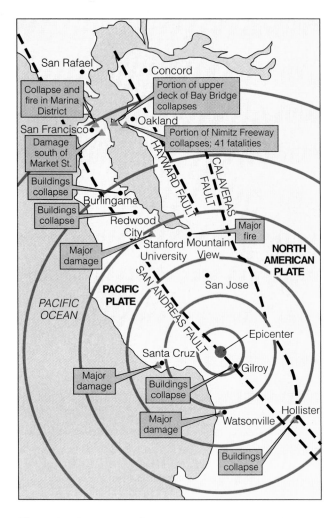

Figure 8-5 At 5:04 P.M., October 17, 1989, a 7.1-magnitude earthquake occurred in northern California along the San Andreas Fault (Figure 8-3). It was the largest earthquake in northern California since 1906, when the great San Francisco quake and the resulting fires destroyed much of San Francisco. In the 1989 quake the most extensive damage was within a radius of 32 kilometers (20 miles) from the epicenter, but seismic waves caused major damage as far away as San Francisco and Oakland. Sixty-seven people were killed, and official damage estimates were as high as $10 billion. This was North America's costliest natural disaster until Hurricane Andrew devastated Florida in 1992 and the Los Angeles earthquake in 1994.

effects on people and structures, such as buildings, bridges, freeway overpasses, dams, and pipelines. Secondary effects of earthquakes include rockslides, urban fires, and flooding due to subsidence of land. Coastal areas can also be severely damaged by large, earthquake-generated water waves, called *tsunamis* (misnamed "tidal waves," even though they have nothing to do with tides), that travel as fast as 950 kilometers (590 miles) per hour. Two of the worst earthquakes, in terms of lives lost, happened in China, with about 830,000 killed in 1556 and 500,000 in 1976.

Loss of life and property from earthquakes can be reduced. To do this we look at historical records and make geologic measurements to locate active fault zones, make maps showing where ground conditions are more subject to shaking (Figure 8-6), establish building codes that regulate the placement and design of buildings in areas of high risk, and, hopefully, learn to predict when and where earthquakes will occur.

Actual prediction of earthquakes will have to be based primarily on collecting data on *precursor phenomena*, events that precede an earthquake, as well as on the pattern and frequency of earthquakes in an area. Precursor phenomena include such characteristics of earth materials as slight tilting of rock; changes in electrical and magnetic properties; the amount of radon (a radioactive gas) dissolved in groundwater; the speed of seismic waves passing through the area from nearby quakes; and even the unusual behavior of animals that may be able to sense an imminent earthquake.

Volcanoes An active **volcano** occurs where magma (molten rock) reaches the earth's surface through a central vent or a long crack (fissure). Such volcanic activity can release *ejecta* (debris ranging from large chunks of lava rock to ash that may be glowing hot), liquid lava, and various gases into the environment.

Some volcanoes, such as those at Mount St. Helens in Washington (Case Study, p. 200), and Mount Pinatubo in the Philippines (which erupted in 1991) have a steep, flaring cone shape and usually erupt explosively. Others, such as those in the islands of Iceland and Hawaii (Figure 7-1), typically erupt more quietly, with lava flowing or spraying into the air.

Prediction of volcanic activity has improved considerably during this century, but every volcano has its own personality. As with earthquake maps, the eruptive history of a volcano or volcanic center gives some indication of where the risks are. Volcanologists are now studying precursor phenomena such as tilting or swelling of the cone, changes in magnetic and thermal properties of the volcano, changes in gas composition, and increased seismic activity.

We tend to think of volcanic activity in negative terms, but it also provides some benefits. One is outstanding scenery in the form of majestic mountains, some lakes (such as Crater Lake in Oregon), and other landforms. Geothermal phenomena, such as geysers and hot springs, have aesthetic value as well as economic value in the tourist and travel industry. Geothermal energy is also an important energy source in some areas and may be more widely used in the future (Section 18-8). Perhaps the most important benefit of volcanism is the highly fertile soils produced by the weathering of lava.

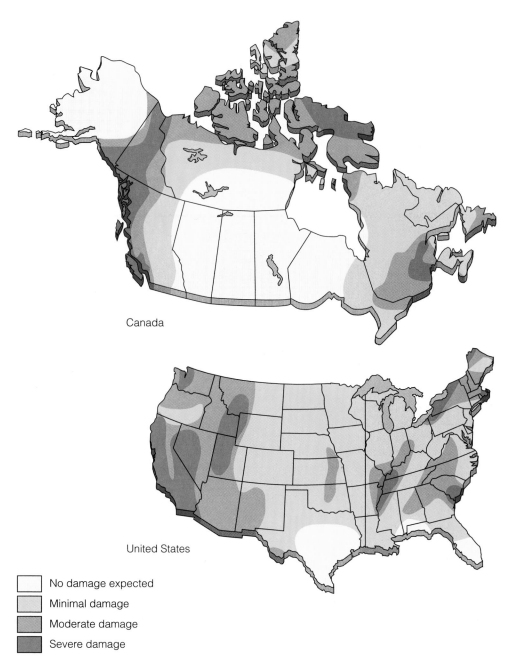

Figure 8-6 Expected damage from earthquakes in Canada and the contiguous United States. Except for a few regions along the Atlantic and the Gulf coasts, virtually every part of the continental United States is subject to some risk from earthquakes, but several areas have a risk of moderate to major damage. This map is based on earthquake records. (Data from U.S. Geological Survey and *Energy, Mines, and Resources Canada, 1976*)

Canada

United States

No damage expected
Minimal damage
Moderate damage
Severe damage

8-4 BIOLOGICAL HAZARDS: DISEASE IN MDCS AND LDCS

Types of Disease A **transmissible disease** is caused by a living organism—often a bacterium, virus, or parasitic worm—and can be spread from one person to another. The infectious agents are spread by air, water, food, body fluids, and, in some cases, insects and other nonhuman carriers (called *vectors*). Examples are sexually transmitted diseases (Connections, p. 202), malaria, schistosomiasis, elephantiasis, sleeping sickness, and measles.

About 40% of the world's people live in malaria-prone tropical and subtropical regions (Figure 8-9). There are 200–300 million new cases each year. Some 1–2 million of these people (some estimate as many as 5 million) die each year—more than half of them children under age 5. Malaria's intermittent symptoms include fever and chills, anemia, an enlarged spleen, severe abdominal pain and headaches, extreme weakness, and greater susceptibility to other diseases.

Malaria is caused by four species of protozoa of the genus *Plasmodium.* Most cases of the disease are transmitted when an uninfected female of any one of 60

A: $11–$50 billion with cleanup costs to be paid by taxpayers

The 1980 Eruption of Mount St. Helens

After 123 years of dormancy, Mount St. Helens near the Washington–Oregon border erupted on May 18, 1980 with an explosive force 1,500 times as powerful as the one that demolished Hiroshima (Figure 8-7). This eruption has been described as the worst volcanic disaster in U.S. history.

Devastation occurred in three semicircular zones north of the volcano. In the direct blast zone (the "tree-removal zone"), extending about 13 kilometers (8 miles), everything was obliterated or carried away. In the "tree-down zone" beyond the direct blast zone to 30 kilometers (19 miles) out, the blast of dense, ash-laden air blew the trees down like matchsticks, so their tops pointed away from the blast. In the "seared zone," 1–2 kilometers (0.6–1.2 miles) beyond the second zone, trees were left standing but were scorched brown.

The explosion also threw ash more than 7 kilometers (4 miles) up into the atmosphere. Several hours after the explosion, the ash darkened the sky enough to trigger automatic streetlights during the morning hours in Yakima and

U.S. Geological Survey

Spokane, Washington. Within two weeks the ash cloud had traveled around the globe and eventually circled the planet several times before the ash settled to the ground.

The heat of the eruption rapidly melted snow and glacial ice on the mountain. Mudflows, mixtures of volcanic debris and water, flowed down the stream valleys, clogging three rivers. Such mudflows commonly flow down steep slopes at speeds of 30–70 kilometers (19–43 miles) per hour; so you can't outrun one, and you may not be able to outdrive some, especially if you are in heavy traffic. The Mount St. Helens eruption did not produce a lava flow, although it could have.

About 60 people died, and several hundred cabins and homes were destroyed or severely damaged. An estimated 7,000 big-game animals (bear, deer, and elk) died, as did all

species of *Anopheles* mosquito bites an infected person, ingests blood that contains the parasite, and then bites an uninfected person. When this happens *Plasmodium* parasites move from the mosquito into the bloodstream, multiply in the liver, and then enter blood cells to continue multiplying (Figure 8-10). Malaria can also be transmitted by blood transfusions or by sharing needles. This cycle repeats itself until immunity develops, treatment is given, or the victim dies.

During the 1950s and 1960s the spread of malaria was sharply curtailed by draining swamplands and marshes, by spraying breeding areas with insecticides, and by using drugs to kill the parasites in the bloodstream. Since 1970, however, malaria has come roaring back. Most of the malaria-carrying *Anopheles* mosquitoes have become genetically resistant to most of

the insecticides used. Worse, the *Plasmodium* parasites have become genetically resistant to the common antimalarial drugs. Irrigation ditches—breeding grounds for mosquitoes—have proliferated, and malaria control budgets have been cut in the mistaken belief that the disease is under control.

Researchers are working to develop new antimalarial drugs and vaccines, as well as biological controls for *Anopheles* mosquitoes. Such approaches are underfunded, however, and are more difficult than originally thought. The World Health Organization estimates that only 3% of the money spent worldwide each year on biomedical research is devoted to malaria and other tropical diseases, even though more people suffer and die worldwide from these diseases than from all others combined.

Q: In tropical and temperate areas, how long does it take to renew 2.54 centimeters (1 inch) of topsoil?

Figure 8-7 Mount St. Helens, a composite volcano in Washington, near the Washington–Oregon border, before (left), during (center), and after (right) its major eruption in May of 1980. It is one in a chain of volcanoes less than 1 million years old that stretches 1,500 kilometers (930 miles) from Lassen Peak in northern California to Mount Garibaldi in British Columbia. As a result of this eruption, much of the northern side of the mountain was blown away, and the altitude of the summit was reduced by about 450 meters (1,475 feet).

birds and most small mammals in the blast area. Salmon hatcheries were damaged, and crops (including alfalfa, apples, potatoes, and wheat) were lost. Many people living in the area lost their jobs.

On the plus side, trace elements that were added to the soil from the ash may benefit agriculture in the long run. Also, volcano visitor centers and other tourism promotions brought new jobs and income. Many

biologists were surprised at how fast various forms of life had begun colonizing many of the most devastated areas by 1990—an instance of primary succession (Figure 5-41) in action.

Prevention offers the best approach to slowing the spread of malaria. Methods include increasing water flow in irrigation systems to prevent mosquito larvae from developing (an expensive and wasteful use of water); using mosquito nets dipped in a nontoxic insecticide (permethrin) in windows and doors of homes; cultivating fish that feed on mosquito larvae; clearing vegetation around houses; and planting trees that soak up water in low-lying marsh areas where mosquitoes thrive.

Diseases such as cardiovascular (heart and blood vessel) disorders, most cancers, diabetes, bronchitis, emphysema, and malnutrition often have multiple (and often unknown) causes. They also tend to develop slowly and progressively over time, are not caused by living organisms, and do not spread from

one person to another. They are classified as **nontransmissible diseases**.

Solutions to Disease in LDCs   Poverty is the underlying cause of lower average life expectancy and higher infant mortality (Figure 6-9) in LDCs, as well as for poor people in MDCs. The overcrowding, unsafe drinking water, poor sanitation, and malnutrition associated with poverty increase the spread of transmissible diseases, which account for about 40% of all deaths in LDCs. The hot, wet climates of tropical and subtropical LDCs also increase the chances of infection, because disease organisms can thrive year-round.

With adequate funding, the health of people in LDCs can be improved dramatically, quickly, and cheaply by providing:

Sex Can Be Hazardous to Your Health

Sexually transmitted diseases (STDs) are passed on during sexual activity. Many STDs can also be transmitted from mother to infant during birth; from one intravenous (IV) drug user to another on shared needles; and by exposure to infected blood.

Worldwide, there are about 250 million new infections and 750,000 deaths from STDs each year. By 2000 the number of deaths from STDs is expected to reach 1.5 million—an average of 4,100 per day. In the United States STDs strike about 12 million people (8 million of them under age 25) per year (Figure 8-8). The number of new reported cases of most STDs has risen every year since 1981.

Since 1981 the *acquired immune deficiency syndrome* or AIDS—a fatal disease that develops from the HIV virus and is transmitted primarily by sexual contact—has become a serious global health threat. The World Health Organization estimated that by August 1993, at least 14 million people worldwide (57% of them in sub-Saharan Africa) had been infected with the HIV virus—many of them unknowingly. By 2000 at least 30 million people are expected to be infected, with 83% of them in Asia and Africa. Heterosexual transmission accounts for about 90% of the new infections worldwide and about 6% in the United States (12% among those ages 13–24).

Within 10 years 95% of those with the virus develop AIDS. There is as yet no cure for AIDS, although drugs may help some infected people live longer. By August 1992 an estimated 2.7 million people had AIDS (69% of them in Africa and 16% in the United States), and 2.5 million had died from the disease, with a death rate of 100,000 per year in 1992. By 2000 as many as 24 million people are expected to have AIDS, and the annual death toll may reach 400,000.

In the United States at least 300,000 and probably 1 million people are infected with HIV, 227,000 have AIDS, and 160,000 have died from the disease. Each day about 150 Americans die from AIDS, with the disease being the sixth leading cause of death among young people ages 15–24.

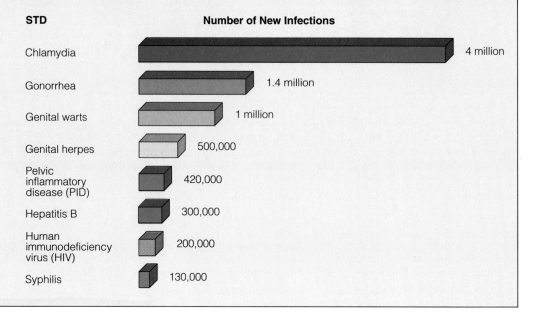

Figure 8-8 Annual number of new infections of sexually transmitted diseases in the United States. (Data from Centers for Disease Control and Prevention)

STD	Number of New Infections
Chlamydia	4 million
Gonorrhea	1.4 million
Genital warts	1 million
Genital herpes	500,000
Pelvic inflammatory disease (PID)	420,000
Hepatitis B	300,000
Human immunodeficiency virus (HIV)	200,000
Syphilis	130,000

- Contraceptives (Figure 6-7), sex education, and family-planning counseling.

- Better nutrition, prenatal care, and birth assistance for pregnant women. At least 500,000 women in LDCs die each year of mostly preventable pregnancy-related causes, compared with 6,000 in MDCs.

- Better nutrition for children.

- Greatly improved postnatal care (including the promotion of breastfeeding) to reduce infant mortality.

- Immunization against tetanus, measles, diphtheria, typhoid, and tuberculosis.

- Oral rehydration for diarrhea victims (a simple solution of water, salt, and sugar).

- Antibiotics for infections.

Q: Worldwide, how much topsoil is eroded each year?

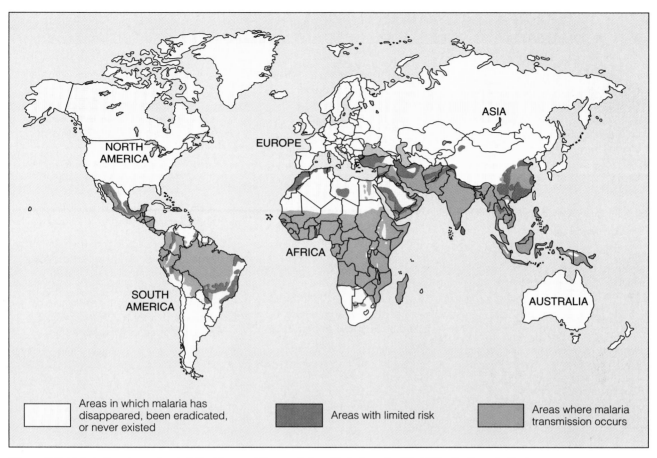

Figure 8-9 Malaria threatens about 40% of the world's population. (Data from the World Health Organization)

Areas in which malaria has disappeared, been eradicated, or never existed	Areas with limited risk	Areas where malaria transmission occurs

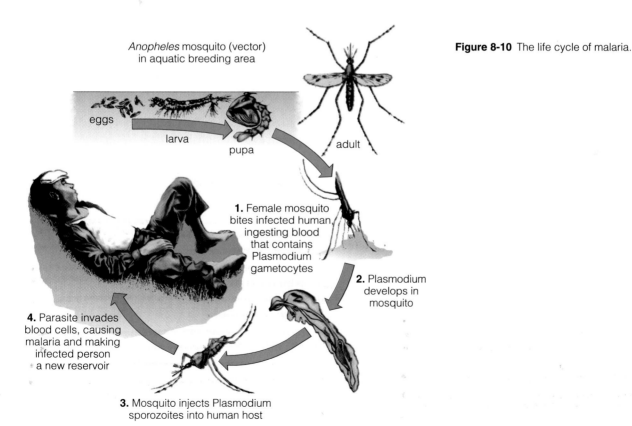

Anopheles mosquito (vector) in aquatic breeding area

Figure 8-10 The life cycle of malaria.

eggs
larva
pupa
adult

1. Female mosquito bites infected human, ingesting blood that contains Plasmodium gametocytes

2. Plasmodium develops in mosquito

3. Mosquito injects Plasmodium sporozoites into human host

4. Parasite invades blood cells, causing malaria and making infected person a new reservoir

A: About 23 billion metric tons (25 billion tons)

Reducing the Toll of Smoking

SOLUTIONS

The good news is that smoking—the single greatest human health hazard—is a preventable hazard. To reduce this hazard the American Medical Association and many health experts advocate:

- Banning all cigarette advertising in the United States.

- Forbidding the sale of cigarettes and other tobacco products to anyone under 21, with strict penalties for violators.

- Banning all cigarette vending machines.

- Classifying nicotine as addictive and dangerous and placing its use in tobacco or other products under the jurisdiction of the Food and Drug Administration.

- Eliminating all federal subsidies to U.S. tobacco farmers and tobacco companies.

- Taxing cigarettes at about $3.00 a pack (instead of the current 56¢, cigarettes are taxed at $3–6 per pack in Norway and at more than $3 per pack in Denmark and Canada). This action would discourage smoking and require that people who choose to smoke pay for the resulting harmful costs now borne by society as a whole—a user-pays approach. Such a program would raise about $50 billion per year in revenue, which could be targeted to help finance a better health care system.

- Forbidding U.S. officials to pressure other governments to import U.S. tobacco or tobacco products. Since 1985 the federal government has threatened trade sanctions against countries that place tariffs and other restrictions on U.S. tobacco products. Thus the U.S. government is coercing other governments into allowing imports of a hazardous (but legal) drug from America, while trying to halt the flow of illegal drugs into the United States.

It is encouraging that the percentage of U.S. adults who smoke dropped from 42% in 1964 to 26% in 1992. Each year in the United States an estimated 1.3 million adults quit smoking, but about 1.1 million teenagers start smoking.

 Clean drinking water and sanitation facilities to the one-third of the world's population that lacks them.

Extending such primary health care to all the world's people would cost an additional $10 billion per year, a mere 4% of what the world spends every year on cigarettes or devotes every four days to military spending.

Diseases in MDCs　As a country industrializes and makes the *demographic transition* (Figure 6-30), it also makes an *epidemiological transition*, in which the infectious diseases of childhood become less important and the chronic diseases of adulthood (heart disease and stroke, cancer, and respiratory conditions) become more important in determining mortality.

Each year about 2.3 million people die in the United States: 37% from heart attacks, 22% from cancer, 7% from strokes, 5% from accidents, and the remaining 29% from a variety of causes. In the United States and other MDCs, most deaths are a result of environmental and lifestyle factors rather than infectious agents invading the body. Except for auto accidents, these deaths result from chronic diseases that take a long time to develop and that have multiple causes. They are largely related to location (urban or rural),

work environment, diet, smoking (p. 192 and Solutions, above), exercise, sexual habits (Connections, p. 202), and the level of use of alcohol or other harmful drugs.

Changing these harmful lifestyle factors could prevent 40–70% of all premature deaths, one-third of all cases of acute disability, and two-thirds of all cases of chronic disability. Currently about 95% of the money spent on health care in the United States is used to *treat* rather than to *prevent* disease.

8-5 RISK ANALYSIS

Estimating Risks　Risk analysis involves identifying hazards and evaluating their associated risks (*risk assessment*); ranking risks (*comparative risk analysis*); determining options and making decisions about reducing or eliminating risks (*risk management*); and informing decision makers and the public about risks (*risk communication*).

Risk assessment involves determining the types of hazards involved, estimating the probability that each hazard will occur, estimating how many people are likely to be exposed to it, and how many may suffer serious harm. Probabilities based on past experience, animal and other tests, and epidemiological studies

　　　　　　Q: How much land has become desertified during the past 50 years?

are used to estimate risks from older technologies and products (Section 8-2). For new technologies and products, much more uncertain statistical probabilities, based on models rather than actual experience, must be calculated. Table 8-1 is an example of *comparative risk analysis*, summarizing the greatest ecological and health risks identified by a panel of scientists acting as advisers to the U.S. Environmental Protection Agency.

The more complex a technological system is and the more people that are needed to design and run it, the harder it is to calculate the risks. The overall reliability of any technological system is the product of two factors:

$$\text{system reliability (\%)} = \text{technology reliability} \times \text{human reliability} \times 100$$

With careful design, quality control, maintenance, and monitoring, a highly complex system such as a nuclear power plant or a space shuttle can achieve a high degree of technology reliability. However, human reliability is almost always much lower than technology reliability and is virtually impossible to predict; to be human is to err.

Suppose, for example, that the technology reliability of a nuclear power plant is 95% (0.95) and that the human reliability is 75% (0.75). Then the overall system reliability is only 71% (0.95 × 0.75 = 71%). Even if we could make the technology 100% reliable (1.0), the overall system reliability would still be only 75% (1.0 × 0.75 = 75%).

This crucial dependence—of even the most carefully designed systems—on unpredictable human reliability helps explain essentially "impossible" tragedies like the Three Mile Island and Chernobyl (p. 516) nuclear power plant accidents and the explosion of the space shuttle *Challenger*.

Poor management, poor training, and poor supervision increase the chances of human errors. Maintenance workers or people who monitor warning panels in the control rooms of nuclear power plants (Figure 8-11) get bored because most of the time nothing goes wrong. They may fall asleep on duty (as has happened in control rooms at several U.S. nuclear plants) or falsify maintenance records because they think the system is safe without their help. They may be distracted by personal problems or illness. And managers may order them to take shortcuts that increase short-term profits or make the managers look more efficient and productive.

One way to make a system more foolproof or "fail-safe" is to move more of the potentially fallible elements from the human side to the technical side. However, chance events such as a lightning bolt can knock out automatic control systems. And no machine or computer program can completely replace human

---

Table 8-1 Greatest Ecological and Health Risks

High-Risk Ecological Problems

Global climate change

Stratospheric ozone depletion

Wildlife habitat alteration and destruction

Species extinction and loss of biodiversity

Medium-Risk Ecological Problems

Acid deposition

Pesticides

Airborne toxic chemicals

Toxic chemicals, nutrients, and sediment in surface waters

Low-Risk Ecological Problems

Oil spills

Groundwater pollution

Radioactive isotopes

Acid runoff to surface waters

Thermal pollution

High-Risk Health Problems

Indoor air pollution

Outdoor air pollution

Worker exposure to industrial or farm chemicals

Pollutants in drinking water

Pesticide residues on food

Toxic chemicals in consumer products

Data from Science Advisory Board, *Reducing Risks*, Washington, DC, Environmental Protection Agency, 1990. Items in each category are not listed in rank order.

---

judgment in assuring that a complex system operates properly and safely. Also, the parts in any automated control system are manufactured, assembled, tested, certified, and maintained by fallible human beings.

Risk-Benefit Analysis The key question is whether the estimated short- and long-term benefits of using a particular technology or product outweigh the estimated short- and long-term risks of other alternatives. One method for making such evaluations is **risk-benefit analysis**. It involves estimating the short- and long-term societal benefits and risks involved and then dividing the benefits by the risks to find a **desirability quotient**:

$$\text{desirability quotient} = \frac{\text{societal benefits}}{\text{societal risks}}$$

Figure 8-11 Control room of a nuclear power plant. Watching these indicators is such a boring job that government investigators have found some operators asleep, and in one case they found the control room empty.

Assuming that benefits and risks can be calculated accurately (a big assumption), here are several possibilities:

- $\text{large desirability quotient} = \dfrac{\text{large societal benefits}}{\text{small societal risks}}$

  Example: *X rays.* Use of X rays (a form of ionizing radiation, Figure 3-4) to detect bone fractures and other medical problems has a large desirability quotient—provided that the dose is no larger than needed, that less harmful procedures are not available, and that doctors do not overprescribe them. Other examples in this category include mining, most dams, and airplane travel. Proponents of nuclear power plants place nuclear technology in this category.

- $\text{small desirability quotient} = \dfrac{\text{large societal benefits}}{\text{much larger societal risks}}$

  Example: *Coal-burning power plants and nuclear power plants* (Sections 19-3 and 19-4, respectively).

- $\text{uncertain desirability quotient} = \dfrac{\text{potentially large benefits}}{\text{unknown risks}}$

  Example: *Genetic engineering* (Pro/Con, next page). Some see biotechnology as a way to increase food supplies, degrade toxic wastes, eliminate certain genetic diseases and afflictions, and make enormous amounts of money. Others fear that, without strict controls, it could do great harm.

Problems with Risk Assessment Calculation of desirability quotients, as well as other ways of evaluating or expressing risk, is extremely difficult, imprecise, and controversial. Here are some of the key problems and issues:

- *Will some technologies benefit one group of people (population A) while imposing a risk on another (population B, often the poor and lower middle class)? Who should decide which population to favor?*

- *Should estimates emphasize short-term risks, or should more weight be put on long-term risks?* Who decides this?

- *Should the primary goal of risk analysis be to determine how much risk is acceptable (the current approach) or to figure out how to do the least damage (a prevention approach)?*

- *Who should do a particular risk-benefit analysis or risk assessment?* Should it be the corporation or government agency that develops or manages the technology, or some independent laboratory or panel of scientists? For an outside evaluation, who chooses the people to do the study? Who pays the bill and thus can influence the outcome by denying the lab, agency, or experts future business?

- *Once a risk-assessment study is done, who reviews the results? Is it a government agency? Independent scientists? The general public?*

Q: How much of the planet's land is threatened by desertification?

Genetic Engineering: Savior or Monster?

PRO/CON

"Genetic engineers" have learned how to splice genes and recombine sequences of existing DNA molecules in organisms to produce DNA with new genetic characteristics (recombinant DNA). This rapidly developing technology excites many scientists and investors. They see it as a way to increase crop and livestock yields, and also as a way to produce, patent, and sell crop varieties with greater resistance to diseases, pests, frost, and drought—and with more protein than existing varieties. They hope to create bacteria that can destroy oil spills, degrade toxic wastes, concentrate metals found in low-grade ores, and serve as biological "factories" for new vaccines, drugs, and therapeutic hormones. Gene therapy, proponents argue, could eliminate certain genetic diseases and other genetic afflictions.

When the U.S. Supreme Court ruled in 1980 that genetically engineered organisms could be patented, investors began pouring billions of dollars into the fledgling biotech industry, with dreams of more billions in profits. Already genetic engineering has produced a drug to reduce heart attack damage and agents to fight diabetes, hemophilia, and some forms of cancer. It has also been used to diagnose AIDS and cancer. In agriculture, gene transfer has been used to develop strawberries that resist frost, as well as smaller cows that produce more milk. Toxin-producing genes have been transferred from bacteria to plants, increasing the plants' immunity to insect attack.

Some people are worried that biotechnology may run amok. Most recognize that it is unrealistic to stop genetic engineering altogether, but they argue that it should be kept under strict control. They believe that we don't understand how nature works well enough to have unregulated control over the genetic characteristics of humans and other species.

Critics also fear that unregulated biotechnology could lead to the development of "superorganisms." For example, bacteria genetically altered to clean up ocean oil spills by degrading the oil would cause havoc if they were able to multiply rapidly and began degrading the world's remaining oil supplies. Genetically engineered organisms might also mutate, change their form and behavior, and then migrate. Unlike defective cars and other products, they couldn't be recalled.

Proponents argue that the risk of this or other biotech catastrophes is small. To have a serious effect, such organisms would have to be outstanding competitors and resistant to predation. In addition, they would have to be capable of becoming dominant in ecosystems and in the ecosphere.

Nevertheless, critics fear that the potential profits from biotechnology are so enormous that, without strict controls, greed—not ecological wisdom and restraint—will prevail. They point to the serious and widespread ecological problems when we have accidentally or deliberately introduced alien organisms into biological communities.

In 1989 a committee of prominent ecologists appointed by the Ecological Society of America warned that the ecological impacts of new combinations of genetic traits from different species would be difficult to predict. Their report called for a case-by-case review of any proposed environmental releases, as well as carefully regulated, small-scale field tests before any bioengineered organism is put into commercial use.

This controversy illustrates the difficulty of balancing the actual and potential benefits of a technology with its actual and potential risks of harm.

■ *Should cumulative impacts of various risks be considered, or should risks be considered separately, as is usually done?* For example, suppose a pesticide is found to have a risk of killing one in a million people, the current EPA limit. Cumulatively, however, effects from 40 such pesticides might kill 40—or even 400—of every 1 million people. Is this acceptable?

Some see risk analysis as a useful and much-needed tool. Others see it as a way to justify premeditated murder in the name of profit. According to hazardous waste expert Peter Montague (Guest Essay, p. 37), "The explicit aim of risk assessment is to convince people that some number of citizens *must* be killed each year to maintain a national lifestyle based on necessities like Saran Wrap, throwaway cameras, and lawns without dandelions." Risk evaluator Mary O'Brien says, "Risk assessment is industry's attempt to make the intolerable appear tolerable."

Such critics accuse industries of favoring risk analysis because so little is known about health risks from pollutants and because existing data are controversial. This allows risk-benefit analysis to be crafted to support almost any conclusion and then be labeled as "scientific decision making." The huge uncertainties in risk analysis also allow industries to delay regulatory decisions for decades by challenging data in the courts.

Despite the inevitable uncertainties involved, proponents argue that risk analysis is a useful way to organize available information, identify significant hazards, focus on areas that need more research, and stimulate people to make more informed decisions about health and environmental goals and priorities. However, at best risk assessments yield only a range of probabilities and uncertainties based on different assumptions—not the precise numbers that decision makers want.

Managing Risk   Once an assessment of risk is made, decisions must be made about what to do about the risk. **Risk management** includes the administrative, political, and economic actions taken to decide whether and how a particular societal risk is to be reduced to a certain level—and at what cost. Risk management involves trying to answer the following questions:

- Which of the vast number of risks facing society should be evaluated and managed with the limited funds available?

- In what sequence or priority should the risks be evaluated and managed?

- How reliable is the risk-benefit analysis or risk assessment carried out for each risk?

- How much risk is acceptable? How safe is safe enough?

- How much money will it take to reduce each risk to an acceptable level?

- How much will each risk be reduced if available funds are limited, as is usually the case?

- How will the risk management plan be communicated to the public, monitored, and enforced?

Risk managers must make difficult decisions based on inadequate and uncertain scientific data. These decisions have potentially grave consequences for human health and the environment, and they also have large economic effects on industry and consumers. Each step in this process involves value judgments and trade-offs to find some reasonable compromise between conflicting political, economic, and environmental interests.

Risk Perception and Communication   Most of us do poorly in assessing the risks from the hazards that surround us, and we tend to be full of contradictions. For example, many people deny or shrug off the high-risk nature of activities such as motorcycling (1 death per 50 participants), smoking (1 death in 600 by age 35 for a pack-a-day smoker), a car crash (1 in 5,000 with a seat belt versus 1 in 2,500 without), parachuting (1 in 4,000), and rock climbing (1 in 5,000). Yet these same people may be terrified about the possibility of dying in a commercial airplane crash (1 in 800,000), in a train crash (1 in 20 million), from a shark attack (1 in 300 million), or from drinking water with the EPA limit of trichloroethylene (1 in 2 billion).

Being bombarded with news about people killed or harmed by various hazards distorts our sense of risk. *The real news each year is that 99% of the people on Earth didn't die.* However, that's not what we see on TV or hear about every day.

The public generally sees a technology or a product as being riskier than experts see it (Table 8-1) when:

- *It is fairly new or complex rather than familiar.* Examples might include genetic engineering or nuclear power as opposed to dams or automobiles.

- *It is mostly involuntary instead of voluntary.* Examples might include nuclear power plants, nuclear weapons, industrial pollution, or food additives, as opposed to smoking, drinking alcohol, or driving.

- *It is viewed as unnecessary rather than as beneficial or necessary.* Examples might include CFC and hydrocarbon propellants in aerosol spray cans or food additives used to increase sales appeal—as opposed to cars or firearms for self-defense.

- *It involves a large, well-publicized death toll from a single catastrophic accident rather than the same or even larger death toll spread out over a longer time.* Examples might include a severe nuclear power plant accident, an industrial explosion, or a plane crash, as opposed to coal-burning power plants, automobiles, smoking, or malnutrition in LDCs.

- *It involves unfair distribution of the risks.* For example, citizens are outraged when government officials decide to put a hazardous-waste landfill or incinerator in or near their neighborhood even when the decision is based on risk-benefit analysis. This is usually seen as politics, not science.

- *It is poorly communicated* (Guest Essay, next page). Does the decision-making agency or company seem trustworthy and concerned or dishonest, unconcerned, and arrogant (Exxon after the Valdez oil spill, the Nuclear Regulatory Agency and the nuclear industry since the Three Mile Island accident)? Does it involve the community in the decision-making process from start to finish and reveal what's going on before the real decisions are made? Does it understand, listen to, and respond to community concerns?

- *It does not involve a sincere search for and evaluation of alternatives.*

Q: What major food-producing country is doing the most to reduce soil erosion?

Public Confidence in Industry and Government: A Crisis in Environmental Risk Communication

Vincent T. Covello

GUEST ESSAY

Vincent T. Covello is professor of environmental sciences in the School of Public Health at Columbia University and Director of Columbia University's Center for Risk Communication in New York City. Before these appointments he was director of the Risk Assessment Program at the National Science Foundation and a senior scientist for the White House Council on Environmental Quality. He is on the editorial board of several scientific journals and is the past president of the Society for Risk Analysis. He has written or edited over 25 books and 75 articles on environmental risk management. Two of his most recent books are Effective Risk Communication *(Plenum, 1989) and* Principles and Methods for Analyzing Health and Environmental Risks *(Council on Environmental Quality, 1988).*

Obtaining reliable scientific information about environmental risk is a matter of intense concern to the public. In the United States the majority of people see industry and government as the most knowledgeable about those risks. Yet at the same time most Americans view industry and government as two of the least trustworthy sources of information about the risks of chemicals, radiation, and potential environmental hazards. Several factors have contributed to this crisis in communication between industry and government and concerned citizens. They include

- *Disagreements among scientific experts.* Because of different assumptions, data, and methods, experts in industry and government often disagree and engage in highly visible debates about the reliability, validity, and interpretation of risk-assessment results. While such debates may advance scientific knowledge, they can undermine confidence in the ability of industry and government.

- *Lack of resources for risk assessment and management.* Most people don't buy explanations that the generation of health and environmental data about risks is constrained by financial, technical, statutory, legal, or other limitations.

- *Lack of adequate coordination among responsible authorities.* Regulatory agencies are rarely required to develop coherent, coordinated, consistent, and interrelated plans, programs, and guidelines for assessing and managing risks. This fragmentation often leads to jurisdictional conflicts about which agency or level of government has the ultimate responsibility for assessing and managing the risk in question, and in many cases to competing estimates of risk.

- *Lack of attention to and priority for risk communication.* Many industry and government officials are poor communicators. Their complex language and jargon-ridden statements often make them look unresponsive, dishonest, or evasive. They aggravate the problem by ignoring the many pitfalls of risk communication, including using quantitative risk numbers in public presentations inappropriately, attacking the credibility of environmental or consumer activists, using risk- and cost-benefit studies inappropriately, and comparing risks for activities the public sees as different.

- *Insensitivity to the information needs and concerns of the public.* Risk experts and the general public frequently look at risks differently. For example, experts tend to focus on the numbers, while laypeople consider a complex array of factors such as voluntariness (choice), familiarity, effects on children, effects on future generations, benefits, origin (natural or from human activities), and fairness.

Public trust and confidence in industry and government will be restored only when people are convinced that industry and government officials share their concerns about environmental risks; that they are scientifically, technologically, and managerially competent; that they are honest, fair, and open; and that they are personally committed to eliminating risks or reducing them to an absolute minimum. Even more basic trust and credibility will be restored only when industry and government officials recognize, accept, and involve the public as a legitimate partner in making decisions about risk.

Such a partnership also places demands on citizens, who must be open-minded and willing to take the time to learn about risk issues. A guiding principle of risk communication in a democracy is that people and communities have a right to participate in decisions that affect their lives, their property, and the things they value. The goal of risk communication in a democracy should be to produce an informed public that is involved, interested, reasonable, thoughtful, solution oriented, and collaborative. It should not be to defuse public concerns or to replace needed action.

Critical Thinking

1. In general do you believe information provided by government or industry officials about environmental risks? Explain why you do or don't. Whom do you trust as sources of such information? Why?

2. Are you willing to learn about and participate in making decisions about environmental risks? Explain. Have you ever done so? Why or why not?

People who believe their lives and the lives of their children are being threatened because they live near an actual or proposed chemical plant, toxic-waste dump, or waste incinerator don't care that experts say the chemical is likely to kill only one in a million people in the general population. Only a few of those million people live or will live near the plant, dump, or incinerator, as they do. And that one might be their child. To them it is a personal threat, not a statistical abstraction.

Solutions: Reducing Health Risks   The most effective and cheapest way to reduce the risks we face from chemical and biological hazards is pollution prevention. We can do this on a personal level by becoming more informed about the relative risks of various hazards and by making lifestyle choices—choosing not to smoke or be around smokers, to practice safe sex, to follow a healthy diet coupled with some exercise, and to protect our eyes and skin from the sun.

We must also work to see that the poor have such choices instead of being forced to face serious hazards in order to survive. At the local, national, and global levels we must become involved politically to ensure that people are not involuntarily exposed to unnecessary health hazards. We must insist that pollution prevention be the centerpiece of all environmental, health, and economic policy and the best way to reduce risks.

Not all waste and pollution can be eliminated.... What is absolutely crucial, however, is to recognize that pollution prevention should be the first choice and the option against which all other options are judged. The burden of proof imposed on individuals, companies, and institutions should be to show that pollution prevention options have been thoroughly examined, evaluated, and used before lesser options are chosen.

JOEL HIRSCHORN

Critical Thinking

1. Do you believe that smoking is the single greatest threat to human health? Explain why you agree or disagree.

2. Should standards for allowed pollution levels be set to protect the most sensitive person or only the average person in a population? Explain. Should we have zero pollution levels for all toxic and hazardous chemicals? Explain.

3. Do you believe that health and safety standards in the workplace should be strengthened and enforced more vigorously—even if this causes a loss of jobs because companies transfer operations to countries with weaker standards? Explain.

4. How would you go about reducing the overall threat to human health from AIDS and other STDs? What things do you do in your own life to reduce your risk of getting a sexually transmitted disease?

5. What restrictions, if any, do you believe should be placed on genetic engineering research and use? How would you enforce any restrictions?

6. Explain why you agree or disagree with each of the following proposals.
   a. All advertising of cigarettes and other tobacco products should be banned.
   b. All smoking should be banned in public buildings and commercial airplanes, buses, subways, and trains.
   c. All government subsidies to tobacco farmers and the tobacco industry should be eliminated.
   d. Cigarettes should be taxed at about $3 per pack so that U.S. smokers alone pay for the health and productivity losses now borne by society as a whole.

7. Do you believe that risk-benefit analysis should be used as the primary method to evaluate and manage risks? Explain.

*8. Conduct a survey to determine the major physical, chemical, and biological hazards to human health at your school. Develop a plan for preventing or reducing these hazards, and present the results of your survey and plan to school officials.

*9. Make a concept map of the key ideas in this chapter using the section heads and subheads and the key terms (shown in boldface type in the chapter). See the inside front cover and Appendix 4 for information on concept maps.

PART III
Resources: Air, Water, Soil, Minerals, and Wastes

I am utterly convinced that most of the great environmental struggles will be either won or lost in the 1990s, and that by the next century it will be too late to act.

THOMAS E. LOVEJOY

9 Air

When Is a Lichen Like a Canary?

Nineteenth-century coal miners took canaries with them into the mines—not for their songs but for that moment when they stopped singing. Then the miners knew the air was bad, and it was time to get out of the mine.

Nowadays we use sophisticated equipment to monitor air quality, but living things like lichens (Figure 9-1) still have a role in warning us of bad air. A lichen consists of fungus and alga living together, usually in a mutually beneficial (mutualistic) partnership.

With more than 20,000 known species, lichens can live almost anywhere—on rocks, trees, bare soil, buildings, gravestones, and even sun-bleached bones—and some survive for more than 4,000 years.

These hearty pioneer species are good air pollution detectors because they are always absorbing air as a source of nourishment. Certain lichen species are sensitive to specific air-polluting chemicals. Old man's beard (*Usnea trichodea*) (Figure 9-1, bottom) and yellow *Evernia* lichens, for example, sicken or die in the presence of too much sulfur dioxide.

Since lichens are widespread, long-lived, and reliably anchored in their spots, they can also be used to track pollution to its source. The scientist who discovered sulfur dioxide pollution on Isle Royale in Lake Superior, where no car or smokestack had ever intruded, used *Evernia* lichens to point the finger northward to coal-burning facilities at Thunder Bay, Canada.

Last, but not least, lichens can replace electronic monitoring stations that would cost more than $100,000 each. This is not so much a triumph of nature over technology but a partnership between the two, for technicians use highly sophisticated methods to analyze lichens for pollution and to measure their rates of photosynthesis.

Because we all must breathe air from a shared global atmospheric commons, air pollution anywhere is a potential threat elsewhere, and in some cases everywhere. Although lichens can alert us to the danger, as with all forms of pollution, the only real solution is prevention.

Figure 9-1 Red and yellow crustose lichens growing on slate rock in the foothills of the Sierra Nevada near Merced, California (top), and Usnea trichodea lichen growing on a branch of larch tree in Gifford Pinchot National Park, Washington. Various species of lichens can be used to detect levels of specific air pollutants and to track down their sources. (Kenneth W. Fink/ Ardea London, top; Milton Rand/Tom Stack & Associates, bottom)

I thought I saw a blue jay this morning. But the smog was so bad that it turned out to be a cardinal holding its breath.

MICHAEL J. COHEN

This chapter is devoted to answering the following questions:

- What layers are found in the atmosphere?
- What are air pollutants, and where do they come from?
- What is smog?
- What is acid deposition?
- What are the harmful effects of air pollutants?
- How can we prevent and control air pollution?

9-1 THE ATMOSPHERE

The Troposphere: Life Giver and Weather
Breeder   We live at the bottom of a "sea" of air called the **atmosphere**. This thin envelope of life-sustaining gases surrounding the earth is divided into several spherical layers characterized by abrupt changes in temperature due to differences in the absorption of incoming solar energy (Figure 9-2).

About 75% of the mass of Earth's air is found in the atmosphere's innermost layer, the **troposphere**, which extends only about 17 kilometers (11 miles) above sea level at the equator and about 8 kilometers (5 miles) over the poles. If the earth were an apple, this lower layer, containing the air we breathe, would be no thicker than the apple's skin. This thin and turbulent layer of rising and falling air currents and winds is the planet's weather breeder.

The composition of the atmosphere has varied considerably throughout Earth's long history (Section 5-5). Today about 99% of the volume of clean, dry air in the troposphere consists of two gases: nitrogen (78%) and oxygen (21%). The remainder has slightly less than 1% argon (Ar), 0.036% carbon dioxide (CO_2), and trace amounts of neon (Ne), helium (He), methane (CH_4), krypton (Kr), hydrogen (H_2), Xenon (Xe), and chlorofluorocarbons (CFCs, put there by human activities). Air in the troposphere also holds water vapor in amounts varying from 0.01% by volume at the frigid poles to 5% in the humid tropics.

Temperature drops with altitude in the troposphere. At the top of this zone there is a sudden reversal, and temperatures start to rise. The boundary where this occurs is called the *tropopause*; it limits mixing between the troposphere and upper layers.

The Stratosphere: Earth's Global Sunscreen
The tropopause marks the end of the troposphere

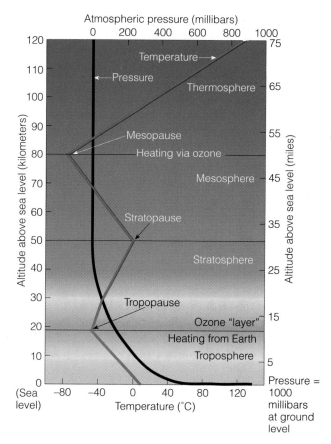

Figure 9-2 Earth's present atmosphere consists of several layers. Most ultraviolet radiation from the sun is absorbed by ozone (O_3) in the stratosphere, most of which is found in the so-called *ozone layer*, between 17 and 26 kilometers (11–16 miles) above sea level.

and the beginning of the **stratosphere**, the atmosphere's second layer, which extends from about 17–48 kilometers (11–30 miles) above Earth's surface (Figure 9-2). Although the stratosphere contains less matter than the troposphere, its composition is similar, with two notable exceptions: Its volume of water vapor is about 1,000 times less, and its volume of ozone (O_3) is about 1,000 times greater.

Stratospheric ozone is produced when some of the oxygen molecules there interact with lightning and solar radiation. Ozone is continuously being formed and destroyed, but as long as the rates of these two reversible processes are equal, the average concentration of ozone in the stratosphere remains constant (although the concentration varies at different altitudes and at different places). The presence of ozone in the stratosphere keeps about 99% of the sun's harmful ultraviolet radiation (especially ultraviolet-B, or UV-B) given off by the sun from reaching the earth's surface. This filtering action helps protect the human body from sunburn, skin and eye cancer, cataracts, and damage to the immune system.

This "global sunscreen" also prevents damage to some plants, aquatic organisms, and other land animals. Furthermore, by keeping most high-energy UV radiation from reaching the troposphere, ozone in the stratosphere prevents much of the oxygen in the troposphere from being converted to ozone, which is harmful to us and other forms of life. The trace amounts of ozone that do form in the troposphere as a component of urban smog damage plants, the respiratory systems of humans and other animals, and materials such as rubber.

Thus our good health, and that of many other species, depends on having enough "good" ozone in the stratosphere and as little "bad" ozone as possible in the troposphere. Unfortunately, our activities are both increasing the amount of harmful ozone in the tropospheric air we must breathe and decreasing the amount of beneficial ozone in the stratosphere.

Air in the stratosphere, unlike that in the troposphere, is calm, with little vertical mixing. Pilots like to fly in this layer because it has so little turbulence and such excellent visibility (due to the almost complete absence of clouds). Flying in the stratosphere also improves fuel efficiency because the thin air offers little resistance to the forward thrust of the plane. And unlike in the troposphere, temperature rises with altitude in the stratosphere until there is another reversal at the *stratopause*, which marks the end of the stratosphere and the beginning of the atmosphere's next layer.

Mesosphere and Thermosphere   Above the stratosphere the temperature begins falling with altitude in a layer called the **mesosphere**, or middle layer (Figure 9-2). At the top of this layer there is another reversal of temperature called the *mesopause*. This marks the beginning of the **thermosphere**, where temperatures rise again. Temperatures are very high in this layer because its gaseous molecules are constantly bombarded by high-energy solar radiation and cosmic rays (Figure 3-4). This input of high-energy electromagnetic radiation also converts the gaseous molecules in this layer to ions.

Connections: Disrupting Earth's Nutrient Cycles   Earth's biogeochemical cycles (Section 4-4) work fine as long as we don't disrupt them by overloading them at certain points or by removing too many vital chemicals at other points. In fact, our activities are disrupting the gaseous parts of some of these cycles.

We produce one-fourth as much CO_2 as the rest of nature does, as we burn up nature's one-time-only deposit of fossil fuels and as we clear forests. This massive human intrusion into the carbon cycle (Figure 4-

29) has the potential to warm the earth and alter global climate and food-producing regions (Section 10-2). We also disrupt natural energy flows (Figure 4-4) by producing massive heat islands and dust domes over urban areas (Figure 6-22).

By burning fossil fuels and using nitrogen fertilizers, we are releasing three times the nitrogen oxides and gaseous ammonia into the atmosphere as are natural processes in the nitrogen cycle (Figure 4-30). In the atmosphere these nitrogen oxides are converted to nitric acid (HNO_3) that returns to the earth, where it can increase the acidity of soils, streams, and lakes.

We are currently releasing twice as much sulfur into the atmosphere as are natural processes (Figure 4-33). Most of this is sulfur dioxide emitted by petroleum refining and by burning coal and oil. In the atmosphere this sulfur dioxide is converted to sulfuric acid (H_2SO_4) and sulphate salts that return to the earth as components of acid deposition.

These are only a few of the chemicals we are spewing into the atmosphere. Small amounts of toxic metals like arsenic, cadmium, and lead are also circulated in the ecosphere in chemical cycles. We now inject twice as much arsenic into the atmosphere as nature does, 7 times as much cadmium, and 17 times as much lead.

9-2   OUTDOOR AIR POLLUTION: POLLUTANTS, SMOG, AND ACID DEPOSITION

Types and Sources of Air Pollution   As clean air in the troposphere (Figure 9-2) moves across the earth's surface, it collects the products of both natural events (dust storms and volcanic eruptions, Figure 8-7) and human activities (emissions from cars and smokestacks). These potential pollutants, called **primary pollutants**, mix with the churning air in the troposphere. Some may react with one another or with the basic components of air to form new pollutants, called **secondary pollutants** (Figure 9-3). Long-lived pollutants travel far before they return to the earth as particles, droplets, or chemicals dissolved in precipitation.

Table 9-1 lists the major classes of pollutants found in outdoor air. Outdoor pollution in industrialized countries comes mostly from five groups of primary pollutants: *carbon oxides* (CO and CO_2), *nitrogen oxides* (mostly NO and NO_2, or NO_x), *sulfur oxides* (SO_2 and SO_3), *volatile organic compounds* (VOCs, mostly hydrocarbons), and *suspended particles*—all produced primarily by combustion of fossil fuels (Figure 3-5).

In MDCs, most pollutants enter the atmosphere from the burning of fossil fuels in both power plants

Q: What is the largest source of solid waste in the United States?

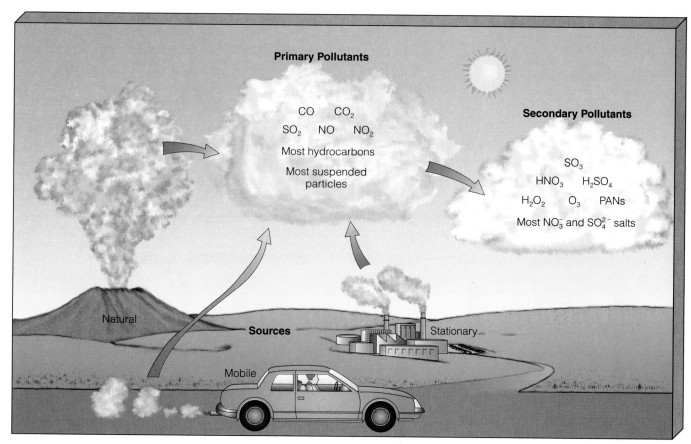

Figure 9-3 Primary and secondary air pollutants.

Table 9-1 Major Classes of Air Pollutants

Class	Examples
Carbon oxides	Carbon monoxide (CO), carbon dioxide (CO_2)
Sulfur oxides	Sulfur dioxide (SO_2), sulfur trioxide (SO_3)
Nitrogen oxides	Nitric oxide (NO), nitrogen dioxide (NO_2), nitrous oxide (N_2O) (NO and NO_2 are often lumped together and labeled as NO_x)
Volatile organic compounds	Methane (CH_4), propane (C_3H_8), benzene (C_6H_6), chlorofluorocarbons (CFCs)
Suspended particles	Solid particles (dust, soot, asbestos, lead, nitrate and sulfate salts), liquid droplets (sulfuric acid, PCBs, dioxins, pesticides)
Photochemical oxidants	Ozone (O_3), peroxyacyl nitrates (PANs), hydrogen peroxide (H_2O_2), aldehydes
Radioactive substances	Radon-222, iodine-131, strontium-90, plutonium-239 (Table 3-1)
Toxic compounds	Trace amounts of at least 600 toxic substances (many of them volatile organic compounds), 60 of them known carcinogens

and factories (*stationary sources*) and in motor vehicles (*mobile sources*) (Figure 9-3). In car-clogged cities like Los Angeles, São Paulo, London, and Mexico City, motor vehicles are responsible for 80–88% of the air pollution.

Connections: Smog Photochemical smog is a mixture of primary and secondary pollutants that forms when some of the primary pollutants interact under the influence of sunlight (Figure 9-4). The resulting mix of more than 100 chemicals is dominated

A: Mining of nonfuel minerals, creating waste equal to 6 times the garbage produced by the U.S. population   CHAPTER 9

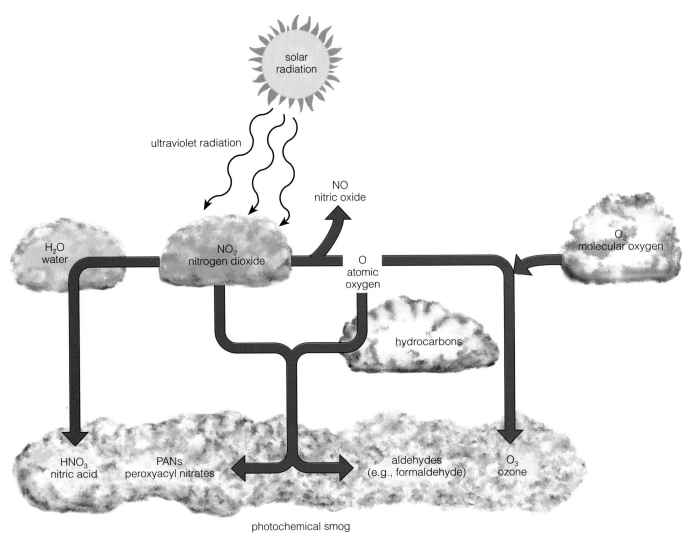

Figure 9-4 Simplified scheme of the formation of photochemical smog. The severity of smog is generally associated with atmospheric concentrations of ozone at ground level.

by ozone, a highly reactive gas that harms most living organisms.

Virtually all modern cities have photochemical smog, but it is much more common in cities with sunny, warm, dry climates and lots of motor vehicles. Los Angeles (Figure 9-5), Denver, Salt Lake City, Sydney, Mexico City (Figure 6-17), and Buenos Aires all have serious photochemical smog problems. The hotter the day, the higher the levels of ozone and other components of photochemical smog. On a sunny day the smog builds up to peak levels by early afternoon, irritating people's eyes and respiratory tracts. People with asthma and other respiratory problems, and healthy people who exercise outdoors between 11 A.M. and 4 P.M., are especially vulnerable.

Thirty years ago cities like London, Chicago, and Pittsburgh burned large amounts of coal and heavy oil (which contain sulfur impurities), in power plants and factories, and for space heating. During winter such cities suffered from **industrial smog**, consisting mostly of a mixture of sulfur dioxide, suspended droplets of sulfuric acid (formed from some of the sulfur dioxide, Figure 9-6), and a variety of suspended solid particles. Today, in most parts of the world coal and heavy oil are burned only in large boilers with reasonably good pollution control or with tall smokestacks; so industrial smog, sometimes called *gray-air smog*, is rarely a problem. However, in China, Ukraine, and some eastern European countries, such as Poland, large quantities of coal are still burned with inadequate pollution controls.

The frequency and severity of smog in an area depend on several things: the local climate and topography, the density of the population, the amount of industry, and the fuels used in industry, heating, and transportation. In areas with high average annual pre-

Q: How much of the solid waste produced in the U.S. is municipal solid waste (garbage)?

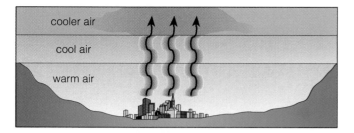

Normal pattern

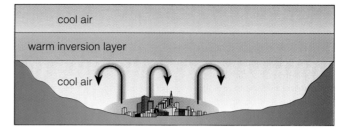

Thermal inversion

Figure 9-5 Thermal inversions trap pollutants in a layer of cool air that cannot rise to carry the pollutants away. The photograph shows a Los Angeles freeway covered with an orange haze of photochemical smog at twilight. Because of its topography Los Angeles has frequent thermal inversions, many of them prolonged during the summer months.

cipitation, rain and snow help cleanse the air of pollutants. Winds also help sweep pollutants away and bring in fresh air, but they may transfer some pollutants to distant areas.

Hills and mountains tend to reduce the flow of air in valleys below and allow pollutant levels to build up at ground level. Buildings in cities generally slow wind speed, thereby reducing dilution and removal of pollutants.

During the day the sun warms the air near the earth's surface. Normally this heated air expands and rises, carrying low-lying pollutants higher into the troposphere. Colder, denser air from surrounding high-pressure areas then sinks into the low-pressure area created when the hot air rises (Figure 9-5). This continual mixing of the air helps keep pollutants from reaching dangerous concentrations near the ground.

Sometimes, however, weather conditions trap a layer of dense, cool air beneath a layer of less dense, warm air in an urban basin or valley—a phenomenon called a **temperature inversion** or a **thermal inversion** (Figure 9-5). In effect, a lid of warm air covers the region and prevents ascending air currents (that would disperse pollutants) from developing. These inversions usually last for only a few hours; but sometimes, when a high-pressure air mass stalls over an area, they last for several days. Then air pollutants at ground level build up to harmful and even lethal concentrations.

The first U.S. air pollution disaster occurred in 1948, when fog laden with sulfur dioxide and suspended particulate matter stagnated for five days over the town of Donora in Pennsylvania's Monongahela valley south of Pittsburgh. About 6,000 of the town's 14,000 inhabitants fell ill, and 20 of them died. This

Figure 9-6 Sulfur dioxide emissions from a coal-burning power plant. Special camera filters revealed these otherwise invisible pollutants. In the atmosphere sulfur dioxide is converted to droplets of sulfuric acid, a major component of acid deposition, which can damage plants and kill fish.

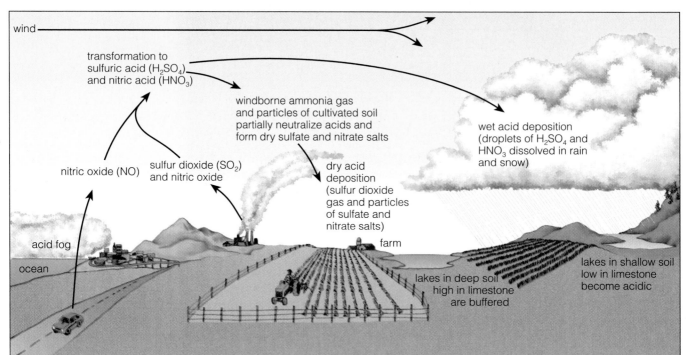

Figure 9-7 Acid deposition consists of acidified rain, snow, dust, or gas with a pH lower than 5.6. The lower the pH level, the greater the acidity of this wet and dry deposition, commonly called acid rain.

killer fog resulted from a combination of mountainous terrain surrounding the valley and stable weather conditions. These conditions trapped and concentrated the deadly pollutants emitted by the community's steel mill, zinc smelter, and sulfuric acid plant. Ther-

mal inversions also enhance the harmful effects of urban heat islands and dust domes that build up over urban areas (Figure 6-22).

A city with several million people and automobiles in an area with a sunny climate, light winds, mountains

Q: How many tires do Americans throw away each year?

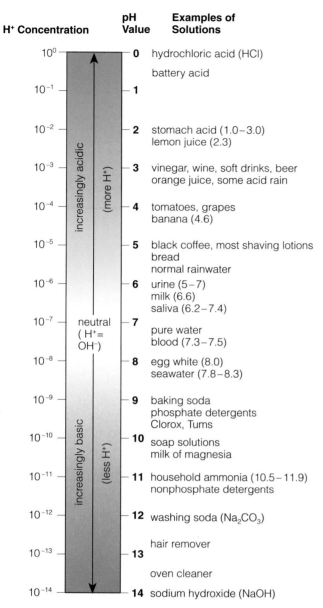

Figure 9-8 The pH scale, used to measure acidity and alkalinity of water solutions. Values shown are approximate. A neutral solution has a pH of 7; one with a pH greater than 7 is basic, or alkaline; and one with a pH less than 7 is acidic. The lower the pH below 7, the more acidic the solution. Each whole-number drop in pH represents a 10-fold increase in acidity.

on three sides, and the ocean on the other has ideal conditions for photochemical smog worsened by frequent thermal inversions. This describes the Los Angeles basin, which has 14 million people, 10.6 million motor vehicles, thousands of factories, and thermal inversions at least half of the year (Figure 9-5). Despite having the world's toughest air pollution control program, Los Angeles is the air pollution capital of the United States. By 2010 the Los Angeles area is expected to have 21 million people and 13 million cars.

Connections: Acid Deposition   To reduce local air pollution (and meet government standards without having to add expensive air pollution control devices), most coal-burning power plants, ore smelters, and other industrial plants began using tall smokestacks to emit sulfur dioxide, suspended particles, and nitrogen oxides above the inversion layer (Figure 9-5, right). As this practice spread in the 1960s and 1970s, pollution in downwind areas began to rise. In addition to smokestack emissions, large quantities of nitrogen oxides are also released by motor vehicles.

As sulfur dioxide and nitrogen oxides are transported as much as 1,000 kilometers (600 miles) by prevailing winds, they form secondary pollutants such as nitric acid vapor, droplets of sulfuric acid, and particles of sulfate and nitrate salts (Figures 9-3 and 9-6). These chemicals descend to the earth's surface in two forms: *wet*, as acidic rain, snow, fog, and cloud vapor, and *dry*, as acidic particles. The resulting mixture is called **acid deposition**, commonly called *acid rain* (Figure 9-7). Because these acidic components remain in the atmosphere for only a few days, acid deposition occurs on a regional, rather than a global, basis.

Acidity of substances in water is commonly expressed in terms of pH (Figure 9-8), with pH values greater than 7 being alkaline and pH values less than 7 being acidic. Natural precipitation is slightly acidic, with a pH of 5.0–5.6. However, typical rain in the eastern United States is about 10 times more acidic, with a pH of 4.3. In some areas it is 100 times more acidic, with a pH of 3—as acidic as vinegar. And some cities and mountaintops downwind from cities are bathed in acid fog as acidic as lemon juice, with a pH of 2.3—about 1,000 times the acidity of normal precipitation.

Acid deposition has a number of harmful effects (especially when the pH falls below 5.1 and below 5.5 for aquatic systems), including the following:

- It damages statues, buildings, metals, and car finishes.

- It can contaminate fish in some lakes with highly toxic methylmercury. Apparently, increased acidity of lakes converts inorganic mercury compounds in lake-bottom sediments into more toxic methylmercury, which is more soluble in the fatty tissue of animals.

- It can damage foliage and weaken trees, especially conifers such as red spruce at high elevations, which are bathed almost continuously in very acidic fog and clouds (Figure 9-9).

- It and other air pollutants can make trees more susceptible to stresses such as cold temperatures, diseases, insects, drought, and fungi that thrive under acidic conditions (Figure 9-9 and p. 211).

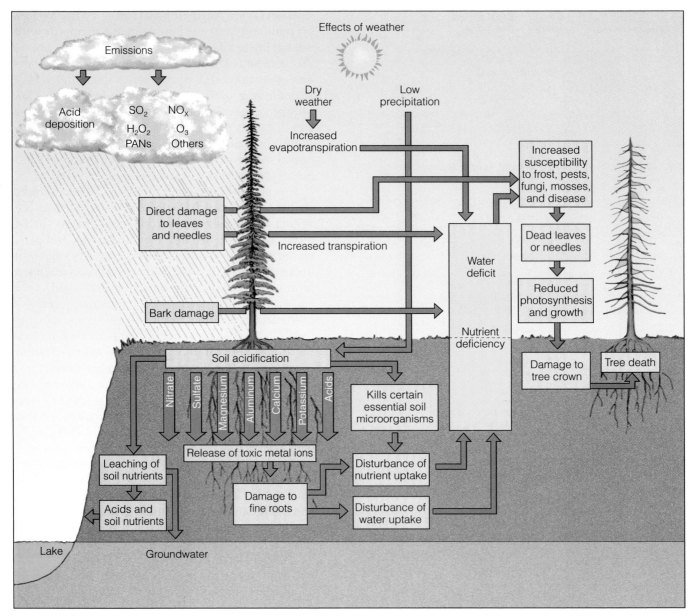

Figure 9-9 Possible or suspected harmful effects of air pollutants on trees.

- It can release soluble aluminum ions, which damage tree roots, from soil (Figure 9-9). When washed into lakes, the released aluminum ions also can kill many kinds of fish by increasing mucus formation that clogs their gills.

- It leads to excessive levels of nitrogen in the soil, which can overstimulate plant growth and increase depletion of other soil nutrients.

- It contributes to human respiratory diseases such as bronchitis and asthma, which can cause premature death.

Acid deposition is a serious regional problem (Figure 9-10) in many areas downwind from coal-burning power plants, smelters, factories, and large urban areas. Vegetation and aquatic life in lakes in areas with thin, acidic soils are especially sensitive to acid deposition (Figure 9-10). They are less vulnerable in other areas where soils and bedrock contain limestone and other alkaline substances that easily dissolve in water and can neutralize acids. However, even in these areas repeated exposure year after year can deplete the acid-neutralizing chemicals in highly alkaline soils.

A large-scale, government-sponsored research study on acid deposition in the United States in the 1980s concluded that the problem was serious but not yet at a crisis stage. Representatives of coal companies, and of industries that burn coal and oil, claim

Q: What percentage of glass beverage bottles in the U.S. are refillable?

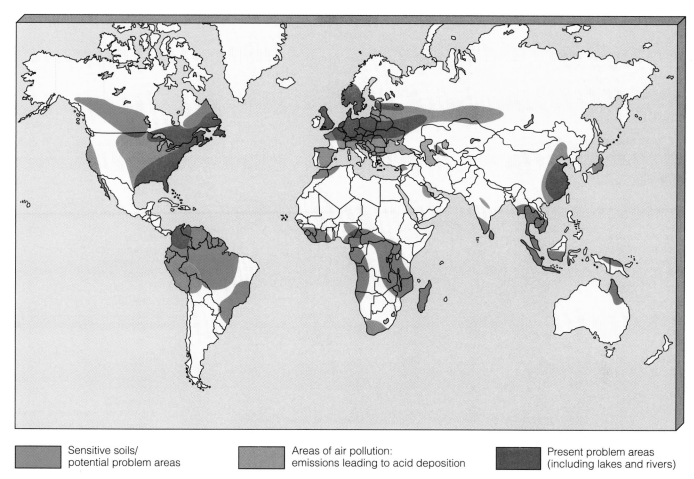

 (legend)

| | Sensitive soils/ potential problem areas | | Areas of air pollution: emissions leading to acid deposition | | Present problem areas (including lakes and rivers) |

Figure 9-10 Regions where acid deposition is now a problem, and regions with potential acid deposition problems. (Data from World Resource Institute and Environmental Protection Agency)

that adding expensive air pollution control equipment or burning low-sulfur coal or oil costs more than the resulting health and environmental benefits are worth. However, a 1990 economic study indicated that the benefits of controlling acid deposition will be worth at least $5 billion (some say $10 billion) per year, about 50% greater than the costs of controlling acid deposition.

A large portion of the acid-producing chemicals generated in one country may be exported to others by prevailing winds. For example, more than three-fourths of the acid deposition in Norway, Switzerland, Austria, Sweden, the Netherlands, and Finland is blown to those countries from industrialized areas of western and eastern Europe.

Chemical detective work indicates that more than half the acid deposition in southeastern Canada and the eastern United States originates from coal- and oil-burning power plants and factories in seven states: Ohio, Indiana, Pennsylvania, Illinois, Missouri, West Virginia, and Tennessee. The large net flow of acid deposition from the United States to Canada strained relations between the two countries in the 1980s. How-

ever, tensions were eased by the Clean Air Act of 1990, which calls for a significant reduction in U.S. emissions of sulfur dioxide and a modest reduction in emissions of nitrogen oxides by the year 2000. Canada has agreed to make comparable reductions in its emissions of these air pollutants. In areas near and downwind from large urban areas, NO_x emissions—mostly from cars—leading to the formation of nitric acid may be the main culprit.

Within a few decades, NO_x and SO_2 emissions from LDCs are expected to outstrip those from MDCs, leading to greatly increased damage from acid deposition over a much wider area, especially where soils are sensitive to acidification.

There is some good news. A 1993 study by the U.S. Geological Survey found that the concentration of sulfate ions—a key component of acid deposition—declined at 26 out of 33 of the U.S. rainwater collection sites between 1980 and 1991. And between 1970 and 1992, emissions of sulfur dioxide in the United States have dropped 30%. The 1990 amendments to the Clean Air Act require the nation to reduce its sulfur dioxide output by 40% of the 1980 level by 2000.

A: 11% (compared to 95% in Finland and 73% in Germany)

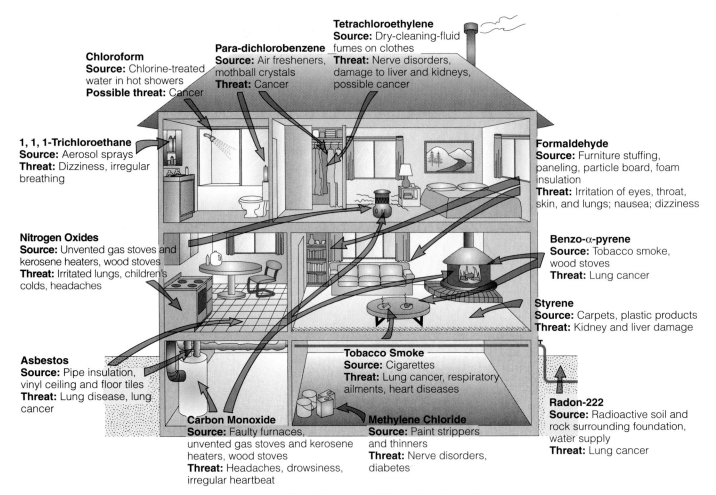

Chloroform
Source: Chlorine-treated water in hot showers
Possible threat: Cancer

Para-dichlorobenzene
Source: Air fresheners, mothball crystals
Threat: Cancer

Tetrachloroethylene
Source: Dry-cleaning-fluid fumes on clothes
Threat: Nerve disorders, damage to liver and kidneys, possible cancer

1, 1, 1-Trichloroethane
Source: Aerosol sprays
Threat: Dizziness, irregular breathing

Formaldehyde
Source: Furniture stuffing, paneling, particle board, foam insulation
Threat: Irritation of eyes, throat, skin, and lungs; nausea; dizziness

Nitrogen Oxides
Source: Unvented gas stoves and kerosene heaters, wood stoves
Threat: Irritated lungs, children's colds, headaches

Benzo-α-pyrene
Source: Tobacco smoke, wood stoves
Threat: Lung cancer

Styrene
Source: Carpets, plastic products
Threat: Kidney and liver damage

Asbestos
Source: Pipe insulation, vinyl ceiling and floor tiles
Threat: Lung disease, lung cancer

Tobacco Smoke
Source: Cigarettes
Threat: Lung cancer, respiratory ailments, heart diseases

Carbon Monoxide
Source: Faulty furnaces, unvented gas stoves and kerosene heaters, wood stoves
Threat: Headaches, drowsiness, irregular heartbeat

Methylene Chloride
Source: Paint strippers and thinners
Threat: Nerve disorders, diabetes

Radon-222
Source: Radioactive soil and rock surrounding foundation, water supply
Threat: Lung cancer

Figure 9-11 Some important indoor air pollutants. (Data from Environmental Protection Agency)

9-3 INDOOR AIR POLLUTION

Types and Sources If you are reading this book indoors, you may be inhaling more air pollutants with each breath than if you had been outside. As many as 150 dangerous chemicals, in concentrations 10–40 times higher than those outdoors, occur in the typical American home (Figure 9-11). A 1993 study found that pollution levels inside cars can be up to 18 times higher than those outside the vehicles.

The health risks from exposure to such chemicals are magnified because people spend 70–98% of their time indoors. In 1990, the EPA placed indoor air pollution at the top of the list of 18 sources of cancer risk. At greatest risk are smokers, the young, the old, the sick, pregnant women, people with respiratory or heart problems, and factory workers.

Pollutants found in buildings produce dizziness, headaches, coughing, sneezing, nausea, burning eyes, chronic fatigue, and flulike symptoms—the "sick building syndrome." According to the EPA, at least one-fifth of all U.S. buildings are considered "sick," at

an estimated cost to the nation of $60 billion per year in absenteeism and reduced productivity. Some indoor pollutants cause disease and premature death. According to the EPA and public health officials, cigarette smoke (p. 192), radioactive radon-222 gas (Case Study, at right), asbestos, and formaldehyde are the four most dangerous indoor air pollutants.

Severe indoor air pollution, especially from particulate matter, occurs inside the dwellings of many poor rural people in LDCs. The burning of wood, dung, and crop residues in open fires or in unvented or poorly vented stoves for cooking and heating (in temperate and cold areas) exposes the people, especially women and young children, to very high levels of indoor air pollution. Partly as a result, respiratory illnesses are a major cause of death and illness in most LDCs.

What Should Be Done About Asbestos?

There is intense controversy over what to do about possible exposure to tiny fibers of asbestos. Unless completely sealed within a product, asbestos easily crumbles into a dust of fibers tiny enough to become

Q: What are the best ways to deal with solid waste?

Is Your Home Contaminated with Radioactive Radon Gas?

Radon-222 is a colorless, odorless, tasteless, naturally occurring radioactive gas produced by the radioactive decay of uranium-238. Small amounts of uranium-238 are found in most soil and rock, but this isotope is much more concentrated in underground deposits of uranium, phosphate, granite, and shale.

When radon gas from such deposits seeps upward through the soil and is released outdoors, it disperses quickly in the atmosphere and decays to harmless levels. However, when the gas is drawn into buildings through cracks, drains, and hollow concrete blocks, or when it seeps into water in underground wells over such deposits, it can build up to high levels (Figure 9-12). Stone and other building materials obtained from radon-rich deposits can also be a source of indoor radon contamination.

Radon-222 gas quickly decays into solid particles of other radioactive elements that can be inhaled, exposing lung tissue to a large amount of ionizing radiation from alpha particles (Figure 3-7). Repeated exposure to these radioactive particles over 20–30 years can cause lung cancer, especially in smokers. According to studies by the EPA and the National Research Council, 7,000–30,000—with 13,600 as a best estimate—of lung cancer deaths each year in the United States may be caused by prolonged exposure to radon or to radon acting together with smoking.

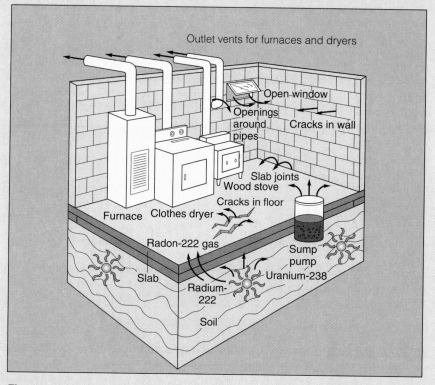

Figure 9-12 Sources and paths of entry for indoor radon-222 gas. (Data from Environmental Protection Agency)

According to the EPA, average radon levels above 4 picocuries (one-trillionth of a curie, used to measure radioactivity) per liter of air in a closed house are considered unsafe. However, some researchers cite evidence suggesting that radon becomes dangerous only if indoor levels exceed 20 picocuries per liter—the level accepted in Canada, Sweden, and Norway.

EPA indoor radon surveys suggest that 4–5 million U.S. homes may have annual radon levels above 4 picocuries per liter of air

and that 50,000–100,000 homes may have levels above 20 picocuries per liter. Unsafe levels can build up easily in a superinsulated or airtight home unless the building has an air-to-air heat exchanger to change indoor air without losing much heat. According to a 1992 EPA survey of 1,000 schools, one school in five has levels of radon gas higher than 4 picocuries per liter, exposing an estimated 11 million students and teachers to this potential health threat.

(continued)

suspended in the air and to be inhaled into the lungs, where they remain for many years. Prolonged exposure to asbestos fibers can cause asbestosis (a chronic lung condition that eventually makes breathing nearly impossible), lung cancer, and mesothelioma (an inoperable cancer of the chest cavity lining). Smokers exposed to asbestos fibers have a much greater chance of

dying from lung cancer than do nonsmokers exposed to such fibers.

Most of these diseases occur in people exposed for years to high levels of asbestos fibers. This group includes asbestos miners, insulators, pipe fitters, shipyard employees, auto mechanics (from brake linings), and workers in asbestos-producing factories. More

If the 4 picocuries per liter standard is adopted (as proposed by the EPA), the cost of testing and correcting the problem could run from $30–$100 billion, with a 15–20% reduction in radon-related death—mostly among smokers. Some researchers, however, argue that it makes more sense to spend perhaps only $500 million to find and fix homes and buildings with radon levels above 20 picocuries per liter until more reliable data are available on the threat from exposure to lower levels of radon.

When groundwater in or near radon-laden rock is withdrawn, heated, and used for showers and for washing clothes and dishes, it releases radon-222. According to the EPA, fewer than 200 people die each year from ingesting or inhaling radon from drinking water. Some scientists argue that there is no direct epidemiological or laboratory animal evidence of cancer's being caused by ingestion of radon in drinking water. They argue that drinking water standards for radon being proposed by the EPA are too stringent and too costly to implement compared to the health risk involved.

Because radon "hot spots" can occur almost anywhere, it's impossible to know which buildings have unsafe levels of radon without carrying out tests. In 1988 the EPA and the U.S. Surgeon General's Office recommended that everyone living in a detached house, a town house, or a mobile home—or on the first three floors of an apartment building—test for radon.* By 1992, only 9% of U.S. households had conducted such tests (at $20–$100 per home). If testing reveals an unacceptable level, you can consult the free EPA publication *Radon Reduction Methods* for ways to reduce radon levels and health risks. Has the building where you live or work been tested for radon?

*For information see "Radon Detectors: How to Find Out If Your House Has a Radon Problem," *Consumer Reports*, July 1987. Ideally, radon detectors should be left in the main living area for a year.

than 350,000 former U.S. workers have come down with asbestos-caused diseases, and each year an estimated 10,000 people—mostly U.S. industrial and construction workers exposed to high levels of asbestos fibers for years—die prematurely from asbestos-related diseases. Asbestos manufacturing companies in the United States have been swamped with health claims from workers, and some have been driven into bankruptcy.

Between 1900 and 1986, asbestos was widely used in the United States. Much of it was sprayed on ceilings and walls of schools and other public and private buildings for fireproofing, soundproofing, insulation of heaters and pipes, and wall and ceiling decoration. The EPA banned those uses in 1974. In 1989 the EPA ordered a ban on almost all remaining uses of asbestos (such as brake linings, roofing shingles, and water pipes) in the United States by 1997. Representatives of the asbestos industry in the United States and Canada (which produces most of the asbestos used in the United States) challenged the ban in court, contending that with proper precautions these asbestos products can be safely used and that the costs of the ban outweigh the benefits. In 1991 a federal appeals court overturned the 1989 EPA ban.

In 1986 Congress passed the Asbestos Hazards Emergency Response Act, which required all schools to have a qualified inspector check for asbestos and submit plans for containment or removal by May 8, 1989. In 1988 the EPA estimated that more than 760,000 buildings—one of every seven commercial and public buildings in the United States (including 30,000 schools)—contain asbestos that has crumbled or could crumble and release fibers. Removal of asbestos from such buildings could cost $50–$150 billion.

Critics of this law contend that the health benefits of asbestos removal from many schools, homes, and other buildings are not worth the costs, unless measurements (not just visual inspection) show that the buildings have a high level of airborne asbestos fibers. They call for sealing, wrapping, and other forms of containment instead of removal of most asbestos, and they point out that improper or unnecessary removal can release more asbestos fibers than sealing off asbestos that is not crumbling. After much controversy and huge expenditures of money on asbestos removal, there is now general agreement that asbestos should not be removed from buildings where it has not been damaged or disturbed; instead it should be sealed or wrapped to minimize release of fibers.

9-4 EFFECTS OF AIR POLLUTION ON LIVING ORGANISMS AND ON MATERIALS

Damage to Human Health Your respiratory system has a number of mechanisms that help protect you from air pollution. For example, hairs in your nose fil-

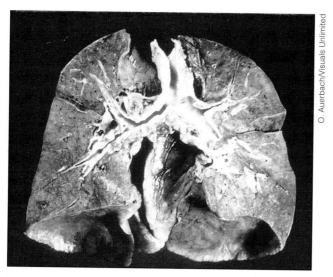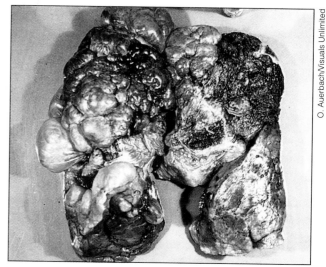

Figure 9-13 Normal appearance of human lungs (left) and appearance of the lungs of a person who died from emphysema (right). Prolonged smoking and exposure to air pollutants can cause emphysema in anyone, but about 2% of emphysema cases are caused by a defective gene that reduces the elasticity of the air sacs in the lungs. Anyone with this hereditary condition, for which testing is available, should not smoke and should not live or work in a highly polluted area.

ter out large particles. Sticky mucus in the lining of your upper respiratory tract captures small particles and dissolves some gaseous pollutants. Sneezing and coughing expel contaminated air and mucus when your respiratory system is irritated by pollutants. The cells of your upper respiratory tract are also lined with hundreds of thousands of tiny, mucus-coated hairlike structures called cilia. They continually wave back and forth, transporting mucus and the pollutants they trap to your throat, from which they are either swallowed or expelled.

Years of smoking and exposure to air pollutants can overload or break down these natural defenses. This causes or contributes to respiratory diseases such as *lung cancer*, *asthma* (typically an allergic reaction causing constriction of the bronchial tubes, resulting in acute shortness of breath), *chronic bronchitis* (damage to the cells lining the bronchial tubes, causing mucus buildup, coughing, and shortness of breath), and *emphysema* (damage to air sacs leading to abnormal dilation of air spaces, loss of lung elasticity, and acute shortness of breath, Figure 9-13).

Fine particles (the extremely small, invisible particles emitted by incinerators, motor vehicles, and power and industrial plants) are especially hazardous because they are small enough to penetrate the body's natural defenses against air pollution. They can also bring with them droplets or other particles of toxic or cancer-causing pollutants that become attached to their surfaces. Eight recent studies of air pollution in U.S. cities have indicated that fine particles not above currently permissable standards are killing up to

60,000 Americans each year. More than a dozen studies have directly or indirectly confirmed this conclusion. Also, there appears to be no threshold level below which the harmful effects of fine particles disappear. Elderly people, infants, pregnant women, and people with heart disease, asthma, or other respiratory diseases are especially vulnerable to air pollution.

No one knows how many people die prematurely from respiratory or cardiac problems caused or aggravated by air pollution. In the United States estimates of annual deaths related to outdoor air pollution range from 7,000 to 180,000 people. If indoor air pollution is included, estimated annual deaths from air pollution in the United States range from 150,000 to 350,000 people. The wide range of these estimates shows how difficult it is to get accurate information concerning deaths attributable to air pollutants, mostly because of the large number of interacting factors affecting human health (Section 8-2). Millions more suffer illness and lose work time; illness effects are especially well documented for ozone. According to the EPA and the American Lung Association, air pollution costs the United States at least $150 billion annually in health care and lost work productivity, with $100 billion of that caused by indoor air pollution.

The World Health Organization estimates that worldwide about 1.3 billion people—one person in four, mostly in LDCs, live in cities where outdoor air is unhealthy to breathe (Spotlight, p. 226). The annual global death toll from air pollution is estimated to be at least four times the annual U.S. death toll from air pollution.

A: Burial and incineration—the two most widely used methods

The World's Most Polluted City

The air pollution capital of the world may be Cubatão, a heavily industrialized city an hour's drive south of São Paulo, Brazil (Figure 6-16). This city of 100,000 people lies in a coastal valley that has frequent thermal inversions that trap polluted air.

Scores of factories spew thousands of tons of pollutants per day into the frequently stagnant air. More babies are born deformed in Cubatão than anywhere else in Latin America. Residents call the area "the valley of death."

In one recent year, 13,000 of the 40,000 people living in the downtown core area suffered from respiratory disease. One resident says, "On some days, if you go outside, you will vomit." The mayor refuses to live in the city.

Most residents would like to live somewhere else, but they need the jobs available in the city and cannot afford to move. The government has begun some long overdue efforts to control air pollution, but it has a long way to go. Meanwhile, the poor continue to pay the price of this form of economic progress: poor health and premature death.

Figure 9-14 Sulfur dioxide and other fumes have killed the forest once found on the land around this nickel smelter, which has operated for several decades at Sudbury, Ontario, Canada.

Damage to Plants Some gaseous pollutants (especially ozone) damage leaves of crop plants and trees directly when they enter leaf pores (Figure 9-9). Chronic exposure of leaves and needles to air pollutants can break down the waxy coating that helps prevent excessive water loss and damage from diseases, pests, drought, and frost. Such exposure also interferes with photosynthesis and plant growth, reduces nutrient uptake, and causes leaves or needles to turn yellow or brown and drop off (Figure 9-9). Spruce, fir, and other conifers, especially at high elevations, are most vulnerable to air pollution because of their long life spans and the year-round exposure of their needles to polluted air.

Prolonged exposure to high levels of multiple air pollutants from smelters (plants that extract metal from ores) can kill all trees and most other vegetation in an area (Figure 9-14). Prolonged exposure to a mix of air pollutants from coal-burning power and industrial plants and from cars can also damage trees and other plants. However, the effects may not become visible for several decades, when large numbers of trees suddenly begin dying off because of depletion of soil nutrients and increased susceptibility to pests, diseases, fungi, and drought (Figure 9-9). This phenomenon, known as

Waldsterben (forest death), has turned whole forests of spruce, fir, and beech into stump-studded meadows. The five European countries with the highest percentages of coniferous forest damage are Poland (75%), Czechoslovakia (71%, photo on p. 211), Greece (64%), Great Britain (64%), and Germany (60%). The overall productivity of European forests has been reduced by about 16% mostly because of air pollution, causing damages of roughly $30 billion per year.

Similar diebacks in the United States have occurred, mostly on high-elevation slopes that face moving air masses and are dominated by red spruce. The most seriously affected areas are in the Appalachian Mountains. Air pollution is also implicated in the recent dieback of sugar maples in Canada and the northeastern United States.

Air pollution, mostly by ozone, also threatens some crops—especially corn, wheat, soybeans, and peanuts—and is reducing U.S. food production by 5–10%. In the United States, estimates of economic losses to agriculture as a result of air pollution range from $1.9 billion to $5.4 billion per year.

Damage to Aquatic Life High acidity (low pH) can severely harm the aquatic life in freshwater lakes that have low alkaline content, or in areas in which surrounding soils have little acid-neutralizing capacity (Figure 9-10). Much of the damage to aquatic life in the Northern Hemisphere is a result of *acid shock*, caused by the sudden runoff of large amounts of highly acidic water and toxic aluminum into lakes and streams when snow melts in the spring or when heavy rains follow a drought. The aluminum leached from the soil and lake sediment kills fish by producing mucus and clogging their gills.

At least 16,000 lakes in Norway and Sweden contain no fish, and 52,000 more lakes have lost most of

Q: What country recycles the largest amount of its municipal solid waste?

Table 9-2 Harmful Effects of Air Pollution on Materials

Material	Effects	Principal Air Pollutants
Stone and concrete	Surface erosion, discoloration, soiling	Sulfur dioxide, sulfuric acid, nitric acid, particulate matter
Metals	Corrosion, tarnishing, loss of strength	Sulfur dioxide, sulfuric acid, nitric acid, particulate matter, hydrogen sulfide
Ceramics and glass	Surface erosion	Hydrogen fluoride, particulate matter
Paints	Surface erosion, discoloration, soiling	Sulfur dioxide, hydrogen sulfide, ozone, particulate matter
Paper	Embrittlement, discoloration	Sulfur dioxide
Rubber	Cracking, loss of strength	Ozone
Leather	Surface deterioration, loss of strength	Sulfur dioxide
Textile	Deterioration, fading, soiling	Sulfur dioxide, nitrogen dioxide, ozone, particulate matter

their acid-neutralizing capacity, because of excess acidity. In Canada, some 14,000 acidified lakes are almost fishless, and 150,000 more are in peril.

In the United States, about 9,000 lakes are threatened with excess acidity, one-third of them seriously. Most are concentrated in the Northeast and the upper Midwest—especially Minnesota, Wisconsin, and the upper Great Lakes—where 80% of the lakes and streams are threatened by excess acidity. Over 200 lakes in New York's Adirondack Mountains are too acidic to support fish.

Acidified lakes can be neutralized by treating them or the surrounding soil with large amounts of limestone, but liming is an expensive and only temporary remedy. Moreover, it can kill some types of plankton and aquatic plants, and it can harm wetland plants that need acidic water. It is also tricky to use.

Damage to Materials Each year air pollutants cause billions of dollars in damage to various materials we use (Table 9-2). The fallout of soot and grit on buildings, cars, and clothing requires costly cleaning. Air pollutants break down exterior paint on cars and houses, and they deteriorate roofing materials. Irreplaceable marble statues, historic buildings, and stained-glass windows throughout the world have been pitted and discolored by air pollutants (Figure 9-15). Damage to buildings in the United States from acid deposition alone is estimated at $5 billion per year.

Figure 9-15 This marble monument on a church in Surrey, England, has been damaged by exposure to acidic air pollutants.

9-5 SOLUTIONS: PREVENTING AND REDUCING AIR POLLUTION

U.S. Air Pollution Legislation In the United States, Congress passed Clean Air acts in 1970, 1977, and 1990, giving the federal government considerable power to control air pollution, with federal regulations enforced by each state. These laws required the EPA to establish *national ambient air quality standards (NAAQS)* for seven outdoor pollutants: suspended particulate matter, sulfur oxides, carbon monoxide, nitrogen oxides, ozone, hydrocarbons, and lead. Each

A: Japan (50%)

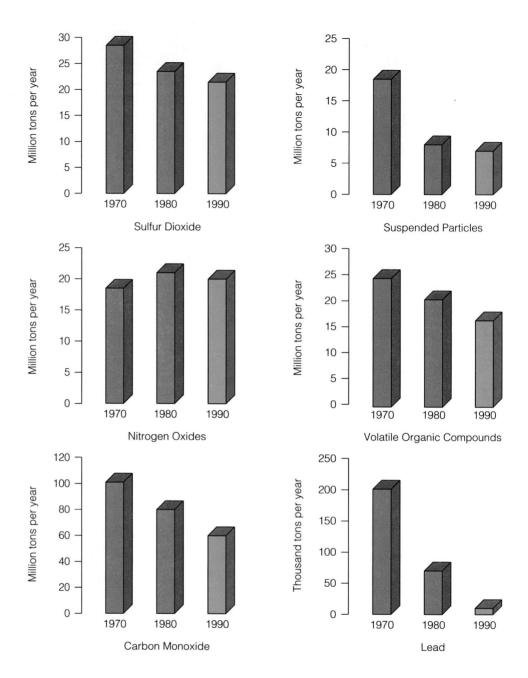

Figure 9-16 Trends in emissions of six major outdoor pollutants in the United States, 1970–90. (Data from Environmental Protection Agency)

standard specifies the maximum allowable level, averaged over a specific time period, for a certain pollutant in outdoor (ambient) air.

Between 1970 and 1990, the United States has significantly reduced levels of five of the six major outdoor air pollutants (Figure 9-16), and these downward trends continue in the early 1990s. Emissions of nitrogen oxides have increased somewhat because of a combination of insufficient automobile emission standards and a growth in both the number of motor vehicles and the distances traveled. These conditions have also led to increases in ozone levels in many major urban areas.

Without the 1970 standards for emissions of pollutants shown in Figure 9-16, air pollution levels would be 130% to 315% higher today. Even so, in 1992 at least 86 million people lived in areas that exceeded at least one air pollution standard. A typical new car in 1992 emitted only about 5% the pollution a new car in 1972 did.

A serious problem is that most U.S. air pollution control laws are based on pollution cleanup rather than pollution prevention. The only air pollutant with a sharp drop in its atmospheric level was lead (Figure 9-16), which was virtually banned in gasoline. This shows the effectiveness of the pollution prevention ap-

Q: What percentage of U.S. municipal solid waste could be recycled, composted, or reused?

A Market Approach to Pollution Control

SOLUTIONS

To help reduce SO_2 emissions, the Clean Air Act of 1990 allows utilities to buy and sell SO_2 pollution rights. With this *emissions trading policy* each utility has a specified limit on its annual SO_2 emissions. A utility that emits less SO_2 than its limit would receive pollution credits. The utility could then use its credits to avoid reductions in SO_2 emissions in some of its other facilities, could bank them for future expansions, or could sell them to other utilities, private citizens, or environmental groups.

Instead of the government dictating how each utility should meet its emissions target, this approach lets the marketplace determine the cheapest, most efficient way to get the job done. If this market-based approach works for reducing SO_2 emissions, it could be applied to other air and water pollutants.

Some environmentalists see this as an improvement over the current regulatory approach. Others argue that it continues to set a bad example by legally sanctioning the right to pollute, especially if the annual legal pollution limits are not lowered as companies find better ways to reduce emissions. Moreover, if in trading pollution rights a *net* reduction in pollution does not result, no real progress has been achieved.

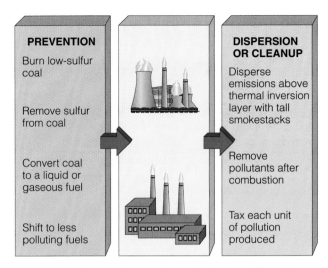

Figure 9-17 Methods for reducing emissions of sulfur oxides, nitrogen oxides, and particulate matter from stationary sources such as coal-burning electric power plants and industrial plants.

proach. If the current 4% annual growth in the distance that vehicles travel in the United States continues, annual emissions of nitrogen dioxide, carbon monoxide, and hydrocarbons will rise about 45% between 1995 and 2009 without stricter standards. Some analysts, however, believe that the marketplace can be used to control air pollution more efficiently than prescribed regulation (Solutions, above).

Environmentalists believe that the Clean Air Act of 1990 will further reduce air pollution in the United States. However, they point to several deficiencies in the law. These include:

- *Failing to sharply increase the fuel efficiency standards for cars and light trucks*, which would cut oil imports and air pollution more quickly and effectively than any other method. It would also save consumers enormous amounts of money.

- *Failing to classify the ash from municipal trash incinerators as hazardous waste*, thus encouraging incineration instead of pollution prevention as a solu-

tion to solid- and hazardous-waste reduction (Chapter 13).

- *Giving municipal trash incinerators 30-year permits*, which locks the nation into hazardous air pollution emissions and toxic waste from incinerators well into the twenty-first century. It also undermines pollution prevention, recycling, and reuse.

- *Setting weak standards for air pollution emissions from incinerators*, thus allowing unnecessary emissions of mercury, lead, dioxins, and other toxic pollutants (Section 13-5).

- *Setting municipal recycling goals at a token 25% (which the law allows the EPA and states to waive) instead of an achievable 60%.* This undermines recycling and reuse and encourages reliance on burying and burning solid and hazardous wastes.

- *Doing essentially nothing to reduce emissions of carbon dioxide and other greenhouse gases* (Section 10-2).

- *Continuing to rely almost entirely on pollution cleanup rather than pollution prevention.*

Solutions to Outdoor Air Pollution   Figure 9-17 summarizes ways to reduce emissions of sulfur oxides, nitrogen oxides, and particulate matter from stationary sources such as electric power plants and industrial plants that burn coal. So far, most of the emphasis has been on dispersing the pollutants with tall smokestacks or adding equipment that removes some of the pollutants after they are produced (Figure 9-18). Environmentalists call for taxes on air pollutant emissions and a shift to prevention methods.

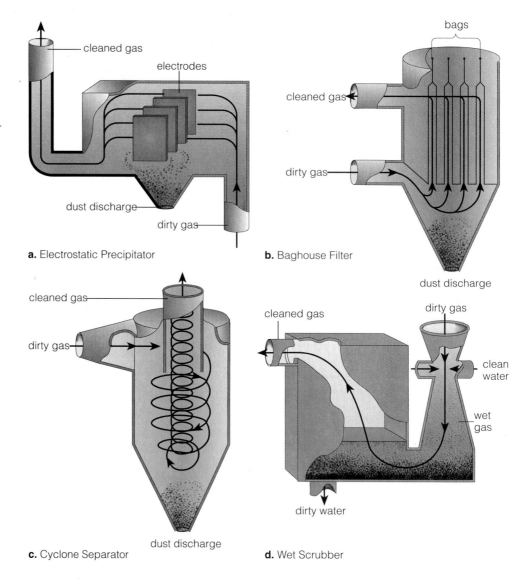

Figure 9-18 Four commonly used methods for removing particulates from the exhaust gases of electric power and industrial plants. Of these, only baghouse filters remove many of the more hazardous fine particles. Also, all produce hazardous materials that must be disposed of safely and—except for cyclone separators—all methods are expensive. The wet scrubber is also used to reduce sulfur dioxide emissions.

a. Electrostatic Precipitator

b. Baghouse Filter

c. Cyclone Separator

d. Wet Scrubber

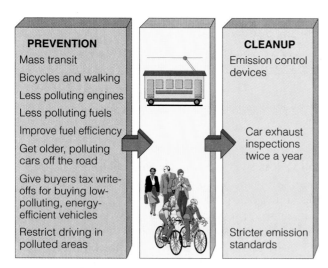

PREVENTION	CLEANUP
Mass transit	Emission control devices
Bicycles and walking	
Less polluting engines	
Less polluting fuels	
Improve fuel efficiency	Car exhaust inspections twice a year
Get older, polluting cars off the road	
Give buyers tax write-offs for buying low-polluting, energy-efficient vehicles	
Restrict driving in polluted areas	Stricter emission standards

Figure 9-19 Methods for reducing emissions from motor vehicles.

Figure 9-19 shows ways to reduce emissions from motor vehicles, the primary culprits in producing photochemical smog (Figure 9-4), which contains damaging ozone. Table 9-3 evaluates alternative fuels.

Despite strict air pollution control laws, one of every four people in the United States is routinely exposed to O_3 concentrations that exceed standards set under the Clean Air Act. However, progress is being made. From 1982 to 1992 overall U.S. smog incidence dropped by 8%, even as population and consumption rose. And during this same period the number of people exposed to unhealthful ozone levels in the Los Angeles air basin was cut in half. Today the motor vehicles and industries operating in California are among the cleanest in the world. In 1992 a new car sold in California emitted one-tenth the pollution emitted by a new car in 1970.

Despite these successes, smog levels in the Los Angeles air basin are too high much of the year (al-

Q: How much of the world's wastepaper was recycled in 1992?

Table 9-3 Evaluation of Alternatives to Gasoline

Advantages	Disadvantages
Compressed Natural Gas	
Fairly abundant domestic and global supplies	Cumbersome fuel tank required
Low emissions of hydrocarbons, CO, and CO_2	Expensive engine modification required ($2,000)
Currently inexpensive	One-fourth the range
Vehicle development advanced	New filling stations required
Reduced engine maintenance	Nonrenewable resource
Well suited for fleet vehicles	
Electricity	
Renewable if not generated from fossil fuels or nuclear power	Limited range and power
Zero vehicle emissions	Batteries expensive
Electric grid in place	Slow refueling (6–8 hours)
Efficient and quiet	Power-plant emissions if generated from coal or oil
Reformulated Gasoline (Oxygenated Fuel)	
No new filling stations required	Nonrenewable resource
Low to moderate emission reduction of CO	Dependence on imported oil perpetuated
No engine modification required	Possible high cost to modify refineries
	No emission reduction of CO_2
	Higher cost
	Water resources contaminated by leakage and spills
Methanol	
High octane	Large fuel tank required
Emission reduction of CO_2 (total amount depends on method of production)	One-half the range
	Corrosive to metal, rubber, plastic
Reduced total air pollution (30–40%)	Increased emissions of potentially carcinogenic formaldehyde
	High CO_2 emissions if generated by coal
	High capital cost to produce
	Hard to start in cold weather
Ethanol	
High octane	Large fuel tank required
Emission reduction of CO_2 (total amount depends on distillation process and efficiency of crop growing)	Much higher cost
	Corn supply limited
Emission reduction of CO	Competition with food growing for cropland
Potentially renewable	Less range
	Smog formation possible
	Corrosive
	Hard to start in cold weather
Solar-Hydrogen	
Renewable if produced using solar energy	Nonrenewable if generated by fossil fuels or nuclear power
Lower flammability	Large fuel tank required
Virtually emission-free	No distribution system in place
Zero emissions of CO_2	Engine redesign required
Nontoxic	Currently expensive

most one of every three days in 1992) and could rise as population and consumption rise. To meet federal air pollution standards, Los Angeles must reduce air pollution further. The 1990 Clean Air Act gives Los Angeles until 2010 to meet federal air pollution standards. However, the law requires the region to make progress toward this goal, and the California Clean Air Act requires the region to reduce emissions by 5% a year until health standards have been met. A 1989 study estimated that meeting federal standards for ozone and particulates alone would provide $9.4 billion in health benefits every year.

In 1989, California's South Coast Air Quality Management District Council proposed a drastic program

What You Can Do About Indoor Air Pollution

INDIVIDUALS MATTER

To reduce your exposure to indoor air pollutants:

■ *Test for radon and take corrective measures as needed* (Case Study, p. 223).

■ *Install air-to-air heat exchangers or regularly ventilate your house by opening windows.*

■ *Test indoor air for formaldehyde at the beginning of the winter heating season when the house is closed up.** The cost is $200–$300.

■ *Don't buy synthetic wall-to-wall carpeting, furniture, and other products containing formaldehyde, and use "low-emitting formaldehyde" or nonformaldehyde building materials.*

■ *Reduce indoor levels of formaldehyde and several other toxic gases by growing house plants.* Examples are the spider or airplane plant (the most effective), golden pothos, syngonium, philodendron (especially the elephant-ear species), chrysanthemum, ligustrum, photina, variegated liriope, aloe vera, ficus (weeping fig), peace lily, and Gerbera daisy. About 20 plants of such species can help clean the air in a typical home. Plants should be potted with a mixture of soil and granular charcoal (which absorbs organic air pollutants).

■ *Test your house or workplace for asbestos fiber levels if it was built before 1980.*† If airborne asbestos levels are too high, hire an independent consultant—not an asbestos-removal firm—to advise you on what to do. (The typical charge is $500 or more, but this could save you asbestos-removal costs of $10,000–$100,000.). Don't buy a pre-1980 house without having its indoor air tested for asbestos.

■ *Don't store gasoline, solvents, or other volatile hazardous chemicals inside a home or attached garage.*

■ *Don't use commercial room deodorizers or air fresheners.*

■ *Don't use aerosol spray products.*

■ *Don't smoke.* If you must smoke, do it outside or in a closed room vented to the outside.

■ *Make sure that wood-burning stoves, fireplaces, and kerosene- and gas-burning heaters are properly installed, vented, and maintained.*

*To locate a testing laboratory in your area, write to Consumer Product Safety Commission, Washington, DC 20207, or call 301-492-6800.

†To get a free list of certified asbestos laboratories that charge $25-$50 to test a sample, send a self-addressed envelope to NIST/ NVLAP, Building 411, Room A124, Gaithersburg, MD 20899, or call the EPA's Toxic Substances Control Hotline at 202-554-1404.

---

to reduce O_3, photochemical smog, and other major air pollutants in the Los Angeles area. This plan would do the following things:

■ *Require 10% of new cars sold in California by 2003 to emit no air pollutants.*

■ *Outlaw drive-through facilities to keep vehicles from idling in lines.*

■ *Substantially raise parking fees and assess high fees for families owning more than one car.*

■ *Strictly control or relocate industrial plants that release large quantities of hydrocarbons and other pollutants.* These facilities include petroleum-refining, dry-cleaning, auto-painting, printing, baking, and trash-burning plants.

■ *Find substitutes for or ban use of consumer products that release hydrocarbons,* including aerosol propellants, paints, household cleaners, and barbecue starter fluids.

■ *Eliminate gasoline-burning engines over two decades by converting trucks, buses, chain saws, outboard motors, and lawn mowers to run on electricity or on alternative fuels.*

■ *Require gas stations to use a hydrocarbon-vapor recovery system on gas pumps and to sell alternative fuels* (Table 9-3).

The plan may be defeated by public opinion when residents begin to feel the economic pinch from such drastic changes. Proponents argue, however, that the economic impact of not carrying out such a program will cost consumers and businesses much more. Such measures are a glimpse of what most cities will have to do as people, cars, and industries proliferate.

Solutions to Indoor Air Pollution For many people, indoor air pollution poses a greater threat to health than does outdoor air pollution. Yet the EPA spends $200 million per year trying to reduce outdoor air pollution and only $5 million a year on indoor air pollution.

To sharply reduce indoor air pollution, it's not necessary to establish mandatory indoor air quality standards and monitor the more than 100 million homes and buildings in the United States. Instead air pollution experts suggest that indoor air pollution reduction can be achieved by several means, including the following:

■ *Modify building codes to prevent radon infiltration, or require use of air-to-air heat exchangers or other devices to change indoor air at regular intervals*

Q: How much of the world's wastepaper could be recycled by 2000?

- *Require exhaust hoods or vent pipes for appliances burning natural gas or another fossil fuel*

- *Set formaldehyde emission standards for building, furniture, and carpet materials*

- *Equip work stations with adjustable fresh air inputs* (much like those for passengers on commercial aircraft)

- *Find substitutes for potentially harmful chemicals in aerosols, cleaning compounds, paints, and other products used indoors* (Table 13-2)

Each of us can reduce our own exposure to indoor air pollutants (Individuals Matter, at left). In LDCs significant reductions in respiratory illnesses would occur if governments gave the poor simple stoves that burn biofuels more efficiently (which would also reduce deforestation) and that are vented outside.

Connections: Protecting the Atmosphere

Environmentalists believe that protecting the atmosphere, and thus the health of people and many other organisms, will require the following significant changes throughout the world:

- *Integrate air pollution, water pollution, energy, land-use, and population regulation policies.*

- *Emphasize pollution prevention rather than pollution control.* Widespread use of solar-produced hydrogen fuel (Section 18-7) would eliminate most air pollution.

- *Improve energy efficiency* (Section 18-2).

- *Reduce use of fossil fuels, especially oil and coal* (Sections 19-1 and 19-3).

- *Shift to renewable energy resources* (Chapter 18).

- *Emphasize distribution of low-emission and more efficiently vented cookstoves in rural areas of LDCs.*

- *Discourage automobile use* (Section 6-4).

- *Increase recycling and reuse, and reduce the production of all forms of waste* (Chapter 13).

- *Develop air quality strategies based on the air flows and pollution sources of an entire region instead of the current piecemeal, city-by-city approach.*

- *Slow population growth* (Section 6-5).

- *Include the social costs of air pollution and other forms of pollution in the market prices of goods and services* (Section 7-2).

As population and consumption rise, we can generate new air pollution faster than we can clean up the old, even in MDCs with strict air pollution control laws. This shows the need for both slowing population growth and relying on pollution prevention.

We are in somewhat the same position in regard to polluted air as the fish are to polluted water.

ALLAN V. KNEESE

Critical Thinking

1. Evaluate the pros and cons of the following statement: "Since we have not proven absolutely that anyone has died or suffered serious disease from nitrogen oxides, present federal emission standards for this pollutant should be relaxed."

2. What topographical and climate factors either increase or help decrease air pollution in your community?

3. Should all tall smokestacks be banned? Explain.

*4. Do buildings in your school contain asbestos? If so, what are the indoor levels? Should this asbestos be removed? What are the current indoor levels of asbestos fibers in any buildings from which asbestos has been removed within the past five years? If indoor asbestos testing has not been done, talk with school officials about having it done.

*5. Have dormitories and other buildings on your campus been tested for radon? If so, what were the results? What has been done about areas with unacceptable levels? If this testing has not been done, talk with school officials about having it done.

*6. Make a concept map of the key ideas in this chapter using the section heads and subheads and the key terms (shown in boldface type in the chapter). See the inside front cover and Appendix 4 for information on concept maps.

10

Climate, Global Warming, and Ozone Loss

2040 A.D.: *Hard Times on Planet Earth**

Mary Wilkins sat in the living room of her underground home in Illinois, which she shared with her daughter Jane and her family (Figure 10-1). It was July 4, 2040—Independence Day. There would be no parade or fireworks today. People didn't stay outside for long now because of the searing heat and intense ultraviolet radiation. With the food riots and martial law in place since 2020 people stayed home.

Many of her friends and millions of other Americans had long ago migrated to Canada to find a cooler climate and more plentiful food supply after America's Midwestern breadbasket and central and southern California were abandoned because of a lack of water and food. Her friend June had recently written wondering where to go now that Canadian farmland was drying up.

Behind her erupted shouts and squeals from her grandchildren, Jessica and Jeffrey, playing Refugees and Border Patrol. It had been a popular game since 2020 when the United States had built a "Great Wall" along its border with Mexico in a mostly vain attempt to keep out millions of Latin Americans trying to find food and work in the north.

Mary sighed as she went to corral her pale and undernourished grandchildren. She felt sorry for them

---

*Compare this fictional worst-case scenario with the hopeful scenario that opens Chapter 2.

and began telling them about the old days—before the Warming and severe ozone depletion—when school was not a TV set with lessons year-round and kids could play outside all day during summer. There were green parks to play in and green trees to climb, swimming pools full of water, and lakes and rivers everywhere. And in winter cold white stuff called snow fell from the sky and could be gathered up in balls to throw at each other.

She also told them that almost everyone had a car. "What's a car?" asked Jeffrey, the oldest child. "Is it like the bus that Mommy rides to work?" "Yes, only much smaller—just for one person or one family," Mary answered. "It could go fast and ran on a fuel called gasoline—much too rare to be used anymore."

"Why did people let things get so bad?" Jessica asked. Mary's eyes filled with tears as she took the child wordlessly in her arms. "Why didn't we listen to the warnings of scientists in the 1980s and 1990s?" she asked herself. Then she looked at Jessica and admitted, "Because we didn't want to believe anything bad could happen."

Although our species has existed for only an eyeblink of Earth's history, evidence indicates that we are altering its atmosphere 10 to 100 times faster than the natural rate of change over the past 10,000 years. Global warming from our binge of fossil-fuel burning and deforestation, and depletion of stratospheric ozone from our use of chlorofluorocarbons and other chemicals, are now threats.

Figure 10-1 This earth-sheltered house in Will County, Illinois, could be like Mary Wilkins's 2040 house. Across the United States about 13,000 families have built earth-sheltered houses. (Courtesy of Pat Armstrong/Visuals Unlimited)

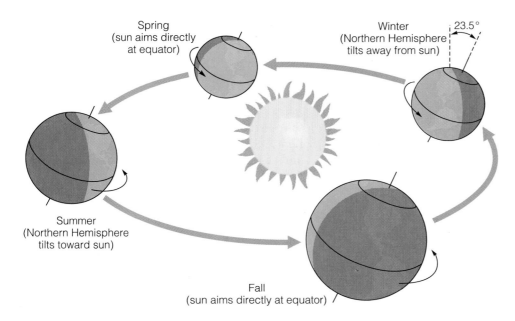

Spring
(sun aims directly
at equator)

Winter
(Northern Hemisphere
tilts away from sun)

23.5°

Summer
(Northern Hemisphere
tilts toward sun)

Fall
(sun aims directly at equator)

Figure 10-2 Seasonal changes in climate (shown here for the Northern Hemisphere only) are caused by variations in the amount of solar energy reaching various areas of the earth. As Earth makes its annual revolution around the sun on an axis tilted about 23.5°, various regions are tipped toward or away from the sun, which causes changes in seasons.

We, humanity, have finally done it: disturbed the environment on a global scale.

THOMAS E. LOVEJOY

This chapter will be devoted to answering the following questions:

■ What key factors determine variations in Earth's climate?

■ Can we really make Earth warmer, and if so, what will a few degrees matter?

■ What can we do about possible global warming?

■ Can we accelerate ozone depletion in the stratosphere, and why should we care?

■ What can we do to slow projected ozone depletion?

10-1 WEATHER AND CLIMATE: A BRIEF INTRODUCTION

What Are Weather and Climate? At every moment at any spot on Earth, the troposphere has a particular set of physical properties including such things as temperature, pressure, humidity, precipitation, sunshine, cloud cover, and wind direction and speed. These short-term properties of the troposphere at a given place and time are what we call **weather**.

Climate is the long-term average weather of an area; it is a region's general pattern of atmospheric or weather conditions, seasonal variations, and weather extremes (such as hurricanes and prolonged drought or rain) over a long period (at least 30 years).

What Factors Influence Climate? The two most important factors determining the climate of an area are temperature and precipitation (Figure 5-2). The temperature and precipitation patterns that lead to different climates are caused mostly by the way air circulates over the earth's surface. Several factors determine these patterns of global air circulation:

■ *Long-term variations in the amount of solar energy striking the earth.*

■ *Uneven heating of the Earth's surface.* Air is heated much more at the equator, where the sun's rays strike directly throughout the year, than at the poles, where sunlight strikes at a glancing angle. These differences help explain why tropical regions near the equator are hot, polar regions are cold, and temperate regions in between generally have intermediate temperatures (Figure 5-2).

■ *The tilt of the earth's axis* (an imaginary line connecting the North and South Poles). Because of this tilt various regions are tipped toward or away from the sun as the earth makes its annual revolution around the sun (Figure 10-2). This creates opposite seasons in the Northern and Southern hemispheres.

■ *Rotation of the earth.* Forces in the atmosphere created by this rotation deflect winds (air masses moving north and south from the equator) to the right in the Northern Hemisphere and to the left in the Southern Hemisphere, in what is called the *Coriolis effect.* The result is six huge convection cells of swirling air masses—three north and three south of the equator (Figure 10-3).

- *Properties of air and water.* Cold air is denser (weighs more per unit volume) than hot air and thus tends to sink through less dense, warmer air; hot air, being less dense, tends to rise. Hot air can also hold more water vapor than cold air. When heated by the sun, ocean water evaporates and removes heat from the oceans to the atmosphere. This moist, hot air expands, becomes less dense, and rises in fairly narrow vortices that spiral upward, creating an area of low pressure at the earth's surface. As this moisture-laden air rises, it cools and releases moisture as condensation (because cold air can hold less water vapor than warm air). The heat released when water vapor condenses radiates into space. The resulting cooler, drier air becomes denser, sinks (subsides), and creates an area of high pressure. As this area flows across the earth's surface, it picks up heat and moisture and begins to rise again. The resulting small and giant convection cells circulate air, heat, and moisture both vertically and from place to place in the troposphere, leading to different climates and patterns of vegetation (Figures 10-4 and 5-5).

- *Ocean currents.* The factors just listed, plus differences in water density, cause warm and cold ocean currents. These currents, along with air masses above, redistribute heat received from the sun (Figure 5-2). Ocean currents, like air currents, redistribute heat and thus influence climate and vegetation, especially near coastal areas. For example, without the warm Gulf Stream, which transports 25 times more water than all the world's rivers, the climate of northwestern Europe would be subarctic. Currents also help mix ocean waters and distribute nutrients and dissolved oxygen needed by aquatic organisms. Changes in prevailing winds, however, can change the temperature of surface waters, weaken or alter ocean currents, and trigger weather changes over large areas (Figure 10-5).

- *Atmospheric composition.* Small amounts of carbon dioxide and water vapor and trace amounts of

Figure 10-3 Formation of prevailing surface winds, which disrupt the general flow of air from the equator to the poles and back to the equator. As Earth rotates, its surface turns faster beneath air masses at the equator and slower beneath those at the poles. This deflects air masses moving north and south to the west or east, creating six huge convection cells in which air swirls upward and then descends toward Earth's surface at different latitudes. The direction of air movement in these cells sets up belts of prevailing winds that distribute air and moisture over Earth's surface. These winds affect the general types of climate found in different areas and also drive the circulation of ocean currents. (Used by permission from Cecie Starr and Ralph Taggart, *Biology: The Unity and Diversity of Life*, 6th ed., Belmont, Calif.: Wadsworth, 1992)

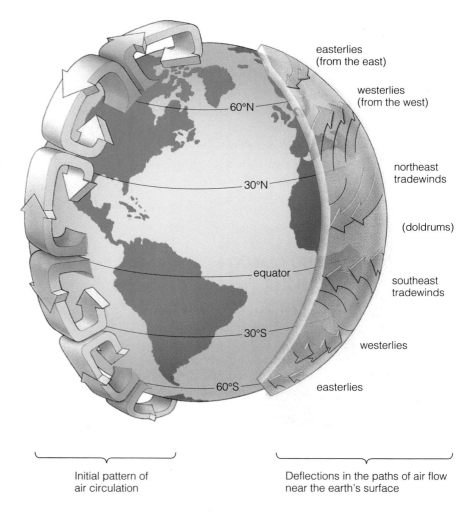

easterlies
(from the east)

westerlies
(from the west)

northeast
tradewinds

(doldrums)

southeast
tradewinds

westerlies

easterlies

60°N

30°N

equator

30°S

60°S

Initial pattern of
air circulation

Deflections in the paths of air flow
near the earth's surface

Q: According to the EPA, how many landfills in the U.S. will eventually leak?

ozone, methane, nitrous oxide, chlorofluorocarbons, and other gases in the troposphere play a key role in determining the earth's average temperatures and thus its climates. Collectively, these gases, known as **greenhouse gases**, act somewhat like the glass panes of a greenhouse (or of a car parked in the sun with its windows rolled up): They allow light, infrared radiation, and some ultraviolet radiation from the sun (Figure 3-4) to pass through the troposphere. The earth's surface absorbs much of this solar energy and degrades it to longer, infrared radiation—that is, heat— which then rises into the troposphere (Figure 4-4). Some of this heat escapes into space; some is absorbed by molecules of greenhouse gases, warming the air; and some radiates back toward the earth's surface. This trapping of heat in the troposphere is called the **greenhouse effect** (Figure

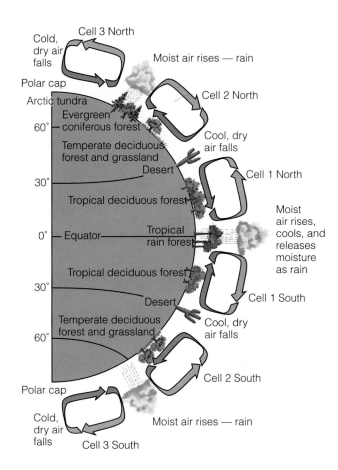

Figure 10-4 (right) Global air circulation and biomes. Heat and moisture are distributed over Earth's surface by vertical convection currents that form into six large convection cells (called Hadley cells) at different latitudes. The direction of air flow and the ascent and descent of air masses in these convection cells determine Earth's general climatic zones. The uneven distribution of heat and moisture over the planet's surface leads to the forests, grasslands, and deserts that make up Earth's biomes.

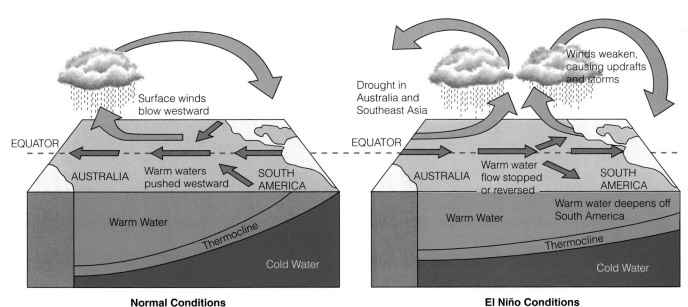

Normal Conditions   **El Niño Conditions**

Figure 10-5 Surface winds blowing westward cause shore upwellings of cold, nutrient-rich bottom water in the tropical Pacific Ocean near the coast of Peru (left). The warm and cold water are separated by a zone of gradual temperature change called the thermocline. Every few years a climate shift known as the *El Niño-Southern Oscillation (ENSO)* disrupts this pattern. Westward surface winds weaken, which depresses the coastal upwellings and warms the surface waters off South America (right). ENSOs typically occur for several months to over a year every three or four years, although the interval has been as long as seven years. A long-lasting ENSO severely disrupts populations of plankton, fish, and seabirds in upwelling areas and can trigger extreme weather changes over at least two-thirds of the globe, especially in the countries along the Pacific and Indian oceans. Some areas receive abnormally high rainfall; others suffer severe droughts.

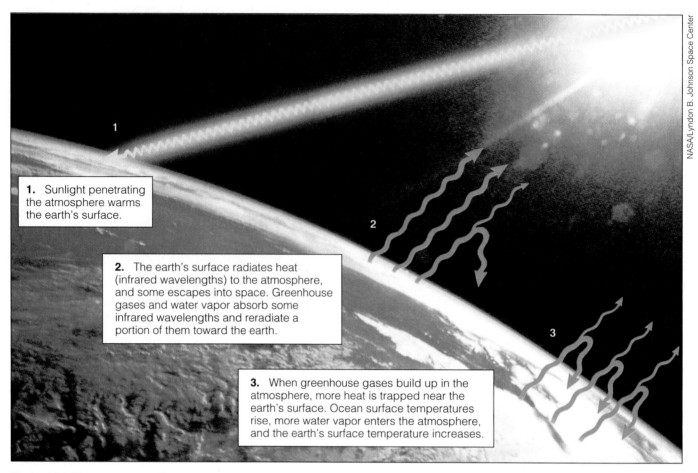

Figure 10-6 The greenhouse effect. Without the atmospheric warming provided by this natural effect, the earth would be a cold and mostly lifeless planet. According to the greenhouse theory, when concentrations of greenhouse gases rise the average temperature of the troposphere rises. (Used by permission from Cecie Starr and Ralph Taggart, *Biology: The Unity and Diversity of Life*, 6th ed., Belmont, Calif.: Wadsworth, 1992)

10-6) and is itself normal and natural. Without greenhouse gases (especially water vapor), the earth would be a cold and lifeless planet with an average surface temperature of –18°C (0°F) instead of its current 15°C (59°F). We and other species currently benefit from a comfortable level of greenhouse gases, with only minor and slow fluctuations, but global warming or cooling over decades instead of hundreds to thousands of years would be disastrous for our species and many others.

10-2 GLOBAL WARMING? OR A LOT OF HOT AIR?

What We Know About the Earth's Climate

The greenhouse effect is among the most widely tested and accepted scientific theories. However, there is much debate over whether human activities are now

or will soon be intensifying the natural greenhouse effect, thereby raising the earth's average temperature (global warming) and changing the climate found in various parts of the world. There is also controversy over how much the temperature might rise and how this rise might affect the climate in different areas. If we are to evaluate the possibility of global warming and its possible effects, we need to look at what we know and don't know about the earth's climate.

In 1990 and 1992, the Intergovernmental Panel on Climate Change (IPCC) published reports by several hundred leading atmospheric scientists on the best available evidence concerning past climate change, the greenhouse effect, and recent changes in global temperatures. Based on the panel's reports and other studies, the following points constitute current scientific consensus on these matters:

- Earth's average surface temperature has fluctuated considerably over geologic time, including several ice ages that covered much of the planet

Q: How much of the hazardous waste produced in the United States is regulated by federal laws?

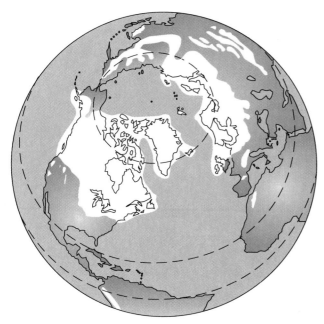

Figure 10-7 The region that was covered by ice at various times during the most recent glaciation, which ended about 10,000 years ago. During the past 2.0–2.5 million years, much of the Northern Hemisphere was covered several times with thick ice sheets. These glacial periods alternated with warmer interglacial periods.

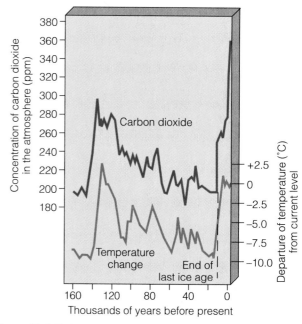

Figure 10-8 Estimated long-term variations in mean global surface temperature and average tropospheric carbon dioxide levels over the past 160,000 years. Since the last great ice age ended about 10,000 years ago, we have enjoyed a warm interglacial period. One factor in the earth's mean surface temperature is the greenhouse effect. Changes in tropospheric levels of carbon dioxide, a major greenhouse gas, correlate closely with changes in the earth's mean surface temperature and thus its climate, although other factors also influence global climate.

with thick ice during the past 800,000 years (Figure 10-7). Each glacial period lasted about 100,000 years and was followed by a warmer interglacial period of 10,000–25,000 years. As the ice melted at the end of the last ice age, average sea levels rose about 100 meters (300 feet), greatly reducing the amount of dry land.

■ For the past 10,000 years we have enjoyed the relative warmth of the latest interglacial period (called the Holocene), during which mean surface temperatures have usually fluctuated only 0.5–1°C (0.9–1.8°F) over 100- to 200-year periods. This relative climatic stability over thousands of years has saved us from drastic changes in the nature of soils and vegetation patterns throughout the world, allowing large increases in food production and thus in population (Figure 2-3). However, even these small temperature changes have led to large migrations of peoples in response to changed agricultural and grazing conditions.

■ Heat trapped by greenhouse gases in the atmosphere is what keeps the planet warm enough to allow us and other species to exist (Figure 10-6 and Connections, p. 117). This idea, proposed by mathematician Jean Fourier more than 100 years ago, has been confirmed by numerous laboratory

experiments and atmospheric measurements, and is one of the most widely accepted scientific theories.

■ Over the past 160,000 years, levels of water vapor in the troposphere have remained fairly constant while those of CO_2 have fluctuated by a factor of two. Estimated changes in the levels of tropospheric CO_2 (Figure 10-8) correlate closely with estimated variations in the earth's mean surface temperature.

■ Measured atmospheric levels of certain greenhouse gases—CO_2, methane, nitrous oxide, and CFCs—have risen in recent decades (Figure 10-9).

■ Most of the increased levels of these greenhouse gases have been caused by human activities: burning fossil fuels, use of CFCs, agriculture, and deforestation.

■ Since 1860, when measurements began, mean global temperature has risen 0.3–0.6°C (0.5–1.1°F) (Figure 10-10).

■ Eight of the 13 years from 1980 to 1992 were among the hottest in the 110-year recorded history of land-surface temperature measurements, and 1990 was the hottest of all.

Figure 10-9 Increases in average concentrations of major greenhouse gases in the troposphere, mostly because of human activities. (Data from Electric Power Research Institute. Adapted and updated by permission from Cecie Starr and Ralph Taggart, *Biology: The Unity and Diversity of Life*, 6th ed., Belmont, Calif.: Wadsworth, 1992)

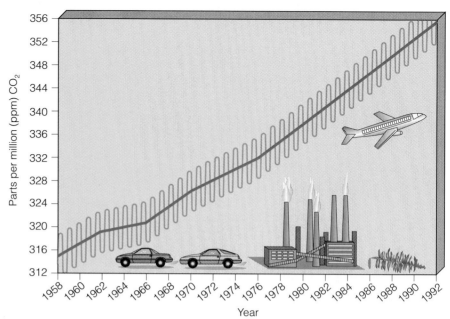

a. Carbon dioxide (CO$_2$) contributes about 55% to global warming from greenhouse gases produced by human activities. Industrial countries account for about 76% of annual emissions. The main sources are fossil-fuel burning (67%) and land clearing and burning (33%). CO$_2$ remains in the atmosphere for 50 to 500 years.

b. Chlorofluorocarbons (CFCs) are believed to be responsible for 24% of the human contribution of greenhouse gases. They also deplete ozone in the stratosphere. The main sources are leaking air conditioners and refrigerators, evaporation of industrial solvents, production of plastic foams, and aerosol propellants. CFCs take 10–20 years to reach the stratosphere and generally trap 1,500–7,000 times as much heat per molecule as CO$_2$ while they are in the troposphere. This heating effect in the troposphere may be partially offset by the cooling caused when CFCs deplete ozone during their 65–110-year stay in the stratosphere.

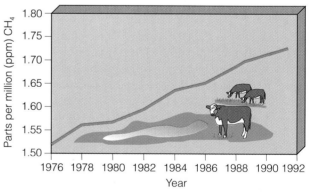

c. Methane (CH$_4$) accounts for about 18% of the increase in greenhouse gases. Methane is produced when anaerobic bacteria break down dead organic matter in moist places that lack oxygen. These areas include swamps and other natural wetlands; rice paddies; landfills; and the intestinal tracts of cattle, sheep, and termites. Production and use of oil and natural gas—and incomplete burning of organic materials (including biomass burning in the tropics)—also are significant sources. CH$_4$ stays in the troposphere for 7–10 years. Each CH$_4$ molecule traps about 25 times as much heat as a CO$_2$ molecule.

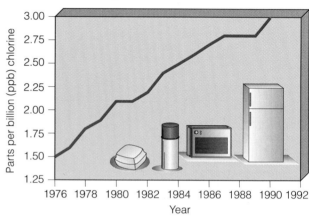

d. Nitrous oxide (N$_2$O) is responsible for 6% of the human input of greenhouse gases. Besides trapping heat in the troposphere, it also depletes ozone in the stratosphere. It is released from nylon production; from burning of biomass and nitrogen-rich fuels (especially coal); and from the breakdown of nitrogen fertilizers in soil, livestock wastes, and nitrate-contaminated groundwater. Its life span in the troposphere is 140–190 years, and it traps about 230 times as much heat per molecule as CO$_2$.

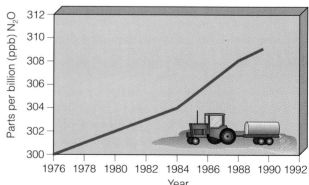

Q: What percentage of U. S. children under age 6 have unsafe levels of lead in their blood?

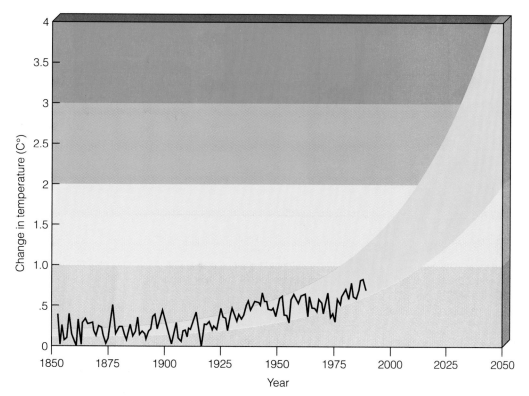

Figure 10-10 Changes in the earth's mean surface temperature between 1860 and 1990 (dark line). The yellow region shows global warming projected by various computer models of the earth's climate systems. Note that the computer projections roughly match the historically recorded change in temperature between 1860 and 1990. All current models suggest that global temperature will rise between now and 2050. Scientists admit that current models could underestimate or overestimate the amount of warming by a factor of two. (Data from National Academy of Sciences and National Center for Atmospheric Research)

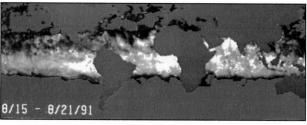

Larry Stowe/National Oceanic Atmospheric Administration

Figure 10-11 Satellite view of sulfur dioxide and sulfuric acid aerosols in the atmosphere 12 weeks after the massive eruption of Mount Pinatubo in the Philippines in June 1991—the largest volcanic eruption of this century. These chemicals are expected to reflect incoming sunlight and to lower average global temperatures 0.6°C (1°F) between 1992 and 1995, until the aerosols return to Earth's surface.

- So far, any temperature changes possibly caused by an enhanced greenhouse effect have been too small to exceed normal short-term variations in mean atmospheric temperature caused by volcanic eruptions (Figures 8-7 and 10-11), air pollution, and other climatic factors.

- Warming or cooling by more than 2°C (4°F) over a few decades (instead of over centuries, as has happened during the last 10,000 years) would be disastrous for Earth's ecosystems and for human economic and social systems. Such rapid climate change would alter conditions faster than some species, especially plants, could adapt or migrate.

These changes might also shift areas where people could grow food. Some areas might become uninhabitable because of drought or because of floods following a rise in average sea levels.

- We don't know enough about how the earth works to make accurate projections about the possible effects of our inputs of greenhouse gases on either global and regional climates or on the biosphere.

Computers As Crystal Balls: Modeling Greenhouse Warming   To project the behavior of climatic and other complex systems, scientists develop mathematical models that simulate such systems and then run them on computers (Figures 6-28 and 6-29). *It is important to understand that such models are projections, not predictions.* They are scenarios of what could happen based on various assumptions and data fed into the model. How well the results correspond to the real world depends both on the design of the model and on the accuracy of the data and the assumptions used.

Current climate models (Figure 10-12) generally agree on how global climate might change within a factor of two (Figure 10-10) but disagree on changes for individual regions. Here are the main projections of the major climate models:

- Earth's mean surface temperature will rise 1.5–5.5°C (2.7–9.9°F) by 2050 if inputs of greenhouse gases continue to rise at the current rate

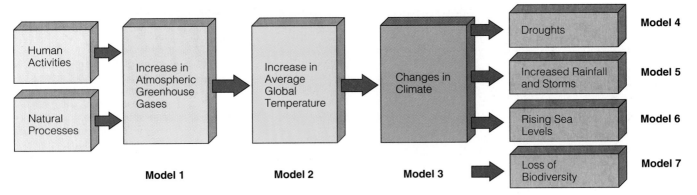

Figure 10-12 Generalized model of the greenhouse effect and its possible consequences. The greenhouse effect itself (model 1) is well established. The increase in temperature resulting from higher concentrations of greenhouse gases (model 2) is projected by crude climate models. These models project that average global temperature should rise 1.5°C–5.5°C (2.7°F–9.9°F) if atmospheric concentrations of greenhouse gases double. Because of uncertainties in the models, however, a particular model's projections could be off by a factor of two in either direction. Current models cannot project the specific climate changes (model 3) and their consequences (models 4–7) in various parts of the world.

(Figure 10-10). Even at the lower value, the earth would be warmer than it has been for 10,000 years (see cartoon, right).

- The Northern Hemisphere will warm more and faster than the Southern Hemisphere, mostly because water takes longer to warm than air and the latter has more ocean (Figure 5-18), which can absorb more heat than land can.

- Temperatures at middle and high latitudes should rise two to three times the global average, whereas temperatures near the equator should rise less than the global average.

- Soil will be drier in some middle latitudes, especially during summers in the Northern Hemisphere.

- More areas will have extreme heat waves and more forest and brush fires.

- The average sea level will rise 2–4 centimeters (0.8–1.6 inches) per decade.

A Cloudy Crystal Ball: What We Don't Know
Because we have only partial knowledge about how the earth's climate system works, our models and scenario projections are flawed, but they are all we have. The following list contains some factors that might dampen or amplify a rise in average atmospheric temperature, determine how fast temperatures might climb (or fall), and influence what the effects might be on various areas:

- *Changes in the amount of solar energy reaching the earth.* Solar output varies by about 0.1%—apparently in 11-year, 80-year, and other cycles—which

can temporarily warm or cool the earth and thus affect the projections of climate models. Two 1992 studies concluded, however, that the projected warming power of greenhouse gases will outweigh the climatic influence of the sun over at least the next 50 years.

- *Where some of the CO_2 we put into the atmosphere goes.* Only about 49% of carbon dioxide released each year by human activities remains in the atmosphere. Models and some measurements indicate that about 29% of our input of CO_2 is absorbed by the oceans. This leaves about 22% of the CO_2 we put into the atmosphere unaccounted for. Researchers believe that it is being absorbed by plants and soil on land, but have not been able to pinpoint where this is happening.

- *Effects of oceans on climate.* The world's oceans could slow global warming by absorbing more heat, but this depends on how long the heat takes to reach deeper layers. We don't know much about this effect, although recent measurements by oceanographers indicate that deep vertical mixing in the ocean occurs extremely slowly (taking up to hundreds of years) in most places because water density increases with depth, inhibiting mixing of different layers. The oceans also help moderate tropospheric temperature by removing about 29% of the excess CO_2 (a major greenhouse gas) we pump into the atmosphere, but we don't know if they can absorb more. If the oceans warm up enough, more CO_2 will bubble out of solution than dissolves (just as in a glass of ginger ale left out in the sun), amplifying and accelerating global warming. On the other hand,

Q: What are the two most desirable ways to deal with hazardous waste?

THIS IS YOUR PLANET.

THIS IS YOUR PLANET ON FOSSIL FUELS.

ANY QUESTIONS?

Matt Wuerker (Color added)

warmer air might speed up photosynthesis by oceanic phytoplankton, which would absorb more CO_2 from the atmosphere and slow global warming. Another possibility is that warmer air will evaporate more water from the oceans and create more clouds. Depending on their type (thick or thin) and altitude, more clouds could contribute to either warming or cooling. We don't know which type might predominate and how this factor would vary in different parts of the world.

■ *Changes in polar ice.* The Greenland and Antarctic ice sheets act like enormous mirrors reflecting sunlight back into space. If warmer temperatures melted some of this ice and exposed darker ground or ocean that would absorb more sunlight, warming would be accelerated. Then more ice would melt, amplifying the rise in atmospheric temperature even more. On the other hand, the early stages of global warming might actually increase the amount of the earth's water stored as ice. Warmer air would carry more water vapor, which could drop more snow on some glaciers, especially the gigantic Antarctic ice sheet. If snow accumulated faster than ice was lost, the ice sheet would grow, reflect more sunlight, and help cool the atmosphere—perhaps ushering in a new ice age within a thousand years.

■ *Air pollution.* Projected global warming might be partially offset by particles and droplets of various air pollutants—released by volcanic eruptions (Figures 8-7 and 10-11) and human activities—because they reflect back some of the incoming sunlight. However, things aren't that simple. Pollutants in the lower troposphere can either warm or cool the air and surface below them, depending on the reflectivity of the underlying surface. These

contradictory and patchy effects, plus improved air pollution control, make it unlikely that air pollutants will counteract any warming very much in the next half century. Even if they did, levels of these pollutants, which already kill hundreds of thousands of people a year and damage vegetation (including food crops), need to be reduced.

■ *Effects of increased CO_2 on photosynthesis.* Some studies suggest that more CO_2 in the atmosphere is likely to increase the rate of photosynthesis, with the increased growth of plants and other producers removing more CO_2 from the atmosphere and slowing global warming. Other studies suggest that this effect varies with different types of plants and in different climate zones. Also, much of the increased plant growth could be offset by plant-eating insects that breed more rapidly and year-round in warmer temperatures.

■ *Methane release from wetlands.* Natural wetlands making up just 5% of Earth's land area are a major source of methane emissions, produced when organic material is decomposed by anaerobic bacteria that thrive in such flooded, oxygen-deficient soils. Some scientists speculate that in a warmer world huge amounts of methane tied up in arctic tundra soils and in muds on the bottom of the Arctic Ocean might be released if the blanket of permafrost covering tundra soils melts and the oceans warm considerably. Because methane is a potent greenhouse gas, this release could greatly amplify global warming. On the other hand, some scientists believe that bacteria in tundra soils would rapidly oxidize the escaping methane to CO_2, a less potent but still important greenhouse gas. A 1993 study found that parts of Alaska's tundra soils are now releasing CO_2—possibly a signal that global warming is taking its toll.

■ *Adequate information on Earth's past climate history.* Recent data from analysis of carbon dioxide and other gases trapped in layers of ice formed over the past 260,000 years suggest that the earth's climate shifted often, drastically, and sometimes surprisingly quickly in the interglacial period preceding the one we live in. The new evidence (which has not been supported by another recent analysis of ancient ice) indicates that average temperatures during the warm interglacial period that began about 125,000 years ago varied as much as 10°C (18°F) in only a decade or two. If these findings are confirmed and also apply to the current interglacial period, fairly small changes in concentrations of greenhouse gases could trigger climate instability with sharp and perhaps rapid shifts in average global temperatures.

A: Don't make them and recycle or reuse them

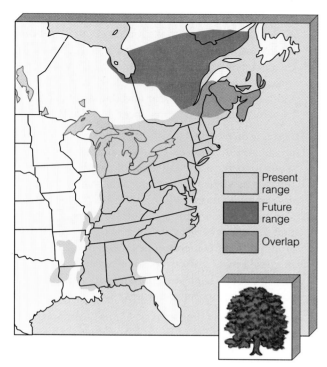

Figure 10-13 According to one projection, if CO$_2$ emissions doubled between 1990 and 2050, beech trees (now common throughout the eastern United States) would be able to survive only in a greatly reduced range in northern Maine and southeastern Canada (shown in orange and red). (Data from Margaret B. Davis and Catherine Zabinski, University of Minnesota)

Because of these and numerous other uncertainties in global climate models, projections made so far might be off by a factor of two in either direction. In other words, global warming during the next 50–100 years could be half the temperature increase projected in Figure 10-10 (the best-case scenario) or double it (the worst-case scenario).

Some Possible Effects of a Warmer World

So what's the big deal? Why should we worry about a possible rise of only a few degrees in the mean surface temperature of the earth? We often have that much change between June and July, or between yesterday and today. The key point is that we are not talking about normal swings in *local weather*, but about a projected *global* change in *average climate* from a thickening blanket of greenhouse gases.

A warmer troposphere would have different consequences for different peoples and species. Some places would get drier, some wetter. Some would get hotter, others cooler. Here are some possible effects of a warmer global climate:

- *Changes in food production.* Food productivity could increase in some areas and drop in others. Past archeological evidence and computer models indicate that climate belts and thus tolerance ranges of plant species (including crops) would shift northward by 100–150 kilometers (60–90 miles) or 150 meters (500 feet) vertically (Figure 5-6) for each 1°C (1.8°F) rise in global temperature. Computer models have projected drops in the global yield of key food crops ranging from 30% to 70%. With current knowledge, we can't predict where changes in crop-growing capacity might occur or how long such changes might last.

- *Reductions in water supplies.* Lakes, streams, and aquifers in some areas that have watered ecosystems, croplands, and cities for centuries could shrink or dry up altogether, forcing entire populations to migrate to areas with adequate water supplies—if they could. We can't say with much certainty where this might happen.

- *Changes in forests.* Forests in temperate and subarctic regions (Figure 5-3) would be forced to move toward the poles or to higher altitudes (Figure 5-6) leaving more grassland and shrubland in their wake. However, tree species can move only through the slow growth of new trees along forest edges—typically about 0.9 kilometer (0.5 mile) per year or 9 kilometers (5 miles) per decade. If climate belts moved faster than this—or if migration were blocked by cities, roads, or other barriers built by people—entire forests of oak, beech (Figure 10-13), and other deciduous trees could die and release CO$_2$ into the atmosphere as they decompose. According to Oregon State University scientists, projected drying from global warming could cause massive fires in up to 90% of North American forests, destroying wildlife habitats and injecting huge amounts of CO$_2$ into the atmosphere.

- *Reductions in biodiversity.* Large-scale forest diebacks would also cause mass extinction of species that couldn't migrate to new areas. And fish would die as temperatures soared in streams and lakes, and as lowered water levels concentrated pesticides. Any shifts in regional climate would threaten many parks, wildlife reserves, wilderness areas, wetlands, and coral reefs, wiping out many current efforts to stem the loss of biodiversity.

- *Rising sea level.* Water expands slightly when heated. This explains why global sea levels would rise if the oceans warmed, just as the fluid in a thermometer rises when heated. If warming at the poles caused ice sheets and glaciers to melt even partially, global sea level would rise even more. About one-third of the world's population and more than a third of the world's economic

Q: What are the three least desirable ways of handling hazardous waste?

infrastructure are concentrated in coastal regions. Thus, even a moderate rise in sea level would flood low-lying areas (many of which contain major cities), as well as lowlands and deltas where crops are grown. It would also destroy wetlands and coral reefs and accelerate coastal erosion. Currently, global sea levels are rising a few centimeters per decade. Scientists project that two-thirds of this rise is the result of global warming. Recent studies indicate the remaining third may be due to cutting down forests (which transfers water in trees and soil to the atmosphere and eventually to the ocean), draining of wetlands, and large-scale pumping of groundwater (which transfers water to Earth's surface where some of it reaches the ocean). One comedian jokes of plans to buy land in Kansas because it will probably become valuable beachfront property; another boasts she isn't worried because she lives in a houseboat—the "Noah strategy."

- *Weather extremes.* In a warmer world, prolonged heat waves and droughts would become the norm in many areas. And as the upper layers of seawater warmed, hurricanes and typhoons would occur more frequently and blow more fiercely.

- *Threats to human health.* A warmer world would disrupt supplies of food and fresh water, displacing millions of people and altering disease patterns in unpredictable ways. The spread of tropical climates from the equator would bring malaria (Section 8-4), encephalitis, yellow fever, dengue fever, and other insect-borne diseases to formerly temperate zones. Sea-level rise could spread infectious disease by flooding sewage and sanitation systems in coastal cities.

10-3 SOLUTIONS: DEALING WITH THE THREAT OF GLOBAL WARMING

More Research or Action? Some analysts believe we should wait until we know more before taking any serious action to deal with the possibility of global warming. However, even with better understanding, which could take decades, our knowledge will be limited because climate is so incredibly complex. Thus scientists will never be able to offer the certainty that some decision makers want before making such tough decisions as phasing out fossil fuels and replacing deforestation with reforestation.

Others urge us to adopt the *precautionary principle*, the idea that when dealing with risky, unpredictable, and often irreversible environmental problems it is

Denial Can Be Deadly

SPOTLIGHT

Many people say we should wait for more information about possible global warming before taking any action. Others contend there is no problem, or they hope that some factor or combination of factors will make the problem go away. Some see such behavior as a form of denial—an unwillingness to face up to potentially harmful behavior. Psychologist Robert Ornstein calls such denial the *boiled frog syndrome*. It's like trying to alert a frog to danger as it sits in a pan of water being very slowly heated on the stove.

If the frog could talk, it might say, "I'm a little warmer, but I'm doing fine." As the water gets hotter, we would warn the frog that it will die, but it might reply, "The temperature has been increasing for a long time, and I'm still alive. Stop worrying."

Eventually the frog dies because it has no experience of the lethal effects of boiling water and thus cannot perceive that its situation is dangerous. Like the frog, we also face a possible future without precedent, and our senses are unable to pick up warnings of impending danger. Unlike the frog, however, we can act to prevent or at least lessen the possible consequences of our actions.

Suppose however, that we, like the frog, continue to deny or not want to think about the possibility of climate change until the earth's mean temperature rises to the point at which it exceeds normal climatic fluctuations (which some climate scientists expect to happen within the next 5–20 years). Reacting then to improve energy efficiency, replace fossil fuels with renewable energy resources, halt deforestation, and start massive reforestation will take another 40–50 years. By then much of the damage will already have been done, and the effects would probably last for hundreds if not thousands of years.

On the other hand, perhaps the threat of global warming will not materialize. Should we take actions now that will cost enormous amounts of money and create political turmoil, all based on a cloudy crystal ball? There are no simple or easy answers.

often wise to take action before there is enough scientific knowledge to justify it (Spotlight, above).

Slowing Possible Global Warming The good news according to Gus Speth, former president of the World Resources Institute, is that "even though climate change threatens to be bigger, more irreversible,

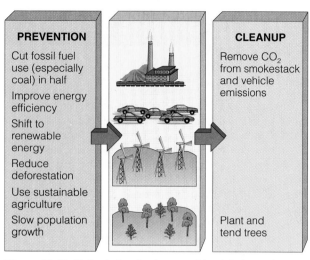

PREVENTION	CLEANUP
Cut fossil fuel use (especially coal) in half	Remove CO$_2$ from smokestack and vehicle emissions
Improve energy efficiency	
Shift to renewable energy	
Reduce deforestation	
Use sustainable agriculture	
Slow population growth	Plant and tend trees

Figure 10-14 Methods for slowing possible global warming.

and more pervasive than other environmental problems, it's also *controllable*, if we act now." Reducing this risk also may not be as costly as some have projected. According to recent studies by the National Academy of Sciences and the Congressional Office of Technology Assessment, the United States could reduce its greenhouse gas emissions by 25–35% of 1990 levels using existing technology at low or no net cost, and by 48% at low-to-moderate cost. A 1993 study estimated that a doubling of current atmospheric levels of greenhouse gases by the middle of the next century could cost the U.S. economy alone about $60 billion per year and up to $335 billion per year, assuming temperatures continued to rise.

Figure 10-14 presents a variety of ways to slow possible global warming; currently, none of these solutions is being vigorously pursued. The quickest, cheapest, and most effective way to reduce emissions of CO$_2$ and other air pollutants over the next two to three decades is to use energy more efficiently (Section 18-2 and Solutions, at right). According to the National Academy of Sciences, this strategy alone could lower U.S. greenhouse gas emissions by 10–40% at no net cost to the economy. Shifting from fossil fuels to renewable energy resources that do not emit CO$_2$ (Chapter 18) could cut projected U.S. CO$_2$ emissions by 8–15% by 2000 and could virtually eliminate them by 2025.

Natural gas (Section 19-2) could be used to help make the transition to an age of energy efficiency and renewable energy. When burned, natural gas emits only half as much CO$_2$ per unit of energy as coal (Figure 19-11) and emits far smaller amounts of most other air pollutants as well. Halting deforestation (Sections 16-2 and 16-3) and switching to Earth-sustaining agriculture (Section 14-4) would reduce CO$_2$ emissions

SOLUTIONS

Energy Efficiency to the Rescue

According to energy expert Amory Lovins (Guest Essay, p. 513), *the major remedies for slowing possible global warming are things we should be doing already, even if there were no threat of global warming.* He argues that if we waste less energy, reduce air pollution by cutting down on our use of fossil fuels, and harvest trees sustainably, we and other forms of life—in this and in future generations—would be better off regardless of whether we can affect global climate.

He also argues that getting countries to sign treaties and to agree to cut back and reallocate their use of fossil fuels in time to reduce serious environmental effects is difficult, if not almost impossible, and very costly. Climate models suggest that CO$_2$ emissions must be reduced by 80% to slow projected global warming to a safe rate; so far countries can't even agree to a 20% reduction.

According to Lovins, improving energy efficiency (Section 18-2) would be the fastest, cheapest, and surest way to slash emissions of CO$_2$ and most other air pollutants within two decades using existing technology. He estimates that doing this would also save the world up to $1 trillion per year in reduced energy costs—as much as the annual global military budget.

Moreover, using energy more efficiently would reduce all forms of pollution, help protect biodiversity, and forestall arguments among governments about how CO$_2$ reductions should be divided up and enforced. This approach would also make the world's supplies of fossil fuel last longer, reduce international tensions over who gets the dwindling oil supplies, and give us more time to phase in alternatives to fossil fuels.

To Lovins and most environmentalists, greatly improving worldwide energy efficiency *now* is a money-saving, life-saving, Earth-saving, win-win offer that we should not refuse, even if climate change is not a possibility.

and help preserve biodiversity. Slowing population growth is also important. If we cut per capita greenhouse gas emissions in half but world population doubles, we're back where we started.

All the food we eat, the houses we live in, any cars we drive, and almost anything we do requires energy (mostly from fossil fuels), and this adds carbon dioxide to the atmosphere. People in LDCs with much lower average resource consumption add much less

Q: How are 71% of official hazardous wastes in the United States handled?

CO_2 per person, but they hope to increase their resource consumption. For example, between 1982 and 1992 emissions of carbon dioxide in China increased 65%, largely due to a sharp rise in coal burning. If projected coal burning occurs, by 2025 China will emit more CO_2 than the current combined total of the United States, Japan, and Canada.

Some analysts have suggested that MDCs and LDCs enter into win-win pacts to reduce the threat of global warming. In this "let's make a deal" strategy, LDCs would agree to stop deforestation, protect biodiversity, slow population growth, enact fairer land distribution policies, and phase out coal burning. In return, MDCs would forgive much of LDCs' foreign debt and help fund the transfer to LDCs of modern energy efficiency, solar energy, pollution control, pollution prevention, sustainable agriculture, and reforestation technologies. MDCs would also agree to make substantial cuts in their use of fossil fuels, abandon use of ozone-depleting chemicals, greatly improve energy efficiency, stop deforestation, shift to sustainable agriculture, and slow their population growth.

Removing significant amounts of CO_2 from exhaust gases is not currently feasible. Available methods can remove only about 30% of the CO_2 and would at least double the cost of electricity. Planting and tending trees is important for restoring deforested and degraded land and for reducing soil erosion but is only a stopgap measure for slowing CO_2 emissions. To absorb the CO_2 we put into the atmosphere, each person in the world would need to plant and tend an average of 1,000 trees every year. Also, if much of the resulting newly grown forests are cleared and burned by us or by massive forest fires caused by global warming—or if much of the new forest died because of drought—most of the CO_2 removed would be released, accelerating global warming.

Some scientists have suggested various "technofixes" for dealing with possible global warming, including (1) fertilizing the oceans with iron to stimulate the growth of marine algae, which could remove more CO_2 through photosynthesis; (2) covering the oceans with white Styrofoam chips to help reflect more energy away from the earth's surface; (3) unfurling gigantic foil-faced sun shields in space to reduce solar input; and (4) injecting sunlight-reflecting sulfate particulates into the stratosphere to cool the earth's surface. All of these schemes are quite expensive and would have unknown—possibly harmful—effects on the earth's ecosystems and climate.

Preparing for Possible Global Warming Some analysts suggest that we begin preparing for the long-term effects of possible global warming. Their suggestions include the following courses of action:

- *Breed food plants that need less water or can thrive in salty water.*
- *Build dikes to protect coastal areas from flooding*, as the Dutch have done for centuries.
- *Move storage tanks of hazardous materials away from coastal areas.*
- *Ban new or rebuilt construction on low-lying coastal areas.*
- *Stockpile 1–5 years' worth of key foods throughout the world as short-term insurance against disruptions in food production.*
- *Expand existing wilderness areas, parks, and wildlife refuges northward in the Northern Hemisphere and southward in the Southern Hemisphere, and create new wildlife reserves in these areas.*
- *Connect wildlife reserves with corridors that would allow mobile species to move with climate change.*
- *Waste less water* (Section 11-5).

All of the measures together for slowing or responding to climate change would cost us money, but the bill would be far less than the $12 trillion we have spent since 1945 to protect us from the possibility of nuclear war. A 1992 study by four environmental groups concluded that aggressive action to lower CO_2 emissions in the United States by 70% over the next 40 years could cost the economy about $2.7 trillion. But it would also save consumers and industry about $5 trillion in fuel and electricity bills, leading to a net saving of $2.3 trillion. To many analysts the possibility of global warming is as serious a threat as the possibility of nuclear war and will require action at international, national, local, and individual levels (Individuals Matter, p. 248).

10-4 OZONE DEPLETION: SERIOUS THREAT OR HOAX?

The Threat: Letting in Deadly Rays Thanks to the evolution of photosynthetic, oxygen-producing bacteria, the earth has had a stratospheric global sunscreen—the ozone layer (Figure 9-2)—for the past 450 million years. Considerable evidence indicates that we are thinning this screen with our recent use of chlorine- and bromine-containing compounds. A few scientists dismiss the threat of ozone depletion, but the overwhelming consensus of researchers in this field is that ozone depletion by chemicals we have released into the atmosphere is a real threat. If the measurements and models (Figure 10-15) used to characterize this change in the chemical makeup of the stratosphere are correct, ozone depletion will have serious long-term

A: Buried in deep wells ponds, pits, or landfills (64%) and incinerated (7%)

What You Can Do to Reduce Global Warming

INDIVIDUALS MATTER

While the world's governments argue over what to do about projected global warming, we can take matters into our own hands:

- *Reduce your use of fossil fuels.* Driving a car that gets at least 15 kilometers per liter (35 miles per gallon), joining a car pool and using mass transit, and walking or bicycling as much as possible will reduce emissions of CO_2 and other air pollutants, will save energy and money, and can improve your health.

- *Use energy-efficient light bulbs, refrigerators, and other appliances.*

- *Use solar energy to heat household space or water as much as possible.*

- *Cool your house by using shade trees and available breezes and by making it energy efficient.*

- *Plant and care for trees to help absorb CO_2.*

- *Lobby for laws aimed at encouraging energy efficiency, reducing deforestation, and curbing emissions of greenhouse gases and other air pollutants.* In 1990, Connecticut was the first state to pass a global warming law. This legislation bans use of electric heating in new or renovated buildings after 1993 unless they meet certain minimum energy standards.

effects on human health, animal life, and the sunlight-driven primary producers (mostly plants) that support the earth's food chains and webs.

Chlorofluorocarbons: From Dream Chemicals to Nightmare Chemicals How did we get into this situation? It started when Thomas Midgley, Jr., a General Motors chemist, discovered the first chlorofluorocarbon (CFC) in 1930, and chemists then made similar compounds to create a family of highly useful CFCs. The two most widely used are CFC-11 (trichlorofluoromethane, CCl_3F) and CFC-12 (dichlorofluoromethane, CCl_2F_2).

These useful, chemically stable, odorless, nonflammable, nontoxic, and noncorrosive compounds seemed to be dream chemicals. Cheap to make, they became popular as coolants in air conditioners and refrigerators, propellants in aerosol spray cans, cleaners for electronic parts such as computer chips, sterilants for hospital instruments, fumigants for granaries and ship cargo holds, and building blocks for the bubbles in Styrofoam (used for insulation and packaging).

But it was too good to be true. In 1974 chemists Sherwood Rowland and Mario Molina made calculations indicating that CFCs were creating a global time bomb by lowering the average concentration of ozone in the stratosphere. They shocked both the scientific community and the $28-billion-per-year industry that makes these chemicals by calling for an immediate ban of CFCs in spray cans.

Here's what Rowland and Molina found: Spray cans, discarded or leaky refrigeration and air-conditioning equipment, and the production and burning of plastic foam products release CFCs into the atmosphere. These molecules are too unreactive to be removed, and (mostly through convection, random drift, and the turbulent mixing of air in the troposphere) they rise slowly into the stratosphere, taking 10–20 years to make the journey. There, under the influence of high-energy ultraviolet (UV) radiation, they break down and release chlorine atoms, which speed up the breakdown of ozone (O_3) into O_2 and O and cause ozone to be destroyed faster than it is formed. Each CFC molecule can last in the stratosphere for 65–110 years. During that time, each chlorine atom in these molecules—like a gaseous Pac-Man—can convert as many as 100,000 molecules of O_3 to O_2. If Rowland and Molina's calculations and later models of this problem (Figure 10-15) are correct, these one-time dream molecules have turned into a nightmare of global ozone terminators.

Although Rowland and Molina warned us of this problem in 1974, it took 15 years of interaction between the scientific and political communities before countries agreed to begin slowly phasing out CFCs.* The CFC industry was a powerful, well-funded adversary with a lot of profits and jobs at stake. But through 15 years of attacks Rowland and Molina held their ground, expanded their research, and relentlessly explained the meaning of their calculations to other scientists, elected officials, and the press.

CFCs are not the only ozone-eaters. A few other chemicals can release highly reactive chlorine and bromine atoms if they end up in the stratosphere and are exposed to intense UV radiation, including: **(1)** bromine-containing compounds called *halons* and *HBFCs* (both used in fire extinguishers) and *methyl bromide* (a widely used pesticide); and **(2)** *carbon tetrachloride* (a cheap, highly toxic solvent) and *methyl chloroform*, or *1,1,1-trichloroethane* (used as a cleaning solvent for clothes and metals and as a propellant in more than 160 consumer products, such as correction fluid, dry-cleaning sprays, spray adhesives, and other

*For a fascinating account of how corporate stalling, politics, economics, and science can interact, see Sharon Roan's *Ozone Crisis: The 15-Year Evolution of a Sudden Global Emergency* (New York: Wiley, 1989).

Q: How many sites in the United States contain potentially hazardous wastes?

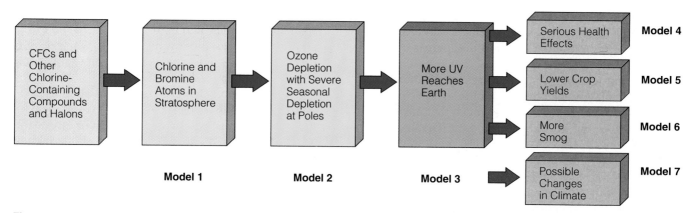

Figure 10-15 Generalized model of ozone layer depletion and its effects. Models 1–3 have been confirmed. The extent of the effects in models 4–7 is still being studied.

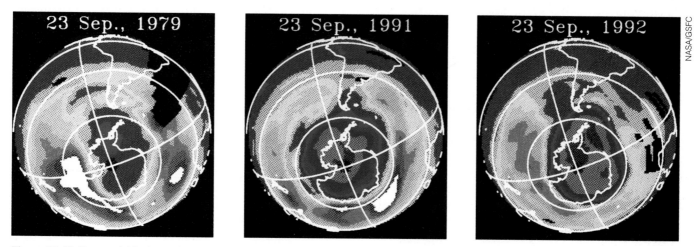

Figure 10-16 Seasonal thinning or loss of ozone (shown by shades of pink), called an ozone hole, in the upper stratosphere over the Antarctic region, as measured by the Nimbus-7 satellite on September 23 of 1979, 1991, and 1992. Since 1987 this area of seasonal thinning (where the normal concentration of ozone has been cut at least in half) has spanned an area larger than the continental United States. In 1992 and again in 1993—aided by sulfuric acid droplets formed as a result of the 1991 eruption of Mount Pinatubo (Figure 10-11)—the size of the hole spread and covered an area three times the area of the continental United States, and ozone levels above the South Pole fell to an all-time low.

aerosols). Substitutes are available for virtually all uses of these chemicals.

Holes in the Ozone Layer and Other Surprises

Each year the news about ozone loss seems to worsen, and sometimes it takes scientists by surprise. The first surprise came in 1985, when researchers analyzing satellite data discovered that 50% of the ozone in the upper stratosphere over Antarctica was being destroyed during the Antarctic spring and early summer (September–December), when sunlight returned after the dark Antarctic winter. This pronounced seasonal loss of ozone had not been predicted by computer models of the stratosphere. Since then, this pronounced seasonal Antarctic *ozone hole*, or *ozone thinning*, has expanded in most years (Figure 10-16). In

both 1992 and 1993 it covered an area three times the size of the continental United States.

Measurements have indicated that CFCs are the primary culprits. Each sunless winter, steady winds blow in a circular pattern over Earth's poles, creating *polar vortices*—huge swirling masses of bitter-cold air that are trapped above the poles for several months until the sun returns. When water droplets in clouds enter these large circling streams of frigid air, they form tiny ice crystals. The surfaces of these ice crystals collect CFCs and other ozone-depleting chemicals in the stratosphere, enabling them to destroy ozone much faster. The return of sunlight two to three months later triggers weeks of ozone depletion before the vortex breaks up. Huge clumps of ozone-depleted air above Antarctica then flow northward and linger for a few

A: 34,000 (and perhaps as many as 425,000)

more weeks over parts of Australia and New Zealand and over the southern tips of South America and Africa, thus raising UV radiation levels in these areas by as much as 20%.

A 1993 study suggests that the seasonal Antarctic ozone hole may be prolonging its life each year by cooling the air inside the vortex so much that in effect it delays the spring warming by ten days. That means ten more days of ice clouds—and ten more days of ozone destruction. If this finding is correct, seasonal ozone depletion above the Antarctic may perpetuate itself, regardless of any change in CFCs or other ozone-depleting chemicals; and it could even get worse because of this self-reinforcing behavior.

In 1988 scientists discovered that similar but much less severe ozone thinning occurs over the North Pole during the Arctic spring and early summer (February– June), with a seasonal ozone loss of 10–25% (compared to 50% or more over much of the Antarctic region). When this mass of air above the Arctic breaks up each spring, masses of ozone-depleted air flow southward to linger over parts of Europe, North America, and Asia. Mostly because these air masses flow alternately over land and water, seasonal ozone loss over the North Pole is much lower than that over the South Pole.

The situation could get much worse. Scientists estimate that ozone losses over northern midlatitudes could be 10–30%, as ozone-destroying chemicals drift slowly into the stratosphere. In 1992 atmospheric scientists warned that if rising levels of greenhouse gases (Figure 10-9) change the climate as projected over the next 50 years, the stratosphere over the Arctic could be altered such that it will experience severe ozone thinning like that now found over Antarctica. This in turn would sharply decrease ozone levels over parts of the northern hemisphere, including the United States—another example of connections.

Is Ozone Depletion a Hoax? Political talk-show host Rush Limbaugh, zoologist and former head of the Atomic Energy Commission Dixy Lee Ray (now deceased), and several articles in the popular press have claimed that ozone depletion by CFCs is a hoax. The evidence for these claims comes mostly from articles and books written by S. Fred Singer (a Ph.D. physicist and climate scientist), Rogelio Maduro (who has a bachelor of science degree in geology and who is an associate editor of a science and technology magazine published by supporters of Lyndon LaRouche—considered an extremist politician by most people), and Ralf Schauerhammer (a German writer).

Ray claimed that CFCs should not reach the stratosphere because they are heavier than air. But the troposphere is turbulent and more like a room full of air stirred and mixed by a fan. Since the 1960s mea-

surements have confirmed the presence of heavier-than-air molecules high in the troposphere and lower stratosphere. And since the mid-1970s CFCs have been detected in thousands of stratospheric air samples collected by balloons and aircraft.

Maduro and Schauerhammer claim that the evidence indicates that volcanoes, seawater, and biomass burning have been releasing far more chlorine into the atmosphere than do the CFCs we have emitted. They argue that this has been going on for billions of years and the ozone layer is still here. Critics also argue that the ozone hole over Antarctica was discovered in 1956, before CFCs were in wide use, and thus is natural phenomenon.

Scientists directly involved in ozone-layer research dispute these claims, describing them as being based on selective use of out-of-date research and on unwarranted extrapolation of questionable data. They point out that most chlorine from natural sources—mostly sodium chloride (NaCl) and hydrogen chloride (HCl) from the evaporation of sea spray—never makes it to the stratosphere because these compounds (unlike CFCs) are soluble in water and get washed out of the lower atmosphere by rain. If sodium chloride from sea spray were making it to the stratosphere, there should be evidence of sodium in the lower stratosphere; measurements show that it is not there. Measurements also indicate that no more than 20% of the chlorine from biomass burning (in the form of methylchloride) is making it to the stratosphere, which is about five times less than the contribution from CFCs.

Researchers also dispute the hypothesis that large quantities of HCl are injected into the stratosphere from volcanic eruptions. Most of this water-soluble HCl is injected into the troposphere and is washed out by rain before it reaches the stratosphere. Measurements show that HCl in the stratosphere increased by less than 10% after the eruption of the Mexican volcano El Chichón in 1982, and it increased even less from the eruption of Mt. Pinatubo in the Philippines in 1991 (Figure 10-11). Singer, whose skepticism about some aspects of ozone depletion models has been cited to bolster the case of those calling the whole thing a hoax, agrees with this scientific consensus and in 1993 stated that "CFCs make the major contribution to stratospheric chlorine." Natural chlorine compounds released by large volcanic eruptions that reach the stratosphere can account for some ozone depletion there. However, such eruptions occur infrequently, and the evidence indicates that currently emissions of CFCs are the primary culprit.

In 1956 British scientist Gordon Dobson measured lower (but not unusually lower) levels of ozone over Antarctica than he expected. He had discovered the strong Antarctic polar vortex that does lead to some seasonal ozone loss through natural causes. Between

Q: How many of the world's estimated 30,000 edible plants feed most of the world's people?

1956 and 1976 this natural pattern of slight ozone loss did not change. However, since 1976 these seasonal losses of ozone have increased dramatically (Figure 10-16) and have been linked by measurements and models to rising levels of CFCs in the stratosphere. Critics of the ozone-depletion idea also point out that the expected increase in UV radiation from ozone loss in the stratosphere has not as yet been detected in urban areas in the United States and most other MDCs. However, in 1993 Canadian researchers reported that measurements made at ground level in Toronto showed that levels of skin-damaging ultraviolet-B increased more than 5% per year during wintertime every year from 1989 to 1993, as stratospheric ozone levels dropped. It is hypothesized that significant ground-level increases in UV radiation have not been observed in many areas (especially in industrialized countries in the Northern hemisphere) because air pollution may be filtering out some of the rays—another example of connections. If such air pollution is decreased (Section 9-5), we may experience a greater threat from increased UV radiation; however, if we don't decrease air pollution, we will continue to suffer from its harmful effects (Section 9-4). Polluting our way out of ozone increases at ground level—and possibly global warming—is hardly a reasonable or desirable way to deal with these potentially serious problems.

Since 1992 Canada has monitored surface ultraviolet radiation and given the public daily forecasts of expected UV levels using a scale of 1 (minimal risk) to 10—something Australia has been doing for several years as well. A similar program began in the United States in 1994.

Possible Consequences of Ozone Loss: Life in the Ultraviolet Zone   Why should we care about ozone loss (see cartoon)? With less ozone in the stratosphere, more biologically harmful UV-B radiation will reach the earth's surface. Evidence indicates that increased UV radiation will give us worse sunburns (caused by UV-B), earlier wrinkles (caused by UV-A), more cataracts (a clouding of the lens that reduces vision and can cause blindness if not corrected—caused by UV-B) and more skin cancers (Connections, p. 252).

Cases of skin cancer (Figure 10-17) and cataracts are soaring in Australia, New Zealand, South Africa, Argentina, and Chile, where the ozone layer is very thin for several months after the masses of ozone-depleted air over the South Pole (Figure 10-16) drift northward.

Australian television stations now broadcast daily UV levels and warnings for fair-skinned Australians—with the world's highest rate of skin cancer—to stay inside during bad spells and to protect themselves from the sun's rays with hats, clothing, and sunscreens when they go out during daytime—required by law for schoolchildren.

Chile, the only populous country located under the Antarctic ozone hole, has experienced a fourfold increase in malignant melanoma since 1980. Levels of skin cancer and cataracts are also increasing rapidly in the United States, presumably mostly from increased exposure to the sun as many people spend more time

A: 30 (with most provided by wheat, rice, corn, and potato)

CONNECTIONS

The Cancer You Are Most Likely to Get

Skin specialists have been warning us about the harmful effects of too much sunlight on skin long before there was any concern about additional exposure to UV radiation from a thinning ozone layer. Animal studies and epidemiological surveys indicate that years of exposure to UV-B ionizing radiation in sunlight is the primary cause of basal-cell and squamous-cell skin cancers (Figure 10-17a and 10-17b), which make up 95% of all skin cancers. Typically there is a 20–30-year lag between UV exposure and development of these cancers.

Caucasian children and adolescents who get only a single severe sunburn double their chance of getting these cancers. Some 90–95% of these types of skin cancer can be cured if detected early enough, although their removal may leave disfiguring scars, and they still kill about 2,300 Americans each year. (I have had three basal-cell cancers on my face because of "catching too many rays" in my younger years. I wish I had known then what I know now).

A third type of skin cancer, *malignant melanoma* (Figure 10-17c), spreads rapidly to other organs and kills one-fifth of its victims (most under age 40) within five years—despite surgery, chemotherapy, and radiation. Each year it kills about 100,000 people (including 6,800 Americans)—mostly Caucasians—but it can often be cured if detected early enough. Evidence suggests that people, especially Caucasians, who get three or more blistering sunburns before age 20 subsequently have five times more risk of contracting malignant melanoma than those who have never had severe sunburns.

Melanoma cases in the United States doubled between 1980 and 1990; it is now the most prevalent type of cancer among women ages 25–29. In 1980, 1 in 250 Americans developed melanoma. By 1991 the figure was 1 in 101, and by 2000 it is projected to be 1 in 75. The EPA estimates 200,000 skin cancer deaths in the United States alone over the next 50 years.

It has been assumed that most malignant melanomas are caused by exposure to UV-B radiation (although genetic factors apparently make some individuals more susceptible). However, some scientists suspect that exposure to UV-A radiation, which is not filtered out by the ozone layer and only partially blocked by most suncreeens, may play a role in causing malignant melanoma.

Virtually anyone can get skin cancer, but people with fair and freckled skin, blonde or red hair, and light eye color run the highest risk. People who spend long hours in the sun or in tanning parlors multiply their chances of developing skin cancer and of having wrinkled, dry skin by age 40. Nor does a dark suntan prevent skin cancer. Dark-skinned people are almost immune to sunburn but do get skin cancer, although at a rate one-tenth that of Caucasians. Outdoor workers are particularly susceptible to skin cancer on the face, neck, hands, and arms.

To protect yourself, the safest course is to stay out of the sun and say no to tanning parlors. When you are in the sun, wear tightly woven protective clothing, a wide-brimmed hat, and sunglasses that protect against UV radiation (ordinary sunglasses may actually harm your eyes by dilating your pupils so that more UV radiation strikes the retina). Unfortunately, the world's poor people can't afford such glasses. People who take antibiotics and women who take birth control pills are more susceptible to UV damage.

Apply sunscreen with a protection factor of 15 or more (25 if you

" I MISS THE OZONE LAYER...."

in the sun and move to areas with warmer climates. Some 600,000 Americans a year develop skin cancer. Any increase in UV radiation because of ozone thinning is expected to worsen this situation.

Other effects from increased UV exposure are **(1)** suppression of the immune system, which would reduce our defenses (regardless of skin pigmentation) against a variety of infectious diseases; **(2)** an increase in eye-burning, highly damaging ozone and acid deposition in the troposphere (Section 9-2); **(3)** lower yields (about a 1% decline for each 3% drop in stratospheric ozone) of crops such as corn, rice, soybeans, cotton, beans, peas, sorghum, and wheat, with estimated losses totaling $2.5 billion per year in the United States alone before the middle of the next cen-

Q: How much of Earth's land area is suitable for cultivation?

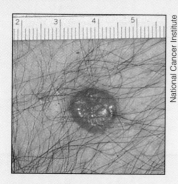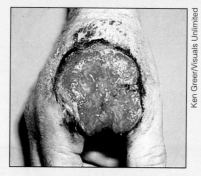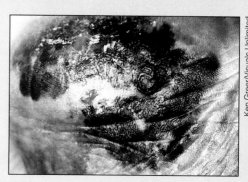

Figure 10-17 Three types of skin cancer: **(a)** basal-cell cancer, **(b)** squamous-cell cancer, and **(c)** malignant melanoma. The incidence of these types of cancer is rising.

have light skin) to all exposed skin, and reapply it after swimming or excessive perspiration. Children who use a sunscreen with a protection factor of 15 anytime they are in the sun from birth to age 18 decrease their chance of skin cancer by 80%; babies under a year old should not be exposed to the sun at all. Because of the possibility that UV-A may play a role in causing malignant melanoma, you may want to seek out newer suncreeens that do a good job of blocking out both UV-A and UV-B radiation. However, a recent study with mice suggests that even these improved sunscreens may not help prevent melanoma.

Before using a particular sunscreen, run a test patch on the skin on the underside of your arm to be sure that you are not allergic to its ingredients or that it doesn't have the reverse effect and amplify sun damage (something I recently learned the hard way). If you like the bronzed look, try some of the improved creams that give you a tanned appearance without spending long hours in the sun; but, again, test them for any allergic reaction first.

Get to know your moles and examine your skin at least once a month. The warning signs of skin cancer are a change in the size, shape, or color of a mole or wart

(the major sign of malignant melanoma, which needs to be treated quickly); sudden appearance of dark spots on the skin; or a sore that keeps oozing, bleeding, and crusting over but does not heal. Be alert for precancerous growths: reddish-brown spots with a scaly crust. If you observe any of these signs, consult a doctor immediately. Are you doing all you can to protect your skin, and perhaps your life, from exposure to UV radiation?

tury; and **(4)** a loss of perhaps $2 billion per year from degradation of paints, plastics, and other polymers in the United States alone.

In a worst-case scenario, most people would have to avoid the sun altogether (see cartoon, far left). Even cattle could graze only at dusk, and farmers and other outdoor workers might need to limit their exposure to the sun to minutes.

10-5 SOLUTIONS: PROTECTING THE OZONE LAYER

Just Say No If the models and measurements of ozone depletion and its possible effects are correct

(Figure 10-15), we know what needs to be done: Stop producing any ozone-depleting chemicals now (global abstinence). After saying no, the models indicate that we still will have to wait 50–100 years for the ozone layer to return to 1985 levels, and another 100–200 years for full recovery.

Substitutes are already available for most uses of CFCs, and others are being developed (Table 10-1 and Individuals Matter, p. 255). CFCs in existing air conditioners, refrigerators, and other products must be recovered and in some cases reused until the substitutes are phased in. This will be expensive, but if ozone depletion should become more severe because of our activities, the ecological, health, and financial costs will be much higher.

A: About 11%

Table 10-1 CFC Substitutes

Types	Pros	Cons
HCFCs (hydrochlorofluorocarbons)	Break down faster (2–20 years). Pose about 90% less danger to ozone layer. Can be used in aerosol sprays, refrigeration, air conditioning, foam, and cleaning agents.	Are greenhouse gases. Will still deplete ozone, especially if used in large quantities. Health effects largely unknown. HCFC-123 causes benign tumors in the pancreas and testes of male rats and may be banned for use in aerosol sprays, foam, and cleaning agents. May lower energy efficiency of appliances.
HFCs (hydrofluorocarbons)	Break down faster (2–20 years). Do not contain ozone-destroying chlorine. Can be used in aerosol sprays, refrigeration, air conditioning, and foam.	Are greenhouse gases. Safety questions about flammability and toxicity still unresolved. May lower energy efficiency of appliances. Production of HFC-134a, a refrigerant substitute, yields an equal amount of methyl chloroform, a serious ozone depleter.
Hydrocarbons (such as propane and butane)	Cheap and readily available. Can be used in aerosol sprays, refrigeration, foam, and cleaning agents.	Can be flammable and poisonous. Some increase ground-level pollution.
Ammonia	Simple alternative for refrigerators; widely used before CFCs.	Toxic if inhaled. Must be handled carefully.
Water and Steam	Effective for some cleaning operations and for sterilizing medical instruments.	Creates polluted water that must be treated. Wastes water unless the used water is cleaned up and reused.
Terpenes (from the rinds of lemons and other citrus fruits)	Effective for cleaning electronic parts.	None.
Helium	Effective coolant for refrigerators, freezers, and air conditioners.	This rare gas may become scarce if use is widespread, but very little coolant is needed per appliance.

Substitutes such as hydrofluorocarbons (HFCs) and hydrochlorofluorocarbons (HCFCs) may help ease the replacement of CFCs for essential uses, but models indicate that these chemicals also can deplete ozone (although at a slower rate) and probably should be banned no later than 2005. Both categories contain greenhouse gases.

Can Technofixes Save Us?　What about a quick fix from technology so we can keep on using CFCs? One suggestion is to collect some of the ozone-laden air at ground level over Los Angeles and other cities and ship it up to the stratosphere. Even if we knew how to do this, the Los Angeles air would dilute the stratospheric ozone concentration, rather than increasing it.

Two atmospheric scientists have speculated that we might inject large quantities of ethane and propane into the stratosphere, where they might react with CFCs to remove the offending chlorine atoms. They estimate this would take only 1,000 jumbo-jet flights over a critical 30-day period every year for several decades. But the scientists proposing this possibility warn that the plan could backfire, accelerate ozone depletion, and have unpredictable effects on climate.

Others have suggested using tens of thousands of lasers to blast CFCs out of the atmosphere before they can reach the stratosphere. However, the energy requirements would be enormous and expensive, and decades of research would be needed to perfect the types of lasers needed—time the models suggest we don't have. And no one can predict the possible effects on climate, birds, or planes.

Some Hopeful Progress　In 1987, 24 nations meeting in Montreal developed a treaty—commonly known as the *Montreal Protocol*—to cut emissions of CFCs (but not other ozone depleters) into the atmosphere by about 35% between 1989 and 2000. After hearing more bad news about ozone depletion, representatives of more than half the world's nations met in Copenhagen in 1992 and agreed to **(1)** phase out production (except for essential uses) of halons by January 1, 1994, and CFCs, carbon tetrachloride, HBFCs (halon substitutes), and

Q: How much of the world's food is produced on irrigated cropland?

Ray Turner and His Refrigerator

INDIVIDUALS MATTER

Ray Turner, an aerospace manager at Hughes Aircraft in California, made an important low-tech, ozone-saving discovery by using his head—and his refrigerator. His concern for the environment led him to look for a cheap and simple substitute for the CFCs used as cleaning agents for removing films of oxidation in the manufacture of most electronic circuit boards at his plant and elsewhere.

He started his search for a low-tech solution to a high-tech problem by looking in his refrigerator for a better circuit board cleaner. He decided to put drops of various substances on a corroded penny to see whether any of them would remove the film of oxidation. Then he used his soldering gun to see if solder would stick to the cleaned surface of the penny, indicating that the film had been cleaned off.

First, he tried vinegar. No luck. Then he tried some ground-up lemon peel, also a failure. Next he tried a drop of lemon juice and watched as the solder took hold. The rest is history.

In the months that followed Turner and a Hughes team perfected his discovery, which is now used on several of the company's military projects. Since it was introduced, the new cleaning technique has reduced circuit board defects by about 75% at Hughes. And Turner got a hefty bonus. Maybe you can find a solution to an environmental problem in your refrigerator.

What You Can Do to Help Protect the Ozone Layer

INDIVIDUALS MATTER

If you believe, as do most scientists, that there is a serious threat to the ozone layer, here are some things you can do:

- *Don't buy products containing CFCs, carbon tetrachloride, or methyl chloroform (1,1,1-trichloroethane on most ingredient labels).* Read labels and seek out substitutes for these products.

- *Don't buy CFC-containing polystyrene foam insulation.* Types of insulation that don't contain CFCs are extended polystyrene (commonly called EPS or beadboard), fiberglass, rock wool, cellulose, and perlite.

- *Don't buy halon fire extinguishers for home use.* Instead, buy those that use dry chemicals (CO_2 extinguishers release this greenhouse gas into the troposphere). If you already have a halon extinguisher, store it until a halon-reclaiming program is developed.

- *Stop using aerosol spray products, except in some necessary medical sprays.* Even those that don't use CFCs and HCFCs (such as Dymel) emit hydrocarbons or other propellant chemicals into the air. Use roll-on and hand-pump products instead.

- *Pressure legislators to ban all CFCs, halons, methyl bromide, carbon tetrachloride, and methyl chloroform by 1996 (with no loopholes)—and HCFCs by 2005 instead of by 2030.*

- *Pressure legislators not to exempt military and space programs from any phaseout of ozone-depleting chemicals.*

- *Buy new refrigerators and freezers that use vacuum insulation (as in Thermos bottles) instead of rigid-foam insulation and that use helium as a coolant.* (Such refrigerators are available from Cryodynamics, 1101 Bristol Road, Mountainside, NJ 07092). China has purchased 9 million of them.

- *If you junk a car, a refrigerator, a freezer, or an air conditioner, make sure the coolant is removed and kept safely for reuse or destruction.*

- *Have car and home air conditioners checked regularly for CFC leaks—and repair them if necessary.*

- *If you buy a car with an air conditioner, look for one that doesn't use CFCs.* These should be available on most new models by 1995.

chloroform by January 1, 1996; **(2)** freeze consumption of HCFCs at 1991 levels by 1996 and eliminate them by 2030; and **(3)** freeze methyl bromide production at 1991 levels by 1995, and phase it out by 2001. LDCs signing the treaty are given a 10-year grace period in meeting these deadlines, but this may be accelerated in 1995.

The agreements reached so far are important examples of global cooperation. Indeed, after reaching a historical high in 1988, global production of CFCs had dropped 50% by 1992—roughly equaling the amount produced in 1970. Some scientists claim that such action is premature because models of ozone depletion and its effects are not reliable enough. But many other scientists believe that ozone loss is a serious threat, that the agreements do not go far enough fast enough, and that we must expand and strengthen our efforts to provide environmental security (Guest Essay, p. 256). Individuals can play a role in reducing the threat of ozone depletion (Individuals Matter, at right).

A: About 33%

Framing a Global Response to Environmental Threats

Jessica Tuchman Mathews

Jessica Tuchman Mathews is vice president of the World Resources Institute, a highly respected center for policy research on global resource and environmental issues. Dr. Mathews has served as director of the Office of Global Issues on the President's National Security Council and on the editorial board of The Washington Post. *In 1989 she published an influential article in* Foreign Affairs *calling for nations to redefine national security in terms of national and global environmental security.*

National security and *national sovereignty* must be redefined during the 1990s to accommodate new global environmental realities, just as they were redefined during the 1970s to accommodate global economic realities.

Intricately interconnected effects of the way we live—the buildup of greenhouse gases, the depletion of the ozone layer, and the loss of tropical forests and species—are shifting the center of gravity in international relations. These phenomena threaten national securities, defy solution by one or a few countries, and render national borders irrelevant. By definition, then, they pose a major challenge to national sovereignty.

One of the few clear things about the post–Cold War era is that national security will increasingly depend on how resource, environmental, and demographic issues are resolved. It is no coincidence that control of Persian Gulf oil was central to this era's first international crisis.

Regional environmental decline is already threatening well-being and thereby political stability in many parts of the world. Eastern Europe's horrendous environmental degradation is undercutting attempts to rebuild its shattered economies. For instance, 95% of the water in Poland's rivers is unfit for human consumption, land is being withdrawn from cultivation because of contamination with toxic heavy metals, and air pollution causes heavy economic losses due to health costs and lost productivity. In LDCs natural resources such as farmland, forests, and fisheries are being ravaged while the number of people these resources must sustain is expected to grow by nearly 1 billion during the 1990s.

The fallout from global environmental trends goes far beyond economic and political arrangements. Unless the community of nations finds ways to reverse these trends, they will eventually shake not just the security of states but also the foundations of life. No one nation can succeed alone.

Dealing with the global environmental problems of tropical deforestation and species loss, climate change, and ozone depletion will require a rising level of collective international cooperation and management. Fortunately, nations are beginning to act as though they understand their mutual interest in cooperation. The most spectacular demonstrations of understanding were the agreements reached by 93 nations in 1990 and 1992 to phase out emissions of most ozone-destroying chemicals.

Turning this mutual interest into effective international management remains an elusive goal. Progress does not lie in a vain attempt to apply uniform environmental standards to nations whose members differ by 100-fold in per capita income and that have vastly different cultures, climates, religions, resources, and attitudes toward nature. Instead, it lies in institutional innovations as sweeping as those that inaugurated the post–World War II period.

The new international system must be designed to catalyze cooperation. Instead of the glacial pace required to negotiate treaties that set particular performance standards, we need fluid international processes that respond quickly to changes in scientific understanding and that set all nations moving in the same direction at whatever pace is realistic for each nation's circumstances.

Scientific theory and economic, political, and environmental concerns are all in a constant state of flux. Only a new institutional agility can keep international environmental governance closely attuned to these changing realities and ensure the best possible outcome.

Critical Thinking

1. Do you agree that national security and national sovereignty must be redefined to include national and global environmental security? Explain.

2. What changes in the current interactions among nations do you believe must be made to catalyze international cooperation on global environmental problems? How would you bring about these changes?

Some people have wondered whether there is intelligent life in other parts of the universe; others wonder whether there is intelligent life on Earth. To them, if we can seriously deal with the interconnected planetary problems of loss of biodiversity (Chapters 16 and 17), possible climate change, and depletion of ozone in the stratosphere, then the answer is a hopeful yes. This means recognizing that *prevention* is the best (and in the long run the least costly) way to deal with global environmental problems. Otherwise, they believe the answer is a tragic no.

The atmosphere is the key symbol of global interdependence. If we can't solve some of our problems in the face of threats to this global commons, then I can't be very optimistic about the future of the world.

Margaret Mead

Critical Thinking

1. What consumption patterns and other features of your lifestyle directly add greenhouse gases to the atmosphere? Which, if any, of those things would you be willing to give up to slow projected global warming and reduce other forms of air pollution?

2. Explain why you agree or disagree with each of the proposals listed in Section 10-3 for
   a. slowing down emissions of greenhouse gases into the atmosphere and
   b. adjusting to the effects of global warming.
   Explain. What effects would carrying out these proposals have on your lifestyle and that of your descendants? What effects might *not* carrying out these actions have?

3. What consumption patterns and other features of your lifestyle directly and indirectly add ozone-depleting chemicals to the atmosphere? Which, if any, of those things would you be willing to give up to slow ozone depletion?

4. Should all uses of CFCs, halon, and other ozone-depleting chemicals be banned in the United States and worldwide? Explain. Suppose this meant that air conditioning (especially in cars and perhaps in buildings) had to be banned, or that it became five times as expensive. Would you still support such a ban?

5. Do people have the right to use the atmosphere as a dumping ground for pollutants? Explain. If not, how would you restrict such activities?

*6. Make a concept map of the key ideas in this chapter using the section heads and subheads and the key terms (shown in boldface type in the chapter). See the inside front cover and Appendix 4 for information on concept maps.

A: 128 (94 at home and 34 abroad)

11 Water

Water Wars in the Middle East

Because of differences in climate some parts of the earth have an abundance of water and other parts have little water (Figures 5-2 and 4-27). As population, agriculture, and industrialization grow, there is increasing competition for water—especially in dry regions.

The next wars in the Middle East may well be fought over water, not oil. Most water in this dry region comes from three shared river basins: the Jordan, the Tigris–Euphrates, and the Nile (Figure 11-1).

Arguments among Ethiopia, Sudan, and Egypt over access to water from the Nile River basin are escalating rapidly. Ethiopia, which controls the headwaters of 80% of the Nile's flow, has plans to divert

more of this water; so does Sudan. This could reduce the amount of water available to water-short Egypt, whose terrain is desert except for the thin strip of irrigated cropland along the Nile and its delta. By 2025 Egypt's population is expected to double—increasing the demand for water. Its options are to go to war with Sudan and Ethiopia to obtain more water, or to slash population growth and improve irrigation efficiency.

There is also fierce competition for water among Jordan, Syria, and Israel, which get most of their water from the Jordan River basin. Israel uses water more efficiently than any country in the world. Nevertheless, it is now using 95% of its potentially renewable supply of fresh water, and the supply is projected to fall 30% short of demand by the year 2000 because of increased immigration. Water also plays a role in the conflict over Israeli-occupied territories of the West Bank. Under the hills of the West Bank lies the Mountain aquifer, a strategic source of water taken over by water-short Israel after the Six-Day War in 1967.

Currently, 88 million people live in the water-short basins of the Tigris and Euphrates rivers and by the year 2020 the population there is projected to almost double to 170 million. Turkey—located at the headwaters of these two rivers (Figure 11-1)—has abundant water. It plans to build 22 dams along the upper Tigris and Euphrates and to construct pipelines to transport and sell water to parched Saudi Arabia and Kuwait, and perhaps Syria, Israel, and Jordan. The greatest threat to Iraq is a cutoff of its water supply by Turkey and Syria.

Clearly, distribution of water resources will be a key issue in any future peace talks in this region. The keys to resolving these water supply problems in this volatile region involve a combination of regional cooperation, slowed population growth, and improved water efficiency.

By the middle of the next century almost twice as many people will be trying to share the same amount of fresh water the earth has today. Already, 1.2 billion people lack access to clean drinking water, 2.2 billion live without sewage systems, and two-thirds of the world's households don't have running water. As fresh water becomes scarcer, access to water resources will become a major economic and political issue.

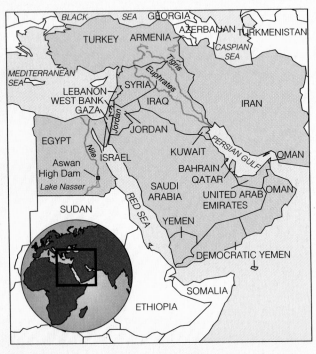

Figure 11-1 Middle Eastern countries have some of the highest population growth rates in the world. Because of their dry climate, food production depends on irrigation. In the 1990s and beyond, conflicts among countries in this region over access to water may overshadow long-standing religious and ethnic clashes, as well as disputes over ownership of oil supplies.

Our liquid planet glows like a soft blue sapphire in the hard-edged darkness of space. There is nothing else like it in the solar system. It is because of water.

JOHN TODD

In this chapter we will answer the following questions:

- What are water's unique physical properties?
- How much fresh water is available to us, and how much of this are we using?
- What are the world's water resource problems?
- How can we get more water, and how can we use it more efficiently?
- What pollutes water, where do the pollutants come from, and what effects do they have?
- What are the water pollution problems of streams, lakes, oceans, and groundwater?
- How can we prevent and reduce water pollution?

11-1 WATER'S IMPORTANCE AND UNIQUE PROPERTIES

We live on the water planet. From space, it is the blue colors of water and the white moisture-laden clouds that distinguish the earth from other planets. Earth's green colors are not possible without water, and its browns show water-short areas that cannot support much vegetation

A precious film of water—most of it salt water—covers about 71% of Earth's surface (Figure 5-18). Earth's organisms are made up mostly of water, or what I like to call *Earth juice*. For example, a tree is about 60% water by weight, and you and most animals are about 65% water.

Fresh water is a vital resource for agriculture, manufacturing, transportation, and countless other human activities. Water also plays a key role in sculpting the earth's surface, moderating climate, and diluting pollutants. In fact, without water, life as we know it could not exist.

Water has many unique—almost magical—properties.

- *Water exists as a liquid over a wide temperature range.* Its high boiling point of 100°C (212°F) and low freezing point of 0°C (32°F) mean that water remains a liquid in most climates on Earth.
- *Liquid water changes temperature very slowly because it can store a large amount of heat without a large change in temperature.* This high heat capacity helps protect living organisms from the shock of abrupt temperature changes. It also moderates Earth's climate and makes water an excellent coolant for car engines, power plants, and other heat-producing industrial processes.
- *It takes a lot of heat to evaporate liquid water.* Water's ability to absorb large amounts of heat as it changes into water vapor—and to release this heat as the vapor condenses back to liquid water—is a primary factor in distributing heat throughout the world.
- *Liquid water can dissolve a variety of compounds.* This enables it to carry dissolved nutrients into the tissues of living organisms, to flush waste products out of those tissues, to serve as an all-purpose cleanser, and to help remove and dilute the water-soluble wastes of civilization. However, water's superiority as a solvent also means that it is easily polluted by water-soluble wastes.
- *Liquid water has strong attractive forces between its molecules.* These forces cause the surface of a liquid to contract (high surface tension) and also create a great capacity to adhere to and coat a solid (high wetting ability). Together these properties allow water to rise through a plant from the roots to the leaves.
- *Most liquids shrink when they freeze, but liquid water expands when it becomes ice.* This means that ice has a lower density (mass per unit of volume) than liquid water, and thus ice floats on water, and bodies of water freeze from the top down instead of from the bottom up. Without this property, lakes and streams in cold climates would freeze solid, and most current forms of aquatic life could not exist. Because water expands on freezing, it can also break pipes, crack engine blocks (which is why we use antifreeze), and fracture streets and rocks.

Water—the lifeblood of the ecosphere—is truly a wondrous substance that connects us to one another, to other forms of life, and to the entire planet. Despite its importance, water is one of the earth's most poorly managed resources. We waste it and pollute it; we also charge too little for making it available, thus encouraging even greater waste and pollution of this vital renewable resource.

11-2 SUPPLY, RENEWAL, AND USE OF WATER RESOURCES

Worldwide Supply, Renewal, and Distribution
Only a tiny fraction of the planet's abundant water is available to us as fresh water. About 97% is found in the oceans and is too salty for drinking, irrigation, or industry (except as a coolant).

The remaining 3% is fresh water. About 2.997% is locked up in ice caps or glaciers or is buried so deep that it costs too much to extract. Only about 0.003% of the earth's total volume of water is easily available to us as soil moisture, exploitable groundwater, water vapor, and lakes and streams. If the world's water supply were only 100 liters (26 gallons), our usable supply of fresh water would be only about 0.003 liter (one-half teaspoon) (Figure 11-2).

Fortunately, the available fresh water amounts to a generous supply that is continuously collected, purified, and distributed in the *hydrologic cycle* (Figure 4-34). This natural recycling and purification process provides plenty of fresh water so long as we don't overload it with slowly degradable and nondegradable wastes or withdraw water from underground supplies faster than it is replenished. Unfortunately we are doing both. Further, usable fresh water is unevenly distributed around the world. Differences in average annual precipitation divide the world into water "haves" and "have-nots."

As population and industrialization increase, water shortages in already water-short regions will intensify, and water wars may erupt (p. 258). Projected global warming (Section 10-2) also might cause changes in rainfall patterns and disrupt water supplies, and we can't predict which areas might be affected.

Surface Water   The fresh water we use comes from surface water and groundwater (Figure 11-3). Precipitation that does not soak into the ground or return to the atmosphere by evaporation or transpiration is called **surface water**. It forms streams, lakes, wetlands, and artificial reservoirs.

Watersheds, also called **drainage basins**, are areas of land that drain into bodies of surface water. Water flowing off the land into these bodies is called **surface runoff**.

Groundwater   Some precipitation infiltrates the ground and fills the pores in soil and rock. The subsurface area where all available soil and rock spaces are filled by water is called the **zone of saturation**, and the water in these pores is called **groundwater** (Figure 11-3). The **water table** is the upper surface of the zone of saturation. It is the fuzzy and fluctuating dividing line between saturated soil and rock (where every available pore is full) and unsaturated (but still wet) rock and soil where the pores can absorb more water.

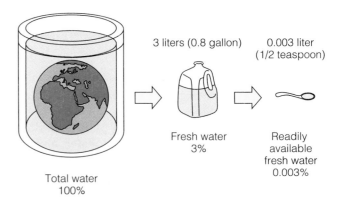

Figure 11-2 Only a tiny fraction of the world's water supply is available as fresh water for human use.

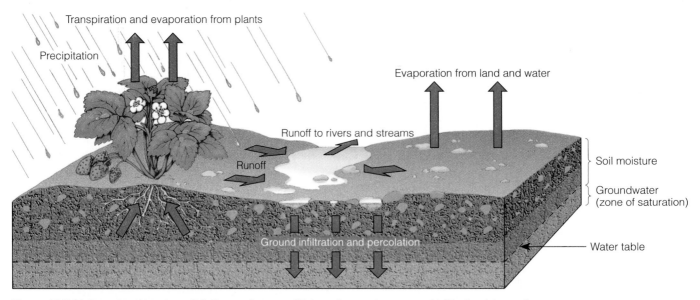

Figure 11-3 Main routes of local precipitation: surface runoff into surface waters, ground infiltration into aquifers, and evaporation and transpiration into the atmosphere.

Q: How much of the world's cropland is used to grow livestock feed?

The water table falls in dry weather and rises in wet weather. Porous, water-saturated layers of sand, gravel, or bed rock through which groundwater flows (and that can yield an economically significant amount of water) are called **aquifers** (Figure 11-4).

Most aquifers are replenished naturally by precipitation, which percolates downward through soil and rock in what is called **natural recharge**. Any area of land through which water passes into an aquifer is called a **recharge area**. Groundwater moves from the recharge area through an aquifer and out to a discharge area (well, spring, lake, geyser, stream, or ocean) as part of the hydrologic cycle.

Normally groundwater moves from points of high elevation and pressure to points of lower elevation and pressure. This movement is quite slow, typically only a meter or so (about 3 feet) per year and rarely more than 0.3 meter (1 foot) in any given day. Thus, most aquifers are somewhat like huge, slow-moving underground lakes.

There is 40 times as much groundwater as there is surface water. However, groundwater is unequally distributed, and only a small amount of it is economically exploitable.

If the withdrawal rate of an aquifer exceeds its natural recharge rate, the water table around the withdrawal well is lowered, creating a waterless volume known as a *cone of depression* (Figure 11-5). Any pollutant discharged onto the land above will be pulled directly into this cone and will pollute water withdrawn by the well.

Some aquifers, called *fossil aquifers*, get very little—if any—recharge. Often found deep underground, they are (on a human time scale) nonrenewable resources. Withdrawals from fossil aquifers amount to "water mining"—and if kept up, will deplete these ancient deposits of liquid Earth capital.

World and U.S. Water Use Two common measures of human water use are withdrawal and consumption. **Water withdrawal** is taking water from a groundwater or surface-water source to a place of use. **Water consumption** occurs when water that has been withdrawn is not returned to the surface water or

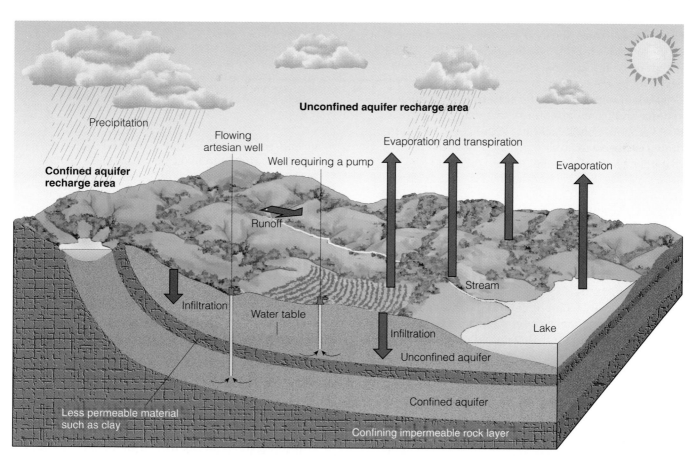

Figure 11-4 The groundwater system. An *unconfined*, or *water table*, *aquifer* forms when groundwater collects above a layer of rock or compacted clay through which water flows very slowly (low permeability). A *confined aquifer* is sandwiched between layers such as clay or shale that have low permeability. Groundwater in this type of aquifer is confined and under pressure.

A: More than 50% (66% in the United States)

Figure 11-5 Drawdown of water table and cone of depression.

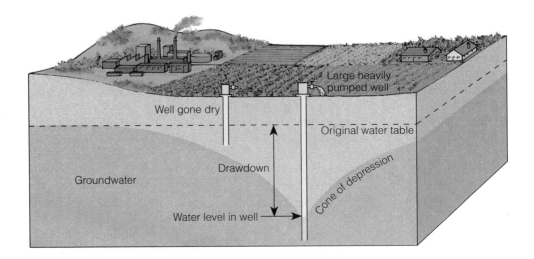

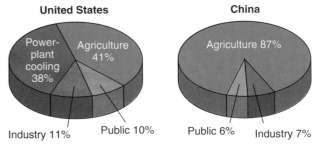

groundwater from which it came—and so cannot be used again in that area. This usually occurs because the water has evaporated or transpired into the atmosphere. Worldwide, about 60% of the water withdrawn is consumed. Once consumed, it is still involved in the global water cycle.

Since 1950 the global rate of water withdrawal has increased almost fivefold, and per capita use has trebled, largely to meet the food and other resource needs of the world's rapidly growing population. Water withdrawal rates are projected to at least double in the next two decades.

Uses of withdrawn water vary from one region to another and from one country to another (Figure 11-6). Averaged globally, about 69% of the water withdrawn each year is used to irrigate 18% of the world's cropland, especially in the former USSR and Mexico. In Asia 82% of the water withdrawn is used to irrigate crops; in Africa the figure is 68%, in the United States 41%, and in Europe 30%. Much irrigation water is wasted, with 70–80% of the water either evaporating or seeping into the ground before reaching crops.

In the western United States, irrigation accounts for about 85% of all water use, and much of this water is used inefficiently. One reason for this is that federal subsidies (mostly federally financed dams and water delivery systems) make water so cheap that Western farmers dependent on this water would lose money by investing in more efficient irrigation.

Worldwide, about 23% of the water withdrawn is used for energy production (oil and gas production and power-plant cooling) and industrial processing, cleaning, and waste removal. The amount of water U.S. industry uses each year to cool, wash, circulate, and manufacture materials is equivalent to 30% of the water in the world's streams. Agricultural and manufactured products both require large amounts of water, much of which could be used more efficient-

Figure 11-6 Use of water in the United States and China. The United States has the world's highest per capita use of water—amounting to an average of 5,300 liters (1,400 gallons) per person every day. (Data from Worldwatch Institute and World Resources Institute)

ly and reused. For example, it takes 380,000 liters (100,000 gallons) to make an automobile, 3,800 liters (1,000 gallons) to produce 454 grams (1 pound) of aluminum, and 3,000 liters (800 gallons) to produce 454 grams (1 pound) of grain-fed beef in a feedlot, where large numbers of cattle are confined to a fairly small area. Together, all U.S. nuclear reactors need to be cooled each year by a volume of water greater than that in Lake Erie.

Domestic and municipal use accounts for about 8% of worldwide withdrawals and about 13–16% of withdrawals in MDCs. As population, urbanization, and industrialization grow, the volume of wastewater needing treatment will increase enormously.

11-3 WATER RESOURCE PROBLEMS

Too Little Fresh Water Droughts—periods in which precipitation is much lower and evaporation is higher than normal—cause more damage and suffering worldwide than any other natural hazard. Since the 1970s drought has killed more than 24,000 people

Q: What percentage of U.S. cropland is used to produce fruits and vegetables?

per year and created swarms of environmental refugees. At least 80 arid and semiarid countries, where nearly 40% of the world's people live, experience years-long droughts.

In water-short areas many women and children must walk long distances each day, carrying heavy jars or cans, to get a meager supply of sometimes contaminated water. Areas likely to face increased water shortages in the 1990s and beyond include northern Africa, parts of India, northern China, much of the Middle East, Mexico, parts of the western United States, Poland, and much of the former USSR.

Reduced precipitation, higher-than-normal temperatures, or both usually trigger a drought; rapid population growth makes it worse. Deforestation (Sections 16-3 and 16-4), overgrazing by livestock (Section 16-5), desertification (Figure 12-25), and replacing diverse natural grasslands (Figure 5-10) with monoculture fields of crops (Figure 5-12) can intensify the effects of drought. Millions of poor people in LDCs have no choice but to try to survive on drought-prone land.

If global warming occurs as projected (Section 10-2), severe droughts may become more frequent in some areas of the world and may jeopardize food production. Some water-starved cities may have to be abandoned.

Water will be the burning foreign policy issue for water-short countries in the 1990s and beyond. Almost 150 of the world's 214 major river systems are shared by two countries, and another 50 are shared by three to ten nations. All of these rivers are potential powder kegs in world geopolitics. Some 40% of the world's population already clashes over water, especially in the Middle East (Figure 11-1). Other intense clashes over water are brewing along the Mekong, which flows through Thailand, Cambodia, Laos, and Vietnam. For over three decades India and Pakistan have been in conflict over water from the Jhelum River, with Pakistanis fearing that Indians might interrupt the flow of water they need. Bangladesh has similar fears about India building dams on the Ganges. Competition between cities and farmers for scarce water is also escalating in regions such as the western United States and China.

Some countries have lots of water, but the largest rivers (which carry most of the runoff) are far from agricultural and population centers where the water is needed. For example, South America has the largest annual water runoff of any continent, but 60% of the runoff flows through the Amazon River in remote areas where few people live (Figure 6-16).

Strategies for capturing some of this runoff and bringing the water to people include building dams and reservoirs and using aqueducts to transport water to other areas. These approaches, however, are expensive and have harmful environmental impacts in addition to their benefits, as discussed in Section 11-4.

Too Much Water Some countries have enough annual precipitation but get most of it at one time of the year. In India, for example, 90% of the annual precipitation falls between June and September, the monsoon season. This prolonged downpour causes floods, waterlogs soils, leaches soil nutrients, and washes away topsoil and crops.

Natural flooding by streams, the most common type of flooding, is caused primarily by heavy rain or rapid melting of snow; this causes water in the stream to overflow the channel in which it normally flows and to cover the adjacent area, called a **floodplain** (Figure 11-7).

Floods are a natural phenomenon and have several benefits. They provide the world's most productive farmland, which is regularly covered with nutrient-rich silt left after flood waters recede. Floods also recharge groundwater under plains and refill wetlands, which help keep rivers flowing during droughts and provide important breeding and feeding grounds for fish and waterfowl.

People have settled on floodplains since the beginnings of agriculture. The soil is fertile, and ample water is available for irrigation. Communities have access to the water for transportation of people or goods, and floodplains are flat sites, suitable for cropland, buildings, highways, and railroads. Floodplains include highly productive wetlands, provide natural flood and erosion control, help maintain high water quality, and contribute to recharging groundwater. However, prolonged rains can cause streams and lakes anywhere to overflow and flood the surrounding floodplain (Figure 11-7, right), but low-lying river basins such as the Ganges River basin in India and Bangladesh are especially vulnerable. Hurricanes and typhoons can also flood low-lying coastal areas.

In the 1970s floods killed more than 4,700 people per year and caused tens of billions of dollars in property damages—a trend that continued (and even worsened) in the 1980s and early 1990s, including massive flooding in the United States in 1993 along the floodplains of the Mississippi, Missouri, and Illinois rivers (Figure 11-8). And in India, for example, flood losses doubled in the 1980s. A 25-year analysis revealed that 39% of the deaths from natural hazards were caused by floods, followed by typhoons and hurricanes (36%), earthquakes (13%), gales and thunderstorms (5%), and volcanic eruptions (2%).

Floods, like droughts, are usually considered to be natural disasters, but human activities have contributed to the sharp rise in flood deaths and damages since the 1960s. The main way humans increase the

A: About 2%

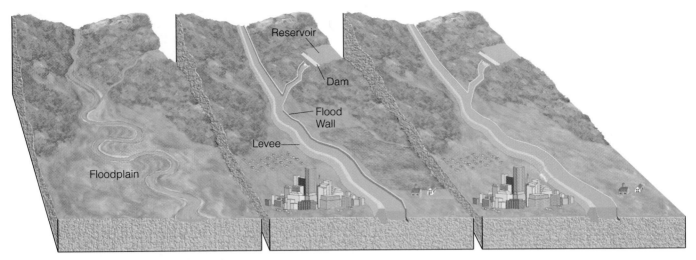

Figure 11-7 Land in a natural floodplain (left) along a river is flooded after prolonged rains. When the floodwaters recede alluvial deposits of silt are left, creating a nutrient-rich soil. To reduce the threat of flooding and thus allow people to live in floodplains in many LDCs (as well as the United States), rivers have been dammed to create reservoirs to store and release water as needed, narrowed and straightened, and equipped with protective levees and walls (middle). However, this can give a false sense of security to floodplain dwellers who actually live in risky areas, and in the long run can greatly increase flood damages. Though dams, levees, and walls prevent flooding in most years, they can be overwhelmed by prolonged rains (right), as happened in the midwestern United States during the summer of 1993.

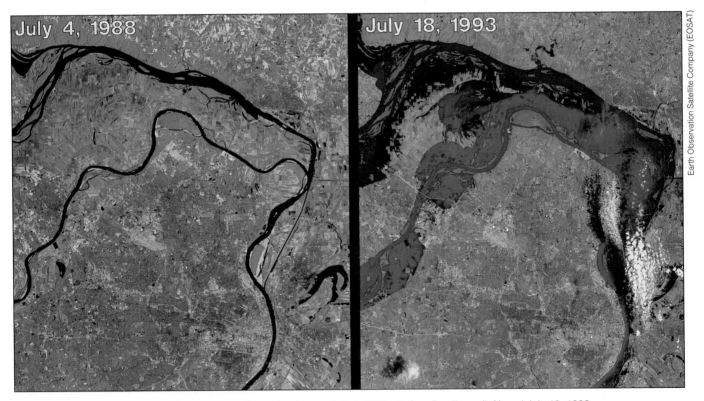

Figure 11-8 Satellite images of the area around St. Louis taken on July 4, 1988—before flooding—(left) and July 18, 1993 (right) after severe flooding from prolonged rains.

probability and severity of flooding is by removing vegetation—through logging (Figure 11-9), timbering operations, overgrazing by livestock, construction, forest fires, and certain mining activities. Vegetation retards surface runoff and increases infiltration; when the vegetation is removed by human activities or natural occurrences, precipitation reaches streams more directly, often with a large load of sediment, which increases the chance of flooding by making a stream more shallow (Connections, p. 266).

264

Q: How many people are underfed and undernourished?

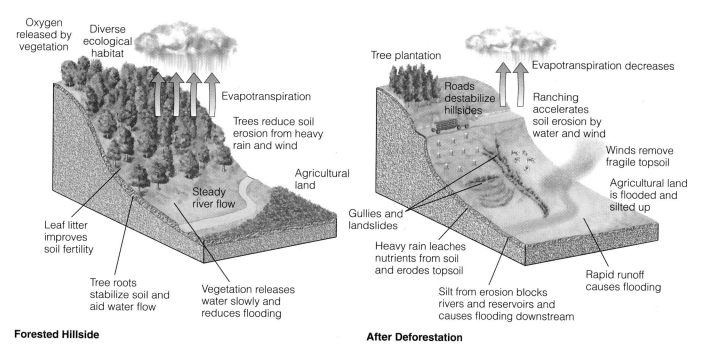

Forested Hillside

Oxygen released by vegetation

Diverse ecological habitat

Evapotranspiration

Trees reduce soil erosion from heavy rain and wind

Agricultural land

Steady river flow

Leaf litter improves soil fertility

Tree roots stabilize soil and aid water flow

Vegetation releases water slowly and reduces flooding

After Deforestation

Tree plantation

Roads destabilize hillsides

Evapotranspiration decreases

Ranching accelerates soil erosion by water and wind

Winds remove fragile topsoil

Agricultural land is flooded and silted up

Gullies and landslides

Heavy rain leaches nutrients from soil and erodes topsoil

Silt from erosion blocks rivers and reservoirs and causes flooding downstream

Rapid runoff causes flooding

Figure 11-9 A mountainside before and after deforestation. When a hillside or mountainside is deforested—for timber and fuelwood, for grazing livestock, or for unsustainable farming—water from rains rushes down denuded slopes, eroding precious topsoil and flooding downstream areas. A 3,000-year-old Chinese proverb says: "To protect your rivers, protect your mountains."

Urbanization (Section 6-3) also increases flooding (even with moderate rainfall) by replacing vegetation and soil with highways, parking lots, and buildings that lead to rapid runoff of rainwater. If sea levels rise during the next century, as projected, many low-lying coastal cities, wetlands, and croplands will be under water.

Solutions: Reducing the Risks of Flooding
Various ways have been developed to reduce the hazard of stream flooding. One controversial way to reduce flooding is *channelization*, in which a section of a stream is deepened, widened, or straightened to allow more rapid runoff (Figure 11-7, middle). This can reduce upstream flooding. However, the increased flow of water can increase upstream bank erosion and increase downstream deposits of sediment and flooding.

Artificial levees and embankments can be built along stream banks to reduce the chances of water flowing over into nearby floodplains (Figure 11-7, middle). By containing and straightening a river, channelization and levees speed up river flow and increase its capacity to do damage downstream. And if the levee breaks or the flood spills over it, floodwater may be trapped between the levee and the valley wall long after the stream discharge has decreased (Figure 11-7, right), as many people in the midwestern United States learned during massive flooding from heavy rains during the summer of 1993 (Figure 11-8). After the flood waters receded, a toxic sludge laced with pesticides, indus-

trial wastes, and raw sewage was left on large areas of land along the floodplains of the Mississippi, Missouri, and Illinois rivers. Severe erosion from the 1993 flooding reduced crop yields on already heavily eroded fields.

A *flood control dam* can be built across a stream to hold back and store flood water and release it more gradually (Figure 11-7, middle). The dam and its reservoir may also provide such secondary benefits as hydroelectric power, water for irrigation, and recreational facilities. But a reservoir can reduce floods only if the water level is kept low. However, it is much more profitable for dam operators to keep water levels high to produce electricity and supply irrigation water. As a result, with prolonged rains the reservoir can overflow or operators release large volumes of water to prevent overflow, worsening the severity of flooding downstream. Also, the reservoir of the dam gradually fills with sediment (that once fertilized downstream floodplains) until it is useless, and it also has other drawbacks (Figure 11-10).

Tragically, some flood control dams have failed for one reason or another, causing catastrophic, unpredicted flooding. According to the Federal Emergency Management Agency the United States has about 1,300 unsafe dams in populated areas. The agency reported that the dam safety programs of most states are inadequate because of weak laws and budget cuts.

Over the last 65 years the U.S. Army Corps of Engineers has spent $25 billion on channelization, dams,

Living Dangerously in Bangladesh

CONNECTIONS

Bangladesh (Figure 6-13) is one of the world's most densely populated countries, with 114 million people packed into an area roughly the size of Wisconsin. Women bear an average of 4.9 children, and the population could reach 211 million by 2025. Bangladesh is also one of the world's poorest countries, with an average per capita income of about $180.

Most of the country consists of floodplains and shifting islands of silt formed by a delta at the mouth of three major rivers (Figure 5-31). Runoff from annual monsoon rains in the Himalaya mountains of India, Nepal, Bhutan, and China flow down the rivers through Bangladesh into the Bay of Bengal.

The people of Bangladesh are used to moderate annual flooding during the summer monsoon season, and they depend on the floodwaters to grow rice, their primary source of food. The annual deposit of Himalayan soil in the delta basin also helps maintain soil fertility.

In the past, great floods occurred every 50 years or so, but during the 1970s and 1980s they came about every 4 years. Bangladesh's flood problems begin in the Himalayan watershed. There, a combination of rapid population growth, deforestation, overgrazing, and unsustainable farming on steep, easily erodible mountain slopes has greatly diminished the ability of the soil to absorb water (Figure 11-9). Instead of being absorbed and released slowly, water from the monsoon rain runs off the denuded Himalayan foothills, carrying vital topsoil with it. This runoff, combined with heavier-than-normal monsoon rains, has caused severe flooding in Bangladesh.

In 1988, for example, a disastrous flood covered two-thirds of the country's land area for three weeks and leveled 2 million homes after the heaviest monsoon rains in 70 years. At least 2,000 people drowned, and 30 million people— 1 in 4—were left homeless. At least a quarter of the country's crops were destroyed, costing at least $1.5

billion and causing thousands of people to die of starvation.

In their struggle to survive, the poor in Bangladesh have cleared many of the country's coastal mangrove forests (Figure 5-24) for fuelwood and for cultivation of crops. This deforestation has led to more severe flooding because these coastal wetlands shelter the low-lying coastal areas from storm surges and cyclones, with fierce winds whipping up waves as high as 9 meters (30 feet).

In 1970 as many as 1 million people drowned in one storm. Another surge killed some 140,000 people in 1991. Flood damages and deaths in areas still protected by mangrove forests are much lower than in areas where the forests have been cleared. This problem can be solved only if Bangladesh, Bhutan, China, India, and Nepal all cooperate in reforestation efforts and flood control measures, and reduce their population growth.

and levees. Yet the financial cost of flood damage has risen steadily during this period. The massive floods in the midwestern United States along the banks of the Mississippi, Missouri, and Kansas rivers during the summer of 1993 took 50 lives, made almost 70,000 people homeless, and caused an estimated $12 billion in damage to property and crops (Figure 11-8). Billions more will have to be spent to repair the more than 800 levees and embankments (of the 1,400 in the nine states affected by the disaster) that were topped or breached. This disaster revealed that dams, levees, and channelization can give a false sense of security, encourage people to settle on floodplains, and worsen the severity of flood damage. Instead of trying to completely manage natural water flows, some countries are realizing that the risks of doing so can outweigh the benefits. In Germany, for example, plans are underway to bulldoze through dikes and allow parts of the floodplains to again flood regularly.

From an environmental viewpoint, *floodplain management* is the best approach to reduce the risks to hu-

mans from flooding. The first step is to construct a *flood-frequency curve*, based on historical records and an examination of vegetation to determine how often, *on average*, a flood of a certain size might occur in a particular area. Using these data, a plan is developed to prohibit certain types of buildings or activities in high-risk areas; to elevate or otherwise flood-proof buildings that are allowed on the legally defined floodplain; and to construct a floodway that allows flood water to flow through the community with minimal damage. Cities can also use floodplains as greenbelts. Such belts can provide urban open space, which—when not flooded—can be used as parks and recreational areas.

The Federal Flood Disaster Protection Act was passed by the Congress in 1973. This act requires that, to be eligible for federal flood insurance, local governments must adopt such floodplain development regulations. It also denies federal funding to proposed construction projects in areas that are officially subject to flooding.

Q: How many people die each year from hunger-related causes?

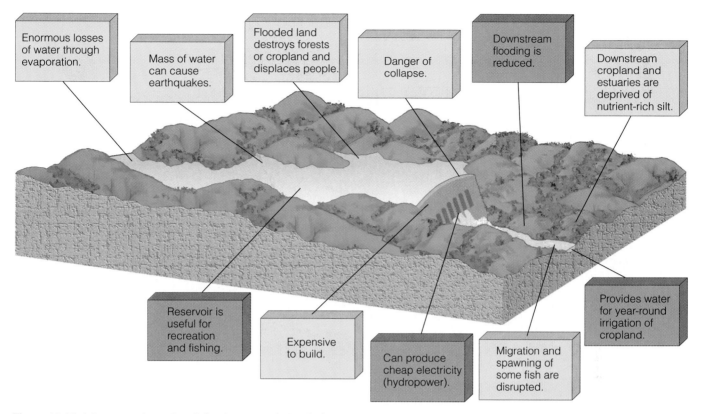

Figure 11-10 Advantages (green) and disadvantages (yellow) of large dams and reservoirs.

Despite these efforts, in 1990 more than 16.8 million households and $758 billion worth of property were located on floodplains in the United States. On the average, floods kill more than 200 people and cause over $4 billion in property losses per year in the United States. California, Florida, Texas, and Louisiana are the four most flood-prone states, but in 1993 land along the Mississippi, Missouri, and Kansas rivers from Illinois to Louisiana was devastated by massive flooding (Figure 11-8).

The federal flood insurance program underwrites $185 billion in policies because private insurance companies are unwilling to fully insure people who live in flood-prone areas against damages. This federal program actually encourages many people to build on floodplains and low-lying coastal areas with a high risk of flooding.

Some economists argue that it would make more economic sense and save taxpayers money if the government would buy the 2% of the country's land that repeatedly floods, instead of continuing to make disaster payments. This would also increase wetlands, allow restoration of wetlands, and ultimately improve water quality. Others believe that people should be free to live along floodplains, coasts, or other flood-prone areas, but that those choosing to do so should accept the risk and not expect to be bailed out by other taxpayers who subsidize federal flood insurance. However, attempts to reduce or eliminate federal flood insurance coverage or to restrict development on floodplains are usually defeated politically by intense protests from property owners who already live in these risky areas.

Water Resources in the United States

Although the United States has plenty of fresh water, much of it is in the wrong place at the wrong time or is contaminated by agriculture and industry. From a human perspective, the eastern states usually have ample precipitation, whereas many of the western states have too little. In the East the largest uses for water are for energy production, cooling, and manufacturing; in the West the largest use by far is for irrigation.

In many parts of the eastern United States the most serious water problems are flooding, occasional urban shortages, and pollution. For example, the 3 million residents of Long Island, New York, get their water from an aquifer that is becoming severely contaminated.

The most serious water problem in the arid and semiarid areas of the western half of the country is a shortage of runoff caused by low precipitation, high evaporation, and recurring prolonged drought. In many areas, water tables are dropping rapidly as

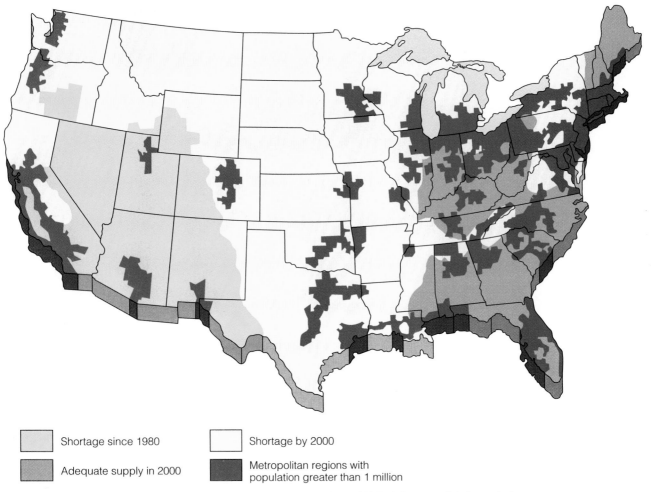

Shortage since 1980	Shortage by 2000
Adequate supply in 2000	Metropolitan regions with population greater than 1 million

Figure 11-11 Current and projected water-deficit regions in the continental United States and their proximity to metropolitan areas having populations greater than 1 million. (Data from U.S. Water Resources Council and U.S. Geological Survey)

farmers and cities deplete groundwater aquifers faster than they are recharged.

Many major urban centers, especially those in the West and Midwest, are located in areas that don't have enough water or are projected to have water shortages by 2000 (Figure 11-11). Experts project that current shortages and conflicts over water supplies will get much worse as more industries and people migrate west and compete with farmers for scarce water. These shortages could worsen even more if climate warms as a result of an enhanced greenhouse effect (Section 10-2). Because water is such a vital resource, the data depicted in Figure 11-11 may be useful in deciding where to live in the coming decades.

Methods for Managing Water Resources

One way to manage water resources is to increase the supply in a particular area by building dams and reservoirs, bringing in surface water from another

area, or tapping groundwater. The other approach is to improve the efficiency of water use.

LDCs rarely have the money to develop the water storage and distribution systems needed to increase their supply. Their people must settle where the water is. In MDCs people tend to live where the climate is favorable and bring in water from another watershed.

11-4 SOLUTIONS: SUPPLYING MORE WATER

Constructing Dams and Reservoirs Rainwater and water from melting snow can be captured and stored in large reservoirs created by damming streams. This water can then be released in a controlled flow as desired to produce hydroelectric power at the dam site, to irrigate land below the dam, to control flooding of land below the reservoir, and to pro-

Q: What human activity has the most harmful overall environmental impact?

vide water carried by aqueduct to towns and cities. Reservoirs are also used for recreation activities, such as swimming, fishing, and boating.

About 13.5% of the electrical power used in the United States is hydroelectric. This is a potentially renewable source of energy as long as drought or long-term changes in climate don't reduce water flow in the dam's basin. Large dams and reservoirs have benefits and drawbacks (Figure 11-10). Building small dams, which have fewer destructive effects than large dams and reservoirs, is a useful way to trap water for irrigation.

Proposed dams in LDCs will cover vast areas and uproot millions of people. Despite protests by more than 50,000 villagers, the Indian government is going ahead with plans to build 30 large dams and thousands of smaller ones along the Namada River and 41 of its tributaries. These projects are small compared to China's Three Gorges project—the world's largest proposed hydroelectric dam and reservoir. When completed, this $11 billion, 18-year project will create a 590-kilometer- (370-mile-) long lake on the Yangtze River. The 1.6-kilometer- (1-mile-) long and 180-meter- (600-foot-) high dam will supply power to industries and 150 million Chinese. It will also help hold back the floodwaters of the Yangtze which have killed millions over the centuries.

On the negative side, it will flood large areas of farmland and 800 existing factories and displace 1.2 million people (the largest resettlement in world history), including two cities each containing 100,000 people. Critics also charge that the dam will transform the Yangtze into a sewer for industrial wastes and possibly trigger landslides and earthquakes. They also believe that the flood-control benefits will be minimal as the river fills with more sediment. With prolonged rains, this could expose half a million people to severe flooding.

Watershed Transfers   Tunnels, aqueducts, and underground pipes can transfer stream runoff collected by building dams and reservoirs from water-rich watersheds to water-poor areas. Three of the world's largest watershed transfer projects are the California Water Project (Case Study, p. 270), the James Bay project in Canada, and the diversion of water from rivers feeding the Aral Sea in Central Asia to irrigate cropland.

The Aral Sea story is a particularly interesting and tragic example of a large-scale water transfer project. The states of Kazakhstan and Uzbekistan in the former USSR have the driest climate in Central Asia. Since 1960 enormous amounts of irrigation water have been diverted from the inland Aral Sea (a huge freshwater lake in Central Asia) and its two feeder rivers to grow cotton and food crops. The irrigation canal, the

Figure 11-12 Once the world's fourth largest freshwater lake, the Aral Sea has been shrinking and getting saltier since 1960 because most of the water from the rivers that replenish it has been diverted to grow cotton and food crops. As the lake shrinks, it leaves a salty desert.

world's longest, stretches over 1,300 kilometers (800 miles).

The diversion (coupled with droughts) has caused a regional ecological disaster, described by one former Soviet official as "ten times worse than the 1986 Chernobyl nuclear power plant accident." The sea's salinity has tripled, its surface area has shrunk by 46% (Figure 11-12), and its volume has decreased by 69%. The two supply rivers are mere trickles. About 30,000 square kilometers (11,600 square miles) of former lake bottom have turned into desert, and the process continues.

All the native fish are gone, devastating the area's fishing industry, which once provided work for more than 60,000 people. Two major fishing towns are now surrounded by a desert containing stranded fishing boats and rusting commercial ships. Roughly half of the area's bird and mammal species have also disappeared.

Salt, dust, and dried pesticide residues have been carried as far as 300 kilometers (190 miles) by the wind. As the salt spreads, it kills crops, trees, and wildlife and destroys pastureland. This phenomenon has added a new term to our vocabulary of environmental ills: *salt rain*.

These changes have also affected the area's already semiarid climate. The once-huge Aral Sea acted as a thermal buffer, moderating the heat of summer

In California the basic water problem is that 75% of the population lives south of Sacramento but 75% of the rain falls north of it. The California Water Project uses a maze of giant dams, pumps, and aqueducts to transport water from water-rich northern California to heavily populated areas of the state and to arid and semiarid agricultural regions (Figure 11-13).

For decades northern and southern Californians have been feuding over how state water should be allocated under this project. Southern Californians say they need more water from the north to support Los Angeles, San Diego, and other growing urban areas, and to grow more crops. Agriculture uses 82% of the water withdrawn in California. Irrigation for just two crops, alfalfa and cotton, uses as much water as the residential needs of all 30 million Californians.

Opponents in the north say that sending more water south would degrade the Sacramento River, threaten fisheries, and reduce the flushing action that helps clean San Francisco Bay of pollutants. They also argue that much of the water already sent south is wasted and that making irrigation just 10% more efficient would provide enough water for domestic and industrial uses in southern California.

To supply agribusiness in California and other western states with cheap water, the Bureau of Reclamation has drained major rivers and lakes, and it has destroyed vast areas of wetland waterfowl habitat and salmon spawning habitat. Environmentalists believe that the federal government should not award new long-term water contracts that give many farmers and ranchers cheap, government-subsidized water for irrigating crops—especially if the irrigation is for grass to feed cows or for "thirsty" crops such as rice, alfalfa, and cotton that could be grown more cheaply in rain-fed areas. They also propose enacting state laws requiring cities wanting more water to pay farmers to install water-saving irrigation technology. Cities could use the water saved.

If water supplies in California were to drop sharply because of projected global warming, water delivered by the state's huge distribution system would plummet as well. Most irrigated agriculture in California would have to be abandoned, and much of the population of southern California might have to move to areas with more water. The six-year drought that northern and southern California experienced between 1986 and 1992 was a small taste of a possible future.

Groundwater is no answer. Throughout most of California it is already being withdrawn faster than it is replenished. Santa Barbara, Los Angeles, and several other southern California cities are planning experimental plants to supply fresh water by removing salt from seawater—an option five to six times more costly than state-provided water. Improving irrigation efficiency and allowing farmers to sell their water allotments are much quicker and cheaper solutions.

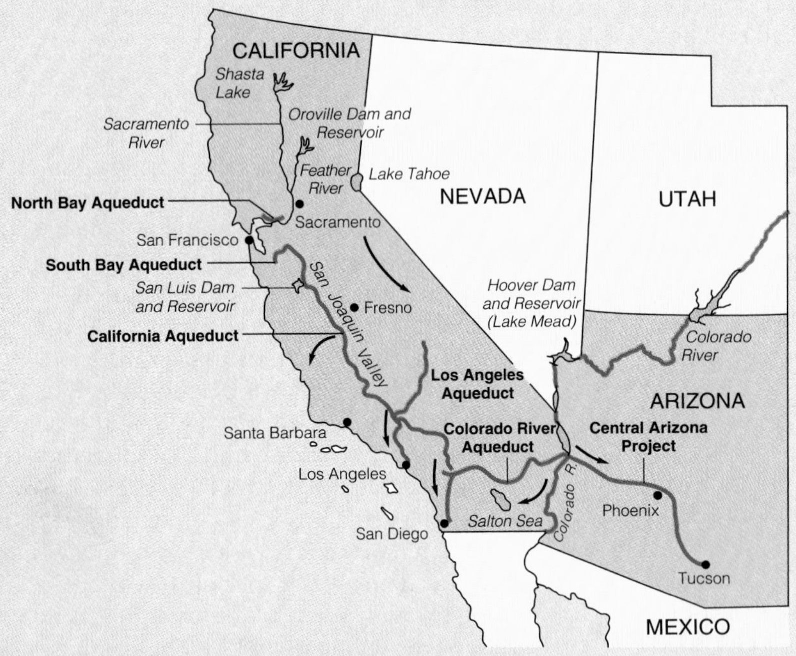

Figure 11-13 California Water Project and Central Arizona Project for large-scale transfer of water from one watershed to another. Arrows show general direction of water flow.

and the extreme cold of winter. Now there is less rain, summers are hotter, winters are colder, and the growing season is shorter. Cotton and crop yields have dropped dramatically.

Local farmers have turned to using herbicides, insecticides, and fertilizers on some crops. Many of these chemicals have percolated downward and accumulated to dangerous levels in the groundwater, from which most of the drinking water comes.

Ways to deal with this problem include **(1)** charging farmers more for irrigation water to reduce waste and encourage a shift to less water-intensive crops; **(2)**

decreasing irrigation water quotas; **(3)** introducing water-saving technologies; **(4)** developing a regional integrated water management plan; **(5)** planting protective forest belts; **(6)** using underground water to supplement irrigation water and lower the water table to reduce waterlogging and salinization; **(7)** improving health services; and **(8)** slowing the area's rapid population growth (3% per year).

In 1992, Kazakhstan, Uzbekistan, Kyrgyzstan, and Turkmenistan, which share the Aral Sea basin, signed an agreement describing how the waters of the two rivers feeding the Aral Sea should be divided, and they created a council to manage the basin's resources. However, even with help from foreign countries, the United Nations, and agencies such as the World Bank, the money needed to save the Aral Sea may not be available.

Another major watershed transfer project is the James Bay Project—a $60-billion, 50-year scheme to harness the wild rivers that flow into the James and Hudson bays in Canada's Quebec province—in order to produce electric power for Canadian and U.S. consumers (Figure 11-14). If completed, it would **(1)** construct 600 dams and dikes that will reverse or alter the flow of 19 giant rivers covering a watershed three times the size of New York State; **(2)** flood 176,000 square kilometers (68,000 square miles) of boreal forest and tundra—an area the size of Washington State or Germany; and **(3)** displace thousands of indigenous Cree and Inuit who have lived sustainably off James Bay by subsistence hunting, fishing, and trapping for 5,000 years.

After 20 years and $16 billion, Phase I has been completed. The second and much larger phase (Figure 11-14) is scheduled to begin soon but is being opposed in court by the Cree, whose ancestral hunting grounds would be flooded. In 1990 the federal court of Canada ordered a full environmental impact assessment for this project, including public hearings. In the summer of 1992, New York State canceled one of its two contracts to buy electricity produced by Phase II—stating that it was uneconomic to buy the power compared to the other options. A court case has been filed against the second contract. It may also be canceled for economic reasons and as a result of pressure from environmentalists. But the battle is not over.

Tapping Groundwater   In the United States, 23% of all fresh water used is groundwater. About half of the country's drinking water (96% in rural areas and 20% in urban areas) and 40% of irrigation water is pumped from aquifers.

Overuse of groundwater can cause or intensify several problems (Figure 11-15): *aquifer depletion, aquifer subsidence* (sinking of land when groundwater

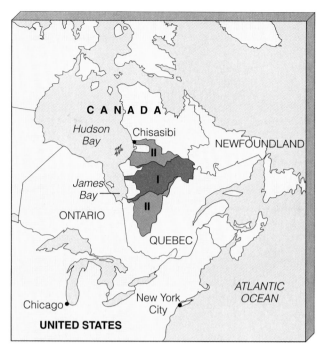

Figure 11-14 The James Bay project in northern Quebec will alter or reverse the flow of 19 major rivers and flood an area the size of Washington State to produce hydropower for consumers in Quebec and the United States, especially in New York State. Phase I of this 50-year project is completed. Phase II is scheduled to begin but is being opposed by the indigenous Cree, whose ancestral hunting grounds would be flooded, and by environmentalists in Canada and the United States.

is withdrawn), and *intrusion of salt water into aquifers.* Groundwater can also become contaminated from industrial and agricultural activities, septic tanks, and other sources. Because groundwater is the source of about 40% of the stream flow in the United States, groundwater depletion also robs streams of water.

Currently, about one-fourth of the groundwater withdrawn in the United States is not replenished. The most serious overdraft is in parts of the huge Ogallala Aquifer, extending from southern South Dakota to northwestern Texas (Case Study, p. 273). Aquifer depletion is also a problem in Saudi Arabia, northern China, Mexico City, Bangkok, and parts of India.

Ways to slow groundwater depletion include **(1)** controlling population growth, **(2)** discontinuing planting of water-thirsty crops in dry areas, **(3)** developing crop strains that require less water and are more resistant to heat stress, and **(4)** wasting less irrigation water.

When fresh water is withdrawn from an aquifer near a coast faster than it is recharged, salt water intrudes into the aquifer (Figure 11-17). Saltwater intrusion threatens to contaminate the drinking water of many towns and cities along the Atlantic and Gulf coasts (Figure 11-15) and in the coastal areas of Israel,

A: About 33% (42% in U.S. coastal waters)

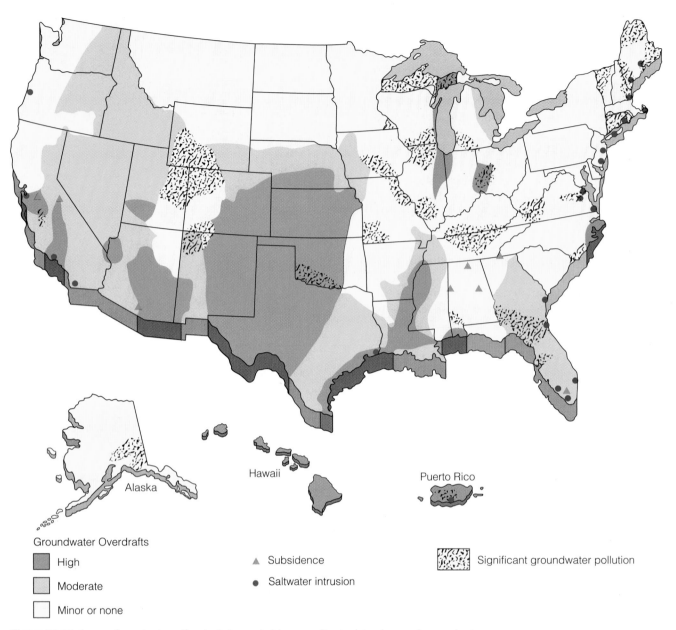

Figure 11-15 Areas of greatest aquifer depletion, subsidence, saltwater intrusion, and groundwater contamination in the United States. (Data from U.S. Water Resources Council and U.S. Geological Survey)

Syria, and the Arabian Gulf states. Another growing problem in the United States and many other MDCs is groundwater contamination, discussed in Section 11-7.

Desalination   Desalination—the removal of dissolved salts from ocean water or from brackish (slightly salty) groundwater—is another way to increase fresh water supplies. Distillation and reverse osmosis are the two most widely used methods. *Distillation* involves heating salt water until it evaporates and condenses as fresh water, leaving salts behind in solid form. In *reverse osmosis* salt water is pumped at high pressure through a thin membrane whose pores

allow water molecules—but not dissolved salts—to pass through.

About 7,500 desalination plants in 120 countries provide about 0.1% of the fresh water used by humans. Desalination plants in the Middle East (especially Saudi Arabia and Kuwait) and North Africa produce about two-thirds of the world's desalinated water.

Desalination, however, has a downside. It uses vast amounts of electricity and therefore costs three to five times more than water from conventional sources. Distributing the water from coastal desalination plants costs even more in terms of the energy needed to

Q: What percentage of U.S. crops are grown using organic methods (no pesticides or commercial fertilizers)?

Mining Groundwater—The Shrinking Ogallala Aquifer

CASE STUDY

Water pumped from the Ogallala Aquifer (Figure 11-16)—the world's largest known aquifer—has helped transform much of a vast prairie that is too dry for rainfall farming into one of the United States' most productive farmlands. Although this aquifer is gigantic, it is essentially a nonrenewable fossil aquifer with an extremely slow recharge rate. Water is being pumped out eight times as fast as natural recharge occurs, mostly for irrigation to supply 15% of the country's corn and wheat, 25% of its cotton, and 40% of its feedlot beef.

The withdrawal rate is 100 times the recharge rate for parts of the aquifer that lie beneath Texas, New Mexico, Oklahoma, and Colorado. Water experts project that at the current rate of withdrawal, one-fourth of the aquifer's original supply will be depleted by 2020—much

sooner in areas where it is shallow; then depleted areas will become a desert. It will take thousands of years to replenish the aquifer. Depletion is encouraged by federal tax laws that allow farmers and ranchers to deduct the cost of drilling equipment and sinking wells.

Long before the water is gone, the high cost of pumping water from a rapidly dropping water table will force many farmers to grow drought-tolerant crops instead of profitable but "thirsty" crops such as cotton and sugar beets. Some farmers will go out of business. Total irrigated area is already declining in five of the seven states using this aquifer because water must be pumped from depths as great as 1,830 meters (6,000 feet). If farmers in the Ogallala region conserve more water and switch to crops that require less water, depletion of the aquifer could be delayed.

Ogallala Aquifer

Figure 11-16 The Ogallala—the world's largest known aquifer. If the water in this aquifer were above ground, it would be enough to cover the entire lower 48 states with 0.5 meter (1.5 feet) of water. This fossil aquifer, which is renewed very slowly, is being depleted to grow crops and raise cattle.

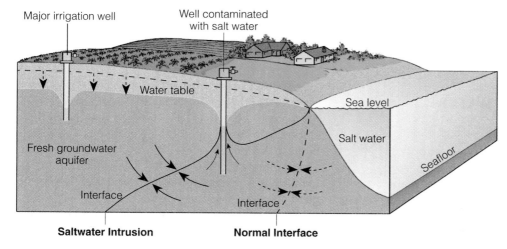

Figure 11-17 Saltwater intrusion along a coastal region. When the water table is lowered, the normal interface (dotted line) between fresh and saline groundwater moves inland (solid line).

pump desalinated water uphill and inland. Moreover, desalination produces large quantities of brine, with high levels of salt and other minerals that must go somewhere. Dumping the concentrated brine in the ocean near the plants might seem to be the logical solution, but this would increase the salt concentration and threaten food resources in estuarine waters. And

if these wastes were dumped on the land, they could contaminate groundwater and surface water. There is no "away."

Desalination can provide fresh water for coastal cities in arid countries, such as sparsely populated Saudi Arabia, where the cost of getting fresh water by any method is high. However, desalinated water will

A: About 3%

probably never be cheap enough to use for irrigating conventional crops or to meet much of the world's demand for fresh water, unless efficient solar-powered methods can be developed. And the problem of what to do with the salt must be solved.

Cloud Seeding and Towing Icebergs   For years several countries, particularly the United States, have been experimenting with seeding clouds with chemicals to produce more rain over dry regions and snow over mountains. Cloud seeding involves injecting a large, suitable cloud with a powdered chemical such as silver iodide, either from a plane or from ground-mounted burners. Small water droplets in the cloud clump together around tiny particles of the chemical (condensation nuclei) and form drops or ice particles large enough to fall to Earth as precipitation.

Unfortunately, cloud seeding is not useful in very dry areas, where it is most needed, because rain clouds are rarely available. Furthermore, widespread cloud seeding would introduce large amounts of the cloud-seeding chemicals into soil and water systems, possibly harming people, wildlife, and agricultural productivity. A final obstacle to cloud seeding is legal disputes over the ownership of water in clouds. For example, during the 1977 drought in the western United States, the attorney general of Idaho accused officials in neighboring Washington of "cloud rustling" and threatened to file suit in federal court.

There also have been proposals to tow massive icebergs to arid coastal areas (such as Saudi Arabia and southern California) and pump the fresh water from the melting bergs ashore. However, the technology for doing this is not available, and the costs may be too high, especially for water-short LDCs.

11-5 SOLUTIONS: USING WATER MORE EFFICIENTLY

Curbing Waste   Increasing the water supply in some areas is important, but soaring population, food needs, and industrialization, along with unpredictable shifts in water supplies, will eventually outstrip this approach. It makes much more sense economically and environmentally to use water more efficiently.

Mohamed El-Ashry of the World Resources Institute estimates that *65–70% of the water people use throughout the world is wasted through evaporation, leaks, and other losses.* The United States—the world's largest user of water—does slightly better but still wastes 50% of the water it withdraws. El-Ashry believes that it is economically and technically feasible to reduce water waste to 15%, thus meeting most of the world's water needs for the foreseeable future.

Conserving water would have many other benefits, including reducing the burden on wastewater plants and septic systems; decreasing pollution of surface water and groundwater; reducing the number of expensive dams and water-transfer projects that destroy wildlife habitats and displace people; slowing depletion of groundwater aquifers; and saving energy and money needed to supply and treat water.

A prime cause of water waste in the United States (and in most countries) is artificially low water prices. Cheap water is the only reason that farmers in Arizona and southern California can grow water-thirsty crops like alfalfa in the middle of the desert. It also enables people in Palm Springs, California, to keep their lawns and 74 golf courses green in a desert area.

Water subsidies are paid for by all taxpayers through higher taxes. Because these external costs don't show up on monthly water bills, consumers have little incentive to use less water or to install water-conserving devices and processes. Raising the price of water to reflect its true cost (Section 7-2) would be a powerful incentive for using water more efficiently.

The federal Bureau of Reclamation supplies one-fourth of the water used to irrigate land in the western United States under long-term contracts (typically 40 years) at greatly subsidized prices. During the 1990s hundreds of these long-term water contracts will come up for renewal. Sharply raising the price of federally subsidized water would encourage investments in improving water efficiency, and many of the West's water supply problems could be eased. Outdated laws governing access to and use of water resources also encourage unnecessary water waste (Spotlight, next page). Such reforms, however, are opposed by politically powerful western states.

Another reason for the water waste in the United States is that the responsibility for water resource management in a particular watershed may be divided among many state and local governments, rather than being handled by one authority. For example, the Chicago metropolitan area has 349 water-supply systems, divided among some 2,000 local units of government over a six-county area.

In sharp contrast is the regional approach to water management used in England and Wales. The British Water Act of 1973 replaced more than 1,600 agencies with 10 regional water authorities based on natural watershed boundaries. Each water authority owns, finances, and manages all water supply and waste treatment facilities in its region. The responsibilities of each authority include water pollution control, water-based recreation, land drainage and flood control, inland navigation, and inland fisheries. Each water authority is managed by a group of elected local officials and a

Q: How much of the food produced in the United States is wasted?

Laws regulating surface-water access and use differ in the eastern and western parts of the United States. In most of the East water use is based on the doctrine of **riparian rights**. Basically this system of water law gives anyone whose land adjoins a flowing stream the right to use water from the stream as long as some is left for downstream landowners. However, as population and water-intensive land uses grow, there often is not enough water to meet the needs of all the people along a stream.

In the arid and semiarid West the riparian system does not work because large amounts of water are needed in areas far from major surface-water sources. In most of this region the principle of **prior appropriation** regulates water use. In this first-come, first-served approach, the first user of water from a stream establishes a legal right for continued use of the amount originally withdrawn. If there is a shortage, later users are cut off in order, one by one, until there is enough water to satisfy the demands of the earlier users. Some states have a combination of riparian and prior appropriation water rights.

To retain their prior appropriation rights, users within a particular state must withdraw a certain amount of water even if they don't need it—a use-it-or-lose-it approach—which discourages farmers from adopting water-conserving irrigation methods. However, this use-it-or-lose-it rule does not apply to water bodies shared by two or more states. Water allocation between states is determined by interstate compacts and court decrees.

Most groundwater use is based on common law, which holds that subsurface water belongs to whoever owns the land above such water. This means that landowners can withdraw as much as they want to use on their land.

When many users tap the same aquifer, that aquifer becomes a common-property resource. The largest users have little incentive to conserve and can deplete the aquifer for everyone, creating another tragedy of the commons (Connections, p. 16).

Environmentalists and many economists call for a change in laws allocating rights to surface and groundwater supplies, with emphasis on *water marketing*. They believe that farmers and other users who save water through conservation or a switch to less water-thirsty crops should be able to sell or lease the water they save to industries and cities rather than losing their rights to this water.

---

smaller number of officials appointed by the national government.

Reducing Irrigation Losses   Irrigation accounts for 69% of water use. Since almost two-thirds of this water is wasted, more efficient use of even a small amount of irrigation water frees water for other uses.

Most irrigation systems distribute water from a groundwater well or a surface canal by downslope or gravity flow through unlined field ditches (Figure 11-18). This method is cheap as long as farmers in water-short areas don't have to pay the real cost of making this water available. However, it delivers far more water than needed for crop growth, with only 50–60% of the water reaching crops because of evaporation, deep percolation (seepage), and runoff.

Farmers can prevent seepage by placing plastic, concrete, or tile liners in irrigation canals. Lasers can also be used as a surveying aid to help level fields so that water gets distributed more evenly. Small check dams of earth and stone can capture runoff from hillsides and channel it to fields. Holding ponds can store rainfall or capture irrigation water for recycling to crops. Reforesting watersheds also leads to a more manageable flow of irrigation water, instead of a devastating flood (Figure 11-9 and Connections, p. 266).

Many farmers served by the dwindling Ogallala Aquifer now use center-pivot sprinkler systems (Figure 11-18), with which 70–80% of the water reaches crops. Some farmers are switching to low-energy precision-application (LEPA) sprinklers. These systems bring 75–85% of the water to crops by spraying it closer to the ground and in larger droplets than does the center-pivot system. They also reduce energy use and costs by 20–30%. However, because of the high initial costs, sprinklers are used on only about 1% of the world's irrigated land.

In the 1960s, highly efficient trickle or drip irrigation systems were developed in arid Israel. A network of perforated piping, installed at or below the ground surface, releases a trickle of water close to plant roots (Figure 11-18). This minimizes evaporation and seepage, and it brings 80–90% of the water to crops. These systems are expensive to install but are economically feasible for high-profit fruit, vegetable, and orchard crops and for home gardens. They would become cost-effective in most areas if water prices reflected the true cost of this resource.

A: About 25%

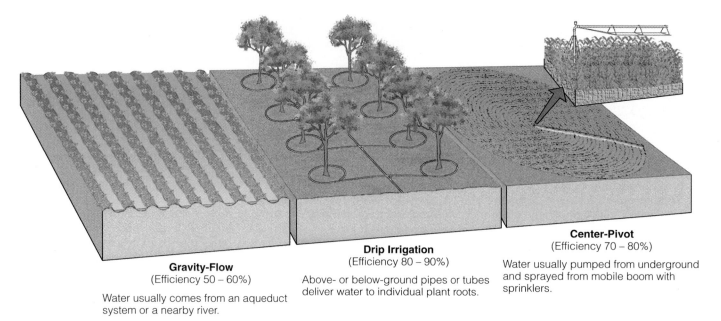

Gravity-Flow
(Efficiency 50 – 60%)

Water usually comes from an aqueduct system or a nearby river.

Drip Irrigation
(Efficiency 80 – 90%)

Above- or below-ground pipes or tubes deliver water to individual plant roots.

Center-Pivot
(Efficiency 70 – 80%)

Water usually pumped from underground and sprayed from mobile boom with sprinklers.

Figure 11-18 Major irrigation systems.

Irrigation efficiency can also be improved by computer-controlled systems that monitor soil moisture and provide water only when necessary. Farmers can switch to more water-efficient, drought-resistant, and salt-tolerant crop varieties. In addition, farmers can use organic farming techniques, which produce higher crop yields per hectare and require only one-fourth of the water and commercial fertilizer of conventional farming. Since 1950 water-short Israel has used many of these techniques to slash irrigation water waste by about 84%—while irrigating 44% more land. However, so long as irrigation water is cheap and plentiful, farmers have little incentive to invest in water-saving techniques.

As fresh water becomes scarce and cities consume water once used for irrigation, carefully treated urban wastewater, which is rich in nitrate and phosphate plant nutrients—could be used for irrigation. For example, Israel now uses 35% of its municipal wastewater, mostly for irrigation, and plans to reuse 80% of this flow by 2000.

Wasting Less Water in Industry   Manufacturing processes can use recycled water or be redesigned to save water. Japan and Israel lead the world in conserving and recycling water in industry. For example, a paper mill in Hadera, Israel, uses one-tenth as much water as most other paper mills do. Manufacturing aluminum from recycled scrap rather than from virgin ores can reduce water needs by 97%.

In the United States, industry is the largest conserver of water. However, the potential for water recycling in U.S. manufacturing has hardly been tapped because the cost of water to many industries is subsi-

dized. A higher, more realistic price would stimulate additional water reuse and conservation in industry.

Wasting Less Water in Homes and Businesses
Flushing toilets, washing hands, and bathing account for about 78% of the water used in a typical U.S. home. In the arid western United States and in dry Australia, lawn and garden watering can take 80% of a household's daily usage. Much of this water is wasted.

Green lawns in an arid or semiarid area can be replaced with vegetation adapted to a dry climate—a form of landscaping called *xeriscaping*, from the Greek word *xeros*, meaning "dry." A xeriscaped yard typically uses 30–80% less water than a conventional one. Drip irrigation (Figure 11-18) can be used to water gardens and other vegetation around homes.

More than half the water supply in Cairo, Lima, Mexico City, and Jakarta disappears before it can be used, mostly from leaks. Leaky pipes, water mains, toilets, bathtubs, and faucets waste 20–35% of water withdrawn from public supplies in the United States.

Many cities offer no incentive to reduce leaks and waste. In New York City, for example, 95% of the residential units don't have water meters. Users are charged flat rates, with the average family paying less than $100 a year for virtually unlimited use of high-quality water. The same is true for one-fifth of all U.S. public water systems. And many apartment dwellers have little incentive to conserve water because their water use is included in their rent.

In Boulder, Colorado, the introduction of water meters reduced water use by more than one-third. Tucson, Arizona, is a desert city that has ordinances that require conserving and reusing water. Tucson now

Q: What percentage of the world's potential food supply is lost to pests?

How to Save Water and Money

- *For existing toilets, reduce the amount of water used per flush* by putting a tall plastic container weighted with a few stones into each tank, or by buying and inserting a toilet dam.

- *Install water-saving toilets that use no more than 6 liters (1.6 gallons) per flush.*

- *Flush toilets only when necessary.* Consider using the advice found on a bathroom wall in a drought-stricken area: "If it's yellow, let it mellow—if it's brown, flush it down."

- *Install water-saving showerheads and flow restrictors on all faucets.*

- *Check frequently for water leaks in toilets and pipes, and repair them promptly.* A toilet must be leaking more than 940 liters (250 gallons) *per day* before you can hear the leak. To test for toilet leaks, add a water-soluble vegetable dye to the water in the tank, but don't

flush. If you have a leak, some color will show up in the bowl's water within a few minutes.

- *Turn off sink faucets while brushing teeth, shaving, or washing.*

- *Wash only full loads of clothes;* if smaller loads must be used, use the lowest possible water level setting.

- *When buying a new washer, choose one that uses the least amount of water and that fills up to different levels for loads of different sizes.* Front-loading clothes models use less water and energy than comparable top-loading models.

- *Use automatic dishwashers for full loads only.* Also, use the short cycle and let dishes air-dry to save energy and money.

- *When washing many dishes by hand, don't let the faucet run.* Instead, use one filled dishpan or sink for washing and another for rinsing.

- *Keep one or more large bottles of water in the refrigerator rather than*

running water from the tap until it gets cold enough for drinking.

- *Don't use a garbage disposal system—a large user of water. Instead, compost your food wastes.*

- *Wash a car from a bucket of soapy water and use the hose for rinsing only. Use a commercial car wash that recycles its water.*

- *Sweep walks and driveways instead of hosing them off.*

- *Reduce evaporation losses by watering lawns and gardens in the early morning or evening, rather than in the heat of midday or when windy.*

- *Use drip irrigation and mulch for gardens and flower beds. Better yet, landscape with native plants adapted to local average annual precipitation so that watering is unnecessary.*

- *To irrigate plants, install a system to capture rainwater or collect, filter, and reuse normally wasted gray water from bathtubs, showers, sinks, and the clothes washer.*

---

consumes half as much water per capita as Las Vegas, a desert city where water conservation is still voluntary. A city might spend billions to build a dam, reservoir, and aqueducts to supply it with water, when spending much less on repairing leaky pipes and providing or subsidizing citizens with low-flow toilets and showerheads could eliminate the need for more water. Each of us can conserve water and in the process can often save money (Individuals Matter, above).

In some parts of the United States systems can be leased that purify and completely recycle wastewater from houses, apartments, or office buildings. Such a system can be installed in a small outside shed and serviced for a monthly fee about equal to that charged by most city water and sewer systems. In Tokyo all the water used in Mitsubishi's 60-story office building is purified for reuse by an automated recycling system. A New Jersey office complex cut water consumption 62% with an on-site treatment and reuse system.

A California water utility gives rebates for water-saving toilets; it also distributed some 35,000 water-saving showerheads, cutting per capita water use 40%

in only one year. In 1988 Massachusetts became the first state to require that all new toilets use no more than 6 liters (1.6 gallons) per flush. Since then 14 other states have followed suit, and most also have water-saving standards for new faucets and showerheads.

Gray water from bathtubs, showers, bathroom sinks, and clothes washers can be collected, stored, carefully treated, and reused for irrigation and other purposes. California has become the first state to legalize reuse of gray water to irrigate landscapes. An estimated 50–75% of the water used by a typical house could be reused as gray water.

11-6 SOLUTIONS: SAVING THE EVERGLADES

Use and Abuse of the Everglades Most people think of the Florida Everglades as a swamp. Actually it is a slow-moving river 80 kilometers (50 miles) wide and generally only 15 centimeters (6 inches) deep that

---

A: About 55% (35% before harvest and 20% after harvest)

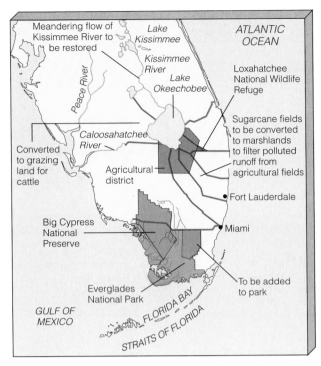

Figure 11-19 The world's largest ecological restoration project is an attempt to undo and redo an engineering project that is destroying Florida's Everglades.

flows south from just below Orlando through Everglades National Park to the estuary of Florida Bay (Figure 11-19).

As the thin layer of water trickles south to the Florida Bay, it creates a vast wetland—the world's largest freshwater marshland—with a variety of habitats that are sanctuaries for a staggering variety of wildlife. It is a haven for 14 endangered or threatened species, including the American alligator (Figure 4-36) and the highly endangered Florida panther (Figure 11-20). Without the Everglades rain distribution and aquifer recharge system, Miami and the rest of south Florida would be uninhabitable.

Since the Civil War half of the original Everglades has been lost to development and water management projects. Today, the Everglades is crisscrossed with 2,250 kilometers (1,400 miles) of canals, levees, spillways, and pumping stations.

The most devastating blow came in the 1960s, when the U.S. Army Corps of Engineers transformed the meandering 103-mile-long Kissimmee River (Figure 11-19) into a straight 84-kilometer (56-mile) canal—called "the dirty ditch" by environmentalists. The canal provided flood control by speeding the flow of water; but in the process it drained the water from large wetlands north of Lake Okeechobee, which farmers turned into cow pastures. Nutrients from cow manure running off the new pastures shot down the straight Kissimmee canal and into Lake Okeechobee,

spawning the explosive growth of algae and other vegetation that deplete the water of oxygen.

Below Lake Okeechobee, vast agricultural fields of sugarcane and vegetables were planted (Figure 11-19). The runoff of phosphorus and other plant nutrients from these fields has stimulated the growth of cattails, which have taken over and displaced saw grass and disrupted the food chains in a vast area of once diverse wetland ecosystems of the Everglades.

To help preserve the lower end of the system, the U.S. government established the Everglades National Park in 1947, which contains about 20% of the remaining Everglades (Figure 11-19). This didn't work because—as environmentalists had predicted—the massive plumbing and land development project to the north cut off much of the water flow needed to sustain the park's wildlife—making it the country's most endangered national park.

Falling water levels also began threatening Miami's water supply as more water was drawn from its aquifer than was replenished. This caused salt water to move inland, irreversibly contaminating supplies of drinking water in some areas of south Florida.

As large volumes of fresh water that once flowed through the park into Florida Bay were diverted for crops and cities, the bay has become saltier and warmer. This and increased nutrient input from cropfields and cities have stimulated the growth of massive algae blooms in the bay, threatening the coral reefs and the fishing industry of the Florida Keys.

Can We Save the Everglades from Ourselves?

By the 1970s state and federal officials recognized that this massive plumbing project—especially converting the winding Kissimmee River to a straight canal—had been a serious ecological blunder. However, it took over 20 years of political haggling before the state and federal government agreed upon a massive ecological project to undo some of the damage. Some if its goals are to restore the curving flow of over half of the Kissimmee River, reclaim large areas of wetlands along the floodplain above and below Lake Okeechobee, remove spillways and levees blocking water flow south of Lake Okeechobee, and add land to Everglades National Park's eastern border (Figure 11-19).

In 1993 Congress authorized the Corps of Engineers to restore the natural contour of 84 kilometers (52 miles) of the Kissimmee. This ambitious project will take more than 15 years and cost at least $588 million (some say the cost will be at least twice this amount), with funds coming from the federal and state governments. If completed, this reengineering project will not restore the Everglades to its natural state, but hopefully it will fix some of what was broken.

This is only part of the cost of possibly saving the Everglades. Another $465 million to $700 million are

Q: What percentage of U.S. crops are lost to pests?

Figure 11-20 The highly endangered Florida panther. Some 30–50 of these animals remain in the Everglades and in Big Cypress National Preserve. These nocturnal carnivores roam up to 50 square miles to hunt deer, wild hogs, and small mammals. To reduce killing of these animals by cars, underpasses have been built along Interstate 75. A captive breeding program has transferred five breeding pairs from the wild to zoos, with the goal of breeding 200 panthers for return to the wild by 2000—if a place can be found to return them to.

to be spent in a 20-year project to reclaim as wetlands land between agricultural areas and the park. These reclaimed marshes would increase the water flow through the Everglades, help filter phosphorus and other nutrient runoff from sugarcane and vegetable fields, and reduce the excessive input of these nutrients into the park and bay.

This plan was developed in the 1980s but has been delayed by court challenges from sugarcane and vegetable growers who didn't want to pick up any of the cost of restoring wetlands their activities had degraded. In 1993, a compromise was reached, with the growers agreeing to pay between $232 million and $322 million of the cleanup costs, with no further financial liability. The growers say this is more than they should have to pay, whereas environmentalists say the agreement lets the growers off too easily. They would end up paying only about two-thirds of the low estimated cost of the project and only 46% of the more realistic $700 million estimate. The rest of the bill will be paid by Florida taxpayers, water users, and the federal government. Expensive and controversial attempts to undo some of the harm we have imposed on the Everglades provide another lesson from nature showing that prevention is the cheapest and best way to go.

11-7 POLLUTION OF STREAMS, LAKES, AND GROUNDWATER

Principal Water Pollutants The most common types of water pollutants are:

- *Disease-causing agents (pathogens).* These include bacteria, viruses, protozoa, and parasitic worms that enter water from domestic sewage and animal wastes (Table 11-1). In LDCs they are the biggest cause of sickness and death, prematurely killing an average of 13,700 people each day, half of them children under age 5. A good indicator of the quality of water for drinking or swimming is the number of colonies of *coliform bacteria* present in a 100-milliliter (0.1-quart) sample of water. The World Health Organization recommends a coliform bacteria count of 0 colonies per 100 milliliters for drinking water, and the EPA recommends a maximum level for swimming water of 200 colonies per 100 milliliters.

- *Oxygen-demanding wastes.* These are organic wastes that can be decomposed by aerobic. (oxygen-requiring) bacteria. Large populations of bacteria supported by these wastes can deplete water of dissolved oxygen (Figure 11-21), causing fish and other forms of oxygen-consuming aquatic life to die. The quantity of oxygen-demanding wastes in water can be determined by measuring the **biological oxygen demand (BOD)**: the amount of dissolved oxygen needed by aerobic decomposers to break down the organic materials in a certain volume of water over a five-day incubation period at 20°C (68°F).

- *Water-soluble inorganic chemicals.* These consist of acids, salts, and compounds of toxic metals such as mercury and lead. High levels of these chemicals can make water unfit to drink, harm fish and

A: About 37% (7% higher than in the 1940s despite a 33-fold increase in pesticide use since then)

Table 11-1 Common Diseases Transmitted to Humans Through Contaminated Drinking Water

Type of Organism	Disease	Effects
Bacteria	Typhoid fever	Diarrhea, severe vomiting, enlarged spleen, inflamed intestine; often fatal if untreated
	Cholera	Diarrhea, severe vomiting, dehydration; often fatal if untreated
	Bacterial dysentery	Diarrhea; rarely fatal except in infants without proper treatment
	Enteritis	Severe stomach pain, nausea, vomiting; rarely fatal
Viruses	Infectious hepatitis	Fever, severe headache, loss of appetite, abdominal pain, jaundice, enlarged liver; rarely fatal but may cause permanent liver damage
Parasitic protozoa	Amoebic dysentery	Severe diarrhea, headache, abdominal pain, chills, fever; if not treated can cause liver abscess, bowel perforation, and death
	Giardia	Diarrhea, abdominal cramps, flatulence, belching, fatigue
Parasitic worms	Schistosomiasis	Abdominal pain, skin rash, anemia, chronic fatigue, and chronic general ill health

other aquatic life, depress crop yields, and accelerate corrosion of equipment that uses the water.

- *Inorganic plant nutrients.* These are water-soluble nitrates and phosphates that can cause excessive growth of algae and other aquatic plants, which then die and decay, depleting water of dissolved oxygen and killing fish. Excessive levels of nitrates in drinking water can reduce the oxygen-carrying capacity of the blood and can kill unborn children and infants, especially those under one year old.

- *Organic chemicals.* These include oil, gasoline, plastics, pesticides, cleaning solvents, detergents, and many other chemicals. They threaten human health and harm fish and other aquatic life.

- *Sediment or suspended matter.* This is insoluble particles of soil and other solids that become suspended in water, mostly when soil is eroded from the land (Figure 5-21). By weight this is by far the biggest water pollutant. Sediment clouds water and reduces photosynthesis; it also disrupts aquatic food webs and carries pesticides, bacteria, and other harmful substances. Sediment that settles out destroys feeding and spawning grounds of fish, and it clogs and fills lakes, artificial reservoirs, stream channels, and harbors.

- *Radioactive isotopes that are water-soluble or capable of being biologically amplified to higher concentrations as they pass through food chains and webs.* Ionizing radiation from such isotopes can cause birth defects, cancer, and genetic damage.

- *Heat absorbed by water used to cool electric power plants.* The resulting rise in water temperature lowers dissolved oxygen content and makes aquatic organisms more vulnerable to disease, parasites, and toxic chemicals.

- *Alien species (genetic pollution).* Each day several thousand species (mostly small ones such as mussels and phytoplankton) are transported in cargo and ballast water in ships to new marine systems. They may then outcompete many native species, reduce biodiversity, and cause economic losses. Alien marine species also spread through canals linking bodies of water and by deliberate introduction to enhance fishery production.

Total damage from water pollution in the United States is estimated to cost $20 billion per year.

Point and Nonpoint Sources of Pollution

Point sources discharge pollutants at specific locations through pipes, ditches, or sewers into bodies of surface water. Examples include factories, sewage treatment plants (which remove some but not all pollutants), active and abandoned underground mines, offshore oil wells, and oil tankers. Because point sources are at specific places (mostly in urban areas), they are fairly easy to identify, monitor, and regulate. In MDCs many industrial discharges are strictly controlled, whereas in most LDCs such discharges are largely uncontrolled.

Nonpoint sources are sources that cannot be traced to any single discharge. They are usually large, poorly defined areas that pollute water by runoff, subsurface flow, or deposition from the atmosphere. Examples include runoff of chemicals into surface water

Q: What percentage of insecticides applied to crops in the United States reach the target pests?

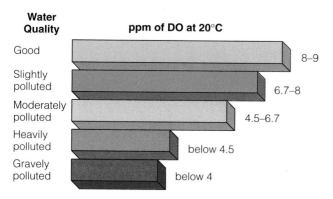

Figure 11-21 Water quality and dissolved oxygen (DO) content in parts per million (ppm) at 20°C (68°F). The solubility of oxygen decreases as the water temperature increases. Only a few species of fish can survive in water with fewer than 4 ppm of dissolved oxygen.

and seepage into the ground from croplands, livestock feedlots, logged forests, streets, lawns, septic tanks, construction sites, parking lots, and roadways.

In the United States, nonpoint pollution from agriculture—mostly in the form of sediment, inorganic fertilizer, manure, salts dissolved in irrigation water, and pesticides—is responsible for an estimated 64% of the total mass of pollutants entering streams and 57% of those entering lakes. Little progress has been made in the control of nonpoint water pollution because of the difficulty and expense of identifying and controlling discharges from so many diffuse sources.

Contaminated Drinking Water　In many parts of the world, water quality has also been degraded. Rivers in eastern Europe, Latin America, and Asia are severely polluted, as are some in MDCs. Aquifers used as sources of drinking water in many MDCs and LDCs are becoming contaminated with pesticides, fertilizers, and hazardous organic chemicals. In China, for example, 41 large cities get their drinking water from polluted groundwater.

According to the World Health Organization, 1.5 billion people don't have a safe supply of drinking water, and 1.7 billion people lack adequate sanitation facilities. At least 5 million people die every year from waterborne diseases that could be prevented by clean drinking water and better sanitation. Most of the 13,700 who die each day from such diseases are children under age 5.

In 1980 the United Nations called for spending $300 billion to supply all of the world's people with clean drinking water and adequate sanitation by 1990. The $30-billion-per-year cost of this program is about what the world spends every 10 days for military purposes. Sadly, only about $1.5 billion per year was actually spent.

Stream Pollution　Flowing streams—including large ones called *rivers*—recover rapidly from degradable, oxygen-demanding wastes and excess heat by a combination of dilution and bacterial decay. This recovery process works so long as streams are not overloaded with these pollutants and so long as their flow is not reduced by drought, damming, or diversion for agriculture and industry. Slowly degradable and nondegradable pollutants are not eliminated by these natural dilution and degradation processes.

This breakdown of degradable wastes by bacteria depletes dissolved oxygen, which reduces or eliminates populations of organisms with high oxygen requirements until the stream is cleansed. The depth and width of the resulting *oxygen sag curve* (Figure 11-22) (and thus the time and distance a stream takes to recover) depend on the stream's volume, flow rate, temperature, and pH level (Figure 9-8), as well as the volume of incoming degradable wastes. Similar oxygen sag curves can be plotted when heated water from power plants is discharged into streams. The types of pollutants, flow rates, dilution capacity, and recovery time vary widely with different river basins; these factors also vary in the three major zones of a river as it flows from its headwaters, to its wider and deeper middle sections, and finally to an ocean or lake (Figure 5-31).

Requiring cities to withdraw their drinking water downstream rather than upstream (as is done now) would dramatically improve water quality as the stream flows toward the sea. Then each city would be forced to clean up its own waste outputs rather than pass them downstream (as is done now). However, upstream users, who have the use of fairly clean water without high cleanup costs, fight this pollution prevention approach.

Water pollution control laws enacted in the 1970s have greatly increased the number and quality of wastewater treatment plants in the United States and in many other MDCs. Laws have also required industries to reduce or eliminate point source discharges into surface waters. These efforts spurred by individuals (Individuals Matter, p. 283) have enabled the United States to hold the line against increased pollution of most of its streams by disease-causing agents and oxygen-demanding wastes, an impressive accomplishment considering the rise in economic activity and population since the laws were passed.

One success story is the cleanup of Ohio's Cuyahoga River, which was so polluted that in 1969 it caught fire as it flowed through the city of Cleveland. That prompted city and state officials to pass laws limiting the discharge of wastes by industries into the river and sewage systems, as well as laws and funds to upgrade sewage treatment facilities. Today the river

A: 1–2% (and often less than 0.1%)

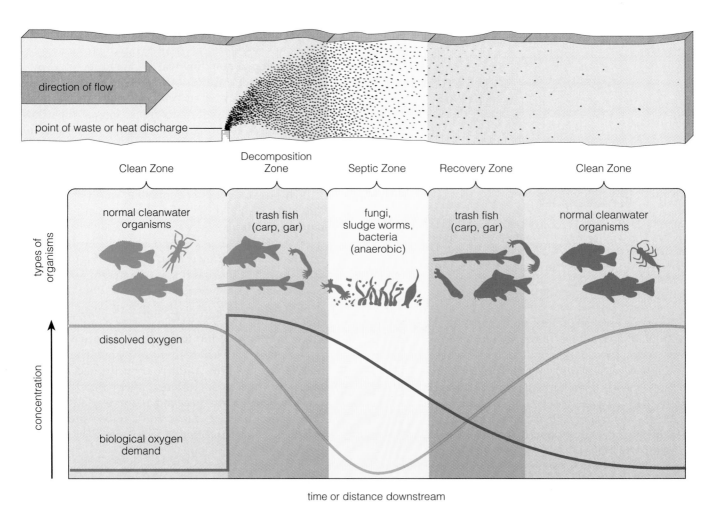

Figure 11-22 Dilution and decay of degradable, oxygen-demanding wastes and heat, showing the oxygen sag curve (orange) and the curve of oxygen demand (blue). Depending on flow rates and the amount of pollutants, streams recover from oxygen-demanding wastes and heat if they are given enough time and are not overloaded.

has made a comeback and is widely used by boaters and anglers.

However, we know relatively little about stream quality because water quality in 64% of the length of U.S. streams has not been measured. And many existing monitoring stations are not located in places that are suitable for assessing the presence or absence of pollutants from their drainage basins, mostly because of a lack of money. Furthermore, even this limited monitoring does not measure toxics and ecological indicators of water quality.

Available data indicate that stream pollution from huge discharges of sewage and industrial wastes is a serious and growing problem in most LDCs (Figure 11-24), where waste treatment is practically nonexistent. Numerous streams in the former USSR and in eastern European countries are severely polluted. Currently, more than two-thirds of India's water resources are polluted. Of the 78 streams monitored in China, 54 are seriously polluted. In Latin America and Africa,

most streams passing through urban or industrial areas are severely polluted.

Lake Pollution   In lakes (and reservoirs), dilution is often less effective than in streams because these bodies of water frequently contain stratified layers (Figure 5-28) that undergo little vertical mixing. Stratification also reduces levels of dissolved oxygen, especially in the bottom layer. In addition, lakes and reservoirs have little flow, further reducing dilution and replenishment of dissolved oxygen. The flushing and changing of water in lakes and large artificial reservoirs can take from 1 to 100 years, compared with several days to several weeks for streams.

Thus, lakes are more vulnerable than streams to contamination by plant nutrients, oil, pesticides, and toxic substances that can destroy bottom life and kill fish. Atmospheric fallout and runoff of acids is a serious problem in lakes vulnerable to acid deposition (Figure 9-10).

Q: Since 1945 how many premature deaths from insect-transmitted diseases have been saved by using insecticides?

Rescuing a River

When Marion Stoddart (Figure 11-23) first moved to Groton, Massachusetts, in the early 1960s the nearby Nashua River was considered one of the nation's filthiest rivers. Dead fish bobbed on its waves, and at times the water was red, green, or blue from pigments discharged by paper mills.

Marion Stoddart was appalled. Instead of thinking nothing could be done, she committed herself to restoring the Nashua and establishing public parklands along its banks. She didn't start by filing lawsuits or organizing demonstrations. Rather, she created a careful cleanup plan and approached state officials with it in 1962. They laughed, but she was not deterred and began practicing the most time-honored skill of politics—one-on-one persuasion. She identified the power brokers in the riverside communities and began to educate them, win them over, and get them to cooperate in cleaning up the river.

She got the state to ban open dumping in the river. When promised federal matching funds for building the treatment plant failed to arrive, Stoddart got 13,000 signatures on a petition to President Nixon. The funds arrived in a hurry.

Stoddart's next success was getting a federal grant to beautify the river. She hired high school dropouts to clear away mounds of debris. When the river cleanup was completed, she persuaded communities along the river to create some 2,400 (6,000 acres) hectares of riverside park and woodlands along both banks.

Now, over two decades later, the Nashua is still clean, and a citizens' group founded by Stoddart keeps watch on water quality. The river's waters support many kinds of fish and other wildlife, and they are used for canoeing (Figure 11-23) and other kinds of recreation.

The project is considered a model for other states and is testimony to what a committed individual can do to change the world by getting people to work together.

For her efforts Stoddart has been named by the UN Environment Programme as an outstanding worldwide worker for the environment. She might say, however, that the naturally blue and canoeable Nashua is itself her best reward.

Seth Resnick

Figure 11-23 Earth citizen Marion Stoddart canoeing down the Nashua River near Groton, Massachusetts. She spent over two decades spearheading successful efforts to have this river cleaned up.

A: About 7 million

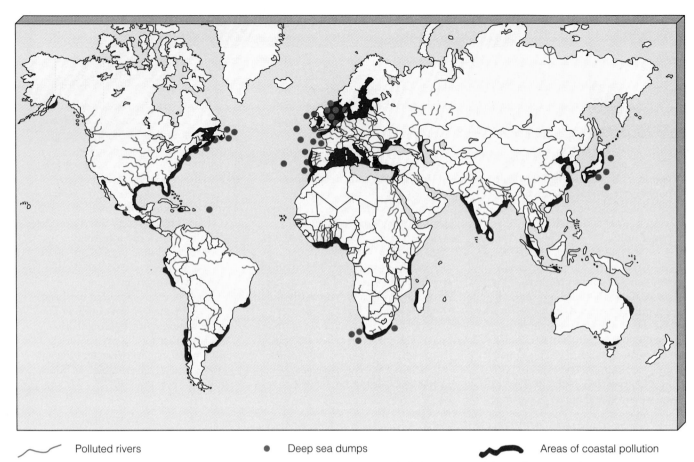

| ~~~ | Polluted rivers | ● | Deep sea dumps | ⬛ | Areas of coastal pollution |

Figure 11-24 Pollution of rivers and the oceans. (Data from United Nations and U.S. Environmental Protection Agency)

In any body of water, some synthetic organic compounds and toxic metals such as lead and mercury are not biodegraded, and others are biodegraded very slowly. Many toxic chemicals also enter lakes and reservoirs from the atmosphere. Concentrations of some chemicals, such as DDT, PCBs (Figure 11-25), some radioactive isotopes, and some mercury compounds can be biologically amplified as they pass through food webs.

Lakes receive inputs of nutrients and silt from the surrounding land basin as a result of natural erosion and runoff. Some of these lakes become more eutrophic over time (Figure 5-29), but others don't because of differences in the surrounding waterbasin. Near urban or agricultural areas the input of nutrients to a lake can be greatly accelerated by human activities, a process known as **cultural eutrophication**. Such a change is caused mostly by nitrate- and phosphate-containing effluents from sewage treatment plants, runoff of fertilizers and animal wastes, and accelerated erosion of nutrient-rich topsoil (Figure 11-26).

During hot weather or drought this nutrient overload produces dense growths of organisms such as algae, cyanobacteria, water hyacinths, and duckweed.

Dissolved oxygen in both the surface layer of water near the shore and in the bottom layer is depleted when large masses of algae die, fall to the bottom, and are decomposed by aerobic bacteria. This oxygen depletion can kill fish and other oxygen-consuming aquatic animals. If excess nutrients continue to flow into a lake, the bottom water becomes foul and almost devoid of animals, as anaerobic bacteria take over and produce smelly decomposition products such as highly toxic hydrogen sulfide and flammable methane.

About one-third of the 100,000 medium-to-large lakes and about 85% of the large lakes near major population centers in the United States suffer from some degree of cultural eutrophication (Case Study, p. 288). A quarter of China's lakes are classified as eutrophic.

The best solution to cultural eutrophication is twofold: to use both prevention methods to reduce the flow of nutrients into lakes and reservoirs, and to use pollution cleanup methods to clean up lakes already suffering from excessive eutrophication. Major prevention methods include advanced waste treatment (Section 11-9); bans or limits on phosphates in household detergents and other cleaning agents; and soil conservation and land-use control to reduce nutrient

Q: Worldwide, how many people are poisoned each year by pesticides?

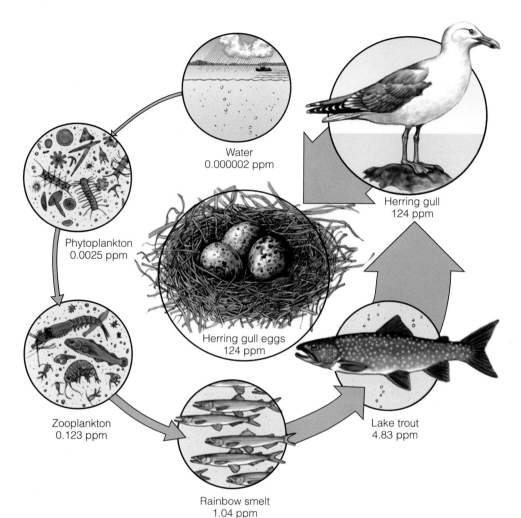

Figure 11-25 Biological amplification of PCBs (polychlorinated biphenyls) in an aquatic food chain in the Great Lakes. Most of the 209 different PCBs are insoluble in water, soluble in fats, and resistant to biological and chemical degradation—properties that result in their bioaccumulation in the tissues of organisms and their biological amplification in food chains and webs. The long-term health effects on people exposed to low levels of PCBs are unknown. However, in laboratory animals high doses of PCBs produce liver and kidney damage, gastric disorders, birth defects, bronchitis, miscarriages, skin lesions, hormonal changes, reduction in penis size, and tumors. Some studies indicate that most of these harmful effects are caused by polychlorinated dibenzofurans (commonly called furans) found as contaminants in some PCBs. In the United States and Canada PCBs have been banned since 1976. Prior to that, however, millions of metric tons of these chemicals were released into the environment, many of them ending up in bottom sediments of lakes, streams, and oceans.

Water
0.000002 ppm

Phytoplankton
0.0025 ppm

Herring gull
124 ppm

Herring gull eggs
124 ppm

Zooplankton
0.123 ppm

Lake trout
4.83 ppm

Rainbow smelt
1.04 ppm

runoff. Major cleanup methods are dredging bottom sediments to remove excess nutrient buildup; removing excess weeds; controlling undesirable plant growth with herbicides and algicides; and pumping air through lakes and reservoirs to avoid oxygen depletion (an expensive and energy-intensive method).

Groundwater Pollution While highly visible oil spills get lots of media attention, a much greater threat to human health is the out-of-sight pollution of groundwater (Figure 11-27), which is a prime source of water for drinking and irrigation. This vital form of Earth capital is easy to deplete and pollute because it is renewed so slowly. Laws protecting groundwater are weak in the United States and nonexistent in most countries.

When groundwater becomes contaminated, it does not cleanse itself of degradable wastes as surface water can if it is not overloaded (Figure 11-22). Because groundwater flows are slow and not turbulent, contaminants are not effectively diluted and dis-

persed. Also, groundwater has much smaller populations of decomposing bacteria than do surface water systems, and its cold temperature slows down decomposition reactions. Thus, it can take hundreds to thousands of years for contaminated groundwater to cleanse itself of degradable wastes—and nondegradable wastes are there permanently.

Results of limited testing of groundwater in the United States are alarming. In a 1982 survey, the EPA found that 45% of the large public water systems served by groundwater were contaminated with synthetic organic chemicals that posed potential health threats. Another EPA survey in 1984 found that two-thirds of the rural household wells tested violated at least one federal health standard for drinking water, usually pesticides or nitrates from fertilizers (which cause a life-threatening blood disorder in infants during their first year). The EPA has documented groundwater contamination by 74 pesticides in 38 states.

Crude estimates indicate that although only 2% by volume of all U.S. groundwater is contaminated,

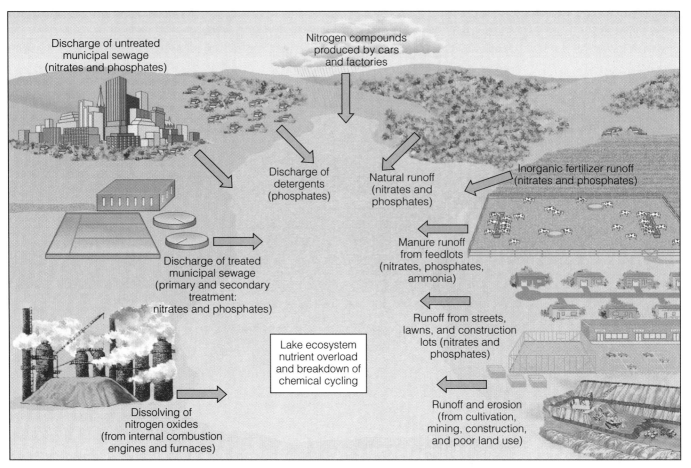

Figure 11-26 Principal sources of nutrient overload, or cultural eutrophication, in lakes. The amount of nutrients from each source varies, depending on the types of human activities taking place in each airshed and watershed. Levels of dissolved oxygen (Figure 11-21) drop when these excess algae and plants die and are decomposed by aerobic bacteria. This can kill fish and other aquatic life, and lower the aesthetic and recreational values of the lake.

up to 25% of usable groundwater is contaminated, and in some areas as much as 75% is contaminated. In New Jersey, for example, every major aquifer is contaminated. In California, pesticides contaminate the drinking water of more than 1 million people. In Florida, where 92% of the residents rely on groundwater for drinking, over 1,000 wells have been closed.

Groundwater can be contaminated from a number of sources, including underground storage tanks, landfills, abandoned hazardous-waste dumps, deep wells used to dispose of liquid hazardous wastes, and industrial-waste storage lagoons located above or near aquifers (Figure 11-27). An EPA survey found that one-third of 26,000 industrial-waste ponds and lagoons have no liners to prevent toxic liquid wastes from seeping into aquifers. One-third of those sites are within 1.6 kilometers (1 mile) of a drinking water well.

The EPA estimates that at least 1 million underground tanks are leaking their contents into groundwater. A slow gasoline leak of just 4 liters (1 gallon) per day can seriously contaminate the water supply for 50,000 people. Such slow leaks usually remain undetected until someone discovers that a well is contaminated.

Determining the extent of a leak can cost $25,000–$250,000. Cleanup costs range from $10,000 for a small spill to $250,000 or more if the chemical reaches an aquifer. Replacing a leaking tank adds an additional $10,000–$60,000. Legal fees and damages to injured parties can run into the millions. Stricter regulations should reduce leaks from new tanks but would do little about the millions of older tanks that are "toxic time bombs." Some analysts call for above-ground storage of hazardous liquids so that leaks can be easily detected and rectified.

11-8 OCEAN POLLUTION

The Ultimate Sink The oceans are the ultimate sink for much of the waste matter we produce. This is summarized in the African proverb, "Water may flow

Q: What is the most serious drawback to using chemicals to control pests (especially insects)?

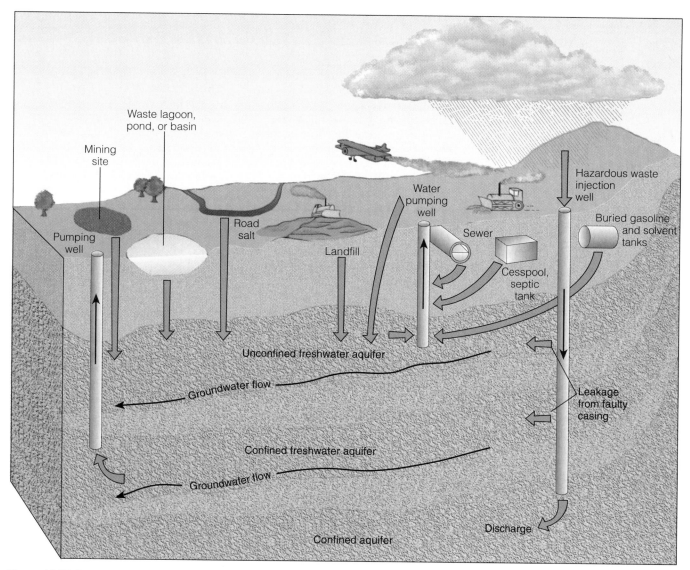

Figure 11-27 Principal sources of groundwater contamination in the United States.

in a thousand channels, but it all returns to the sea." About three-quarters of the total pollution load entering the oceans worldwide comes from human activities on land.

Oceans can dilute, disperse, and degrade large amounts of raw sewage, sewage sludge, and oil—and some types of industrial waste, especially in deepwater areas. Marine life has also proved to be more resilient than some scientists had expected, leading some of them to suggest that it is generally safer to dump sewage sludge and most other hazardous wastes into the deep ocean than to bury them on land or burn them in incinerators.

Other scientists dispute this idea, pointing out that we know less about the deep ocean than we do about outer space. They add that dumping waste in the ocean would delay urgently needed pollution pre-

vention and promote further degradation of this vital part of Earth's life-support system. Marine explorer Jacques Cousteau has warned that "the very survival of the human species depends upon the maintenance of an ocean clean and alive, spreading all around the world. The ocean is our planet's life belt."

Overwhelming Coastal Areas: Living Near the Edge Coastal areas—especially wetlands and estuaries (Figures 5-21 and 5-23), coral reefs (Figures 5-20 and 5-22), and mangrove swamps (Figure 5-24)—bear the brunt of our enormous inputs of wastes into the ocean (Figure 11-24). This is not surprising, for half the world's population lives on or within 100 kilometers (160 miles) from the coast (Figure 5-27), nearly one-fifth of the world's people live in coastal cities, and coastal populations are growing more rapidly than

A: Development of genetic resistance to the chemicals by target pests

The five interconnected Great Lakes contain at least 95% of the surface fresh water in the United States and 20% of the world's fresh surface water (Figure 11-28). The Great Lakes basin is home for about 35 million people, making up about 30% of the Canadian population and 13% of the U.S. population. About 40% of U.S. industry and half of Canadian industry are located in this watershed. Great Lakes tourism generates $16 billion annually, with $4 billion of that from sport fishing.

Despite their enormous size, these lakes are vulnerable to pollution from point and nonpoint sources because less than 1% of the water entering the Great Lakes flows out to the St. Lawrence River each year. In addition to land runoff these lakes also receive large quantities of acids, pesticides, and other toxic chemicals by deposition from the atmosphere—often blown in from hundreds or thousands of kilometers away.

By the 1960s many areas of the Great Lakes were suffering from severe cultural eutrophication, huge fish kills, and contamination from bacteria and other wastes. The impact on Lake Erie was particularly intense because it is the shallowest of the Great Lakes, it has the smallest volume of water, its drainage basin is heavily industrialized, and it has the largest human population. Many bathing beaches had to be closed, and by 1970 the lake had lost nearly all its native fish.

Since 1972 a $20-billion pollution control program, carried out by Canada and the United States, has greatly decreased levels of phosphates, coliform bacteria, and many toxic industrial chemicals in the Great Lakes. Algal blooms have also decreased, dissolved oxygen levels and sport and commercial fishing have increased, and most swimming beaches have been reopened.

These improvements were brought about mainly by new or upgraded sewage treatment plants and improved treatment of industrial wastes. Also, phosphate detergents, household cleaners, and water conditioners were banned or their phosphate levels were lowered in many areas of the Great Lakes drainage basin.

The most serious pollution problem today is contamination from toxic wastes flowing into the lakes (especially Lake Erie and Lake Ontario) from land runoff, streams, and atmospheric deposition (Figure 11-28). Toxic chemicals such as PCBs have built up in food chains and webs (Figure 11-25) and have contaminated many types of fish caught by anglers and depleted populations of birds, river otters, and other animals feeding on contaminated fish. A survey by Wisconsin biologists revealed that one in four fish taken from the Great Lakes is unsafe for consumption.

In 1991 the U.S. government passed a law requiring accelerated cleanup of the lakes, especially of 42 toxic hot spots, and an immediate reduction in emissions of toxic air pollutants in the region. However, meeting these goals may be delayed by lack of federal and state funds.

Environmentalists call for a ban on the use of chlorine as a bleach in the pulp and paper industry around the Great Lakes, a ban on all new incinerators in the area, and an immediate ban on toxic discharge into the lakes of 70 toxic chemicals that threaten human health and wildlife.

Pollution is not the only problem. Since the 1800s the Great Lakes have been invaded by numerous alien species that have sharply reduced populations of commercial and sport fish, and caused other problems. Canals built in the early 1800s allowed the lakes to be invaded by the sea lamprey, a parasite that attaches itself to the body of soft-skinned fish and sucks out blood and other body fluids (Figure 4-41). Between the 1920s and the mid-1950s a combination of large populations of sea lampreys and overfishing devastated populations of game and commercial fish. In 1954 a poison was found that could kill the larvae of sea lampreys. By 1962 the poison had caused a sharp drop in sea lamprey populations, and a $10-million-per-year program has kept them under control.

Since the 1980s, however, the sea lamprey has been breeding in large rivers near the lakes, which are hard to treat with poisons. Unless other control methods are developed, the sea lamprey may again decimate populations of desirable game and commercial fish in the Great Lakes.

In 1986 larvae of an alien species—the *zebra mussel*—arrived in water discharged from a European ship near Detroit. With no known natural enemies, these tiny mussels have run amok. They deplete the food supply for other lake species, clog irrigation pipes, shut down water intake systems for power plants and city water supplies, foul beaches, and grow in huge masses on boat hulls, piers, and other surfaces.

These invaders cost the Great Lakes basin at least $500 million per year. It is estimated that the annual costs could reach $5 billion by 2000.

global population. In Southeast Asia, over two-thirds of the people live on or near the coastlines.

In most coastal LDCs and in some coastal MDCs, untreated municipal sewage and industrial wastes are often dumped into the sea without treatment. In the United States, about 35% of all municipal sewage ends up virtually untreated in marine waters. Most U.S. harbors and bays are badly polluted from municipal sewage, industrial wastes, and oil. San Francisco Bay, the largest estuary in the western United States, is

Q: What percentage of food bought in U.S. supermarkets has pesticide residue levels above the legal limit?

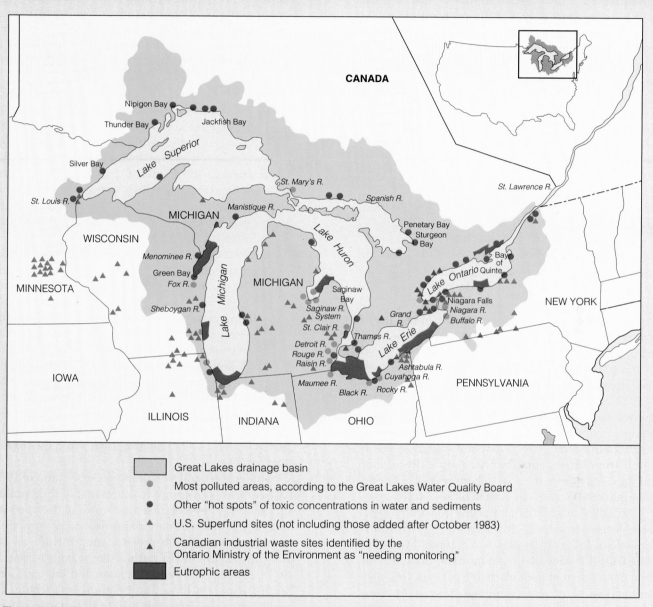

Figure 11-28 The Great Lakes basin. (Data from Environmental Protection Agency)

Even worse, the zebra mussel is expected to spread uncontrollably and dramatically alter most freshwater communities throughout the United States and southern Canada by 2000.

There is even worse news. In 1991 another larger and potentially more destructive species—the *quagga mussel*—invaded the Great Lakes, probably brought in by a Russian freighter. It can survive at greater depths and tolerate more extreme temperatures than the zebra mussel. There is concern that it may eventually colonize areas such as the Chesapeake Bay and waterways in parts of Florida.

overrun by alien species and can no longer support commercial fishing. Also, during last 140 years 60% of its water area has been filled in and converted to land.

Each year fully one-third of the area of U.S. coastal waters around the lower 48 states are closed to shellfish harvesters because of pollution and habitat disruption. In 1992 there were more than 2,600 beach closings in 22 coastal states, mostly because of bacterial contamination from inadequate and overloaded sewage treatment systems. Many more would be

A: 1–3%

closed if their waters were tested regularly. Coastal waters of beaches near San Diego, California, are being contaminated with massive inputs of raw sewage from Mexico. Most of this pollution comes from Tijuana, Mexico, where pollution controls are poorly enforced. The problem is expected to get much worse as Tijuana's population of more than 1 million is expected to reach 2 million by the year 2000.

The Chesapeake Bay is an example of an estuary in trouble because of human activities. It is the largest estuary in the United States, and one of the world's most productive. It is the largest source of oysters in the United States and the largest producer of blue crab in the world. Between 1940 and 1993 the number of people living in the Chesapeake Bay area grew from 3.7 million to 15 million, and by 2000 it may reach 18 million.

The estuary receives wastes from point and nonpoint sources scattered throughout a huge drainage basin that includes 9 large rivers and 141 smaller streams and creeks in parts of six states (Figure 11-29). The bay has become a huge pollution sink because it is quite shallow—with an average depth of less than 7 meters (23 feet)—and because only 1% of the waste entering it is flushed into the Atlantic Ocean.

Levels of phosphate and nitrate plant nutrients have risen sharply in many parts of the bay, causing algal blooms and oxygen depletion (Figure 11-29). Studies have shown that point sources, primarily sewage treatment plants, contribute about 60% by weight of the phosphates. Nonpoint sources—mostly runoff from urban, suburban, and agricultural land and deposition from the atmosphere (Figure 9-7)—are the origin of about 60% by weight of the nitrates.

Air pollutants account for nearly 30% of the nitrogen entering the estuary. In addition, large quantities of pesticides run off cropland and urban lawns; and industries discharge large amounts of toxic wastes, often in violation of their discharge permits. Commercial harvests of oysters, crabs, and several important fish have fallen sharply since 1960 because of a combination of overfishing, pollution, and disease.

Since 1983 more than $700 million in federal and state funds have been spent on a Chesapeake Bay cleanup program that will ultimately cost several billion dollars. Since 1987 nitrogen and phosphorus from nonpoint sources dropped about 7%, but goals for the year 2000 are unlikely to be met. To add to its problems, the bay will soon be invaded by zebra and quagga mussels now causing massive damage to the Great Lakes (Case Study, p. 288). Halting the deterioration of this vital estuary will require the prolonged, cooperative efforts of citizens, officials, and industries throughout its entire watershed with much greater emphasis on pollution prevention.

Ocean Dumping Dumping of industrial waste off U.S. coasts has stopped, although it still takes place in a number of other MDCs and some LDCs. However, barges and ships legally dump large quantities of **dredge spoils** (materials, often laden with toxic metals, scraped from the bottoms of harbors and rivers to maintain shipping channels) off the Atlantic, Pacific, and Gulf coasts at 110 sites (Figure 11-24).

In addition, many countries, including Great Britain, dump into the ocean large quantities of **sewage sludge**—a gooey mixture of toxic chemicals, infectious agents, and settled solids removed from wastewater at sewage treatment plants. This practice was banned in the United States as of 1992. Some elected officials and scientists oppose this ban, however, arguing that ocean disposal, especially in the deep ocean, is safer and cheaper than land dumping and incineration.

Fifty countries with at least 80% of the world's merchant fleet have agreed not to dump sewage and garbage at sea, but this agreement is hard to enforce and is often violated. Most ship owners save money by dumping at sea and risking small fines if caught. Each year as many as 2 million seabirds and more than 100,000 marine animals (including whales, seals, dolphins, sea lions, and sea turtles) die when they ingest or become entangled in plastic cups, bags, six-pack yokes, broken sections of fishing nets, ropes, and other debris dumped into the sea and discarded on beaches. Some ships also incinerate their wastes and illegally dump the ash at sea, releasing toxins into the marine environment.

Under the London Dumping Convention of 1972, 100 countries agreed not to dump highly toxic pollutants and high-level radioactive waste in the open sea beyond the limits of national jurisdiction. And since 1983 these nations observed a moratorium on the dumping of low-level radioactive wastes at sea. In 1994 this became a permanent ban, but France, Great Britain, Russia, China, and Belgium may legally exempt themselves from this ban. In 1992 it was learned that for decades the former Soviet Union had been dumping large quantities of high- and low-level radioactive wastes into the Arctic Ocean and tributaries that flow into this ocean.

Oil Pollution *Crude petroleum* (oil as it comes out of the ground) and *refined petroleum* (fuel oil, gasoline, and other processed petroleum products) are accidentally or deliberately released into the environment from a number of sources.

It is encouraging that between 1981 and 1989 oil pollution from ships dropped by 60%, according to a study made by the U.S. National Academy of Sciences. Tanker accidents (Case Study, p. 292) and blowouts

Q: How many cancer deaths in the United States are caused by exposure to pesticide residues in foods?

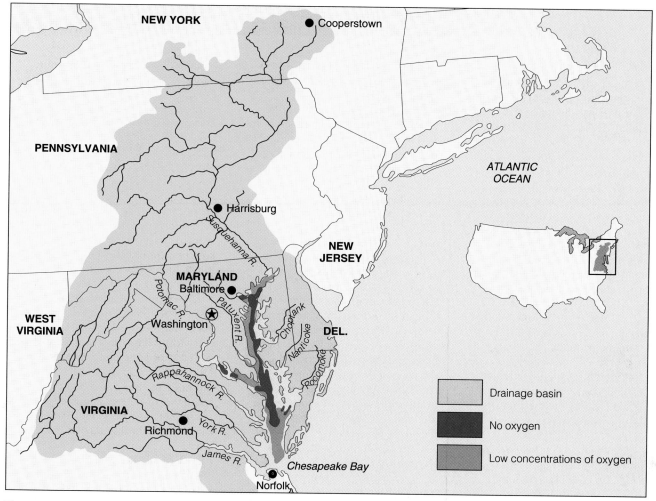

Figure 11-29 Chesapeake Bay. The largest estuary in the United States is severely degraded as a result of water pollution from point and nonpoint sources in six states, as well as from deposition of pollutants from the atmosphere.

Legend:
- Drainage basin
- No oxygen
- Low concentrations of oxygen

(oil escaping under high pressure from a borehole in the ocean floor) at offshore drilling rigs get most of the publicity, but more oil is released by normal operation of offshore wells, washing tankers and releasing the oily water, and from pipeline and storage tank leaks. A 1993 Friends of the Earth study estimated that each year U.S. oil companies unnecessarily spill, leak, or waste oil equal to that held by 1,000 *Exxon Valdez* tankers, or more oil than Australia uses.

Although natural oil seeps also release large amounts of oil into the ocean at some sites, most ocean oil pollution comes from activities on land. Almost half (some experts estimate 90%) of the oil reaching the oceans is waste oil dumped onto the land or into sewers by cities, individuals, and industries. Each year oil equal to 20 times the amount spilled by the *Exxon Valdez* is improperly disposed of by U.S. citizens changing their own motor oil. Worldwide, about 10% of the oil that reaches the ocean comes from the atmosphere, mostly from oil fire smoke.

The effects of oil on ocean ecosystems depend on a number of factors: type of oil (crude or refined), amount released, distance of release from shore, time of year, weather conditions, average water temperature, and ocean currents. Research shows that most forms of marine life recover from exposure to large amounts of crude oil within three years. However, recovery from exposure to refined oil, especially in estuaries, may take 10 years or longer. A recent study also showed that diesel oil spilled at sea becomes more toxic to marine life as it ages. The effects of spills in cold waters (such as Alaska's Prince William Sound and Antarctic waters) and in shallow enclosed gulfs and bays (such as the Persian Gulf) generally are more damaging and last longer.

Oil slicks that wash onto beaches can have a serious economic impact on coastal residents, who lose income from fishing and tourist activities. Oil-polluted beaches washed by strong waves or currents are cleaned up after about a year, but beaches in sheltered

A: 4,000–20,000 per year

The Valdez Oil Spill

Crude oil from Alaska's North Slope fields near Prudhoe Bay is carried by pipeline to the port of Valdez and then shipped by tanker to the West Coast (Figure 11-30). In the early 1970s environmentalists had predicted that a large, damaging spill might occur in these waters, made treacherous by icebergs, submerged reefs, and violent storms. On March 24, 1989, the *Exxon Valdez*, a tanker more than three football fields long, went off course in a 16-kilometer-wide (11-mile) channel in Prince William Sound near Valdez, Alaska. It hit submerged rocks on a reef, creating the worst oil spill ever in U.S. waters.

The rapidly spreading oil slick coated more than 1,600 kilometers (1,000 miles) of shoreline, almost the length of the shoreline between New Jersey and South Carolina. The oil killed between 300,000 and 645,000 birds (including 144 bald eagles), up to 5,500 sea otters, 30 seals, 23 whales, and unknown numbers of fish. The real toll on wildlife will never be known, however, because most of the dead animals sank and decomposed without being counted.

Since the spill Exxon has spent $2.5 billion on the actual cleanup. In 1991, Exxon pleaded guilty to federal felony and misdemeanor charges and agreed to pay the federal government and the state of Alaska $1 billion in fines and civil damages. After tax write-offs and inflation adjustments, Exxon will end up paying about $500 million in fines, with the rest being absorbed by taxpayers through lost tax revenue. Exxon still faces some $59 billion in lawsuits from the Alaskan fishing industry, landowners, cannery workers, Native Americans, and other injured parties. By 1993 none of the money paid by Exxon to Alaska had been used either for substantive ecological restoration or for acquisition and protection of vulnerable forests and other habitats in the Valdez area.

This multibillion-dollar accident might have been prevented if Exxon had spent $22.5 million to fit the tanker with a double hull. In the early 1970s, then Interior Secretary Rogers Morton told Congress that all oil tankers using Alaskan waters would have double hulls, but under pressure from oil companies the requirement was later dropped. Today, virtually all merchant ships have double hulls—except oil tankers. Legislation passed since the spill requires all new tankers to have double hulls and requires that existing large single-hulled oil tankers be phased out between 1995 and 2005. However, the oil industry is working to have these and other stricter requirements enacted since the spill weakened as the public memory of the spill fades.

Others must share the blame for this tragedy. State officials had been lax in monitoring Alyeska, a company formed by the seven oil companies extracting oil from Alaska's North Slope, and the Coast Guard did not effectively monitor tanker traffic because of inadequate radar equipment and personnel. American consumers also get some of the blame. Their unnecessarily wasteful use of oil and gasoline (Section 18-2) is the driving force behind the search for more domestic oil without employing adequate environmental safeguards.

---

areas remain contaminated for several years. Estuaries and salt marshes (Figure 5-23) suffer the most damage and cannot effectively be cleaned up.

11-9 SOLUTIONS: PREVENTING AND REDUCING WATER POLLUTION

Nonpoint-Source Pollution The leading nonpoint source of water pollution is agriculture. Farmers can sharply reduce fertilizer runoff into surface waters and leaching into aquifers by using moderate amounts of fertilizer—and by using none at all on steeply sloped land. They can use slow-release fertilizers and alternate their plantings between row crops and soybeans or other nitrogen-fixing plants to reduce the need for fertilizer. Farmers can also be required to plant buffer zones of permanent vegetation between cultivated fields and nearby surface water.

Farmers can also reduce pesticide runoff and leaching by applying pesticides only when needed. They can reduce the need for pesticides by using biological control or integrated pest management (Section 15-5). Nonfarm uses of inorganic fertilizers and pesticides—on golf courses, yards, and public lands, for example—could also be sharply reduced.

Livestock growers can control runoff and infiltration of manure from feedlots and barnyards by managing animal density, by planting buffers, and by not locating feedlots on land that slopes toward nearby surface water. Diverting the runoff into detention basins would allow this nutrient-rich water to be pumped out and applied as fertilizer to cropland or forestland.

The 1990 Farm Bill encourages farmers to voluntarily participate in programs that reduce use of pesticides and fertilizers and control agricultural runoff. However, the financial incentives for these programs

Q: What percentage of U.S. lawns are treated with pesticides?

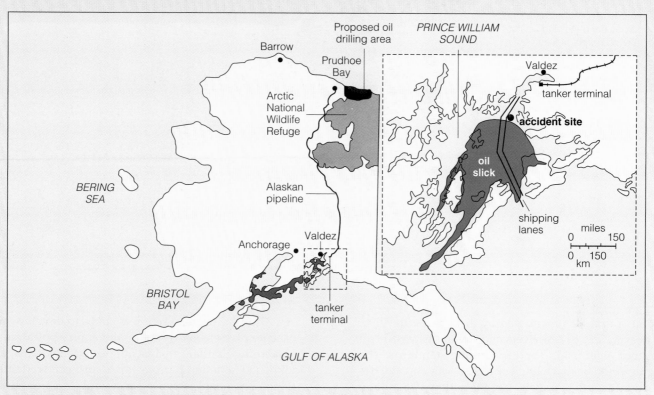

Figure 11-30 Site of the oil spill in Alaska's Prince William Sound from the tanker *Exxon Valdez* on March 24, 1989.

This spill highlighted the importance of pollution prevention. Even with the best technology and a fast response by well-trained people, scientists estimate that no more than 11–15% of the oil from a major spill can be recovered. This spill also shows that we cannot truly fix an ecosystem we have damaged.

are much lower than those of other U.S. Department of Agriculture programs that encourage land exploitation and intensive use of pesticides and fertilizers to promote high yields.

Critical watersheds should also be reforested. Besides reducing water pollution from sediments, reforestation would reduce soil erosion and the severity of flooding (Connections, p. 266); it would also help slow projected global warming (Section 10-2) and loss of the earth's vital biodiversity (Chapters 16 and 17).

Point-Source Pollution: The Legal Approach

In many LDCs and in some MDCs, sewage and waterborne industrial wastes are discharged without treatment into the nearest waterway or into wastewater lagoons. Worldwide, some 1.7 billion people in LDCs don't have sanitary ways to dispose of their sewage. In Latin America, less than 2% of urban sewage is treated. Only 15% of the urban wastewater in China

receives treatment. Treatment facilities in India cover less than a third of the urban population.

In MDCs, most wastes from point sources are purified to varying degrees. The Federal Water Pollution Control Act of 1972 (passed when Congress overrode then President Richard Nixon's veto and renamed the Clean Water Act of 1977 when it was amended) and the 1987 Water Quality Act form the basis of U.S. efforts to control pollution of the country's surface waters. The main goals of the Clean Water Act were to make all U.S. surface waters safe for fishing and swimming by 1983 and to restore and maintain the chemical, physical, and biological integrity of the nation's waters. Progress has been made, but these goals have not been met.

Between 1972 and 1992, U.S. taxpayers and the private sector have spent more than $541 billion on water pollution control—nearly all of it on end-of-pipe controls on municipal and industrial discharges from

A: About 40%

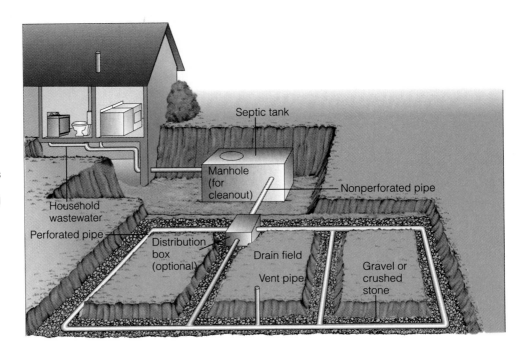

Figure 11-31 Septic tank system used for disposal of domestic sewage and wastewater in rural and suburban areas. This system traps greases and large solids and discharges the remaining wastes over a large drainage field. As these wastes percolate downward, the soil filters out some potential pollutants, and soil bacteria decompose biodegradable materials. To be effective, septic tank systems must be properly installed in soils with adequate drainage, not placed too close together or too near well sites, and pumped out when the settling tank becomes full.

point sources, as mandated by these laws. By 1990, however, 37% of U.S. rivers and streams assessed still did not meet fishing and swimming standards. The figure could be much higher because water quality has been measured in only 36% of the nation's 2.8 million kilometers (1.8 million miles) of rivers.

These acts require the EPA to establish *national effluent standards* and to set up a nationwide system for monitoring water quality. The effluent standards limit the amounts of certain conventional and toxic water pollutants that can be discharged into surface waters from factories, sewage treatment plants, and other point sources. Each point-source discharger must get a permit specifying the amount of each pollutant that a facility can discharge. However, a 1993 study found that about 18% of the USA's 7,000 major industries have found it cheaper to pay repeated fines for violating their permits by dumping wastes into waterways than to eliminate such pollution.

The original 1972 act emphasized pollution prevention. It forbade the discharge of *any* toxic pollutants into U.S. waters by 1985. This requirement has not been enforced. Although toxic inputs have been reduced in many areas, industry still releases an average of 164 thousand metric tons (180 thousand tons) of toxic substances into U.S. surface waters per year.

Another requirement of the 1972 act was that then-clean surface waters in the United States be kept clean. Protecting existing clean waters from pollution was left up to the states. Faced with cleaning up already-polluted waters, money-short state governments have not implemented this requirement. Some clean waters have been protected, but many have become dirtier.

The 1972 act also requires states to develop and execute plans to control nonpoint source pollution—something that largely has not been done. It also established a federal program to protect wetlands, which has been partially successful. Most environmentalists believe that the original goals of the 1972 Clean Water Act (up for renewal in 1994), requiring pollution prevention and protecting existing clean waters from being polluted, are sound. To them the problem is that we need to get more serious about implementing these requirements, which have been on the books for 25 years. They agree that doing this will be costly, but they point out that not doing it will in the long run be much more costly.

Environmentalists call for this law to be strengthened by:

- Increasing funding and the authority to control nonpoint sources of pollution

- Strengthening programs to prevent and control toxic water pollution, including phasing out use of certain toxic discharges (such as many organic chemicals containing chlorine)

- Providing more funding and authority for watershed planning

- Expanding the ability of citizens to bring lawsuits to see that water pollution laws are enforced

Point-Source Pollution: The Technological Approach In rural and suburban areas with suitable soils, sewage from each house is usually discharged into a **septic tank** (Figure 11-31). About 25% of all homes in the United States are served by septic tanks.

Q: What is the best way to control pests?

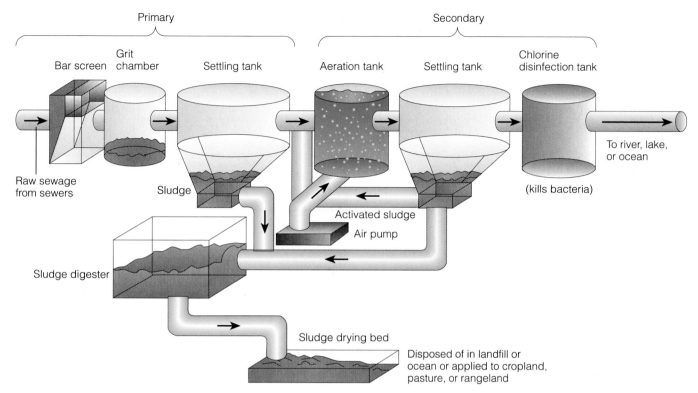

Figure 11-32 Primary and secondary sewage treatment.

A septic tank needs to be cleaned out every three to five years by a reputable contractor so that it won't contribute to groundwater pollution. Many do-it-yourself septic tank cleaners contain toxic chemicals that can kill bacteria important to sewage decomposition in the system.

In urban areas, most waterborne wastes from homes, businesses, factories, and storm runoff flow through a network of sewer pipes to wastewater treatment plants. Some cities have separate lines for stormwater runoff, but in 1,200 U.S. cities the lines for these two systems are combined because it is cheaper. When rains cause combined sewer systems to overflow, they discharge untreated sewage directly into surface waters.

When sewage reaches a treatment plant, it can undergo up to three levels of purification, depending on the type of plant and the degree of purity desired. **Primary sewage treatment** is a mechanical process that uses screens to filter out debris such as sticks, stones, and rags. Then suspended solids settle out as sludge in a settling tank (Figure 11-32).

Secondary sewage treatment is a biological process in which aerobic bacteria are used to remove up to 90% of biodegradable, oxygen-demanding organic wastes (Figure 11-32). Some plants use *trickling filters*, in which aerobic bacteria degrade sewage as it seeps through a bed of crushed stones covered with bacteria and protozoa. Others use an *activated sludge*

process, in which the sewage is pumped into a large tank and mixed for several hours with bacteria-rich sludge and air bubbles to increase degradation by microorganisms. The water then goes to a sedimentation tank, where most of the suspended solids and microorganisms settle out as sludge. The sludge produced from either primary or secondary treatment is broken down in an anaerobic digester and incinerated, dumped in the ocean or a landfill, or applied to land as fertilizer. After secondary treatment, however, wastewater still contains about 3–5% by weight of the oxygen-demanding wastes, 3% of the suspended solids, 50% of the nitrogen (mostly as nitrates), 70% of the phosphorus (mostly as phosphates), and 30% of most toxic metal compounds and synthetic organic chemicals. Virtually none of any long-lived radioactive isotopes or persistent organic substances such as pesticides is removed.

As a result of the Clean Water Act, most U.S. cities have secondary sewage treatment plants. In 1989, however, the EPA found that more than 66% of sewage treatment plants have water quality or public-health problems, and studies by the General Accounting Office have shown that most industries sometimes violate regulations. Also, 500 cities have failed to meet federal standards for sewage treatment plants, and 34 East Coast cities simply screen out large floating objects from their sewage before discharging it into coastal waters.

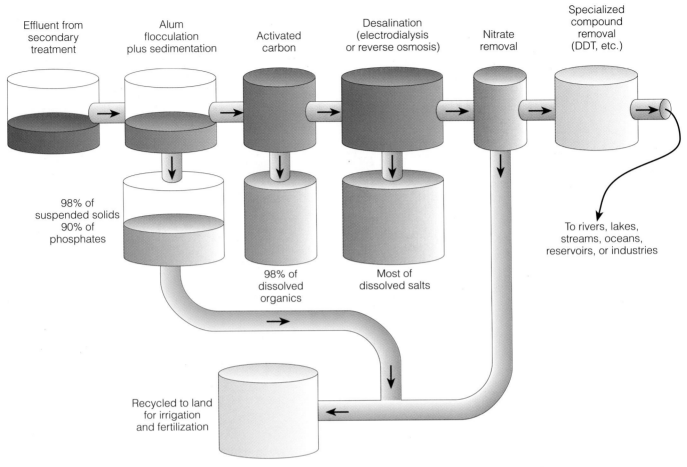

Figure 11-33 Advanced sewage treatment. Often only one or two of these processes are used to remove specific pollutants in a particular area. This expensive method is not widely used.

Labels in figure:
Effluent from secondary treatment
Alum flocculation plus sedimentation
Activated carbon
Desalination (electrodialysis or reverse osmosis)
Nitrate removal
Specialized compound removal (DDT, etc.)
98% of suspended solids 90% of phosphates
98% of dissolved organics
Most of dissolved salts
To rivers, lakes, streams, oceans, reservoirs, or industries
Recycled to land for irrigation and fertilization

Advanced sewage treatment is a series of specialized chemical and physical processes that remove specific pollutants left in the water after primary and secondary treatment (Figure 11-33). Types of advanced treatment vary depending on the specific contaminants to be removed. Without advanced treatment, sewage treatment plant effluents contain enough nitrates and phosphates to contribute to accelerated eutrophication of lakes, slow-moving streams, and coastal waters (Figure 11-26). Advanced treatment is rarely used because such plants typically cost twice as much to build and four times as much to operate as secondary plants. However, despite the cost, advanced treatment is used for more than a third of the population in Finland, the former West Germany, Switzerland, and Sweden, and to a lesser degree in Denmark and Norway.

Before water is discharged after primary, secondary, or advanced (tertiary) treatment, it is bleached to remove water coloration and disinfected to kill disease-carrying bacteria and some, but not all, viruses. The usual method for doing this is *chlorination*. However, chlorine can react with organic materials in water to form small amounts of chlorinated hydrocarbons,

some of which cause cancers in test animals. Disinfectants such as ozone and ultraviolet light are being used in some places, but they cost more than chlorination.

Sewage treatment produces a toxic gooey sludge that must be disposed of or recycled to the land as fertilizer. About 54% by weight of all municipal sludge produced in the United States is applied to farmland, forests, highway medians, and degraded land as fertilizer, and 9% is composted. The rest is dumped in conventional landfills (where it can contaminate groundwater) or incinerated (which can pollute the air with traces of toxic chemicals, and the resulting toxic ash is usually buried in a landfill that will eventually leak).

Sewage sludge—sometimes called "black gold" or "Metrogro"—is rich in humus and improves soil structure more than commercial inorganic fertilizers at a much lower cost. Before it is applied to land, sewage sludge can be heated to kill harmful bacteria, as is done in Switzerland and parts of Germany; it can also be treated to remove toxic metals and organic chemicals before application, but that can be expensive. The best and cheapest solution is to prevent these toxics from reaching sewage treatment plants. However, untreated sludge can be applied to land not used for

Q: What percentage of the USDA's research and education budget is spent on integrated pest management?

crops or livestock. Examples include forests, surface-mined land, golf courses, lawns, cemeteries, and highway medians.

It is encouraging that some communities and individuals are seeking better ways to purify contaminated water by working with nature (Solutions, p. 298).

Protecting Drinking Water   Treatment of water for drinking by urban residents is much like wastewater treatment. Areas that depend on surface water usually store it in a reservoir for several days to improve clarity and taste by allowing the dissolved oxygen content to increase and suspended matter to settle out. The water is then pumped to a purification plant, where it is treated to meet government drinking water standards. Usually, it is run through sand filters, then through activated charcoal, and then it is disinfected. In areas with very pure sources of groundwater, little, if any, treatment is necessary.

Only about 54 countries, most of them in North America and Europe, have safe drinking water standards. The Safe Drinking Water Act of 1974 requires the EPA to establish national drinking water standards, called *maximum contaminant levels*, for any pollutants that may have adverse effects on human health. This act has helped improve drinking water in much of the United States, but there is still a long way to go. At least 700 potential pollutants are found in municipal drinking water supplies. Of the ones that have been tested, 97 cause cancers, 82 cause mutations, 28 are toxic, and 23 promote tumors in test animals. By the end of 1995 the EPA must set maximum contaminant levels for 108 contaminants.

Privately owned wells in suburban and rural areas are not required to meet federal drinking water standards. The biggest reasons are the cost of testing each well regularly (at least $1,000) and ideological opposition to mandatory testing and compliance by some homeowners.

In 1993 a study by the Natural Resources Defense Council found that 43% of all U.S. drinking water systems violated one or more drinking water standards during 1991. They found 250,000 violations affecting more than 120 million people and blamed dirty drinking water for an estimated 900,000 illnesses and 900 deaths per year. In most cases, people were not notified when their drinking water was contaminated, and state and federal regulators acted on just 3,900 (1.6%) of the 250,000 violations. Most of these problems are the result of inadequate monitoring of water quality (especially in small public and private systems), inadequate treatment, or slow enforcement. For only half-a-cent more a day from all water users, high-quality monitoring of drinking water throughout the United States could be provided. Currently, more money is spent on military bands each year than on the EPA's enforcement of the Safe Drinking Water Act.

Contaminated wells and concern about possible contamination of public drinking water supplies have created a boom in the number of U.S. citizens drinking bottled water (at costs about 1,500 times more than that of tap water) or of those adding water purification devices to their home systems. This has created enormous profits for both legitimate companies and con artists in these businesses. Many bottled-water drinkers are getting ripped off. More than one-third of the bottled water comes from the same sources used to supply tap water, which is regulated much more strictly than bottled water.

To be safe, consumers should purchase bottled water only from companies that have their water frequently tested and certified, ideally by EPA-certified laboratories. Before buying bottled water the consumer should determine whether the bottler belongs to the International Bottled Water Association (IBWA) and adheres to its testing requirements. The IBWA requires its members to test for 181 contaminants, and it sends an inspector from the National Sanitation Foundation, a private lab, to bottling plants annually to check all pertinent records and make sure the plant is run cleanly.

Protecting Coastal Waters   The key to protecting oceans is to reduce the flow of pollution from the land worldwide, especially in the coastal zone. Worldwide, some 33% of all pollutants entering the oceans come from air emissions from land-based sources and 44% from streams emptying into the oceans. The most important suggestions for preventing excessive pollution of coastal waters and for cleaning them up include the following:

Prevention

- *Greatly reducing the discharge of toxic pollutants into coastal waters from both industrial facilities and municipal sewage treatment plants.*

- *Greatly reducing all discharges of raw sewage from sewer-line overflows by requiring separate storm and sewer lines in cities.*

- *Banning all ocean dumping of sewage sludge and hazardous dredged materials.*

- *Enacting and enforcing laws and land-use practices that sharply reduce runoff from nonpoint sources in coastal areas.*

- *Protecting sensitive marine areas from all development by designating them as ocean sanctuaries.*

- *Regulating coastal development to minimize its environmental impact, and eliminating subsidies and tax incentives that encourage harmful coastal development.*

A: 1%

An exciting low-tech, low-cost alternative to expensive waste treatment plants is to create an artificial wetland, as the residents of Arcata, California, did. In this coastal town of 15,000, some 63 hectares (155 acres) of wetlands have been created on land that was once a dump between the town and adjacent Humboldt Bay (Figure 11-34). The project was completed in 1974 for $3 million less than the estimated cost of a conventional treatment plant.

Here's how it works: First, sewage is held in sedimentation tanks, where the solids settle out. This resulting sludge is removed and processed for use as fertilizer. The liquid is pumped into oxidation ponds, where the wastes are broken down by bacteria. After a month or so, the water is released into the artificial marshes, where it is further filtered and cleansed by plants and bacteria. Although the water is clean enough to discharge directly into the bay, state law requires that it first be chlorinated. So the town chlorinates the water—and then dechlorinates it—before sending it into the bay, where oyster beds thrive. Some water from the marshes is piped into the city's salmon hatchery.

The marshes and lagoons are an Audubon Society bird sanctuary and provide habitats for thousands of seabirds and marine animals. The treatment center is a city park and attracts many tourists. The town even celebrates its natural sewage treatment system with an annual "Flush with Pride" festival. Over 150 cities and towns in the United States now use natural and artificial wetlands for treating sewage.

Can you use natural processes for treating wastewater if there isn't a wetland available, or enough land on which to build one? According to ecologist John Todd, you can: Set up a greenhouse lagoon and use sunshine the way nature does (Figure 11-35). The process begins when sewage flows into a greenhouse containing rows of large aquarium tanks covered with plants such as water hyacinths, cattails, and bul- rushes. In these tanks algae and microorganisms decompose wastes into nutrients absorbed by the plants. The decomposition is speeded up by sunlight streaming into the greenhouse, and toxic metals are absorbed into the tissues of trees that will be transplanted outside. Then the water passes through an artificial marsh of sand, gravel, and bulrush plants that filters out algae and organic waste. Next the water flows into aquarium tanks, where snails and zooplankton consume microorganisms and are themselves consumed by crayfish, tilapia, and other fish that can be eaten or sold as bait. After 10 days the now-clear water flows into a second artificial marsh for final filtering and cleansing. When working properly, such solar-aquatic treatment systems have produced water fit for drinking.

These natural alternatives to building expensive treatment plants may not solve the waste problems of large cities. But they can help, and they are an attractive alternative for small towns, the edges of urban areas, and rural areas.

- *Prohibiting oil drilling in ecologically sensitive offshore and nearshore areas.*

- *Collecting used oils and greases from service stations and other sources, and reprocessing them for reuse.* Currently more than 90% of the used oil collected for "recycling" is burned as fuel, releasing lead, chromium, arsenic, and other toxic pollutants into the air.

- *Requiring all existing oil tankers to have double hulls, double bottoms, or other oil-spill prevention measures by 1998.* A recent study by the National Academy of Sciences and the Coast Guard concluded that double hulls are the most cost-effective way to help prevent tanker oil spills.

- *Greatly increasing the financial liability of oil companies for cleaning up oil spills, thus encouraging pollution prevention.*

- *Routing oil tankers as far as possible from sensitive coastal areas.*

- *Having Coast Guard or other vessels guide tankers out of all harbors and enclosed sounds and bays.*

- *Banning the rinsing of empty oil tanker holds and the dumping of oily ballast water into the sea.*

- *Banning discharge of garbage from vessels into the sea, and levying large fines on violators.*

- *Adopting a nationwide tracking program to ensure that medical waste is safely disposed of.*

Cleanup

- *Greatly improving oil-spill cleanup capabilities.* However, according to a 1990 report by the Office of Technology Assessment, there is little chance that large spills can be effectively contained or cleaned up.

- *Upgrading all coastal sewage treatment plants to at least secondary treatment, or developing alternative methods for sewage treatment (Solutions, above).*

Q: What percentage of Earth's land area is covered by forests?

Figure 11-34 Marsh sewage treatment area in Arcata, California.

Figure 11-35 At the Providence, Rhode Island, Solar Sewage Plant, biologist John Todd is demonstrating how ecological waste engineering in a greenhouse can be used to purify wastewater. Todd and others are carrying out research to perfect such solar aquatic systems based on working with nature.

Protecting Groundwater Pumping polluted groundwater to the surface, cleaning it up, and returning it to the aquifer is usually prohibitively expensive—$5–10 million or more for a single aquifer. Recent attempts to pump and treat contaminated aquifers show that it may take decades, even hundreds of years, of pumping before all of the contamination is forced to the surface. Thus preventing contamination is the only effective way to protect groundwater resources. Water pollution experts suggest that this could be accomplished by the following means:

- *Banning virtually all disposal of hazardous wastes in sanitary landfills and deep injection wells* (Figure 11-27)

- *Monitoring aquifers near existing sanitary and hazardous-waste landfills, underground tanks, and other potential sources of groundwater contamination* (Figure 11-27)

- *Controlling application of pesticides and fertilizers by farmers and homeowners much more strictly*

- *Requiring that people who use private wells for drinking water have their water tested once a year*

- *Establishing pollution standards for groundwater*

- *Emphasizing above-ground storage of hazardous liquids—an in-sight-in-mind approach that allows rapid detection and collection of leaking materials*

Sustainable Water Use Sustainable use of Earth's water resources involves developing an integrated approach to managing water resources and water pollution throughout each watershed. It also means reducing or eliminating water subsidies so that its market price more closely reflects water's true cost.

Once we stop overloading aquatic systems with pollutants, they recover amazingly fast. Doing this requires that we shift from pollution cleanup to pollution prevention. To make such a shift we must truly

A: About 34% (6% with tropical forests)

What You Can Do to Reduce Water Pollution

INDIVIDUALS MATTER

- *Use manure or compost instead of commercial inorganic fertilizers to fertilize garden and yard plants.*

- *Use biological methods or integrated pest management to control garden, yard, and household pests (Section 15-5).*

- *Use low-phosphate, phosphate-free, or biodegradable dishwashing liquid, laundry detergent, and shampoo.*

- *Don't use water fresheners in toilets.*

- *Use less harmful substances instead of commercial chemicals for most household cleaners (Table 13-2).*

- *Don't pour pesticides, paints, solvents, oil, antifreeze, or other products containing harmful chemicals down the drain or on the ground. Contact your local health department about disposal.*

- *If you get water from a private well or suspect that municipal water is contaminated, have it tested by an EPA-certified laboratory for lead, nitrates, trihalomethanes, radon, volatile organic compounds, and pesticides.*

- *If you have a septic tank, have it cleaned out every three to five years by a reputable contractor so that it won't contribute to groundwater pollution.*

- *Get to know your local bodies of water and form community watchdog groups to help monitor, protect, and restore them.*

- *Support efforts to clean up riverfronts and harbors.*

accept that the environment—air, water, soil, life—is an interconnected whole. Without an integrated approach to all forms of pollution, we will continue to shift environmental problems from one part of the environment to another.

Individuals can do their part by reducing water waste (Individuals Matter, p. 277) and preventing water pollution (Individuals Matter, above).

It is not until the well runs dry, that we know the worth of water.

BENJAMIN FRANKLIN

Critical Thinking

1. How do human activities increase the harmful effects of prolonged drought? How can these effects be reduced?

2. How do human activities contribute to flooding and flood damage? How can these effects be reduced?

3. Explain how dams and reservoirs can cause more flood damage than they prevent. Should all proposed large dam and reservoir projects be scrapped? Explain.

4. Should prices of water for all uses in the United States be raised sharply to encourage water conservation? Explain. What effects might this have on the economy, on you, on the poor, and on the environment?

5. List 10 major ways to conserve water on a personal level. Which, if any, of these practices do you now use or intend to use?

6. Why is dilution not always the solution to water pollution? Give examples and conditions for which this solution is, and is not, applicable.

7. How can a stream cleanse itself of oxygen-demanding wastes? Under what conditions will this natural cleansing system fail?

8. Should all dumping of wastes in the ocean be banned? Explain. If so, where would you put the wastes instead? What exceptions would you permit, and why?

9. Should the injection of hazardous wastes into deep-underground wells be banned? Explain. What would you do with these wastes?

*10. In your community:
   a. What are the major sources of the water supply?
   b. How is water use divided among agricultural, industrial, power-plant cooling, and public uses? Who are the biggest consumers of water?
   c. What has happened to water prices during the past 20 years? Are they too low to encourage water conservation and reuse?
   d. What water supply problems are projected?
   e. How is water being wasted?

*11. In your community:
   a. What are the principal nonpoint sources of contamination of surface water and groundwater?
   b. What is the source of drinking water?
   c. How is drinking water treated?
   d. What contaminants are tested for?
   e. Has drinking water been analyzed recently for the presence of synthetic organic chemicals, especially chlorinated hydrocarbons? If so, were any found, and are they being removed?
   f. How many times during each of the past five years have levels of tested contaminants violated federal standards? Was the public notified about the violations?

*12. Make a concept map of the key ideas in this chapter using the section heads and subheads and the key terms (shown in boldface type in the chapter). See the inside front cover and Appendix 4 for information on concept maps.

12 Minerals and Soil

The Great Terrain Robbery

Want to get rich at taxpayers' expense? You can if you know how to make use of a little known mining law passed in 1872 to encourage mining of gold, silver, lead, copper, uranium, and other hard-rock minerals on public lands.

Under this 1872 law, any person or corporation can assume legal ownership of any public land not classified as wilderness or park simply by "patenting" it. This involves declaring their belief that the land contains valuable hard-rock minerals, spending $500 to improve the land for mineral development, filing a claim, and then paying the federal government $6–$12 per hectare ($2.50–$5.00 an acre) for the land. So far public lands with a total area about the size of Connecticut have been transferred to private interests at such bargain basement prices.

In 1993, for example, the Manville Corporation patented about 809 hectares (2,000 acres) of federal land in Montana for $10,000 that contains an estimated $32 billion worth of platinum and palladium. In the same year, Secretary of Interior Bruce Babbitt protested that this antiquated mining law may force him to sell federal land containing an estimated $10 billion worth of gold to a foreign corporation for only $10,000.

Mining companies operating under this law—almost half of them owned mostly by foreign corporations—annually remove mineral resources worth $4–$6 billion on land they have bought at absurdly low prices. They pay no royalties, which should bring in $200–$500 million per year to the U.S. Treasury.

The 1872 law does not even require that patented property be mined. Land speculators have often purchased such properties at 1872 prices and then sold them for thousands of times what they paid. In 1986, for example, a mining company paid the government $42,500 to buy (patent) 7,000 hectares (17,000 acres) of oil shale land in Colorado. A few weeks later, the patent holders sold the land to major oil companies for $37 million. Other owners have developed sites as casinos, homes, ski resorts, golf courses, and vacation-home developments. According to Senator Dale Bumpers, "The 1872 mining law is a license to steal and the biggest scam in America."

There is also no provision in the 1872 law requiring reclamation of damaged land. Miners can ravage the land and dump toxic chemicals with little likelihood of having to reclaim the disrupted land and water (Figure 12-1). At least 50 hazardous-waste sites on the EPA's Superfund list of the nation's worst dumps are from hard-rock mining. Estimated cleanup costs of all land damaged by hard-rock mining on existing or former public lands—to be paid by the government—range between $11 billion and $50 billion, depending on whether groundwater and toxic waste impacts are included. It's common for a company to mine a site, abandon it, file for bankruptcy, and leave the public with the cleanup bill. Environmentalists have been trying, with little success, to have this law revised to protect taxpayers and the environment.

Figure 12-1 Contamination of Bear Trap Creek in Montana from gold mining. Gold mining can contaminate water with highly toxic cyanide or mercury used to extract gold from its ore. Sulfur in gold ore is also converted by air and water to sulfuric acid, which releases toxic metals such as cadmium and copper into streams and groundwater. (Courtesy of Bryan Peterson)

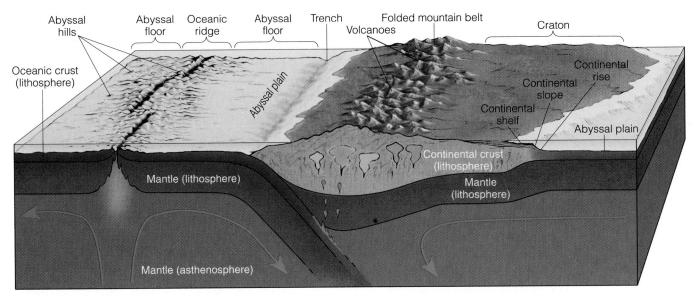

Abyssal hills · Abyssal floor · Oceanic ridge · Abyssal floor · Trench · Volcanoes · Folded mountain belt · Craton · Oceanic crust (lithosphere) · Abyssal plain · Continental rise · Continental slope · Continental shelf · Abyssal plain · Continental crust (lithosphere) · Mantle (lithosphere) · Mantle (lithosphere) · Mantle (asthenosphere)

Figure 12-2 Major features of the earth's crust and upper mantle. The lithosphere is the rigid, brittle outer zone, composed of the crust and outermost mantle. The asthenosphere is a zone in the mantle that can be deformed by heat and pressure like most forms of plastic.

Below that thin layer comprising the delicate organism known as the soil is a planet as lifeless as the moon.

G. Y. JACKS AND R. O. WHYTE

In this chapter we will consider the following questions:

- What are the major geologic processes occurring on and in the earth?

- How does the rock cycle recycle earth materials and concentrate resources?

- What are the environmental impacts of extracting and using mineral resources?

- How fast are nonfuel mineral supplies being used up?

- How can we increase supplies of key minerals?

- What is soil, and what types are best for crops?

- Why should we worry about soil erosion?

- How can we reduce erosion of topsoil?

12-1 GEOLOGIC PROCESSES*

Earth's Structure Over billions of years, Earth's interior has separated into three major, concentric zones, which geologists identify as the core, the mantle, and the crust (Figure 4-2).

The inner zone or **core** is composed mostly of iron and perhaps some nickel. The core has a solid inner part, surrounded by a liquid core of molten material.

---

*The primary author of Sections 12-1 and 12-2 is **Kenneth J. Van Dellen**, professor of geology and environmental science, Macomb Community College, with assistance from G. Tyler Miller, Jr.

Earth's core is surrounded by a thick, solid zone called the **mantle**. This largest zone of Earth's interior is rich in the elements iron (its major constituent), silicon, oxygen, and magnesium. Most of the mantle is solid rock, but under its rigid outermost part there is a zone of very hot, partly melted rock that flows like soft plastic. This plastic region of the mantle is called the *asthenosphere*.

The outer and thinnest zone of the earth is called the **crust**. It consists of the *continental crust*, which underlies the continents (including the continental shelves extending into the oceans), and the *oceanic crust*, which underlies the ocean basins (Figure 12-2).

Although we tend to think of Earth's crust, mantle, and core as fairly static and unchanging, they are constantly changing by geologic processes taking place within the earth and on the earth's surface, most over thousands to millions of years.

Internal Processes: Plate Tectonics Geologic changes originating from within the earth are called *internal processes*. Generally they build up the planet's surface. Heat from the earth's interior provides the energy for these processes.

A map of Earth's earthquakes and volcanoes shows that most of these phenomena occur along certain lines or belts on the earth's surface (Figure 12-3a). The areas of the earth outlined by these major belts are called **plates** (Figure 12-3b). They are about 100 kilometers (60

Figure 12-3 (right) Earthquakes and volcano sites are distributed mostly in bands along the planet's surface **(a)**. These bands correspond to various types of lithospheric plate boundaries **(b)**, shown in Figure 12-4.

Q: At current loss rates, when will most remaining tropical forests be gone?

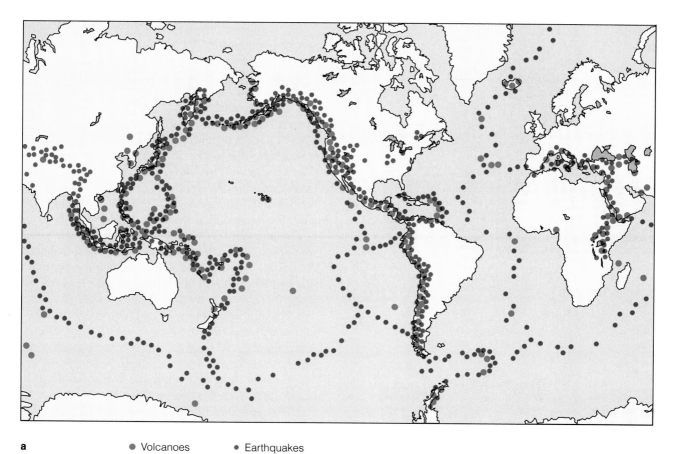

a   • Volcanoes   • Earthquakes

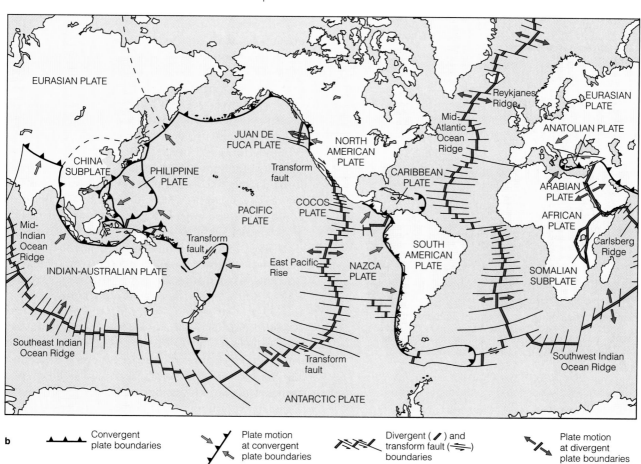

b

| ▲▲▲ Convergent plate boundaries | ↗ Plate motion at convergent plate boundaries | Divergent (╱) and transform fault (═) boundaries | ↕ Plate motion at divergent plate boundaries |

EURASIAN PLATE · **CHINA SUBPLATE** · **PHILIPPINE PLATE** · **JUAN DE FUCA PLATE** · Transform fault · **NORTH AMERICAN PLATE** · Reykjanes Ridge · Mid-Atlantic Ocean Ridge · **EURASIAN PLATE** · **ANATOLIAN PLATE** · **CARIBBEAN PLATE** · **ARABIAN PLATE** · **AFRICAN PLATE** · **PACIFIC PLATE** · **COCOS PLATE** · Mid-Indian Ocean Ridge · **INDIAN-AUSTRALIAN PLATE** · Transform fault · East Pacific Rise · **NAZCA PLATE** · **SOUTH AMERICAN PLATE** · Carlsberg Ridge · **SOMALIAN SUBPLATE** · Southeast Indian Ocean Ridge · Transform fault · Southwest Indian Ocean Ridge · **ANTARCTIC PLATE**

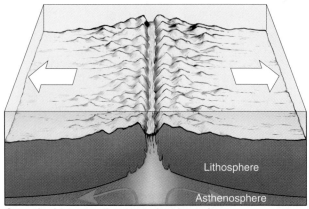

Oceanic ridge at a divergent plate boundary

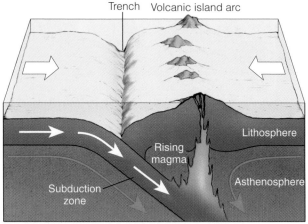

Trench and volcanic island arc at a convergent plate boundary

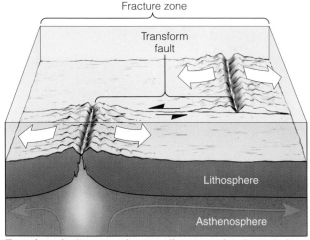

Transform fault connecting two divergent plate boundaries

Figure 12-4 Types of boundaries between Earth's lithospheric plates. All three boundaries occur both in oceans and on continents.

miles) thick and are composed of the crust and the rigid, outermost part of the mantle above the asthenosphere. This combination is called the **lithosphere**. These plates move constantly, carried by the slowly flowing asthenosphere like large pieces of ice floating on the sur-

face of a lake during the spring breakup. Some plates move faster than others, but a typical speed is about the rate at which fingernails grow.

The theory explaining the movements of the plates and the processes that occur at their boundaries is called **plate tectonics.** Throughout Earth's history continents have split and joined as plates have drifted thousands of kilometers back and forth across the planet's surface. For example, geologic evidence indicates that about 225 million years ago the continents we now call Europe, Asia, North America, and South America were combined as one continent. This gigantic continental mass, named *Pangaea*, broke into pieces that gradually drifted apart as a result of movements of Earth's plates, and the continents eventually reached their present positions. In the process the Atlantic Ocean was formed.

The movement of the lithospheric plates is important to us for several reasons. Plate motion produces mountains (including volcanoes), the oceanic ridge system, trenches, and other features of Earth's surface. Plate movements and interactions also concentrate many of the minerals we extract and use.

The theory of plate tectonics also helps explain how certain patterns of biological evolution occurred. By reconstructing how continents have drifted around for millions of years, we can trace how life-forms migrated from one area to another when continents that are now far apart were joined together.

Lithospheric plates have three types of boundaries: divergent, convergent, and transform fault (Figure 12-4). At a **divergent plate boundary** the plates move apart in opposite directions (←|→). They occur mostly where hot and partially molten rock material (magma) pushes up between two plates, cools, and solidifies to form new ocean floor that pushes the plates apart. Mountain chains, called *oceanic ridges*, are formed along the ocean floor where the two plates diverge. Divergent plate boundaries can also occur on continents when great blocks of crust are pushed up and moved apart, forming a "rift" valley between them into which water will flow if there is a link to the ocean.

Where tops of adjacent convection cells of magma flow toward each other, the plates are pushed together (→|←), producing a **convergent plate boundary** (Figure 12-4). At most convergent plate boundaries oceanic lithosphere is carried downward (subducted) under the island arc or the continent at a **subduction zone**. An oceanic trench ordinarily forms at the boundary between the two converging plates; the deepest trenches descend lower below sea level than Mount Everest rises above it. As it descends, some of the lithospheric material melts, producing volcanoes, and the rest is incorporated into the mantle. Stresses in the plate undergoing subduction cause earthquakes at convergent plate boundaries (Figure 12-3a).

Q: Worldwide, how many trees are planted for each ten cut?

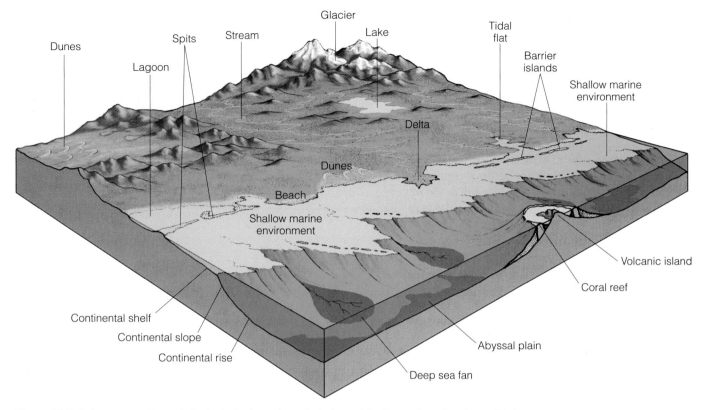

Figure 12-5 Solar energy—through the hydrologic cycle and wind—and the force of gravity, along with the activities of organisms such as reef-building corals (Figure 5-20), have produced a variety of landforms and sedimentary environments.

The third type of plate boundary, called a **transform fault**, occurs where plates move in opposite but parallel directions along a fracture (fault) in the lithosphere (Figure 12-4). Like the other types of plate boundaries, most transform faults are on the ocean floor. California's San Andreas Fault (Figure 8-3) is one of the exceptions.

External Processes: Erosion and Weathering

Geological changes based directly or indirectly on energy from the sun and on gravity, instead of on heat in the earth's interior, are called *external processes*. Whereas internal processes generally build up the earth's surface, external processes tend to lower it.

Erosion is the process or group of processes by which loosened material (as well as material not yet separated) is dissolved, loosened, or worn away from one part of the earth's surface and deposited in other places. Streams are the most important agent of erosion, operating everywhere on Earth except in the polar regions. They produce ordinary valleys and canyons, and the resulting sediments may form deltas where streams flow into lakes and oceans (Figures 5-31 and 12-5).

Loosened material that can be eroded is usually produced by **weathering**. Weathering can occur as a result of mechanical processes, chemical processes, or both. In *mechanical weathering* a large rock mass is broken into smaller fragments of the original material, similar to the results you would get using a hammer to break a rock into small fragments. The most important agent of mechanical weathering is *frost wedging*, in which water collects in pores and cracks of rock, expands upon freezing, and splits off pieces of the rock.

In *chemical weathering* a mass of rock is decomposed by one or more chemical reactions (usually with oxygen, carbon dioxide, or water), resulting in products that are chemically different from the original material. The products usually include both solid and dissolved components.

Disintegration of rock by mechanical weathering accelerates chemical weathering by increasing the surface area that can be attacked by chemical weathering agents. This is similar to the way granulated sugar dissolves much faster than a large chunk of sugar. Chemical weathering is also aided by higher temperatures and precipitation. It occurs most rapidly in the tropics and next most rapidly in temperate climates. Weathering is responsible for the development of soil, as discussed in Section 12-6. Human activities, particularly those that destroy the vegetation, accelerate erosion, as discussed in Section 12-7.

12-2 MINERALS, ROCKS, AND THE ROCK CYCLE

Minerals, Rocks, and Ores Earth's crust, which is still forming in various places, is composed of minerals and rocks. It is the source of virtually all the nonrenewable resources we use—fossil fuels, metallic minerals, and nonmetallic minerals (Figure 1-9). It is also the source of soil and of the elements that make up our bodies and those of other living organisms.

A **mineral** is an element or an inorganic compound that occurs naturally and is solid. Some minerals consist of a single element, such as gold, silver, diamond (carbon), and sulfur. However, most of the over 2,000 identified minerals occur as inorganic compounds formed by various combinations of the eight elements that make up 98.5% by weight of Earth's crust (Figure 12-6).

Rock is any material that makes up a large, natural, continuous part of Earth's crust. Some kinds of rock, such as limestone (calcium carbonate, or $CaCO_3$) and quartzite (silicon dioxide, or SiO_2), contain only one mineral, but most rocks consist of two or more minerals.

Some metal resources (Figure 1-9) seem more abundant than they really are in the earth's crust because slow-acting, infrequent, or localized processes have selectively concentrated them into ores. An **ore** is a metal-yielding material that can be economically extracted at a given time. Copper, for example, makes up 0.0058% by weight of the earth's crust, but, to be profitable, copper ore must contain at least 0.5% copper; thus the concentration of copper in ore is at least 86 times (0.5/0.0058) its average crustal abundance. To be profitable, the gold in gold ore must be concentrated 1,600 times its crustal average, and mercury an astonishing 100,000 times.

This limited and uneven concentration of nonrenewable metal resources raises serious questions about the wisdom of extracting concentrated deposits and scattering them all over the countryside in landfills, junkyards, or refuse dumps. Instead, we should think of discarded items made of these nonrenewable materials as potential resources to be recycled and reused to reduce energy use, extraction of virgin minerals, pollution, and waste.

Rock Types and the Rock Cycle Geologic processes constantly redistribute the chemical elements within and at the surface of the earth. Based on the way it forms, rock is placed in three broad classes: igneous, sedimentary, or metamorphic.

Igneous rock can form below the earth's surface—as well as on it—when magma wells up from the upper mantle or deep crust, cools, and hardens into

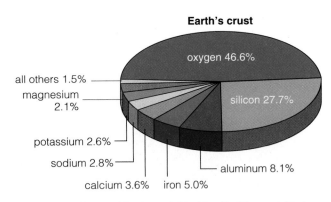

Figure 12-6 Composition by weight of the Earth's crust. Various combinations of only eight elements make up the bulk of most minerals.

rock. Although often covered by sedimentary rocks or soil, igneous rocks form the bulk of the earth's crust. They also are the main source of many nonfuel mineral resources. Granite and its relatives are used for monuments and as decorative stone in buildings, basalt as crushed stone where gravel is scarce, and volcanic rocks in landscaping. Many of the popular gemstones, such as diamond, tourmaline, garnet, ruby, and sapphire, are part of igneous rocks.

Sedimentary rock forms from sediment. Most such rocks are formed when preexisting rocks are weathered and eroded into small pieces, transported from their sources, and deposited in a body of surface water. As these deposited layers become buried and compacted, the resulting pressure causes their particles to bond together to form sedimentary rocks such as sandstone and shale.

Some sedimentary rocks, such as dolomite and limestone, are formed from the compacted shells, skeletons, and other remains of dead organisms. Lignite and bituminous coal are sedimentary rocks derived from plant remains (Figure 12-7).

Besides making up much of the planet's scenic landscape, some sedimentary rocks are important resources. Limestone, for example, is used as crushed stone, as building stone, as flux in blast furnaces for smelting iron ore, and with shale for making Portland cement.

Metamorphic rock is produced when a preexisting rock is subjected to high temperatures (which may cause it to melt partially), high pressures, chemically active fluids, or a combination of those agents. Anthracite (Figure 12-7), slate (used for roofs and floors of buildings), and marble (formed from limestone and used in buildings, sculptures, and monuments) are economically important metamorphic rocks. Talc, asbestos, graphite, titanium, and some gems are also found in metamorphic rocks.

Q: What percentage of the earth's species live in tropical forests?

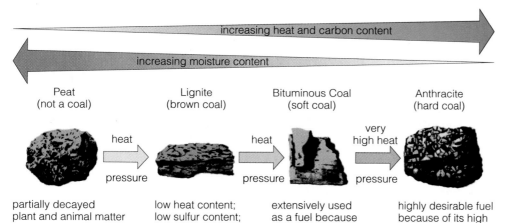

increasing heat and carbon content

increasing moisture content

Peat
(not a coal)

Lignite
(brown coal)

Bituminous Coal
(soft coal)

Anthracite
(hard coal)

heat

pressure

heat

pressure

very
high heat

pressure

partially decayed
plant and animal matter
in swamps and bogs;
low heat content

low heat content;
low sulfur content;
limited supplies in
most areas

extensively used
as a fuel because
of its high heat content
and large supplies;
normally has a
high sulfur content

highly desirable fuel
because of its high
heat content and
low sulfur content;
supplies are limited
in most areas

Figure 12-7 Stages in the formation of coal over millions of years. Peat is a soil material made of moist, partially decomposed organic matter and is not classified as coal. Lignite and bituminous coal are sedimentary rocks, and anthracite is a metamorphic rock.

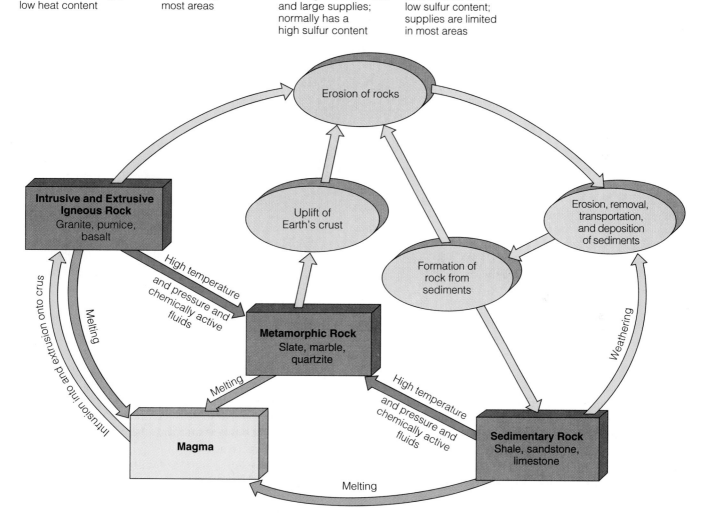

Figure 12-8 The rock cycle, the slowest of Earth's cyclic processes. Earth's materials are recycled over millions of years by three processes: melting, erosion, and metamorphism. These processes produce igneous, sedimentary, and metamorphic rocks, respectively. Rock of any of the three classes can be converted to rock of either of the other two classes or can even be recycled within its own class.

Rocks are constantly being exposed to various physical and chemical conditions that over time can change them. The interaction of processes that change rocks from one type to another is called the **rock cycle** (Figure 12-8). Recycling material over millions of years, this slowest of Earth's cyclic processes is responsible for concentrating mineral resources on which humans depend (Figure 1-9). Most of these resources are nonrenewable on a human time scale because of the slowness of the rock cycle.

Figure 12-9 This open-pit copper mine in Bingham, Utah, the largest human-made hole in the world, is 4.0 kilometers (2.5 miles) in diameter and 0.8 kilometer (0.5 mile) deep. The amount of material removed from this mine is seven times the amount moved to build the Panama Canal.

Don Green/Kennecott Copper Corporation (now owned by British Petroleum)

12-3 ENVIRONMENTAL IMPACTS OF EXTRACTING AND USING MINERAL RESOURCES

Locating and Extracting Crustal Resources

Mining companies use several methods to find promising mineral deposits. Geological information about plate tectonics (Figure 12-3) and mineral formation suggests areas for closer study. Aerial photos and satellite images sometimes reveal rock formations associated with certain minerals. Other instruments on planes and satellites can detect mineral deposits by effects on Earth's magnetic or gravitational fields.

After profitable deposits of minerals are located, deep deposits are removed by **subsurface mining** and shallow deposits by **surface mining**. Subsurface mining disturbs less than one-tenth as much land as surface mining and usually produces less waste material. However, it leaves much of the resource in the ground and is more dangerous and expensive than surface mining. For example, roofs and walls of underground mines collapse, trapping and killing miners; explosions of dust and natural gas injure or kill them; and prolonged inhalation of mining dust causes lung diseases.

In surface mining, mechanized equipment strips away the **overburden** of soil and rock, and usually discards it as a waste material called **spoils**. Surface mining extracts about 90% by weight of the mineral and rock resources and more than 60% by weight of the coal in the United States.

The type of surface mining used depends on the resource being sought and on the local topography. In **open-pit mining**, machines dig holes and remove ores such as iron and copper (Figure 12-9). This method is also used for sand and gravel, and for building stone such as limestone, sandstone, slate, granite, and marble. Another form of surface mining is **dredging**, in which chain buckets and draglines scrape up underwater mineral deposits.

Strip mining is surface mining in which bulldozers, power shovels, or stripping wheels remove the overburden in strips. It is used mostly for removing coal and some phosphate rock.

Surface mining is used to extract almost two-thirds of the coal used in the United States. Most coal that is surface-mined is removed by area strip mining or contour strip mining, depending on the terrain.

Area strip mining is used where the terrain is fairly flat. An earthmover strips away the overburden, and then a power shovel digs a cut to remove a mineral deposit, such as coal. After the mineral is removed, the trench is filled with overburden, and a new cut is made

Q: What percentage of tropical forest plants have been studied for their possible use as human resources?

Figure 12-10 Effects of area strip mining of coal near Mulla, Colorado. Restoration of newly strip-mined areas is now required in the United States, but many previously mined areas have not been restored.

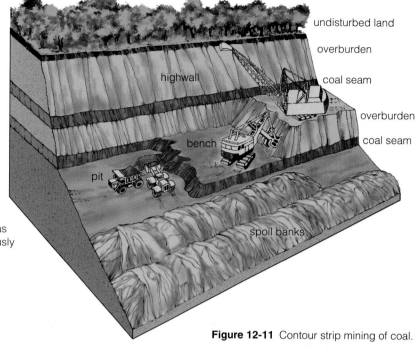

Figure 12-11 Contour strip mining of coal.

Figure 12-12 Grantsville, Maryland. With the land returned to its original contour and grass planted to hold the soil in place, it is hard to tell that this was once a site of surface coal mining. However, about three-fourths of the coal that can be surface-mined in the United States is in the West, in arid and semiarid regions, where the climate and the soil usually prevent full restoration.

parallel to the previous one. This process is repeated for the entire deposit. If the land is not restored, this type of mining leaves a wavy series of highly erodible hills of rubble called *spoil banks* (Figure 12-10).

Contour strip mining is used in hilly or mountainous terrain. A power shovel cuts a series of terraces into the side of a hill (Figure 12-11). An earthmover removes the overburden and a power shovel extracts the coal, with the overburden from each new terrace dumped onto the one below. Unless the land is restored, a wall of dirt is left in front of a highly erodible bank of soil and rock called a *highwall*. Sometimes giant augers are used to drill horizontally into a hillside to extract underground coal.

In the United States contour strip mining for coal is used mostly in the mountainous Appalachian region. If the land is not restored (Figure 12-12), this type of surface mining has a devastating impact on the land.

Subsurface mining is used to remove coal too deep to be extracted by surface mining. Miners dig a deep vertical shaft, blast subsurface tunnels and rooms to get to the deposit, and haul the coal or ore to the surface. In the *room-and-pillar method* as much as half of the coal is left in place as pillars to prevent the mine from collapsing. In the *longwall method* a narrow tunnel is dug and then supported by movable metal pillars. After a cutting machine has removed the coal or ore from part of the mineral seam, the roof supports are moved forward, allowing the earth behind the supports to collapse. No tunnels are left behind after the mining operation has been completed.

Environmental Impacts The mining, processing, and use of crustal resources require enormous amounts of energy and often cause land disturbance, erosion, and air and water pollution (Figure 12-13).

Mining can affect the environment in several ways. Most noticeable are scarring and disruption of the land surface (Figures 12-9, 12-10, and 12-11) and the ugliness of spoil heaps and tailings. Land above underground mines collapses or subsides, causing roads to buckle, houses to tilt, railroad tracks to bend, sewer

A: About 1%

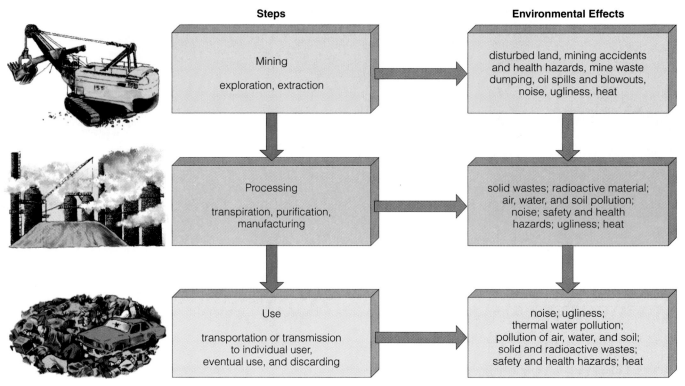

Steps | **Environmental Effects**

Mining

exploration, extraction

disturbed land, mining accidents and health hazards, mine waste dumping, oil spills and blowouts, noise, ugliness, heat

Processing

transpiration, purification, manufacturing

solid wastes; radioactive material; air, water, and soil pollution; noise; safety and health hazards; ugliness; heat

Use

transportation or transmission to individual user, eventual use, and discarding

noise; ugliness; thermal water pollution; pollution of air, water, and soil; solid and radioactive wastes; safety and health hazards; heat

Figure 12-13 Some harmful environmental effects of resource extraction, processing, and use. The energy used to carry out each step causes additional pollution and environmental degradation. This harm could be minimized by requiring that the full costs of the pollution and environmental degradation caused by mining, processing, and manufacturing companies be included in the price of their products. Many of these "external" costs are now passed on to society as a whole in the form of poorer health, increased health and insurance costs, and increased taxes to deal with pollution and environmental degradation (Section 7-2).

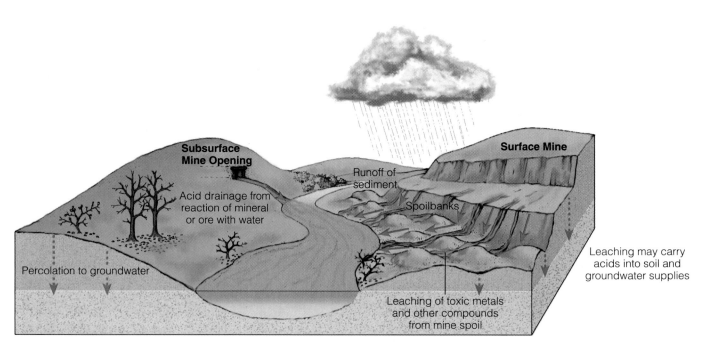

Figure 12-14 Degradation and pollution of a stream and groundwater by runoff of acids—called *acid mine drainage*—and toxic chemicals from surface and subsurface mining operations. These substances can kill fish and other aquatic life. In the United States acid mine drainage has damaged over 26,000 kilometers (16,000 miles) of streams, mostly in Appalachia and in the West.

Q: How much of the world's area of tropical forests is managed sustainably?

SPOTLIGHT

The Environment and the New Gold Rush

For thousands of years, gold has been a source of wealth and power because it is rare, can be hammered or drawn into almost endless shapes, does not corrode or tarnish, and resists chemical attack.

Only gold's high market price makes extracting it pay off. The metal occurs in such minute quantities in the earth's crust (with gold ore averaging only about 0.00033% gold) that miners must extract and process massive quantities of soil and rock.

Mostly because of higher gold prices and cheaper extraction technologies, the amount of gold mined between 1980 and 1992 almost doubled. This new gold rush is great news for owners and investors of gold mining companies, but it is bad news for the environment. Extracting the gold needed just to make a typical pair of wedding bands creates enough waste to fill a large car. In the United States alone, the waste produced by gold mining operations each year would fill a bumper-to-bumper convoy of

dump trucks that would wrap around the equator.

Solid waste is only the beginning. Gold mining can also severely pollute surface water and groundwater. In Australia and North America, a new mining technology—called *cyanide heap leaching*—has been developed. It is cheap enough to allow mining companies to level entire mountains containing very low-grade gold ore. To extract the gold, miners spray a cyanide solution (which reacts with gold) onto huge open-air piles of crushed ore. They then collect the solution in ponds, recirculate it a number of times, and extract gold from it.

Unfortunately, cyanide is extremely toxic and can be harmful or lethal to people, plants, and wildlife—especially to birds and mammals drawn to cyanide collection ponds as a source of water. Cyanide collection ponds can also leak or overflow, posing threats to underground drinking water supplies and to wildlife (especially fish) in lakes and streams (Figure 12-1). Although cyanide in soils and surface waters normally breaks down

fairly rapidly, it can contaminate groundwater for long periods.

The gold rush of the 1980s has also caused millions of miners in various Latin American (especially Brazil, Figure 6-16), Asian, and African LDCs—many of them the landless poor—to stream into tropical forests and other areas in search of gold.

These small-scale miners use destructive mining techniques such as digging large pits by hand, river dredging, and hydraulic mining (a technique, outlawed in the United States since 1884, in which water jets are used to wash entire hillsides into sluice boxes). Highly toxic mercury is usually used to extract the gold from the other materials. In the process, much of the mercury contaminates water supplies and gets into fish consumed by people living in mining areas.

To environmentalists, the only solution to this problem is for governments to more closely regulate gold and other forms of mining and to make mining companies pay for the harmful environmental effects they cause (p. 301 and Solutions, p. 315).

lines to crack, gas mains to break, and groundwater systems to be disrupted. In addition, spoil heaps and tailings can be eroded by wind and water. The air can be contaminated with dust and toxic substances, and water pollution is a serious concern (Figure 12-1 and Spotlight, above).

Acid mine drainage occurs when aerobic bacteria produce sulfuric acid from iron sulfide minerals in spoil from coal mines and some ore mines. Rainwater seeping through the mine or mine wastes may carry the acid to nearby streams, destroying aquatic life and contaminating surface water supplies (Figure 12-14). It may also infiltrate the ground and contaminate groundwater. Other harmful materials running off (or dissolved from underground mines or aboveground mining wastes) are radioactive uranium compounds and compounds of toxic metals such as lead, arsenic, or cadmium.

After extraction from the ground, many resources must be separated from other matter, a process that can pollute the air and water. Ore, for example, typically contains two parts: the ore mineral, which contains the desired metal, and the *gangue*, which is the waste mineral material. *Beneficiation*, or separation in a mill of the ore mineral from the gangue, produces solid and often hazardous waste called **tailings**.

Most ore minerals do not consist of pure metal; so **smelting** is done to separate the metal from the other elements in the ore mineral. Without effective pollution control equipment, smelters emit enormous quantities of air pollutants, which damage vegetation and soils in the surrounding area. Pollutants include sulfur dioxide, soot, and tiny particles of arsenic, cadmium, lead, and other toxic elements and compounds found in many ores. For example, decades of uncontrolled sulfur dioxide emissions from copper-smelting

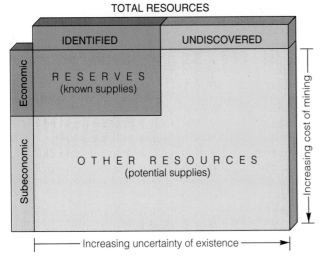

TOTAL RESOURCES

Figure 12-15 General classification of mineral resources by the U.S. Geological Survey. Note that this is not an area graph depicting abundance of reserves relative to other resources.

operations near Copperhill and Ducktown, Tennessee, killed all vegetation over a wide area around the smelter. Another dead zone has been created around the Sudbury, Ontario, nickel smelter in Canada (Figure 9-14). New zones without vegetation are forming in parts of eastern Europe, the former Soviet Union, and Chile. Smelters also cause water pollution and produce liquid and solid hazardous wastes that must be disposed of safely.

12-4 WILL THERE BE ENOUGH MINERALS?

Availability: Geology and Economics

We know how to find and extract more than 100 nonrenewable minerals from the earth's crust. We convert these raw materials into many everyday items we use and then discard, reuse, or recycle (Figure 7-4).

The U.S. Geological Survey divides mineral resources into two broad categories, *identified* and *undiscovered*, based on degree of geologic understanding and certainty that the resource exists (Figure 12-15). **Identified resources** are deposits of a particular mineral resource that have a known location, quantity, and quality or that are estimated from direct geological evidence and measurements. **Reserves** are identified resources that can be extracted economically at present prices with current mining technology.

Undiscovered resources are potential supplies of a particular mineral resource that are believed to exist on the basis of geologic knowledge and theory, though specific locations, quality, and amounts are

unknown. **Other resources** (Figure 12-15) are identified and unidentified resources that are not classified as reserves.

Worldwide the demand for mineral commodities is soaring because both population and per capita consumption are rising (Figure 1-12). Concentrations of many nonrenewable mineral resources that formed millions of years ago are being depleted in decades.

Most published estimates of particular mineral resources refer to *reserves* (Figure 12-15). Reserves can be increased when exploration finds previously undiscovered economic-grade mineral resources. They can also be increased when extraction of identified subeconomic-grade mineral resources becomes profitable because of new technology or higher prices.

The future supply of such resources depends on two factors: **(1)** the actual or potential supply and **(2)** the rate at which that supply is being used.

We never completely run out of any mineral. However, a mineral becomes *economically depleted* when finding, extracting, transporting, and processing the remaining deposits cost more than the results are worth. At that point we have four choices: recycle or reuse existing supplies, waste or use less, find a substitute, or do without.

Depletion time is the time it takes to use a certain portion—usually 80%—of the reserves of a mineral at a given rate of use. When experts disagree about depletion times, they are using different assumptions about supply and rate of use (Figure 12-16).

The shortest depletion time assumes no recycling or reuse and no increase in reserves (curve A, Figure 12-16). A longer depletion time assumes that recycling will stretch existing reserves and that better mining technology, higher prices, and new discoveries will, say, double the reserves (curve B, Figure 12-16). An even longer depletion time assumes that new discoveries will expand reserves, say, five- or tenfold, and that recycling, reuse, and reduced consumption will extend supplies (curve C, Figure 12-16). Finding a substitute for a resource dictates a whole new set of depletion curves for the new resource.

Some minerals are more important than others. Minerals essential to the economy of a country are called **critical minerals**, and those necessary for national defense are called **strategic minerals**. The definition of *critical* or *strategic* may vary from country to country or from time to time.

Who Has the World's Nonfuel Mineral Resources?
Nonfuel mineral resources are unevenly distributed in the world. The former USSR, the United States, Canada, Australia, and South Africa supply most of the world's 20 most important nonfuel minerals.

Q: How many people cannot find or buy enough fuelwood to meet their basic needs?

No industrialized country is self-sufficient in mineral resources, although the former Soviet Union came close and was a major exporter of critical and strategic minerals. Its breakup means that some of its countries have ample minerals and others do not. By contrast, Japan—in addition to lacking coal, oil, and timber resources—has virtually no metals. Japan depends on resource imports, which it upgrades to finished products and then sells abroad to buy the resources it needs to sustain its economy. Most western European countries depend heavily on minerals from Africa.

Because of mineral and energy wealth, the United States became the world's richest and most powerful nation in less than 200 years. However, this meteoric rise had a price: the rapid depletion of many of its energy (especially oil) and nonfuel mineral resources (such as lead, aluminum ore, and iron ore).

The United States will never again be self-sufficient in oil or in many key metals. Massive resource consumption and unsustainable depletion of much of its domestic nonrenewable Earth capital have also cost the country millions of jobs. Even though it is the world's largest producer of nonfuel minerals, the United States must import 50% or more of 24 of its 42 most important nonfuel minerals. Some are imported because they are used faster than they can be produced from domestic supplies; others because foreign ore deposits are of a higher grade and are cheaper to extract than remaining U.S. reserves.

Figure 12-17 shows U.S. reserves and major foreign sources for 20 important nonfuel minerals, to the year 2000. Most U.S. mineral imports come from reliable and politically stable countries. However, experts are concerned about four strategic minerals—manganese, cobalt, platinum, and chromium—for which the United States has little or no reserves and depends on imports from potentially unstable countries in the former USSR and Africa (South Africa, Zambia, Zaire). As the American Geological Institute notes, "Without manganese, chromium, platinum, and cobalt, there can be no automobiles, no airplanes, no jet engines, no satellites, and no sophisticated weapons—not even home appliances."

The United States stockpiles critical and strategic minerals to cushion against short-term supply interruptions and price jumps. These supplies are supposed to last through a three-year conventional war (minus the amounts available from domestic sources and secure foreign sources), but most of them would not.

Are There Supply Limits?   World mineral use increased tenfold from 1750 to 1900. And since 1900 it has jumped at least thirteenfold. Experts disagree about how long affordable supplies of key nonfuel minerals will last. Geologists and environmentalists tend to see

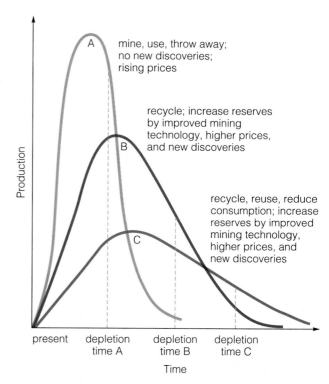

Figure 12-16 Depletion curves for a nonrenewable resource, such as aluminum or copper, using three sets of assumptions. Dashed vertical lines show when 80% depletion occurs.

Earth's supply of minerals as finite because of their uneven distribution. Also, the two laws of energy set minimum grades of ore that can be processed without spending more money than they are worth or causing unacceptable environmental damage (Figure 12-13). Many economists, by contrast, tend to see mineral supplies as essentially infinite because we can improve technologies for finding and processing minerals—or find substitutes (Guest Essay, p. 22).

Are Rich Nations Exploiting Mineral Supplies in LDCs?   Many LDCs fear that MDCs will gobble up most of the world's resources for their own economic growth, before these poorer countries can develop. To LDCs and to some resource analysts, asking whether we will have enough affordable minerals really means asking whether MDCs will have enough. The mineral resource needs of the LDCs, which have 78% of the world's people but now use only about 20% of the mineral resources, are largely ignored.

MDCs argue that when they buy nonfuel minerals from LDCs, they are helping fund the LDCs' economic development. Multinational companies point out that mines they develop in LDCs mean jobs and income for those countries.

LDCs respond that they typically get such low prices for their resources that their long-term economic development is jeopardized. MDC-based companies

A: About 1.5 billion—roughly one of every four persons on Earth

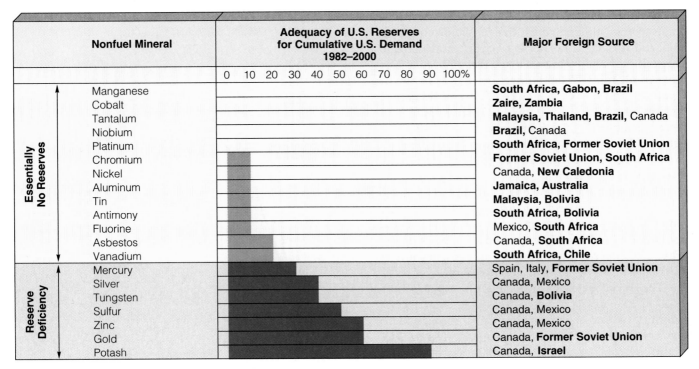

Figure 12-17 Evaluation of the supply of selected nonfuel minerals in the United States, 1982–2000, and major foreign sources of these minerals. Foreign sources subject to potential interruption of supply by political, economic, or military disruption are shown in boldface print. (Data from U.S. Geological Survey)

end up with most of the profits and, in extracting minerals, often cause severe environmental degradation and pollution in LDCs. Moreover, economic growth in mineral-exporting LDCs is stunted when they borrow money from MDCs to buy expensive imported products. Low mineral prices also encourage resource waste in MDCs, which increases the rates at which supplies of key nonrenewable resources are depleted.

Politics also plays a key role in mineral prices. If an LDC opposes a mineral-importing MDC's policies, the MDC may cut off imports from that LDC (assuming other sources are available) or stop its foreign aid. An LDC that does not succumb to such "economic blackmail" faces economic ruin.

Currently LDCs owe MDCs over $1.4 trillion, and each year, they owe $43 billion in interest alone. To correct this situation, LDCs have called for economic changes that would shift what they believe is a fairer share of the world's wealth to them. Their proposals include:

- Much more aid from MDCs to LDCs

- Forgiveness of some of their debt to MDCs

- Removal of trade barriers in MDCs against some LDC products

- Higher prices for minerals, timber, and other LDC exports to MDCs

- Greater LDC influence over the decisions of international lending institutions, such as the World Bank and the International Monetary Fund

So far MDCs have generally opposed these ideas, arguing that it would cost them jobs and reduce their economic growth. What is your position on this issue?

Are There Environmental Limits? Some environmentalists and resource experts believe that the greatest danger from high levels of resource consumption may not be the exhaustion of resources but the damage that their extraction and processing do to the environment (Figure 12-13).

The minerals industry accounts for 5–10% of world energy use, making it a major contributor to air and water pollution and to greenhouse gases. As more remote, deeper deposits are mined, even more energy will be needed to dig bigger holes (Figure 12-9) and transport the metal ores and extracted coal, oil, and natural gas farther.

The grade of an ore—its percentage of metal content—largely determines the environmental impact of metal mining. Generally we use the more accessible and higher-grade ores first. As they are depleted, it takes more money, energy, water, and other materials to exploit lower-grade ores. Environmental impacts increase accordingly (Figure 12-13).

Q: By 2000 how many people may not be able to get enough fuelwood?

Reforming the 1872 Mining Law

To environmentalists the 1872 mining law (p. 301) is a glaring example of welfare for the rich at the expense of all taxpayers, but have been unable to have this law revised. They would (1) end the practice of granting title to public lands for actual or imaginary hard-rock mining; (2) allow carefully regulated hard-rock mineral extraction under a lease not to exceed 20 years; (3) require mining companies to pay rents that cover at least the costs of administration, roads, and other government subsidies related to any mining operation during the term of the lease; (4) have mining companies pay a royalty of no less than 12.5% on any hard-rock minerals they extract; (5) make mining companies responsible for restoring the land and cleaning up environmental damage caused by their activities; and (6) require the companies to post a performance bond up front to cover the likely cost of environmental damage and restoration.

Environmentalists point out that mining is the only natural resource industry in the United States that by law can buy public lands and the only resource industry that pays no rents or royalties for resource extraction. In contrast oil, gas, and coal companies can only lease—not buy—public lands, must pay a 12.5% royalty on all energy resources they extract, and are responsible for environmental cleanup and land restoration.

In 1993 the Clinton administration asked for a similar 12.5% royalty on hard-rock minerals extracted from public lands along with strict environmental standards and provisions for cleanup of abandoned sites. However, the proposal was blocked by powerful western senators who receive large election campaign contributions from mining companies, ranchers, and others who benefit from cheap access to resources on public lands at taxpayers' expense. A royalty may be imposed because of public pressure. However, if mining companies get their way, they will still be able to buy public land at bargain prices and have to pay only a token royalty of about 1–2% by imposing such fees on the value of ores at the mouth of a mine (instead of after the ore has been processed) and by allowing them to deduct mining costs from royalties.

Mining companies claim that charging royalties for minerals taken from public lands and making them responsible for cleanup and restoration will force them to do their mining in other countries, which would cost American jobs and reduce tax revenues. However, environmentalists counter that this is merely a smoke screen tactic and that mining companies (like energy companies) would still make a reasonable profit on minerals they get from public lands.

They also point out that Canada, Australia, South Africa and other countries that are major extractors of hard-rock minerals don't sell public lands to mining companies and require the companies to pay rent on any public land they lease and royalties on the minerals they extract. What do you think should be done?

12-5 INCREASING MINERAL RESOURCE SUPPLIES

Economics and Resource Supply Geologic processes determine how and where a mineral resource is concentrated in the earth (Section 12-1). Economics determines what part of the known supply will be used (Figures 12-15 and 12-16).

According to standard economic theory, in a competitive free market a plentiful resource is cheap because supply exceeds demand (Figure 7-2); but when a resource becomes scarce its price rises—stimulating exploration and development of better mining technology. Rising prices also make it profitable to mine ores of ever-lower grades and will motivate the search for substitutes. However, this theory may no longer apply to MDCs. In the mixed economic systems (Section 7-1) of such countries, industry and government control supply, demand, and prices of minerals to such a large extent that a truly competitive free market does not exist.

Most mineral prices are artificially low because countries subsidize development of their domestic mineral resources. In the United States mining companies get depletion allowances amounting to 5–22% of their gross income, depending on the mineral. The companies can also deduct much of their cost for finding and developing mineral deposits. Moreover, the U.S. mining industry gets another big federal subsidy through virtual giveaways of federal lands and hard-rock minerals (p. 301 and Solutions, above).

Over the last decade these mining subsidies have cost U.S. taxpayers $5 billion ($560 million in 1992 alone). The U.S. Treasury could gain about $2 billion over the next five years if the hard-rock mineral industry were taxed on the same basis as other industries. Taxing rather than subsidizing the extraction of non-fuel mineral resources would create incentives for

Mining with Microbes

SOLUTIONS

One emerging prospect for improving mining technology is the use of microorganisms for in-place (*in situ*) mining, which would remove desired metals from ores while leaving the surrounding environment relatively undisturbed.

Once an ore deposit had been identified and deemed economically viable, wells would be drilled into it and the ore fractured. Then the ore would be inoculated with either natural or genetically engineered bacteria to extract the desired metal. Next the ore would be flooded with water, which would be pumped to the surface, where the desired metals would be removed. Thus metal production would become essentially biological, with little energy input compared with current extraction technologies.

In 1990 a gold-mining plant using microbes went into operation in Colorado. It uses *Thiobacillus ferroxidans*, a natural bacterium, to extract gold that is embedded in iron sulfide ore and that is uneconomical to extract by conventional methods. The bacteria consume the sulfur and iron in the ore, leaving gold in a form that is easily extracted by current technology. The plant is expected to produce gold at about $240 per ounce, well below the current market price.

Microbiological processing of ores, however, is slow: It can take decades to remove the same amount of material that conventional methods can remove within months or years. So far biological methods are economically feasible only with low-grade ore (such as gold), for which conventional techniques are too expensive.

Some people have expressed concern about potential harm from genetically engineered organisms (Pro/Con, p. 207). However, proponents of mining biotechnology point out that these organisms get their energy from chemical sources, not from living or formerly living organisms, as do genetically engineered organisms used in agriculture and medicine. Consequently, they argue, the risks are several orders of magnitude lower than those from other genetically engineered organisms.

more efficient resource use, promote waste reduction and pollution prevention, encourage recycling and reuse, and provide governments with revenue.

Another problem is that the cost of nonfuel mineral resources is only a small part of the final cost of goods. Thus scarcity of minerals does not raise the market prices of products very much; so industries and consumers have no incentive to reduce demand for products in time to avoid economic depletion of the minerals.

Low mineral prices, made possible by ignoring the harmful environmental costs of mining and processing (Figure 12-13), encourage waste, pollution, and environmental damage. Mining companies and manufacturers have little incentive to reduce resource waste and pollution as long as they can pass many of the harmful environmental costs of their production on to society. Most environmentalists and many economists believe that until we have full-cost pricing for resources and manufactured items that includes their environmental and social costs (Section 7-2), we will continue depleting and wasting nonrenewable mineral resources (Solutions, p. 315).

An economic factor limiting production of nonfuel minerals is investment capital. With today's fluctuating mineral markets and rising costs, investors are wary of tying up large sums for long periods with no assurance of a reasonable return.

Finding New Land-Based Mineral Deposits

Geologic exploration guided by better knowledge, satellite surveys, and other new techniques will increase present reserves of most minerals. Although most of the easily accessible, high-grade deposits are already known, new discoveries will be found, mostly in unexplored areas of LDCs.

Exploring for new resources, however, takes lots of capital and is a risky venture. Typically, if geologists identify 10,000 possible deposits of a given resource, only 1,000 sites are worth exploring; only 100 justify drilling, trenching, or tunneling; and only 1 will become a producing mine or well. Even if large new supplies are found, no nonrenewable mineral supply can stand up to continued exponential growth in its use.

Improving Mining Technology and Mining Low-Grade Ore

Some analysts assert that all we need to do to increase supplies of any mineral is to extract lower grades of ore. They point to new earth-moving equipment, new purifying techniques for removing impurities, and other technological advances during the past few decades.

For example, in 1900 the average copper ore mined in the United States was about 5% copper by weight; today it is 0.5%—and copper costs less (adjusted for inflation). Technological improvements also increased world copper reserves 500% between 1950 and 1980. Future advances may let us use even lower-grade ores of some metals (Solutions, at left).

Several factors limit the mining of lower-grade ores, however. As ever-poorer ores are mined, we reach a point where it costs more to mine and process

Q: What percentage of the original old-growth forests in the United States have been cut?

such resources than they are worth, unless we have a virtually inexhaustible source of cheap energy. Availability of fresh water also may limit the supply of some mineral resources, because large amounts of water are needed to extract and process most minerals. Many mineral-rich areas lack fresh water.

Finally, exploitation of lower-grade ores may be limited by the environmental impact of waste material produced during mining and processing (Figures 12-9 and 12-13). At some point the costs of land restoration and pollution control exceed the current value of the minerals, unless we continue to pass these harmful costs on to society and to future generations.

Mining the Oceans   Ocean mineral resources are found in three areas: seawater, sediments and deposits on the shallow continental shelf (Figure 12-5), and sediments and nodules on the deep-ocean floor. Most of the chemical elements found in seawater occur in such low concentrations that recovering them takes more energy and money than they are worth. Only magnesium, bromine, and sodium chloride are abundant enough to be extracted profitably at present prices with current technology.

Deposits of minerals (mostly sediments) along the continental shelf and near shorelines are already significant sources of sand, gravel, phosphates, and nine other nonfuel mineral resources. Offshore wells also supply large amounts of oil and natural gas.

The deep-ocean floor at various sites may be a future source of manganese and other metals. For example, at a few sites manganese-rich nodules have been found in large quantities. These cherry- to potato-sized rocks contain 30–40% by weight manganese, used in certain steel alloys. They also contain small amounts of other strategically important metals, such as nickel, copper, and cobalt. These nodules might be sucked up from the ocean floor by pipe or scooped up by a continuous cable with buckets to a mining ship. However, most of these nodule beds occur in international waters. Their development has been put off indefinitely because of squabbles over who owns them.

Iron, manganese, copper, and zinc also occur in sulfide deposits around vents found at certain locations on the deep-ocean floor, where hot gases and mineral-laden steam are released from fissures in the earth's crust. However, concentrations of metals in most of these deposits are too low to be valuable mineral resources.

Environmentalists recognize that seabed mining would probably cause less harm than mining on land. They are concerned, however, that removing seabed mineral deposits and dumping back unwanted material will stir up ocean sediments, which could destroy some seafloor organisms and have unknown effects on poorly understood ocean food webs. Surface waters

The Materials Revolution

SOLUTIONS

Scientists and engineers are rapidly developing new materials—in particular, ceramics and plastics—as replacements for metals. Ceramics have many advantages over conventional metals. They are harder, stronger, lighter, and longer-lasting than many metals. Also, they withstand intense heat and do not corrode. Because they can burn fuel at higher temperatures than metal engines, ceramic engines can boost vehicle fuel efficiency by 30–40%. We also have ceramic knives, scissors, batteries, fishhooks, and artificial limbs.

Within a few decades we may have high-temperature ceramic superconductors in which electricity flows without resistance. That may lead to faster computers, more efficient power transmission, and affordable electromagnets for propelling magnetic levitation trains (Figure 6-20). To date Japanese scientists have filed more patent applications for such superconductors than the rest of the world combined.

Plastics also have advantages over many metals. High-strength plastics and composite materials strengthened by lightweight carbon and glass fibers are likely to transform the automobile and aerospace industries. They cost less to produce than metals because they require less energy, don't need painting, and can easily be molded into any shape.

Many cars now have plastic body parts, which reduce weight and boost fuel economy. Planes and cars made almost entirely of plastics—held together by new superglues—may be common in the next century. New plastics and gels are also being developed to provide superinsulation without taking up much space.

The materials revolution is also transforming medicine. So-called biomaterials, made of new plastics, ceramics, glass composites, and alloys, are being used in artificial skin, arteries, organs, and joints.

might also be polluted by the discharge of sediments from mining ships and rigs.

Finding Substitutes   Some people believe that even if supplies of key minerals become very expensive or scarce, human ingenuity will find substitutes. They point to the current materials revolution in which silicon and other abundant elements are being substituted for scarce metals in many uses (Solutions, above).

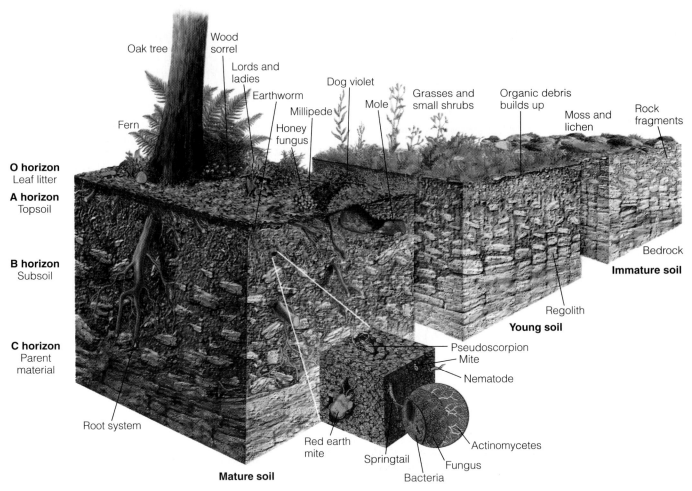

O horizon Leaf litter
A horizon Topsoil
B horizon Subsoil
C horizon Parent material

Oak tree
Wood sorrel
Lords and ladies
Earthworm
Dog violet
Millipede
Mole
Grasses and small shrubs
Organic debris builds up
Rock fragments
Fern
Honey fungus
Moss and lichen
Root system
Red earth mite
Springtail
Bacteria
Fungus
Actinomycetes
Nematode
Mite
Pseudoscorpion
Regolith
Bedrock
Mature soil
Young soil
Immature soil

Figure 12-18 Formation and generalized profile of soils. Horizons, or layers, vary in number, composition, and thickness, depending on the type of soil. (Used by permission of Macmillan Publishing Company from Derek Elsom, *Earth*, New York: Macmillan, 1992. Copyright ©1992 by Marshall Editions Developments Limited)

Substitutes can undoubtedly be found for many scarce mineral resources, but the search is costly, and phasing a substitute into a complex manufacturing process takes time. Also, while a vanishing mineral is being replaced, people and businesses dependent on it may suffer economic hardships. Moreover, finding substitutes for some key materials—such as helium, phosphorus for phosphate fertilizers, manganese for making steel, and copper for wiring motors and generators—may be hard or impossible. Finally, some substitutes are inferior to the minerals they replace. For example, aluminum could replace copper in electrical wiring, but producing aluminum takes much more energy than producing copper. Aluminum wiring is also more of a fire hazard than copper wiring.

12-6 SOIL: THE BASE OF LIFE

Soil Layers, Components, and Types   Unless you are a farmer or a gardener, you probably think of soil as dirt—as something you don't want on your

hands, clothes, or carpet. Yet, your life and the lives of most other organisms depend on soil, especially topsoil. In addition to food, soil provides us with wood, paper, cotton, and medicines, and it helps purify the water we drink and decompose and recycle biodegradable wastes. And soil nutrients eroded naturally from land help support food webs in aquatic systems. Yet since the beginnings of agriculture we have abused this vital, potentially renewable resource. Entire civilizations have collapsed because they mismanaged the topsoil that supported their populations.

Pick up a handful of soil and notice how it feels and looks. **Soil** is a complex mixture of inorganic materials (clay, silt, pebbles, and sand), decaying organic matter, water, air, and billions of living organisms. Soil forms when formerly living matter decays, when solid rock weathers and crumbles, and when sediments resulting from erosion are deposited.

The living organisms (mostly decomposers) in soils and the plants, animals, and microorganisms supported by soils make up the bulk of the earth's biodiversity.

Q: How much of all U.S. land consists of public lands?

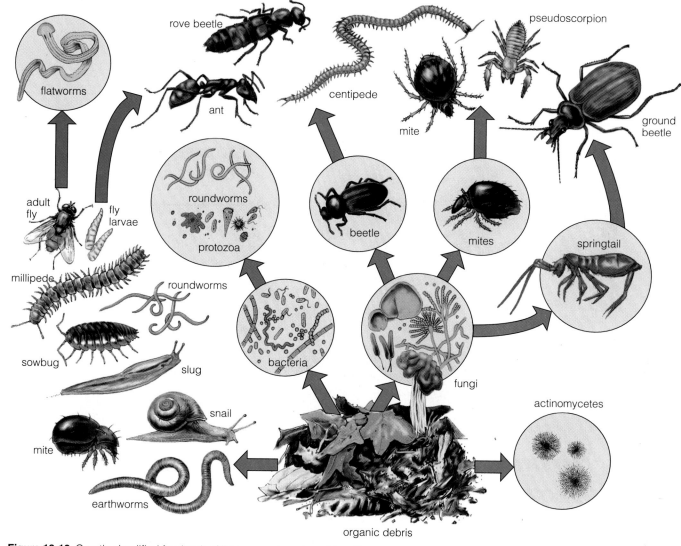

Figure 12-19 Greatly simplified food web of living organisms found in soil.

Mature soils are arranged in a series of zones called **soil horizons**, each with a distinct texture and composition that varies with different types of soils. A cross-sectional view of the horizons in a soil is called a **soil profile**. Most mature soils have at least three of the possible horizons (Figure 12-18).

The top layer, the *surface-litter layer* or *O-horizon*, consists mostly of freshly fallen and partially decomposed leaves, twigs, animal waste, fungi, and other organic materials. Normally it is brown or black in color. The *topsoil layer*, or *A-horizon*, is a porous mixture of partially decomposed organic matter (humus) and some inorganic mineral particles. Usually it is darker and looser than deeper layers. The roots of most plants and most of a soil's organic matter are concentrated in these two upper layers. As long as these layers are anchored by vegetation, soil stores water and releases it in a nourishing trickle instead of a devastating flood.

The two top layers of most well-developed soils teem with bacteria, fungi, earthworms, and small insects (Figure 12-18). These layers are also home for burrowing animals such as moles and gophers. These soil-dwellers interact in complex food webs (Figure 12-19).

Bacteria and other decomposer microorganisms are found by the billions in every handful of topsoil. They recycle the nutrients we and other land organisms need by breaking down some of the complex organic compounds in the upper soil into simpler inorganic compounds soluble in soil water. Soil moisture carrying these dissolved nutrients is drawn up by the roots of plants and transported through stems and into leaves (Figure 12-20).

Some organic litter in the two top layers is broken down into a sticky, brown residue of partially decomposed organic material called **humus**. Because humus

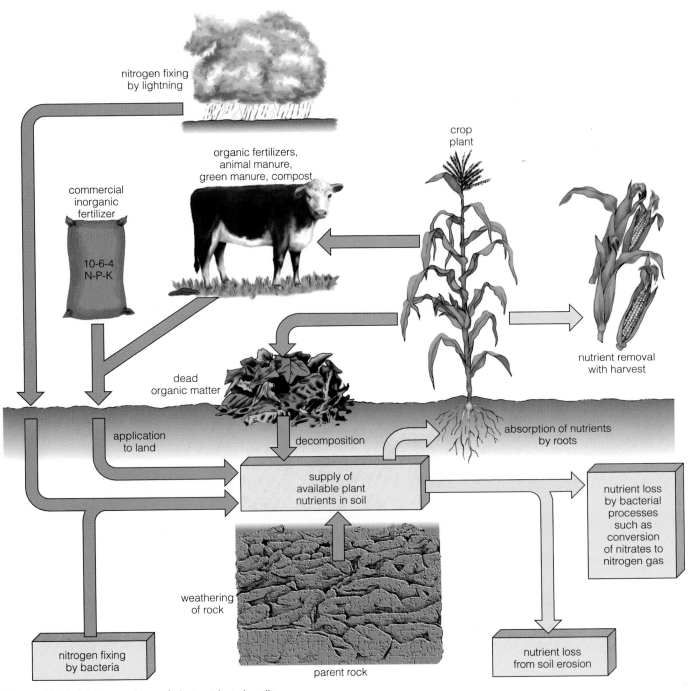

nitrogen fixing
by lightning

commercial
inorganic
fertilizer

10-6-4
N-P-K

organic fertilizers,
animal manure,
green manure, compost

crop
plant

nutrient removal
with harvest

dead
organic matter

application
to land

decomposition

absorption of nutrients
by roots

supply of
available plant
nutrients in soil

nutrient loss
by bacterial
processes
such as
conversion
of nitrates to
nitrogen gas

weathering
of rock

parent rock

nitrogen fixing
by bacteria

nutrient loss
from soil erosion

Figure 12-20 Addition and loss of plant nutrients in soils.

is only slightly soluble in water, most of it stays in the topsoil layer. A fertile soil, producing high crop yields, has a thick topsoil layer with lots of humus.

Humus is an important soil material. It coats the sand, silt, and clay particles in topsoil and binds them together into clumps, giving a soil its *structure*. Humus also helps topsoil hold water and nutrients taken up by plant roots. Particles of humus and of clay tend to have a negative electrical charge on their surfaces. This allows them to attract positively charged nutrient ions

such as potassium (K^+), calcium (Ca^{2+}), and ammonium (NH_4^+) strongly enough to keep them from being carried away as rainwater percolates downward through the topsoil.

Humus also provides spaces for the growth of nutrient-absorbing root hairs and a class of fungi, known as mycorrhizae, that are the mutualistic partners of some trees and other plants.

Color tells us a lot about how useful a soil is for growing crops. For example, dark-brown or black top-

Q: Where is most of the federally owned and managed land in the United States?

soil is nitrogen-rich and high in organic matter. Gray, bright yellow, or red topsoils are low in organic matter and will need nitrogen fertilizer to support most crops.

The *B-horizon (subsoil)* and the *C-horizon (parent material)* contain most of a soil's inorganic matter. It is mostly broken-down rock, a varying mixture of sand, silt, clay, and gravel. The C-horizon lies on a base of unweathered parent rock called bed rock.

The spaces, or pores, between the solid organic and inorganic particles in the upper and lower soil layers contain varying amounts of air (mostly nitrogen and oxygen gas) and water. Plant roots need oxygen for respiration in their cells.

Some of the rain falling on the soil percolates through the soil layers and occupies many of the pores. This downward movement of water through soil is called **infiltration**. As the water seeps down, it dissolves and picks up various soil components in upper layers and carries them to lower layers—a process called **leaching**.

Soils develop and mature slowly (Figure 12-18). One maturation process is **humification**—in which organic matter in the upper soil layers becomes humus. In lower layers a soil matures through **mineralization**, in which decomposers turn organic materials into inorganic ones.

Mature soils vary widely from biome to biome in color, content, pore space, acidity (pH, Figure 9-8), and depth. Five important soil types, each with a distinct profile, are shown in Figure 12-21. Most of the world's crops are grown on soils exposed when grasslands (Figure 5-10) and deciduous forests (Figure 5-15) are cleared.

Soil Texture and Porosity   Soils vary in their content of *clay* (very fine particles), *silt* (fine particles), *sand* (medium-size particles), and *gravel* (coarse to very coarse particles). The relative amounts of the different sizes and types of mineral particles determine **soil texture**, as summarized in Figure 12-22. Soils containing a mixture of clay, sand, silt, and humus are called **loams**.

To get an idea of a soil's texture, take a small amount of topsoil, moisten it, and rub it between your fingers and thumb. A gritty feel means that it contains a lot of sand. A sticky feel means a high clay content, and you should be able to roll it into a clump. Silt-laden soil feels smooth like flour. A loam topsoil, best suited for plant growth, has a texture between these extremes—a crumbly, spongy feeling with many of its particles clumped loosely together.

Soil texture helps determine **soil porosity**: a measure of the volume of pores or spaces per volume of soil and the average distances between those spaces. A porous soil (with many pores) can hold more water and air than a less porous soil. The average size of the spaces or pores in a soil determines **soil permeability**: the rate at which water and air move from upper to lower soil layers. Soil porosity is also influenced by **soil structure**: how soil particles are organized and clumped together.

Soil texture, porosity, and permeability determine a soil's *water-holding capacity*, *aeration* (or *oxygen content*, the ability of air to move through the soil), and *workability* (how easily it can be cultivated). Table 12-1 compares the main physical and chemical properties of sand, clay, silt, and loam soils.

Loams are the best soils for growing most crops because they hold lots of water but not too tightly for plant roots to absorb. Sandy soils are easy to work, but water flows rapidly through them They are useful for growing irrigated crops or those with low water requirements, such as peanuts and strawberries.

The particles in clay soils are very small and easily compacted. When these soils get wet, they form large, dense clumps, explaining why wet clay can be molded into bricks and pottery. Clay soils are more porous and have a greater water-holding capacity than sandy soils, but the pore spaces are so small that these soils have a low permeability. Because little water can infiltrate to lower levels, the upper layers can easily become too waterlogged for most crops.

Soil Acidity (pH)   The acidity or basicity (alkalinity) of a soil is another factor determining the types of plants it can support. Acidity and basicity of substances in water solution are commonly expressed in terms of **pH** (Figure 9-8).

Soils vary in acidity, and the pH of a soil influences the uptake of soil nutrients by plants. Plants vary in the pH ranges they can tolerate. When soils are too acidic, the acids can be partially neutralized by an alkaline substance such as lime. Because lime speeds up the decomposition of organic matter in the soil, however, manure or another organic fertilizer should also be added to maintain soil fertility.

In dry regions such as much of the western and southwestern United States, calcium and other alkaline compounds are not leached away by rain. Soils in such areas may be too alkaline (pH above 7.5) for some crops. If drainage is good, irrigation can leach the alkaline compounds away. Adding sulfur, which is gradually converted into sulfuric acid by soil bacteria, also reduces soil alkalinity.

The burning of fossil fuels, especially coal, releases sulfur dioxide and nitrogen oxides, which form acidic compounds in the atmosphere. These compounds fall back to the earth as *acid deposition* (Figure 9-7). As acidic rain or melted acidic snow infiltrates the soil, the hydrogen ions (H^+) in the acids are attracted to particles of minerals and humus in the topsoil layer,

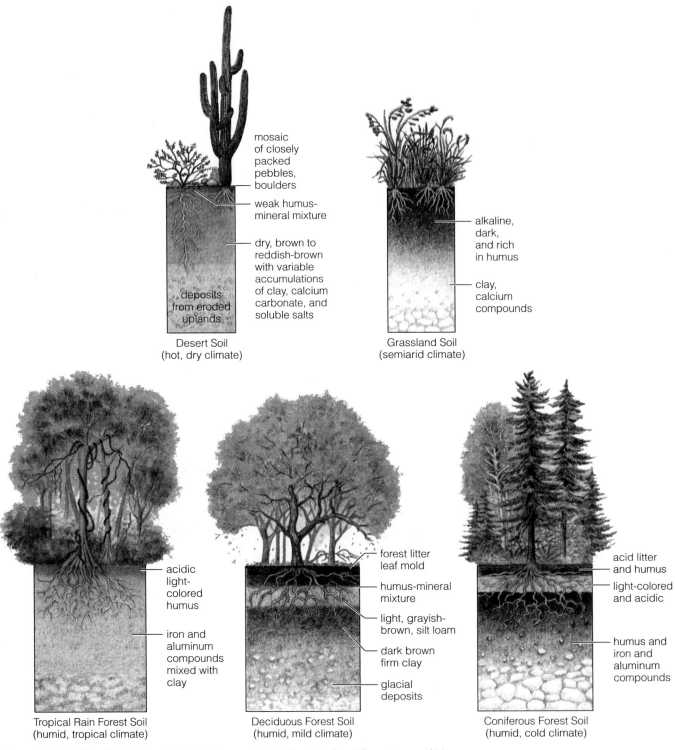

mosaic of closely packed pebbles, boulders

weak humus-mineral mixture

dry, brown to reddish-brown with variable accumulations of clay, calcium carbonate, and soluble salts

deposits from eroded uplands

Desert Soil
(hot, dry climate)

alkaline, dark, and rich in humus

clay, calcium compounds

Grassland Soil
(semiarid climate)

acidic light-colored humus

iron and aluminum compounds mixed with clay

Tropical Rain Forest Soil
(humid, tropical climate)

forest litter leaf mold

humus-mineral mixture

light, grayish-brown, silt loam

dark brown firm clay

glacial deposits

Deciduous Forest Soil
(humid, mild climate)

acid litter and humus

light-colored and acidic

humus and iron and aluminum compounds

Coniferous Forest Soil
(humid, cold climate)

Figure 12-21 Soil profiles of the principal soil types typically found in five different types of biomes.

displacing some of the potassium (K^+), calcium (Ca^{2+}), magnesium (Mg^{2+}), and ammonium (NH_4^+) ions that were attached to those particles. The resulting loss of soil fertility can reduce crops and tree growth and make then more vulnerable to drought, disease, and pests (Figure 9-9).

12-7 SOIL EROSION

Natural and Human-Accelerated Erosion

Soil erosion is the movement of soil components, especially surface-litter and topsoil, from one place to another. The two main agents of erosion are flowing water

Q: What percentage of timber in the United States is harvested by clear-cutting?

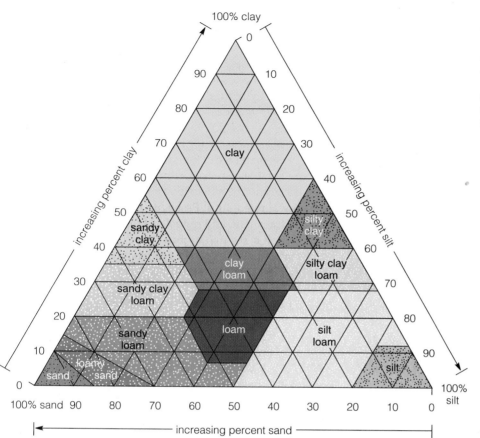

Figure 12-22 Soil texture depends on the percentages of clay, silt, and sand particles in the soil. Soil texture affects soil porosity—the average number and spacing of pores in a volume of soil. Loams—roughly equal mixtures of clay, sand, silt, and humus—are the best soils for growing most crops. (Data from Soil Conservation Service)

Table 12-1 Useful Properties of Soils with Different Textures

Soil Texture	Nutrient-Holding Capacity	Water-Infiltration Capacity	Water-Holding Capacity	Aeration	Workability
Clay	Good	Poor	Good	Poor	Poor
Silt	Medium	Medium	Medium	Medium	Medium
Sand	Poor	Good	Poor	Good	Good
Loam	Medium	Medium	Medium	Medium	Medium

(Figure 12-23) and wind (Figure 4-1). Some soil erosion is natural (Figure 12-5). In undisturbed vegetated ecosystems, the roots of plants help anchor the soil, and usually soil is not lost faster than it forms. However, farming, logging, building, overgrazing by livestock, off-road vehicles, fire, and other activities that destroy plant cover leave soil vulnerable to erosion.

Although wind causes some erosion, most is caused by moving water. Soil scientists distinguish between three types of water erosion. *Sheet erosion* occurs when surface water moves down a slope or across a field in a wide flow and peels off uniform sheets, or layers, of soil. Because the topsoil disappears evenly, sheet erosion may not be noticeable until much damage has been done. In *rill erosion* the surface water forms fast-flowing little rivulets that cut small channels in the soil (Figure 12-23). In *gully erosion* rivulets of fast-flowing water join together and with each succeeding rain cut the channels wider and deeper until they become ditches or gullies (Figure 12-23). Gully erosion usually happens on steep slopes where all or most vegetation has been removed (Figures 1-3 and 11-9).

A: About 66% (33% in national forests)

Losing topsoil makes a soil less fertile and less able to hold water. The resulting sediment, the largest source of water pollution, clogs irrigation ditches, boat channels, reservoirs, and lakes. The water is cloudy and tastes bad, fish die, and flood risk increases as a stream fills with sediment (Figure 11-7). Rivers running brown with silt show Earth capital hemorrhaging from the land (Figure 5-22).

Soil, especially topsoil, is classified as a potentially renewable resource because it is continuously regenerated by natural processes. However, in tropical and temperate areas it takes 200–1,000 years for 2.54 centimeters (1 inch) of new topsoil to form, depending on climate and soil type. If topsoil erodes faster than it forms on a piece of land, the soil there becomes a nonrenewable resource. Annual erosion rates for farmland throughout the world are 7–100 times the natural renewal rate (Guest Essay, p. 333). Soil erosion is milder on forestland and rangeland than on cropland, but forest soil takes two to three times longer to restore itself than does cropland. Construction sites usually have the highest erosion rates by far.

The World Situation Today topsoil is eroding faster than it forms on about one-third of the world's cropland. In some countries more than half the land is affected, including Nepal (95%), Peru (95%), Turkey (95%) Lesotho (88%), Madagascar (79%), and Ethiopia (53%). In Africa soil erosion has increased 20-fold in the last three decades.

Figure 12-23 Rill and gully erosion of vital topsoil from irrigated cropland in Arizona.

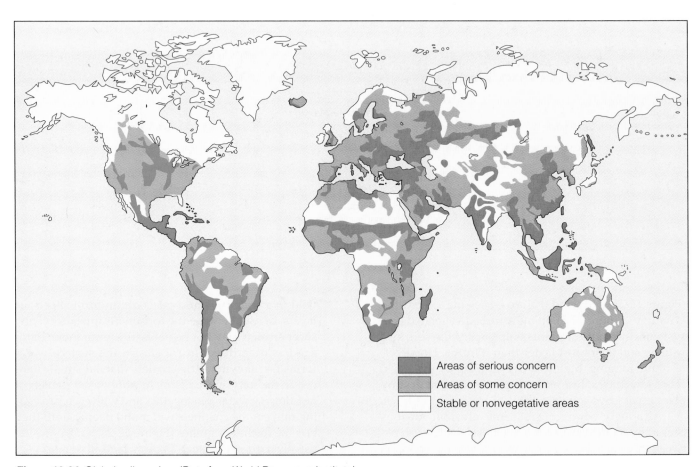

Figure 12-24 Global soil erosion. (Data from World Resources Institute)

Q: How much of the U.S. Forest Service budget is devoted to timber sales?

A 1992 study by the World Resources Institute found that soil on more than 12 million square kilometers (5 million square miles) of land—an area the size of China and India combined—had been seriously eroded since 1945 (Figure 12-24). The study also found that 89,000 square kilometers (34,000 square miles) of land scattered across the globe was too eroded to grow crops anymore.

Two-thirds of the seriously degraded lands are in Asia and Africa; Central America and the United States have each lost 25% of their productive cropland. Overgrazing is the worst culprit, accounting for 35% of the damage, with the heaviest losses in Africa and Australia. Deforestation causes 30% of Earth's severely eroded land and is most prevalent in Asia and South America (Section 16-4). Unsustainable methods of farming account for 28% of such erosion, with two-thirds of the damage found in North America.

Each year we must feed 90 million more people with an estimated 24 billion metric tons (26 billion tons) less topsoil. The topsoil that each year washes and blows into the world's streams, lakes, and oceans would fill a train of freight cars long enough to encir-cle the planet 150 times. At that rate the world is losing about 7% of its topsoil from potential cropland each decade—a serious problem for farmers and eventually for all of us. The situation is worsening as many farmers in LDCs plow marginal lands to survive.

According to a 1993 report by the United Nations Food and Agriculture Organization (FAO), farmers worldwide must adopt better farming practices. Otherwise, high-quality topsoil will disappear from an estimated 140 million hectares (345 million acres)—equal to the area of Alaska—in 20 years.

Spreading Desertification   **Desertification** is a process whereby the productive potential of arid or semiarid land falls by 10% or more, and this drop is caused mostly by human activities. *Moderate desertification* is a 10–25% drop in productivity, and *severe desertification* is a 25–50% drop. *Very severe desertification* is a drop of 50% or more, usually creating huge gullies and sand dunes.

Desertification is a serious and growing problem in many parts of the world (Figure 12-25). The regions most affected by desertification are all cattle-producing

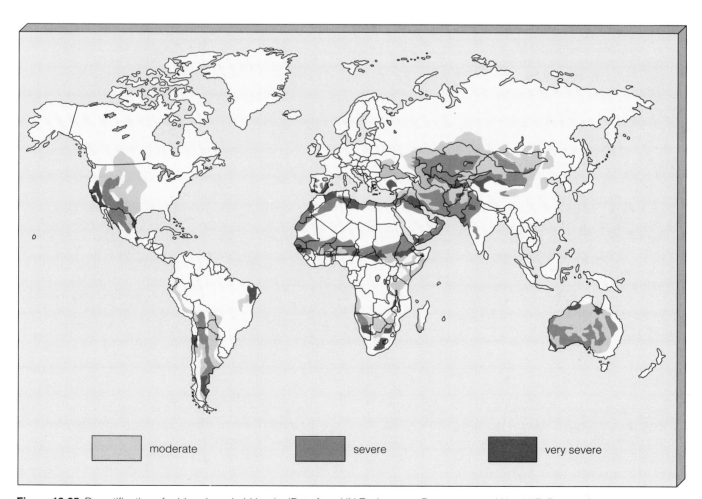

Figure 12-25 Desertification of arid and semiarid lands. (Data from UN Environment Programme and Harold E. Dregnue)

Desertification and Alien Species in Australia

CASE STUDY

Within 200 years after English settlers arrived in Australia in the eighteenth century, they had destroyed most of the country's forest cover, eliminated or endangered 70% of its mammal species, caused widespread erosion and salting of the soil (leading to desertification throughout much of the country, Figures 12-25 and 12-26), and decimated the once-sizable aboriginal population. Topsoil is now eroding 50 times faster than it did prior to 1788.

In 1859 Thomas Austin, longing for home pastimes, imported and released 24 English rabbits into the Australian countryside for hunting. Without any natural enemies the rabbits multiplied so rapidly that within 10 years they numbered in the millions, and like hordes of locusts they began munching their way across the continent. They ate crops, competed with livestock for grass, and destroyed wild vegetation, displacing and endangering many native animal species.

Australians tried without success to contain this plague of rabbits until the 1950s, when they introduced a virus that killed off 99% of the rabbits. Now, however, the rabbits are back. Through natural selection the remaining 1% became resistant, began multiplying, and now number more than 200 million, causing about $75 million of damages per year to grazing land. Farmers and ranchers attack the rabbits with guns, dogs, poison, and dynamite, but they keep multiplying. Australian scientists are using genetic engineering in a desper-

Figure 12-26 Desertification in this arid outback region of Australia was caused by cattle overgrazing the vegetation. Livestock cropped plants so short that the vegetation died; trampling kept new seedlings from growing. Overgrazing is the biggest cause of desertification.

ate attempt to develop stronger viruses before the country is again overrun.

In the 1930s someone came up with the bright idea of importing the South American cane toad to help control beetles that were devastating the sugarcane crop. Unfortunately, these plump 0.5-kilogram (1-pound) toads stayed on the ground, while the sugarcane beetles flew above; so they never met. Then the toads began multiplying. They now occupy 40% of Queensland and are moving into New South Wales and the Northern Territory.

The poisonous toads are almost impossible to kill. Toads that are shot recover and hop away. The ones that are run over by a car simply swallow their squashed guts and limp away.

Australians are now trying to reverse some of the consequences of their centuries of destruction and thoughtless introductions of alien species. The government has abolished tax advantages for clearing land and replaced them with incentives for planting trees, preferably the native species they destroyed. They have a daunting task.

---

areas and include sub-Saharan Africa (between North Africa's barren Sahara and the plant-rich land to its south), the Middle East, western Asia, parts of Central and South America, the western half of the United States, and Australia (Case Study, above).

Moderate desertification can go unrecognized. For example, overgrazing has reduced the productivity of

much of the grassland in the western United States. Yet most of the residents do not realize they live in a moderately desertified area.

Practices that leave topsoil vulnerable to desertification include overgrazing (Figure 12-26); deforestation without reforestation (Section 16-4); surface mining without land reclamation (Figures 12-9, 12-10, and

Q: Between 1978 and 1992, how much money did the Forest Service lose on timber sales?

12-11); irrigation techniques that lead to increased erosion, salt buildup, and waterlogged soil; farming on land with unsuitable terrain or soils; and soil compaction by farm machinery and cattle hoofs.

These destructive practices are linked to rapid population growth, high human and livestock densities, poverty, and poor land management. The consequences of desertification include worsening drought, famine, declining living standards, and swelling numbers of environmental refugees whose land is too eroded to grow crops or feed livestock.

It is estimated that 810 million hectares (2 billion acres)—an area the size of Brazil and 12 times the size of Texas—have become desertified during the past 50 years. The UN Environment Programme estimates that worldwide 63% of rangelands, 60% of rain-fed croplands, and 30% of irrigated croplands are threatened by desertification. The total area of this threatened land is 33 million square kilometers (13 million square miles)—about the size of North and South America combined. If present trends continue, desertification could threaten the livelihoods of 1.2 billion people by 2000.

Every year an estimated 60,000 square kilometers (23,000 square miles—an area the size of West Virginia) of new desert are formed, and another 210,000 square kilometers (81,000 square miles—an area the size of Kansas) lose so much soil and fertility that they are no longer worth farming or grazing.

The most effective way to slow the march of desertification is to drastically reduce overgrazing, deforestation (Sections 16-3 and 16-4), and the destructive forms of planting, irrigation, and mining that are to blame. In addition, reforestation programs will anchor soil and hold water while providing fuelwood, slowing desertification, and reducing the threat of global warming (Section 10-2).

The total cost of such prevention and rehabilitation would be about $141 billion, only 3.5 times the estimated $42 billion annual loss in agricultural productivity from desertified land. Thus, once this potential productivity is restored, the cost of the program could be recouped in three or four years. So far, however, only about one-tenth of the needed money has been provided.

The U.S. Situation   Erosion is not limited to LDCs. Vanishing topsoil and creeping desertification have become serious problems in parts of the United States (Figures 12-24 and 12-25). According to the Soil Conservation Service, about one-third of the nation's original prime topsoil has been washed or blown into streams, lakes, and oceans, mostly as a result of overcultivation, overgrazing, and deforestation.

In the 1930s the United States learned a harsh lesson about soil when much of the soil in several mid-

Figure 12-27 The Dust Bowl of the Great Plains, where a combination of periodic severe drought and poor soil conservation practices led to severe erosion of topsoil by wind in the 1930s.

western states was lost through a combination of poor cultivation practices and drought. Windy and dry, the vast grasslands of the Great Plains stretch across 10 states, from Texas through Montana and the Dakotas. Before settlers began grazing livestock and planting crops there in the 1870s, the deep and tangled root systems of native prairie grasses anchored the fertile topsoil firmly in place (Figure 12-21). Plowing the prairie tore up these roots, and the agricultural crops the settlers planted annually in their place had less extensive root systems.

After each harvest the land was plowed and left bare for several months, exposing it to the plains winds. Overgrazing also destroyed large expanses of grass, denuding the ground. The stage was set for severe wind erosion and crop failures, needing only a long drought to raise the curtain.

Such a drought arrived, lasting from 1926 to 1934. In the 1930s, dust clouds created by hot, dry windstorms darkened the sky at midday in some areas. Rabbits and birds choked to death on the dust. During May 1934 the entire eastern United States was blanketed with a cloud of topsoil blown off the Great Plains as far as 2,400 kilometers (1,500 miles) away. Journalists began calling the Great Plains the Dust Bowl (Figure 12-27).

Cropland equal in area to Connecticut and Maryland combined was stripped of topsoil, and an area the size of New Mexico was severely eroded. Thousands of displaced farm families from Oklahoma, Texas, Kansas, and other states migrated to California or to the industrial cities of the Midwest and East. Most found no jobs because the country was in the midst of the Great Depression.

In that memorable May of 1934, Hugh Bennett of the U.S. Department of Agriculture (USDA) was pleading before a congressional hearing in Washington for new programs to protect the country's topsoil.

A: $4.2 billion (some say $7 billion)

Lawmakers took action when Great Plains dust began seeping into the hearing room.

In 1935 the United States established the Soil Conservation Service (SCS) under the Department of Agriculture. With Bennett as its first head, the SCS began promoting good conservation practices, first in the Great Plains states and later in every state. Soil conservation districts were formed throughout the country, and farmers and ranchers were given technical assistance in setting up soil conservation programs. But these heroic efforts have not stopped human-accelerated erosion in the Great Plains. The basic problem is that much of the region is better suited for moderate grazing than for farming.

Depletion of groundwater in the massive Ogallala aquifer underlying the Great Plains (Figure 11-16) is proceeding much faster than the water is being replenished; this also threatens farming and ranching in parts of the Dust Bowl area. If Earth warms as projected (Figure 10-10), the region could become even drier, and farming might have to be abandoned.

Today, soil on cultivated land in the United States is eroding about 16 times faster than it can form. And erosion rates are even higher in heavily farmed regions, such as the Great Plains, which has lost one-third or more of its topsoil in the 150 years since it was first plowed. Parts of the western rangelands and the Great Plains are rapidly becoming deserts from overcultivation, overgrazing, and depletion of groundwater used for irrigation—a new Dust Bowl waiting to happen. Some of the country's most productive agricultural lands, such as those in Iowa, have lost about half their topsoil. California's soil is eroding 80 times faster than it can be formed.

Enough topsoil erodes away each day in the United States to fill a line of dump trucks 5,600 kilometers (3,500 miles) long. About 86% of it comes from land used to graze cattle or to raise crops to feed cattle. The other 14% of eroded soil comes from land used to raise crops for human consumption. David Pimentel (Guest Essay, p. 333) estimates that the direct and indirect costs of soil erosion and runoff in the United States exceed $25 billion per year—an average loss of $2.9 million per hour!

12-8 SOLUTIONS: SOIL CONSERVATION

Conservation Tillage Soil conservation involves reducing soil erosion and restoring soil fertility. Most methods used to control soil erosion involve keeping the soil covered with vegetation (Figure 12-28).

In **conventional-tillage farming** the land is plowed, and the soil is broken up and smoothed to make a planting surface. In areas such as the midwestern United States (Figure 12-27) harsh winters prevent plowing just before the spring growing season. Thus cropfields are often plowed in the fall. This bares the soil during the winter and early spring months, leaving it vulnerable to erosion.

To reduce erosion, many U.S. farmers are trying **conservation-tillage farming** (or *minimum-tillage* or *no-till farming*. The idea is to disturb the soil as little as possible while planting crops. With minimum tillage, special tillers break up and loosen the subsurface soil without turning over the topsoil, previous crop residues, and any cover vegetation. In no-till farming special planting machines inject seeds, fertilizers, and weed-killers (herbicides) into slits made in the unplowed soil.

Besides reducing soil erosion, conservation tillage saves fuel, cuts costs, holds more water in the soil, keeps the soil from getting packed down, and allows more crops to be grown during a season (multiple cropping). Yields are at least as high as those from conventional tillage. At first, conservation tillage was thought to require more herbicides. However, a 1990 U.S. Department of Agriculture (USDA) study of corn production in the United States found no real difference in levels of herbicide use between conventional and conservation-tillage systems.

Conservation tillage is now used on about one-third of U.S. croplands and is projected to be used on over half of them by the year 2000. The USDA estimates that using conservation tillage on 80% of U.S. cropland would reduce soil erosion by at least half. So far, the practice is not widely used in other parts of the world.

Terracing, Contour Farming, Strip Cropping, and Alley Cropping For hundreds of years farmers have used various methods to reduce soil erosion (Figure 12-28). **Terracing** can be used on steeper slopes. The slope is converted into a series of broad, nearly level terraces that run across the land contour with short vertical drops from one terrace to another (Figure 12-28a). Terracing retains water for crops at each level and cuts soil erosion by controlling runoff. In areas of high rainfall, diversion ditches must be built behind each terrace to permit adequate drainage.

In mountainous areas, such as the Himalaya on the border between India and Tibet and the Andes near the west coast of South America, farmers have traditionally built elaborate systems of terraces. Terracing allowed them to cultivate steeply sloping land that would otherwise quickly lose its topsoil.

Today, however, some of these slopes are being farmed without terraces, leaving the land too poor after 10–40 years to grow crops or generate new forest. Although most poor farmers know the risk of not

Q: How much of U.S. public rangeland is in unsatisfactory (fair or poor) condition?

Prato/Bruce Coleman Ltd.

a. Terracing in Bali

Soil Conservation Service

b. Contour planting and strip cropping in Illinois

P. A. Sanchez/North Carolina State University

c. Alley cropping in Peru

Figure 12-28 Soil conservation methods.

Soil Conservation Service

d. Windbreaks in South Dakota

terracing, many have too little time and too few workers to build terraces. They must plant crops or starve. The resultant loss of protective vegetation and topsoil also greatly intensifies flooding below these watersheds, as in Bangladesh (Figure 11-9 and Connections, p. 266).

Soil erosion can be reduced 30–50% on gently sloping land by means of **contour farming**: plowing and planting crops in rows across, rather than up and down, the sloped contour of the land (Figure 12-28b). Each row planted horizontally—along the contour of the land—acts as a small dam to help hold soil and slow the runoff of water.

In **strip cropping**, a row crop like corn is alternated in strips with a soil-saving cover crop, such as a grass or a grass-legume mixture, that completely covers the soil and thus reduces erosion (Figure 12-28b). The strips of cover crop trap soil that erodes from the row crop. The cover crops catch and reduce water runoff and also

help prevent the spread of pests and plant diseases from one strip to another. In addition, they help restore soil fertility if nitrogen-fixing legumes, such as soybeans or alfalfa, are planted in some of the strips.

Erosion can also be reduced by **alley cropping**, or **agroforestry**, a form of *intercropping* in which several crops are planted together in strips or alleys between trees and shrubs that can provide fruit or fuelwood (Figure 12-28c). The trees provide shade (which reduces water loss by evaporation) and help to retain soil moisture and release it slowly. The tree and shrub trimmings can be used as mulch (green manure) for the crops and as fodder for livestock.

Gully Reclamation, Windbreaks, Land Classification, and PAM Water runoff quickly creates gullies in sloping bare land (Figures 1-3 and 12-23). Such land can be restored by **gully reclamation**. Small gullies can be seeded with quick-growing

Slowing Erosion in the United States

The 1985 Farm Act established a strategy to reduce soil erosion in the United States. In the first phase of this program farmers are given a subsidy for highly erodible land they take out of production and replant with soil-saving grass or trees for 10 years. The land in such a *conservation reserve* cannot be farmed, grazed, or cut for hay. Farmers who violate their contracts must pay back all subsidies with interest.

By 1992, more than 14 million hectares (35 million acres) of land had been placed in the conservation reserve, cutting soil erosion on U.S. cropland by almost one-third. If the program is expanded and adequately enforced, it could cut soil losses on U.S. cropland by 80%.

The second phase of the program required all farmers with highly erodible land to develop Soil Conservation Service (SCS)-approved five-year soil-conservation plans for their entire farms by the end of 1990. Farmers not implementing their plans by 1995 can lose eligibility for government subsidies and loans. A third provision of the Farm Act authorizes the government to forgive all or part of farmers' debts to the Farmers Home Administration if they agree not to farm highly erodible cropland or wetlands for 50 years. The farmers are required to plant trees or grass on this land or to convert it back into wetland.

In 1987, however, the SCS eased the standards that farmers' soil-conservation plans must meet to keep them eligible for other subsidies. Environmentalists have also accused the SCS of laxity in enforcing the Farm Act's "swampbuster" provisions, which deny federal funds to farmers who drain or destroy wetlands on their property. Despite some weaknesses the 1985 Farm Act makes the United States the first major food-producing country to make soil conservation a national priority.

plants such as oats, barley, and wheat for the first season, whereas deeper gullies can be dammed to collect silt and gradually fill in the channels. Fast-growing shrubs, vines, and trees can also be planted to stabilize the soil, and channels can be built to divert water from the gully and prevent further erosion.

Wind erosion can be reduced by **windbreaks**, or **shelterbelts**: long rows of trees planted so that they partially block the wind (Figure 12-28d). Windbreaks are especially effective if uncultivated land is kept covered with vegetation; they also help retain soil moisture, supply some wood for fuel, and provide habitats for birds, pest-eating and pollinating insects, and other animals. Unfortunately, many of the windbreaks planted in the upper Great Plains after the 1930s Dust Bowl disaster (Figure 12-27) have been cut down to make way for large irrigation systems and farm machinery.

Land can be evaluated with the goal of identifying easily erodible (marginal) land that should neither be planted in crops nor cleared of vegetation. In the United States, the Soil Conservation Service (SCS) has set up a land-use classification system. The SCS basically relies on voluntary compliance with its guidelines in the almost 3,000 local and state soil- and water-conservation districts it has established, and it provides technical and economic assistance through local district offices.

Recently a chemical, polyacrylamide (PAM), has been used to sharply reduce erosion of some irrigated fields at moderate cost in most soils (except desert soils). Tests show that adding just ten parts per million (ppm) of this white crystal to water during the first hour of irrigation can reduce erosion by 70–99%. After that, irrigation can continue for 12–24 hours without further treatment. The chemical works best when negatively charged PAM crystals are used on soils with an overall negative charge. It is speculated that the negatively charged PAM particles may bind to the positively charged clay particles in soils and thus reduce erosion by increasing the cohesiveness of surface soil particles.

Of the world's major food-producing countries, only the United States is reducing some of its soil losses (Solutions, at left). Even so, effective soil conservation is practiced on only about half of all U.S. agricultural land, and on less than half of the country's most erodible cropland.

Maintaining and Restoring Soil Fertility

Fertilizers partially restore plant nutrients lost by erosion, crop harvesting (Figure 12-20), and leaching when water flows through soil layers. Farmers can use **organic fertilizer** from plant and animal materials and **commercial inorganic fertilizer** produced from various minerals.

Three basic types of organic fertilizer are animal manure, green manure, and compost. **Animal manure** includes the dung and urine of cattle, horses, poultry, and other farm animals. It improves soil structure, adds organic nitrogen, and stimulates beneficial soil bacteria and fungi. Despite its effectiveness the use of animal manure in the United States has decreased. One reason is that separate farms for growing crops and raising animals have replaced most mixed animal-raising and crop-farming operations. Animal manure is available at feedlots near urban areas, but transporting it to distant rural crop-growing areas usually costs too much.

Thus much of this valuable resource is wasted and can end up polluting nearby bodies of water (Figure 11-26). In addition, tractors and other motorized farm machinery have replaced horses and other draft animals that naturally added manure to the soil.

Green manure is fresh or growing green vegetation plowed into the soil to increase the organic matter and humus available to the next crop. It may consist of weeds in an uncultivated field, grasses and clover in a field previously used for pasture, or legumes such as alfalfa or soybeans grown to build up soil nitrogen.

Compost is a rich natural fertilizer and soil conditioner. It aerates soil, improves its ability to retain water and nutrients, helps prevent erosion, and prevents nutrients from being wasted in landfills. Farmers and homeowners produce it by piling up alternating layers of carbohydrate-rich plant wastes (such as grass clippings, leaves, weeds, hay, straw, and sawdust), kitchen scraps (such as vegetable remains and egg shells), animal manure, and topsoil (Figure 12-29). This mixture provides a home for microorganisms that aid the decomposition of the plant and manure layers.

Another method for conserving soil nutrients is **crop rotation**. Corn, tobacco, and cotton can deplete the topsoil of nutrients (especially nitrogen) if planted on the same land several years in a row. One year farmers using crop rotation plant areas or strips with such nutrient-depleting crops; the next year they plant the same areas with legumes, whose root nodules (Figure 4-31) add nitrogen to the soil, or with crops such as oats, barley, rye, or sorghum. This method helps restore soil nutrients and reduces erosion by keeping the soil covered with vegetation.

Will Inorganic Fertilizers and Irrigation Save the Soil? Today, especially in the United States and other MDCs, farmers rely on *commercial inorganic fertilizers* containing nitrogen (as ammonium ions, nitrate ions, or urea), phosphorus (as phosphate ions), and potassium (as potassium ions). Other plant nutrients may also be present in low or trace amounts.

Inorganic commercial fertilizers are easily transported, stored, and applied. Worldwide, their use increased about ninefold between 1950 and 1984, but since then it has increased little and has actually declined slightly since 1988. Today, the additional food they help produce feeds one of every three people in the world. Without them, world food output would plummet an estimated 40%.

Commercial inorganic fertilizers have some disadvantages, however. They do not add humus to the soil. Unless animal manure and green manure are also added, the soil's content of organic matter—and thus its ability to hold water—will decrease, and the soil will become compacted and less suitable for crop

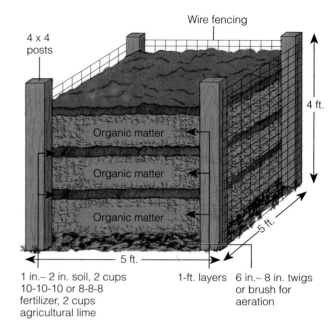

Figure 12-29 A simple home compost bin can be used to produce a mulch for garden and yard plants. You can also create an odorless indoor compost pile in a plastic bag or garbage container.

growth. By decreasing the soil's porosity, inorganic fertilizers also lower the oxygen content of soil and keep added fertilizer from being taken up as efficiently. In addition, most commercial fertilizers supply only 2 or 3 of the 20-odd nutrients needed by plants.

The widespread use of commercial inorganic fertilizers, especially on sloped land near streams and lakes, also causes water pollution as some of the nutrients in the fertilizers are washed into nearby bodies of water; there the resulting cultural eutrophication (Figure 11-26) causes algae blooms that use up oxygen dissolved in the water, thereby killing fish. Rainwater seeping through the soil can also leach nitrates in commercial fertilizers into groundwater. Drinking water drawn from wells containing high levels of nitrate ions can be toxic, especially for infants.

The approximately 18% of the world's cropland that is now irrigated (Figure 11-18) produces about one-third of the world's food. Irrigated land can produce crop yields that are two to three times greater than those from rain-watering, but irrigation has its downside. Irrigation water contains dissolved salts. In dry climates, much of the water in this saline solution evaporates, leaving its salts (such as sodium chloride) in the topsoil. The accumulation of these salts, called **salinization** (Figure 12-30), stunts crop growth, lowers yields, and eventually kills crop plants and ruins the land.

It is estimated that salinization is reducing yields on one-fourth of the world's irrigated cropland. Worldwide, 50–65% of all currently irrigated cropland

What You Can Do to Help Protect the Soil

- When building a home, save all the trees possible. Require the contractor to disturb as little soil as possible, to set up barriers that catch any soil eroded during construction, and to save and replace any topsoil removed instead of hauling it off and selling it. Plant dis-turbed areas with fast-growing native ground cover (preferably not grass) immediately after construction is completed.

- Landscape the area not used for gardening with a mix of wildflowers, herbs (for cooking and for repelling insects), low-growing ground cover, small bushes, and other forms of vegetation natural to the area. This biologically diverse type of yard saves water, energy, and money, and it reduces infestation of mosquitoes and other damaging insects by providing a diversity of habitats for their natural predators.

- Set up a compost bin (Figure 12-29) and use it to produce mulch and soil conditioner for yard and garden plants.

Figure 12-30 Salinization and waterlogging of soil on irrigated land without adequate drainage lead to decreased crop yields.

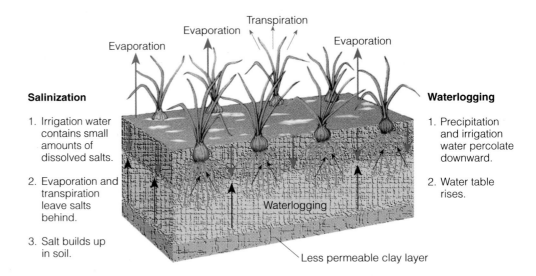

Salinization

1. Irrigation water contains small amounts of dissolved salts.

2. Evaporation and transpiration leave salts behind.

3. Salt builds up in soil.

Waterlogging

1. Precipitation and irrigation water percolate downward.

2. Water table rises.

Less permeable clay layer

will probably have undergone reduced productivity from salting by 2000.

Salts can be flushed out of soil by applying much more irrigation water than is needed for crop growth, but this practice increases pumping and crop-production costs, and it wastes enormous amounts of water. Heavily salinized soil can also be renewed by taking the land out of production for two to five years, installing an underground network of perforated drainage pipes, and flushing the soil with large quantities of low-salt water. This costly scheme, however, only slows the salt buildup; it does not stop the process. Flushing salts from the soil also makes downstream irrigation water saltier, unless the saline water can be drained into evaporation ponds rather than returned to the stream or canal.

Another problem with irrigation is **waterlogging** (Figure 12-30). Farmers often apply heavy amounts of irrigation water to leach salts deeper into the soil. Without adequate drainage, however, water accumulates underground, gradually raising the water table.

Saline water then envelops the roots of plants and kills them. At least one-tenth of all irrigated land worldwide suffers from waterlogging, and the problem is getting worse.

Responsibility for reducing soil erosion should not be limited to farmers. At least 40% of soil erosion in the United States is caused by timber cutting, overgrazing, mining, and urban development carried out without proper regard for soil conservation.

Because soil is the base of life, human-accelerated soil depletion must be slowed and replaced with soil rebuilding and conservation so that these vital, slowly renewable resources are used sustainably. Each of us has a role to play in seeing that soil resources are used sustainably (Individuals Matter, above).

At some point, either the loss of topsoil from the world's croplands will have to be checked by effective soil conservation practices, or the growth in the world's population will be checked by hunger and malnutrition.

LESTER R. BROWN

Land Degradation and Environmental Resources

David Pimentel

GUEST ESSAY

David Pimentel is professor of insect ecology and agricultural sciences in the College of Agriculture and Life Sciences at Cornell University. He has published over 350 scientific papers and 12 books on environmental topics including land degradation, agricultural pollution and energy use, biomass energy, and pesticides. He was one of the first ecologists to employ an interdisciplinary, holistic approach in investigating complex environmental problems.

At a time when the world's human population is rapidly expanding and its need for more land to produce food, fiber, and fuelwood is also escalating, valuable land is being degraded through erosion and other means at an alarming rate. Soil degradation is of great concern because soil reformation is extremely slow. Worldwide annual erosion rates for agricultural land are about 20–100 times the average of 500 years (with a range of 220 to 1,000 years) required to renew 2.5 centimeters (1 inch) of soil in tropical and temperate areas—a renewal rate of about 1 metric ton of topsoil per hectare of land per year.

Erosion rates vary in different regions because of topography, rainfall, wind intensity, and the type of agricultural practices used. In China, for example, the average annual soil loss is reported to be about 40 metric tons per hectare (18 tons per acre) while the U.S. average is 18 metric tons per hectare (8 tons per acre). In states like Iowa and Missouri, however, annual soil erosion averages more than 35 metric tons per hectare (16 tons per acre).

Worldwide, about 10 million hectares (25 million acres) of land—about the size of Virginia—are abandoned for crop production each year because of high erosion rates plus waterlogging of soils, salinization, and other forms of soil degradation. In addition, according to the UN Environment Programme, crop production becomes uneconomical on about 20 million hectares (49 million acres) each year because soil quality has been severely degraded.

Soil erosion also occurs in forestlands but is not as severe as that in the more exposed soil of agricultural land. However, soil erosion in managed forests is a primary concern because the soil reformation rate in forests is about two to three times longer than that in agricultural land. To compound this erosion problem, at least 24 million hectares (59 million acres) of forest are being cleared each year throughout the world, with most of this land used to grow food and graze cattle.

The effects of agriculture and forestry are interrelated in other ways. Deforestation reduces fuelwood supplies and forces the poor in LDCs to substitute crop residue and manure for fuelwood. When these plant and animal wastes are burned instead of being returned to the land

as ground cover and organic fertilizer, erosion is intensified and productivity of the land is decreased. These factors, in turn, increase pressure to convert more forestland into agricultural land, further intensifying soil erosion.

One reason that soil erosion is not of high-priority concern among many governments and farmers is that it usually occurs so slowly that its cumulative effects may take decades to become apparent. For example, the removal of 1 millimeter (1/25 inch) of soil is so small that it goes undetected. But over a 25-year period the loss would be 25 millimeters (1 inch)—taking about 500 years to replace by natural processes.

Besides reduced soil depth, soil erosion leads to reduced crop productivity because of losses of water, organic matter, and soil nutrients. Water is the primary limiting factor for all natural and agricultural plants and trees. A 50% reduction of soil organic matter on a plot of land has been found to reduce corn yields as much as 25%.

When soil erodes, vital plant nutrients such as nitrogen, phosphorus, potassium, and calcium are also lost. With U.S. annual cropland erosion rates of about 18 metric tons per hectare (8 tons per acre), an estimated $18 billion of plant nutrients are lost annually. Using fertilizers to replace these nutrients substantially adds to the cost of crop production.

Some analysts who are unaware of the numerous and complex effects of soil erosion have falsely concluded that the damages are relatively minor. For example, they report that soil loss causes an annual reduction in crop productivity of only 0.1–0.5% in the United States. However, we need to consider all the ecological effects caused by erosion, including reductions in soil depth, in availability of water for crops, and in soil organic matter and nutrients. When this is done, agronomists and ecologists report a 15–30% reduction in crop productivity—a key factor in increased levels of costly fertilizer and declining yields on some land despite high levels of fertilization. Because fertilizers are not a substitute for fertile soil, they can be applied only up to certain levels before crop yields begin to decline.

Reduced agricultural productivity is only one of the effects of soil erosion. In the United States water runoff is responsible for transporting about 3 billion metric tons (3.3 billion tons) of sediment (about 60% from agricultural land) each year to waterways in the lower 48 states. Off-site damages to U.S. water storage capacity, wildlife, and navigable waterways from these sediments cost an estimated $6 billion each year. Dredging sediments from U.S. streams, harbors, and reservoirs alone costs about $570 million each year. About 25% of new water storage capacity in U.S. reservoirs is built solely to compensate for sediment buildup.

(continued)

When soil sediments that include pesticides and other agricultural chemicals are carried into streams, lakes, and reservoirs, fish production is adversely affected. These contaminated sediments interfere with fish spawning, increase predation on fish, and destroy fisheries in estuarine and coastal areas.

Increased erosion and water runoff on mountain slopes flood agricultural land in the valleys below, further decreasing agricultural productivity. Eroded land also does not hold water very well, again decreasing crop productivity. This effect is magnified in the 80 countries (with nearly 40% of the world's population) that experience frequent droughts. The rapid growth in the world's population, accompanied by the need for more crops and a projected doubling of water needs in the next 20 years, will only intensify water shortages, particularly if soil erosion is not contained.

Thus soil erosion is one of the world's critical problems and, if not slowed, will seriously reduce agricultural and forestry production and degrade the quality of aquatic ecosystems. Solutions are not particularly difficult but are often not implemented because erosion occurs so gradually that we fail to acknowledge its cumulative impact until damage is irreversible. Many farmers have also been conditioned to believe that losses in soil fertility can be remedied by applying more fertilizer or by using more fossil-fuel energy.

The principal way to control soil erosion and its accompanying runoff of sediment is to maintain adequate vegetative coverage on soils [methods discussed in Section 12-8]. These methods are also cost-effective, especially when off-site costs of erosion are included. Scientists, policymakers, and agriculturists need to work together to implement soil and water conservation practices before world soils lose most of their productivity.

Critical Thinking

1. Some analysts contend that average soil erosion rates in the United States and the world are low and that the soil erosion problem can easily be solved with improved agricultural technology such as no-till cultivation and increased use of commercial inorganic fertilizers. Do you agree or disagree with this position? Explain.

2. What specific things do you believe elected officials should do to decrease soil erosion and the resulting water pollution by sediment in the United States?

Critical Thinking

1. Explain what would happen if plate tectonics stopped. Explain what would happen if erosion and mass wasting stopped.

2. Explain why you agree or disagree with each of the following propositions:
   a. The competitive free market will control the supply and demand of mineral resources.
   b. New discoveries will provide all the raw materials we need.
   c. The ocean will supply all the mineral resources we need.
   d. We will not run out of key mineral resources because we can always mine lower-grade deposits.
   e. When a mineral resource gets scarce, we can always find a substitute.
   f. When a nonrenewable resource gets scarce, all we have to do is recycle or reuse it.

3. Explain why you support or oppose:
   a. Eliminating all tax breaks and depletion allowances for extraction of virgin resources by mining industries
   b. Stopping the grant of title to public lands for actual or imaginary hard-rock mining (p. 301)
   c. Requiring mining companies to pay a royalty of 12.5% on any hard-rock minerals they extract from federal land

   d. Making hard-rock mining companies responsible for restoring the land and cleaning up environmental damage caused by their activities

4. Use the second law of energy (Section 3-6) to analyze the physical and economic feasibility of the following:
   a. Extracting most minerals dissolved in seawater
   b. Recycling minerals that are widely dispersed
   c. Mining increasingly lower-grade mineral deposits
   d. Using inexhaustible solar energy to mine minerals
   e. Continuing to mine, use, and recycle minerals at increasing rates

5. Why should everyone, not just farmers, be concerned with soil conservation?

6. What are the main advantages and disadvantages of commercial inorganic fertilizers? Why should both inorganic and organic fertilizers be used?

*7. As a class project evaluate soil erosion on your school grounds. Develop a soil conservation plan for your school and present it to school officials.

*8. What mineral resources are extracted in your local area? What mining methods are used? Do laws require restoration of the landscape after mining? If so, how stringently are those laws enforced?

*9. Make a concept map of the key ideas in this chapter using the section heads and subheads and the key terms. See the inside front cover and Appendix 4 for information on concept maps.

13 Wastes: Reduction and Prevention

There Is No "Away": The Love Canal Tragedy

Between 1942 and 1953 Hooker Chemicals and Plastics Corporation (now owned by Occidental Chemical) sealed its chemical wastes into steel drums and dumped them into an old canal excavation (called Love Canal after its builder, William Love) near Niagara Falls, N.Y. Several federal agencies (especially the army) also dumped wastes at the site.

In 1953 Hooker Chemicals covered the dump site with clay and topsoil, and under threat of condemnation reluctantly sold it to the Niagara Falls school board for $1. The company inserted in the deed a disclaimer saying it was not legally liable for any injury caused by the wastes. In 1957 Hooker warned the school board not to disturb the clay cap because of damage from toxic wastes.

By 1959 an elementary school, playing fields, and 949 homes had been built in the Love Canal area (Figure 13-1). Between 1971 and 1977 the residents of Love Canal discovered that "out of sight, out of mind" can be hazardous to your health and peace of mind. Toxic industrial wastes buried decades earlier bubbled to the surface, found their way into groundwater, and ended up in storm sewers, gardens, basements of homes and the school next to the canal, and the school playground.

In 1978, after considerable pressure from alarmed residents and unfavorable publicity, the state acted. It closed the school, permanently relocated the 236 families whose homes were closest to the dump, and fenced off the area around the canal. Two years later, after protests from outraged families still living fairly close to the landfill, President Jimmy Carter declared Love Canal a federal disaster area and had the remaining families permanently relocated.

The school and 239 homes within a block and a half of the canal were torn down. All remaining homes, except 60 whose owners decided to stay, were purchased by the state. The dump site was covered with a new clay cap and surrounded by a drain system that pumps leaking wastes to a new treatment plant. By 1991 the total cost for cleanup and relocation had reached $275 million.

Because of the difficulty in linking exposure to a variety of chemicals to specific health effects (Section 8-2), possible long-term health effects of exposure to hazardous chemicals on Love Canal residents remain controversial and unknown. However, the psychological damage to the evacuated families is enormous. For the rest of their lives they will worry about the possible effects of the chemicals on themselves and their children and grandchildren.

In June 1990 the EPA declared the area—renamed Black Creek Village—safe and allowed state officials to begin selling some of the 236 dilapidated and boarded-up houses at 10–20% below market value. Yet, the dump has not been cleaned up but only fitted with a drainage system, and even the EPA acknowledges that it will leak again sooner or later.

The Love Canal incident is a vivid reminder that we can never really throw anything away, that wastes don't stay put, and that preventing pollution is much safer, easier, and cheaper than trying to clean it up.

Figure 13-1 The Love Canal housing development near Niagara Falls, New York, was built near a hazardous-waste dump site. This photo shows the area when it was abandoned in 1980. In 1990 the EPA declared the site safe, and people were allowed to buy the remaining houses and move back into the area despite protests from environmentalists. (NY State Department of Environmental Conservation)

Solid wastes are only raw materials we're too stupid to use.
ARTHUR C. CLARKE

This chapter is devoted to answering the following questions:

- What is solid waste, and how much is produced?
- What can we do about solid waste?
- What is hazardous waste, and how much is produced?
- What can we do about hazardous waste?
- What is done with hazardous waste in the United States?
- How can we make the transition to a low-waste society?

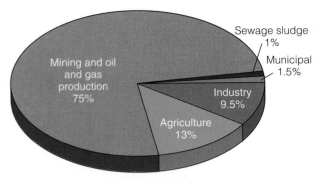

Figure 13-2 Sources of the 10 billion metric tons (11 billion tons) of solid waste produced each year in the United States. Some 65 times as much solid waste is produced by mining and industrial activities as by household garbage. (Data from Environmental Protection Agency and U.S. Bureau of Mines)

13-1 WASTING RESOURCES: SOLID WASTE AND THE THROWAWAY APPROACH

Sources of Solid Waste With only 4.7% of the world's population, the United States produces 33% of the earth's **solid waste**: any unwanted or discarded material that is not a liquid or a gas. The United States generates about 10 billion metric tons (11 billion tons) of solid waste per year—an average of 40 metric tons (44 tons) per person! While garbage produced directly by households and businesses is a significant problem, about 98.5% of the solid waste in the United States comes from mining, oil and natural gas production, agriculture, and industrial activities (Figure 13-2). Although individuals don't generate this waste directly, it is produced indirectly through the products they consume.

Most mining waste is left piled near mine sites and can pollute the air, surface water, and groundwater. Although mining waste is the single largest category of U.S. solid waste, the EPA has done little to regulate its disposal, mostly because Congress has exempted mining wastes from regulation as a hazardous waste. In LDCs there is even less regulation of mining procedures and wastes.

Industrial solid waste includes scrap metal, plastics, paper, fly ash (removed by air pollution control equipment—Figure 9-18—in industrial and electrical power plants), and sludge from industrial waste treatment plants. Most of it is buried or incinerated at the plant site where it was produced.

The remaining 1.5% of solid waste produced in the United States is **municipal solid waste** (MSW) from homes and businesses in or near urban areas. An estimated 182 million metric tons (200 million tons) of

municipal solid waste—often referred to as garbage—was produced in the United States in 1993. This waste would fill a bumper-to-bumper convoy of garbage trucks encircling the earth almost six times. It amounts to 700 kilograms (1,540 pounds) per person annually—two to three times that in most other MDCs (except Australia, New Zealand, Canada, France, Finland, Norway, and Denmark) and many times that in LDCs.

About 17% of these potentially usable resources is recycled or composted (up from 7% in 1960). The other 83% is hauled away and either dumped in about 5,300 landfills (72%) or burned in 179 incinerators and waste-to-energy plants (11%) at a cost of about $30 billion per year (projected to rise to $45 billion by 1995 and to $75 billion by 2000). By 1995, the EPA estimates that 58% of U.S. municipal solid waste will be sent to landfills, 17% burned, and 25% recycled or composted.

Litter is also a source of solid waste. For example, helium-filled balloons are often released into the atmosphere at ball games and parties. When the helium escapes or the balloons burst, they become litter. Fish, turtles, seals, whales, and other aquatic animals die when they ingest balloons that fall into oceans and lakes. The use of these balloons implies that it is acceptable to litter, to waste helium (a scarce resource) and energy (used to separate helium from air), and to kill wildlife (even if we switched to balloons that biodegraded within six weeks). Mass releases of helium balloons are now banned in Florida, Connecticut, Tennessee, and Virginia.

What It Means to Live in a Throwaway Society With only 4.7% of the world's population, U.S. consumers throw away:

- Enough aluminum to rebuild the country's entire commercial airline fleet every three months.

Q: How much of all U.S. land area is protected as wilderness?

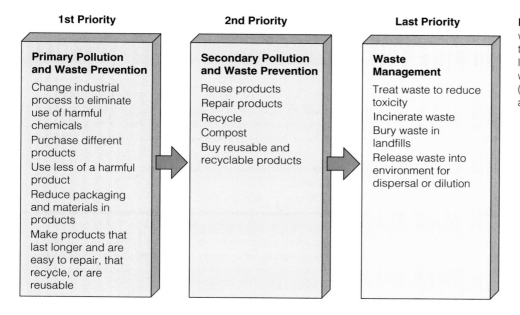

1st Priority	2nd Priority	Last Priority
Primary Pollution and Waste Prevention Change industrial process to eliminate use of harmful chemicals Purchase different products Use less of a harmful product Reduce packaging and materials in products Make products that last longer and are easy to repair, that recycle, or are reusable	**Secondary Pollution and Waste Prevention** Reuse products Repair products Recycle Compost Buy reusable and recyclable products	**Waste Management** Treat waste to reduce toxicity Incinerate waste Bury waste in landfills Release waste into environment for dispersal or dilution

Figure 13-3 Priorities for dealing with material use and waste. So far these priorities are not being followed, with most efforts devoted to waste management (bury or burn). (Environmental Protection Agency and National Academy of Sciences)

- Enough glass bottles to fill the two 412-meter-high (1,350-foot) towers of the New York World Trade Center every two weeks.

- Enough tires each year to encircle the earth almost three times.

- Enough disposable plates and cups each year to serve everyone in the world six meals.

- About 18 billion disposable diapers per year, which if linked end to end would reach to the moon and back seven times.

- About 2 billion disposable razors, 1.6 billion throwaway ballpoint pens, and 500 million disposable cigarette lighters each year.

- About 2.5 million nonreturnable plastic bottles each hour.

- Some 14 billion catalogs (an average of 54 per American) and 38 billion pieces of junk mail per year.

And this is only part of the 1.5% of solid waste labeled as municipal in Figure 13-2!

13-2 SOLUTIONS: REDUCING AND REUSING SOLID WASTE

What Are Our Options? There are two ways to deal with the mountains of solid waste we produce: *waste management* and *pollution (waste) prevention*. Waste management is a *throwaway* or *high-waste approach* that encourages waste production (Figure 3-12) and then attempts to manage the wastes in ways that will reduce

environmental harm—mostly by burying them or burning them.

Sooner or later, however, even the best-designed waste incinerators release some toxic substances into the atmosphere and leave a toxic residue that must be disposed of—usually in landfill. Furthermore, even the best-designed landfills eventually leak wastes into groundwater. And eventually we run out of affordable or politically acceptable sites for landfills and incinerators.

No approach to dealing with solid waste is more widely praised and less widely used than producing less garbage. The basic problem is that modern economic systems give higher rewards to those who produce waste than to those who try to use resources more efficiently. We give timber, mining, and energy companies tax write-offs and other subsidies to cut trees and to find and mine copper, oil, coal, and uranium (p. 301). At the same time we seldom subsidize companies and businesses that recycle copper or paper, use oil or coal more efficiently, or develop renewable alternatives to fossil fuels. That tilts the economic playing field against waste prevention and in favor of waste production.

Preventing pollution and waste is a *low-waste approach*, which views solid wastes as resources or wasted solids that we should be recycling, reusing, or not using in the first place (Figures 3-13 and 13-3). The low-waste approach involves teaching people to see trash cans and dumpsters as resource containers, and trash as concentrated urban ore that needs to be mined for useful materials for recycling or reuse. With this approach, the economic system is used to discourage waste production and encourage waste prevention especially by shifting to full-cost pricing (Section 7-2 and Guest Essay, p. 338).

A: 4% (only 1.8 % in the lower 48 states)

We Have Been Asking the Wrong Questions About Wastes

Lois Marie Gibbs

GUEST ESSAY

In 1977 Lois Marie Gibbs was a housewife living nearing the Love Canal, New York, toxic dump site (p. 335). She had never engaged in any sort of political action until toxic chemicals began oozing from the dump site into front yards and basements. Then she organized her neighborhood and became the president and major strategist for the Love Canal Homeowners Association. This dedicated grass-roots political action by "amateurs" brought hazardous-waste issues to national prominence and spurred passage of the federal Superfund legislation to help clean up abandoned hazardous-waste sites. Lois Gibbs then moved to Washington, D.C., and formed Citizens' Clearinghouse for Hazardous Wastes, an organization that has helped over 7,000 community grass-roots organizations protect themselves from hazardous wastes. Her story is told in her autobiography, Love Canal: My Story *(State University of New York Press, 1982), and was also the subject of a CBS movie,* Lois Gibbs: The Love Canal, *which aired in 1982. She is an inspiring example of what an ordinary citizen can do to change the world.*

Just about everyone knows our environment is in danger. One of the most serious threats is the massive amount of waste put into the air, water, and ground every year. All across the United States and around the world, there are thousands of places that have been, and continue to be, polluted by toxic chemicals, radioactive waste, and just plain garbage.

For generations the main question people have asked is, "Where do we put all this waste? It's got to go somewhere." That is the wrong question, as has been shown by a series of experiments in waste disposal and by the simple fact that there is no "away" [Section 3-4].

We tried dumping our waste in the oceans. That was wrong. We tried injecting it into deep, underground wells. That was wrong. We've been trying to build landfills that don't leak. That doesn't work. We've been trying to get rid of waste by burning it in high-tech incinerators. That only produces different types of pollution, such as air pollution and toxic ash. We've tried a broad range of "pollution" controls. But all that does is allow legalized, high-tech pollution. Even recycling, which is a very good thing to do, suffers from the same problem as all the other methods: It addresses waste *after* it has been produced.

For many years people have been assuming that "it's got to go somewhere," but now many people, especially young people, are starting to ask, "Why?" Why do we produce so much waste? Why do we need products and services that have so many toxic by-products? Why can't industry change the way it makes things so that it stops producing so much waste?

These are the *right* questions. When you start asking them, you start getting answers that lead to *pollution prevention* and *waste reduction* instead of simply *pollution control* and *waste management*. People, young and old, who care about pollution prevention are challenging our use and disposal of enormous amounts of polystyrene (Styrofoam) plastic each year. They are challenging companies to stop making products with gases that destroy the ozone layer [Section 10-4] and contribute to the threatening possibility of global warming [Section 10-2]. They are asking why so many goods are wrapped in excessive, throwaway packaging. They are challenging companies that sell pesticides, cleaning fluids, batteries, and other hazardous products to either remove the toxics from those products or take them back for recovery or recycling, rather than disposing of them in the environment. They are demanding alternatives to throwaway materials in general.

Since 1988 hundreds of student groups have contacted my organization to get help and advice in taking these effective types of actions. Many of these groups begin by working to get polystyrene food packaging out of their school cafeterias and out of local fast-food restaurants. Oregon students even took legal action to get rid of cups and plates made from bleached paper, because the

According to the U.S. National Academy of Sciences, this prevention approach should have the following hierarchy of goals: **(1)** *Reduce* waste and pollution by preventing its creation; **(2)** *Reuse* as many things as possible; **(3)** *Recycle and compost* as much waste as possible; **(4)** *Incinerate or treat* waste that can't be reduced, reused, recycled, or composted; and **(5)** *Bury* what is left in state-of-the-art landfills after the first four goals have been met.

Experience in the Netherlands and various communities in the United States indicates that 60–80% of municipal solid waste could be composted (40%) or recycled (20–40%), leaving only 20–40% for incineration or burial. Although, technically, 80% of municipal solid waste could be composted or recycled, the actual rate will depend on the economic rewards built into recycling systems. We are slowly undergoing a shift in how we think about garbage. The old mentality is that recycling is what we save and separate from garbage. The new thinking is that garbage is what's left over after we reduce, reuse, and recycle.

Reducing Waste and Pollution: The Best Choice

Cutting waste and preventing pollution generally save

paper contains the deadly poison dioxin. They were asking the right questions—and getting the right answer—when they demanded the school systems switch to reusable cups, plates, and utensils.

Dozens of student groups have joined with local environmental and grass-roots organizations in their communities to get toxic-waste sites cleaned up or to stop new toxic-waste sites, radioactive waste sites, or waste incinerators from being built.

Waste issues are not simply environmental issues. They are all tied up with economics. Our economy is geared to producing and then disposing of waste. *Somebody* is making money from every scrap of waste and has a vested interest in leaving things the way they are. Environmentalists and industry officials constantly argue about what's called "cost-benefit analysis" [Section 7-2]. Simply stated, this poses the question of whether the benefit of controlling pollution or waste will be greater than the cost. This is another example of the wrong question. The right question is, "Who will benefit and who will pay the cost?"

Waste issues are also issues of *justice* and *fairness.* Again, there's a lot of debate between industry officials and environmentalists, especially those in federal and state environmental agencies, about what they call "acceptable risk." Simply stated, that means industry officials and environmentalists will decide people's exposure to toxic chemicals. Unfortunately they hardly ever ask the people who are actually going to be exposed how they feel about it. Instead, industry officials debate and ask one another how much people will be exposed to. Again, this is the wrong question. It's simply not fair to expose people to chemical poisons without their consent.

Risk analysts often say, "But there's only a one in a million chance of increased death from this toxic chemical." That may be true. But suppose I took a pistol and went to the edge of your neighborhood and began shooting. There's probably only a one in a million chance that I'd hit somebody. But would you give me permission,

would you give me a license, to do that? As long as we don't stand up for our rights and demand that "bullets" in the form of hazardous chemicals not be "fired" in our neighborhoods, we are giving environmental regulators and waste producers a license to kill a certain number of us without our even being consulted.

When you study environmental issues, remember that they are not abstract issues that only happen somewhere else. We *all* have to live, breathe, and survive in this environment. We have all learned that decisions made for us or by us in the past have come back to haunt us. Likewise, today's decisions will affect all of us tomorrow and far into the future.

From my personal experience I know that decisions made to dump wastes at Love Canal and in thousands of other places in the past were not made simply on the basis of the best available scientific knowledge. The same holds true for decisions made about how to manage the wastes we produce today and to produce less waste.

Instead, the world we live in is shaped by decisions based on money and power. If you really want to understand what's behind any given environmental issue, the first question you should ask is, "Who stands to profit from this?" Then ask, "Who is going to pay the price?" You can then identify both sides of the issue and decide whether you want to be part of the problem or part of the solution.

Critical Thinking

1. What changes would you be willing to make in your own lifestyle to prevent pollution and reduce waste?

2. What political and economic changes, if any, do you believe must be made so that we shift from a waste production and waste management society to a pollution prevention and waste reduction society? What actions, if any, are you taking to bring about such social changes?

---

more energy and virgin resources than recycling does, and they reduce the environmental impacts of extracting, processing, and using resources (Figure 12-13). Ways to reduce waste include:

- *Decreasing consumption.* This begins when we ask whether we really need (as opposed to want) a particular product.

- *Using less material per product* (lighter but still strong and safe cars, for example, Solutions, p. 317).

- *Redesigning manufacturing processes to use fewer resources and produce less waste.* For example in the 1970s Campbell Soup redesigned its cans to use 30% less steel.

- *Making and using products that last longer and that are easy to repair, reuse, or recycle.* Several European auto manufacturers, for example, are designing their cars for easy disassembly, reuse, and recycling of various parts, and they are trying to minimize the use of nonrecyclable or hazardous materials. In 1993, IBM introduced its PS/2E personal computer—also known as the

Figure 13-4 Energy used to make a 400-milliliter (12-fluid-ounce) beverage container. (Data from Argonne National Laboratory)

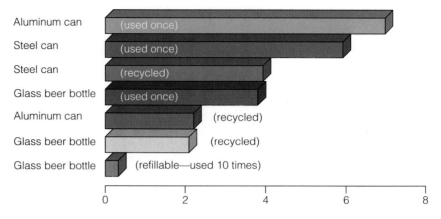

Energy Used (thousand BTUs)

"green machine"—which contains 25% recycled plastic and is designed for easy disassembly and reuse or recycling of most of its parts.

- *Cutting down on unnecessary packaging.* In the United States, packaging accounts for 50% of all paper produced, 90% of all glass, 11% of all aluminum, and 3% of all energy used. Packaging makes up about 50% by volume and 30% by weight of municipal solid waste. Nearly $1 of every $10 spent for food in the United States goes for throwaway packaging. A 1992 study concluded that lightweight packaging material (including plastics and paper) has less overall environmental impact in most products than heavy types of packaging such as glass and steel. Some U.S. manufacturers have reduced the weight of some of their packaging bottles and cartons by 10–30%, but much more needs to be done.

- *Eliminating or sharply reducing the use or production of toxic materials such as arsenic, cadmium, chromium, lead, mercury, dioxins, pesticides, and chlorine-containing compounds.*

- *Reusing products.* The good news is that an array of strategies for waste reduction is already known or being developed. And with incentives, taxes, and full-cost pricing (Table 7-1), which would make the economic playing field more level, they could be phased in within 10–20 years.

Reuse: The Next Best Choice Reuse—a form of waste reduction—extends resource supplies and reduces energy use and pollution even more than recycling. A popular bumper sticker reads: "Recyclers do it more than once." A better version might be, "Recyclers do it more than once, but reusers do it the most." One example is the refillable glass beverage bottle, which can be used 50 times or more and is the most energy-efficient beverage container on the market

(Figure 13-4). Collected and filled at local bottling plants, they reduce transportation and energy costs, create local jobs, and keep more of the money people spend circulating in the local economy. Moreover, studies by Coca-Cola and PepsiCo of Canada show that 0.5-liter (16-ounce) bottles of their soft drinks cost one-third less in refillable bottles.

In 1964, 89% of all soft drinks and 50% of all beer in the United States was sold in refillable glass bottles. Refillable glass bottles now make up only about 6% of the beer and soft drink market, and only 10 states even have refillable bottles.

Denmark has led the way by banning all beverage containers that can't be reused. To encourage use of refillable glass bottles, Ecuador has a beverage container deposit fee that is 50% higher than the cost of the drink. This has been so successful that bottles as old as 10 years continue to circulate. Sorting is not a problem because only two sizes of glass bottles are allowed. In Finland 95% of the soft drink, beer, wine, and spirits containers are refillable, and in Germany 73% are refillable.

Another reusable container is the metal or plastic lunch box that most workers and schoolchildren once used. Today many people carry their lunches in throwaway paper or plastic bags. At work people can have their own reusable glasses, cups, dishes, utensils, cloth napkins, and towels, and they can use washable cloth handkerchiefs instead of throwaway paper tissues. In Germany, outdoor festivals increasingly use washable plates, glasses, and cutlery. Plastic containers with tops are reusable for lunch items and refrigerator leftovers. They can be used in place of throwaway plastic wrap and aluminum foil (most of which is not recycled). This and most forms of reuse also save money.

Let's say you've just paid for your groceries and you're offered a choice between plastic or paper bags. Which do you choose? The answer is, neither. Both are

Q: How much of the medicines sold in the world have active ingredients extracted from wildlife (mostly plants)?

environmentally harmful, and the question of which is the more damaging has no clear-cut answer. On the one hand, plastic bags are made from nonrenewable fossil fuels, degrade slowly in landfills, and can harm wildlife if swallowed; and producing them pollutes the environment.

On the other hand, the brown paper bags in most supermarkets are made from virgin paper. Even those that supposedly contain some recycled fiber use mostly pre-consumer paper waste. Papermaking uses trees, pollutes the air and water, and releases toxic dioxins. On a weight basis plastic bags take less energy to make and transport than paper bags, and they use less landfill space. Paper bags, however, break down faster in landfills and produce fewer toxic substances. Moreover, unlike the oil used to make plastic bags, the wood used to make paper bags is a potentially renewable resource. Also, the technology for recycling paper is more advanced than that for plastics, and recycled paper can be used in a greater variety of products.

Don't fall for labels claiming paper and plastic bags are recyclable. Just about anything is in theory recyclable. What counts is whether it is, in fact, recycled.

So don't choose between paper and plastic. Instead, bring your own *reusable* canvas or string containers when you shop for groceries or other items (Figure 13-5), or save and reuse any paper or plastic bags you get. Several of these reusable bags can be folded up and kept in a handbag, pocket, or car.

What should we do about diapers? About 95% of the 18 billion disposable diapers discarded each year in the United States end up in landfills, where they remain entombed for centuries. Even the new biodegradable diapers take 100 years to break down in a landfill. Production of disposable diapers uses trees, consumes plastic resources (about a third of each diaper is plastic), and creates air and water pollution. Disposable diapers, however, make up only about 1.2% by volume of the typical landfill's contents, compared with paper waste (40–50%) and construction debris (20–30%).

Cloth diapers can be washed and reused 80–200 times and then retired into lint-free rags, which keeps the roughly 10,000 disposable diapers the average baby uses before becoming toilet trained from reaching landfills. Cloth diapers also save trees and money. For example, the disposable diapers needed for one baby cost about $1,533. A cloth diaper service costs about $975, and washing cloth diapers at home costs about $283.

Does the choice seem simple? Consider this: Laundering cloth diapers produces 9 times as much air pollution and 10 times as much water pollution as disposable diapers do in their lifetime. Over their lifetimes, cloth diapers also consume six times more water and three times more energy than disposables.

Figure 13-5 An example of good Earth-keeping. A reusable string bag (or a cloth or canvas bag) can be used to carry groceries and other purchases. This eliminates the need for throwaway paper and plastic bags, both of which are environmentally harmful even if they are recycled.

So the choice between disposable diapers and reusable cloth diapers is not clear-cut. Many people opt for cheaper cloth diapers and use disposable ones for trips or at child-care centers that won't allow use of cloth diapers.

Another big solid-waste problem is created by used tires. There are now 2–4 billion used tires heaped in landfills, old mines, abandoned houses, and other dump sites throughout the United States. Each year the pile grows by about 250 million tires (about 47 million of them suitable for retreading). About 80% of the tires thrown away each year are disposed of in landfills (where they tend to float or rise to the surface and pierce landfill cover), stockpiled, or dumped illegally. Tire dumps are fire hazards and breeding grounds for mosquitoes. A fire in a tire dump is extremely hard to put out and produces air pollution and toxic runoff that can pollute nearby surface water and groundwater.

Instead of being dumped, used tires can be burned to produce electricity and pulverized to make resins for products ranging from car bumpers, garbage cans,

Sailing Through Life in an Earthship

When is a ship not a ship? When it's an Earthship. Because then it's a passive solar house, with three thick walls (made from dirt-packed used tires) partly buried in the ground and the other wall facing the sun, with a glass wall to capture solar energy (Figure 13-6). Though they are rooted in the earth, Earthship architect Michael Reynolds sees such houses as ships designed to "sail" through good times and bad.

Earthship houses can be built for as low as $2 a square meter ($20 a square foot). You could build one yourself, using things that someone else has thrown away—tires, which most tire dealers are glad to give away or even pay you to haul off, and aluminum cans. Once the

house is up, depending on the features you choose to build in, you can pay little for electricity, heating, cooling, and water-heating. You can even grow your own food in the greenhouse created by the sun-facing glass wall, which also captures solar energy for heating.

To build the walls, discarded steel-belted radial tires are filled with dirt and layered in staggered courses like bricks. Dirt (from excavated soil) is packed tight inside (usually with a sledgehammer) and between the tires to make the wall rock solid; then it's finished with a coat of plaster or adobe.

Once heated by the sun, these thick and strong walls have so much thermal mass that they stay warm for a long time, steadily radiating that warmth into the spaces they enclose. Their huge thermal

mass allows them to be built in areas without enough sunlight for conventional solar homes. If, as Reynolds prefers, the whole house is partly underground on the three sides not facing the sun, an Earthship can maintain a steady mean temperature of 18–21°C (65–70°F) in such outside temperature extremes as those in Taos, New Mexico, ranging from 38°C (100°F) in summer to –34°C (–30°F) in winter.

Some of the inner walls can be built of another type of solid waste that is often thrown away—aluminum beverage cans—embedded in cement mortar. Like the tires, they are packed with dirt to provide thermal mass. To Reynolds, living in an Earthship is a great way to live more gently on the earth—and also save money.

Figure 13-6 Earthship house in Taos, New Mexico, during construction (left) and completed (right).

and doormats to road-building materials. By 1994 federal law requires that 5% of U.S. roads built with federal funds use asphalt with recycled rubber. By 1997 the figure rises to 20%, resulting in the recycling of about 100 million tires per year. At a plant near St. Louis, Missouri, Illinois Power company is burning about 7.5 million tires per year instead of coal. This cuts the company's sulfur dioxide emmissions in half, produces 50% more energy, and reduces costs by at least $1 million per year. Discarded tires can also be

reused for the foundations and walls of low-cost passive solar homes (Solutions, above). And some worn-out tires have been reused to build artificial reefs to attract fish.

One of the best ways to reduce the output of discarded tires is to extend the life of tires—a form of waste reduction and pollution prevention. During the past 40 years improvements in tire manufacturing have more than doubled the useful life of tires to about 64,000 kilometers (40,000 miles). Some tires are

Q: What percentage of the world's estimated plant species have been evaluated for their medical uses?

now being produced with an average life of 97,000 kilometers (60,000 miles), and researchers believe this could be extended to at least 160,000 kilometers (100,000 miles).

Another way to reuse wastes is to establish community and industry *exchange programs.* In Wellesley, Massachusetts, people can get rid of the items they no longer want and exchange them for items they want or need at little or no cost at the town landfill and recycling center. Businesses and industries can cooperate in publishing catalogs or on-line computer inventories of chemicals or products to exchange with other businesses in a community or region.

13-3 SOLUTIONS: RECYCLING SOLID WASTE

Recycling Organic Wastes: Community Composting Composting occurs when microorganisms (mostly fungi and aerobic bacteria) present in soil break down organic matter such as leaves, food wastes, paper, and wood in the presence of oxygen to produce **compost,** a sweet-smelling, dark-brown humus-like material that is rich in organic matter and soil nutrients. Biodegradable wastes—including solid waste from slaughterhouses and food-processing plants, kitchen and yard waste, manure from animal feedlots, paper too contaminated to be recycled, wood, and municipal sewage sludge—make up about 35% by weight of the municipal solid waste output. Composting can take place in backyards or in centralized community facilities.

To compost such wastes we mix them with soil, put the mixture into a pile or container, stir it occasionally, and let the stuff rot for several months. Heat, generated by microbial decomposition, increases inside the pile. The pile is periodically turned and mixed to ensure that the temperature is high enough throughout to kill pathogens and weed seeds but not hot enough to kill the decomposing microbes. Although compostable wastes decompose naturally, controlling factors such as moisture, oxygen (to reduce odors from anaerobic decompostion), temperature, nutrients, and particle size can speed up the process and reduce odors and other problems.

The resulting compost can be used as an organic soil fertilizer or conditioner, as topsoil, as a landfill cover, and as fertilizer for golf courses, parks, forests, roadway medians, and the grounds around public buildings. Compost can also be used to help restore eroded soil on hillsides and along highways and to help restore strip-mined land, overgrazed areas, and eroded cropland. Studies have shown that up to 20–40% of municipal solid waste could be composted, especially in areas where landfill costs are high.

To be successful, a composting program must overcome siting problems (no one wants to live near a giant compost pile or plant), control odors, and exclude toxic materials than can contaminate the compost and make it unsuitable for use as a fertilizer on crops and lawns. Large-scale composting is widely used in many European countries, including the Netherlands, Germany, France, Sweden, and Italy, and in states such as Minnesota, Pennsylvania, New Jersey, Massachusetts, and Oregon.

Odors can be controlled by enclosing the facilities and filtering the air inside, but residents near large composting plants still complain of unacceptable odors. The odor problem can be reduced by creating municipal compost operations near existing landfills or at other isolated sites. However, research indicates that the best way to control odor is to decompose biodegradable wastes in a closed metal container in which air is recirculated to give precise control of oxygen and temperature. This composting technology—sometimes called the "Dutch tunnel"—has been used successfully in the Netherlands for 20 years to overcome the odor problem. It is now being introduced into the United States and several other MDCs.

There are two types of large-scale composting systems: mixed waste and source separation. In 1993 there were 22 mixed solid waste (MSW) composting plants operating and another 31 in planning stages in the United States. However, such facilities, which bring in mixed garbage and then sort and compost the organic materials, often fail. Sorting is expensive and much of the compostable organic matter is contaminated with hazardous wastes such as household cleaners, leaking batteries, and motor oil. The result is often contaminated compost that is hard to sell and compost piles that can contaminate groundwater with toxic chemicals. Instead of the mixed-waste approach, a number of composting experts urge communities to establish source-separation programs. With this approach compostable organic wastes are not mixed with other solid waste and are collected in separate containers or compartmentalized garbage trucks. Such systems are less costly, produce more marketable compost, and don't tie up large amounts of capital in building and maintaining large and complex mixed-waste composting plants.

Households can use backyard compost bins (Figure 12-29) or indoor versions for food and yard wastes. Seattle, for example, promotes backyard composting by using a network of volunteer composting experts to help people get started. Apartment dwellers can compost by using indoor bins in which a special type of earthworm converts food waste into humus. By 1993, about 5% of U.S. yard waste in almost 2,300 cities was composted, with Minnesota, New Jersey, Wisconsin, and Michigan leading the way.

Recycling Other Wastes   There are two types of recycling: primary and secondary. The most desirable type is *primary,* or *closed-loop, recycling,* in which products are recycled to produce new products of the same type—newspaper into newspaper and aluminum cans into aluminum cans, for example. The less desirable type is *secondary,* or *open-loop, recycling,* in which waste materials are converted into different products. This does not reduce the use of resources as much as the first type of recycling. For example, primary recycling reduces the use of virgin materials in making a product by 20–90%; secondary recycling reduces it by 25% at most.

In 1992 about 17% of municipal solid waste in the United States was recycled or composted, and a 30% recovery rate seems within reach by 2000. Japan already recycles 40% of its municipal solid waste.

Many environmentalists believe we can do much better. Recent pilot studies in several U.S. communities show that a 60–80% recycling and composting rate is possible. By 1992, ten states were recycling 20–29% of their municipal solid waste and five states 30–38% (Minnesota recycling the most). However, 21 states were recycling less than 11% (with North Carolina, Hawaii, and Wyoming tying for last place at 4%). Sixteen states have adopted goals of recycling half or more of their municipal solid waste by 2000. However, these goals may not be met unless equally strong efforts are made to increase the demand for recycled materials and products.

Centralized Recycling of Mixed Wastes   Large-scale recycling can be accomplished by collecting mixed urban waste and bringing it to centralized *materials-recovery facilities (MRFs)* (rhymes with *smurf*). There, machines shred and automatically separate the mixed waste to recover glass (which can be melted and converted to new bottles or to fiberglass insulation), iron, aluminum, and other valuable materials (Figure 13-7). These materials are then sold to manufacturers as raw materials, and the remaining paper, plastics, and other combustible wastes are recycled or burned. The resulting heat produces steam or electricity to run the recovery plant or to be sold to nearby industries or homes.

Thwarted by widespread citizen opposition to incinerators and landfills, large waste management companies in the United States believed they could make big profits in recycling by building large MRFs. By 1993 the United States had more than 220 materials-recovery facilities and at least 60 more in the planning stages.

However, such plants are expensive to build and maintain. Furthermore, once trash is mixed it takes a lot of money and energy to separate it. Thus, MRFs must have a large input of garbage to make them financially successful, giving their owners a vested interest in having us produce more and more trash—the reverse of what prominent scientists believe we should be doing (Figure 13-3). These facilities also can emit toxic air pollutants and produce a toxic ash that must be disposed of safely. Other problems include health and accident threats to workers in poorly designed and managed plants and increased truck traffic, odor, and noise.

According to the National Solid Wastes Management Association, the average processing cost at MRFs in the United States is $50 per ton, while the average value of the extracted recyclables is about $30 per ton. To avoid losing money MRF operators have entered into long-term contracts that require communities to make up the difference or they charge fees (called tipping fees) for each ton of garbage or recyclables dropped off at the plant.

Separating Wastes for Recycling   It makes more sense economically and environmentally for households and businesses to separate trash into recyclable and reusable categories before it is picked up. With this approach homes and businesses put different kinds of waste materials—usually glass, paper, metals, and plastics—into separate containers. Compartmentalized city collection trucks, private haulers, or volunteer recycling organizations pick up the segregated wastes and sell them to scrap dealers, compost plants, and manufacturers. Another alternative is the establishment of a network of drop-off centers, buy-back centers, and deposit/refund programs where individuals deliver and either sell or donate their separated recyclables.

The source-separation approach produces little air and water pollution, reduces litter, and has low start-up costs and moderate operating costs (although collection costs are usually higher than for MRFs). It also saves more energy and provides more jobs for unskilled workers than centralized MRFs, and it creates three to six times more jobs per unit of material than landfilling or incineration. Recyclables are also cleaner and can usually be sold for a higher price. Another advantage is that collecting aluminum, paper, glass, plastics, and other materials for recycling is an important source of income for volunteer service organizations and for many people, especially the homeless and the poor in MDCs and LDCs. Source separation also educates people about the need for waste reduction, reuse,

Q: How many of Earth's species are believed to become extinct each year because of human activities?

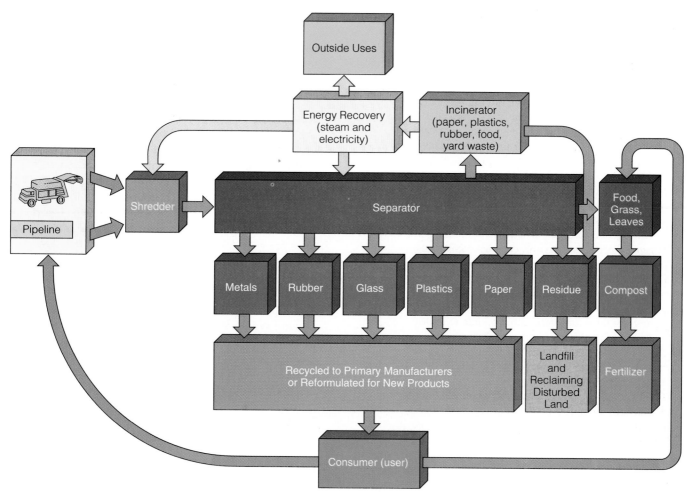

Figure 13-7 Generalized materials-recovery facility used to sort mixed wastes for recycling and burning to produce energy. Because such plants depend on high volumes of trash to be economical, they discourage reuse and waste reduction.

and recycling instead of encouraging them to go on producing large amounts of waste that are magically whisked off to MRFs, landfills, or incinerators—the out-of-sight, out-of-mind approach to waste outputs.

Ironically, many small- and medium-scale source-separation operations—pioneers in the recycling business—are being squeezed out by large waste management companies operating MRFs. In some communities, elected officials have signed long-term (20–30-year) contracts giving large waste management companies exclusive ownership over all garbage. The result is that small-scale recyclers have been sued for recycling materials that large waste management companies claim they own. The fight used to be over what to do with garbage. Now, as garbage is becoming more valuable, the fight is over who owns it.

Recycling Aluminum   Worldwide, the recycling rate for aluminum in 1991 was about 33% (30% in the

United States). Recycling aluminum produces 95% less air pollution and 97% less water pollution, and it requires 95% less energy than mining and processing aluminum ore.

In 1992, 68% (compared to 15% in 1972) of aluminum beverage cans in the United States were recycled at more than 10,000 recycling centers set up by the aluminum industry, other private interests, and local governments. People who returned the cans got about a penny per can for their efforts, earning about $900 million. Within six weeks the average recycled aluminum can had been melted down and was back on the market as a new can. Makers of aluminum cans saved $566 million and the energy equivalent of 20 million barrels of oil. (The electricity used to produce one aluminum can from virgin ore would keep a 100-watt light bulb burning for 100 hours.)

Despite this progress, about 32% of the 92.4 billion aluminum cans produced in 1992 in the United States

Recycled Paper Hype and Trade-offs

True or false: Buying recycled paper products will reduce solid waste. The answer is, not necessarily. Buying recycled paper products can save trees (grown mostly in plantations) and energy and reduce pollution, but it does not necessarily reduce solid waste. Only products made from *post-consumer waste*—intercepted on its way from consumer to the landfill—will do that.

Most recycled paper is actually made from *pre-consumer waste*—scraps and cuttings recovered from paper and printing plants. Since the paper industry has always recycled this waste, it has never contributed to landfill problems. Now this paper is labeled "recycled" as a marketing ploy, giving the false impression that people who buy such products (often at higher prices) are helping the solid-waste problem. Most "recycled" paper has no more than 50% recycled fibers, with only 10% from post-consumer waste.

This book is printed on acid-free recycled paper containing the highest content of recycled fiber available and that was not prohibitively expensive. We go further than simply using the best available recycled paper. Each year the publisher and I donate money to organizations that protect existing tropical forests and to tree-planting organizations, so that at least two trees are planted and tended for each tree used in printing this book. I also see that 50 trees are planted and tended for each tree that I use (in the form of paper) in researching and writing this book. To save paper and reduce pollution, I sent the manuscript for this book to the publisher on small computer discs.

were still thrown away. If these cans were laid end-to-end, they would wrap around the earth more than 120 times. These discarded cans contain more aluminum than most countries use for all purposes, and each discarded aluminum can represents almost indestructible solid waste.

Recycling aluminum cans is great, but many environmentalists believe that aluminum cans are a glaring example of an unnecessary item that could be replaced by refillable glass bottles—a switch from recycling to reuse that also creates jobs in local communities. One way to encourage this change would be to place a heavy tax on nonrefillable containers (aluminum, glass, or plastic) and no tax on reusable glass bottles. Many environmentalists also believe that unnecessary use of aluminum could be reduced by not

giving the aluminum industry huge electric power subsidies, which hide the high environmental cost of using this metal. Meanwhile, they urge consumers not to buy beverages in aluminum cans and to buy reusable glass containers when they are available.

Recycling Wastepaper Overpackaging—including double packages and oversized containers—is a major contributor to paper use and waste. Junk mail also wastes enormous amounts of paper. Each year the U.S. work force throws away enough office and writing paper to build a 4-meter- (12-foot-) high wall stretching from New York City to Los Angeles. Each American throws away paper equivalent to an average of four trees per year, for a total of more than 1 billion trees.

Environmentalists estimate that at least 50% of the world's wastepaper (mostly newspapers, corrugated board and paperboard, office paper, and computer and copier paper) could be recycled by the year 2000. During World War II, when recycling was a national priority, the United States recycled about 45% of its wastepaper, compared to 39% in 1992 (up from 25% in 1989). Some other countries do better; including the Netherlands (53%), Japan (50%), Mexico (45%), Germany (41%), and Sweden (40%). Recently a process has been developed that converts old newspapers, magazines, cardboard, junk mail, and telephone books into a lightweight, sturdy, honeycombed panel (called Gridcore) that can be used as a substitute for wood and plywood.

Recycling the Sunday newspapers in the U.S. would save 500,000 trees per week. In 1992, about 43% of all newspapers published in the United States were recycled, but only about 10% of U.S. newspapers were printed on paper using recycled fiber.

Apart from saving trees (most of which are grown on tree farms), recycling (and reusing) paper has a number of benefits. It

- saves energy because it takes 30–64% less energy to produce the same weight of recycled paper as to make the paper from trees.

- reduces air pollution from pulp mills by 74–95%.

- lowers water pollution by 35%.

- helps prevent groundwater contamination from the toxic ink left after paper rots in landfills over a 30–60-year period.

- conserves large quantities of water.

- saves landfill space.

- creates five times more jobs than harvesting trees for pulp.

- can save money. In 1988, for example, American Telephone and Telegraph earned more than

Q: What is the greatest threat to wild species?

$485,000 in revenue and saved $1.3 million in disposal costs by collecting and recycling high-grade office paper.

Separating paper from other waste materials is a key to increased recycling. Otherwise, paper becomes so contaminated that wastepaper dealers won't buy it.

In the United States tax subsidies and other financial incentives make it cheaper to make paper from trees than from recycled wastepaper—except in Florida, where virgin newsprint is taxed. Also, 11 states have passed laws requiring newspaper publishers to increase their content of recycled paper. This has encouraged the paper industry to invest $3.7 billion between 1993 and 1998 in new mills designed to use recycled paper.

Eliminating or reducing tax subsidies on virgin paper and shifting them to recycled paper would make it cheaper and increase demand. One way to increase demand would be to require that federal and state governments use recycled paper products as much as possible. Such a federal law exists, needing only to have its many loopholes closed.

In 1994 President Bill Clinton issued an executive order requiring all federal agencies to purchase paper with at least 20% post-consumer recycled content by the end of 1994 and 30% by the end of 1998. The order also mandates government purchase of recycled oil and retreaded tires. This order should spare more than a million trees annually and increase the demand for recycled paper. However, its effectiveness may be reduced because of a sizeable loophole that allows sawdust and several other wood-waste materials to be counted as post-consumer paper waste.

Even with the best of intentions, governments, businesses, and individuals may be fooled by claims about recycled paper products. Moreover, as usual there are trade-offs (Spotlight, p. 346).

Recycling Plastics   Plastics are made from petrochemicals (chemicals produced from oil). The plastics industry is among the leading producers of hazardous waste. Plastics now account for about 8% by weight and 20% by volume of municipal solid wastes in the United States and about 60% of the debris found on U.S. beaches. In landfills toxic cadmium and lead compounds used as binders can leach out of plastics and ooze into groundwater and surface water. When plastics are thrown away as litter, they can harm animals that swallow or get entangled in them (Figure 13-8).

Most plastics used today are nondegradable or take 200–400 years to degrade. Even biodegradable plastics take decades to partially decompose in landfills because of a lack of oxygen and moisture.

Figure 13-8 Before this discarded piece of plastic was removed, this Hawaiian monk seal was slowly starving to death. Each year plastic waste dumped from ships and left as litter on beaches threatens the lives of millions of marine animals and seabirds that ingest such debris, become entangled in it, or choke on it.

In theory most plastics could be recycled; in 1992, however, about 2.5% by weight of all plastic wastes and 5% of plastic packaging (including 14% of plastic bottles) used in the United States were recycled. Plastics are more difficult to recycle than glass or aluminum because there are so many different types of plastics. Thus for recycling to work, most plastic trash must be sorted into different categories—a costly procedure. And many of the recycled products are difficult to sell because they cost more and look worse than those made from virgin plastics. As long as the true environmental costs of producing, using, and throwing away plastic are not included in product prices, it will usually be cheaper to make new plastic than to recycle it (Section 7-2).

Another possibility is to convert mixed plastic waste into fuel oil. Currently, the cost of doing this is about $28 per barrel of the synthetic oil, compared with the $16- to $20-per-barrel price of conventional oil.

The $140-billion-per-year U.S. plastics industry has established the Partnership for Plastics Progress (also known as P3) and the American Plastics Council. Critics argue that the main purpose of these public relations organizations is to keep us buying more plastic containers, utensils, and other mostly throwaway items and to fight any laws banning or limiting the use of throwaway plastic items.

Consumers should also be aware that some of the plastic bags and bottles dropped off at recycling centers don't get recycled, because of a lack of market demand. Some get dumped in landfills. And a study by Greenpeace found that some of the plastics accepted by the plastics industry for recycling are exported and dumped in LDCs.

A: Loss and disturbance of habitat by human activities

Environmentalists recognize the importance of plastics in many products, but many of them think that their widespread use in throwaway beverage and food containers should be sharply reduced and replaced with less harmful and wasteful alternatives, such as reusable glass bottles.

Recycling: The German Experience In 1991 Germany enacted the world's toughest packaging law, designed to reduce the amount of waste being landfilled or incinerated. The goal is to recycle or reuse 65% of the nation's packaging by 1995, including 90% of metals and glass, and 80% of paper, board, and plastics in both domestic and imported goods. Product distributors must take back their boxes and other containers for reuse or recycling. And incineration of packaging, even if it is used to generate power, is not allowed.

To implement the system over 600 German manufacturers and distributors formed and pay fees to a nonprofit company to collect, sort, and reprocess the packaging discards (coded with a green dot) of member firms.

There have been problems, including the following:

■ Failure to create a market for recycled packaging materials equal to the supply. This has led to a glut of collected but unrecycled materials (especially plastics) crowding storage facilities.

■ A large supply of recycled packaging material being sold throughout Europe, overwhelming markets for recycled materials and drastically lowering the prices recycling programs in these countries used to get.

■ High costs—about twice as much per ton as collecting and disposing of other German household waste. However, these costs don't include the savings in the reduced use of virgin materials and reduced pollution.

Despite the problems, most Germans support the system. And the program has prompted four out of five German manufacturers to reduce their use of packaging, especially difficult-to-recycle plastics. The Netherlands and France have adopted similar recycling programs.

Does Recycling Make Economic Sense? The answer is yes and no, depending on different ways of looking at the economic and environmental costs and benefits of recycling. Critics contend that recycling

■ has become almost a religion that should not be criticized regardless of how much it costs com-

munities and diverts funds that could otherwise be used for schools, reducing crime, or other purposes.

■ does not make sense if it costs more to recycle materials than to send them to a landfill or incinerator—as is the case in many areas.

■ is not needed in most areas to compensate for a lack of landfill space. Most areas in the United States are not running out of landfill space.

■ is not needed because of a threat to human health because properly operated state-of-the-art landfills and incinerators do not pose serious threats to human health.

■ may make economic sense for valuable and easy to recycle materials (such as aluminum, mercury, and steel) but not for renewable paper (made from tree farms), cheap or plentiful resources (glass from silica), and most plastics (which are very expensive to recycle).

Many communities are faced with a glut of collected materials that can't be sold profitably and are losing money on their recycling programs. Yet, because of the popularity of recycling, officials are reluctant to reduce or abandon such programs.

Proponents contend that recycling

■ should not be judged on whether it pays for itself or makes a profit. Landfilling and incineration of wastes are not profitable for communities but are considered services that citizens should pay for.

■ should be evaluated by including the "avoided" costs of disposal—the savings earned by *not* having to dispose of garbage in landfills and incinerators.

■ includes benefits in reducing pollution and environmental degradation associated with saving energy and using virgin resources (Figure 12-13) that are not reflected in the market prices of most goods and services because of a lack of full-cost pricing (Section 7-2). If such external costs (which people end up paying for indirectly) were included, the large economic and environmental benefits of recycling compared to burying and burning wastes would be obvious.

■ would cost less if more emphasis was placed on creating a demand for goods made from recycled materials.

Obstacles to Reuse and Recycling Several factors hinder recycling (and reuse) in the United States. One is that many of the environmental and health

Q: What is the best way to prevent wildlife extinction from human activities?

costs of items are not reflected in their market prices, so consumers have little incentive to recycle, reuse, or reduce their use of throwaway products.

Another problem is that the logging and mining (p. 301), industries get huge tax breaks, depletion allowances, cheap access to public lands, and other subsidies to encourage them to extract virgin resources as quickly as possible. By contrast, recycling industries get few tax breaks or other subsidies. Finally, the lack of large, steady markets for recycled materials (mostly because they lack the tax breaks given to the manufacturers using virgin materials) makes recycling a risky boom-and-bust business that attracts little investment capital.

Another problem is that some communities and businesses have leaped into recycling with little emphasis on creating a demand for recycled products. By neglecting this fundamental of economics, the supply of waste materials collected for recycling has often exceeded the demand, lowering the prices paid for such materials. To help correct this situation, in 1992 the National Recycling Coalition established a campaign to get businesses to buy recycled material and to pump up the demand for products made from recycled materials. Consumers also increase demand when they buy goods made from recycled materials, especially if they are based on primary recycling and are made from the highest feasible amount of post-consumer waste.

13-4 LAST RESORTS: BURNING OR BURYING SOLID WASTE

Incineration Some solid waste is burned in *conventional incinerators,* but most is burned in *trash-to-energy incinerators* that also produce steam or electricity, which can be sold or used to run the incinerator. Most are *mass-burn incinerators,* which burn mixed trash without separating out hazardous materials (such as batteries) and noncombustible materials that can interfere with combustion conditions and pollute the air. Denmark and Sweden burn 50% of their solid waste to produce energy, compared with 11% in the United States.

Incinerating solid waste kills germs and reduces the volume of waste going to landfills by about 60% (not 90% as usually cited). However, incinerators are costly to build, operate, and maintain, create very few long-term jobs and are plagued by heavy truck traffic. Unlike landfills, which tend to be located in remote areas, incinerators are often sited close to or within communities to facilitate the transport of garbage. Moreover, even with advanced air pollution control devices (mostly a scrubber followed by a baghouse filter or electrostatic precipitator, Figure 9-18), incinerators release toxic substances such as dioxins (Section 13-5), mercury (mostly from batteries), easily inhaled fine particles of heavy metals such as lead (from batteries and plastics), and cadmium (from plastics) into the atmosphere. Animal testing has shown that these substances can cause cancers and nervous-system disorders. Without continuous maintenance and good operator training and supervision, the air pollution control equipment on incinerators often fails, so that emission standards are exceeded.

Incinerators also produce residues of toxic *fly ash* (lightweight particles removed from smokestack emissions by air pollution control devices) and less toxic *bottom ash;* approximately 10% is fly ash, and 90% bottom ash. Usually the two types of ash are mixed and disposed of in ordinary landfills. Although the amount of material to be buried is greatly reduced, its toxicity is increased. Incinerator ash, especially fly ash, usually contains toxic metals such as cadmium, lead, mercury, arsenic, beryllium, zinc, and copper as well as trace amounts of dioxins.

Ways to deal with incinerator ash include **(1)** not mixing usually more toxic fly ash with bottom ash; **(2)** disposing all ash in specially designed hazardous waste landfills; **(3)** chemically or physically treating the ash to reduce its toxicity before disposal; and **(4)** not burning materials containing toxic heavy metals.

Environmentalists have pushed Congress and the EPA to classify incinerator ash as hazardous waste, disposable only in landfills designed to handle hazardous waste, as is done in Japan. To date, however, no action has been taken, largely because waste management companies claim it would make incineration too expensive. For some reason, they also oppose the other options for dealing with incinerator ash—insisting that it is not a health threat. Environmentalists counter that if the companies can't properly dispose of the toxic ash they produce, they shouldn't be in the incineration business.

Newer Japanese incinerators are more strictly controlled than those built in the United States. In Japan hazardous wastes and unburnable materials that would pollute the air are removed before wastes are incinerated, which also greatly reduces the amount and toxicity of the remaining ash. This is rarely done in the United States.

Furthermore, to protect Japanese workers, bottom ash and fly ash are removed by conveyor belt and transported to carefully designed and monitored hazardous-waste landfills in sealed trucks. Before disposal the ash is often solidified in cement blocks. In the United States most incinerator ash is simply dumped into conventional landfills.

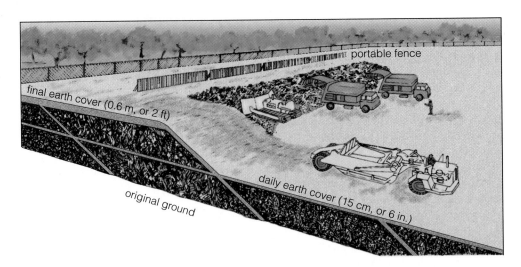

Figure 13-9 A sanitary landfill. Wastes are spread in a thin layer and then compacted with a bulldozer. A scraper (foreground) covers the wastes with a fresh layer of clay or plastic foam at the end of each day. Portable fences catch and hold windblown debris.

In Japan violations of air standards are punishable by large fines, plant closings, and—in some cases—jail sentences for company officials. In the United States, monitoring is not as strict, and punishment for violations is much less severe. Japanese incinerator workers must have an engineering degree. They spend 6–18 months learning how the incinerator works and undergo closely supervised on-site training. U.S. incinerator workers need no degrees and get far less training.

Incineration in the United States has also been plagued by faulty equipment and human errors that have exposed workers and people in surrounding areas to dangerous levels of air pollution. Thanks mostly to Japanese and German technology some of the newer incinerators being built in the United States are safer.

About 40% of Japan's incinerators use *fluidized-bed incineration*. This technology has a higher combustion efficiency and lower emissions of nitrogen oxides, sulfur dioxide, and dioxins than other commercial incinerator designs. However, this technology has not been promoted in the United States because it is more expensive, with the first such plant scheduled to open in Cook County, Illinois, by 1995.

In 1993 there were 179 incinerators (145 of them trash-to-energy incinerators) plus more than 3,100 medical waste incinerators operating in the United States. These trash-to-energy incinerators produced enough electricity to power 1.3 million homes. Since 1985, however, over 73 new incinerator projects have been blocked, delayed, or canceled because of public opposition and high costs. Of the 70 plants still in the planning stage, most face stiff opposition, and many may not be built as communities discover that recycling, reuse, composting, and waste reduction are cheaper and safer alternatives that can handle 60–80% of municipal waste.

Most environmentalists oppose heavy dependence on incinerators because it encourages people to con-

tinue tossing away paper, plastics, and other burnable materials. Many existing incinerators have 20- to 30-year contracts with cities to supply them with a certain volume of trash, which makes it hard for cities to switch to large-scale recycling, composting, reuse, and pollution prevention—a claim disputed by the waste incinerator industry. In 1992, Rhode Island became the first state to ban solid-waste incineration because of its threats to the health and safety of Rhode Islanders, especially children, and its unacceptably high cost. In 1993, West Virginia enacted a similar ban. Sweden banned the construction of new incinerators in 1985.

In 1993, the *Wall Street Journal* warned its readers that incinerators for municipal trash are financial disasters for local governments. They require residential and commercial customers—as well as taxpayers—to pay hundreds of millions of dollars a year over and above the going market rate for trash. The basic problem is that in the 1980s waste management companies lured local officials into signing long-term contracts requiring the community to supply the incinerator with a fixed amount of trash or pay a cash penalty, even if the community can find a cheaper way of dealing with its garbage.

Garbage Graveyards In 1993 about 72% by weight of the municipal solid waste in the United States was buried in sanitary landfills, compared with 98% in Australia, 93% in Canada, 90% in Great Britain, 54% in France, 44% in Sweden, 18% in Switzerland, and 17% in Japan. A **sanitary landfill** is a garbage graveyard in which wastes are spread out in thin layers, compacted, and covered daily with a fresh layer of clay or plastic foam (Figure 13-9).

Modern state-of-the-art landfills are lined with clay and plastic before being filled with garbage (Figure 13-10). The site must be geologically suitable. The bottom is covered with an impermeable liner usually

Q: How much of the earth's land surface has been set aside to protect wildlife?

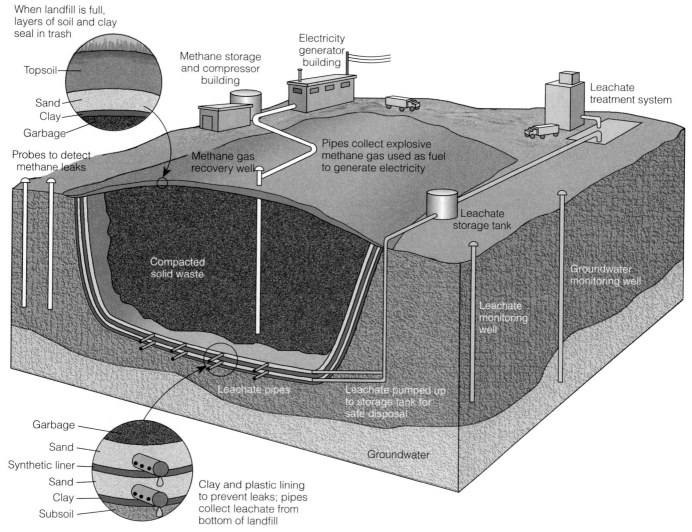

When landfill is full, layers of soil and clay seal in trash

Topsoil
Sand
Clay
Garbage

Probes to detect methane leaks

Methane storage and compressor building

Electricity generator building

Methane gas recovery well

Leachate treatment system

Pipes collect explosive methane gas used as fuel to generate electricity

Compacted solid waste

Leachate storage tank

Groundwater monitoring well

Leachate monitoring well

Leachate pipes

Leachate pumped up to storage tank for safe disposal

Groundwater

Garbage
Sand
Synthetic liner
Sand
Clay
Subsoil

Clay and plastic lining to prevent leaks; pipes collect leachate from bottom of landfill

Figure 13-10 A state-of-the-art sanitary landfill designed to eliminate or minimize environmental problems that plague older landfills. Only a few municipal landfills in the United States have such state-of-the-art design, and 85% of U.S. landfills are unlined. Furthermore, even state-of-the-art landfills will eventually leak, passing contamination and cleanup costs on to the next generation.

made of several layers of clay, thick plastic, and sand. This liner collects leachate (rainwater that is contaminated as it percolates down through the solid waste) and is supposed to keep it from leaking into groundwater. Collected leachate ("garbage juice") is pumped from the bottom of the landfill, stored in tanks, and sent either to a regular sewage treatment plant or to an on-site treatment plant. When the landfill is full, it is covered with clay, sand, gravel, and topsoil to prevent water from seeping in. Several wells are drilled around the landfill to monitor any leakage of leachate into nearby groundwater. Methane gas produced by anaerobic decomposition in the sealed landfill is collected and burned to produce steam or electricity. Since 1993, all landfills in the United States must meet such standards, forcing operators to either upgrade or close their operations.

Sanitary landfills offer certain benefits. No air-polluting open burning is allowed. Odor is seldom a problem, and rodents and insects cannot thrive. Sanitary landfills should be located so as to reduce water pollution from leaching, but that is not always done. Moreover, a sanitary landfill can be put into operation quickly, has low operating costs, and can handle a huge amount of solid waste. And after a landfill has been filled, the land can be graded, planted, and used as a park, a golf course, a ski hill, an athletic field, a wildlife area, or some other recreation area.

However, landfills also have drawbacks. While they operate they cause traffic, noise, and dust. Most also emit toxic gases. In addition, paper and other biodegradable wastes break down very slowly in today's compacted and water- and oxygen-deficient landfills. For example, newspapers dug up from some landfills are still readable after 30 or 40 years, and hot dogs, carrots, and chickens that have been dug up after 10 years have not rotted (Figure 13-11). Biodegradable plastics also take decades to decompose in landfills.

The underground anaerobic decomposition of organic wastes at landfills produces explosive methane gas, toxic hydrogen sulfide gas, and smog-forming volatile organic compounds that escape into the air. Landfills can be equipped with vent pipes to collect these gases, and the collected methane can be burned to produce steam or electricity (Figure 13-10). A single large landfill can provide enough methane to meet the energy needs of 10,000 homes. Besides saving energy, using the methane gas from all large landfills worldwide would lower atmospheric emissions of methane and help reduce projected global warming from greenhouse gases (Figure 10-9).

Contamination of groundwater and nearby surface water from *leachate* that seeps from the bottom of unlined landfills or cracks in the lining of lined landfills is another serious problem. Even when leachate is collected it is rarely treated to render it harmless. Some 86% of the U.S. landfills studied have contaminated groundwater. Once groundwater is contaminated it is extremely difficult—often impossible—to clean up.

To lower possible environmental and health risks from leachate, all U.S. landfills operating after 1993 have had to install a groundwater monitoring system and set aside money to pay for groundwater cleanup; but this does not apply to the thousands of closed landfills. Modern double-lined landfills (Figure 13-10)

required in the United States after 1996 delay the release of toxic leachate into groundwater below landfills, but they do not prevent it. These landfills are designed to accept waste for 10–40 years, and current EPA regulations require owners to maintain and monitor them for at least 30 years after they are closed. However, they could begin to leak after this period, passing the health risks and costs of contamination on to the next generation.

Landfill consultant G. Fred Lee believes that the only solution to the leachate problem is to apply clean water continuously to landfills and collect and treat the resulting leachate in carefully designed and monitored facilities. After 10-20 years of such washing he

Figure 13-11 Hot dogs after 10 years in a sanitary landfill. The idea that a landfill is a large compost pile in which things biodegrade fairly rapidly is a myth. Decomposition in modern landfills is slow because the garbage is tightly packed and exposed to little moisture and essentially no light.

Figure 13-12 There is no "away." In 1987 the barge *Mobro* shown here tried to dump 2,900 metric tons (3,190 tons) of garbage from Islip, Long Island. It was refused permission to unload in North Carolina, Florida, Louisiana, the Bahamas, Mexico, and Honduras. After 164 days and a 9,700-kilometer (6,000-mile) journey, it came back to New York City, where it was barred from docking. After staying in the harbor for three months, its garbage was incinerated in Brooklyn; the resulting 364 metric tons (400 tons) of ash was shipped back to Islip for burial in a local landfill. The publicity from this event catalyzed Islip into developing a recycling program, and by 1989 the town was recycling 35% of its solid waste. This has saved the community $2 million a year and extended the life of its landfill.

Q: How many unwanted dogs and cats are killed by U.S. animal shelters each year because of pet overpopulation?

contends that there should be little potential for pollution of groundwater left in the landfill. He argues that we can either pay more money now to prevent water pollution from landfills or pass the problem on to the future generations whose drinking water may become polluted by landfill leakage.

Land for landfills is scarce in some urban areas, but most areas in the United States are not running out of landfill space—although landfills are vigorously opposed by residents in most communities. The number of landfills in the United States dropped from nearly 20,000 in 1979 to 8,000 in 1988 and about 5,300 in 1993. This sharp decline is the result of landfills either reaching their capacity or closing because they don't meet more stringent state or federal design and operating standards. A further drop is expected because after 1996 only more expensive state-of-the-art landfills will be allowed to operate (Figure 13-10). However, the country's overall landfill capacity is not declining, because most of the smaller local landfills are being replaced by larger local and regional landfills.

Some cities without enough landfill space are shipping their trash to other states or other countries, especially LDCs. However, a few of these states and some LDCs are refusing to accept it (Figure 13-12). Landfills also deprive present and future generations of valuable resources and encourage waste production instead of pollution prevention and waste reduction.

13-5 HAZARDOUS WASTE: TYPES AND PRODUCTION

What Is Hazardous Waste, and How Much Is Produced? According to the Environmental Protection Agency, **hazardous waste** is any discarded material that: **(1)** contains one or more of 39 toxic, carcinogenic, mutagenic, or teratogenic compounds at levels that exceed established limits (Section 8-2); **(2)** is flammable; **(3)** is reactive or unstable enough to explode or release toxic fumes; or **(4)** is capable of corroding metal containers such as tanks, drums, and barrels.

However, this narrow official definition of hazardous wastes (mandated by Congress) does not include:

- *Radioactive wastes* (Section 19-4).

- *Hazardous and toxic materials discarded by households* (Table 13-1).

- *Mining wastes* (Figure 13-2).

- *Oil-and gas-drilling wastes,* routinely discharged into surface waters or dumped into unlined pits and landfills.

Table 13-1 Common Household Toxic and Hazardous Materials

Cleaning Products

Disinfectants

Drain, toilet, and window cleaners

Oven cleaners

Bleach and ammonia

Cleaning solvents and spot removers

Septic tank cleaners

Paint and Building Products

Latex and oil-based paints

Paint thinners, solvents, and strippers

Stains, varnishes, and lacquers

Wood preservatives

Acids for etching and rust removal

Asphalt and roof tar

Gardening and Pest Control Products

Pesticide sprays and dusts

Weed killers

Ant and rodent killers

Flea powder

Automotive Products

Gasoline

Used motor oil

Antifreeze

Battery acid

Solvents

Brake and transmission fluid

Rust inhibitor and rust remover

General Products

Dry-cell batteries (mercury and cadmium)

Artist paints and inks

Glues and cements

- *Liquid waste containing organic hydrocarbon compounds* (80% of all liquid hazardous waste). The EPA allows it to be burned as fuel in cement kilns and industrial furnaces with little regulation (although this may change).

A: 15 million—an average of 41,000 per day

- *Cement kiln dust* produced when liquid hazardous wastes are burned in the kilns—a practice classified as recycling by the EPA but called dangerous "sham recycling" by environmentalists.

- *Municipal incinerator ash*, which if classified as hazardous waste would be so expensive to ship to and bury in special landfills that the whole waste incineration industry would collapse.

- *Wastes from the thousands of small businesses and factories that generate less than 100 kilograms (220 pounds) of hazardous waste per month.*

- *Waste generated by the military*, except at 116 sites so toxic that they are on the EPA's list of priority sites to be cleaned up. U.S. military installations produce more hazardous waste each year (about a ton per minute) than the top five U.S. chemical companies combined. Studies by the Department of Defense have identified over 17,482 contaminated sites at 1,855 military bases, in every state.

Environmentalists call these omissions "linguistic detoxification" designed to save industries and the government money and mislead the public. They urge that all excluded categories be designated hazardous waste. This would quickly shift the emphasis and innovation from waste management and pollution control to waste reduction and pollution prevention and save lots of money.

The EPA estimates that at least 5.5 billion metric tons (6 billion tons) of hazardous waste are produced each year in the United States—an average of 21 metric tons (23 tons) per person. However, because only 6%, or 350 million metric tons (385 million tons), of the total is *legally* defined as hazardous waste, *94% of the country's hazardous waste is not regulated by hazardous-waste laws.*

Lead: Poisoning Children   One example of a toxic waste is lead. Atmospheric emissions of lead from human sources are 28 times greater than those from natural sources. And emissions of lead into the soil and aquatic ecosystems from human activities are almost three times those emitted into the atmosphere.

We take in small amounts of lead in the air we breathe, the food we eat, and the water we drink. Once lead enters the blood only about 10% is excreted. The rest is stored in the bones for decades. Even at very low levels it damages the central nervous system, especially in young children. Pregnant women can also transfer dangerous levels of lead to their unborn children.

Each year 12,000–16,000 American children (mostly poor and nonwhite) under age 9 are treated for acute lead poisoning, and about 200 die. About 30% of the survivors suffer from palsy, partial paralysis, blindness, and mental retardation.

Children under age 6 with levels greater than 10 micrograms of lead in each deciliter (about half a cup) of blood—the current maximum legal standard—are especially vulnerable. Several studies indicate that exposure even to this level in early childhood can reduce IQ levels by 5–10% and cause behavior problems. To achieve this blood level, a child would need to ingest lead equal in weight to that in only one-third of a granule of sugar a day—something easily done by touching soil or house dust contaminated with lead and then sucking the thumb. In the United States it is estimated that more than 6 million preschoolers and 400,000 pregnant women have blood levels of lead above the current standard. Some scientists believe that there is no minimum threshold for safe exposure of children and fetuses to lead, while others say the harmful effects of lead appear only when blood levels in young children and fetuses are over 20 micrograms per deciliter.

According to a 1986 EPA study, *at least one in six children in the United States under age 6 has unsafe blood levels of lead that may retard its mental, physical, and emotional development.* This epidemic affects children of every socioeconomic background, but those in poor families and minority groups suffer most.

The greatest sources of lead in the United States are:

- *Lead particles injected into the atmosphere that settle onto the soil and become outdoor or indoor dust.* Children ingest this lead as they play in contaminated soil or dust on carpeting, toys, or the floor, and then put their thumbs or hands in their mouths. The major sources of atmospheric lead particles today are solid-waste and hazardous-waste incinerators, lead smelters, furnaces burning used motor oil, and battery manufacturing plants. Leaded gasoline has been phased out, but the massive amounts of indestructible lead particles that fell out of the atmosphere for 50 years before the ban contaminate land almost everywhere. In countries that have not banned leaded gasoline, it is the largest atmospheric source of lead.

- *Interior paint in 52%, or 57 million, of the houses built before 1978, when use of lead compounds in interior and exterior paint was banned.* These houses are a major source of lead poisoning for children ages 1–3, who inhale lead dust from cracking and peeling paint or ingest it by sucking their thumbs, putting contaminated toys in their mouths, or gnawing on window sills or furniture. People living in houses or apartments built before 1980 should have samples of the paint analyzed for lead by the local health department or by a private testing laboratory (cost $100–$450). Correcting this problem can cost $2,000–$10,000 per

Q: How much of the commercial energy used in the United States is wasted?

home—unaffordable for many people, including the 2 million impoverished families living in older houses and apartments with deteriorating lead-based paint. The estimated total cost for removing this hazard is $10 billion.*

- *Groundwater contaminated by lead leached from landfills.* More than 3,000 sources of community drinking water are believed to be contaminated in this way.

- *Drinking water contaminated by plumbing containing lead pipes or lead solder.* According to the EPA nearly one in eight Americans in 819 tested communities drinks tap water containing unsafe levels of lead that is leached by acidic or soft water from solder and connectors used with copper piping or from lead pipes used in plumbing systems. Homeowners with copper pipes or joints in their homes, or in homes built before 1930 with plumbing made of lead, should have the local water department or a private laboratory (cost $20–$100) test their tap water for lead. Before buying or renting an existing house or apartment, prospective buyers or renters should have its water (that has been standing in pipes for at least 12 hours) tested for lead. In 1991 the EPA ordered removal of lead from municipal drinking water systems but gave the country's largest municipalities up to 21 years to do the job.

- *Lead solder used to seal the seams on food cans.* This applies especially to acidic foods such as tomatoes and citrus juices. This type of solder has been sharply reduced in U.S. food cans but may be found in cans of imported foods.

- *Imported cups, plates, pitchers, leaded glass crystal, and other items used to cook, store, or serve food, especially acidic foods and hot liquids and foods.* Before using such items, test them for lead content.

- *Vegetables and fruits grown on soil contaminated by lead.* This applies especially to cropland or home gardens near highways, incinerators, and smelters. Careful washing should remove at least half of this lead.

- *Burning comic strips, Christmas wrapping paper, or painted wood, in wood stoves and fireplaces.*

According to the Centers for Disease Control and Prevention and the Department of Health and Human

---

*If you find lead in your home, send a postcard to U.S. Consumer Product Safety Commission, Washington, DC 20207, and ask for the free pamphlet, *What You Should Know About the Lead-Based Paint in Your Home.* Two home kits for testing paint for lead are sold by HybriVet Systems (800-262-LEAD) and Frandon Enterprises (800-359-9000).

Protecting Children from Lead Poisoning

SOLUTIONS

Ways to protect children from lead poisoning include:

- Setting lead standards to protect children and fetuses

- Requiring that all U.S. children be tested for lead by age 1, and establishing a lead-screening program to test all children under age 6

- Eliminating leaded paint and contaminated dust in housing

- Testing all community sources of drinking water, especially in schools and homes, for lead contamination, and removing the contamination or providing alternate sources of drinking water

- Banning the use of lead solder in plumbing pipes and in food cans, and removing lead from municipal drinking water systems within 7 years instead of the current 20 years

- Making sure children wash their hands thoroughly before eating

- Requiring that all ceramicware (whether produced domestically or imported) used to cook, store, or serve food be lead-free

- Banning incineration of municipal solid waste and hazardous waste—the largest new source of lead

- Mounting a global campaign to reduce lead poisoning in LDCs

- Banning leaded gasoline throughout the world

Doing these things will cost lots of money (an estimated $50 billion in the United States). But health officials say the alternative is to keep poisoning and mentally handicapping large numbers of children. What do you think should be done?

Services, *lead is the number one environmental health threat to children in the United States* and should be a matter of the highest priority (Solutions, above).

Dioxins: Unraveling a Complex Health Threat
Dioxins are a family of 75 different chlorinated hydrocarbon compounds formed as by-products in chemical reactions involving chlorine and hydrocarbons, usually at high temperatures. One of these compounds, TCDD, sometimes simply called dioxin, is the most harmful and most widely studied.

TCDD and other dioxins result from the burning of chlorine-containing wastes in municipal and

---

84% (43% of this energy is unnecessarily wasted)

hazardous-waste incinerators and cement kilns; from chlorine-bleaching of bleach pulp and paper; and from the manufacture of certain herbicides (for example, 2,4,5-T), many plastics (for example, PVC), and many chlorinated hydrocarbon chemicals. Dioxins persist in the environment, especially in soil and human fatty tissue, and they can apparently be biologically amplified to higher levels in food webs.

Researchers studying the health effects of TCDD in the early 1970s were puzzled because it produced different effects in different species, as well as different effects in the same species at various doses. Also, epidemiological studies of people exposed to TCDD found a variety of health effects, including various cancers, but no consistent effects.

In 1991 representatives of the paper industry claimed that the inconsistent results of health studies exonerated TCDD and other dioxins, and they pushed the EPA for a reassessment of the health risks of dioxin. This strategy may have backfired when the preliminary results of EPA's reevaluation released in 1992 indicated that dioxin was an even greater health threat than previously thought. This new review showed that TCDD does cause cancer in humans, but unlike most carcinogens it does not damage DNA. Instead, TCDD promotes cancer by activating DNA already damaged by other carcinogens. This explains why researchers found a variety of cancers rather than a single type, as is usually the case with most carcinogens.

This review also revealed that TCDD's immunological, developmental, and neurological effects at very low exposure levels may pose an even greater threat to human health than its cancer-promoting ability. Studies on mice revealed that minuscule doses of TCDD disrupted and suppressed the immune system—making the test animals (and presumably humans) more vulnerable to a variety of diseases. In 1993 scientists conducting research on rhesus monkeys strongly linked exposure to TCDD as a contributor to endometriosis, a painful disease that affects an estimated 1 out of 10 women and can cause reproductive problems.

Equally worrisome, TCDD was found to be an environmental hormone that imitates naturally occurring sex and growth hormones. Male mice and rats exposed to low levels of TCDD had delayed sexual development, feminization, and reduced sex drive and sperm counts—a result described by one researcher as chemical castration.

Tainted foods such as fish, mother's milk, milk from cows eating grass growing in contaminated soil, and crops grown in contaminated soil are the major sources of exposure to dioxins for the general population. Americans routinely eat food containing 7 to 120

times more TCDD than the EPA considers safe, depending on whose data is accepted regarding dioxin contamination in the food chain. Industries producing dioxins as by-products face an avalanche of lawsuits and possible bankruptcy as long as the EPA classifies low-level exposures to dioxin as a serious human health threat. This classification also threatens to halt the burning of hazardous wastes in incinerators and cement kilns, which the EPA has sanctioned as an important way to deal with such wastes. With so much at stake, there is intense industry pressure to discredit and water down this health reassessment of dioxins. Since 1992 industry has mounted a full-scale public relations campaign to convince the public that dioxin and most other toxic chemicals are not serious threats and that we are wasting billions of dollars regulating these chemicals and cleaning up toxic waste dumps.

According to environmentalists, this campaign is a form of linguistic detoxification. They point to growing scientific evidence that dioxin and many other toxic chemicals are more harmful than previously thought, not only in causing cancers but also in disrupting the endocrine, immune, and nervous systems.

To reduce the health threats from dioxin, environmentalists call for the following actions:

- Ban the use of chlorine for bleaching wood pulp and substitute oxygen or other nonchlorine processes. Sweden, for example, has banned the sale of chlorine-bleached disposable diapers. Both Austria and Sweden use unbleached (slightly brown) coffee filters, toilet paper, milk cartons, and other paper products. By the early 1990s, the German paper industry had achieved totally chlorine-free paper production (which is 30% cheaper than using chlorine to bleach paper), and the rest of Europe is moving in this direction.

- Ban all incineration of chlorine-containing chemicals.

- Phase out the use of chlorine in manufacturing, except for a few essential uses such as the manufacture of certain drugs containing chlorine compounds.

Such proposals are vigorously opposed by paper and other chlorine-using industries—which produce 72 billion dollars worth of chlorine per year in the United States—as being unnecessary and too costly. Environmentalists point out the demand for chlorine-free paper and other products is rising and that companies making such products can tap into a growing market. In 1994 the Clinton administration unveiled a clean water plan that would eliminate most uses of chlorine.

Q: How much of the energy input of an incandescent light bulb is converted to light?

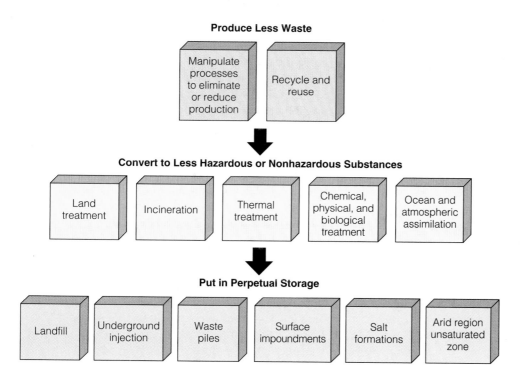

Produce Less Waste

Manipulate processes to eliminate or reduce production

Recycle and reuse

Convert to Less Hazardous or Nonhazardous Substances

Land treatment | Incineration | Thermal treatment | Chemical, physical, and biological treatment | Ocean and atmospheric assimilation

Put in Perpetual Storage

Landfill | Underground injection | Waste piles | Surface impoundments | Salt formations | Arid region unsaturated zone

Figure 13-13 Priorities for dealing with hazardous waste. (National Academy of Sciences)

13-6 SOLUTIONS: DEALING WITH HAZARDOUS WASTE

What Are Our Options? There are five basic options for dealing with hazardous wastes: **(1)** Don't make them in the first place (pollution prevention); **(2)** recycle or reuse them (this is pollution prevention if done within production processes or on site, but waste management otherwise); **(3)** detoxify them; **(4)** burn them, and **(5)** hide them by putting them into a deep well, pond, pit, or landfill, or by dumping them into the ocean.

Pollution Prevention, Recycling, and Reuse
Despite much talk about preventing pollution, the order of priorities for dealing with hazardous waste in the United States is the reverse of what many prominent scientists say it should be (Figures 13-13 and 13-14). Prevention—the most desirable option—involves substituting safer chemicals, reformulating products, modifying production processes, improving operations and maintenance, and practicing closed-loop recycling and reuse of wastes on site (Solutions, p. 358).

Some people are using less hazardous (and usually cheaper) cleaning products (Table 13-2) and are using pesticides and other hazardous chemicals only when absolutely necessary and in the smallest amount possible. Three inexpensive chemicals—baking soda, vinegar, and borax—can be used for most cleaning

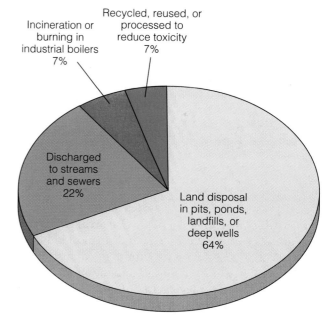

Incineration or burning in industrial boilers 7%

Recycled, reused, or processed to reduce toxicity 7%

Discharged to streams and sewers 22%

Land disposal in pits, ponds, landfills, or deep wells 64%

Figure 13-14 Management of hazardous waste in the United States. (Data from Worldwatch Institute)

and clothes bleaching. Baking soda can also be used as a deodorant and a toothpaste.

No country has an effective pollution prevention program for hazardous waste, but countries like Denmark, the Netherlands, Germany, and Sweden are all

far ahead of the United States in this area. For example, in 1993 Dutch government and chemical industry officials entered into a pact with the goal of cutting emissions of key toxic chemicals into the air and water by 80–97% from 1985 levels by 2010. And in 1992 thirteen European nations agreed in principle to eliminate all discharges and emissions of chemicals that are toxic, persistent, or likely to bioaccumulate in food chains and webs (Figure 11-25). In short, these nations made a binding commitment to try to achieve "zero discharge" of persistent toxic substances.

Effective pollution prevention requires assuming that any waste or pollutant is potentially harmful unless shown otherwise. This *precautionary principle* is the inverse of the waste management approach, in which wastes are assumed to be benign until shown to be harmful.

After preventing pollution, the next most desirable options are recycling and reuse (Figure 13-13), currently applied to only 7% of legally regulated U.S. hazardous waste (Figure 13-14). Yet the EPA devotes less than 1% of its waste management budget to prevention, reuse, and recycling of hazardous waste.

Detoxification　The next priority in hazardous-waste management is to convert any remaining waste into less hazardous or nonhazardous materials (Figure 13-13). Conversion methods include spreading biodegradable wastes on the land; using heat, chemical, or physical methods (or natural or bioengineered bacteria) to break them down; and burning them in incinerators.

Denmark has the most comprehensive and effective hazardous-waste detoxification program. Each municipality has at least one facility that accepts paints, solvents, and other hazardous wastes from households. Toxic waste from industries is delivered to 21 transfer stations throughout the country. All waste is then transferred to a large treatment facility where about 75% of the waste is detoxified and the rest is buried in a carefully designed and monitored landfill.

Biological treatment of hazardous waste, or *bioremediation*, may be the wave of the future. In this process, bacteria secrete enzymes that break down large complex molecules into smaller molecules they can absorb. The end result is cell mass and carbon dioxide. If toxin-munching bacteria can be found or engineered for specific hazardous chemicals, these substances can be fed to them at less than half the cost of disposal in landfills, and only one-third the cost of on-site incineration. However, releasing genetically engineered microorganisms into the environment is expensive and controversial (Pro/Con, p. 207). Studies indicate that most wastes can be digested better and more cheaply by naturally occurring microbes.

Incineration　According to the EPA, in 1993 the United States had 184 incinerators, 128 industrial boilers and furnaces, and 43 cement kilns burning hazardous waste. In 1980, the EPA exempted cement kilns and other industrial boilers and furnaces that burn hazardous waste as fuel from restrictions imposed on commercial incinerators burning hazardous waste.

The EPA estimates that 60% of U.S. hazardous waste could be incinerated. With proper air pollution controls and highly trained personnel, the agency considers incineration as a potentially safe, if expensive, disposal method. However, most environmentalists and some EPA scientists disagree.

Generators of hazardous waste like incinerators. Incineration generally is affordable, and anything can be burned legally. Burning gets rid of the waste and

Table 13-2 Alternatives to Some Common Household Chemicals

Chemical	Alternative	Chemical	Alternative
Deodorant	Sprinkle baking soda on a damp wash cloth and wipe skin.	General surface cleaner	Mixture of vinegar, salt, and water.
Oven cleaner	Baking soda and water paste, scouring pad.	Bleach	Baking soda or borax.
Toothpaste	Baking soda.	Mildew remover	Mix ½ cup vinegar, ½ cup borax, and warm water.
Drain cleaner	Pour ½ cup salt down drain, followed by boiling water; or pour 1 handful baking soda and ½ cup white vinegar and cover tightly for one minute.	Disinfectant and general cleaner	Mix ½ cup borax in 1 gallon hot water.
Window cleaner	Add 2 teaspons white vinegar to 1 quart warm water.	Furniture or floor polish	Mix ½ cup lemon juice and 1 cup vegetable or olive oil.
Toilet bowl, tub, and tile cleaner	Mix a paste of borax and water; rub on and let set one hour before scrubbing. Can also scrub with baking soda and a brush.	Carpet and rug shampoos	Sprinkle on cornstarch, baking soda, or borax and vacuum.
Floor cleaner	Add ½ cup vinegar to a bucket of hot water; sprinkle a sponge with borax for tough spots.	Detergents and detergent boosters	Washing soda or borax and soap powder.
Shoe polish	Polish with inside of a banana peel, then buff.	Spray starch	In a spray bottle, mix 1 tablespoon cornstarch in a pint of water.
Silver polish	Clean with baking soda and warm water.	Fabric softener	Add 1 cup white vinegar or ¼ cup baking soda to final rinse.
Air freshener	Set vinegar out in an open dish. Use an opened box of baking soda in closed areas such as refrigerators and closets. To scent the air, use pine boughs or make sachets of herbs and flowers.	Dishwasher soap	1 part borax and 1 part washing soda.
		Pesticides (indoor and outdoor)	Use natural biological controls. (Section 15-5)

of any legal liability at the same time. Once wastes are mixed and burned it is virtually impossible to trace the resulting toxic ash or air pollution to any one customer of the incinerator company. And the EPA likes hazardous-waste incineration because it gives the appearance of solving the hazardous-waste problem in a way favored by industry.

However, after making an extensive study of hazardous-waste incineration, chemist Peter Montague (Guest Essay, p. 37), an expert in this field, has concluded that it is an out-of-control technology that should be banned. He and other environmentalists point out that all incinerators release toxic air pollution (especially very harmful fine particles of metals such as lead and mercury that cannot be removed by scrubbers and other devices); create new toxic air pollutants like dioxins; and leave a highly toxic ash to be disposed of in landfills that even the EPA says will eventually leak.

Furthermore, a 1992 memo by EPA's Director of Solid Waste admitted that no U.S. hazardous-waste incinerator can destroy 99.9999% of the most hazardous chemicals, as required by law. Technically, all U.S. hazardous-waste incinerators violate federal law and should be shut down. The EPA, however, continues to allow them to operate.

Most of these pollutants move downwind from the incinerators, where people can breathe them or eat food contaminated by them. In 1993 several epidemiological studies linked respiratory and nerve disorder problems to people working at or living near hazardous-waste incinerators. Even a videotape produced by Keep America Beautiful (the voice of the waste management industry) admits that incinerators aren't safe enough.

According to EPA hazardous-waste expert William Sanjour, EPA incinerator regulations don't work because

the regulations require no monitoring of the outside air in the vicinity of the incinerator. Because operators maintain the records, they can easily cheat.... Government inspectors are poorly trained and have low morale and high turnover.... Government inspectors typically work from nine to five Monday through Friday. So if there is anything particularly nasty to burn, it will be done at night or on weekends. When complaints come in, ... the inspector may visit the plant but rarely finds anything. The enforcement officials tend to view the incinerator operator as their

Is Deep-Well Disposal of Hazardous Waste a Good Idea?

With deep-well disposal, liquid hazardous wastes are pumped under pressure through a pipe into dry, porous geologic formations or into fracture zones of rock far beneath aquifers (Figure 11-4) tapped for drinking and irrigation water. In theory these liquids soak into the porous rock material and are isolated from overlying groundwater by essentially impermeable layers of rock.

This method is simple and cheap. Also, it is less visible (because it is usually done on company land) and is less carefully regulated than other disposal methods. Its use is increasing rapidly as other methods are legally restricted or become too expensive.

If sites are chosen according to the best geological and seismic data, deep wells may be a reasonably safe way of disposing of fairly dilute solutions of organic and inorganic waste. With proper site selection and care, it may be safer than incineration. Also, if some use eventually were found for the waste, it could be pumped back to the surface.

However, the Office of Technology Assessment and many environmentalists believe that current regulations—for geologic evaluation, long-term monitoring, and long-term liability if wells contaminate groundwater—are inadequate and may allow injected wastes to:

- Spill or leak at the surface and leach into groundwater
- Escape into groundwater from corroded pipe casing or leaking seals in the well
- Migrate down or horizontally from the porous layer of rock to aquifers (either through existing fractures or through new ones caused by earthquakes, or even by stresses from the introduction of the wastes)

Until this method is more carefully evaluated and regulated, most environmentalists believe that its use should not be allowed to increase.

---

client and the public as a nuisance.... There is no reward to inspectors for finding serious violations.

In 1993, EPA head Carol Browner declared her intention to get all hazardous-waste burning facilities under full permits and rigorous controls as soon as possible. While this is being done, the EPA plans to impose both tougher standards on existing facilities and a freeze on all new burning. Environmentalists applaud these goals but fear that the powerful waste management industry (with many former EPA officials in high positions) will find ways to slow, weaken, or overturn such actions.

Land Disposal Most U.S. hazardous waste is disposed of by deep-well injections (Pro/Con, at left), surface impoundments and state-of-the-art landfills (Figure 13-10). Ponds, pits, or lagoons (Figure 11-27) used to store hazardous waste are supposed to be sealed with a plastic liner on the bottom. Solid wastes settle to the bottom and accumulate, while water and other volatile compounds evaporate into the atmosphere. According to the EPA, however, 70% of these storage basins have no liners, and as many as 90% may threaten groundwater. Eventually all liners leak, and waste will percolate into groundwater. Major storms or hurricanes can cause overflows. Moreover, volatile compounds, such as hazardous organic solvents, can evaporate into the atmosphere and eventually contaminate surface water and groundwater in other locations.

About 5% of the legally regulated hazardous-waste produced in the United States is concentrated, put into drums, and buried, either in one of 21 specially designed and monitored commercial hazardous waste landfills (Figure 13-10) or in one of 35 landfills run by companies to handle their own waste. Sweden goes further and buries its concentrated hazardous wastes in underground vaults (Figure 13-15). Ideally such landfills should be located in geologically and environmentally secure places, and carefully monitored for leaks.

However, both the EPA and the U.S. Office of Technology Assessment have concluded that even the best-designed landfill will eventually leak because the liners leak. They can be ripped or punctured during installation or by burrowing animals or dissolved by chemical solvents. Hazardous-waste engineer Peter Montague (Guest Essay, p.37) examined four hazardous-waste landfills equipped with the latest synthetic plastic liners and found they all leaked within one year.

When current and future commercial hazardous-waste landfills do leak and threaten water supplies, many of their operators will declare bankruptcy. Then the EPA will put the landfills on the Superfund list, and taxpayers will pick up the tab for cleaning them up. As EPA hazardous waste expert William Sanjour points out,

The real cost of dumping is not borne by the producer of the waste or the disposer, but by the people whose

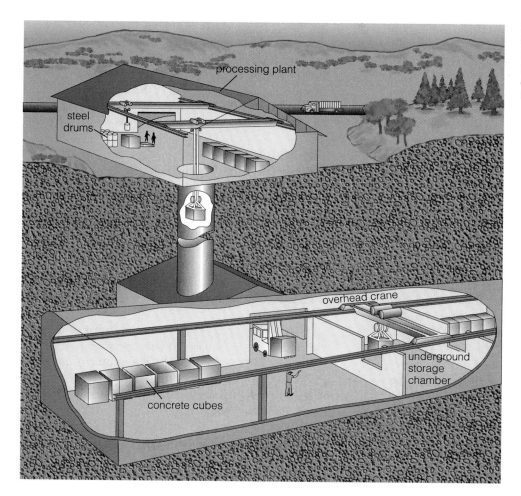

Figure 13-15 Swedish method for handling hazardous waste. Hazardous materials are placed in drums, which are embedded in concrete cubes and then stored in an underground vault.

health and property values are destroyed when the wastes migrate onto their property and by the taxpayers who pay to clean it up.... It is better for liners to leak sooner rather than later, because then there will be responsible parties that they can get to clean it up. Liners don't protect communities. They protect the people who put the waste there and the politicians who let them put the waste there, because they are long since gone when the problem comes up.

Some engineers and environmentalists have proposed storing hazardous wastes above ground in large, two-story, reinforced-concrete buildings until better technologies are developed. The first floor would contain no wastes but would have inspection walkways so people could check for leaks from above. Any leachate would be collected, treated, solidified, and returned to the storage building. Such buildings would last for many decades, perhaps as long as a century. Proponents believe that this *in-sight* approach would be cheaper and safer than *out-of-sight* landfills or incinerators for many hazardous wastes.

There is also growing concern about accidents during some of the more than 500,000 shipments of hazardous wastes (mostly to landfills and incinerators) in the United States each year. Between 1980 and 1990, for example, there were 13,476 toxic-chemical accidents, causing 309 deaths, over 11,000 injuries, and evacuation of over 500,000 people. Few communities have the equipment and trained personnel to deal adequately with hazardous-waste spills.

13-7 HAZARDOUS-WASTE REGULATION IN THE UNITED STATES

Resource Conservation and Recovery Act

In 1976 the U.S. Congress passed the Resource Conservation and Recovery Act (RCRA, pronounced "rick-ra"), amending it in 1984. This law requires the EPA to identify hazardous wastes and set standards for their management, and it provides guidelines and financial aid for states to establish waste management

A: About 10% (with the other 90% given off to the environment as waste heat)

programs. The law also requires all firms that store, treat, or dispose of more than 100 kilograms (220 pounds) of hazardous wastes per month to have a permit stating how such wastes are to be managed.

To reduce illegal dumping, hazardous-waste producers granted disposal permits by the EPA must use a "cradle-to-grave" system to keep track of waste transferred from point of origin to approved off-site disposal facilities. However, the EPA and state regulatory agencies do not have enough people to review the documentation of more than 750,000 hazardous-waste generators and 15,000 haulers each year, let alone detect and prosecute offenders.

If caught, violators are subject to large fines. However, environmentalists argue that the fines are still too low—sending polluters the clear message that crime pays.

Operators of EPA-licensed hazardous-waste landfills must prevent leakage, use at least three wells to monitor the quality of groundwater around the sites, and report any contamination to the EPA. When a landfill reaches capacity and is closed, the operators must cover it with a leakproof cap and monitor the nearby groundwater for 30 years; they are financially responsible for cleanup and damages from leaks for 30 years. Environmentalists consider that provision a serious weakness in the law because leaks from most landfills may not be detected or revealed to the public until after 30 years, passing the hazards and the cleanup costs on to succeeding generations.

Recycled chemical wastes are exempted from control under RCRA. Using this loophole, the EPA (under pressure from producers and handlers of hazardous waste) allows liquid hazardous wastes to be mixed with fuel and burned in industrial boilers, industrial furnaces, and cement kilns, and calls it "recycling." Because these combustion facilities don't have to meet the permit requirements and emission standards of EPA-licensed hazardous-waste incinerators, this increasingly common practice pollutes the air with toxic metals and other hazardous chemicals. And the resulting toxic ash can be mixed with cement (which can then be used in the walls of buildings and in pipes used to deliver drinking water) instead of having to be disposed of in EPA-licensed hazardous-waste landfills. RCRA also allows hazardous wastes to be "recycled" into pesticides as "inert" ingredients.

In 1992 the EPA proposed to exempt any waste containing toxins below certain concentrations from regulation. It did not, however, require any industry that exempted its waste on this basis to produce laboratory analyses or data to support its claim. It is estimated that this rule, based on EPA risk-benefit analysis (Section 8-5), would exempt up to 66% of presently defined hazardous waste from federal regulation, pol-

lute the drinking water of at least 13,200 people getting their water from wells within 1.6 kilometers (1 mile) of landfills receiving exempt waste, and create as many as 1,681 new Superfund sites requiring expensive cleanup. According to the EPA the financial benefits to industries from this rule outweigh its estimated harmful effects. Environmentalists disagree.

Superfund The 1980 Comprehensive Environmental Response, Compensation, and Liability Act is commonly known as the Superfund program. This law (plus amendments in 1986 and 1990) established a $16.3-billion fund financed jointly by federal and state governments and by taxes on chemical and petrochemical industries. The purpose of the Superfund is to identify and clean up abandoned hazardous-waste dump sites such as Love Canal (p. 335) and leaking underground tanks that threaten human health and the environment.

To keep taxpayers from footing most of the bill, cleanups were to be based on the "polluter pays" principle. The EPA was to locate dangerous dump sites, find the culprits, use Superfund money for the cleanup, and then sue the liable parties to recover the cleanup costs, thereby replenishing the fund for cleanups at other sites.

To implement the "polluter pays" principle, the Superfund legislation says that all the polluters of a site are subject to strict, joint, and several liability. This means that each individual polluter can be held liable for the entire cost of cleaning up a site if the other parties can't be found or have gone bankrupt. Such strict liability may seem too harsh, but writers of the legislation realized that any other liability scheme wouldn't work. If EPA had to bring an enforcement action against every party liable for a dump site, the agency would be overwhelmed with lawsuits, and the pace of cleanup would be much slower.

It's not surprising that the companies who created the dumps have vigorously opposed the "polluter pays" principle and favor a "public pays" approach. They argue that the public should pay most of the costs because the public created the pollution by demanding and buying the products sold by the polluting companies (Spotlight, p. 363).

The EPA has identified 34,000 potentially hazardous waste sites (plus 17,482 more at military bases) but has stopped looking for new ones, even though the General Accounting Office estimates that there are between 103,000 and 425,000 such sites. So far the EPA has placed almost 1,300 sites on a National Priority List for cleanup because they threaten nearby populations.

By early 1994, after spending more than $15 billion (polluters $7 billion), the EPA had declared only 214 sites clean or stabilized and had removed only 40

Q: What is the most inefficient and costly way to produce electricity for heating an interior space or water?

Superfund started off with noble goals. Today, however, the program is one that almost everyone loves to hate.

What went wrong? First, during most of the 1980s Superfund was administered by people who had an ideological interest in seeing it fail. During the early 1980s many high-level EPA officials corrupted the program by developing deals and regulations favorable to the polluting companies (the director of the program went to jail and the head of the EPA resigned after the resulting scandal).

Second, the polluters developed a strategy for doing away with the "polluter pays" principle at the heart of the program and shifting most of the bill to the taxpayers. This strategy—which is working—has three components:

- Deny responsibility (stonewalling), in order to tie the EPA up in expensive legal suits for years.

- Sue local governments and small businesses to make them responsible for cleanup, both as a delaying tactic and to turn local governments and small businesses into opponents of Superfund's strict liability requirements.

- Mount a public relations campaign declaring that toxic dumps pose little threat, that cleanup is too expensive compared to the risks involved, and that Superfund is ineffective, wasteful, and unfair.

A common criticism of Superfund is that it has been ineffective in cleaning up hazardous-waste sites, with only about 214 of almost 1,300 sites cleaned up between 1980 and 1994. EPA officials, however, contend that this is a misleading indicator of the program's effectiveness. Cleaning up a site generally requires that polluted groundwater

be cleaned up until it meets federal and state drinking-water standards. Because this pump-and-treat remedy usually takes decades, hundreds of sites being treated this way cannot now be listed as clean.

The EPA also points out that the "polluter pays" principle in Superfund has been effective in making illegal dump sites virtually a thing of the past—an important form of pollution prevention for the future. It has also forced waste producers who are fearful of future liability claims both to reduce their production of such waste and to recycle much more of what they do generate.

Superfund has also been criticized as being wasteful, with more money being spent on legal fees and consultants than on cleanup. Although this is true, the EPA points out that this has happened mostly because insurance companies and major polluters have preferred to spend millions of dollars on lawsuits in order to put off spending hundreds of millions of dollars on cleanups.

A third criticism of Superfund is that its strict liability scheme is unfair. The EPA contends that this has been greatly exaggerated by polluters wanting either to weaken or to get rid of the "polluter pays" principle based on strict liability. After many minor parties such as pizza parlors and doughnut shops were subjected to lawsuits by insurance companies and major polluters, the EPA resolved the liability of almost 5,300 minor parties between 1986 and 1992. And Carol Browner, the new EPA head, is pushing for acceleration of this process.

Industry opponents of Superfund have also proposed that cleanups be tied to the future intended use of a site. Environmentalists oppose this idea for two reasons. First, poisons don't always stay inside a site's boundary, as illustrated by the Love Canal (p. 335) and other incidents. Second, many

of these wastes will be around for hundreds to thousands of years. With growing population and industrialization, there is no way to know how a site might be used during this time frame.

Waste management experts on both sides of this issue have proposed that the Superfund program be improved by:

- Not removing or weakening its "polluter pays" principle based on strict liability. Opponents of strict liability, however, have proposed that all parties should pay only their fair share of the cleanup (as determined by a panel of administrative judges). Superfund money would pay for the shares of liable parties that are insolvent, defunct, or unknown. Weakening this provision by having the EPA identify and assign proportionate costs to all parties will reduce the number of sites cleaned up, increase court challenges to EPA decisions, shift much of the financial responsibility to taxpayers instead of the parties who created the dumps, and deplete the fund with administrative costs instead of using it for cleanup.

- Retaining the goal of permanent cleanup but ranking sites in three general categories: **(1)** those requiring immediate full cleanup; **(2)** those considered to be a serious threat but that are not located near concentrations of people or endangered ecosystems (these sites would receive emergency cleanup and then be isolated by barriers and signs, with fuller cleanup to come later); and **(3)** lower-risk sites requiring only stabilization (capping and containment) and monitoring unless wastes start migrating into the air or water or until better and more cost-effective cleanup methods are developed.

(continued)

- Establishing and using standardized remedies for sites with similar problems.

- Using more Superfund money to fund research on better and more cost-effective ways to clean up sites and to establish demonstration sites to test new cleanup technologies.

- Accelerating the process for dealing with minor waste contributors to sites.

- Reducing some of the cleanup costs for polluters or their insurance companies who agree to binding arbitration instead of lengthy and costly litigation. Such a provision would have to be monitored carefully to be sure that big polluters are not let off too easily relative to their waste contribution.

- Having the U.S. Army Corps of Engineers assist the EPA in

cleanups paid for by Superfund money.

- Allowing real citizen participation in the process. Too often the people affected by risks from Superfund sites are not consulted until the EPA and polluters have already worked out a deal.

---

from the priority list. It takes an average of 9 years from the time a site is identified until it is cleaned up, with an average cleanup cost of $25 million per site. Only $2.4 billion was spent on site-specific activities, with the rest used for outside consultants, administration, management, and litigation. And according to a 1989 report by the Office of Technology Assessment (OTA), about 75% of the cleanups are unlikely to work over the long term. Also, the EPA has recovered less than one-fifth of the $4.3 billion that was supposed to be paid by polluters. And the statue of limitations (6 years from the start of cleanup) has run out for many polluters.

The OTA and the Waste Management Research Institute estimate that the final list could include at least 10,000 priority sites, with cleanup costs of $750 million to $1 trillion, not counting legal fees. Cleaning up toxic military dumps will cost another $100–$200 billion and take at least 30 years; and cleaning up contaminated Department of Energy sites used to make nuclear weapons will cost an additional $100–$400 billion and take 30–50 years. Indeed, full cleanup of many sites may not be scientifically or economically feasible. It is hard to imagine a more convincing reason for emphasizing pollution prevention (Figure 13-13).

Are the fears of the more than 40 million people living near identified hazardous-waste sites justified? In 1993 researchers found that a higher number of children are born with birth defects when their mothers live near hazardous waste dumps. However, according to a 1992 study by the National Academy of Sciences, we don't know much about the effects of living near such dumps. The study concluded that the federal government has (1) no comprehensive inventory of waste sites; (2) no program for discovering new sites; (3) insufficient data for determining safe

exposure levels; (4) questionable methods for assessing the public health danger at Superfund and other hazardous-waste sites; (5) an inadequate system for identifying sites that require immediate action; and (6) no cost-effective methods to clean up sites.

Meanwhile real people are living near thousands of real hazardous-waste dumps. These victims are trapped in a toxic nightmare that fills them with fear and that has made any property they own essentially worthless.

Grass-Roots Action Studies show that incinerators, landfills, or treatment plants for hazardous wastes have traditionally been located in communities populated by African Americans, Asian Americans, Hispanics, and poor whites. Such actions have been condemned as a mixture of environmental racism and economic discrimination (Guest Essay, p. 367).

Now people of color, the poor, and middle-class whites have joined together in a loose-knit coalition known as the grass-roots movement for *environmental justice*. By 1993 this network had grown to 350 national, regional, and local groups. Here are some examples of such injustice uncovered by a variety of studies:

- 60% of black and Hispanic Americans live in communities with uncontrolled toxic waste sites, compared with 50% of the general population.

- Minorities suffer greater exposure to air pollution than do whites.

- Farm workers—more than 80% of them minorities—suffer the most from pesticide poisoning.

- A 1992 investigation by the *National Law Journal* found that between 1985 and 1991 violators of

Q: What are the two most energy-efficient ways to heat interior space?

Figure 13-16 The Mississippi River between Baton Rouge and New Orleans, Louisiana, is lined with oil refineries and petrochemical plants. Along this corridor, known as "Cancer Alley" because of its abnormally high cancer rates, tons of carcinogenic and mutagenic chemicals leak into groundwater or are discharged into the river. In 1988 an environmental alliance of black and white residents protested chemical dumping in their communities and groundwater by marching the 137 kilometers (85 miles) from Baton Rouge to New Orleans.

federal hazardous-waste laws that the EPA took to court received fines that were 500% higher if they polluted white communities than if they violated the law in minority communities.

- The same study showed that abandoned toxic waste sites in minority areas took 20% longer to be placed on the Superfund priority cleanup list than those in white communities.

- A 1988 study found that in families with annual incomes less than $6,000-$15,000 the percentage of black children with unacceptably high blood lead levels was about twice that of white children. And in families with annual incomes greater than $15,000 high lead levels were found in 38% of black children, compared to 12% of white children.

In 1990 leaders of the environmental justice movement sent an open letter to the "Big 10" environmental groups, calling on them to integrate their staffs and memberships and focus more on the problem of environmental injustice. By 1993 most organizations had responded to varying degrees to such criticisms and had initiated new studies and environmental programs to help minority communities.

This coalition offers the following guidelines for achieving environmental justice for all:

- *Don't compromise our children's futures by cutting deals with polluters and regulators.* Environmental justice should not be bought or sold.

- *Hold polluters—and elected officials who go along with them—personally accountable, because what they are doing is wrong.*

- *Don't fall for the argument that protesters against hazardous-waste landfills, incinerators, and injection wells are holding up progress in dealing with hazardous wastes.* Instead, recognize that the best way to deal with waste and pollution is not to produce so much of it (Guest Essays, pp. 37 and 338). After that has been done we can decide what to do with what is left, as suggested by the National Academy of Sciences (Figure 13-13).

- *Oppose all hazardous-waste landfills, deep-disposal wells, and incinerators.* This will sharply raise the cost of dealing with hazardous materials, discourage location of such facilities in poor neighborhoods often populated by minorities (Figure 13-16 and Guest Essay, p. 367), and encourage waste producers and elected officials to get serious about pollution prevention. The goal of politically powerful waste management companies is to have us produce more and more hazardous (and nonhazardous) waste so they can make higher profits. In 1992 the total revenues in the burgeoning hazardous-waste business were $18 billion (compared to $0.5 billion in 1977), and by 2000 they are expected to reach $43 billion.

- *Recognize that there is no such thing as "safe" disposal of hazardous waste.* For such materials the goal should be "Not in Anyone's Backyard" (NIABY) or "Not on Planet Earth" (NOPE).

Ash to Cash: The International Hazardous-Waste Trade

To save money and to escape regulations and local opposition, cities and waste disposal companies in the United States and other MDCs legally ship vast quantities of hazardous waste to other countries. Most legal U.S. exports of hazardous wastes go to Canada and Mexico, but at least nine African countries have also accepted them.

These shipments can take place without EPA approval because U.S. hazardous-waste laws allow exports for "recycling." Sometimes exported wastes labeled as materials to be recycled are dumped after reaching their destination.

U.S. companies are also exporting hazardous waste and jobs by moving highly polluting smelters and manufacturing plants to countries with weak or poorly enforced pollution control laws—and with workers willing to work for lower wages under dangerous conditions. A glaring example is the growing number of U.S. and other foreign factories located along Mexico's northern border. Mexico has some strong environmental laws, but enforcement is lax. Most host countries, hungry for jobs and foreign capital, turn a blind eye to unsafe and polluting practices.

There is also a growing illegal trade in hazardous wastes across international borders. There are too few customs inspectors, and they are not trained to detect such shipments. Hazardous wastes have also been mixed with wood chips or sawdust and shipped as burnable material.

Waste disposal firms can charge high prices for picking up hazardous wastes. If they can then dispose of them—legally or illegally—at low costs, they pocket huge profits. Officials of poor LDCs find it hard to resist the income (often in the form of bribes) from receiving these wastes.

Currently at least 83 countries have banned imports of hazardous waste, and some have adopted a "return to sender" policy when illegal waste shipments are discovered. Environmentalists and some members of Congress call for the United States to ban all exports of hazardous waste (including radioactive waste). They would also ban exports of pesticides and drugs not approved for use in the United States; and they would classify violations of these bans as criminal acts. They argue that exporting hazardous wastes to other countries (or to other states) encourages the throwaway mentality and discourages pollution prevention. Also, exports of toxic waste may come back to haunt the exporters. For instance, they may contaminate soil or fertilizer used to grow food that is imported by the exporting countries.

In 1989, 105 countries meeting in Basel, Switzerland, drew up the Basel Convention, which establishes principles to be enforced by international law that would control shipments of toxic waste across national borders. However, the United States has refused to sign the convention, and some countries and international law experts say that the pact is too vague and full of loopholes to stop the international trade in hazardous wastes.

An effective U.S. or worldwide ban on all hazardous waste exports would help but would not end illegal trade in these wastes. The potential profits are simply too great. The only real solution to the hazardous-waste problem is to stop most of it from ever being produced.

- *Ban release of any toxic chemical that is persistent in any medium (water, air, sediment, soil) or that bioaccumulates in living things* (Figure 11-25). In 1992, 13 European nations agreed in principle to work toward achieving this goal.

- *Pressure elected officials to pass legislation requiring that unwanted industries and waste facilities be distributed more widely instead of being concentrated in poor and working-class neighborhoods, many populated mostly by minorities.*

- *Ban all hazardous-waste exports from one country to another* (Spotlight, above).

It is encouraging that in 1994 President Bill Clinton issued an executive order requiring the EPA to reduce environmental injustice. An Office of Environmental Equity has been established to incorporate environmental justice in every rule, regulation, and proposal the EPA issues.

Making the Transition to a Low-Waste Society

So far we have had our priorities for dealing with solid waste (Figure 13-3) and hazardous waste (Figure 13-13) backwards from what scientists tell us are the best ways to deal with them. According to the strategy proposed by the National Academy of Sciences, we need to replace the "two Bs" of waste management—Burn or Bury—with the "three Rs" of Earth care: Reduce, Reuse, Recycle. This invokes thinking of wastes as wasted resources. We will always produce some wastes, but we can produce much less.

Q: What is the most energy-efficient fuel for powering a motor vehicle?

Robert D. Bullard

Robert D. Bullard is a professor of sociology at the University of California, Riverside. For more than a decade he has worked on and conducted research in the areas of urban land use, housing, community development, industrial facility siting, and environmental justice. His scholarship and activism have made him one of the leading experts on environmental racism—the systematic selection of communities of color for waste facilities and polluting industries. He is the author of four books and more than three dozen articles, monographs, and scholarly papers that address equity concerns. His book, Dumping in Dixie: Race, Class, and Environmental Quality *(Westview Press, 1990), has become a standard text in the field. His most recent book is* Confronting Environmental Racism *(South End Press, 1993).*

Despite widespread media coverage and volumes written on the U.S. environmental movement, environmentalism and social justice have seldom been linked. Nevertheless, an environmental revolution is now taking shape in the United States that combines the environmental and social justice movements into one framework.

People of color (African Americans, Latinos, Asians, Pacific Islanders, and Native Americans), working-class people, and poor people in the United States suffer disproportionately from industrial toxins, dirty air and drinking water, unsafe work conditions, and the location of noxious facilities such as municipal landfills, incinerators, and toxic waste dumps. Despite the government's attempts to level the playing field, all communities are not created equal.

The environmental justice movement attempts to dismantle exclusionary zoning ordinances, discriminatory land-use practices, differential enforcement of environmental regulations, disparate siting of risky technologies, and the dumping of toxic waste on the poor and people of color in the United States and in LDCs.

All communities are not treated as equals when it comes to resolving environmental and public health concerns, either. Over 300,000 farm workers (over 90% of whom are people of color) and their children are poisoned by pesticides sprayed on crops in the United States. Some 3–4 million children (many of them African Americans or Latinos living in the inner city) are poisoned by lead-based paint in old buildings, lead-soldered pipes and water mains, lead-tainted soil contaminated by industry, and air pollutants from smelters. Lead poisoning is considered the number one environmental health problem facing children in the United States. Yet little has been done over the past 20 years to rid the nation of this preventable childhood hazard.

All communities do not bear the same burden or reap the same benefits from industrial expansion. This is true

in the case of the mostly African American Emelle, Alabama (home of the nation's largest hazardous-waste landfill); Navajo lands in Arizona where uranium is mined; and the 2,000 factories known as *maquiladores*, located just across the U.S. border in Mexico.

Communities, states, and regions that contain hazardous-waste disposal facilities (importers) receive far fewer economic benefits (jobs) than the geographic locations that generate the wastes (exporters). Nationally 60% of African Americans and 50% of Latinos live in communities with at least one uncontrolled toxic-waste site. Three of the five largest hazardous-waste landfills are located in communities that are predominantly African American or Latino.

The marginal status of many people of color in the United States makes them prime actors in the movement for environmental and social justice. For example, the organizing theme of the 1991 First National People of Color Environmental Summit, held in Washington, D.C., was justice, fairness, and equity. More than 650 delegates from all 50 states, as well as Puerto Rico, Mexico, Chile, Colombia, and the Marshall Islands, participated in this historic four-day gathering.

Environmental justice does not stop at the U.S. borders. Environmental injustices exist from the *favelas* of Rio de Janeiro [Figure 7-9] to the shantytowns of Johannesburg. Members of the environmental justice movement are also questioning the wasteful and nonsustainable development models being exported to the developing world.

It is no mystery why grass-roots environmental justice groups in Louisiana's "Cancer Alley," [Figure 13-16], Chicago's southside, and Los Angeles's East and South Central neighborhoods are attacking the institutions they blame for their underdevelopment, disenfranchisement, and poisoning. Some people see these threats to their communities as a form of genocide.

Grass-roots leaders are demanding justice. Residents of communities such as West Dallas and Texarkana (Texas), West Harlem (New York), Rosebud (South Dakota), Kettleman City (California), and Sunrise, Lions, and Wallace (Louisiana) see their struggle for environmental justice as a life-and-death matter. Unfortunately their stories of environmental racism are not piped into the nation's living rooms during the nightly news, nor are they blasted across the front pages of national newspapers and magazines. To a large extent the communities that are the victims of environmental injustice remain "invisible" to the larger society.

The environmental justice movement is led, planned, and to a large extent funded by individuals who are not part of the established environmental community or the "Big 10" environmental organizations. Most environ-

(continued)

mental justice groups are small and operate with resources generated from the local community.

For too long these groups and their leaders have been "invisible" and their stories muted. This is changing as these grass-roots groups are forcing their issues onto the nation's environmental agenda.

The United States has a long way to go in achieving environmental justice for all its citizens. The membership of decision-making boards and commissions still does not reflect the racial, ethnic, and cultural diversity of the country. And token inclusion of persons of color on boards and commissions does not necessarily mean that their voices will be heard or their cultures respected. The ultimate goal of any inclusion strategy should be to de-

mocratize the decision-making process and empower disenfranchised people to speak and do for themselves.

Critical Thinking

1. Does your lifestyle and political involvement help promote or reduce environmental racism in society as a whole and in the community where you live?

2. How would you go about helping prevent polluting factories and hazardous-waste facilities from being located in or near communities made up largely of people of color, working-class people, and poor people?

What You Can Do: Reduce Waste and Save Money

INDIVIDUALS MATTER

- *Buy less by asking yourself whether you really need a particular item.*

- *Buy things that are reusable or recyclable, and be sure to reuse and recycle them.*

- *Buy beverages in refillable glass containers instead of cans or throwaway bottles.*

- *Use plastic or metal lunch boxes and metal or plastic garbage containers without throwaway plastic liners (unless such liners are required for garbage collection).*

- *Carry sandwiches and store food in the refrigerator in reusable containers instead of wrapping them in aluminum foil or plastic wrap.*

- *Use rechargeable batteries and recycle them when their useful life is over.* In 1993, Rayovac began selling mercury-free, rechargeable alkaline batteries that outperform conventional nickel-cadmium rechargeable batteries.

- *Carry groceries and other items in a reusable basket, a canvas or string bag, or a small cart.*

- *Use sponges and washable cloth napkins, dish towels, and handkerchiefs instead of paper ones.*

- *Don't use throwaway paper and plastic plates and cups, eating utensils, and other disposable items when reusable or refillable versions are available.*

- *Buy recycled goods, especially those made by primary recycling, and then recycle them.*

- *Reduce the amount of junk mail you get.* This can be accomplished at no charge by writing to Mail Preference Service, Direct Marketing Association, 11 West 42nd St., P.O. Box 3681, New York, NY 10163-3861, or by calling 212-768-7277 and asking that your name not be sold to large mailing-list companies. Of the junk mail you do receive, recycle as much of the paper as possible.

- *Buy products in concentrated form whenever possible.*

- *Choose items that have the least packaging or, better yet, no packaging ("nude products").*

- *Don't buy helium-filled balloons, and urge elected officials and school*

administrators to ban balloon releases except for atmospheric research and monitoring.

- *Compost your yard and food wastes, and lobby local officials to set up a community composting program.*

- *Use pesticides and other hazardous chemicals (Table 13-1) only when absolutely necessary, and in the smallest amount possible.*

- *Use less hazardous (and usually cheaper) cleaning products (Table 13-2).*

- *Do not flush hazardous chemicals down the toilet, pour them down the drain, bury them, throw them away in the garbage, or dump them down storm drains.* Consult your local health department or environmental agency for safe disposal methods.

- *Support legislation that would encourage pollution prevention and waste reduction.*

Q: How much money could be saved if the world got serious about improving energy efficiency?

Table 13-3 Three Systems for Handling Discarded Materials

Item	For a High-Waste Throwaway System	For a Moderate-Waste Resource Recovery and Recycling System	For a Low-Waste Sustainable-Earth System
Glass bottles	Dump or bury	Grind and remelt; remanufacture; convert into building materials	Ban all nonreturnable bottles; reuse bottles
Bimetallic "tin" cans	Dump or bury	Sort, remelt	Limit or ban production; use returnable bottles
Aluminum cans	Dump or bury	Sort, remelt	Limit or ban production; use returnable bottles
Cars	Dump	Sort, remelt	Sort, remelt; tax cars getting less than 17 kilometers per liter (40 miles per gallon)
Metal objects	Dump or bury	Sort, remelt	Sort, remelt; tax items lasting less than 10 years
Tires	Dump, burn, or bury	Grind and revulcanize or use in road construction; incinerate to generate heat or electricity	Recap usable tires; tax or ban all tires not usable for at least 96,000 kilometers (60,000 miles)
Paper	Dump, burn, or bury	Incinerate to generate heat	Compost or recycle; tax all throwaway items; eliminate overpackaging
Plastics	Dump, burn, or bury	Incinerate to generate heat or electricity	Limit production; use returnable glass bottles instead of plastic containers; tax throwaway items and packaging
Yard wastes	Dump, burn, or bury	Incinerate to generate heat or electricity	Compost; return to soil as fertilizer; use as animal feed

With this new way of thinking about waste the most important question is, How can we produce less waste—and, for especially hazardous substances, no waste—(Individuals Matter, p. 368)? Table 13-3 compares the throwaway resource system of the United States, a resource recovery and recycling system, and a sustainable-Earth (or low-waste) resource system.

Making the transition to a low-waste society will not be easy and is very controversial, but in the long run environmentalists argue that it will provide more economic and environmental benefits than not doing it. Environmentalists challenge us to set the following goals:

- *Cut industrial hazardous-waste production 50% over 1990 levels by 2000 and 80% by 2010.*

- *Reuse and recycle (including composting) 60% of municipal solid waste by 2000 and 80% by 2010.*

To prevent pollution and reduce waste, environmentalists urge us to understand and live by four key principles: **(1)** Everything is connected; **(2)** there is no "away" for the wastes we produce; **(3)** dilution is not the solution to most pollution; and **(4)** the best and cheapest way to deal with waste is not to produce so much.

The presumption should be that any waste or pollutant is potentially harmful and preventable.... With a pollution prevention strategy, human intelligence and creativity as well as science and technology can focus on preventing, eliminating, or reducing the production of all wastes and pollutants.

JOEL HIRSCHORN

Critical Thinking

1. Explain why you support or oppose the following:
   a. Passing a national beverage-container deposit law
   b. Requiring that all beverage containers be reusable
   c. Requiring all households and businesses to sort recyclable materials for curbside pickup in separate containers
   d. Requiring consumers to pay for plastic or paper bags at grocery and other stores to encourage the use of reusable shopping bags

2. Keep a list for a week of the solid waste you throw away. What percentage is materials that could be

recycled, reused, or burned for energy? What percentage of the items could you have done without?

3. Would you oppose having a hazardous-waste landfill, a waste-treatment plant, a deep injection well, or an incinerator in your community? Explain. If you oppose these disposal facilities, how should the hazardous waste generated in your community and your state be managed?

4. Give your reasons for agreeing or disagreeing with each of the following proposals for dealing with hazardous waste:
   a. Reduce the production of hazardous waste and encourage recycling and reuse of hazardous materials by levying a tax or fee on producers for each unit of waste generated.
   b. Ban all land disposal of hazardous waste to encourage recycling, reuse, and treatment and to protect groundwater from contamination.
   c. Provide low-interest loans, tax breaks, and other financial incentives to encourage industries producing hazardous waste to recycle, reuse, treat, destroy, and reduce generation of such waste.
   d. Ban the shipment of hazardous waste from the United States to any other country.
   e. Ban the shipment of hazardous waste from one state to another.

*5. What hazardous wastes are produced at your school? In your community? What happens to these wastes?

*6. Make a concept map of the key ideas in this chapter using the section heads and subheads and the key terms (shown in boldface type in the chapter). See the inside front cover and Appendix 4 for information on concept maps.

PART IV

Biodiversity: Living Resources

It is the responsibility of all who are alive today to accept the trusteeship of wildlife and to hand on to posterity, as a source of wonder and interest, knowledge, and enjoyment, the entire wealth of diverse animals and plants. This generation has no right by selfishness, wanton or intentional destruction, or neglect, to rob future generations of this rich heritage. Extermination of other creatures is a disgrace to humankind.

WORLD WILDLIFE CHARTER

14 Food Resources

Perennial Crops on the Kansas Prairie

When you think about farms in Kansas, you probably picture endless fields of wheat or corn plowed up and planted each year. By 2040 this picture might change, thanks to pioneering work at the nonprofit Land Institute near Salina, Kansas (Figure 14-1).

The institute, founded by Wes and Dana Jackson, is experimenting with an ecological approach to agriculture on the midwestern prairie, based on planting a mix of *perennial* grasses, legumes, sunflowers, and grain crops in the same field (polyculture); because they are perennial, they don't have to be replanted each year like traditional food crops. The institute's goal is to raise food by mimicking many of the natural conditions of the prairie.

By eliminating yearly soil preparation and planting, perennial polyculture requires much less labor than conventional monoculture and diversified organic farms growing annual crops. Perennial polyculture rewards farmers more for their wits, creative thinking, and land care efforts, and less for routine drudgery and labor.

If the institute and similar groups doing such Earth-sustaining research succeed, within a few decades, many people may be eating food made from perennials such as *Maximilian sunflower* (which produces seeds with as much protein as soybeans), *eastern gamma grass* (a relative of corn with three times as much protein as corn and twice as much as wheat), *Illinois bundleflower* (a wild nitrogen-producing legume that can enrich the soil and whose seeds may serve as livestock feed), and *giant wild rye* (once eaten by Mongols in Siberia).

These discoveries will come none too soon. To feed the 8.5 billion people projected by 2025, we must produce and distribute as much food during the next 30 years as was produced in all the years since agriculture began about 10,000 years ago.

Figure 14-1 The Land Institute in Salina, Kansas. It is a farm, prairie laboratory, and school dedicated to changing the way we grow food by substituting a diverse mixture of edible perennial plants for traditional annual monoculture crops. (Terry Evans)

The discussion in this chapter answers several general questions:

- How is the world's food produced?
- What are the world's food problems?
- What can we do to solve the world's food problems?
- How can we design and shift to sustainable-Earth agricultural systems?

14-1 HOW FOOD IS PRODUCED

Plants and Animals That Feed the World

The species of plants and animals (species diversity) and the varieties of various plants and animals (genetic diversity) that provide us with food are an important part of the planet's biodiversity. Biologists estimate that the earth has perhaps 30,000 plant species with parts that people can eat. However, just 15 plants and 8 animal species supply 90% of our food. Four crops—wheat, rice, corn, and potato—make up more of the world's total food production than all other crops combined.

Grains provide about half the world's calories, with two out of three people eating mainly a vegetarian diet—mostly because they can't afford meat. As incomes rise, people consume even more grain, but indirectly—in the form of meat, eggs, milk, cheese, and other products of domesticated livestock. Although only about one-third of the world's people can afford to eat meat, more than half of the world's cropland (and almost two-thirds of the cropland in the United States) is used to produce livestock feed to supply these individuals with meat. In addition, one-third of the world's fish catch is converted into fish meal to feed livestock consumed by meat eaters in MDCs.

Types of Food Production

There are two major types of agricultural systems: industrialized and traditional. **Industrialized agriculture** uses large amounts of fossil-fuel energy, water, commercial fertilizers, and pesticides to produce huge quantities of one crop or animal for sale. Practiced on about 25% of all cropland, mostly in MDCs, industrialized agriculture has spread since the mid-1960s to some LDCs (Figure 14-2). **Plantation agriculture**, a form of industrialized agriculture

mostly in tropical LDCs, grows cash crops such as bananas, coffee, and cacao, mostly for sale to MDCs.

Traditional agriculture consists of two main types. **Traditional subsistence agriculture** typically produces only enough crops or livestock for a farm family's survival; in good years there may be a surplus to sell or to put aside for hard times. Subsistence farmers use human labor and draft animals. Examples of this type of agriculture include shifting cultivation in tropical forests (Figure 2-5) and nomadic livestock herding. With **traditional intensive agriculture** farmers increase their inputs of human and draft labor, fertilizer, and water to get a higher yield per area of cultivated land to produce enough food to feed their families and perhaps a surplus for sale. These forms of traditional agriculture are practiced by about 2.7 billion people—almost half the people on Earth—in LDCs. Figure 14-3 show the relative inputs of land, human and animal labor, fossil-fuel energy, and capital needed to produce one unit of food energy by various types of food production.

Industrialized Agriculture and Green Revolutions

Farmers can produce more either by farming more land or by getting higher yields per unit of area from existing cropland. Since 1950 most of the increase in global food production has come from raising the yield per unit of area in a process called a **green revolution**. This process involves planting monocultures of genetically improved plant varieties and lavishing fertilizer, pesticides, and water on them. This approach dramatically increased crop yields in most MDCs between 1950 and 1970 in what is considered the *first green revolution* (Figure 14-4, p. 376).

A *second green revolution* has been taking place since 1967, when fast-growing dwarf varieties of rice and wheat, specially bred for tropical and subtropical climates, were introduced into several LDCs (Figure 14-4). With enough fertilizer, water, and pesticides, yields of these new plants can be two to five times those of traditional wheat and rice varieties (Figure 14-5, p. 376). And fast growth allows farmers to grow two or even three crops a year (multiple cropping) on the same land parcel.

Nearly 90% of the increase in world grain output in the 1960s, about 70% in the 1970s, and 80% of that in the 1980s resulted from this second green revolution. In the 1990s at least 80% of any increase is expected to come from green-revolution techniques.

These increases depend heavily on fossil fuels to run machinery, produce and apply inorganic fertilizers and pesticides, and pump water for irrigation. Since 1950 agricultural use of fossil fuels has increased fourfold, the number of tractors has quadrupled, irrigated area has tripled, use of commercial fertilizer has

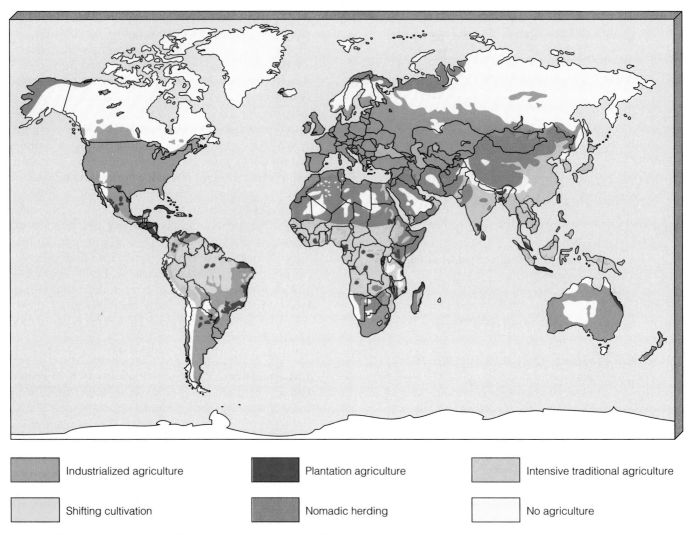

	Industrialized agriculture		Plantation agriculture		Intensive traditional agriculture
	Shifting cultivation		Nomadic herding		No agriculture

Figure 14-2 Generalized location of the world's principal types of food production.

risen 9-fold, and use of pesticides has risen 32-fold. All told, green-revolution agriculture now uses about 8% of the world's oil output.

These high inputs of energy, water, fertilizer, and pesticides have yielded dramatic results, but at some point additional inputs become useless because no more output can be squeezed from the land. In fact, yields may even start dropping because the soil erodes, loses fertility, and becomes salty and waterlogged; because underground water supplies become depleted; and because populations of rapidly breeding pests develop genetic immunity to widely used pesticides.

Since 1940 U.S. farmers have more than doubled crop production without cultivating more land. They have accomplished this through industrialized agriculture, using green revolution techniques in a favorable climate on some of the world's most fertile and productive soils.

Farming has become *agribusiness* as big companies and larger family-owned farms have taken control of most U.S. food production. In 1992, only 1.8% (4.6 million people—down from 23 million in 1950 and 6 million in 1980) of the U.S. population lived on the country's 2.1 million farms. And only about 650,000 Americans worked full-time at farming in 1992. However, about 23 million people—9% of the population—are involved in the U.S. agricultural system, from growing and processing food to distributing it to selling it at the supermarket. In terms of total annual sales, agriculture is the biggest industry in the United States—bigger than the automotive, steel, and housing industries combined. It generates about 18% of the country's gross national product and 19% of all jobs in the private sector, employing more people than any other industry.

U.S. farmland, called the "breadbasket" of the world, produces half the world's grain exports. In

Q: What country has the highest industrial energy efficiency?

1992, each U.S. farmer fed and clothed 130 people (96 at home and 34 abroad), up from 58 people in 1976.

The industrialization of agriculture was made possible by the availability of cheap energy. Most of this energy comes from oil, followed by natural gas used for drying and producing inorganic fertilizers. Agriculture consumes about 17% of all commercial energy used in the United States each year (Figure 14-6).

Most plant crops in the United States provide more food energy than the energy used to grow them. However, raising livestock requires much more fossil-fuel energy than the animals provide in food energy. If we include crops and livestock, U.S. farms currently use about 3 units of fossil-fuel energy to produce 1 unit of food energy. Indeed, if all people in the United States were vegetarians, the country's oil imports could be cut by 60%.

Energy efficiency is much worse if we look at the whole U.S. food system. Considering the energy used to grow, store, process, package, transport, refrigerate, and cook all plant and animal food, *an average of about 10 units of nonrenewable fossil-fuel energy are needed to put 1 unit of food energy on the table.* By comparison, every unit of energy from the human labor of subsistence farmers provides at least 1 unit of food energy and, with traditional intensive farming, up to 10 units of food energy.

Examples of Traditional Agriculture Farmers in LDCs grow about 20% of the world's food on about 75% of its cultivated land. Many traditional farmers simultaneously grow several crops on the same plot, a method called **interplanting**. This biological diversity reduces the chances of losing most or all of their year's food supply to pests, flooding, drought, or other disasters. Common interplanting strategies include

- **Polyvarietal cultivation**, in which a plot is planted with several varieties of the same crop.

- **Intercropping**, in which two or more different crops are grown at the same time on a plot—for example, a carbohydrate-rich grain that uses soil nitrogen alongside a protein-rich legume that puts it back.

- **Agroforestry**, or **alley cropping**, a variation of intercropping in which crops and trees are planted together. For example, a grain or legume crop is planted around fruit-bearing orchard trees, or in rows between fast-growing trees or shrubs that can be used for fuelwood or for adding nitrogen to the soil (Figure 12-28c).

- **Polyculture**, a more complex form of intercropping in which many different plants maturing at various times are planted together. If cultivated

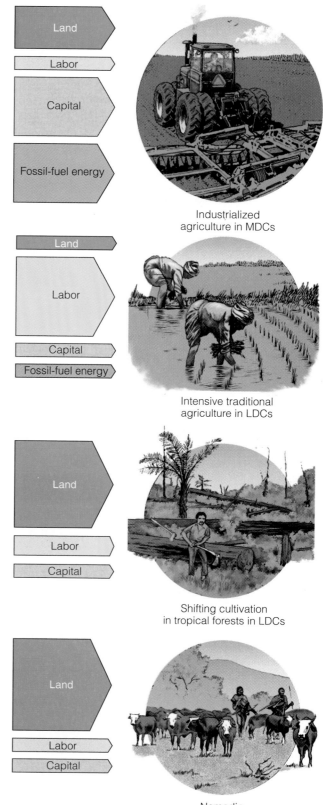

Figure 14-3 Relative inputs of land, labor, capital, and fossil-fuel energy to the principal agricultural systems. An average of 60% of the people in LDCs are involved directly in producing food, compared with only 8% in MDCs (2% in the United States).

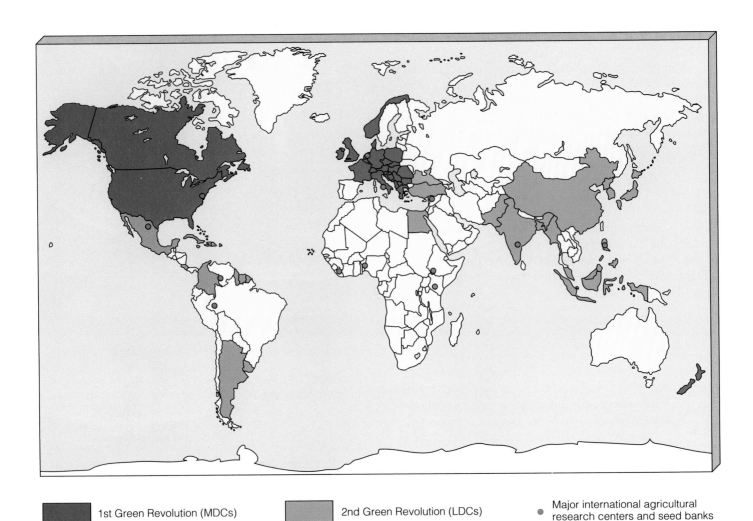

■ 1st Green Revolution (MDCs)	■ 2nd Green Revolution (LDCs)	● Major international agricultural research centers and seed banks

Figure 14-4 Countries whose crop yields per unit of land area increased during the two green revolutions. The first took place in MDCs between 1950 and 1970, and the second has occurred since 1967 in LDCs with enough rainfall or irrigation capacity. Thirteen agricultural research centers and genetic storage banks play a key role in developing high-yield crop varieties.

Figure 14-5 Two parent strains of rice—PETA from Indonesia (center) and DGWG from China (right)—were crossbred to yield IR-8 (left), a new high-yield, semi-dwarf variety of rice used in the second green revolution. The shorter and stiffer stalks of the new varieties allow them to support larger heads of grain without toppling over.

International Rice Research Institute, Manila

Q: How much of the heat in U.S. homes and other buildings escapes through closed windows?

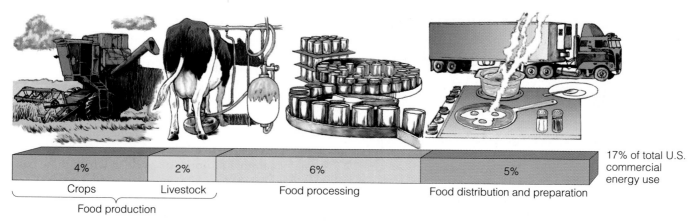

4%	2%	6%	5%	17% of total U.S. commercial energy use
Crops	Livestock	Food processing	Food distribution and preparation	

Food production

Figure 14-6 Commercial energy use by the U.S. industrialized agriculture system. About 20% of the total energy used directly on farms to produce crops is for pumping irrigation water. On average, a piece of food eaten in the United States has traveled 2,100 kilometers (1,300 miles). Processing food also requires large amounts of energy. For example, supplying orange juice takes four times more energy than providing fresh oranges that contain the same amount of juice.

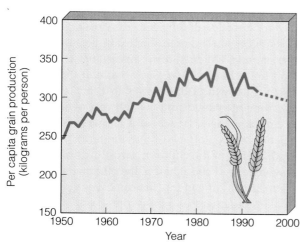

Figure 14-7 World grain production per person, 1950–93, with projection to 2000. (Data from U.S. Department of Agriculture and Worldwatch Institute)

properly, these plots can provide food, medicines, fuel, and natural pesticides and fertilizers on a sustainable basis.

14-2 WORLD FOOD PROBLEMS

Good and Bad News About Food Production

Between 1950 and 1984, world grain production almost tripled, and per capita production rose by 40% (Figure 14-7), reducing hunger and malnutrition around the world. During the same period average food prices adjusted for inflation dropped by 25%, and the amount of food traded in the world market quadrupled.

Despite these impressive achievements in food production, population growth is outstripping food production where 2 billion people live. Global grain production increased at a record 3% per year from 1950 to the peak year of 1984. Since then, however, it has risen barely 1% per year—lower than the rate of population growth. As a result, per capita global grain production has declined by roughly 1% per year since 1984 —an 11% drop between 1984 and 1993 (Figure 14-7). Since 1978, grain production has lagged behind population growth in 69 of the 102 LDCs for which data are available. In 22 African countries, per capita food production has dropped 28% since 1960 and may drop another 30% during the next 25 years. More than 100 countries now regularly import food from the United States, Canada, Australia, Argentina, western Europe, and a few other surplus producers.

Other trends besides population growth reduce per capita availability of food. Since 1978, the irrigated area per person has dropped by 7%; fertilizer use per capita has not increased since 1984; and the global fish catch per person has dropped since 1989. Rangelands, a major source of animal protein mostly from beef and mutton, are suffering from extensive overgrazing on every continent. Like oceanic fisheries, they may be at or near their maximum sustainable yield. If so, rangeland production of beef and mutton may increase little, if at all, in the future—leading to a decline in per capita supply.

In other words, we are now in a situation where population growth is outstripping food production. Prominent scientists (Spotlight, p. 6) have doubts about the ability of new food production technologies to keep up with current levels of population growth, mostly because of the harmful environmental effects of agriculture. If this is correct, we are faced with two options: reduced food consumption and rising death rates among the population (especially the poor) or an all-out effort to slow population growth.

A: About 33%—an energy loss equal to all the oil flowing through the Alaska pipeline each year

Nutritional Deficiency Diseases

The two most common nutritional-deficiency diseases are marasmus and kwashiorkor. **Marasmus** (from the Greek, "to waste away") occurs when a diet is low in both calories and protein. Most victims are nursing infants of malnourished mothers, or children who do not get enough food after weaning. A marasmic child has a thin body, a bloated belly, wide eyes, and an old-looking face (Figure 1-7). If the child is treated in time with a balanced diet, most of these effects can be reversed.

Kwashiorkor (meaning "displaced child" in a West African dialect) is a severe protein deficiency occurring in infants and children ages 1–3 years, usually after the arrival of a new baby deprives them of breast milk. The displaced child's diet changes to grain or sweet potatoes, which provide enough calories but not enough protein. Such children are lethargic and irritable and have a bloated abdomen. They suffer from diarrhea, lose their hair, and may have liver damage. If caught soon enough, most of the effects can be cured with a balanced diet. Otherwise, even if they survive, their growth will be stunted, and they may be mentally retarded.

Nutritional Deficiencies People who cannot grow or buy enough food to meet their basic energy needs suffer from **undernutrition**. To maintain good health and to resist disease, however, people need not only a certain number of calories but also food with the proper amounts of protein (from animal or plant sources), carbohydrates, fats, vitamins, and minerals. People who are forced to live on a low-protein, high-carbohydrate diet consisting only of grains such as wheat, rice, or corn often suffer from **malnutrition**—deficiencies of protein and other key nutrients. Many of the world's desperately poor people, especially children (Figure 1-7), suffer from both undernutrition and malnutrition (Spotlight, above). According to the U.N. Food and Agriculture Organization, the percentage of people in developing regions without enough food to maintain normal body weight and engage in light activity is 33% in Africa, 19% in the Far East, and 13% in Latin America.

According to the World Health Organization, about 1.3 billion people—one out of four, and one in three children—are underfed and undernourished (a low estimate is around 0.8 billion). Each year 40 million people—half of them children under age 5—die prematurely from undernutrition, malnutrition, or normally nonfatal infections and diseases worsened by malnutrition. Some put the annual death toll at 20 million, while others say it is 60 million. With any of these estimates, we have a tragic situation.

Chronically undernourished and malnourished individuals are disease-prone and too weak to work productively or think clearly. As a result, their children are also underfed and malnourished. If these children survive to adulthood, many are locked in a tragic malnutrition-poverty cycle in which these conditions are often passed on to succeeding generations (Figure 14-8).

Each of us must have a small daily intake of vitamins that cannot be made in the human body. Although balanced diets, vitamin-fortified foods, and vitamin supplements have slashed the number of vitamin-deficiency diseases in MDCs, millions of cases occur each year in LDCs. For example, each year more than 500,000 children in LDCs are partially or totally blinded because their diet lacks vitamin A.

Other nutritional-deficiency diseases are caused by the lack of certain minerals. For example, too little iron causes anemia, which in turn causes fatigue, makes infection more likely, increases a woman's chances of dying in childbirth, and increases an infant's chances of dying from infection during its first year of life. In tropical regions of Asia, Africa, and Latin America, iron-deficiency anemia affects about one-tenth of the men, more than half of the children, two-thirds of the pregnant women, and about half of the other women.

Too little iodine in the diet can cause goiter, an abnormal enlargement of the thyroid gland in the neck, which leads to deafness if untreated. It affects up to 80% of the population in the mountainous areas of Latin America, Asia, and Africa, where soils are deficient in iodine and people have no access to iodine-rich seafood. However, children don't have to die because of malnutrition (Solutions, p. 380).

While 15% of the people in LDCs suffer from severe undernutrition and malnutrition, about 15% of the people in MDCs—including at least 34 million Americans—suffer from **overnutrition**. This is an excessive intake of food, especially fats, that can cause obesity (excess body fat) in people who do not suffer from physiological disorders that promote obesity.

Overnutrition is associated with at least two-thirds of the deaths in the United States each year. A study of thousands of Chinese villagers indicates that the healthiest diet for humans is nearly vegetarian, with only 10–15% of calories coming from fat, in contrast to the typical meat-based diet in which 40% of the calories come from fats (Solutions, p. 194).

There is also concern over possible harmful effects from some of the chemicals, called **food additives**, that are added to processed foods for sale in grocery stores and restaurants. Additives retard spoilage, enhance

Q: What is the best way to save oil, slow global warming, and reduce air pollution?

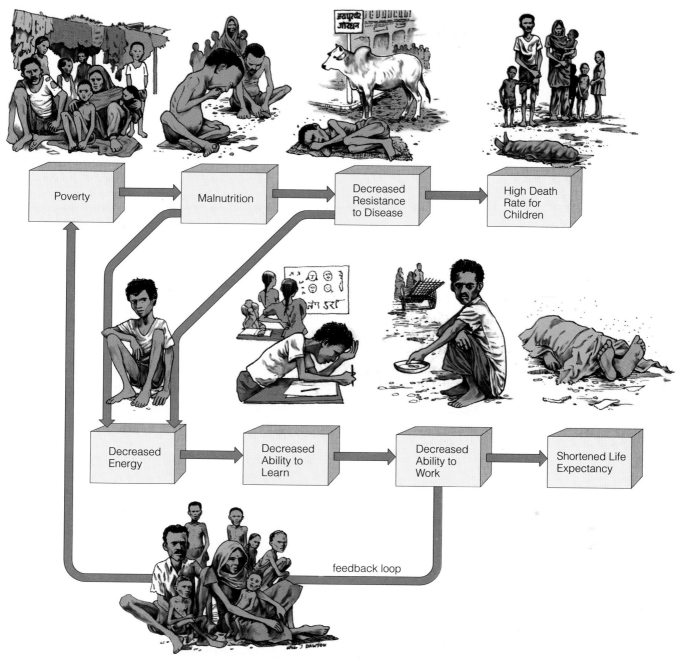

Figure 14-8 Interactions among poverty, malnutrition, and disease can form a tragic cycle that tends to perpetuate such conditions in succeeding generations of families.

flavor, color, and texture, or provide missing vitamins or other nutrients. The presence of synthetic chemical additives does not necessarily mean that a food is harmful, however, and the fact that a food is completely natural is no guarantee that it is safe. A number of natural or totally unprocessed foods contain potentially harmful toxic substances.

Food Supply and Distribution   The good news is that we produce more than enough food to meet the basic needs of every person on Earth. Indeed, if dis-

tributed equally, the grain currently produced world-wide would be enough to give 6 billion people—the projected world population in 1998—a meatless subsistence diet. The bad news is that food is not distributed equally among the world's people because of differences in soil, climate, political and economic power, and average per capita income throughout the world.

By contrast, if everyone ate the diet typical of a person in an MDC, with 30–40% of the calories coming from animal products, the current world agricultural system would support only 2.5 billion people—

A: Improve the energy efficiency of motor vehicles

Saving Children

SOLUTIONS

Officials of the United Nations Children's Fund (UNICEF) estimate that between half and two-thirds of childhood deaths from nutrition-related causes could be prevented at an average annual cost of only $5–$10 per child—10–19¢ per week. This life-saving program would involve the following simple measures:

- Immunizing children against childhood diseases such as measles
- Encouraging breastfeeding
- Preventing dehydration from diarrhea by giving infants a solution of a fistful of sugar and a pinch of salt in a glass of water
- Preventing blindness by giving people a vitamin A capsule twice a year at a cost of about 75¢ per person
- Providing family-planning services to help mothers space births at least two years apart
- Increasing education for women, with emphasis on nutrition, sterilization of drinking water, and child care

less than half the present population and only one-fourth of the 10 billion people projected sometime in the next century (Figure 1-1).

Increases in global food production and food production per person often hide wide differences in food supply and quality among and within countries. For example, nearly half of India's population is too poor to buy or grow enough food to meet basic needs, and an estimated two-thirds of its land is threatened by erosion, water shortages, and salinization. This, coupled with current population growth of 17 million per year, means that India might again suffer from famine in the 1990s and beyond.

MDCs also have pockets of poverty, hunger, and malnutrition. For example, a study by Tufts University researchers found that in 1991 at least 20 million people (12 million children and 8 million adults)—1 out of every 11 Americans and 1 out of 5 American children under age 8—were undernourished, malnourished, or both, mostly because of cuts in food stamps and other forms of government aid since 1980.

Environmental Effects of Producing Food

Agriculture—both industrialized and traditional—has a greater impact on air, soil, and water resources than

any other human activity, as discussed throughout this book. These problems include the following:

Soil Degradation

- Erosion (Figures 12-23 and 12-24). About 85% of U.S. topsoil loss is caused by livestock overgrazing.
- Salinization and waterlogging of heavily irrigated soils (Figure 12-30).
- Desertification, caused by cultivation of marginal land in arid and semiarid climates with unsuitable soil or terrain, as well as by overgrazing, deforestation, and failure to use soil conservation techniques (Figures 12-25 and 12-26).

Water Use and Depletion

- Massive use of water, often in water-short areas, to irrigate 18% of the world's cropland.
- Groundwater depletion by excessive withdrawals for irrigation (Figures 11-15 and 11-16).

Pollution

- Air and water pollution from extraction, processing, transportation, and combustion of fossil fuels used in industrialized agriculture (Figures 9-3, 10-9, 11-26, 12-13).
- Air and water pollution from droplets of pesticides sprayed from planes or ground sprayers (Section 15-3).
- Pollution of streams, lakes, and estuaries, and killing of fish and shellfish by pesticide runoff.
- Pollution of groundwater caused by leaching of water-soluble pesticides, nitrates from commercial inorganic fertilizers, and salts from irrigation water (Figure 11-15).
- Overfertilization of lakes and slow-moving rivers caused by runoff of nitrates and phosphates in commercial inorganic fertilizers, livestock wastes, and food-processing wastes (Figure 11-26). Livestock in the United States produce 21 times more excrement than is produced by the country's human population. Only about half of this livestock waste is recycled to the soil as organic fertilizer.
- Sediment pollution of surface waters caused by erosion and runoff from farm fields, overgrazed rangeland, deforested land, and animal feedlots (Section 11-7 and Figure 14-9).

Biodiversity Loss

- Loss of genetic diversity in plants caused by clearing biologically diverse grasslands (Figure 5-10) and forests (Figures 5-14 and 5-15) and often

Figure 14-9 Huge cattle feedlot near Coalinga, California. Most steers in the United States feed on the open range or on pasturelands for a year. Then they are brought to feedlots and fed for about 100 days on grain (mostly corn) laced with antibiotics (to prevent disease in the crowded conditions)—to fatten them up before slaughter. Chickens and pigs may be kept in feedlots from birth to death. These feedlots increase production efficiency, but they also produce huge concentrations of animal wastes. Without proper controls, these wastes can pollute groundwater and contribute to cultural eutrophication of nearby lakes and slow-moving streams.

Gene Daniels/National Archives/EPA Documerica

replacing them with monocultures (Figures 5-12 and 5-16).

- Endangerment and extinction of wildlife from loss of habitat when grasslands and forests are cleared and wetlands are drained for farming (Section 17-2).

Human Health Threats

- Nitrates in drinking water and pesticides in drinking water, food, and the atmosphere (Section 11-7).

- Human and animal wastes discharged or washed into irrigation ditches and sources of drinking water (Section 11-7).

- Pesticide residues in food. Meat accounts for 55% of such residues in the U.S. diet, compared to 6% from vegetables, 4% from fruits, and 1% from grains.

14-3   SOLUTIONS TO WORLD FOOD PROBLEMS

Increasing Crop Yields   Agricultural experts expect most future increases in food production to come from increased yields per hectare on existing cropland, from improved strains of plants, and from expansion of green-revolution technology to other parts of the world. Scientists are working to create new green revolutions—actually *gene revolutions*—by using genetic engineering and other forms of biotechnology (Pro/Con, p. 207). Over the next 20–40 years they hope to breed high-yield plant strains that are more resistant to insects and disease; thrive on less fertilizer; make their own nitrogen fertilizer (as do legumes, Figure 4-31); do well in slightly salty soils; can withstand drought; and can use solar energy more efficiently during photosynthesis. Even only occasional break-

throughs could generate enormous increases in global crop production before the middle of the next century.

However, several factors have limited the success of the green and gene revolutions so far—and may continue to do so:

- Without huge amounts of fertilizer and water, most green-revolution crop varieties produce yields that are no higher (and are often lower) than those from traditional strains; this is why the second green revolution has not spread to many arid and semiarid areas (Figure 14-4).

- Without ample water, good soil and weather, new genetically engineered crop strains could fail.

- Continuing to increase inputs of fertilizer, water, and pesticides eventually produces no additional increase in crop yields as the J-shaped curve of increased crop productivity slows down, reaches its limits, levels off, and becomes an S-shaped curve. Grain yields per hectare are still increasing in almost every country, but at a much slower rate. Worldwide, such yields dropped from an annual 2.3% increase between 1950 and 1989 to a mere 1% annual increase between 1984 and 1993.

- Without careful land use and environmental controls (Sections 6-4 and 12-8), degradation of water and soil can limit the success of green and gene revolutions.

- The cost of genetically engineered crop strains is too high for most of the world's subsistence farmers in LDCs.

- The severe and increasing loss of Earth's biodiversity—from deforestation, destruction and degradation of other ecosystems, and replacement of a diverse mixture of natural vegetation with monoculture crops—limits the potential of future green and gene revolutions (Connections, p. 383).

A: About 17% (9% in the United States)—mostly hydropower and biomass.

Figure 14-10 The winged bean, a protein-rich annual plant from the Philippines, is only one of many unfamiliar plants that could become important sources of food and fuel. Its edible winged pods, spinachlike leaves, tendrils, and seeds contain as much protein as soybeans, and its edible roots contain more than four times the protein of potatoes. Its seeds can be ground into flour or used to make a caffeine-free beverage that tastes like coffee. Indeed, this plant yields so many different edible parts that it has been called a "supermarket on a stalk." Because of nitrogen-fixing nodules in its roots, this fast-growing plant needs little fertilizer.

Figure 14-11 In South Africa, "Mopani"—emperor moth larvae—are among several insects eaten. Kalahari Desert dwellers eat cockroaches. Lightly toasted butterflies are a favorite food in Bali. French-fried ants are sold on the streets of Bogota, Colombia, and Malaysians love deep-fried grasshoppers. Most of these insects are 58–78% protein by weight—three to four times as protein-rich as beef, fish, or eggs.

Developing New Food Sources Some analysts recommend greatly increased cultivation of various nontraditional plants to supplement or replace such staples as wheat, rice, and corn. One of many possibilities is the winged bean, a protein-rich legume now common only in New Guinea and Southeast Asia (Figure 14-10). Insects are also important potential sources of protein, vitamins, and minerals (Figure 14-11). The problem is getting farmers to cultivate such crops and convincing consumers to try new foods.

Most crops we depend on are tropical annuals. Each year the land is cleared of all vegetation, dug up, and planted with their seeds. David Pimentel (Guest Essay, p. 333) and plant scientists at the Land Institute in Salina, Kansas, believe that we should rely more on polycultures of perennial crops, which are better adapted to regional soil and climate conditions than most annuals, which must be replanted each year (p. 372). This strategy would eliminate the need to till soil each year, greatly reducing use of energy; it would also save water and reduce soil erosion and sediment water pollution.

If not overharvested, certain wild animals could be an important source of food. Prolific Amazon river turtles, for example, are used as a source of protein by local people. Another delicacy is the green iguana, "the chicken of the trees." If managed properly, these large, tasty lizards yield up to ten times as much meat as cattle on the same amount of land. In Indonesia the piglike babirusa is an important source of meat.

Cultivating More Land Humankind grows crops on about 11% of Earth's ice-free land area and pastures

Q: What is the largest untapped energy source in the United States?

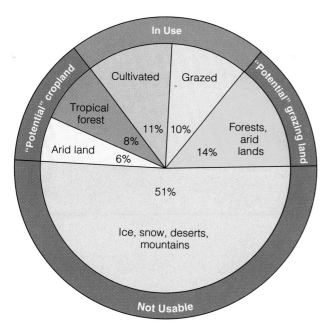

Figure 14-12 Classification of the earth's land. Theoretically we could double the amount of cropland by clearing tropical forests and irrigating arid lands. However, converting this marginal land into cropland would destroy valuable forest resources, reduce the earth's biodiversity, and cause other serious environmental problems—usually without being cost-effective.

livestock on nearly 25%. The rest is either covered by ice, is too dry or too wet, too hot or too cold, too steep, has unsuitable soils, or is inhabited by people.

Theoretically, the world's cropland could be more than doubled by clearing tropical forests and irrigating arid land (Figure 14-12). About 83% of the world's potential new cropland is in the remote rain forests of South America and Africa, primarily in Brazil (Figure 6-16) and Zaire. Clearing rain forests to grow crops and graze livestock, however, has disastrous consequences, as discussed in Section 16-4. And in Africa, potential cropland in savanna (Figure 5-9) and other semiarid land cannot be used for farming or livestock grazing because it is infested by 22 species of the tsetse fly. Its bite can infect people with incurable sleeping sickness and can transmit a fatal disease to livestock.

Researchers hope to develop new methods of intensive cultivation in tropical areas. But some scientists argue that it makes more ecological and economic sense to combine the ancient method of shifting cultivation (followed by fallow periods that are long enough to restore soil fertility, Figure 2-5) with various forms of interplanting. Scientists also recommend plantation cultivation of rubber trees, oil palms, and banana trees, which are adapted to tropical climates and soils.

Much of the world's potentially cultivable land lies in dry areas, especially in Australia and Africa.

Loss of Genetic Diversity

CONNECTIONS

The UN Food and Agriculture Organization estimates that by the year 2000, two-thirds of all seed planted in LDCs will be of uniform strains. This genetic uniformity increases the vulnerability of food crops to pests and diseases. This, plus widespread species extinction, severely limits the potential of future green and gene revolutions.

In the mid-1970s, for example, a valuable wild corn species was barely saved from extinction. When this strain was discovered, only a few thousand stalks survived in three tiny patches in south central Mexico that were about to be cleared by squatter cultivators and commercial loggers. This wild species is the only known perennial strain of corn. Crossbreeding it with commercial varieties could reduce the need for yearly plowing and sowing, which would reduce soil erosion, water use, and energy use. Even more important, this wild corn has a built-in genetic resistance to four of the eight major corn viruses. Furthermore, this wild corn grows in cooler and damper habitats than established commercial strains. Overall the genetic benefits from this wild plant could total several billion dollars per year.

The loss of genetic diversity is a global problem. In Indonesia, for example, 1,500 local varieties of rice have disappeared in the past 15 years. In India, which once had 30,000 varieties of rice, more than 75% of the rice production now comes from 10 varieties. In other words we are rapidly shrinking the world's genetic "library" just when we need it more than ever.

Wild varieties of the world's most important plants can be collected and stored in gene banks, agricultural research centers (Figure 14-4), and botanical gardens; however, space and money severely limit the number of species that can be preserved there. Moreover, many cannot be stored successfully in gene banks; and power failures, fires, or unintentional disposal of seeds can cause irreversible losses. Also, stored plant species stop evolving and thus are less fit for reintroduction to their native habitats, which may have changed in the meantime.

Because of these limitations, ecologists and plant scientists warn that the only effective way to preserve the genetic diversity of most plant and animal species is to protect representative ecosystems throughout the world from agriculture and other forms of development.

A: Renewable (92%), compared with coal (5%), oil (2.5%), and uranium (0.5%)

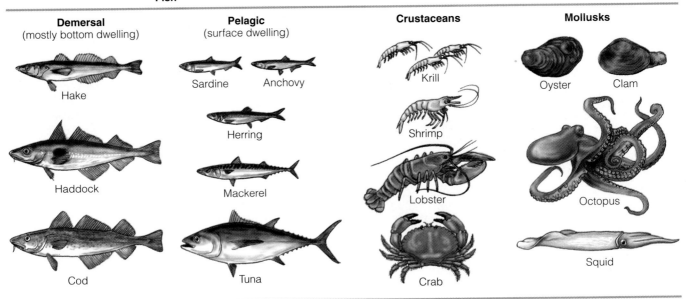

Fish **Shellfish**

Demersal
(mostly bottom dwelling)

Hake

Haddock

Cod

Pelagic
(surface dwelling)

Sardine   Anchovy

Herring

Mackerel

Tuna

Crustaceans

Krill

Shrimp

Lobster

Crab

Mollusks

Oyster   Clam

Octopus

Squid

Figure 14-13 Some major types of commercially harvested marine fish and shellfish.

Large-scale irrigation in these areas would be very expensive, requiring large inputs of fossil fuel to pump water long distances. Irrigation systems would deplete groundwater supplies, and the land would need constant and expensive maintenance against erosion, groundwater contamination, salinization, and waterlogging. Expanding wetland production of rice could accelerate projected global warming by increasing atmospheric methane emissions (Figure 10-9)—another example of connections in nature.

Thus much of the new cropland that could be developed would be on marginal land requiring expensive inputs of fertilizer, water, and energy. Furthermore, these possible increases in cropland would not offset the projected loss of almost one-third of today's cultivated cropland from erosion, overgrazing, waterlogging, salinization, mining, and urbanization.

Pollution is also reducing yields on existing cropland. In the United States ozone pollution in the troposphere reduced harvests of crops by at least 5% during the 1980s. Other air pollutants, such as sulfur dioxide and nitrogen oxides, have also damaged crops. Yields of some crops and populations of marine phytoplankton that support fish and shellfish used as food have been reduced by depletion of ozone in the stratosphere (Figure 10-15). And if the global climate changes, as some scientists project, areas where crops can be grown will shift, which will disrupt food production (Section 10-2). Thus, it is unlikely that there will be a major expansion of cropland that is economically profitable and environmentally sustainable.

Finally, if current population projections are accurate, the global average of 0.28 hectare (0.69 acre) of cropland per capita in 1992 is expected to decline to 0.17 hectare (0.42 acre) by 2025.

Catching More Fish Concentrations of particular aquatic species suitable for commercial harvesting in a given ocean area or inland body of water are called **fisheries**. Worldwide, people get an average of 20% of their animal protein directly from fish and shellfish, and another 5% indirectly from livestock fed with fish meal. In most Asian coastal and island regions, fish and shellfish supply 30–90% of people's animal protein.

About 86% of the annual commercial catch of fish and shellfish comes from the ocean. Ninety-nine percent of this catch is taken from plankton-rich waters (mostly estuaries and upwellings) within 370 kilometers (200 nautical miles) of the coast. However, this vital coastal zone is being disrupted and polluted at an alarming rate (Figure 11-24, Section 11-8, and Case Study, p. 104).

Some commercially important marine species of fish and shellfish are shown in Figure 14-13. Various methods used to harvest such species are shown in Figure 14-14. However, some of these methods also entrap and kill large numbers of other nontarget species such as seals, dolphins, sea turtles, small whales, and other marine species. One of the greatest threats to marine biodiversity is *drift-net fishing* (Figure 14-14). These monster nets drift in the water and catch fish when their gills become entangled in the nylon mesh. Each net descends as much as 15 meters (50 feet) deep and is up to 65 kilometers (40 miles) long. Almost anything that comes in contact with these nearly invisible "curtains of death" becomes entangled. Between 1980 and

Q: How much of U.S. and world energy needs could be provided by renewable resources by 2030 or sooner?

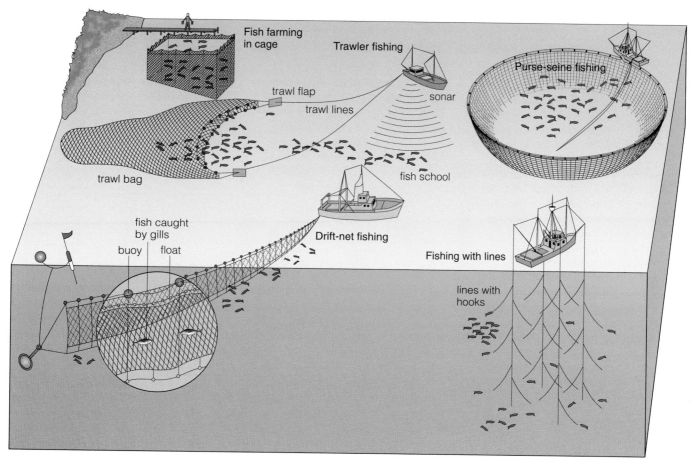

Figure 14-14 Major commercial fishing methods used to harvest marine species.

1991, an estimated 1,800 fishing vessels from Japan, South Korea, and Taiwan used drift-net fishing in international waters. Each night of the fishing season, the fleets' nets were long enough to more than encircle the world.

Every country that has used drift nets in its own waters has eventually banned them. In 1990 the UN General Assembly called for a moratorium of drift-net fishing in international waters after June 1992. The ban, however, has numerous loopholes and no effective mechanism for monitoring, enforcement, and punishment over the earth's vast ocean area, and compliance is voluntary.

Environmentalists believe that the only effective way to reduce drift-net fishing and tuna caught by purse-seine fishing is to mount U.S. and global boycotts of fish caught in these ways This tactic, led by the Earth Island Institute, caused companies selling canned tuna in the United States to stop buying tuna caught by purse-seine or drift-net methods. Consumer power works.

Between 1950 and 1989, the weight of the commercial fish catch increased 4.6-fold, which doubled the seafood catch per person (Figure 14-15). However, the total fish catch declined 7% between 1989 and 1993.

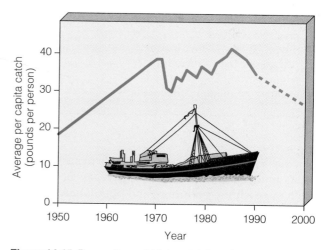

Figure 14-15 Per capita world fish catch has declined since the mid-1980s and is projected to continue dropping through the end of this century. (Data from United Nations and Worldwatch Institute)

Although the total fish catch has grown in most years, the per capita fish catch has declined in most years since 1970 because the human population has grown at a faster rate than the fish catch (Figure 14-15). United Nations and other marine biologists estimate

A: 50–80%

Where Have All the Anchovies Gone?

In 1953 Peru began fishing for anchovies off its western coast. The size of the fishing fleet increased rapidly. The tiny fish were processed into fish meal and sold to MDCs as livestock feed. Between 1965 and 1971 Peruvian anchovies made up about 20% of the world's commercial fish catch.

The bonanza was short-lived, however. Biologists with the UN Food and Agriculture Organization warned that during seven of the eight years between 1964 and 1971 the anchovy harvest exceeded the estimated sustainable yield. Peruvian fishery officials ignored those warnings. The enormous anchovy populations depend on an upwelling of cold, nutrient-rich bottom water near the Peruvian coast (Figure 10-5, left). However, at unpredictable intervals the productivity of these upwellings drops sharply because of a natural weather change called the El-Niño-Southern Oscillation, or ENSO (Figure 10-5, right). The anchovies, as well as other fish, seabirds, and marine mammals in food webs based on phytoplankton, then die back.

Disaster then struck in 1972 when a strong ENSO arrived. The anchovy population, already decimated by overfishing, could not recover from the effects of the ENSO, and the annual yield plummeted (Figure 14-16). By putting short-term profits above a long-term anchovy fishery, Peru lost a major source of income and jobs, and had to increase its foreign debt. Since 1983, the Peruvian anchovy fishery has made only a slight recovery (Figure 14-16).

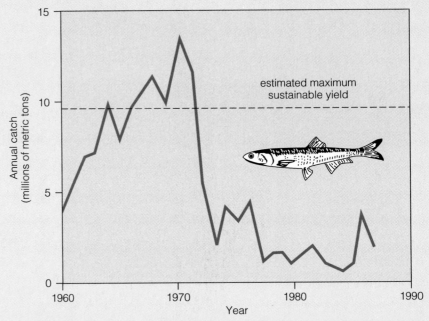

Figure 14-16 Peruvian anchovy catch, showing the combined effects of overfishing and a natural climate change—known as the El-Niño-Southern Oscillation—that occurs every few years. About one-third of the annual marine catch is used to feed animals and fertilize croplands. Scientists estimate that the sustainable yield of the world's marine fishery is 100 million metric tons (110 million tons), which may be reached or exceeded soon. If this yield is exceeded, key fish stocks will be depleted and yields will drop sharply. (Data from UN Food and Agriculture Organization)

that oceanic fisheries may not be able to sustain harvests higher than those in recent years. Because of such overfishing, pollution, and population growth, the per capita world catch is projected to return to the 1960 level by 2000. The fish catch, however, could be increased by cutting waste. Currently, one-fifth of the annual catch is wasted, mainly from throwing back potentially useful fish taken along with desired species. More refrigerated storage at sea to prevent spoilage would also increase the catch.

Fish are potentially renewable resources—as long as the annual harvest leaves enough breeding stock to renew the species for the next year. Ideally, an annual **sustainable yield** figure—the size of the annual catch that could be harvested indefinitely without a decline in the population of a species—should be established for each species to avoid depleting the stock. However, determining sustainable yields is difficult. Counting aquatic populations isn't easy because they disappear beneath the water. And sustainable yield values shift from year to year because of changes in climate, pollution, and other factors. Many marine scientists believe that the annual harvest for each species should be based on an **optimum yield**—the catch that can be economically harvested on a sustained basis; it is usually less than the sustainable yield.

Overfishing is taking so many fish that too little breeding stock is left to maintain numbers—that is, when the sustainable yield is exceeded. Prolonged overfishing leads to **commercial extinction**: So few of a species is left that it's no longer profitable to hunt them. Fishing fleets then move to a new species or to a

Q: How much of the world's electricity could solar cells supply by 2050?

new region, hoping that the overfished species will eventually recover.

Since the early 1980s overfishing has caused declines in the yields of nearly one-third of the world's fisheries and the collapse of 42 valuable fisheries, including 42% of the species commercially fished in U.S. coastal waters. Examples include cod and herring, Atlantic salmon, red snapper, Atlantic bluefin tuna, Alaskan king crab, and Peruvian anchovy (Connections, at left). Overfishing could be reduced by eliminating or sharply lowering government subsidies to the fishing industry. For example, in 1992 the commercial fishing industry spent about $124 billion to catch $70 billion of marine fish worldwide. Much of the $54 billion difference was made up by government subsidies.

Raising More Fish: Aquaculture Aquaculture, in which fish and shellfish are raised for food, supplies about 12% of the world's commercial fish harvest. It is expanding rapidly and by 2005 may account for one-third of the world's fish production. There are two basic types of aquaculture. **Fish farming** involves cultivating fish in a controlled environment, usually a pond, and harvesting them when they reach the desired size. **Fish ranching** involves holding species in captivity for the first few years of their lives—usually in fenced-in areas or floating cages in coastal lagoons and estuaries (Figure 14-14)—and then harvesting the adults when they return to spawn. Ranching is useful for anadromous species—which spend part of their lives in the ocean and part in freshwater streams—such as ocean trout and salmon (Figure 14-17).

Species cultivated in LDCs include carp, tilapia, milkfish, clams, and oysters, which feed on phytoplankton and other aquatic plants. These are usually raised in small freshwater ponds or underwater cages. In MDCs aquaculture is used mostly to stock lakes and streams with game fish and to raise expensive fish and shellfish such as oysters, catfish, crayfish, rainbow trout, shrimp, and salmon.

Aquaculture has several advantages. First, it can produce high yields. Second, little fuel is needed, so yields and profits are not closely tied to the price of oil, as they are in commercial marine fishing. And third, aquaculture is usually labor-intensive and can provide much-needed jobs in LDCs. Experts project that freshwater and saltwater aquaculture production could also be doubled during the next 10 years. If so, it could help stave off seafood shortages and keep seafood prices down, as the world's per capita commercial fish catch declines (Figure 14-15).

There are problems, however. For one thing, large-scale aquaculture requires considerable capital and scientific knowledge, which are in short supply in LDCs. For another, scooping out huge ponds for fish and shrimp farming in some LDCs has destroyed ecologically important mangrove forests (Figure 5-24). Also, fish in aquaculture ponds can be killed by pesticide runoff from nearby croplands; and bacterial and viral infections of aquatic species can also limit yields. Antifoulant chemicals—used to keep nets and cages free of marine life—can be toxic to nearby marine animals. Escaped farm-raised fish may also breed with wild fish, and degrade the genetic stock of such species. Finally, without adequate pollution control, waste outputs from shrimp farming and other large-scale aquaculture operations can contaminate nearby estuaries, surface water, and groundwater and eliminate some native aquatic species. For example, an 8-hectare (20-acre) salmon farm in the United States produces as much organic waste as 10,000 people.

Government Agricultural Policies Agriculture is a risky business. Whether farmers have a good or a bad year is determined by factors over which they have little control—weather, crop prices, crop pests and disease, interest rates, and the global market. Because of the need for reliable food supplies despite variability, most governments provide various forms of assistance to farmers.

Governments have several choices. They can

- *Keep food prices artificially low.* This makes consumers happy but means that farmers may not be able to make a living.

- *Give farmers subsidies to keep them in business and encourage them to increase food production.* In MDCs government price supports and other subsidies for agriculture total more than $300 billion per year ($17–22 billion per year in the United States). However, if government subsidies are too generous and the weather is good, farmers may produce more food than can be sold. Food prices and profits then drop because of the surplus. Large amounts of food become available for export or food aid to LDCs, depressing world food prices. The low prices reduce the financial incentive for farmers to increase domestic food production. Moreover, the taxes citizens in MDCs pay to provide agricultural supports more than offset the lower food prices they enjoy.

- *Eliminate most or all price controls and subsidies,* allowing market competition to be the primary factor determining food prices and thus the amount of food produced.

The U.S. agricultural system produces so much food that the government pays farmers not to produce food on one-fourth of U.S. cropland, or it buys up and stores unneeded crops. Such subsidies can waste taxpayer dollars and serve as a form of welfare for

A: About 30% (50% in the United States)

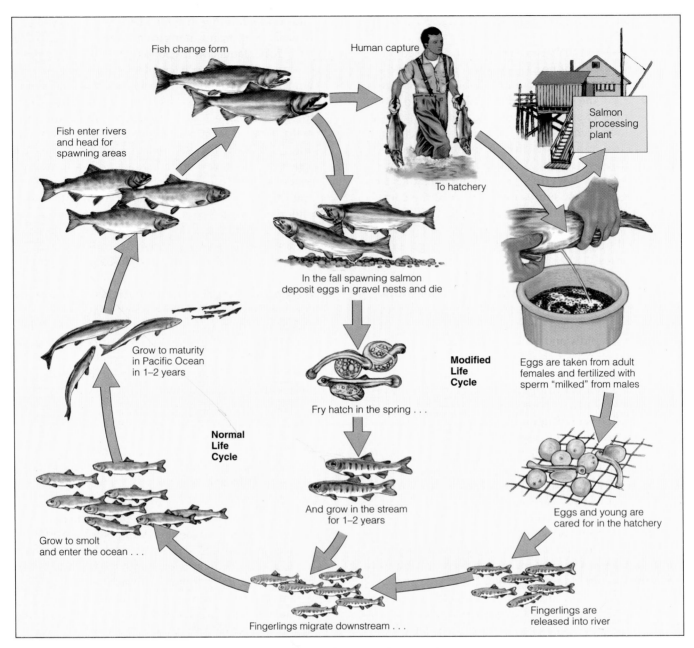

Figure 14-17 Normal (left) and modified (right) life cycle of an anadromous salmon species.

wealthy farmers. In 1989, for example, nearly 60% of federal farm subsidies went to the wealthiest 25% of U.S. farms.

Some analysts call for eliminating all federal subsidies over, say, five years and letting farmers respond to market demand. Only those who were good farmers and financial managers would be able to stay in business. However, such analysts urge that any phaseout of farm subsidies in the United States (or in any other country) should be coupled with increased aid for the poor, who would suffer the most from any increase in food prices.

Instead of eliminating all subsidies, many environmentalists believe that we should refocus them by rewarding farmers and ranchers who protect the soil, conserve water, reforest degraded land, protect and restore wetlands, and conserve wildlife. Those who didn't would receive no subsidies. Current subsidies in most countries favor high-input industrial agriculture and certain crops. Some analysts urge that the economic playing field be made more level by providing subsidies for farmers practicing low-input sustainable agriculture and by making subsidies independent of which crops a farmer grows.

Q: How much of the world's electricity could wind power supply by 2050?

Environmentalists also warn that the current treaties designed to deregulate global trade in agricultural and related products could doom most small- and medium-scale farming operations around the world, many of which practice sustainable diversified farming. They would be replaced by larger-scale monoculture farming and feedlots that are much more environmentally destructive. These trade agreements could also reduce or eliminate all farm subsidies, weaken environmental and health safety standards, and undermine local, state, and national authority. However, eliminating trade barriers (with adequate environmental protection) could help LDCs, which now lose an estimated $100 billion in agricultural sales because of such barriers by MDCs.

International Relief Between 1945 and 1985 the United States was the world's largest donor of nonmilitary foreign aid to LDCs; since 1986 Japan has taken over that role. This aid is used mostly for agriculture and rural development, food relief, population planning, health care, and economic development.

Besides helping other countries, foreign aid stimulates economic growth and provides jobs in the donor country. For example, 70¢ of every dollar the United States gives directly to other countries is used to purchase American goods and services. Today 21 of the 50 largest buyers of U.S. farm goods are countries that once received free U.S. food. Food relief, however is a controversial form of aid (Pro/Con, at right).

Land Reform An important step in reducing world hunger, malnutrition, poverty, and land degradation is land reform. This usually means giving the landless rural poor in LDCs either ownership or free use of enough land to produce their food and, ideally, enough surplus to provide some income. So far China and Taiwan have had the most successful land reform programs.

The world's most unequal land distribution is in Latin America, where 7% of the population owns 93% of the farmland. Most of this land is used for export crops such as sugar, tea, coffee, bananas, or beef, or is left idle on huge estates. Of course, land reform is difficult to institute in countries where government leaders are unduly influenced by wealthy and powerful landowners.

14-4 SOLUTIONS: SUSTAINABLE FOOD PRODUCTION

Guidelines for Sustainable Agricultural Systems To many environmentalists, the key to reducing world hunger, poverty, and the harmful environmental impacts of both industrialized and traditional

Food Relief

PRO/CON

Most people view food relief as a humanitarian effort to prevent people from dying prematurely. However, some analysts contend that giving food to starving people in countries with high population growth rates does more harm than good in the long run. By not helping people grow their own food, they argue, food relief can condemn even greater numbers to premature death from starvation and disease in the future.

Biologist Garrett Hardin (Guest Essay, p. 160) has suggested that we use the concept of *lifeboat ethics* to decide which countries get food relief. His basic premise is that there are already too many people in the lifeboat we call Earth. Thus, if food aid is given to countries that are not reducing their populations, this simply adds more people to an already-overcrowded lifeboat. Sooner or later the boat will sink and most of the passengers will drown.

Large amounts of food relief can also depress local food prices, decrease food production, and stimulate mass migration from farms to already-overburdened cities. In addition, food relief discourages the local and national governments from investing in rural agricultural development to enable their countries to grow enough food for their populations.

Another problem is that much food aid does not reach hunger victims. Transportation networks and storage facilities are inadequate, so that some of the food rots or is devoured by pests before it can reach the hungry. Also, typically, some of the food is stolen by officials and sold for personal profit. And some must often be given to officials as bribes for approving the unloading and transporting of the remaining food to the hungry.

Critics of food relief are not against foreign aid. Instead, they believe that such aid should be given to help countries control population growth and grow enough food to feed their population by using sustainable agricultural methods (Section 14-4). Temporary food relief, they believe, should be given only when there is a complete breakdown of an area's food supply because of natural disaster. What do you think?

agriculture is to develop a variety of **sustainable-Earth agricultural systems**. This approach involves combining the wisdom of traditional agricultural systems with new techniques that take advantage of local climates, soils, resources, and cultural systems (Guest Essay, p. 391). Here are some general guidelines:

What You Can Do to Promote Sustainable-Earth Agriculture

INDIVIDUALS MATTER

- *Waste less food.* An estimated 25% of all food produced in the United States is wasted; it rots in the supermarket or refrigerator, or it is scraped off the plate and into the garbage in households and restaurants.

- *Eat lower on the food chain.* This can be done by reducing or eliminating meat consumption to reduce its environmental impact.

- *Don't feed your dog or cat canned meat products.* Balanced grain pet foods are available and are better for your pet.

- *Reduce the use of pesticides on agricultural products by asking grocery stores to stock fresh produce and meat produced by organic methods.** About 0.5% of U.S. farmers grow about 3% of the country's crops using organic methods.

- *Grow some of your own food using organic farming techniques and drip irrigation to water your crops* (Figure 11-18).

- *Think globally, eat locally.* Whenever possible eat food that is locally grown and in season. This supports your local economy, gives you more influence over how the food is grown (organic or conventional methods), saves energy from having to transport food over long distances, and reduces the use of fossil fuels and pollution. If you deal directly with local farmers, you can also save money.

- *Pressure elected officials to develop and encourage sustainable-Earth agricultural systems in the United States and throughout the world.*

*A Consumer's Organic Mail-Order Directory of farmers and wholesalers who sell organically grown food by mail can be obtained for $9.95 (plus $2.50 postage) from California Action Network, P.O. Box 464, Davis, CA 95617.

- *Neither rob the soil of nutrients nor waste water, and return whatever is taken from the earth.* This means minimizing soil erosion by a variety of methods (Section 12-8); not cultivating easily erodible land; preventing overgrazing by livestock; and using organic fertilizers, crop rotation, and intercropping to increase the organic content of soils. It also means raising water prices to encourage water conservation; using irrigation systems that minimize water waste, salinization, and waterlogging (Figure 12-30); and not trying to grow water-thirsty crops in arid and semiarid areas.

- *Encourage systems featuring a diverse mix of crops and livestock, instead of depending too much on pro-* duction of a single crop or livestock type. Emphasize increased use of polyculture and other forms of intercropping. A small but growing number of organic farmers in MDCs are returning to diversified farming.

- *Rely as much as possible on locally available, renewable biological resources, and use them in ways that preserve their renewability.* Examples include using organic fertilizers from animal and crop wastes (green manure and compost); planting fast-growing trees to supply fuelwood and add nitrogen to the soil; building simple devices for capturing and storing rainwater for irrigating crops; and cultivating crops adapted to local growing conditions.

- *Greatly reduce the use of fossil fuels in agriculture by using locally available renewable energy such as sun, wind, and flowing water, and by using more organic fertilizer instead of commercial inorganic fertilizer.*

- *Emphasize biological pest control and integrated pest management instead of overuse of chemical pesticides* (Section 15-5).

- *Provide economic incentives for farmers using sustainable-Earth agricultural systems.*

- *Encourage local people to grow food for local people— and let them plant what they want, instead of encouraging export of cash crops that reduces food available to local people.*

Making the Transition to Sustainable Agriculture in MDCs In MDCs such as the United States, a shift to sustainable-Earth agriculture will not be easy, for it will be opposed by agribusiness, by successful farmers with large investments in industrialized agriculture, and by specialized farmers unwilling to learn the demanding art of farming sustainably. Without proper education, it might also be resisted by consumers who might not be willing to pay higher prices for food. They are already paying these costs in the form of higher taxes and insurance and health costs, but these costs related to growing food are hidden and don't appear in market prices (Figure 7-2).

However, environmentalists believe that this shift could be brought about over a 10- to 20-year period by the following means:

- *Greatly increase government support of research and development of sustainable-Earth agricultural methods and equipment.* Currently, only about 1% of the Department of Agriculture's annual research budget is used for this purpose.

- *Set up demonstration projects in each county so that farmers can see how sustainable systems work.*

- *Establish college curricula for sustainable-Earth agriculture.*

Q: What is the most promising fuel for replacing oil and other fossil fuels?

Mazunte: A Farming and Fishing Ecological Reserve in Mexico

Alberto Ruz Buenfil

GUEST ESSAY

Alberto Ruz Buenfil, an international environmental activist, writer, and performer, is the founder of a land-based ecovillage in the mountains of Mexico, called Huehuecoytl. His articles on ecology and alternative living have appeared in publications from the United States, Canada, Mexico, Japan, and Europe. His book, Rainbow Nation Without Borders: Toward an Ecotopian Millennium *(New Mexico: Bear, 1991) has been published in English, Italian, and Spanish.*

The Pacific coasts of southern Mexico, especially the shores of the state of Oaxaca, are the main sites for turtle nesting, reproduction, and conservation. They also contain some of Mexico's last reserves of wetlands.

Only recently have we begun recognizing the ecological values of wetlands, deltas, and coastal ecosystems [Section 5-2]. They provide habitats for a rich diversity of wildlife, maintain water supplies, protect shorelines from erosion, and play a role in regulating the global climate.

Previously, places like swamps, marshes, and bogs have been considered wastelands to be drained, filled, and turned into "productive land," especially for constructing and developing urban and tourist areas. During the last few decades the coasts of Oaxaca have not escaped such exploitation, especially after business interests discovered the paradiselike beaches of Mazunte, Zipolite, San Agustinilo, Puerto Angel, and Puerto Escondido, and the magnificent bays of Hautulco.

The villages of Mazunte and San Agustinilo were founded in the late 1960s, basically to bring in cheap labor from neighboring indigenous villages and to provide workers for a slaughterhouse producing products from various species of turtles nesting in the area.

Nearly 200 families of indigenous farmers became fishermen and employees for the new turtle meat factory, which was fully operational in the 1970s and 1980s. According to some of these workers nearly 2,000 turtles were killed and quartered every day during those years, with no law to protect them. At night, dozens of poachers came to collect turtle eggs from their nests.

In the 1980s this situation came to the attention of two of the first environmental organizations to speak up in defense of species, forests, and natural resources in Mexico. They began denouncing the massacre of the area's turtles and pointed out the danger that some of the species might be exterminated. With support from other international organizations they campaigned for almost 10 years, until a 1990 presidential decree made it illegal to exploit the turtles and led to the closing of San Agustinilo's turtle slaughterhouse.

The Mexican government provided some funds, boats, and freezers to compensate for the loss of jobs. However, only about 5% of the indigenous population benefited from this compensation. Since then most of the people have been living on the verge of starvation. What had seemed to be an important environmental victory turned into a nightmare for a large population of indigenous people. Understandably these people had no use for ecologists or environmentalists.

In 1990 a group called ECOSOLAR A.C. began efforts to change this situation by implementing a plan for sustainable development of the coast of Oaxaca. They were successful in obtaining funding for this project from different national and international institutions. By 1992 the members of this small but effective group had been able to:

- Make a detailed study of the bioregion, which is being used to define the possible uses of different areas with the participation of the local people

(continued)

- *Establish training programs in sustainable-Earth agriculture for farmers, county farm agents, and Department of Agriculture personnel.*

- *Give subsidies and tax breaks to farmers using sustainable agriculture and to agribusiness companies that develop products for this type of farming.* Iowa, for example, taxes fertilizers and pesticides; it then uses the revenues to support research into and development of sustainable agriculture. Minnesota provides low-interest loans for farmers engaged in sustainable agriculture. Austria, Denmark, Finland, Germany, Norway, and Sweden offer 3–5-year subsidies for farmers converting to sustainable agriculture.

- *Educate the public to understand the hidden environmental and health costs they are paying for food, and gradually incorporate these costs into market prices* (Section 7-2).

Each of us has a role to play in bringing about a shift from unsustainable to sustainable agriculture at the local, national, and global levels (Individuals Matter, p. 390).

The need to bring birthrates well below death rates, increase food production while protecting the environment, and distribute food to all who need it is the greatest challenge our species has ever faced.

PAUL AND ANNE EHRLICH

- Create a system of credits to help native inhabitants build better houses and small family-run restaurants, and manufacture hammocks for rent or sale to visitors.

- Begin building systems for drainage, water collection, and latrines using low-impact technology and local materials and workers, as well as home nurseries for local seeds, reforestation, and wildlife preservation

- Work with the community to promote Mazunte as a center for ecotourism, a place where visitors can experience unique ecosystems containing alligators, turtles, and hundreds of species of birds and fishes

In only two years the native inhabitants of Mazunte and other neighboring communities have completely changed their opinion and perspective about ecology and environmentalists. In May 1992 Mazunte hosted the second annual gathering of "Earth Keepers," involving nearly 150 representatives of 35 organizations from 20 different countries.

For one week these specialists shared their practical knowledge with the local people. They used their health skills to help the community set up an alternative clinic of healing arts; their skills in permaculture (a form of sustainable-Earth agriculture) and organic agriculture to improve local hatcheries and home nurseries; and their skills in ecotechnology to build biodigestors for producing natural fertilizers from biomass and a village recycling center located at the school. In addition, artistic and cultural activities took place every night in the Center for Biological Investigations, which the people of Mazunte want to turn into Mexico's first Marine Turtles Museum. Run by local people, it will attract and educate visitors from around the world.

A few days after the event concluded, the village of Mazunte called a general meeting attended by 150 heads of family to discuss how to protect turtle-nesting shore areas. Out of that meeting came a "Declaration of Mazunte," requesting that competent higher authorities and the president of Mexico put an immediate end to such destruction, which violates the earlier presidential decree forbidding the annihilation of turtles in Mexico.

The community went further and declared that their village and neighboring environments be considered Mexico's first *Farming and Fishing Reserve*. Its goals would be to protect the area's forests, water sources, wetlands, wildlife, shores, beaches, and scenic places, and to "establish new forms of relationship between humans and nature, for the well-being of today's and tomorrow's generations." This declaration has been presented to the government of Mexico and to many national and international organizations.

Mazunte is taking the lead in showing how cooperation between local people and environmental experts can lead to ecologically sustainable communities that benefit local people and wildlife. This model can show farmers and indigenous communities everywhere how they can live sustainably on Earth and turn things around in a short time. It is a message of hope and empowerment for people seeking a better world for themselves and others.

Critical Thinking

1. What lessons have you learned from this essay that could be applied to your own life?

2. Could the rapid change toward sustainability brought about by environmentalists and local people in Mazunte be accomplished in your own community?

Critical Thinking

1. What are the biggest advantages and disadvantages of (a) labor-intensive subsistence agriculture, (b) energy-intensive industrialized agriculture, and (c) sustainable-Earth agriculture?

2. Summarize the advantages and limitations of each of the following proposals for increasing world food supplies and reducing hunger over the next 30 years: (a) cultivating more land by clearing tropical forests and irrigating arid lands, (b) catching more fish in the open sea, (c) producing more fish and shellfish with aquaculture, and (d) increasing the yield per area of cropland.

3. Should price supports and other federal subsidies paid to U.S. farmers out of tax revenues be eliminated? Explain. Try to consult one or more farmers in answering this question.

4. Is sending food to famine victims helpful or harmful? Explain. Are there any conditions you would attach to sending such aid? Explain.

5. Should tax breaks and subsidies be used to encourage more U.S. farmers to switch to sustainable-Earth farming? Explain.

*6. If possible, visit a nearby conventional industrialized farm and an organic farm. Compare soil erosion and other forms of land degradation, use and costs of energy, use and costs of pesticides and inorganic fertilizer, use and costs of natural pest control and organic fertilizer, yields per hectare for the same crops, and overall profit per hectare for the same crops.

*7. Make a concept map of the key ideas in this chapter using the section heads and subheads and the key terms (shown in boldface type in the chapter). See the inside front cover and Appendix 4 for information on concept maps.

15 Protecting Food Resources: Pesticides and Pest Control

Along Came a Spider

The longest war in human history is our war against insects. This war was declared about 10,000 years ago, when we first got serious about agriculture, and we are no closer to winning today than we were then.

In 1962 biologist Rachel Carson warned against relying on synthetic chemicals to kill insects and other species we deem pests. Chinese farmers have recently decided that it's time to change strategies. Instead of spraying their rice and cotton fields with poison, they began to build little straw huts here and there around the fields in the fall.

If this sounds crazy, it was crazy like a fox. These farmers were giving aid and comfort to insects' worst enemy, one that has hunted them for millions of years: spiders. Protected from the worst of the cold by the huts, far more hibernating spiders would awaken the next spring. Ravenous after their winter fast, they would scuttle off into the fields to stalk their insect prey.

Even without human help, the world's 30,000 known species of spiders kill far more insects every year than insecticides do (Figure 15-1). A typical acre of meadow or woods contains an estimated 50,000 to 2 million spiders, each devouring hundreds of insects per year.

Entomologist Willard H. Whitcomb found that leaving strips of weeds around cotton and soybean fields provides the kind of undergrowth favored by insect-eating wolf spiders (Figure 15-1, right). He also sings the praises of one type of banana spider, which lives in warm climates and can keep a house clear of cockroaches.

The idea of encouraging populations of spiders in fields, forests, and even houses scares most people because spiders have bad reputations. Although a few species of spiders (such as the black widow and brown recluse) are dangerous to people, the vast majority—including the ferocious-looking wolf spider—are harmless to humans. Even the giant tarantula rarely bites people, and its venom is too weak to harm us and other large mammals.

As biologist Thomas Eisner puts it, "Bugs are not going to inherit the earth. They own it now. So we might as well make peace with the landlord." As we seek new ways to coexist with the real rulers of the planet, we would do well to be sure that spiders are in our corner.

Figure 15-1 Spiders are insects' worst enemies. Most spiders, like the two species shown here, are harmless to humans. (James C. Cokendolpher, left; Dan Kline/Visuals Unlimited, right)

A weed is a plant whose virtues have not yet been discovered.
RALPH WALDO EMERSON

This chapter will answer the following questions:

- What types of pesticides are used?

- What are the pros and cons of using chemicals to kill insects and weeds?

- How well is pesticide usage regulated in the United States?

- What alternatives are there to using pesticides?

15-1 PESTICIDES: TYPES AND USES

Population Control in Nature A **pest** is any species that competes with us for food, invades lawns, destroys wood in houses, spreads disease, or is simply a nuisance. In diverse ecosystems, populations of species—including the less than 1% we classify as pests—are kept in control by their natural enemies (predators, parasites, and disease organisms)—another crucial type of Earth capital. When we simplify ecosystems we upset these natural checks and balances that keep any one species from taking over for very long. Then we must devise ways to protect our monoculture crops, tree farms, and lawns from pests that nature once controlled at no charge.

We have done this primarily by developing a variety of **pesticides** (or *biocides*)—chemicals to kill organisms we consider undesirable. Common types of pesticides include **insecticides** (insect-killers), **herbicides** (weed-killers), **fungicides** (fungus-killers), **nematocides** (roundworm-killers), and **rodenticides** (rat- and mouse-killers).

Humans didn't invent the use of chemicals to repel or kill other species. Plants have been developing chemicals to ward off or poison herbivores that feed on them for 225 million years. This is a never-ending, ever-changing process: Herbivores overcome various plant defenses through natural selection; then the plants use natural selection to develop new defenses.

The result of these dynamic interactions between predator and prey species is what biologists call *coevolution* (Section 5-6). The lesson to be learned from nature is that long-term survival requires a kaleidoscope of strategies and the ability to change from one to another as needed.

First-Generation Pesticides and Repellents
As the human population grew and agriculture spread, people began looking for better ways to protect their crops, mostly by using chemicals to kill or repel insects. For example, sulfur was used as an insecticide well before 500 B.C. By the fifteenth century A.D. toxic compounds of arsenic, lead, and mercury were being applied to crops as insecticides. This approach was abandoned in the late 1920s when the increasing number of human poisonings and fatalities encouraged a search for less toxic substitutes. However, traces of these nondegradable toxic metals are still being taken up by tobacco, vegetables, and other crops grown on soil dosed with them long ago.

In the seventeenth century nicotine sulfate, extracted from tobacco leaves, came into use as an insecticide. In the mid-1800s two more natural pesticides were introduced. One was pyrethrum, obtained from the heads of chrysanthemum flowers; the other was rotenone, from the root of the derris plant and other tropical forest legumes. These *first-generation pesticides* were mainly natural substances, weapons borrowed from plants that had been at war with insects for eons.

In addition to protecting crops, people have taken another lesson from nature and used chemicals (mostly produced by plants) to repel or kill household pest insects. They also save money and reduce health hazards associated with using commercial insecticides. For example:

- *Ants* can be persuaded to leave within about four days by sprinkling repellents such as red or cayenne pepper (a favorite I use successfully each spring and summer), crushed mint leaves, or boric acid (with an anticaking agent) along their trails inside a house, and by wiping off countertops with vinegar (also works for me).

- *Mosquitoes* can be repelled by planting basil outside windows and doors and rubbing a bit of vinegar on exposed skin. Mosquito attacks can also be reduced by not using scented soaps or wearing perfumes, colognes, and other scented products outdoors during mosquito season.

- *Roaches* can be killed by sprinkling boric acid under sinks and ranges, behind refrigerators, and in cabinets, closets, and other dark, warm places—or by establishing populations of banana spiders. Roaches can also be trapped by greasing the inner neck of a bottle baited with a raw potato or stale beer.

- *Flies* can be repelled by planting sweet basil and tansy (a common herb) near doorways and patios and by hanging a series of polyethylene strips in front of entry doors (like the ones you see on some grocery store coolers). Nontoxic flypaper can be made by applying honey to strips of yellow paper and hanging it from the ceiling in the center of rooms.

Q: What percentage of Earth's proven reserves of oil are in OPEC countries?

- *Fleas* can be kept off pets by using green dye or flea-repellent soaps; by feeding them brewer's yeast or vitamin B; by using flea powders made from eucalyptus, sage, tobacco, wormwood, bay leaf, or vetiver; or by dipping or shampooing pets in a mixture of water and essential oils such as citronella, cedarwood, eucalyptus, pennyroyal, orange, sassafras, geranium, clove, or mint. Recently, researchers have invented an effective trap that uses green-yellow light to attract fleas to an adhesive-coated surface.

Second-Generation Pesticides   A major pest control revolution began in 1939, when entomologist Paul Mueller discovered that DDT (dichlorodiphenyl-trichloroethane), a chemical known since 1874, was a potent insecticide. DDT soon became the world's most-used pesticide, and Mueller received the Nobel Prize in 1948. DDT was the first of the so-called *second-generation pesticides*. Since 1945 chemists have developed many of these synthetic organic chemicals for use in crop fields, homes, gardens, tree farms, lawns, and golf courses.

Worldwide, about 2.3 million metric tons (2.5 million tons) are used yearly—0.45 kilogram (1 pound) for each person on Earth. About 75% of these chemicals are used in MDCs, but use in LDCs is soaring.

In the United States about 570 biologically active (pest-killing) ingredients and 1,820 inert (non-pest-killing) ingredients are mixed to make some 25,000 different pesticide products. Since 1964 pesticide use in the United States has almost doubled, now exceeding 910 million kilograms (2 billion pounds) per year, or about 4 kilograms (8 pounds) per person per year—about eight times the global average. Four crops—corn, cotton, wheat, and soybeans—account for about 70% of the insecticides and 80% of the herbicides used on crops in the United States.

About 30% of the pesticides used in the United States are used to rid houses, gardens, lawns, parks, playing fields, swimming pools, and golf courses of unwanted pests. The average U.S. homeowner applies three to six times more pesticide per hectare than do farmers. Pesticides are added to products as diverse as paints, some shampoos, carpets, mattresses, and contact lenses.

Some pesticides, called *broad-spectrum* agents, are toxic to many species. Others, called *selective* or *narrow-spectrum* agents, are effective against a narrowly defined group of organisms. Pesticides vary in their *persistence*, the length of time they stay deadly and unchanged in the environment (Table 15-1). Most organophosphates (except malathion) are highly toxic to humans and other animals, and they account for most human pesticide poisonings and deaths.

15-2   THE CASE FOR PESTICIDES

Proponents of pesticides believe that their benefits outweigh their harmful effects. They point out the following benefits:

- *Pesticides save human lives.* Since 1945 DDT and other chlorinated hydrocarbon and organophosphate insecticides have probably prevented the premature deaths of at least 7 million people from insect-transmitted diseases such as malaria (carried by the *Anopheles* mosquito, Figure 8-10), bubonic plague (rat fleas), typhus (body lice and fleas), and sleeping sickness (tsetse fly).

- *Pesticides increase food supplies and lower food costs.* About 55% of the world's potential human food supply is lost to pests before (35%) or after (20%) harvest. In the United States, 37% of the potential food supply is destroyed by pests before and after harvest (13% by insects, 12% by plant pathogens, and 12% by weeds). Without pesticides these losses might be worse, and food prices would rise (by 30–50% in the United States, according to pesticide company officials).

- *Pesticides increase profits for farmers.* Pesticide companies estimate that every $1 spent on pesticides leads to an increase in U.S. crop yields worth approximately $4, but environmentalists estimate that the benefit drops to about $2 if the harmful effects of pesticides are included.

- *Pesticides work faster and better than alternatives.* Pesticides can control most pests quickly and at a reasonable cost. They have a long shelf life, are easily shipped and applied, and are safe when handled properly. When genetic resistance occurs, farmers can use stronger doses or switch to other pesticides. Most chemical pesticides kill more than one pest in the same crop. By contrast, nonchemical methods are generally targeted at a single pest species.

- *The health risks of pesticides are insignificant compared with their health and other benefits.* According to Elizabeth Whelan, director of the American Council on Science and Health (ACSH), which presents the position of the pesticide industry, "the reality is that pesticides, when used in the approved regulatory manner, pose no risk to either farm workers or consumers." Pesticide proponents consider the pesticide health-scare news stories to be distorted science and irresponsible reporting, and they point out that about 99.99% of the pesticides we eat are natural chemicals produced by plants. They urge that U.S. regulatory

Table 15-1 Major Types of Pesticides

Type	Examples	Persistence	Biologically Amplified?
Insecticides			
Chlorinated hydrocarbons	DDT, aldrin, dieldrin, toxaphene, lindane, chlordane, methoxychlor, mirex	High (2–15 years)	Yes (Figure 15-4)
Organophosphates	Malathion, parathion, diazinon, TEPP, DDVP, mevingphos	Low to moderate (1–12 weeks), but some can last several years	No
Carbamates	Aldicarb, carbaryl (Sevin), propoxur, maneb, zineb	Low (days to weeks)	No
Botanicals	Rotenone, pyrethrum, and camphor extracted from plants, synthetic pyrethroids (variations of pyrethrum) and rotenoids (variations of rotenone)	Low (days to weeks)	No
Microbotanicals	Various bacteria, fungi, protozoans	Low (days to weeks)	No
Herbicides			
Contact chemicals	Atrazine, simazine, paraquat	Low (days to weeks)	No
Systemic chemicals	2,4-D, 2,4,5-T, Silvex, diruon, daminozide (Alar), alachlor (Lasso), glyphosate (Roundup)	Mostly low (days to weeks)	No
Soil sterilants	Trifualin, diphenamid, dalapon, butylate	Low (days)	No
Fungicides			
Various chemicals	Captan, pentachlorphenol, zeneb, methyl bromide, carbon bisulfide	Mostly low (days)	No
Fumigants			
Various chemicals	Carbon tetrachloride, ethylene dibromide, methyl bromide	Mostly high	Yes (for most)

agencies shift focus from synthetic pesticides to natural carcinogens found in various foods.

- *Safer and more effective pesticides are being developed.* Company scientists are developing pesticides, such as botanicals and microbotanicals (Table 15-1), that are safer to users and less damaging to the environment. Genetic engineering also holds promise (Pro/Con, p. 207). However, total research and development and government-approval costs for a new pesticide have risen from $6 million in 1976 to between $80 and $120 million today. As a result, new chemicals are being developed only for crops such as wheat, corn, and soybeans with large markets. Yet, the need to develop new and safer pesticides is often highest in small-acreage crops such as fruits, nuts, and vegetables which are consumed in much larger amounts.

- *Many new pesticides are used at very low rates per acre compared to older products.* For example, application amounts per acre for many new herbicides are 100 times lower than for older ones.

Scientists continue to search for the ideal pest-killing chemical. It would

- *Kill only the target pest*
- *Harm no other species*
- *Disappear or break down into something harmless after doing its job*
- *Not cause genetic resistance in target organisms*
- *Be cheaper than doing nothing*

Unfortunately, no known pesticide meets all these criteria, and most don't even come close.

15-3 THE CASE AGAINST PESTICIDES

Genetic Resistance and Killing of Natural Pest Enemies Opponents of widespread use of pesticides believe that the harmful effects outweigh the benefits. The biggest problem is the development of

Q: What percentage of Earth's proven reserves of oil are in the United States?

U.S. Department of Agriculture

Figure 15-2 In the cotton fields of the southern United States boll weevils, which lay thousands of eggs and produce a new generation every 21 days, can produce as many as six generations in one growing season. Attempts to control the cotton boll weevil account for at least 25% of the insecticides used in the United States. However, farmers are now increasing their use of natural predators and other biological methods to control this major pest.

genetic resistance to pesticides by pest organisms. Insects breed rapidly (Figure 15-2), and within 5–10 years (much sooner in tropical areas) they can develop immunity through natural selection to the chemicals we throw at them (Connections, at right). Just when we think we've killed them off with our arsenal of chemicals, they come back stronger than before.

Since 1950 more than 500 major insect pests have developed genetic resistance to one or more insecticides; by 2000 virtually all major insect pest species will probably show some genetic resistance. Nearly 275 of the 500-odd major weed species are resistant to one or more herbicides. Pesticide use has also led to genetic resistance in 10 species of rodents (mostly rats). Because of genetic resistance, most widely used insecticides no longer protect people from insect-transmitted diseases in many parts of the world, leading to even more serious incidences of diseases such as malaria (Section 8-4).

Another problem is that broad-spectrum insecticides kill natural predators and parasites that may have been maintaining a pest species at a reasonable level. With wolf spiders (Figure 15-1, right), wasps, predatory beetles, and other natural enemies out of the way, a rapidly reproducing insect pest species can make a strong comeback only days or weeks after initially being controlled. Wiping out natural predators can also unleash new pests whose populations the predators had previously held in check and cause other unexpected effects (Connections, p. 127). Natural predators can also develop genetic resistance to pesticides, but most predators can't reproduce as fast as their insect prey.

CONNECTIONS

The Pesticide Treadmill

When genetic resistance develops, pesticide sales representatives usually recommend more frequent applications, stronger doses, or a switch to new (usually more expensive) chemicals to keep the resistant species under control. This can put farmers on a **pesticide treadmill**, whereby they pay more and more for a pest control program that does less and less good.

A 1989 study by David Pimentel (Guest Essay, p. 333), an expert in insect ecology, based on data from more than 300 agricultural scientists and economists, concluded that

- Although the use of synthetic pesticides has increased 33-fold since 1942, the U.S. loses more of its crops to pests today (37%) than in the 1940s (31%). Losses attributed to insects almost doubled from 7% to 13% despite a 10-fold increase in the use of synthetic insecticides; losses to plant diseases rose from 10% to 12%, and losses to weeds dropped from 14% to 12%.

- The estimated environmental, health, and social costs of pesticide use in the United States range from $4 billion to $10 billion per year.

- Alternative pest control practices (Section 15-5) could halve the use of chemical pesticides on 40 major crops in the United States without reducing crop yields.

- A 50% cut in pesticide use in the United States would cause retail food prices to rise by only about 0.2% but would raise average income for farmers about 9%.

Mobility and Threats to Wildlife   According to the U.S. Department of Agriculture no more than 2% (and often less than 0.1%) of the insecticides applied to crops by aerial spraying (Figure 15-3) or ground spraying actually reaches the target pests—and less than 5% of herbicides applied to crops reaches the target weeds. Pesticides that miss their target pests end up in the air, surface water, groundwater, bottom sediments, food, and nontarget organisms, including humans and wildlife. Pesticide waste and mobility can be reduced by using recirculating sprayers, covering spray booms to reduce drift, and using rope-wick applicators (which deliver herbicides directly to weeds and reduce herbicide use by 90%).

During the 1950s and 1960s populations of fish-eating birds such as the osprey, cormorant, brown pelican, and bald eagle plummeted. Research indicated

A: 4%. The U.S. uses about 30% of global oil production each year.

Figure 15-3 Crop duster spraying an insecticide on grapevines south of Fresno, California. Aircraft are used to apply about 65% of the pesticides used on cropland in the United States. Only 0.1–2% of insecticides reach the target pests.

National Archives/EPA Documerica

Recently the EPA found DDT in 99% of the freshwater fish it tested.

Each year some 20% of U.S. honeybee colonies are wiped out by pesticides, and another 15% are damaged, costing farmers at least $206 million per year from reduced pollination of vital crops. Pesticide runoff from cropland is a leading cause of fish kills worldwide. According to the U.S. Fish and Wildlife Service, pesticides menace about 20% of the endangered and threatened species in the United States.

Threats to Human Health The World Health Organization estimates that at least 1 million people (including 313,000 farm workers in the United States) are accidentally poisoned by pesticides each year; 4,000–20,000 of them die. At least half of them—and 90% of those killed—are farm workers in LDCs, where educational levels are low, warnings are few, and pesticide regulations lax or nonexistent. The actual number of pesticide-related illnesses and deaths among farm workers in the United States and throughout the world is probably greatly underestimated because of poor records, lack of doctors and disease reporting in rural areas, and faulty diagnoses.

A 1993 study by the National Academy of Sciences concluded that the legal limits for pesticides in food may need to be reduced by up to 1,000 times to protect children, who are more vulnerable to such chemicals than adults. According to the *worst-case estimate* in a 1987 study by the National Academy of Sciences, exposure to pesticides in food causes 4,000–20,000 cases of cancer per year in the United States. Just 10 chemicals accounted for 80% of this estimated risk. Tomatoes alone (treated with four of the most dangerous chemicals) accounted for 15% of the estimated total cancer risk from pesticides in food. In 1987 the EPA ranked pesticide residues in foods as the third most serious environmental health threat in the United States in terms of cancer risk.

A 1993 study of Missouri children showed statistically significant correlations between childhood brain cancer and use of various pesticides in the home, including flea and tick collars, no-pest strips, and chemicals used to control pests such as roaches, ants, spiders, mosquitoes, and termites. Another 1993 study by the Environmental Working Group suggested that more than one-third of a child's lifetime exposure to cancer risk from some pesticides accumulates by age 5. According to this study, by their first birthday the average American child's exposure to some carcinogenic pesticides exceeds the government's lifetime acceptable-cancer risk threshold of one malignancy in every million individuals.

Most environmentalists oppose the proposal by the Clinton administration to weaken the Delaney

that a chemical derived from DDT, biologically amplified in food webs (Figure 15-4), made their eggshells so fragile that the birds could not reproduce. Also hard-hit were such predatory birds as the prairie falcon, sparrow hawk, and peregrine falcon (Figure 15-5), which help control rabbits, ground squirrels, and other crop-eaters. Since the U.S. ban on DDT in 1972, most of these species have made a comeback. In 1980, however, DDT levels were again rising in species such as the peregrine falcon and the osprey. These species may be picking up biologically amplified DDT and other banned pesticides in Latin America, where they winter. In those countries the use of such chemicals is still legal. Illegal use of DDT and other banned pesticides in the United States may also play a role. DDT and other long-lived pesticides used in LDCs can also be blown by winds as far away as India and deposited from the atmosphere onto U.S. land and surface water.

Q: How much of the oil used in the United States is imported?

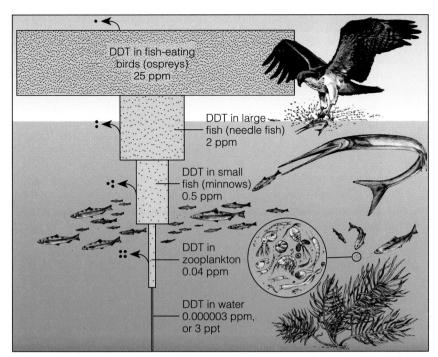

Figure 15-4 DDT concentration in the fatty tissues of organisms was biologically amplified about 10 million times in this food chain of an estuary near Long Island Sound. If each phytoplankton organism in such a food chain retains one unit of DDT from the water, a small fish eating thousands of zooplankton (which feed on the phytoplankton) will store thousands of units of DDT in its fatty tissue. Then a large fish that eats 10 of the smaller fish will receive and store tens of thousands of units, and a bird (or a human) that eats several large fish will ingest hundreds of thousands of units. Dots represent DDT, and arrows show small losses of DDT through respiration and excretion.

Within figure:
DDT in fish-eating birds (ospreys) 25 ppm

DDT in large fish (needle fish) 2 ppm

DDT in small fish (minnows) 0.5 ppm

DDT in zooplankton 0.04 ppm

DDT in water 0.000003 ppm, or 3 ppt

clause in an existing law by replacing zero tolerance of cancer-causing pesticides in processed food with a "one-in-a-million" risk standard per chemical (Spotlight, p. 400). They argue that exposure to a single chemical on one type of food is just one of many exposures consumers get. Instead cumulative exposure to a number of potentially carcinogenic peptides on a variety of crops can build up to an unacceptable level and should be a major factor in deciding acceptable exposure levels, especially for children.

Some scientists are becoming increasingly concerned about possible genetic mutations, birth defects, nervous-system disorders, and effects on the immune and endocrine systems from long-term exposure to low levels of various pesticides. Very little research has been conducted on these effects, and they are not covered adequately in current pesticide health-safety laws.

Accidents and unsafe practices in pesticide plants can expose workers, their families, and sometimes the general public to harmful levels of pesticides or raw chemicals used in their manufacture. For example, in 1984 the world's worst industrial accident occurred at a Union Carbide pesticide plant in Bhopal, India. A highly toxic chemical used in the manufacture of carbamate pesticides leaked from an underground storage tank, killing 5,100 people (2,600 from direct exposure to the hazardous chemical and another 2,500 from the aftereffects). It is estimated that another 10,000 (Union Carbide) to 200,000 (Indian officials) people received serious injuries such as blindness or lung damage.

Hans Reinhard/Bruce Coleman Ltd.

Figure 15-5 The peregrine falcon is endangered in the United States, mostly because of DDT. The insecticide caused their young to die before hatching because the eggshells were too thin to protect them. In 1975 only about 120 peregrine falcons were left in the lower 48 states; today there are about 1,400. Most of them were bred in captivity and then released into the wild in a $2.7-million-per-year recovery program. However, some of these birds are illegally shot or captured for sale on the black market. Peregrine falcons also die because of loss of habitat in their South American wintering grounds.

15-4 PESTICIDE REGULATION IN THE UNITED STATES: IS THE PUBLIC ADEQUATELY PROTECTED?

Because of the potentially harmful effects of pesticides on wildlife and people, Congress passed the Federal Insecticide, Fungicide, and Rodenticide Act (FIFRA)

The Delaney Clause

In 1958 an amendment known as the Delaney clause (after Representative James Delaney), was added to the U.S. food and drug laws. It absolutely prohibits using any additive in foods (including pesticide residues) that has been shown to cause cancer in test animals or people, with no consideration of economic or health benefits or risks. It is one of the first examples of pollution prevention. It is a precautionary or "better-safe-than-sorry" approach that says no deaths will be allowed because of substances added to or finding their way into foods people eat. Since 1958 the food and pesticide industries have fought hard to have this law repealed or changed to allow balancing of health and economic benefits and risks.

Despite its absolute nature, the Food and Drug Administration and the EPA have rarely used the Delaney clause to ban a food additive or pesticide that ends up as a food residue—choosing mostly to ignore or find ways around the law. This ended in 1992 when the U.S. Supreme Court ruled that food-crop residues of pesticides shown to cause cancer in test animals must be banned from use. This ruling means that the EPA would have to ban about 50 pesticides whose active ingredients have been shown to cause cancer in test animals—or it would have to ask Congress to weaken the law.

Food and pesticide producers and the EPA (with the support of President Bill Clinton) want to amend the law to allow continued use of pesticides that pose a small risk, such as less than one chance of a cancer in 1 million people—the current EPA standard. Proponents of replacing the zero carcinogen standard with a "one-in-a-million" risk standard point out humans ingest about 10,000 times as much of natural pesticides produced by plants as defenses against insects as they do of synthetic pesticides. Many of these natural pesticides produce cancer in rodents. If the Delaney clause is sound legislation, why isn't it also applied to natural carcinogens?

Advocates of substituting the negligible-risk standard also point out that the Delaney clause was introduced at a time when only parts per million levels of pesticide residues in food could be detected. Today modern analytical techniques can detect pesticide levels in the parts per quintillion range. They argue that in most cases such levels pose a negligible (if any) health risk to humans and ask why a pesticide should be banned if it is found in foods only at such trace levels.

Most environmentalists believe that the Delaney clause is an excellent example of a pollution prevention or zero-carcinogens policy that genuinely protects the public and that should not be weakened. They argue that exposure to a cancer-causing pesticide on one crop is just one of many exposures consumers get and that the cumulative exposure to such chemicals raises the total risk considerably and should be taken into account.

Because scientists have no effective and affordable way to study the effects of multiple exposures to different pesticides, supporters of the Delaney clause argue that retaining the zero-tolerance principles is the best way to protect human health. Several studies also indicate that permitted residues are 100–500 times what is safe for children—a position generally supported in a 1993 study by the National Academy of Sciences.

Environmentalists also point out that current technologies for estimating the health risks of pesticides—especially for children—are inadequate because of a lack of data and the inherent limitations of risk analysis. Indeed, instead of revoking or weakening the Delaney clause, some health scientists believe it should be strengthened to ban substances in food that have been shown to cause mutations (mutagens), birth defects (teratogens), or nerve-system damage in test animals. What do you think?

---

in 1972. This law, which was amended in 1975, 1978, and 1988, requires that all commercial pesticides be approved for general or restricted use by the EPA. However, the pesticide companies—not the EPA—evaluate the biologically active ingredients in their own products.

Since 1972 the EPA has used this law to ban the use, except for emergencies, of over 50 previously approved pesticides because of potential health hazards. The banned chemicals include most chlorinated hydrocarbon insecticides, several carbamates and organophosphates, and the herbicides 2,4,5-T and Silvex (Table 15-1).

The FIFRA act required the EPA to reevaluate the more than 600 active ingredients approved for use in pre-1972 pesticide products to determine whether any of them caused cancer, birth defects, or other health risks. However, by 1993—21 years after Congress ordered the EPA to evaluate these chemicals—far less than 10% of these active ingredients have been fully tested and evaluated for potential health problems. At this rate it will take over 200 years to evaluate the remaining chemicals.

According to the National Academy of Sciences, federal laws regulating the use of pesticides in the United States are inadequate and poorly enforced by

Q: Will the United States ever again be self-sufficient in oil?

The EPA can ban a chemical immediately in an emergency. Until 1990, however, the law required the EPA to use its already severely limited funds to compensate pesticide manufacturers for their remaining inventory and for all storage and disposal. Because compensation costs for a single chemical could exceed the agency's annual pesticide budget, the only economically feasible solution was for the EPA to allow existing stocks of a dangerous chemical to be sold. One of the 1988 amendments to FIFRA shifted some of the costs of storing and disposing of banned pesticides from the EPA to the manufacturers.

Environmentalists welcome the 1988 amendments, but they point out that the law still has many weaknesses and loopholes. One loophole allows the sale in the United States of insecticides (especially dicofol and chlorobenilate) containing as much as 15% DDT by weight, classified as an impurity. These products, along with others smuggled in and blown in (mostly from Mexico), are believed to be responsible for increased DDT levels since 1980 in some vulnerable forms of wildlife (Figure 15-5) and in some fruits and vegetables grown and sold in the United States (especially in California). Also, this law is the only major environmental statute that does not provide for citizen suits against the EPA for not enforcing the law, an essential tool to ensure government compliance.

According to the FDA, about 1% of the food purchased in the United States has levels above the legal limit for one or more pesticides. Each year FDA inspectors check less than 1% (about 12,000 samples) of domestic and imported food for pesticide contamination (Connections, at left). Moreover, 5 of the 12 FDA regional laboratories cannot detect residues of 80% of the pesticides used in agriculture today, and some foreign food suppliers use pesticides that the FDA cannot detect. Furthermore, the FDA's turnaround time for food analysis is so long that half of the food has been sold and eaten before contamination is detected. Even when contaminated food is found, the growers and importers are rarely penalized.

The 1993 study of pesticide safety by the National Academy of Sciences also urged the government to:

- Make health, not agricultural, impacts the primary consideration for setting limits on pesticide levels allowed in food.

- Collect more and better data on exposure to pesticides for different groups, including farm workers, adults, and children.

- Develop new and better test procedures for evaluating the toxicity of pesticides, especially for children.

both the EPA and the Food and Drug Administration (FDA). The Academy study also concluded that up to 98% of the potential risk of developing cancer from pesticide residues on food grown in the United States would be eliminated if EPA standards were as strict for pre-1972 pesticides as they are now.

In addition, many of the 1,820 so-called inert or biologically inactive ingredients in pesticides clearly can harm humans as well as wildlife. In fact, because inert ingredients make up 80–99% by weight of a pesticide product, they can pose a higher health risk than some active ingredients.

Since 1987 the EPA has been evaluating these inert ingredients and thus far has labeled 100 of them "of known or potential toxicological concern" but has not banned their use. However, by 1991 the EPA had no toxicity information on 80% of the inert ingredients used in pesticides.

FIFRA allows the EPA to leave inadequately tested pesticides on the market and to license new chemicals without full health and safety data. It also gives the EPA unlimited time to remove a chemical even when its health and environmental risks are shown to outweigh its economic benefits. The built-in appeals and other procedures often keep a dangerous chemical on the market for up to 10 years.

Figure 15-6 Genetic engineering against pest damage. Both tomato plants were exposed to destructive caterpillars. The normal plant's leaves are almost gone (left), while the genetically altered plant (right) shows little damage.

Figure 15-7 Biological control of pests. An adult convergent ladybug (right) is consuming an aphid (left).

- In evaluating the health risks of pesticide exposure, consider cumulative exposures of all pesticides in food and water—especially for children—instead of basing regulations on exposure to a single pesticide.

Most environmentalists oppose the Clinton administration's support for weakening the Delaney clause. However, they support the administration's proposals to ban the sale abroad of pesticides that are outlawed in the United States, impose stricter fines for violations of pesticide laws, and require that the EPA set for foods pesticide tolerance levels that protect infants and children.

15-5 SOLUTIONS: OTHER WAYS TO CONTROL PESTS

Cultivation Practices   Chemicals are not the only answer to pests. Other strategies include the following:

- *Changing the type of crop planted in a field each year* (crop rotation).

- *Planting rows of hedges or trees in and around crop fields.* These plantings hinder insect invasions, provide habitats for their natural enemies, and also reduce erosion of soil by wind (Figure 12-28d).

- *Adjusting planting times so that major insect pests either starve or get eaten by their natural predators.*

- *Growing crops in areas where their major pests do not exist.*

- *Switching from vulnerable monocultures to modernized versions of intercropping, agroforestry, and poly-culture that use plant diversity to reduce losses to pests.*

- *Removing diseased or infected plants and stalks and other crop residues that harbor pests.*

- *Using photodegradable plastic to keep weeds from sprouting between crop rows.*

- *Using vacuum machines that gently remove harmful bugs from plants.*

Unfortunately, many U.S. farmers—in order to increase profits, qualify for government subsidies, or avoid bankruptcy—feel they cannot use these cultivation methods.

Building In Resistance   Plants and animals that are genetically resistant to certain pest insects, fungi, and diseases can be developed. However, resistant varieties usually take a long time (10–20 years) and lots of money to develop by conventional methods. Moreover, insects and plant diseases can develop new strains that attack the once-resistant varieties, forcing scientists to continually develop new resistant strains. Genetic engineering is now being used to speed the process (Figure 15-6 and Pro/Con, p. 207).

Also, some resistance factors bred or engineered into crops can be toxic to beneficial insects, humans, and other animals. For example, certain alkaloids bred into potatoes to defend against the Colorado potato beetle are hazardous to humans. So far, traditional crop-breeding methods are largely unregulated.

Using Natural Enemies   Predators (Figures 15-1 and 15-7), parasites, and pathogens (disease-causing bacteria and viruses) can be encouraged or imported to regulate pest populations. More than 300 biological pest control projects worldwide have been successful, especially in China. In Nigeria crop-duster planes release parasitic wasps instead of pesticides to fight the cassava mealybug. Farmers get a $178 return for every $1 they spend on the wasps. In the United States, natural enemies have been used to control more than 70 insect pests, and biological control is catching on as more farmers seek alternatives to chemical warfare.

Other examples of biological control include the use of

- *Guard dogs or llamas to protect livestock from predators.*

- *Geese for weeding orchards, eating fallen and rotting fruit (often a source of pest problems), and controlling grass in gardens and nurseries. Geese also warn of approaching predators or people by honking loudly.*

- *Chickens to control insects and weeds after plants are well established.*

- *Birds to eat insects.* Farmers and homeowners can provide habitats and nesting sites that attract woodpeckers, purple martins, chickadees, barn swallows, nuthatches, and other insect-eating species.

- *Spiders to eat insects* (Figure 15-1).

- *Plants that naturally produce chemicals toxic to their weed competitors or that repel or poison their insect pests.* For example, certain varieties of barley, wheat, rye, sorghum, and Sudan grass will suppress weeds in gardens or orchards. Peppermint can be planted around houses to repel ants (it is also a natural mouth freshener and a cooking spice).

Biological control has several advantages. It homes in on the target species and is nontoxic to other species, including people. Once a population of natural predators or parasites is established, pest control can often be self-perpetuating. Development of genetic resistance is minimized because pest and predator species interact and change together (coevolution). In the United States biological control has saved farmers an average of $25 for every $1 invested.

However, years of research may be needed, both to understand how a particular pest interacts with its various enemies and also to choose the best biological control agent. Biological agents can't always be mass-produced, and farmers find them slower-acting and more difficult to apply than pesticides. Also, biological agents must be protected from pesticides sprayed in nearby fields. Once released into the environment, biological control agents can multiply, cause unpredictable harmful ecological effects, and be impossible

Figure 15-8 Infestation of a steer by screwworm fly larvae. A fully grown steer can be killed in 10 days from thousands of maggots feeding on a single wound.

to recapture. Some may even become pests themselves; others (such as praying mantises) devour beneficial as well as pest insects. Indeed, species introduced for biological pest control have been strongly implicated in the extinction of nearly 100 insect species worldwide.

Using Biopesticides Botanicals such as synthetic pyrethroids (Table 15-1) are an increasingly popular method of pest control, and scientists are busy looking for new plant toxins to synthesize for mass production. Microbes are also being drafted for insect wars. For example, *Bacillus thuringensis (Bt)* toxin is a registered pesticide sold commercially as a dry powder. One of the thousands of strains of this common soil bacterium will kill a specific pest. The bad news is that genetic resistance is already developing to some *Bt* toxins.

Using Birth Control Males of some insect pest species can be raised in the laboratory, sterilized by radiation or chemicals, and then released in hordes in an infested area to mate unsuccessfully with fertile wild females. This technique works best if the females mate only once; if the infested area is isolated so that it can't be repopulated with nonsterilized males; and if the insect pest population has already been reduced to a fairly low level by weather, pesticides, or other factors. Success is also increased if only the "sexiest"—the loudest, fastest, and largest—males are sterilized. Problems with this approach include high costs, the difficulties involved in knowing the mating times and behaviors of each target insect, releasing enough sterile males to do the job, and the relatively few species for which this strategy works.

The Department of Agriculture used the sterile-male approach to essentially eliminate the screwworm fly, a major livestock pest (Figure 15-8), from the

A: About the same in both cases

Figure 15-9 A lemon infested with red scale mites. Pheromones are now being used to help control populations of red scale mites.

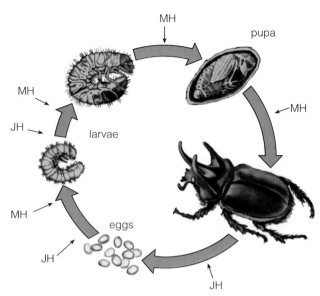

Figure 15-10 For normal growth, development, and reproduction, certain juvenile hormones (JH) and molting hormones (MH) must be present at genetically determined stages in the typical life cycle of an insect. If applied at the right time, synthetic hormones can be used to disrupt the life cycle of insect pests.

southeastern states between 1962 and 1971. In 1972, however, the pest made a dramatic comeback, infesting 100,000 cattle and causing serious losses until 1976, when a new strain of sterile males was developed and released to bring the situation under temporary control. To prevent resurgences of this pest, new strains of sterile male flies have to be released every few years.

Using Insect Sex Attractants In many insect species, a female that is ready to mate releases a minute amount of a chemical sex attractant called a *pheromone*. Pheromones, whether extracted from insects or synthesized in the laboratory, can be used to lure pests into traps or to attract their natural predators into crop fields (usually the more effective approach). Worldwide more than 50 companies sell about 250 pheromones to control pests (Figure 15-9).

These chemicals attract only one species, work in trace amounts, have little chance of causing genetic resistance, and are not harmful to nontarget species. However, it is costly and time-consuming to identify, isolate, and produce the specific sex attractant for each pest or natural predator species.

Using Insect Hormones Hormones are chemicals produced by an organism and sent through its bloodstream to control its growth and development. Each

step in the insect life cycle is regulated by the timely natural release of juvenile hormones (JH) and molting hormones (MH) (Figure 15-10). These chemicals can either be extracted from insects or synthesized in the laboratory, used to disrupt an insect's normal life cycle, and thus cause the insect to die before it can reach maturity and reproduce (Figure 15-11).

Insect hormones have the same advantages as sex attractants, but they take weeks to kill an insect, are often ineffective with a large infestation of insects, and sometimes break down before they can act. They must also be applied at exactly the right time in the target insect's life cycle. Moreover, they sometimes affect the target's predators and other nonpest species, and they can kill crustaceans if they get into aquatic ecosystems. Finally, like sex attractants they are difficult and costly to produce.

Zapping Pests with Hot Water Some farmers have begun using the "Aqua Heat" machine that sprays boiling water on crops to kill both weeds and insects. Water is boiled and drawn from a large stainless steel tank mounted on a tractor and sprayed on crops using a 2-meter- (6-foot-) long boom. So far the system has worked well on cotton, alfalfa, and potato fields sprayed before the growing season and in citrus groves in Florida, where the machine was invented. Operating costs are equal to or lower than those of chemical pesticide application.

Q: How long will the world's oil reserves last at the current consumption rate?

Figure 15-11 Chemical hormones can prevent insects from maturing completely and make it impossible for them to reproduce. The stunted hornworm (left) was fed a compound that prevents its larvae from producing molting hormones (MH); a normal tobacco hornworm is shown on the right.

Zapping Foods with Radiation Certain foods can be exposed after harvest to gamma rays (Figure 3-4) emitted by radioactive isotopes, in order to extend their shelf life—and to kill insects, parasitic worms (such as trichinae in pork), and bacteria (such as salmonellae, which infect 51,000 Americans and kill 2,000 each year). A food does not become radioactive when it is irradiated, just as being exposed to X rays does not make the body radioactive. According to the FDA and the World Health Organization, over 1,000 studies show that foods exposed to low doses of ionizing radiation are safe for human consumption. Currently, 37 countries (8 of them in western Europe) allow irradiation of one or more food items.

However, critics contend that irradiating food destroys some of its vitamins and other nutrients and that the irradiation process may form trace amounts of toxic, possibly carcinogenic, compounds in food.

These opponents say more studies need to be done to ensure the safety and wholesomeness of irradiated food and that it is too soon to see possible long-term effects, which might not show up for 30–40 years. They also argue that Americans want fresh, wholesome food, not old, possibly less nutritious food made to appear fresh and healthy by irradiation. Proponents respond that irradiation of food is likely to lower health hazards to people and reduce pesticide use, and that its potential benefits greatly exceed the risks. Meanwhile, New York, New Jersey, and Maine have prohibited the sale and distribution of irradiated food, as have Germany, Austria, Denmark, Sweden, Switzerland, Sudan, Singapore, Australia, and New Zealand.

Using Integrated Pest Management An increasing number of pest control experts believe that the best way to control crop pests is a carefully designed **integrated pest management (IPM)** program. In this approach, each crop and its pests are evaluated as part of an ecological system. Then a control program is developed that includes a mix of cultivation and biological and chemical methods in proper sequence and with the proper timing.

The overall aim of IPM is not eradication of pest populations, but rather maintenance at just below economically damaging levels (Figure 15-12). Fields are carefully monitored, and when a damaging level of pests is reached farmers first use biological and cultivation controls, including vacuuming up harmful bugs. Small amounts of insecticides (mostly botanicals or microbotanicals) are applied when absolutely necessary, and different chemicals are used to slow development of genetic resistance.

In 1986 the Indonesian government banned the use of 57 pesticides on rice and launched a nationwide program to switch to integrated pest management. The results were dramatic: Between 1987 and 1992, pesticide use dropped by 65%, rice production rose by 15%, and the country now saves about $120 million per year on pesticides—enough to cover the cost of its IPM program.

The experiences of countries such as China, Brazil, Indonesia, and the United States have shown that a well-designed IPM program can reduce pesticide use and pest control costs by 50–90%. IPM can also reduce preharvest pest-induced crop losses by 50%. It can improve crop yields, reduce inputs of fertilizer and irrigation water, and slow the development of genetic resistance because pests are zapped less often and with lower doses of pesticides. Thus IPM is an important form of pollution prevention that reduces risks to wildlife and human health.

However, IPM requires expert knowledge about each pest-crop situation, and it is slower-acting than

A: About 42 years (undiscovered oil might add another 40 years)

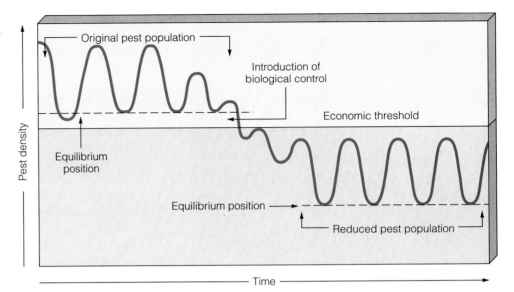

Figure 15-12 The goal of biological control and integrated pest management is to keep each pest population just below the size at which it causes economic loss.

conventional pesticides. Moreover, methods developed for a given crop in one area may not apply to another area with slightly different growing conditions. And although long-term costs are typically lower than the costs of using conventional pesticides, initial costs may be higher.

Ironically, the sharp cost increases for developing new pesticides (mostly because of stricter regulation to protect human health) is hindering the effectiveness of IPM. Because of high costs and low sales volume, pesticide companies have canceled registrations of many existing pesticides and stopped developing new pesticides for low-acreage crops such as fruits and vegetables which make up much of consumers' diets. Low doses of such chemicals are a key ingredient of many IPM programs.

Widespread use of IPM is also hindered by government subsidies of conventional chemical pesticides and by opposition from agricultural chemical companies, whose sales would drop sharply. In addition, farmers get most of their information about pest control from pesticide salespeople and from U.S. Department of Agriculture (USDA) county farm agents, few of whom have adequate training in IPM.

Despite its potential, only about 1% of the USDA's research and education budget is spent on IPM. Environmentalists urge the USDA to promote integrated pest management by **(1)** adding a 2% sales tax on pesticides to fund IPM research and education; **(2)** emphasizing government-sponsored research for development of pesticides and other methods of pest control on selected, small-acreage fruit and vegetable crops. **(3)** setting up a federally supported IPM demonstration project on at least one farm in every county; **(4)** training USDA field personnel and county farm agents

in IPM so they can help farmers use this alternative; **(5)** providing federal and state subsidies and perhaps government-backed crop-loss insurance to farmers who use IPM or other approved alternatives to pesticides; and **(6)** gradually phasing out subsidies to farmers who depend almost entirely on pesticides, once effective IPM methods have been developed for major pest species.

Each of us has a role to play in encouraging the use of IPM and other alternatives to excessive reliance on pesticides and in reducing our exposure to pesticides (Individuals Matter, next page).

We need to recognize that pest control is basically an ecological, not a chemical problem.

ROBERT L. RUDD

Critical Thinking

1. Should DDT and other pesticides be banned from use in malaria control efforts throughout the world? Explain. What are the alternatives?

2. Environmentalists argue that because essentially all pesticides eventually fail, their use should be phased out, and farmers should be given economic incentives for switching to integrated pest management. Explain your position.

3. How can the use of insecticides increase the number of insect pest problems?

4. Debate the following resolution: Because DDT and the other banned chlorinated hydrocarbon pesticides pose no demonstrable threat to human health and have saved millions of lives, they should again be approved for use in the United States.

Q: How long could *all* known and projected U.S. oil deposits supply the world and the U.S. at current consumption?

INDIVIDUALS MATTER

What You Can Do About Pesticide Use

- *Give up the idea that the only good bug is a dead bug.* Recognize that insects such as spiders (p. 393) keep most of the populations of insects we consider to be pests in control and that full-scale chemical warfare on insect pests wipes out many of our insect allies.

- *Buy organically grown produce that has not been treated with synthetic fertilizers, pesticides, or growth regulators; or, grow some of your own produce this way.*

- *Don't insist on perfect-looking fruits and vegetables.* These are more likely to contain high levels of pesticide residues.

- *Get rid of most pesticide residues by carefully washing and scrubbing all* fresh produce, discarding the outer leaves of lettuce and cabbage, and peeling thick-skinned fruits.

- *Use pesticides in your home only when absolutely necessary, and use them in the smallest amount possible.* Try natural alternatives first (p. 394).

- *Dispose of unused pesticides safely.* Contact your local health department or environmental agency for safe disposal methods.

- *Don't become obsessed with having the perfect lawn.* About 40% of U.S. lawns are treated with pesticides, typically at levels three to six times higher per acre than farmland. These chemicals can cause headaches, dizziness, nausea, and eye trouble—and more acute effects in sensitive individuals, including children who play on treated lawns and in parks. Of the 40 pesticides commonly used by the lawn-care industry, 12 are suspected human carcinogens, 10 may cause birth defects, 3 can affect reproduction, 9 can damage liver and kidneys, 20 can cause short-term nervous-system damage, and 29 cause rashes or skin disease.

- *If you hire a lawn-care company, use one that relies only on organic methods.* And get its claims in writing.

- *Urge elected officials to promote integrated pest management, strengthen pesticide laws to protect human health and the environment from the harmful effects of pesticides, and ban exports of pesticides not approved for use in the United States.*

5. Should certain types of foods used in the United States be irradiated? Explain.

6. What changes, if any, do you believe should be made in the Federal Insecticide, Fungicide, and Rodenticide Act regulating pesticide use in the United States?

7. Should U.S. companies be allowed to export to other countries pesticides that have been banned or severely restricted in the United States? Explain.

*8. How are bugs and weeds controlled in your yard and garden? On the grounds of your school and the public schools, parks, and playgrounds where you live? Consider organizing an effort to have integrated pest management and organic fertilizers used on school and public grounds. Do the same thing for your yard and garden.

*9. Make a concept map of the key ideas in this chapter using the section heads and subheads and the key terms (shown in boldface type in the chapter). See the inside front cover and Appendix 4 for information on concept maps.

16 Biodiversity: Sustaining Ecosystems

How Farmers and Loud Monkeys Saved a Forest

It's early morning in a tropical forest in the Central American country of Belize (Figure 6-16). Suddenly, loud roars that trail off into wheezing moans—territorial calls of black howler monkeys (Figure 16-1)—wake up everyone in or near the wildlife sanctuary by the Belize River.

This species is the centerpiece of an experiment integrating ecology and economics by allowing local villagers to make money by helping sustain the forest and its wildlife. The project is the brainchild of an American biologist, Robert Horwich. In 1985 he suggested that villagers establish a sanctuary that would benefit the local black howlers and themselves. He proposed that the farmers leave thin strips of forest along the edges of their fields to provide food for the howlers, who feed on leaves, flowers, and fruits as they travel among the treetops.

To date, more than 100 farmers have participated, and the 47-square-kilometer (18-square-mile) sanctu-ary is now home for an estimated 1,100 black howlers. The idea has spread to seven other villages.

Now, as many as 6,000 ecotourists visit the sanctuary each year to catch glimpses of its loud monkeys and other wildlife. Villagers serve as tour guides, cook meals for the visitors, and lodge tourists overnight in their spare rooms.

Forests, rangelands, parks, and wilderness are key land and biological resources that are coming under increasing stress from population growth and economic development. Learning how to use these potentially renewable forms of Earth capital sustainably—and helping heal those we have degraded—are urgent priorities.

Figure 16-1 Black howler monkey in a tropical forest in Belize. This is one of six howler monkey species in Latin America whose populations have declined as tropical forests in countries such as Guatemala and Mexico have been cleared. The black howler is one of the more than two-thirds of the world's 150 known species of primates threatened with extinction. (Carol Farnetti/Planet Earth Pictures)

Forests precede civilizations, deserts follow them.

FRANÇOIS-AUGUSTE-RENÉ DE CHATEAUBRIAND

In this chapter we will answer the following questions:

■ What are the major types of forests, and why are they such important ecosystems?

■ How should forest resources be managed and conserved?

■ How are forests managed and conserved in the United States?

■ Why are tropical deforestation and fuelwood shortages serious problems, and what can be done about them?

■ Why are rangelands important, and how should they be managed?

■ What problems do parks face, and how should they be managed?

■ Why is wilderness important, and how much should be preserved?

16-1 THE IMPORTANCE OF ECOLOGICAL DIVERSITY

Sustaining Ecosystems   In Chapter 4 you learned that there are three components of the planet's biodiversity: **(1)** *genetic diversity* (variability in the genetic makeup among individuals within a single species), **(2)** *species diversity* (the variety of species on Earth and in different habitats of the planet), and **(3)** *ecological diversity* (the variety of forests, deserts, grasslands, streams, lakes, oceans, and other biological communities that interact with one another and with their nonliving environments).

Because biodiversity is a vital part of the Earth capital that sustains all life, preserving the planet's genes, species, and ecosystems should be among our most important priorities. One way to do this is to protect species from sharp population declines and premature extinctions that result from human activities, as discussed in the next chapter. However, most wildlife biologists believe that the best way to protect species diversity is to sustain and protect the earth's ecosystems that serve as habitats for them, as discussed in this chapter.

This scientific approach recognizes that saving wildlife means saving the places where they live. It is also based on Aldo Leopold's ethical principle that something is right when it tends to maintain Earth's life-support systems for us and other species, and wrong when it doesn't.

Public Lands in the United States: This Land Is Your Land   No nation has set aside so much of its land—about 42%—for public use, enjoyment, and wildlife as the United States. Almost one-third of the country's land is managed by the federal government; 73% of this public land is in Alaska, and another 22% is in the western states. In the western states, 60% of all land is public land owned by the nation as a whole. These public lands can be classified according to how they are used as:

Multiple-Use Lands

National Forest System   These 156 forests (Figure 16-2) and 19 grasslands are managed by the Forest Service. Except for wilderness areas (15%) this land is managed using two principles: **(1)** The principle of sustainable yield states that a potentially renewable resource should not be harvested or used faster than it is replenished; **(2)** the principle of multiple use allows a variety of uses on the same land at the same time.

Today national forests are used for timbering (the dominant use in most cases), mining, grazing, farming, oil and gas extraction, recreation, sport hunting, sport and commercial fishing, and conservation of watershed, soil, and wildlife resources. Off-road vehicles are usually restricted to designated routes.

National Resource Lands   These grasslands, prairies, deserts, scrub forests, and other open spaces in the western states and Alaska are managed by the Bureau of Land Management under the principle of multiple use. Emphasis is on providing a secure domestic supply of energy and strategic minerals and on preserving rangelands for livestock grazing under a permit system. Some of these lands that have not been disturbed by roads are being evaluated for designation as wilderness areas.

Moderately Restricted-Use Lands

National Wildlife Refuges   These 503 refuges (Figure 16-2) and other ranges are managed by the Fish and Wildlife Service. About 24% of this land is designated as wilderness. Most refuges protect habitats and breeding areas for waterfowl and big game to provide a harvestable supply for hunters. A few protect specific endangered species from extinction.

These lands are not officially managed under the principle of multiple use. Nevertheless, sport hunting, trapping, sport and commercial fishing, oil and gas development, mining (old claims only), logging, grazing, some military activities, and farming are permitted as long as the Department of the Interior finds such uses compatible with the purposes of each unit.

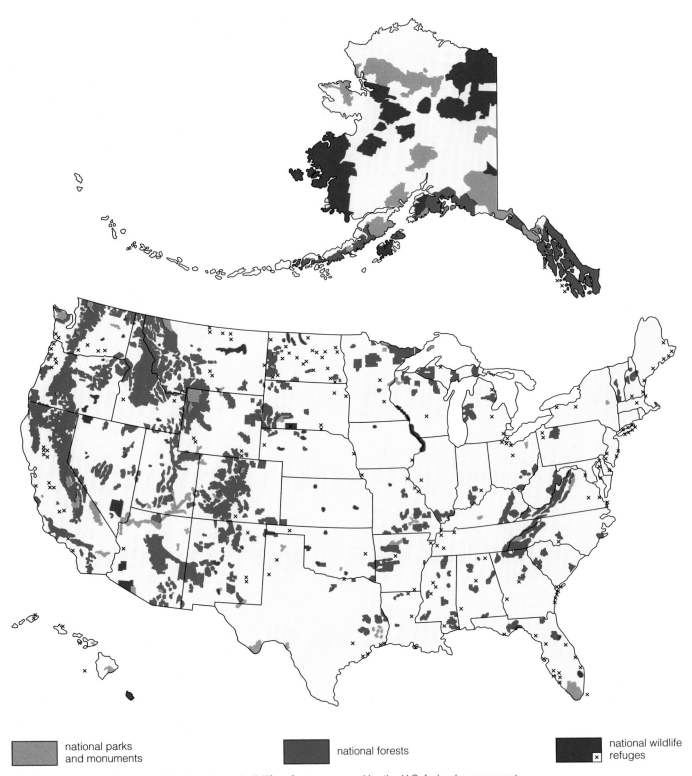

Figure 16-2 National forests, national parks, and wildlife refuges managed by the U.S. federal government. (Data from U.S. Geological Survey)

national parks and monuments

national forests

national wildlife refuges

Q: How much new oil must be discovered and developed to continue using oil at the current rate?

Restricted-Use Lands

National Park System These 367 units include 50 major parks (mostly in the West) and 309 national recreation areas, monuments, memorials, battlefields, historic sites, parkways, trails, rivers, seashores, and lakeshores (Figure 16-2). All are managed by the National Park Service. Its goals are to preserve scenic and unique natural landscapes; preserve and interpret the country's historic and cultural heritage; protect wildlife habitats and wilderness areas within the parks; and provide certain types of recreation.

National parks may be used only for camping, hiking, sport fishing, and boating. Motor vehicles are permitted only on roads. In national recreation areas these same activities plus sport hunting, mining, and oil and gas drilling are allowed. About 49% of National Park System is designated as wilderness.

National Wilderness Preservation System These 474 roadless areas lie within the national parks, national wildlife refuges, and national forests. They are managed by the National Park Service, the Fish and Wildlife Service, and the Forest Service, respectively. These areas are to be preserved essentially untouched "for the use and enjoyment of the American people in such a manner as will leave them unimpaired for future use and enjoyment as wilderness."

Wilderness areas are open only for recreational activities such as hiking, sport fishing, camping, nonmotorized boating, and, in some areas, sport hunting and horseback riding. Roads, logging, grazing, mining, commercial activities, and buildings are banned, except where they predate the wilderness designation. Motorized vehicles and boats are banned except for emergencies, but aircraft may land in Alaskan wilderness.

Between 1970 and 1992, the area of land in all public land systems except the national forests increased significantly (2.7-fold in the National Park System, 3-fold in the National Wildlife Refuge System, and 9-fold in the National Wilderness Preservation System). Most of the additions, by President Carter just before he left office in 1980, are in Alaska. Since then little land has been added to the systems.

Federally administered public lands contain much of the country's commercial timber (40%), grazing land (54%), and energy resources (especially shale oil, uranium, coal, and geothermal energy)—and most of its copper, silver, asbestos, lead, molybdenum, beryllium, phosphate, and potash. For over a century private individuals and corporations have exploited many of these resources, often at below-market prices.

How Should Public Lands Be Managed? Since 1901 conservationists have been split into two major schools of thought on how U.S. public lands should be used and managed. *Preservationists* have sought to protect large areas from mining, logging, and other forms of resource extraction so that these lands can be enjoyed today and passed on unspoiled to future generations. For the most part, preservationists have fought a losing battle. Members of the *wise-use* school see public lands as resources to be used wisely to enhance economic growth and national strength; they advocate efficient and scientific management of these lands to produce sustainable yields of potentially renewable resources (such as trees and grasses for livestock grazing) and to enable extraction of nonrenewable mineral and energy resources.

Despite their basic differences, both schools of early conservationists opposed delivering these public resources into the hands of a few for profit. Both groups have been disappointed. Since 1910 development rights to public lands have routinely been sold at below-market prices to large corporate farms, ranches, mining companies, and timber companies (p. 301). Taxpayers have subsidized this use of resources by absorbing the loss of potential revenue and paying for most of the resulting damage. In recent years, the total costs of subsidies (in public funds spent and taxes and user fees not collected) given to mining, logging, and grazing interests on public lands has exceeded $1 billion per year.

Since the early 1900s a third group of people has attempted to convince the federal government to give or sell most of the nation's public lands to states or to private interests so they can be more fully exploited for economic growth. In the early 1930s, President Herbert Hoover supported this policy, but the Great Depression of the 1930s made the cost of owning such lands unattractive to state governments and private investors.

In the late 1970s a coalition of ranchers, miners, loggers, developers, farmers, politicians, off-road vehicle users, and other users of public lands launched a political campaign known as the *sagebrush rebellion*, with the goal of turning over most western public lands to the states or to private interests. This idea was strongly supported by President Ronald Reagan (a declared "sagebrush rebel") in the 1980s, but it was thwarted by strong opposition in Congress, by public outrage, and by legal challenges from environmental and conservation organizations, whose memberships soared in this period.

In 1988, several hundred local and regional grassroots groups (many financed mostly by developers and by timber, mining, oil, coal, and ranching interests) formed a national coalition with the goals of destroying the environmental movement in the United States, increasing resource exploitation of public lands at low

---

A: A supply equal to the reserve in Saudi Arabia (the world's largest) every 10 years

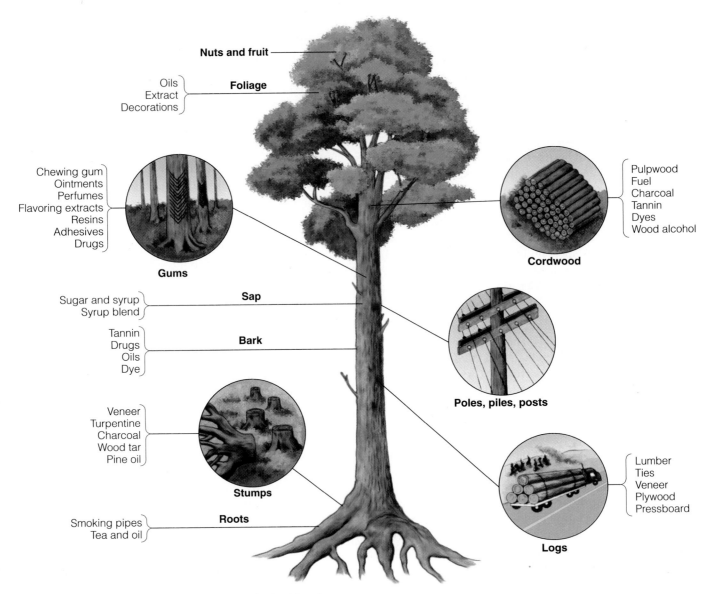

Figure 16-3 Some of the many useful products obtained from trees.

prices, and turning most of these lands over to private enterprise. They try to confuse the public by calling themselves the "wise use movement." This movement is strongly opposed by environmentalists—(including members of the preservationist and original wise-use schools of thought), (Spotlight, p. 187).

16-2 SUSTAINING AND MANAGING FORESTS

Types of Forests There are three general types of forests, depending primarily on climate: tropical, temperate, and polar (Figure 5-5). Since agriculture began about 10,000 years ago, human activities have re-

duced Earth's forest cover by at least one-third, from about 34% of the world's land area to 26%—and only 12% consists of intact forest ecosystems (Figure 5-3). If used sustainably—with the rate of cutting and degradation not exceeding the rate of renewal, and with emphasis on protecting biodiversity—forests are renewable resources. However, forests are disappearing almost everywhere, although losses in Europe and North America have been partially offset by new forest growth.

Old-growth forests are virgin (uncut) forests and second-growth forests that have not been seriously disturbed for several hundred years. They contain massive trees that are hundreds or even thousands of years old. Examples include forests of Douglas fir, western hemlock, giant sequoia (Figure 4-10), and coastal red-

Q: What would happen if oil's harmful effects were included in its price and government subsidies were removed?

woods in the western United States; loblolly pine in the Southeast; and 60% of the world's tropical forests.

Old-growth forests provide ecological niches for a variety of wildlife species (Figure 4-38). These forests also have large numbers of standing dead trees (snags) and fallen logs (boles), which are habitats for a variety of species. Decay of this dead vegetation returns plant nutrients to the soil (Figures 4-17 and 12-18).

Second-growth forests are stands of trees resulting from secondary ecological succession after cutting (Figure 5-42). Most forests in the United States and other temperate regions are second-growth forests that grew back after virgin forests were logged or farms were abandoned. Some second-growth stands have remained undisturbed long enough to become old-growth forests, but many are *tree farms*—managed tracts with emphasis on growing uniform trees of one species (Figure 5-16) that are harvested as soon as they become commercially valuable. About 40% of tropical forests are second-growth forests.

Commercial and Ecological Importance of Forests   Forests give us lumber for housing, biomass for fuelwood, pulp for paper, medicines, and many other products worth more than $300 billion a year (Figure 16-3). Many forestlands are also used for mining, grazing livestock, and recreation. Three countries—the United States, the former Soviet Union, and Canada—supply 53% of the world's commercial timber.

Forested watersheds act as giant sponges, slowing down runoff and absorbing and holding water that recharges springs, streams, and groundwater. Thus they regulate the flow of water from mountain highlands to croplands and urban areas, and they help control soil erosion, reduce flooding, and reduce the amount of sediment washing into streams, lakes, and reservoirs (Figure 11-9 and Connections, p. 266).

Forests also influence local, regional, and global climate. For example, 50–80% of the moisture in the air above tropical forests comes from trees via transpiration and evaporation. If large areas of these lush forests are cleared, average annual precipitation drops, the region's climate gets hotter and drier, and soils become depleted of already-scarce nutrients, baked, and washed away.

Forests are also vital to the global carbon cycle (Figure 4-29) and act as a brake on a possible runaway greenhouse effect (Figures 10-6 and 10-10). And they provide habitats for more wildlife species than any other biome, making them the planet's major reservoir of biodiversity. They also buffer us against noise, absorb air pollutants, and nourish the human spirit.

According to one calculation a typical tree provides $196,250 worth of ecological benefits in the form of oxygen, air cleaning, soil fertility and erosion con-

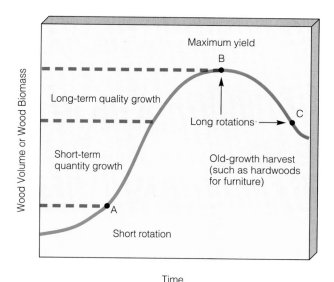

Figure 16-4 Rotation cycle of forest management.

trol, water recycling and humidity control, and wildlife habitats. Sold as timber the same tree is worth only about $590. Even if such estimates are off by a factor of one hundred, the long-term ecological benefits of a tree far exceed its short-term economic benefits. As long as the lasting and renewable ecological benefits of forests are undervalued in the marketplace, we will continue to exploit these forests, and their long-term ecological services, for short-term economic gain.

Types of Forest Management   The total volume of wood produced by a particular stand of forest varies as it goes through different stages of growth and ecological succession (Figure 16-4). If the goal is to produce fuelwood or fiber for paper production in the shortest time, the forest is usually harvested on a short rotation cycle, before the growth rate peaks (point A of Figure 16-4). Harvesting at the peak growth rate gives the maximum yield of wood per unit of time (point B of Figure 16-4). If the goal is high-quality wood for fine furniture or veneer, managers use longer rotations to develop larger, older-growth trees (point C of Figure 16-4).

There are two basic forest management systems: even-aged management and uneven-aged. With **even-aged management**, trees in a given stand are maintained at about the same age and size. Even-aged management begins with the cutting of all or most trees from an area. Then the site is replanted all at once.

Many important tree species that need ample sunlight to grow can be grown only in even-age stands. Most even-age stands in the United States are of mixed tree species and are frequently managed to produce high-quality trees on long rotations. Indeed, the majority of second-growth forests in the United States are even-aged and contain a mixture of tree species.

A: It would be too expensive to use and would be phased out.

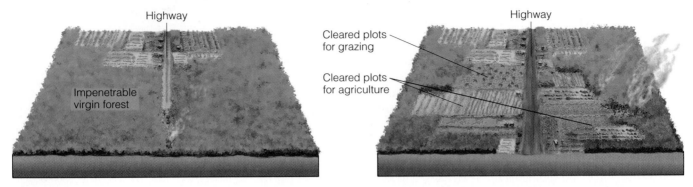

Highway

Impenetrable
virgin forest

Highway

Cleared plots
for grazing

Cleared plots
for agriculture

Figure 16-5 Building roads into previously inaccessible tropical (and other) forests paves the way to destruction, fragmentation, and degradation.

With even-age tree plantations, growers often emphasize single species (monocultures) of fast-growing softwoods to get the best return on their investment in the shortest time. Crossbreeding and genetic engineering can improve both the quality and the quantity of tree-farm wood. Once the trees in such a tree farm (Figure 5-16) reach maturity, the entire stand is harvested, and the area is replanted.

With **uneven-aged management**, trees in a given stand are maintained at many ages and sizes to foster natural regeneration. Here the goals are biological diversity, long-term production of high-quality timber, a reasonable economic return, and multiple use. Mature trees are selectively cut, with clear-cutting used only on small patches of species that benefit from it.

Road Building and Tree Harvesting Logging roads make timber accessible. Unhappily, that's not all. They cause erosion and sediment pollution of waterways, and they expose forests to exotic pests, diseases, and alien wildlife. Their most serious impact, however, is the chain of events they start (Figure 16-5). In many LDCs they open up once-impenetrable forests to farmers, miners, and ranchers (who cut, degrade, damage, or flood large areas of trees), and to hunters (who can deplete wild animal species). A network of roads can lead to severe habitat fragmentation and loss of biodiversity (Figure 16-6).

Once loggers can reach a forest, they use various methods for harvesting the trees (Figure 16-7 and Spotlight, p. 416). The method used depends on the tree species being harvested, the nature of the site, whether the stand consists of trees of the same age or of mixed ages, and the objectives and resources of the owner. Some tree species, for example, grow best in full or moderate sunlight in large clearings. Such sun-loving species are usually harvested by shelterwood cutting, seed-tree cutting, or clear-cutting (Figure 16-7). The problem is that timber companies have a built-in economic incentive to use large-scale clear-cutting,

NASA/Goddard Space Flight Center

Figure 16-6 Computer-enhanced satellite image of a 1,300-square-kilometer (500-square-mile) area of Mount Hood National Forest in Oregon in 1991. Old-growth forest patches are dark red; clear-cut old-growth forest regenerating as tree farms are lighter red; and recently clear-cut areas are blue-green and white. This area has been extensively cleared and fragmented into vulnerable patches.

often on species that could be harvested by less environmentally destructive methods. While some logging operations can be destructive, many loggers carry out their work to minimize environmental damage. They also help maintain diverse and bountiful forests by

Q: How long will proven reserves of natural gas last at current consumption rates?

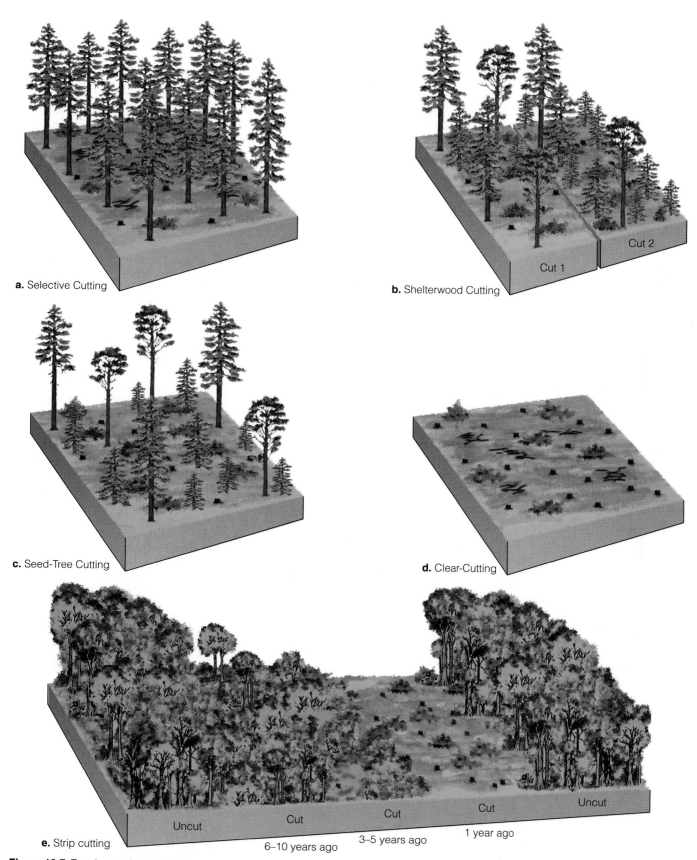

a. Selective Cutting

b. Shelterwood Cutting

Cut 1

Cut 2

c. Seed-Tree Cutting

d. Clear-Cutting

e. Strip cutting

Uncut

Cut
6–10 years ago

Cut
3–5 years ago

Cut
1 year ago

Uncut

Figure 16-7 Tree harvesting methods.

A: About 80 years for the world and 60 years for the United States

Ways to Harvest Trees

Selective Cutting

Intermediate-aged or mature trees in an uneven-aged forest are cut singly or in small groups, creating gaps not much larger than those from natural treefall (Figure 16-7a). This reduces crowding, encourages the growth of younger trees, maintains an uneven-aged stand with trees of different species and sizes, and allows trees to grow back naturally. If done properly it also helps protect the site from soil erosion and wind damage. However, it's costly unless the trees removed are quite valuable; and maintaining a good mixture of tree ages, species, and sizes takes planning and skill. An unsound type of selective cutting is *high grading*, or *creaming*, which removes the most valuable trees. This practice, common in many tropical forests, ends up injuring one-third to two-thirds of the remaining trees when they are knocked over by logging equipment and when the large target trees are felled and removed.

Shelterwood Cutting

All mature trees are removed in a series of cuttings stretched out over about 10 years (Figure 16-7b). This technique can be applied to even- or uneven-aged stands. The first cut removes most mature canopy trees, unwanted tree species, and diseased, defective, and dying trees. A second cut removes more canopy trees. A third cut removes the remaining mature trees, and the even-aged stand of young trees then grows to maturity. This method leaves a fairly natural-looking forest that can be used for a variety of purposes. It also helps reduce soil erosion and provides a good habitat for wildlife.

Seed-Tree Cutting

Nearly all of a stand's trees are harvested in one cutting, leaving a few uniformly distributed seed-producing trees to regenerate a new crop (Figure 16-7c). After the new trees have become established, the seed trees may be harvested. By allowing several species to grow at once, seed-tree cutting leaves an aesthetically pleasing forest, useful for recreation, deer hunting, erosion control, and wildlife conservation. Leaving the best trees for seed can also lead to genetic improvement in the new stand.

Clear-Cutting

All trees from a given area are removed in a single cutting. The clear-cut area may be a whole stand (Figure 16-7d), a strip, or a series of patches. After all trees are cut, the site is reforested naturally from seed released by the harvest, or artificially as foresters broadcast seed over the site or plant seedlings raised in a nursery. It requires much less skill and planning than other harvesting methods and usually gives timber companies the maximum economic return. On the negative side, clear-cutting leaves ugly, unnatural forest openings (Figure 16-7d) and eliminates any potential recreational value. It also destroys wildlife habitats and thus reduces biodiversity. Environmental degradation from clear-cutting above ground is obvious, but equally serious damage goes on underground in the soil, from the loss of fungi, worms, bacteria, and other microbes that nourish plants with water and minerals and help protect them from disease (Figures 12-18 and 12-19). Furthermore, trees in stands bordering clear-cut areas are more vulnerable to windstorms, and large-scale clear-cutting on steep slopes leads to severe soil erosion, sediment water pollution, and flooding (Figure 11-9). Clear-cutting—if done carefully and responsibly—is often the best way to harvest tree farms and stands of some tree species that require full or moderate sunlight for growth. However, for economic reasons it is often done irresponsibly and used on species that don't require this method.

Stripcutting

This variation of clear-cutting can allow a sustainable timber yield from forests without the widespread destruction often associated with conventional clear-cutting. A strip of trees is clear-cut along the contour of the land, with the corridor narrow enough to allow natural regeneration within a few years (Figure 16-7e). After regeneration, another strip is cut above the first, and so on. This allows a forest area to be clear-cut in narrow strips over several decades with minimal damage.

---

planting trees on land previously used for agricultural purposes.

Protecting Forests from Pathogens and Insects

In a healthy and diverse forest, diseases and insect populations are usually controlled by interactions with other species. Thus, they rarely get out of control and seldom destroy many trees. However, a biologically simplified tree farm is vulnerable to attack by pathogens, especially parasitic fungi (such as chestnut blight and Dutch elm disease) and insects (such as bark beetles, the gypsy moth, and spruce budworm).

Biodiversity is the best and cheapest defense against tree diseases and insects. Other methods include banning imported timber that might introduce harmful new parasites, removing infected trees or clear-

Q: How long will the world's proven reserves of coal last at current consumption rates?

a. Surface fire

Figure 16-8 Two types of forest fires. Occasional surface fires, like this one in Florida's Ocala National Forest **(a)** burn deadwood, undergrowth, and leaf litter. This helps prevent more destructive crown fires in some types of forests, such as the one **(b)** in Yellowstone National Park during the summer of 1988. Wildlife that can't escape are killed and wildlife habitats destroyed. Severe erosion can also occur.

b. Crown fire

cutting infected areas and burning all debris, treating diseased trees with antibiotics, developing disease-resistant tree species, applying pesticides (Section 15-1), and using integrated pest management (Section 15-5).

A New Look at Fire   Occasional natural fires set by lightning are an important part of the ecological cycle of many forests. Some species actually need occasional fires. For example, the seeds of the giant sequoia (Figure 4-10) and the jack pine are released or germinate only after being exposed to intense heat.

Forest ecosystems can be affected by different types of fires (Figure 16-8). Some, called *surface fires*, usually burn only undergrowth and leaf litter on the forest floor (Figure 16-8a). These fires kill seedlings and small trees but spare most mature trees. Most wild animals can escape.

In forests where ground litter accumulates rapidly, a surface fire every five or so years burns away flammable material and helps prevent more destructive fires. Surface fires also release valuable mineral nutrients tied up in slowly decomposing litter and undergrowth, increase the activity of underground nitrogen-fixing bacteria, stimulate the germination of certain tree seeds, and help control pathogens and insects. Some wildlife species—deer, moose, elk, muskrat, woodcock, and quail, for example—depend on occasional surface fires to maintain their habitats and to

provide food in the form of vegetation that sprouts after fires.

Some extremely hot fires, called *crown fires* (Figure 16-8b), may start on the ground but eventually burn whole trees and leap from treetop to treetop. They usually occur in forests where all fire has been prevented for several decades, allowing dead wood, leaves, and other flammable ground litter to build up. These rapidly burning fires can destroy most vegetation, kill wildlife, and lead to accelerated erosion.

Sometimes surface fires go underground and burn partially decayed leaves or peat (Figure 12-7). Such *ground fires* most common in northern peat bogs, may smolder for days or weeks before being detected and are difficult to extinguish.

Protecting forest resources from fire involves four approaches: prevention, prescribed burning, presuppression, and suppression. Methods of *prevention* of forest fires include requiring burning permits and closing all or parts of a forest to travel and camping during periods of drought and high fire danger. The most important—and the cheapest—prevention method, however, is education. The Smokey-the-Bear educational campaign of the Forest Service and the National Advertising Council, for example, has prevented countless forest fires in the United States, saving many lives and avoiding billions of dollars in losses. However, ecologists contend that by allowing litter buildup in

A: About 220 years (65 years if use increases 2% a year)

some forests, prevention increases the likelihood of highly destructive crown fires. For that reason many fires in national parks and wilderness areas are now allowed to burn as part of the natural ecological cycle of succession and regeneration (Figure 5-42).

Prescribed burning in some forests can effectively prevent crown fires by reducing litter buildup. These surface fires are also used to control outbreaks of tree disease and pests. These fires are started only by well-trained personnel when weather and forest conditions are ideal for control and proper intensity of burning. Prescribed fires are also timed to keep levels of air pollution as low as possible.

Presuppression involves trying to detect a fire at an early stage and reducing its spread and damage. For example, helicopters and small airplanes can be used to detect small fires before they get out of control. And firefighters can clear vegetation to form firebreaks to reduce the spread of fires.

Once a wildfire starts, fire fighters have a number of methods for fire *suppression*. They use specially designed bulldozers and breaker plows to establish firebreaks. They pump water onto the fire from tank trucks and drop water or fire-retarding chemicals from aircraft. And controlled backfires can be set to create burned areas to confine fires.

Protecting Forests from Air Pollution and Climate Change   Forests at high elevations and forests downwind from urban and industrial centers are exposed to a variety of air pollutants that can harm trees, especially conifers. Besides doing direct harm, prolonged exposure to multiple air pollutants makes trees much more vulnerable to drought, diseases, and insects (Figure 9-9).The solution is to slash emissions of the offending pollutants from coal-burning power plants, industrial plants, and motor vehicles (Section 9-5).

In coming decades an even greater threat to forests, especially temperate and boreal forests, may come from regional climate changes brought about by projected global warming (Figure 10-10). Possible ways to deal with projected global warming are discussed in Section 10-3.

Solutions: Sustainable-Earth Forestry   To timber companies, sustainable forestry means getting a sustainable yield of commercial timber in as short a time as possible. This often means clearing diverse forests and replacing them with intensively managed tree farms (Figure 5-16).

To environmentalists this does not qualify as sustainable use (Pro/Con, at right). They call for widespread use of sustainable-Earth forestry, which recog-

PRO/CON

Monocultures or Mixed Cultures?

Most commercial foresters believe that tree farms are the best way to meet the increasing demand for wood and wood products and to increase short-term profits for timber companies. They contend that intensively managed forests can be harvested and regenerated in ways that conserve these potentially renewable resources for future generations. In addition, high yields from tree farms can reduce pressure to clear old-growth forests. Planting a tree farm is also the quickest way to reforest damaged land, preventing soil erosion and desertification.

However, the German experience between 1840 and 1918 shows that widespread monoculture forestry is not the way to go. Around 1840 German foresters decided to clear-cut diverse natural forests and replace them with pine and spruce plantations to increase the output of wood per hectare. These trees were harvested every 20–30 years to provide low-quality timber and pulp. Deciduous hardwood and true fir species nearly became extinct.

Yields increased for a time, but the monoculture plantations depleted the soil of plant nutrients. After the second or third generation the yield of the pine and spruce stands dropped and the general quality of the wood fell off. Many trees were stunted or fell victim to pests, diseases, or wind.

After 1918 German foresters shifted to a more natural type of forestry management. Monoculture stands were replaced by even- and uneven-aged mixtures of commercially valuable species. Instead of clear-cutting, loggers selectively harvested only certain economically important trees in a stand. The result was increased timber production and an improvement in soil quality. Today German foresters are appalled to see the United States and other countries making the same mistakes their predecessors made in sacrificing long-term, sustainable productivity for short-term economic gain.

nizes that a biologically diverse forest ecosystem is the best protection against erosion, flooding, sediment water pollution, loss of biodiversity, and tree loss from fire, wind, insects, and diseases. Such sustainable-Earth forest management emphasizes:

- *Recycling more paper to reduce the harvest of pulp-wood trees* (Section 13-3)

Q: How long will proven reserves of coal in the United States last at current consumption rates?

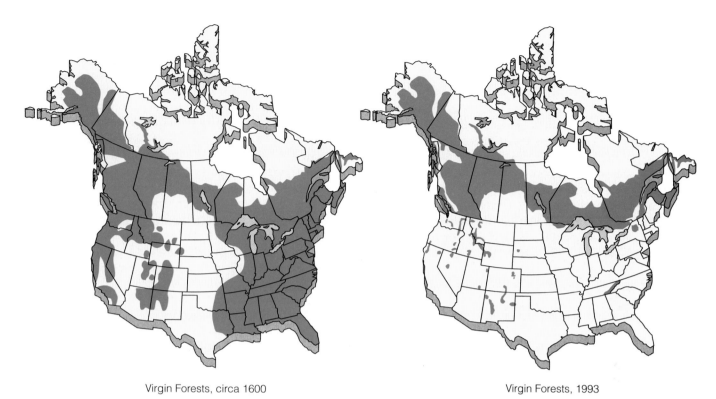

Virgin Forests, circa 1600   Virgin Forests, 1993

Figure 16-9 Vanishing old-growth forests in the United States and Canada. Since about 1600, an estimated 90–95% of the virgin forests that once covered much of the lower 48 states have been cleared away; most of the remaining old-growth forests in the lower 48 states are on public lands. In the Pacific Northwest about 80% of this forestland in 12 national forests is slated for logging. Some analysts believe that this estimate is too high. Regardless of the exact numbers there is no dispute that there has been a massive loss of virgin forests in the United States during the past 395 years. About 60% of old-growth forests in western Canada have been cleared, and much of what remains is slated for cutting. (Data from the Wilderness Society and the U.S. Forest Service, and *Atlas Historique du Canada*, Vol. 1)

- *Growing more timber on long rotations*, generally about 100–200 years, depending on the species and the soil quality (point C, Figure 16-4)

- *Practicing selective cutting of individual trees or small groups of most tree species* (Figure 16-7a)

- *Minimizing fragmentation of remaining larger blocks of forest* (Figure 16-6)

- *Using road building and logging methods that minimize soil erosion and compaction*

- *Practicing stripcutting (Figure 16-7e) instead of conventional clear-cutting and banning all clear-cutting on land that slopes more than 15–20°*

- *Leaving standing dead trees (snags) and fallen timber (boles) to maintain diverse wildlife habitats and to be recycled as nutrients* (Figure 4-17)

Sustainable-Earth forestry does not mean that tree farms or even-aged management should never be used, but it does mean that their use should be limited, especially in old-growth forests.

16-3 FOREST MANAGEMENT AND CONSERVATION IN THE UNITED STATES

U.S. Forest Resources   Though forests cover about one-third of the lower 48 states (Figures 5-3 and 5-4), most of their virgin forests have been cut (Figure 16-9), and what remains is threatened. U.S. forests provide habitats for more than 80% of the country's wildlife species and are a prime setting for outdoor recreation.

Nearly two-thirds of this forestland is capable of producing commercially valuable timber. Since 1950 the United States has met the demand for wood and wood products without serious depletion of its commercial forestlands.

For three centuries the United States was self-sufficient in wood. Since 1940, however, the country has been a net importer of wood, and the gap is widening. This is happening even though the United States cuts more wood than any other country and is the world's largest timber exporter. The primary reason for this import–export gap is not higher per capita consumption of wood (which is half of what it was in 1900), but growth in population (which has tripled since 1900).

Managing U.S. National Forests   About 22% of the commercial forest acreage in the United States is located within the 156 national forests managed by the U.S. Forest Service (Figure 16-2). These forestlands serve as grazing lands for more than 3 million cattle and sheep each year, support multimillion-dollar mining operations, contain a network of roads almost nine times longer than the entire U.S. interstate highway system (and long enough to circle the earth 14 times), receive more recreational visits than any other federal public lands, and supply about 14% of the nation's timber.

The Forest Service is required by law to manage national forests according to the principles of sustained yield and multiple use, a nearly impossible task. For example, timber company officials complain that they aren't allowed to buy and cut enough timber on public lands, especially in remaining old-growth forests in California and the Pacific Northwest (Figures 16-6 and 16-9).

Environmentalists, on the other hand, charge that the Forest Service has allowed timber harvesting to become the dominant use in most national forests. They point out that 70% of the Forest Service budget is devoted directly or indirectly to the sale of timber, and that at current rates of timber removal—the equivalent of about 129 football fields a day—all unprotected ancient forests on public lands in western Washington and Oregon (Figures 16-6 and 16-9) will be gone by the year 2023. A drive through most national forests might give the illusion of traveling through a primeval forest because the Forest Service leaves thin buffers of uncut trees—called "beauty strips"—along the roads. But a flight over these forests will reveal that many of the trees have been clear-cut (Figure 16-6).

The agency keeps most of the money it makes on timber sales, while any losses are passed on to taxpayers. Because logging increases its budget, the Forest Service has a powerful built-in incentive to encourage timber sales. Local county officials also exert tremendous pressure on both members of Congress and Forest Service officials to keep the timber harvests high because counties get 25% of the gross receipts from national forests within their boundaries.

Environmentalists and the General Accounting Office have accused the Forest Service of poor financial management of public forests. By law, the Forest Service must sell timber for no less than the cost of reforesting the land from which it was harvested. However, the cost of access roads is not included in this price but is provided as a subsidy to logging companies. Log-

ging companies also get the timber itself for less than they would normally pay a private landowner.

Studies have shown that between 1978 and 1992, national forests lost at least $4.2 billion (some sources say $7 billion) from timber sales. With interest, this added at least $5.9 billion to the national debt. In most years, the Forest Service loses $200–$250 million ($352 million in 1992, according to a Wilderness Society estimate) because normally only 17 of the 120 national forests open to logging make money. Timber company officials , however, argue that being able to get timber from federal lands fairly cheaply benefits taxpayers by keeping lumber prices down. However, below-cost timber prices discourage investments to produce additional timber in private lands and put more pressure for timber cutting in national and state forests.

Instead of providing economic stability for local industries and communities, logging in national forests tends to have the opposite effect. Communities relying on national-forest timber experience boom-and-bust cycles as the timber is depleted.

The Controversy over Old-Growth Deforestation in the Pacific Northwest   To officials of timber companies, the giant living trees and rotting dead trees in old-growth forests are valuable resources that should be harvested for profit and to provide jobs, not locked up to please environmentalists. They point out that the timber industry annually pumps millions of dollars into the Pacific Northwest's economy and provides jobs for about 100,000 loggers and millworkers. Timber officials claim that protecting large areas of remaining old-growth forests on public lands will cost as many as 60,000 jobs and hurt the economy of logging and milling towns throughout the Pacific Northwest; by contrast, the 1993 Forest Ecosystem Management Assessment Team estimated the job loss at 6,000–13,000.

To environmentalists, remaining ancient forests on the nation's public lands are a treasure whose ecological, scientific, aesthetic, and recreational values far exceed the economic value of cutting them down for short-term economic gain. The fate of these forests is a national issue because these forests are owned by all U.S. citizens, not just the timber industry or the residents of a region. It is also a global issue because these forests are important reservoirs of irreplaceable biodiversity. Also, the way the United States treats its few remaining old-growth forests sets a precedent for other nations' (especially LDCs') treatment of their old-growth forests, wetlands, coral reefs, and other ecosystems.

Environmentalists point out that protecting old-growth forests on public lands is not the main cause of

Q: What would happen if coal's harmful effects were included in its market price and government subsidies were removed?

past and projected job losses in the timber industry in this region. Other factors include automation, export of raw logs overseas (depriving U.S. millworkers of jobs while providing jobs for millworkers in Japan, China, and South Korea), and timber imports from Canada. Loggers, millworkers, and store owners who live in these communities are caught in the middle, pawns in a high-stakes game of corporate profit.

The threatened northern spotted owl (Figure 16-10) has become a symbol in the struggle between environmentalists and timber company officials over the fate of unprotected old-growth forests on public lands in the Pacific Northwest This owl lives almost exclusively in 200-year-old Douglas fir forests in western Oregon and Washington and northern California, mostly in 17 national forests and 5 Bureau of Land Management parcels. The species is vulnerable to extinction because of its low reproductive rates and the low survival rates of juveniles through their first five years. Only 2,000–3,600 pairs remain.

In July 1990 the U.S. Fish and Wildlife Service added the spotted owl to the federal list of threatened species. This requires that its habitat be protected from logging or other practices that would decrease its chances of survival.

This decision is being vigorously fought by the timber industry. They hope to persuade Congress to revise the Endangered Species Act to allow for economic considerations. They also contend that the owls do not require old-growth forest and can adapt to younger second-growth forests.

A major problem is that the media and many politicians and citizens discuss this and other complex environmental problems on a simplistic we-versus-them basis. This disguises the fact that *the controversy over cutting of ancient forests in the Pacific Northwest isn't an owl-versus-jobs issue.* The owl and other threatened species in these forests are merely symbols of the broader clash between timber company owners who want to clear-cut most remaining old-growth stands in the national forests and environmentalists who want to protect them—or at least allow only sustainable harvesting in some areas using selection cutting.

The Endangered Species Act is the best (and only) tool environmentalists have to help them achieve this broader goal of protecting biodiversity by protecting fast-disappearing old-growth forest habitats. The truth is that both owls and humans are utterly dependent on healthy, diverse ecosystems. Timber jobs are disappearing in the Northwest for the same reasons the owls are: The ancient forests they depend on are almost gone. For example, between 1978 and 1987 the seven largest U.S. lumber and wood products companies reduced their timber harvesting in the Northwest by 35%, while increasing their production in the South by 121%. As pri-

Figure 16-10 The threatened northern spotted owl lives secretively in old-growth forests of the Pacific Northwest. Environmentalists have used the Endangered Species Act as a tool to help achieve the much wider goal of preserving biodiversity by preventing further destruction of America's endangered old-growth forests.

vate and public forests in the South are depleted or declared off limits, timber companies are increasing operations in the Northeast.

Most environmentalists believe that supporting sustainable use of public forests—based on allowing limited selective cutting, replanting and restoring cleared areas, diversifying the economy, and encouraging tourism—is the best way logging-based communities can remain economically and ecologically healthy.

Another part of the solution to this dilemma is to recognize that the owls, loggers, and environmentalists are not the problem. *We are all the problem.* We buy wood that is harvested from old-growth forests at such a low price that sustainable logging is not economically feasible. The marketplace is not indicating the real costs of destroying and degrading our forests because we don't insist that the prices of wood and wood products include their full short- and long-term environmental and social costs. Until we change the market system to include these real costs, we will continue to deplete Earth capital and eliminate potentially sustainable jobs.

Solutions: Reducing Consumption and Greening Production of Wood and Paper Products

The greatest threats to U.S. National Forests and most forests throughout the world are rapid growth in demand for forest products because of a combination of growing population and affluence (Figure 1-12) and wasteful practices in making and using wood and

A: It would be too expensive to use and would be phased out

Kenaf: A Substitute for Paper

Could we do away with the use of trees to make paper? Not completely, but a woody annual plant called kenaf could sharply reduce the use of trees for papermaking.

The tall, bamboo-like stalks of this plant grow to heights of 4.6 meters (15 feet). Its yield per acre of fiber suitable to make paper is about the same as pine trees. However, kenaf takes only 150 days to mature, compared to 30-60 years for pine trees.

Few herbicides are needed to cultivate kenaf because it grows faster than most weeds. And not many insecticides are needed be-cause its outer fiber covering is nearly insect-proof.

It doesn't take as many chemicals to break down kenaf into fibers used for paper because its fibers are not dense like wood. Another environmental plus is that its fibers can be bleached in a single-stage process using hydrogen peroxide, rather than chlorine bleach. No chlorine means no toxic dioxin residues in pulp mill effluent (Section 13-5).

In many ways, kenaf paper is better than tree paper. Printers say it runs on presses better than regular paper. And kenaf newspaper is also stronger than tree paper, requires less ink, doesn't yellow, has less print rub-off, and it has an at-tractive off-white color. Kenaf fibers can also be used to make a soil potting mix and an oil spill clean-up product, which absorbs nine times its weight in oil and costs less than other oil-absorbing materials.

There are some drawbacks, but they can be overcome. Because kenaf paper is not slick, its high friction makes it not as effective as conventional paper in copiers and fax machines. Some type of coating could eliminate this problem. Currently kenaf paper costs twice as much as either virgin or recycled paper stock, but as demand increases and producers cut production costs its price should come down.

---

paper materials. According to the Worldwatch Institute and forestry analysts, up to 60% of the wood now being consumed in the United States could be saved through the following conservation and recycling measures:

- *Increasing wood and paper product manufacturing efficiency using existing technology.*

- *Reducing construction waste to levels found in Finland.*

- *Reducing the use of disposable products and conserving paper to the levels found in Norway.*

- *Doubling the use of recycled paper (Section 13-3).*

- *Increasing the demand for paper products using the maximum feasible content of fibers made from postconsumer paper waste (Spotlight, p. 346).*

- *Increasing the use of lumber substitutes* such as boards made from recycled plastic, panels made from ryegrass straw, houses made from straw (Solutions, p. 498), walls made from waste wood fiber mixed with some sand and cement, and Spaceboard paneling made from recycled paper as a substitute for plywood.

- *Greatly increasing the use of kenaf instead of trees to make paper (Solutions, above).*

Solutions: Reforming Federal Forest Management
Some forestry experts and environmentalists have suggested several ways to reduce overexploitation of publicly owned timber resources and provide true multiple use of national forests as required by law:

- *Adopt a national forest management policy that gives first priority to protecting biodiversity, with emphasis on protecting remaining old-growth forests.*

- *Scrap existing individual Forest Service management plans for each national forest and prepare ecosystem plans for whole regions that integrate the uses of all public lands in each region.*

- *Urge elected officials to ban all timber cutting in national forests and to fund the Forest Service completely from recreational user fees.* The Forest Service estimates that recreational user fees (as low as $3 per day) would generate three times what it earns from timber sales.

- *Until a total ban is enacted, reduce the current annual harvest of timber from national forests by one-half,* instead of doubling it as proposed by the timber industry.

- *Preserve at least 50% of remaining old-growth timber in any national forest.*

- *Allow individuals or groups to buy conservation easements that prevent timber harvesting on designated areas of public old-growth forests.* In such conservation-for-tax-relief swaps, purchasers would be allowed tax breaks for the funds they put up.

- *Build no more roads in national forests.*

Q: How much of the world's electricity is supplied by nuclear power?

- *Require that timber from national forests be sold at a price that includes the costs of road building, site preparation, and site regeneration, and that all timber sales in national forests yield a profit for taxpayers.*

- *Don't use money from timber sales in national forests to supplement the Forest Service budget,* which encourages overexploitation of timber resources.

- *Eliminate the provision that returns 25% of gross receipts from national forests to counties containing the forests, or base such returns on recreational user-fee receipts only.*

- *Require use of sustainable-forestry methods* (p. 418).

- *Close loopholes in and strictly enforce the ban on exporting unprocessed logs (but not processed lumber)* from the Pacific Northwest, and expand this ban to the entire United States in order to preserve lumber mill jobs.

- *Build new sawmills or retrofit existing ones so they can cut smaller trees* and not be dependent on harvesting large, old-growth trees whose supply will be diminishing. Several sawmill owners in Oregon and Washington are prospering and have created new jobs after investing in state-of-the-art mills for cutting small diameter logs. Instead of relying on clear-cutting, such mills can get most of the wood they need by selective thinning of forests and tree farms. This improves forest health and reduces potential damage from fires and insects.

- *Give logging and milling towns grants and interest-free loans* to spur economic diversification and to help retrain displaced loggers and millworkers.

- *Provide dislocated timber workers and their families with financial assistance* for housing, job retraining, job searching, health insurance, and extended unemployment benefits.

- *Provide federal funds for extensive reforestation and restoration of cut and degraded areas of national forests* to help renew these ecosystems and furnish alternative jobs for unemployed loggers and millworkers.

Timber company officials vigorously oppose most of these proposals, claiming they would cause economic disruption in their industry and in logging communities and raise the price of timber for consumers. Environmentalists argue that taxpayers are paying higher prices than they think for timber when their tax dollars are used to subsidize logging in national forests. They contend that including these and the harmful environmental costs of unsustainable timber cutting would promote more sustainable use of these resources, help sustain logging communities, and protect biodiversity.

In April, 1993, the Clinton administration held a forest summit meeting in Portland, Oregon, to hear both sides of the old-growth controversy. Several months later the administration released a compromise plan that pleased almost no one. Timber company officials said it was too favorable to environmentalists and environmentalists complained that it favored timber interests.

Environmentalists praised the proposals that would set up a reserve system that includes many ancient-forest stands, attempt to protect watersheds and riparian systems, eliminate $100 million in tax loopholes that encourage U.S. exports of raw logs (a leading cause of mill closings and job losses in the timber industry), and provide $1.2 billion in economic assistance to timber-dependent communities. However, environmentalists criticized the plan for:

- allowing thinning and salvaging cuts (which past experience shows can be used as a disguise to harvest timber that should not be cut) in the proposed riparian and old-growth reserves instead of banning all logging in the reserves.

- cutting back on salmon protection (with the Pacific Northwest salmon fisheries providing more jobs than the region's timber industry) by narrowing the buffer zones around salmon streams where logging is prohibited.

- allowing an annual level of timber cutting in the area's national forests that is considered unsustainable—a claim disputed by the timber industry.

16-4 COMBATING TROPICAL DEFORESTATION AND THE FUELWOOD CRISIS

The Loss of Tropical Forests   Tropical forests, which cover about 6% of the earth's land area, grow near the equator in Latin America, Africa, and Asia (Figure 5-3). They include rainforests, moist deciduous forests, dry and very dry deciduous forests, and forests on hills and mountains. About 56% of the world's tropical forests have already been cleared or damaged. According to a 1993 UN Food and Agricultural Organization (FAO) report, the annual rate of loss rose almost 40% between 1980 and 1990 (Figures 16-11 and 16-12). Satellite scans and ground-level surveys indicate that these forests are vanishing rapidly, at a rate of at least 154,000 square kilometers (59,000 square miles) per year—equivalent to about 34 city

A: About 17% (5% of the world's total commercial energy use)

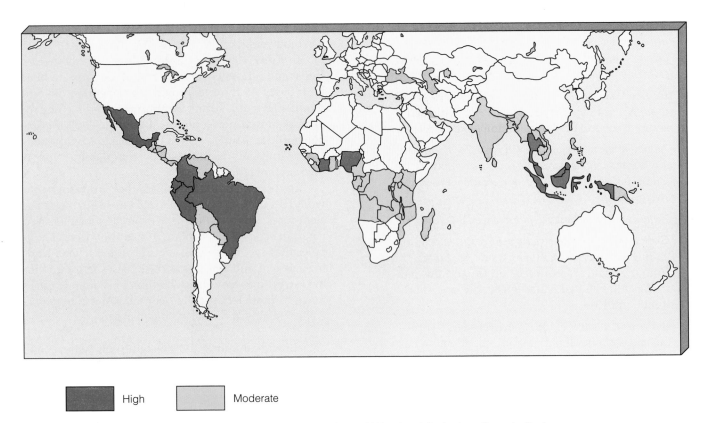

High   Moderate

Figure 16-11 Countries rapidly losing their tropical forests. (Data from UN Food and Agriculture Organization)

Figure 16-12 Photo taken from the space shuttle *Discovery* in September 1988 shows smoke from fires burning cleared areas of tropical rain forests in South America's Amazon basin. The white "clouds" are plumes of smoke, covering an area about three times the size of Texas. These fires may have accounted for 10% of all carbon dioxide entering the atmosphere in 1988. Mostly because of international pressure and an economic slowdown, the amount of forestland cleared in Brazil has dropped somewhat since 1990. Even so, Brazil accounts for a larger area of deforestation each year than any of the other 62 tropical countries.

NASA

Q: Has a scientifically and politically acceptable method for the long-term disposal of nuclear waste been developed?

Scenes from the year 2000:
The last elephant hides in the
last rain forest.

blocks per minute, or almost two football fields per second. It's estimated that an equivalent area of these forests is damaged every year. Such estimates are crude and could be off by as much as a factor of two either way. Regardless, past and current tropical deforestation represent a massive loss of biodiversity.

About 40% of this deforestation is taking place in South America (especially in the vast Amazon Basin, Figure 6-16). However, the rates of tropical deforestation in Southeast Asia and in Central America are about 2.7 times higher than in South America. Haiti has lost an estimated 98% of its original forest cover, the Philippines 97%, and Madagascar 84%.

Reforestation in the tropics scarcely deserves the name, with only one tree planted for every 10 trees cut; in Africa the ratio is 1 to 29. If the current rate of loss continues, all remaining tropical forests (except for a few preserved but still vulnerable patches) will be gone within 30 to 50 years, and much sooner in some areas (see cartoon).

Environmentalists consider the plight of tropical forests to be one of the world's most serious environmental problems (Guest Essay, p. 446). These forests are home to at least 50% (some estimate 90%) of Earth's total stock of species (Figures 1-10 and 16-13)—most of them still unknown and unnamed.

Why Should We Care About Tropical Forests?

Tropical forests touch the daily lives of everyone on Earth through the products and ecological services they provide. These forests supply half of the world's annual harvest of hardwood and hundreds of food products (including coffee, tea, cocoa, spices, nuts, chocolate, and tropical fruits), and materials such as natural latex rubber (Figure 16-14), resins, dyes, and essential oils that can be harvested sustainably. A 1988

Figure 16-13 Tropical forests are the planet's largest storehouse of biological diversity. Two of the species found there are the red uakari monkey (top) from the Peruvian portion of the Amazon Basin and the keel-billed toucan from Belize in Central America (bottom). Most tropical forest species have specialized niches (Figure 4-38). This makes them highly vulnerable to extinction when their forest habitats are cleared or damaged.

Figure 16-14 Rubber tapping is a potentially sustainable use of tropical forests. A diagonal groove is cut through the bark of a live rubber tree, and milky liquid latex trickles into a collecting cup. The latex, which is about 30% rubber, is processed to make rubber, and the scars heal without killing the tree. About 300,000 rubber tappers living in the Amazon Basin also gather Brazil nuts, fruits, and fibers in the forests, and they cultivate small plots near their homes. This photo, taken in 1987, shows Chico Mendes, leader of 70,000 Amazon rubber tappers in his Brazilian home state of Acre, making a cut on a rubber tree. On December 22, 1988, he was murdered near his home in Xapuri by ranchers who opposed his internationally known efforts to protect Brazil's rain forests from land-clearing, colonization, and other destructive forms of development. In 1991 his assassins were convicted of murder, but they escaped from jail in 1993.

Figure 16-15 Rosy periwinkle found in the threatened tropical forests of Madagascar. Two compounds extracted from this plant have been used to cure most victims of two deadly cancers—lymphocytic leukemia (which used to be an almost certain death sentence for children) and Hodgkin's disease (mostly affecting young adults). Annual income from the sale of these two drugs exceeds $80 million—none of which is returned to Madagascar. Only a tiny fraction of tropical plants has been studied for such potential uses, and many will become extinct before we can study them.

study by a team of scientists showed that sustainable harvesting of such nonwood products as nuts, fruits, herbs, spices, oils, medicines, and latex rubber in Amazon rain forests over 50 years would generate twice as much revenue per hectare as timber production and three times as much as cattle ranching.

The active ingredients for 25% of the world's prescription drugs are substances derived from plants, most of which grow in tropical rain forests. Some tropical plants are used directly as medicines, especially in LDCs. However, in MDCs the medically active ingredients in such plants are identified and then synthesized using modern chemistry. This is usually cheaper and reduces the need to deplete tropical forests of such plants.

Such drugs are used in birth control pills, tranquilizers, muscle relaxers, and life-saving drugs for treating malaria, leukemia and Hodgkin's disease (Figure 16-15), testicular and lung cancer, heart disease, high blood pressure, multiple sclerosis, venereal warts, and many other diseases. Commercial sales of drugs with active ingredients derived from forests total an estimated $100 billion per year worldwide and $15 billion per year in the United States. Seventy percent of the 3,000 plants identified by the National Cancer Institute (NCI) as sources of cancer-fighting chemicals come from tropical forests. While you are reading this page, a plant species that could cure a type of cancer, AIDS, or some other deadly disease might be wiped out forever. And some tropical tree species can be used for many purposes (Spotlight, next page).

Most of the original strains of rice, wheat, and corn that supply more than half of the world's food were developed from wild tropical plants. Botanists believe that tens of thousands of strains of plants with potential food value await discovery in tropical forests.

Despite this immense potential, less than 1% of the estimated 125,000 flowering plant species in the world's tropical forests (and less than 3% of the world's 220,000 such species) have been examined closely for their possible use as human resources. Biol-

Q: How many sites in the United States are contaminated with radioactive materials?

A Tree for All Seasons

What if there were a single plant that could quickly reforest bare land, provide fuelwood and lumber in dry areas, produce alternatives to toxic pesticides, treat numerous diseases, and help control population growth? There is—the neem tree, a relative of mahogany. This remarkable tropical species, native to India and Burma, is ideal for reforestation because it can grow 9 meters (30 feet) tall in only six years! And it grows fastest on poor soil in semiarid lands in Africa and Arabia. This drought-resistant tree can provide an abundance of fuelwood, lumber, and lamp oil.

It's also a natural pesticide—chemicals in its leaves can repel or kill over 200 insect species, including termites. And neem seeds and leaves have relieved so many different fevers, infections, and pains that the tree has been called a "village pharmacy." And it's a contraceptive—Neem-seed oil evidently acts as a strong spermicide. Researchers are now trying to use a compound extracted from this oil to create a male birth control pill.

Since 1985 numerous U.S. patents have been awarded to U.S. and Japanese forms for neem-based products. Farmers and scientists in India have criticized this as taking genetic information without giving India a share of the profits made from use of this indigenous knowledge.

ogist E. O. Wilson warns that destroying these forests and the species they contain for short-term economic gain is like throwing away a wrapped present or burning down an ancient library before you read the books.

To most economists and investors, untouched rain forests have economic potential and need to be developed. But others believe that destruction of these reservoirs of biological and cultural diversity must be slowed considerably or stopped and that ways must be found to use parts of them sustainably—as many indigenous peoples have done for thousands of years. Otherwise, within a few decades we will see the premature extinction of numerous tribal cultures (Connections, p. 428) and millions of wild species. In addition, the Environmental Policy Institute estimates that unless destruction of tropical forests stops, the resulting flooding and loss of topsoil will cause as many as a billion people to starve during the next 30 years.

Causes of Tropical Deforestation The major underlying causes of the current massive destruction and degradation of tropical forests are:

- *Population growth and poverty,* which combine to drive subsistence farmers and the landless poor to tropical forests to try to grow enough food to survive.

- *Massive foreign debt and policies of governments and international development and lending agencies that encourage rapid depletion of resources to stimulate short-term economic growth.* LDCs are encouraged to borrow huge sums of money from MDCs to finance economic growth. To pay the interest on their debts, these countries often sell off their forests, minerals, oil, and other resources—mostly to MDCs—at low prices dictated by the international marketplace.

- *Government subsidies that accelerate deforestation by making timber cheap relative to its full ecological cost* (Section 7-2).

The process of degrading a tropical forest begins with a road (Figures 16-5 and 16-17). Once the forest becomes accessible it can be increasingly cleared and degraded by:

- *Unsustainable small-scale farming.* Colonists follow logging roads into the forest to plant crops on small cleared plots, to build homes, and to try to survive. With little experience, many of these newcomers cut and burn too much forest to grow crops and don't allow depleted soils to recover (Figure 2-5), destroying large tracts of forest.

- *Cattle ranching.* Cattle ranches are often established on exhausted and abandoned cropland, often aided by government subsidies to make ranching profitable. When torrential rains and overgrazing turn the soils into eroded wastelands, ranchers move to another area and repeat the destructive process known as *shifting ranching.* In Brazil's Amazon Basin grazing cattle (often on land cleared and abandoned by poor farmers) allows ranchers (supported by government subsidies averaging $5.6 million per ranch) to claim large tracts of land and the mineral rights below it. Then they sell both land and minerals for quick profits—a process called *ghost ranching.* At least half of the 600 large ranches in the Brazilian Amazon Basin have never sent a cow to market, and about 30% are now abandoned. All told, cattle ranching has been responsible for about 60% of the deforestation in this basin. Between 1965 and 1983, Central America lost two-thirds of its tropical forestland (Figure 16-11), much of it cleared to raise beef for export to the United States, Canada, and western Europe. The true cost of a quarter-pound hamburger made from cattle grazing on

Cultural Extinction

Many tribal or indigenous peoples living in tropical forests have used those forests sustainably for centuries. They are stewards of 99% of Earth's genetic resources, and their homelands are sanctuaries for more threatened and endangered species than all the world's official wildlife reserves.

Many of these peoples in tropical forests and other biomes, however, are being driven from their homelands by commercial resource extractors and landless peasants. They are forced to adopt new ways while reeling from shock and hunger, and they are often killed by foreign diseases to which they have no immunity. Those who resist are often killed by ranchers, miners, and settlers.

About 500 years ago some 6 million tribal people lived in the Amazon Basin (Figure 6-16). Today there are only about 200,000, and many of them are threatened. In remote Amazon Basin rain forests of Brazil and Venezuela, the Yanomami—a Stone Age tribe now reduced to fewer than 20,000 people (9,000 in Brazil)—have been struggling for 20 years to preserve their land and way of life (Figure 16-16). Since 1985, 45,000–100,000 gold miners have invaded Yanomami territory in Brazil, bringing malaria and other diseases that killed some of the tribespeople. The miners carved more than 100 airstrips out of the rain forests, cleared and gouged out large tracts of land, turned streams into sewers, and poisoned nearby soil, water, and food webs with toxic mercury used to extract gold from gold ore (Spotlight, p. 311). Since 1974 more than 1,500 Yanomami have been massacred, mostly by invading miners.

Figure 16-16 These are members of the Yanomami tribe in a tropical forest in Brazil's Amazon Basin. Many of Earth's remaining tribal peoples, representing 5,000 cultures, are vanishing as their lands are taken over for economic development.

In 1991 the governments of Brazil and Venezuela recognized an area of Amazon Basin forest about the size of South Dakota as the permanent homeland of the beleaguered Yanomami people. The governments, however, have not adequately protected the Yanomami land from thousands of new gold miners and other illegal intruders.

Environmentalists and others, led by organizations such as Survival International and Cultural Survival, call upon governments to protect and honor the rights and Earth wisdom of remaining indigenous peoples to live on their lands without intrusion and resource extraction. They urge us to heed the words of Argentina's Guarani holyman Pae Antonio, "When the Indians vanish, the rest will follow."

Q: How much will it cost U.S. taxpayers to clean up contaminated nuclear weapons production facilities"?

Figure 16-17 Computer-enhanced satellite image of a 1,300-square-kilometer (500-square-mile) area of tropical rain forest north of Manus in Brazil's Amazon Basin in 1991. Remaining old-growth forest is dark red; regenerating forest is orange or light red; and recently deforested areas are blue-green or white. Most of the same area of old-growth forest in the United States shown in Figure 16-6 is much more extensively cut and fragmented.

land that was once tropical forest is the destruction of 5 square meters (54 square feet) of the forest, an area roughly the size of a small kitchen. Although beef eaters in MDCs are responsible for some of this deforestation, Latin Americans consume about 70% of their own beef production.

- *Commercial logging.* Since 1950, the consumption of tropical lumber has risen 15-fold, with Japan now accounting for 60% of annual consumption. Other leading importers of tropical hardwoods are the United States and Great Britain. Most cleared tropical forests are not replanted, and degraded ones are rarely restored because few if any of the costs of environmental degradation are included in the prices charged loggers. As tropical timber in Asia is depleted in the 1990s, cutting will shift to Latin America and Africa. The World Bank estimates that by 2000 only 10 of the

33 countries now exporting tropical timber will have any left to export. Although timber exports to MDCs contribute to tropical forest depletion and degradation, over 80% of the trees cut in LDCs are used at home.

- *Raising cash crops.* Immense plantations grow crops such as sugarcane, bananas, tea, and coffee, mostly for export to MDCs (Figure 14-2).

- *Mining operations.* Much of the extracted minerals, such as iron ore and bauxite (aluminum ore), are exported to MDCs.

- *Oil drilling and extraction.* Most of the oil extracted from Amazonian forest areas goes to MDCs (about half to the United States). Large amounts of oil have oozed into the forest and waterways from leaky pipelines.

- *Damming rivers and flooding large areas of forest.* Water allowed to flow past the dams is used mostly to produce electricity (hydropower) often used in mining and smelting minerals exported to MDCs. Several large dams in the Amazon Basin have flooded large areas of forest, and the 79 hydroelectric dams planned there over the next 20 years will flood an area equal to that of the state of Georgia—and displace many indigenous people (Figure 11-10).

Solutions: Reducing Tropical Deforestation
Environmentalists have suggested the following ways to reduce tropical deforestation:

- *Use remote-sensing satellites to find out how much of the world is covered with forest and how much has been deforested.* This could be done for about what the world spends for military purposes *every three minutes.*

- *Use economic indicators that include both the environmental services provided by forests and their depletion in their estimated value* (Figure 7-5).

- *Phase in full-cost ecological pricing for timber and other forest products. Do this by phasing out government subsidies that encourage deforestation and phasing in tariffs, use taxes, user fees, and subsidies that favor sustainable forestry* (Table 7-1). As long as the lasting and renewable ecological benefits of forests are undervalued in the marketplace, we will continue to destroy and degrade these forests, and their long-term ecological services, for short-term economic gain.

- *Establish a mandatory international system to identify and label tropical (and other) timber grown and harvested sustainably.* So far, only 0.1% of the world's

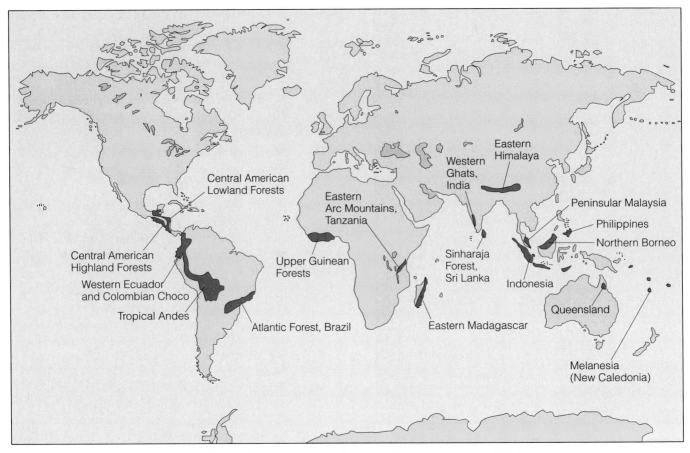

Figure 16-18 Tropical forest "hot spots." These 16 areas, believed to be unusually rich storehouses of biodiversity, are being deforested rapidly. (Data from Norman Myers and Conservation International)

tropical forests are managed sustainably, with the annual harvest not exceeding annual regrowth.

- *Reform tropical timber-cutting regulations and practices.* New logging contracts would charge more for timber-cutting concessions, make the contracts longer (70 years) to encourage conservation and reforestation, and require companies to post adequate bonds for reforestation and restoration.

- *Fully fund the Rapid Assessment Program (RAP),* which sends biologists to assess the biodiversity of "hot spots"—forests and other habitats that are both rich in unique species and in imminent danger—with the goal of channeling funds and efforts toward immediate protection of these valuable ecosystems (Figure 16-18).

- *Use debt-for-nature swaps and conservation easements to encourage countries to protect tropical forests or other valuable natural systems.* In a debt-for-nature swap, participating tropical countries act as custodians for protected forest reserves in return for foreign aid or debt relief (Case Study, next page). With conservation easements, a country, a private

organization, or a group of countries compensates individual countries for protecting selected forest areas. Currently less than 5% of the world's tropical forests are part of parks and preserves, and many of these are protected only on paper.

- *Help settlers learn how to practice small-scale sustainable agriculture* (Figure 2-5).

- *Stop funding tree and crop plantations, ranches, roads, and destructive types of tourism on any land now covered by old-growth tropical forests.*

- *Concentrate farming, tree and crop plantations, and ranching activities on cleared tropical forest areas that are in various stages of secondary ecological succession* (Figure 5-42).

- *Set aside large protected areas for indigenous tribal peoples* (Connections, p. 428).

- *Pressure banks and international lending agencies (controlled by MDCs) not to lend money for environmentally destructive projects*—especially road building (Figure 16-5)—involving old-growth tropical forests.

Debt-for-Nature in Bolivia

In 1984 biologist Thomas Lovejoy suggested that debtor nations willing to protect part of their natural resources should be rewarded. In the **debt-for-nature swaps** Lovejoy proposed that a certain amount of foreign debt be cancelled in exchange for spending funds on better natural resource management. Typically, a conservation organization buys a certain amount of a country's debt from a bank at a discount rate and negotiates the swap. A government or private agency must then agree to supervise the program.

In 1987 Conservation International purchased $650,000 of Bolivia's $5.7-billion national debt from the Citibank affiliate in Bolivia for $100,000. In exchange for not having to pay back this part of its debt, the Bolivian government agreed to expand and protect 1.5 million hectares (3.7 million acres) of tropical forest around its existing Beni Biosphere Reserve in the Amazon Basin—containing some of the world's largest reserves of mahogany and cedar—from harmful forms of development. The government was to legally protect the reserve and create a $250,000 fund, with the interest to be used to manage the reserve.

The plan is supposed to be a model of sustainable economic development (Figure 16-19). A virgin tropical forest is to be set aside as a biological reserve. It is to be surrounded by a protective buffer of savanna used for sustainable grazing of livestock. Controlled commercial logging—as well as hunting and fishing by local natives—would be permitted in some parts of the forest but not in the mountain area above the tract, to protect the area's watershed and to prevent erosion.

As of 1993, six years after the agreement was signed, however, the Bolivian government still had not provided legal protection for the reserve. It also waited until April

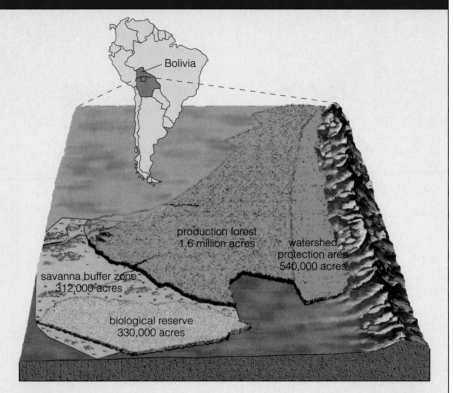

Figure 16-19 Sustainable development in Bolivia. A U.S. conservation organization arranged a debt-for-nature swap to help protect this land from destructive development.

1989 to contribute only $100,000 to the reserve management fund. Meanwhile, with government approval, timber companies have cut thousands of mahogany trees from the area, with most of this lumber exported to the United States. Also, the area's 5,000 native inhabitants were not consulted about the swap plan, even though they are involved in a land-ownership dispute with logging companies.

One lesson learned from this first debt-for-nature swap is that legislative and budget requirements must be met before the swap is executed. Another is that such swaps need to be carefully monitored by environmental organizations to be sure that paper proposals for sustainable development are not disguises for eventual unsustainable development.

Debt-for-nature swaps are a fine idea if carried out properly. However, they put only a small dent in the debt and environmental problems of LDCs. By 1992, 21 debt-for-

nature swaps had eliminated more than $100 million of debt for 11 countries at a cost of $17 million, and they had generated over $65 million for conservation efforts in these countries. However, this amounts to only 0.008% of the $800-billion foreign debt of tropical countries.

Critics also point out that such swaps legitimize the LDCs' massive debt at a time when environmentalists and debtor governments are urging that much of this debt be forgiven in return for carefully monitored and enforced agreements to reduce poverty, redistribute land, control population growth, and protect biodiversity and indigenous peoples in priority areas. Many of the debt swaps also support international (mostly corporate) control over debtor countries' development, while giving banks in MDCs a way to make good on part of essentially bad loans.

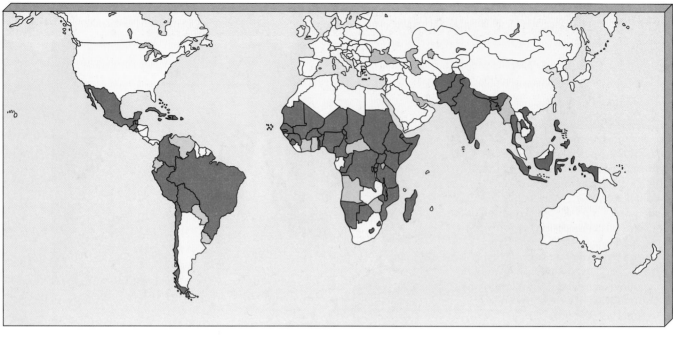

| | Acute scarcity and depletion in 1985 | | Deficits and scarcity by 2000 |

Figure 16-20 Scarcity of fuelwood, 1985 and 2000 (projected). (Data from UN Food and Agriculture Organization)

- *Pressure elected officials to vote down or modify parts of the international General Agreement on Tariffs and Trade (GATT) that weaken environmental protection in the name of free trade.* For example, unless this trade agreement is changed to strengthen environmental safeguards, a country wanting to stop exporting its timber or other resources or to ban the import of beef produced on cleared tropical forests could be accused of violating the principles of free trade.

- *Reduce poverty and the flow of the landless poor to tropical forests by slowing population growth.*

- *Reforest and rehabilitate degraded tropical forests and watersheds.*

Ecological Protection in Costa Rica   Costa Rica (Figure 6-16), smaller in area than West Virginia, was once almost completely covered with tropical forests. Between 1963 and 1983, however, politically powerful ranching families cleared much of the country's forests to graze cattle, with most of the beef exported to the United States and western Europe. By 1983 only 17% of the country's original tropical forest remained, and soil erosion was rampant.

Despite widespread degradation, tiny Costa Rica is a "superpower" of biodiversity, with an estimated 500,000 species of plants and animals. A single park in

Costa Rica is home for more bird species than all of North America.

Another bright note is that in the mid-1970s Costa Rica established a system of national parks and reserves that now protects 12% of its land (6% of it in reserves for indigenous peoples), compared to only 1.8% protected in the lower 48 states of the United States. One reason for this was the establishment in 1963 of the Organization of Tropical Studies (OTS), a consortium of 44 U.S. and Costa Rican universities, with the goal of promoting research and education in tropical ecology. The resulting infusion of several thousand scientists has helped Costa Rica's leaders and people appreciate its great biodiversity. It also led to the establishment of the National Biodiversity Institute (INBio) in 1989, a private nonprofit organization set up by biologists to survey and catalog the country's biodiversity. The program trains students, teachers, bus drivers, and other ordinary people in the art of collecting and identifying species, and in the process it provides them with income.

This strategy has paid off. Today revenue from tourism (almost two-thirds of it from ecotourists) is the country's largest source of outside income. Costa Rica plans to protect 17% of its land from exploitation.

The Fuelwood Crisis in LDCs   By 1985 about 1.5 billion people—almost one out of every three persons

Q: Does using nuclear power add carbon dioxide to the atmosphere?

Figure 16-21 Making fuel briquettes from cow dung in India. As fuelwood becomes scarce, more people collect and burn dung, depriving the soil of an important source of plant nutrients.

on Earth—in 63 LDCs either could not get enough fuelwood to meet their basic needs or were forced to meet their needs by consuming wood faster than it was being replenished (Figure 16-20). The UN Food and Agriculture Organization projects a fuelwood crisis by the end of this century for 2.7 billion people in 77 LDCs.

Besides deforestation and accelerated soil erosion, fuelwood scarcity has other harmful effects. It places an additional burden on the poor, especially women, who must often walk long distances to gather fuel. Buying fuelwood or charcoal can take 40% of a poor family's meager income. And poor families who can't get enough fuelwood often burn dried animal dung and crop residues for cooking and heating (Figure 16-21). As a result, these natural fertilizers never reach the soil. Cropland productivity is reduced, the land is degraded even more, and hunger and malnutrition increase.

Solutions: Reducing the Fuelwood Crisis

LDCs can reduce the severity of the fuelwood crisis by planting more fast-growing fuelwood trees such as leucaenas and acacias, burning wood more efficiently, and switching to other fuels. Fast-growing tree species used to establish fuelwood plantations, however, must be selected carefully to prevent harm to local ecosystems.

Eucalyptus trees, for example, are being used to reforest areas threatened by desertification and to establish fuelwood plantations in some parts of the world. Because these species grow fast even on poor soils, this might seem like a good idea.

Environmentalists see it as an ecological disaster. In their native Australia these trees thrive in areas with good rainfall. When planted in arid areas, however, the trees suck up so much of the scarce soil water that most other plants can't grow. Then farmers don't have

fodder to feed their livestock, and groundwater is not replenished. The eucalyptus trees also deplete the soil of nutrients and produce toxic compounds that accumulate in the soil because of low rainfall. In Karnata, India, villagers became so enraged over a government-sponsored project to plant these trees that they uprooted the saplings.

Experience has shown that planting projects are most successful when local people, especially women (who are a major force for environmental improvement in many LDCs), are involved in their planning and implementation (Solutions, p. 434). Programs work best when village farmers own the land or are given ownership of any trees they grow on village land. This gives them a strong incentive to plant and protect trees for their own use and for sale.

Another promising method is to encourage villagers to use the sun-dried roots of various common gourds and squashes as cooking fuel. These rootfuel plants, which regenerate themselves each year, produce large quantities of burnable biomass per unit of area on dry deforested lands. They also help reduce soil erosion and produce an edible seed that is high in protein.

New, more efficient, less polluting stoves can make use of locally available materials and provide both heat and light, like the open fires they replace—and at the same time reduce indoor air pollution, a major health threat. During the 1980s over 64 million efficient cookstoves were distributed in rural areas in LDCs, especially in Africa and India. Cheap and easily made solar ovens that capture sunlight can also be used to reduce wood use and air pollution.

Despite encouraging success in some countries (such as China, Nepal, Senegal, and South Korea) most LDCs suffering from fuelwood shortages have inadequate forestry policies and budgets, and they lack trained foresters. Such countries are cutting trees for fuelwood and forest products 10–20 times faster than new trees are being planted.

16-5 MANAGING AND SUSTAINING RANGELANDS

The World's Rangeland Resources Almost half of the earth's ice-free land is **rangeland**: land that supplies forage or vegetation (grasses, grasslike plants, and shrubs) for grazing (grass-eating) and browsing (shrub-eating) animals, and that is not intensively managed. Most rangelands are grasslands in areas too dry for unirrigated crops (Figure 5-3).

About 42% of the world's rangeland is used for grazing livestock. Much of the rest is too dry, cold, or remote from population centers to be grazed by large numbers of livestock. In the United States about 34%

SOLUTIONS

The Chipko Movement

In the late fifteenth century Jambeshwar, the son of a village leader in northern India, developed 29 principles for living and founded a Hindu sect called the Bishnois, based on a religious duty to protect trees and wild animals. In 1730, when the Maharajah of Jodhpur in northern India ordered that the few remaining trees in the area be cut down, the Bishnois forbade it. Women rushed in and hugged the trees to protect them, but the Maharajah's minister ordered the work to proceed anyway. According to legend 363 Bishnois women died on that day.

In 1973 some women in the Himalayan village of Gopeshwar in northern India started the modern Chipko (an Indian word for "hug" or "cling to") movement to protect the remaining trees in a nearby forest from being cut down to make tennis rackets for export. It began when Chandi Prasad Bhatt, a village leader, urged villagers to run into the forest ahead of loggers and hold on to and protect the trees (Figure 16-22), thus reviving the tradition started several hundred years earlier by the Bishnois.

When the loggers came, local women, children, and men rushed into the forests, flung their arms around trees, and dared the loggers to let the axes fall on their backs. They were successful, and their action inspired women in other Himalayan villages to protect their forests. As a result, the commercial cutting of timber in the hills of the Indian state of Uttar Pradesh has been banned.

Since 1973 the Chipko movement has widened its efforts. Now the women—who still guard trees from loggers—also plant trees, prepare village forestry plans, and build walls to stop soil erosion. Recently villagers have focused on opposing the planting of fast-growing pine and eucalyptus plan-

Figure 16-22 A member of the Chipko movement protecting a tree in northern India from being cut down for export. This is another example of a successful grassroots group dedicated to helping sustain the earth. Women are the driving force in this group.

tations being proposed by development agencies such as the World Bank. Chipko activist Sunderlal Bahugana explains the objections:

Cruel trees like pine and eucalyptus take up a lot of water from the soil and give nothing back. The ground in pine plantations is barren like a desert. What the villagers need are deciduous trees such as walnut and oak that supply what I call the five F's—food in the form of nuts and berries, firewood, fodder from leaves, fibre to make rope or baskets, and rich soil from fertilizer.

The spreading actions of such Earth citizens should inspire us to help sustain the earth. Working with the earth will come mostly from the bottom up by the actions of ordinary people, not from the top down.

Q: Nuclear weapons existing today could kill everyone in the world how many times?

of the total land area is rangeland. Most of this is short-grass prairies in the arid and semiarid western half of the country (Figure 5-11).

Rangeland Vegetation and Livestock   Most rangeland grasses have deep, complex root systems (Figure 12-21) that not only anchor the plants but also sustain them through several seasons. If the leaf tip of most plants is eaten, the leaf stops growing, but the blades of rangeland grass grow from the base, not the tip. When the upper half of the shoot and blade of grass is eaten, the plant can grow back quickly. As long as only its upper half is eaten, and its lower half remains, rangeland grass is a renewable resource that can be grazed again and again. Range plants do vary in their ability to recover, however, and grazing changes the balance of plant species in grassland communities.

In addition to vast numbers of wild herbivores, the world has about 10 billion domesticated animals. About 3 billion of these are *ruminants*—mostly cattle (1.5 billion), sheep, and goats—which can digest the cellulose in grasses. Three-fourths of these forage on rangeland vegetation before being slaughtered for meat, while the rest are fattened on grain in feedlots (Figure 14-9) before slaughter. Seven billion pigs, chickens, and other livestock are *nonruminants*. They cannot feed on rangeland vegetation and eat mostly cereal grains grown on cropland.

Each type of grassland has a ruminant **carrying capacity**: the maximum number of ruminants that can graze a given area without destroying the metabolic reserve needed for grass renewal. Carrying capacity is influenced by season, range conditions, climatic conditions, past grazing use, soil type, kinds of grazing animals, and amount of grazing.

Overgrazing occurs when too many animals graze too long and exceed the carrying capacity of a grassland area. It lowers the productivity of vegetation and changes the number and types of plants in an area. Large populations of wild ruminants can overgraze rangeland in prolonged dry periods, but most overgrazing is caused by excessive numbers of domestic livestock feeding too long in a particular area (Figure 12-26).

Heavy overgrazing compacts the soil, which diminishes its capacity to hold water and to regenerate itself. Overgrazing converts continuous grass cover into patches of grass and thus exposes the soil to erosion, especially by wind. Then woody shrubs such as mesquite and prickly cactus invade and take over. Figure 16-23 compares normally grazed and severely overgrazed grassland. Overgrazing is the major cause of desertification in arid and semiarid lands (Figure 12-25).

Some ecologists have suggested that wild herbivores, such as eland, oryx, and Grant's gazelle, could

Figure 16-23 Rangeland: overgrazed (left) and lightly grazed (right).

be raised on ranches on grasslands such as tropical savanna (Figure 5-9). Because many wild herbivores have a more diversified diet than cattle, they can make more efficient use of available vegetation and thus reduce the potential for overgrazing. In addition, they need less water than cattle and are more resistant than cattle to animal diseases found in savanna grasslands. These animals also handle much of their own predator control because of long evolutionary experience with lions, leopards, cheetahs, and wild dogs.

Since 1978 David Hopcraft has carried out a successful game ranching experiment on the Athi Plains near Nairobi, Kenya. The ranch is stocked with various native grazers and browsers, including antelope, zebras, giraffes, and ostriches. Cattle, once raised for comparison purposes, are being phased out and may be replaced with native Cape buffalo. The yield of meat from the native herbivores has been rising steadily, the condition of the range has improved, and costs are much lower than those for raising cattle in the same region.

Game ranching of native deer and antelope might also help reverse overgrazing and desertification in much of the western United States, especially with government subsidies to replace those now given for raising cattle.

Condition of the World's Rangelands   *Range condition* is usually classified as excellent (more than 80% of its potential forage production), good

(50–80%), fair (21–49%), and poor (less than 21%). Except in North America no comprehensive survey of rangeland conditions has been done. In 1990, 50% of public rangeland in the United States was rated as being in unsatisfactory (fair or poor) condition, compared with 84% in 1936. This is a considerable improvement but there's still a long way to go.

Environmentalists also point out that overall estimates of rangeland condition obscure severe damage to certain heavily grazed areas, especially vital **riparian zones**, thin strips of lush vegetation along streams. They help prevent floods and help keep streams from drying out during droughts by storing and releasing water slowly from spring runoff and summer storms. They also provide habitats, food, water, and shade for wildlife, acting as centers of biodiversity in the arid and semiarid western lands. Studies indicate that 65–75% of the wildlife in the West is totally dependent on riparian habitats.

Because cattle need lots of water, they tend to concentrate around riparian zones and feed there until the grass and shrubs are gone. The result: Riparian vegetation is destroyed by trampling and by overgrazing. A 1988 General Accounting Office report concluded that "poorly managed livestock grazing is the major cause of degraded riparian habitat on federal rangelands." Arizona and New Mexico have already lost 90% of their riparian areas, primarily to grazing.

Despite the threat to soil and wildlife, little has been done to protect riparian zones on public land from livestock or to repair the damage. Ranching interests are influential, and there has been little pressure from the public (mostly because of unawareness of the problem) to protect these vital areas on public lands.

Managing Rangelands   The primary goal of rangeland management is to maximize livestock productivity without overgrazing rangeland vegetation. The most widely used method to prevent overgrazing is to control the *stocking rate*—the number of each kind of animal placed on a given area—so it doesn't exceed carrying capacity. Determining the carrying capacity of a range site is difficult and costly. And as noted, carrying capacity changes because of drought, invasions by new species, and other environmental factors.

Not only the numbers, but also the distribution of grazing animals over a rangeland, must be controlled to prevent overgrazing. Ranchers can control distribution by fencing off damaged rangeland and riparian zones, rotating livestock from one grazing area to another, providing supplemental feeding at selected sites, and locating water holes and salt blocks in strategic places.

A more expensive and less widely used method of rangeland management is to suppress the growth of unwanted plants (mostly invaders) by herbicide spraying, mechanical removal, or controlled burning. A cheaper and more effective way to remove unwanted vegetation is controlled, short-term trampling by large numbers of livestock.

Growth of desirable vegetation can be increased by seeding and applying fertilizer, but this method usually costs too much. Reseeding is an excellent way to restore severely degraded rangeland, however.

Another aspect of range management for livestock is predator control. For decades, hundreds of thousands of predators have been shot, trapped, and poisoned by ranchers, farmers, hunters, and federal predator control officials. Federally subsidized predator control programs reduced gray wolf and grizzly bear populations to the point where they are endangered species. Now more federal dollars are being spent by the U.S. Fish and Wildlife Service to protect and help revive them. Experience has shown that killing predators is an expensive and often short-lived solution—and one that can make matters even worse (Connections, next page).

Controversy over Livestock and U.S. Public Rangeland   About 29,000 U.S. ranchers hold what are essentially lifetime permits to graze about 4 million livestock (3 million of them cattle) on Bureau of Land Management (BLM) and National Forest Service rangelands in 16 western states. About 10% of these permits are held by small livestock operators. The other 90% belong to ranchers with large livestock operations and to corporations such as Metropolitan Life Insurance Company, Union Oil, Getty Oil, and the Vail Ski Corporation.

Permit holders pay the federal government a grazing fee for this privilege. Since 1981 grazing fees on public rangeland have been set by Congress at only one-fourth to one-eighth the going rate for leasing comparable private land. This means that taxpayers give the 2% of U.S. ranchers with federal grazing permits subsidies amounting to about $140–$150 million a year—the difference between the fees collected and the actual value of the grazing on this land.

The economic value of a permit is included in the overall worth of the ranches. It can be used as collateral for a loan and is usually automatically renewed every 10 years—allowing permit holders in effect to treat federal land as part of their ranches. It's not surprising that politically influential permit holders have fought so hard to block any change in this system.

The public subsidy does not end with low grazing fees, however. When other costs are factored in—

Q: Can switching to increased use of nuclear power in the United States save much oil?

Coyotes (*Canis latrans*) are smaller relatives of wolves. These omnivores live on a diet of small rodents (especially mice and rabbits), animal carcasses, and various plants. They live in small packs, but unlike wolves they usually hunt alone or in pairs. They are prolific and highly adaptable animals.

A controversial issue in the western United States is the effect coyotes have on livestock, especially sheep. On one hand, sheep ranchers claim that coyotes kill large numbers of sheep on the open range and should be exterminated (Figure 16-24). On the other hand, many wildlife experts maintain that although coyotes kill some sheep, their net effect is to increase rangeland vegetation for livestock and wild herbivores.

The ranchers' view has prevailed, and since 1940 western ranchers, the U.S. Department of the Interior, and the Department of Agriculture have waged a controversial war against livestock predators. Elimination of wolves (Connections, p. 442) allowed the coyote population to expand in numbers and range. Because coyotes are so prolific, adaptable, and crafty, environmentalists contend that any attempt to control them is doomed to failure. They also argue that it would cost taxpayers much less to pay ranchers for each sheep or goat killed by a coyote than to spend $30 million per year on predator control, with 60% of the funds coming from the federal government and the rest from county and state governments and livestock associations.

Some range ecologists argue that the coyote may even benefit some grazing animals by helping to control populations of smaller herbivores (especially rabbits) that compete with large herbivores such as cattle or deer. Removing them or reducing their numbers eliminates or reduces these ecological connections. Unlike deer and cattle, sheep (especially lambs) are small enough to be killed by coyotes. The question is whether the increase in sheep production from improved vegetation is greater than the number of sheep killed by the coyote. We don't know.

Environmentalists suggest that a combination of fences, repellents, and trained guard dogs be used to keep predators away. With these methods, for example, sheep producers in Kansas have one of the country's lowest rates of livestock losses to coyotes.

In 1986 Department of Agriculture researchers reported that predation can be sharply reduced by penning young lambs and cattle together for 30 days and then allowing them to graze together on the same range. When predators attack, cattle butt and kick them. In the process, they protect both themselves and the sheep. Llamas and donkeys, also tough fighters against predators, can be used in the same way to protect sheep.

(continued)

Figure 16-24 For decades ranchers in the western United States have waged war on the coyote because of its reputation for killing livestock, as shown by these coyote carcasses lashed to a fence in California. Each year thousands of coyotes are poisoned, trapped, or shot in the western United States. This predator control program for the benefit of a few ranchers is paid for by all U.S. taxpayers.

Environmentalists point out that most coyotes and other predators do not prey on livestock. Instead of trying to kill all predators, they suggest concentrating on killing or removing the rogue individuals.

Some environmentalists call for abolishing the federal predator control program. Others, however, call for much greater control over the program, including reducing kills of many species such as coyotes, banning the federal program from accepting funds from livestock associations and county governments (which increases economic and political pressure to kill more predators), and imposing much stricter penalties and much larger fines on ranchers and hunters who illegally kill predators. These environmentalists contend that without the federal program, ranchers and hunters would take matters into their own hands and kill much larger numbers of predators. Despite opposition from environmentalists the federal predator control program continues to receive funding. And it is not as carefully supervised and regulated as environmentalists would like because of the political power of ranchers and farmers who see the coyote and other predators as threats to their livelihood.

---

fencing, water pipelines, stock ponds, weed control, livestock predator control, clearing of undesirable vegetation, planting of grass for livestock, erosion, lowered recreational values, loss of biodiversity—it's estimated that taxpayers are providing 29,000 ranchers with an annual subsidy of about $2 billion (an average of $67,000 per rancher) to produce only 3% of the country's beef. In 1991, one rancher received subsidies worth almost $950,000. This explains why many critics charge that the public-lands grazing programs are little more than "rancher welfare," mostly for well-to-do ranchers who hold 90% of the permits.

Some environmentalists believe that all commercial grazing of livestock on public lands should be phased out over 10–15 years. They contend that the water-poor western range is not a very good place to raise livestock (cattle and sheep), which require a lot of water. Forage is produced at a slow rate, droughts are frequent, and animals can be grazed only part of the year.

Ranchers counter that meat produced on rangeland that is not otherwise usable for food production represents a net increase in available food. They point out that when livestock are fattened on corn and other grains, this uses land that could be producing more grains for direct human consumption. Furthermore, if ranchers with permits can no longer use public rangeland, many may go out of business and sell their land to developers—destroying most of the land's ecological functions for wildlife.

The Takings Issue: To Take or Not To Take

Since 1991 there has been growing controversy over what is called the *takings* issue. One part of the takings movement involves the rights of private individuals and corporations that have the privilege of using resources on public lands. It began in 1991 when Nevada rancher Wayne Hays filed a $28 million suit against the Forest Service because it reduced or suspended portions of his grazing permit, citing declining range conditions and violation of the terms of the permit.

Hays argues that a permit to graze federal lands is a private property right—not a privilege, as the government maintains. If it is a private property right, Hays contends, the government cannot reduce or take away a grazing permit unless the permit holder is compensated for resulting losses. He bases this argument on the Fifth Amendment of the U.S. Constitution, which provides that private property shall not be taken for public use without just compensation.

Hays's suit has spurred a whole movement—the County Movement—that would establish the legal precedence of county land-use plans and ordinances over federal laws that affect private property rights on public land within county borders. If successful, such legislation at the state level would severely limit the federal government's ability to enforce grazing, logging, mining, wildlife protection, and health laws—by requiring both lengthy evaluations (of possible financial losses from restrictions on public land use) and compensation to parties for any losses incurred. In 1992 Alaska's governor sued the U.S. government for $29 billion for depriving the state of potential revenue from mineral extraction on public lands.

If regulatory agencies were required to compensate parties for restricting resource use on public lands, most agency budgets would be devoted to such compensation. This would leave the agencies with little money for enforcing laws passed by Congress.

Most legal authorities believe that the legal principles underlying the takings issue have little legal merit. These experts point to the property clause of the Constitution, which says, "Congress shall have the power to dispose of all needful rules and regulations respect-

Q: What would happen if nuclear power's harmful effects were included in its price and all subsidies were removed?

ing the territory of or other property belonging to the United States."

In 1993, environmental groups helped defeat takings bills in 23 of the 30 states in which they were introduced. However, the fight will continue, with new bills being introduced each year and with hundreds of takings suits being filed in the courts.

Even if the takings issue is eventually decided in favor of the federal government, the effect of the spreading takings movement is to hinder government agencies from enforcing laws mandated by Congress. An agency so hindered can be tied up in court for years while hundreds of takings suits proceed through the court system.

A different controversy in the takings issue focuses on government action that affects private—not public—lands. It involves the issue of whether private landowners are entitled to compensation for loss of rights in how they can use their property as a result of government action—something called *regulatory taking*. The Fifth Amendment of the U.S. Constitution gives the government the power, known as eminent domain, to force a citizen to sell property needed for a public good. If your land is needed for a road, for example, the government can take your land but you must be paid its fair market value in exchange for your loss of its use.

The law is clear when a physical taking is involved but not so clear with regulatory taking. The legal question is whether the Constitution requires the government to compensate you if it doesn't take your property but reduces its value by not allowing you to do certain things with it. For example, you might not be allowed to build on some or all of your property because it is a wetland protected by law. Or you may not be able to harvest trees on your land because it is a habitat for an endangered species. As environmental regulations increase and greater emphasis is placed on ecosystem protection, the already raging controversy over regulatory taking will grow and affect more property owners.

Most private landowners believe they should receive just compensation for financial losses imposed on them by regulatory taking. But such compensation could cripple the ability of government at the federal and state levels to protect the public good by enforcing environmental laws by imposing massive costs. At the same time, not compensating owners of private property for loss in value because of environmental laws could create a massive backlash against such laws—something that is already beginning to happen.

In 1993 more than 20 state legislatures considered bills addressing this issue. Some are trying to develop *assessment laws* that require the state to determine when a regulation that destroyed the value of private prop-

erty requires compensation. Other states are considering *compensation laws* that would require payment only if the damage reaches a certain threshold—for example, by reducing a property's value by say 40% or more.

Government officials and many environmentalists are looking for solutions to this problem without having to spend large sums of increasingly scarce public funds. One possibility is for the government to sell some public lands not highly valued because of their biological diversity or other ecological values and use the proceeds to purchase conservation easements that would prohibit certain types of activities such as mining or logging on ecologically valuable private lands. Another suggestion is that the government compensate landowners whose land it wishes to preserve by trading land with them.

Solutions: Sustainable Management of Public Rangeland

Some environmentalists agree that with proper management ranching on western rangeland is a potentially sustainable operation and that encouraging sustainable ranching practices keeps the land from being converted to developments. They call for curbing overgrazing and destruction of vital riparian zones on western public rangeland by:

- *Excluding livestock grazing from riparian areas.*

- *Banning or sharply reducing livestock grazing on rangeland in poor condition until it recovers.*

- *Ending federal predator control measures and instead paying ranchers directly for any livestock shown to be killed by predators. Studies show that such a program would save taxpayers money.*

- *Greatly increasing funds for restoration of degraded rangeland and riparian areas.*

- *Sharply raising grazing fees to a fair market value.*

- *Giving family ranchers with small ranches grazing fee discounts. This would help them stay in business and keep their land from being converted to real estate developments. Ranchers with large operations who artificially divide their holdings to qualify would not get discounts.*

- *Giving rebates or other incentives to ranchers who take extra steps to improve or repair range conditions on public land.*

- *Abolishing grazing advisory boards. These boards gives ranchers undue influence on federal officials and lead to about 96% of the allotted funds being spent on grazing instead of for other uses. Other multiple-use public lands have no such boards.*

- *Replacing the current noncompetitive grazing permit system with a competitive bidding system. The*

A: It would be too expensive to use and would be phased out.

The Eco-Rancher

Wyoming rancher Jack Turnell is a new breed of cowpuncher who gets along with environmentalists and talks about riparian ecology and biodiversity as fluently as he talks about cattle: "I guess I have learned how to bridge the gap between the environmentalists, the bureaucracies, and the ranching industry."

Turnell grazes cattle on his 32,000-hectare (80,000-acre) ranch south of Cody, Wyoming, and on 16,000 hectares (40,000 acres) of Forest Service land where he has grazing rights. For the first decade after he took over the ranch he punched cows the conventional way. Since then he's made some changes.

Turnell disagrees with the proposals by environmentalists to raise grazing fees, and to remove sheep and cattle from public rangeland. He believes that if ranchers are kicked off the public range, ranches like his will be sold to developers and chopped up into vacation sites, irreversibly destroying the range for wildlife and livestock alike.

At the same time, he believes that ranches can be operated in more ecologically sustainable ways. To demonstrate this, Turnell began systematically rotating his cows away from the riparian areas, gave up most uses of fertilizers and pesticides, and crossed his Hereford and Angus cows with a French breed that does not like to congregate around water. Most of his ranching decisions are made in consultation with range and wildlife scientists, and changes in range condition are carefully monitored with photographs.

The results have been impressive. Riparian areas on the ranch and Forest Service land are lined with willows and other plant life, providing lush habitat for an expanding population of wildlife including pronghorn antelope, deer, moose, elk, bear, and mountain lions. And this "eco-rancher" now makes more money because the better-quality grass puts more meat on his cattle. He frequently talks to other ranchers about sustainable range management; some of them probably think he has been chewing locoweed.

existing system gives ranchers who obtain essentially lifetime permits an unfair economic advantage over ranchers who can't get such permits, which amount to government subsidies. Competitive bidding would let free enterprise work and might allow conservation and wildlife groups to obtain grazing permits they could use to protect such lands from overgrazing.

- *Subsidizing experimental programs of game ranching for native deer and antelope.*

The problem is that western ranchers wield enough political power to see that measures like those just listed are not enacted by elected officials. For example, in 1993 the Clinton administration proposed a modest increase in grazing fees. However, western senators were able to delay its implementation and hope to prevent it from going into effect. Still, some ranchers have demonstrated that rangeland can be grazed sustainably (Individuals Matter, at left).

16-6 MANAGING AND SUSTAINING NATIONAL PARKS

Parks Around the World   Today, over 1,100 national parks larger than 1,000 hectares each (2,500 acres) are located in more than 120 countries. Together they cover an area equal to that of Alaska, Texas, and California combined. This important achievement in global conservation was spurred by the creation of the first public national park system in the United States in 1912.

The U.S. National Park System is dominated by 50 national parks, most of them in the West (Figure 16-2). These repositories of majestic beauty and biodiversity, sometimes called America's crown jewels, are supplemented by state, county, and city parks. Most state parks are located near urban areas and thus are used more heavily than national parks. Nature walks, guided tours, and other educational services by U.S. Park Service employees have given many visitors a better appreciation for and understanding of nature.

The Plight of Parks Today   Parks everywhere are under siege. In LDCs, parks are often invaded by local people who desperately need wood, cropland, and other resources. Poachers kill animals to get and sell rhino horns, elephant tusks, and furs. Park services in LDCs typically have too little money and staff to fight these invasions, either by force or by education. Also, most of the world's national parks are too small to sustain many of the larger animal species.

Popularity is the biggest problem of national and state parks in the United States (and other MDCs). Because of increased numbers of sufficiently affluent people, roads, and cars, annual recreational visits to National Park System units increased more than 12-fold

Q: Is nuclear fusion the answer to our energy problems?

(and visits to state parks 7-fold) between 1950 and 1993. The number of visits is expected to double by 2020.

During the peak summer season, the most popular national and state parks are often choked with cars and trailers, and they are plagued by noise, traffic jams, litter, vandalism, poaching, deteriorating trails, polluted water, and crime. Many visitors to heavily used parks leave the city to commune with nature, only to find the parks noisier, more congested, and more stressful than where they came from.

Park Service rangers now spend an increasing amount of their time on law enforcement instead of on resource conservation, management, and education; many even wear body armor. Since 1976 the number of federal park rangers—about 3,200—has not changed, while the number of visitors to park units has risen by 75 million and is expected to rise by another 162 million in the next 20 years. Currently, there is one ranger for every 84,200 visitors to the national parks. Many overworked rangers, earning an average salary of less than $20,000 and living in substandard housing, are leaving for better-paying jobs.

Wolves (Connections, p. 442), bears, and other large predators in and near various parks have all but vanished because of excessive hunting, poisoning by ranchers and federal officials, and the limited size of most parks. As a result, populations of species these predators once helped control have exploded, destroying vegetation and crowding out other species.

Alien species have moved or have been introduced into parks. Wild boars (imported to North Carolina in 1912 for hunting) are threatening vegetation in part of the Great Smoky Mountains National Park. The Brazilian pepper tree has invaded Florida's Everglades National Park. Mountain goats in Washington's Olympic National Park trample native vegetation and accelerate soil erosion. At the same time that some species have moved into parks, other native species of animals and plants (including many threatened or endangered species) are being killed or removed illegally in almost half of U.S. national parks.

The greatest threat to many parks today, however, is posed by nearby human activities. Wildlife and recreational values are threatened by mining, logging, grazing, coal-burning power plants, water diversion, and urban development. Polluted air drifts hundreds of kilometers to kill trees in California's Sequoia National Park and blur the awesome vistas at Arizona's Grand Canyon. According to the National Park Service, air pollution affects scenic views in national parks more than 90% of the time.

But that's not all. Mountains of trash wash ashore daily at Padre Island National Seashore in Texas. Water use in Las Vegas threatens to shut down geysers in the Death Valley National Monument. And unless a massive ecological restoration project works, Florida's Everglades National Park may dry up (Figure 11-19).

Solutions: Better Park Management in LDCs

Some park managers, especially in LDCs, are developing integrated management plans that combine conservation practices with sustainable development of resources in the park and surrounding areas (Figure 16-19). In such plans the inner core and especially vulnerable areas of each park are to be protected from development and treated as wilderness. Restricted numbers of people are allowed to use these areas only for hiking, nature study, ecological research, and other nondestructive recreational and educational activities. In buffer areas surrounding the core, controlled commercial logging, sustainable grazing by livestock, and sustainable hunting and fishing by local people are allowed. Money spent by park visitors adds to local income. By involving local people in developing park management and restoration plans, managers and conservationists seek to help them see the park as a vital resource they need to protect and sustain rather than ruin (Solutions, p.444).

Integrated park management plans look good on paper, but they cannot be carried out without adequate funding and the support of nearby land owners and users. Moreover, in some cases the protected inner core may be too small to sustain some of the park's larger animal species.

Solutions: An Agenda for U.S. National Parks

In the United States national parks are managed under the principle of natural regulation—as if they are wilderness ecosystems that, if left alone, will sustain themselves. Many ecologists consider this a misguided policy. Most parks are far too small to even come close to sustaining themselves, and even the biggest ones cannot be isolated from the harmful effects caused by human activities in nearby areas.

The National Park Service has two goals that increasingly conflict: (1) to preserve nature in parks, and (2) to make nature more available to the public. The Park Service must accomplish these goals with a small budget at a time when park usage and external threats to the parks are increasing.

In 1992 federal auditors criticized the National Park Service for not doing a good job of protecting the natural resources in its parks. The report pointed out that the Park Service spends 92% of its budget on visitor services and only 8% on protecting natural resources. Numerous other studies carried out over three decades—including one in 1992 by the National

---

A: No. It cannot become a significant producer of electricity until 2100, if then.

CONNECTIONS

At one time the gray wolf, *Canis lupis* (Figure 16-25), ranged over most of North America. Between 1850 and 1900, some 2 million wolves were shot, trapped, poisoned—and even drenched with gasoline and set afire—by ranchers, hunters, and government employees. The idea was to make the West and the Great Plains safe for livestock and for big-game animals prized by hunters.

By the 1960s the gray wolf was found mostly in Alaska and Canada, with a few hundred left in Minnesota and about 50 in Montana as well. The species is now listed as endangered in all 48 lower states except Minnesota, whose 1,550–1,750 wolves are considered threatened.

Ecologists now recognize the important role these predators once played in parts of the West and the Great Plains. These wolves culled herds of bison, elk, and mule deer; without predators, such species can proliferate and devastate vegetation, thus threatening the niches of other forms of wildlife. The wolves killed only for food, usually taking weaker animals and thereby strengthening the genetic pool of the survivors.

In 1987 the U.S. Fish and Wildlife Service proposed that wolves be reintroduced into the Yellowstone ecosystem, which includes two national parks, seven national forests, and other federal and state lands in Wyoming, Idaho, and Montana. The plan calls for gradually introducing 10 breeding packs, each containing about 100 wolves, in three recovery areas, beginning in 1994. Each year 30 wolves would be transplanted from Canada, with 15 going to Yellowstone National Park and 15 to central Idaho.

This proposal to reestablish ecological connections that humans had eliminated brought outraged howls—both from ranchers (who feared the wolves would attack their cattle and sheep) and from hunters (who feared that the wolves would kill too many big-game animals). One enraged rancher said that the idea was "like reintroducing smallpox."

A 1992 economic study estimated that returning wolves to the Yellowstone ecosystem would bring about $18 million into the local economy the first year and about $100 million over 20 years. And National Park Service officials promised to trap or shoot any wolves that killed livestock outside the recovery areas and to reimburse ranchers for livestock verified as having been killed by wolves from a private $100,000 fund established by Defenders of Wildlife. However, these promises fell on deaf ears, and these ranchers and hunters have worked hard to delay or defeat the plan in Congress.

The wolves may have a better chance to survive if enough of them migrate to the Yellowstone ecosystem on their own. Then, as an endangered species, they will be fully protected, at least legally. By contrast, if they are relocated to the Yellowstone ecosystem, they will be treated as an "experimental" population; then if they venture outside the recovery areas, they can be killed by ranchers and hunters. For this reason, some environmentalists believe that the wolves should be reintroduced but given full protection as endangered species. Some ranchers and hunters say that, either way, they'll take care of the wolves quietly—what they call the "shoot, shovel, and shut up" solution. In 1993 a single gray wolf found in the Yellowstone ecosystem just outside the southern boundary of Yellowstone National Park was shot and killed by a hunter who said he thought it was a coyote. Do you think the gray wolf should be reintroduced to the Yellowstone ecosystem?

Academy of Sciences—have criticized the Park Service's lack of scientific research.

In 1988 the Wilderness Society and the National Parks and Conservation Association suggested the following blueprint for the future of the U.S. national park system:

- *Educate the public about the urgent need to protect, mend, and expand the system.*

- *Significantly increase the number and pay of park rangers.*

- *Acquire new parkland near threatened areas and add at least 75 new parks within the next decade.*

- *Locate most commercial park facilities (such as restaurants, hotels, and shops) outside park boundaries.*

- *Raise the franchise and facilities rental fees charged to private concessionaires who operate lodging, food, and recreation services inside national parks to at least 22% of their gross receipts.* In 1992 private concessionaires in national parks reported gross receipts of $638 million but returned only 3.1% of that amount to the government in fees. Many large concessionaires have long-term contracts by which they pay the government as little as 0.75% of their gross receipts.

Figure 16-25 The gray wolf is an endangered species in the lower 48 states (except Minnesota, where it is listed as threatened). Efforts to return this species to its former habitat in the Yellowstone National Park area are vigorously opposed by ranchers (who fear the wolves will kill some of their sheep) and by hunters (who fear the wolves will kill big-game animals).

- *Halt concessionaire ownership of facilities in national parks, which makes buying buildings back very expensive.*

- *Wherever feasible, place visitor parking areas outside the park boundary.* Use low-polluting vehicles to transport visitors to and from parking areas and within the park.

- *Greatly expand the National Park Service budget for maintenance and for science and conservation programs.* Currently only 1% of national park funding goes for environmental research. The national parks face a $2.3-billion backlog of repairs.

- *Require the Park Service, Forest Service, and Bureau of Land Management to develop integrated management plans so that activities in nearby national forests and rangelands don't degrade national parklands and wilderness areas within the parks.*

16-7 PROTECTING AND MANAGING WILDERNESS

Why Preserve Wilderness? Protecting undeveloped lands from exploitation is an important way of preserving ecological diversity. According to the

Biocultural Restoration in Costa Rica

SOLUTIONS

In a rugged mountainous region covered with tropical rain forest lies a centerpiece of Costa Rica's efforts to preserve its biodiversity—the Guanacaste National Park, which has been designated an international biosphere reserve. In the park's lowlands a small tropical seasonal forest is being restored and relinked to the rain forest on adjacent mountain slopes.

Daniel Janzen, professor of biology at the University of Pennsylvania, has helped galvanize international support and has raised more than $10 million for this restoration project, the world's largest. Janzen is a leader in the growing field of rehabilitation and restoration of degraded ecosystems. He and Roger Gamez also set up INBio to catalog the country's biodiversity.

Janzen's vision is to make the nearly 40,000 people who live near the park an essential part of the restoration of the degraded forest—a concept he calls *biocultural restoration*. By actively participating in the project, local residents reap enormous educational, economic, and environmental benefits. Local farmers have been hired to sow large areas with tree seeds and to plant seedlings started in Janzen's lab.

Students in grade schools, high schools, and universities study the park's ecology in the classroom and then go on annual field trips to the park itself. There are educational programs for civic groups and tourists from Costa Rica and elsewhere. These visitors and their activities will stimulate the local economy. The project will also serve as a training ground in tropical forest restoration for scientists from all over the world.

Janzen recognizes that in 20 to 40 years today's children will be running the park and the local political system. If they understand the importance of their local environment, they are more likely to protect and sustain its biological resources. He understands that education, awareness, and involvement—not guards and fences—are the best way to protect ecosystems from unsustainable use.

Wilderness Act of 1964, **wilderness** consists of those areas "where the earth and its community of life are untrammeled by man, where man himself is a visitor who does not remain." President Theodore Roosevelt summarized what we should do with wilderness: "Leave it as it is. You cannot improve it."

The Wilderness Society estimates that a wilderness area should contain at least 400,000 hectares (1 million acres); otherwise it can be affected by air, water, and noise pollution from nearby mining, oil and natural gas drilling, timber cutting, industry, and urban development. Environmentalists urge that the world's remaining wilderness be protected by law everywhere, focusing first on the most endangered spots in the most wilderness-rich and biodiverse countries (Figure 16-26).

We need wild places where we can experience the beauty of nature and observe natural biological diversity. We need places where we can enhance our mental and physical health by getting away from noise, stress, and large numbers of people. Wilderness preservationist John Muir advised:

Climb the mountains and get their good tidings. Nature's peace will flow into you as the sunshine into the trees. The winds will blow their freshness into you, and the storms their energy, while cares will drop off like autumn leaves.

Even those who never use the wilderness may want to know it is there, a feeling expressed by novelist Wallace Stegner:

Save a piece of country … and it does not matter in the slightest that only a few people every year will go into it. This is precisely its value.… We simply need that wild country available to us, even if we never do more than drive to its edge and look in. For it can be a means of reassuring ourselves of our sanity as creatures, a part of the geography of hope.

Wilderness areas provide recreation for growing numbers of people. Wilderness also has important ecological values. It provides undisturbed habitats for wild plants and animals, protects diverse biomes from damage, and provides a laboratory in which we can discover more about how nature works. Wilderness is a biodiversity bank and an eco-insurance policy. In the words of Henry David Thoreau, "In wildness is the preservation of the world."

How Much Wilderness Should Be Protected?

Most wildlife biologists believe that the best way to preserve biodiversity is through a worldwide network of reserves, parks, wildlife sanctuaries, and other protected areas. As of 1993 there were about 7,000 nature reserves, parks, and other protected areas throughout the world, occupying 4.9% of the earth's land surface. That is an important beginning, but environmentalists say that to help protect biodiversity from being depleted by human activities a minimum of 10% of the globe's land area must be protected. Moreover, many existing reserves are too small to provide any real protection for the populations of wild species that live on them.

Q: What are proven reserves of a nonrenewable resource such as oil or copper?

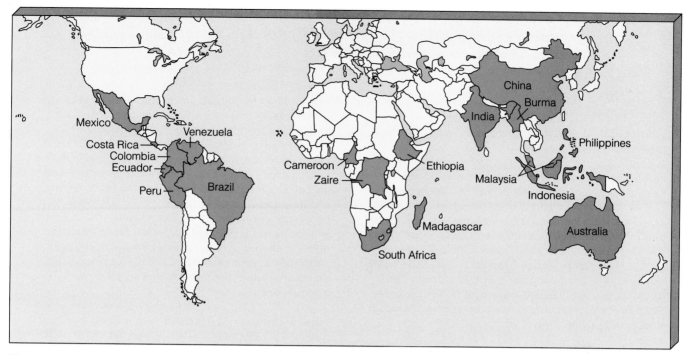

Figure 16-26 Earth's megadiversity countries. Environmentalists believe that efforts to preserve repositories of biodiversity should be concentrated in these species-rich countries. (Data from Conservation International and the World Wildlife Fund)

In 1981 UNESCO proposed that at least one, and ideally five or more, *biosphere reserves* be set up in each of Earth's 193 biogeographical zones. Each reserve should be large enough to prevent gradual species loss and should be designed to combine both conservation and sustainable use of natural resources. To date more than 300 biosphere reserves have been established in 76 countries.

A well-designed biosphere reserve has three zones: **(1)** a *core area* containing an important ecosystem that has had little, if any, disturbance from human activities; **(2)** a *buffer zone*, where activities and uses are managed in ways that help protect the core; and **(3)** a second *buffer* or *transition zone*, which combines conservation and sustainable forestry, grazing, agriculture, and recreation (Figure 16-19). Buffer zones can also be used for education and research.

Research indicates that in some areas several medium-sized interconnected reserves may offer wildlife more protection from extinction than a single large one. If the population of a species is wiped out in one reserve due to fire, epidemic, or some other disaster, the species might still survive in the other reserves. Conservation biologists also suggest establishing protected corridors between reserves to help support more species and allow migrations when environmental conditions in a reserve deteriorate.

So far most biosphere reserves fall short of the bioregional ideal. Most were imposed on existing na-

tional parks and reserves with little emphasis on expanding them to encompass bioregions. Also, too little funding has been provided for their protection and management.

Conservationist Norman Myers (Guest Essay, p. 446) proposes extending all key parks and reserves northward in the Northern Hemisphere (and southward in the Southern Hemisphere), and providing a network of corridors to allow migration of animal species and some plant species if global warming occurs (Section 10-2). Otherwise, he argues, existing parks and reserves will become death traps instead of sanctuaries.

An international fund to help LDCs protect and manage bioregions and biosphere reserves would cost $100 million per year—about what the world's nations spend on arms every 90 minutes. Because there won't be enough money to protect most of the world's biodiversity, environmentalists believe that efforts should be focused on the megadiversity countries (Figure 16-26). In the United States, environmentalists urge Congress to pass an Endangered Ecosystems or Biodiversity Protection Act as an important step toward surveying and preserving the country's biodiversity.

Solutions: The National Wilderness Preservation System In the United States, preservationists have been trying to keep wild areas from being developed since 1900. On the whole they have fought a losing

Tropical Forests and Their Species: Going, Going, Going...?

Norman Myers

GUEST ESSAY *Norman Myers is an international consultant in environment and development with emphasis on conservation of wildlife species and tropical forests. He has served as a consultant for many development agencies and research organizations, including the U.S. National Academy of Sciences, the World Bank, the Organization for Economic Cooperation and Development, various UN agencies, and the World Resources Institute. Among his recent publications (see Further Readings) are* Conversion of Tropical Moist Forests *(1980),* A Wealth of Wild Species *(1983),* The Primary Source: Tropical Forests and Our Future *(1984),* The Gaia Atlas of Future Worlds *(1990),* The Gaia Atlas of Planet Management *(1992),* Ultimate Security *(1993), and* Scarcity or Abundance: A Debate on the Environment *(1994).*

Tropical forests still cover an area roughly equivalent to the "lower 48" United States. Climatic and biological data suggest they could once have covered an area at least twice as large. So we have already lost half of them, mostly in the recent past.

Worse, remote-sensing surveys show that we are now destroying the forests at a rate of at least 1.25% a year, and we are grossly degrading them at a rate of at least another 1.25% a year—and both rates are accelerating rapidly. Unless we act now to halt this loss, within another few decades at most there could be little left except perhaps a block in central Africa and another in the western Amazon Basin. Even those remnants may not survive the combined pressures of population growth and land hunger beyond the middle of the next century.

This means that we are imposing one of the most broad-scale and impoverishing impacts that the biosphere has ever suffered in its 3.8 billion years of existence. Tropical forests represent the greatest celebration of nature to appear on the planet since the first flickerings of life. They are exceptionally complex ecologically, and they are remarkably rich biotically. Although they now cover only about 6% of Earth's land surface, they still are home for half, and perhaps three-quarters or more, of all species of plant and animal life. Thus elimination of these forests is by far the leading factor in the mass extinction of species that appears likely over the next few decades.

Already we are losing several species every day because of clearing and degradation of tropical forests. The time will surely come, and come soon, when we will be losing many thousands every year. The implications are profound, whether they be scientific, aesthetic, ethical—or simply economic. In medicine alone we benefit from myriad drugs and pharmaceuticals derived from tropical forest plants. The commercial value of these products worldwide can be reckoned at $40 billion each year.

By way of example, the rosy periwinkle [Figure 16-15] from Madagascar's tropical forests has produced two potent drugs used to treat Hodgkin's disease, leukemia, and other blood cancers. Madagascar has—or used to have—an estimated 10,000 plant species, 8,000 of which could be found nowhere else; but Madagascar has lost 93% of its virgin tropical forest. The U.S. National Cancer

battle. Not until 1964 did Congress pass the Wilderness Act, which allowed the government to protect undeveloped tracts of public land from development as part of the National Wilderness Preservation System.

Only 4% of U.S. land area is protected as wilderness, with almost three-fourths of it in Alaska. Only 1.8% of the land area of the lower 48 states is protected, most of it in the West. Almost half of the wilderness east of the Mississippi is in the threatened Florida Everglades National Park (Figure 11-19) and Minnesota's Boundary Waters Canoe Area. Of the 413 wilderness areas in the lower 48 states, only four are larger than 400,000 hectares. Furthermore, the present wilderness preservation system includes only 81 of the country's 233 distinct ecosystems. Like the national parks, most wilderness areas in the lower 48 states are habitat islands in a sea of development.

There remain almost 40 million hectares (100 million acres) of public lands that could qualify for designation as wilderness. Some environmentalists want all of this land protected as wilderness and advocate vigorous efforts to rehabilitate other lands to enlarge existing wilderness areas. Such efforts are strongly opposed by timber, mining, ranching, energy, and other interests who want either to extract resources from these and most other public lands or to convert them to private ownership (Spotlight, p. 187).

Wilderness recovery areas could be created by closing roads in large areas of public lands, restoring habitats, allowing natural fires to burn, and reintroducing species that have been driven from such areas. However, resource developers lobby elected officials and government agencies to build roads in national forests (and other areas being evaluated for inclusion in the

Q: When did humans start shifting from hunting and gathering to agriculture?

Institute estimates that many plants in tropical forests might have potential against various cancers, provided pharmacologists can get to them before they are eliminated by chain saws and bulldozers.

We benefit in still other ways from tropical forests. One critical environmental service is the "sponge effect," by which the forests soak up rainfall during the wet season and then release it in regular amounts throughout the dry season. When tree cover is removed and this watershed function is impaired, the result is an annual regime of floods followed by droughts, which destroys property and reduces agricultural productivity. There is also concern that if tropical deforestation continues unabated, it could trigger local, regional, or even global changes in climate. Such climatic upheavals would affect the lives of billions of people, if not the whole of humankind.

All this raises important questions about our role in the biosphere and our relations with the natural world around us. As we proceed on our disruptive way in tropical forests, we—political leaders and the general public alike—give scarcely a moment's thought to what we are doing. We are deciding the fate of the world's tropical forests unwittingly, yet effectively and increasingly.

The resulting shift in evolution's course, stemming from the elimination of tropical forests, will rank as one of the greatest biological upheavals since the dawn of life. It will equal, in scale and significance, the development of aerobic respiration, the emergence of flowering plants, and the arrival of limbed animals, which took place over eons of time. Whereas those were enriching

disruptions in the course of life on this planet, the loss of biodiversity associated with the destruction and degradation of tropical forests will be almost entirely an impoverishing phenomenon brought about entirely by human actions. And it will all have occurred within the twinkling of a geologic eye.

In short our intervention in tropical forests should be viewed as one of the most challenging problems that humankind has ever encountered. After all, we are the first species ever to be able to look upon nature's work and to decide whether we should consciously eliminate it or leave much of it untouched.

So the decline of tropical forests is one of the great "sleeper" issues of our time—an issue with far more sweeping consequences than it might initially seem. Yet we can still save much of these forests and the species they contain. Should we not consider ourselves fortunate that we alone among all generations are being given the chance to preserve tropical forests as the most exuberant expression of nature in the biosphere—and thereby to support the right to life of many of our fellow species and their capacity to undergo further evolution without human interference?

Critical Thinking

1. Should MDCs provide most of the money to preserve remaining tropical forests in LDCs? Explain.

2. What can you do to help preserve some of the world's tropical forests? Which, if any, of these actions do you plan to carry out?

---

wilderness system) so that they can't be designated as wilderness, and they strongly oppose the idea of wilderness recovery areas.

Solutions: Wilderness Management To protect the most popular areas from damage, wilderness managers must designate areas where camping is allowed and limit the number of people hiking or camping at any one time. Managers have increased the number of wilderness rangers to patrol vulnerable areas, and they have enlisted volunteers to pick up trash discarded by thoughtless users.

Historian and wilderness expert Roderick Nash suggests that wilderness areas be divided into three categories. The easily accessible, popular areas would be intensively managed and have trails, bridges, hiker's huts, outhouses, assigned campsites, and exten-

sive ranger patrols. Large, remote wilderness areas would be used only by people who get a permit by demonstrating their wilderness skills. The third category—biologically unique areas–would be left undisturbed as gene pools of plant and animal species, with no human entry allowed.

National Wild and Scenic Rivers System

In 1968 Congress passed the National Wild and Scenic Rivers Act. It allows rivers and river segments with outstanding scenic, recreational, geological, wildlife, historical, or cultural values to be protected in the National Wild and Scenic Rivers System. These waterways are to be kept free of development. They may not be widened, straightened, dredged, filled, or dammed along the designated lengths. The only activities allowed are camping, swimming, nonmotorized boating,

Figure 16-27 Wangari Maathai—the first Kenyan woman to earn a Ph.D. (in anatomy) and head a department (veterinary medicine) at the University of Nairobi—organized the internationally acclaimed Green Belt Movement in 1977. The goal of this widely regarded women's self-help community action group is to plant a tree for each of Kenya's 28 million people. She recruited 50,000 women to establish tree nurseries and to help farmers raise tree seedlings. Members of the group get a small fee for each tree that survives. By 1990 more than 10 million trees had been planted. The success of this project has sparked the creation of similar programs in more than a dozen other African countries. She and members of this group are true Earth heroes.

sport hunting, and sport and commercial fishing. New mining claims are permitted in some areas, however.

Currently only 0.3% of the country's 6 million kilometers (3.5 million miles) of waterways in about 150 stretches are protected by the Wild and Scenic River System. In contrast, 17% of the country's wild river length has been tamed and drowned by dams (Figure 11-10). Environmentalists have urged Congress to add 1,500 additional eligible river segments to the system by 2000—a goal that is vigorously opposed by some local communities and anti-environmental groups. If that goal is achieved, about 2% of the country's river systems would be protected. Environmentalists also urge that a permanent federal administrative body be established to manage the Wild and Scenic Rivers System and that states develop their own wild and scenic river programs.

National Trails System In 1968 Congress passed the National Trails Act, which protects scenic and historic hiking trails in the National Trails System. However, no designated scenic or historic trail is complete because the Trails System has a low priority and gets little funding.

To fulfill the promise of the Trails Act, environmentalists propose that the Trails System be managed by a single agency. Armed with a comprehensive study of present and future trail needs, the agency should plan to develop a countrywide network of trails in the next decade.

Change from the Bottom Up People throughout the world are working to protect forests and other endangered ecosystems (p. 434). Penan tribespeople in Sarawak, Malaysia, have joined forces with environmentalists to halt destructive logging, which has reduced their population from 10,000 to less than 500. In Brazil 500 conservation organizations have formed a coalition to preserve the country's remaining tropical forests. In the United States, members of Earth First! have perched in the tops of giant Douglas firs and lain in front of logging trucks and bulldozers to prevent the felling of ancient trees in national forests.

In Kenya, Wangari Maathai started the Green Belt Movement, a national self-help community action effort by 50,000 women farmers and half a million schoolchildren to plant trees for firewood and to help hold the soil in place (Figure 16-27). This inspiring leader and Earth hero has said:

> I don't really know why I care so much. I just have something inside me that tells me that there is a problem and I have got to do something about it. And I'm sure it's the same voice that is speaking to everyone on this planet, at least everybody who seems to be concerned about the fate of the world, the fate of this planet.

For her heroic efforts to sustain the earth and fight for human rights, the Kenyan government closed down her Green Belt Movement offices (she moved the headquarters into her home) and jailed her twice. In 1992 she was severely beaten by police while leading a hunger strike pressing for release of political prisoners.

Sustaining existing forests, rangelands, wilderness, and parks, and rehabilitating damaged areas are important tasks. This will cost a great deal of money, but the long-term costs of not doing these things will be much higher.

We abuse land because we regard it as a commodity belonging to us. When we see land as a community to which we belong, we may begin to use it with love and respect.

ALDO LEOPOLD

Critical Thinking

1. Should private companies cutting timber from national forests continue to be subsidized by federal payments for reforestation and for building and maintaining access roads ? Explain.

2. Should all cutting on remaining old-growth forests in U.S. national forestlands be banned? Explain.

Q: In terms of resource use and environmental impact, what is the most overpopulated country in the world?

3. Explain why you agree or disagree with each of the proposals listed on pp. 422–423 for reforming federal forest management in the United States.

4. What difference could the loss of essentially all the remaining virgin tropical forests and old-growth forests in North America have on your life and on the lives of your descendants?

5. Explain why you agree or disagree with each of the proposals listed, starting on p. 429, concerning protection of the world's tropical forests.

6. Explain why you agree or disagree with the proposals on pp. 439–440 for providing more sustainable use of western public rangeland in the United States.

7. Should cattle be banned from grazing on western public rangeland? Explain.

8. Should trail bikes, dune buggies, and other off-road vehicles be banned from public rangeland to reduce damage to vegetation and soil? Explain. Should such a ban also include national forests, national wildlife refuges, and national parks? Explain.

9. **a.** Should individuals and corporations with grazing or mineral rights on public lands in the United States be compensated financially if these rights are reduced or taken away by the federal government? Explain.

   b. Should individuals and corporations be compensated financially if they are prevented from using or from extracting timber or other resources from land they own because the land is classified by the federal or state governments as protected wetlands or habitats for endangered or threatened wildlife species? Explain.

10. Explain why you agree or disagree with each of the proposals listed on pp. 442–443 concerning the U.S. national park system.

11. Should more wilderness areas and wild and scenic rivers be preserved in the United States, especially in the lower 48 states? Explain.

*12. Make a concept map of the key ideas in this chapter using the section heads and subheads and the key terms (shown in boldface type in the chapter). See the inside front cover and Appendix 4 for information on concept maps.

17 Biodiversity: Sustaining Wild Species

Figure 17-1 Passenger pigeons, extinct in the wild since 1900. The last known passenger pigeon died in the Cincinnati Zoo in 1914. (John James Audubon/The New-York Historical Society)

The Passenger Pigeon: Gone Forever

In the early 1800s bird expert Alexander Wilson watched a single migrating flock of passenger pigeons darken the sky for over four hours. He estimated that this flock was more than 2 billion birds strong, 386 kilometers (240 miles) long, and 1.6 kilometers (1 mile) wide.

By 1914 the passenger pigeon (Figure 17-1) had disappeared forever. How could the species that was once the most common bird in North America become extinct in only a few decades?

The answer is humans. The main reasons for the extinction of this species were uncontrolled commercial hunting and loss of the bird's habitat and food supply as forests were cleared for farms and cities.

Passenger pigeons were good to eat, their feathers made good pillows, and their bones were used for fertilizer. They were easy to kill because they flew in gigantic flocks and nested in long, narrow colonies.

Beginning in 1858 passenger pigeon hunting became a big business. Shotguns, traps, artillery, and even dynamite were used. Birds were suffocated by burning grass or sulfur below their roosts. Live birds were used as targets in shooting galleries. In 1878 one professional pigeon trapper made $60,000 by killing 3 million birds at their nesting grounds near Petoskey, Michigan.

By the early 1880s commercial hunting had ceased because only a few thousand birds were left. At that point recovery of the species was doomed because the females laid only one egg per nest. On March 24, 1900, a young boy in Ohio shot the last known wild passenger pigeon. The last passenger pigeon on Earth, a hen named Martha after Martha Washington, died in the Cincinnati Zoo in 1914. Her stuffed body is now on view at the National Museum of Natural History in Washington, D.C.

Sooner or later all species become extinct, but humans have become a primary factor in the premature extinction of more and more species as we march relentlessly across the globe. Every day at least 10 and perhaps as many as 150 species become extinct because of our activities. The hemorrhage of biodiversity may soon reach several hundred species per day.

The last word in ignorance is the person who says of an animal or plant: "What good is it?"... If the land mechanism as a whole is good, then every part of it is good, whether we understand it or not.... Harmony with land is like harmony with a friend; you cannot cherish his right hand and chop off his left.

ALDO LEOPOLD

This chapter will be devoted to answering the following questions:

- Why should we care about wildlife?
- What human activities and natural traits endanger wildlife?
- How can we prevent premature extinction of species?
- Can game animals be managed sustainably?
- Can freshwater and marine fish be managed sustainably?

17-1 WHY PRESERVE WILD SPECIES?

Millions of species have vanished over Earth's long history, so why should we worry about losing a few more? Does it matter that the passenger pigeon (Figure 17-1), the California condor (Figure 17-2), the black rhinoceros (Figure 4-41), the northern spotted owl (Figure 16-10), the loggerhead turtle (Figure 17-3), or some unknown plant or insect in a tropical forest becomes extinct mostly because of human activities?

The answer is yes. Here are several reasons for protecting wild species from premature extinction as a result of human activities (and thereby allowing them to play their roles in the ongoing saga of evolution):

- *Economic importance.* About 10% of the U.S. gross national product comes directly from use of **wildlife resources**, wild species that are actually or potentially useful to human beings. Some 90% of today's food crops were domesticated from wild tropical plants (Figure 14-5). Agricultural scientists and genetic engineers will need to use existing wild plant species, most of them still unknown, both as sources of food (Figure 14-10) and to develop new crop strains (Section 14-3). Wild animal species are a largely untapped food source. Wild plants and plants domesticated from wild species supply rubber (Figure 16-14), oils, dyes, fiber, paper, lumber, and other useful products (Figure 16-3). Nitrogen-fixing microbes in the soil and in plants' root nodules (Figure 4-31) supply nitrogen to grow food crops worth almost $50 billion per year worldwide ($7 billion in the U.S.). Pollination by birds and insects is essential to

Figure 17-2 California condor. The few California condors still in the wild were collected in 1985 to become part of a $1.5-million-per-year captive breeding program. By January 1994 there were 72 of these birds in the Los Angeles Zoo and the San Diego Wild Animal Park, and 5 of 8 captive-bred birds released to a natural habitat north of Los Angeles had survived. This species is especially vulnerable to extinction because of its low reproduction rate, the long period (seven years) it needs to reach reproductive age, the failure of parents (scared away from the nest by noise or human activities) to hatch chicks, and the need for a large, undisturbed habitat.

Figure 17-3 Endangered loggerhead turtle lying on a beach at Bahia State, Brazil. Of the world's 270 known turtle species, 42% are rare or are threatened with extinction.

Figure 17-4 The nine-banded armadillo is being used to study leprosy and to prepare a vaccine for it.

many food crops, including 40 U.S. crops valued at approximately $30 billion per year.

- *Medical importance.* About 80% of the world's population relies on plants or plant extracts for medicines. Prescription and nonprescription drugs worth $100 billion per year—roughly half of the medicines used in the world, and 25% of those used in the United States—have active ingredients extracted from wild species (Figure 16-15). Only about 5,000 of the estimated 250,000 plant species have been studied thoroughly for their possible medical uses. In addition, over 3,000 antibiotics—including penicillin and tetracycline—are derived from microorganisms. Many animal species are used to test drugs (Figure 17-4), vaccines, chemical toxicity (Section 8-2), and surgical procedures, and in studies of human health and disease. Under intense pressure from animal rights groups, scientists are using alternatives such as cell and tissue cultures, simulated tissues and body fluids, bacteria, and computer-generated models and simulations. However, they caution that alternative techniques cannot replace all animal research.

- *Aesthetic and recreational importance.* Wild plants and animals are a source of beauty, wonder, joy, and recreational pleasure for many people (Figure 17-5). Wild **game species** provide recreation in the form of hunting, fishing, observation, and photography. Wildlife tourism, sometimes called *ecotourism*, generates as much as $30 billion in revenues each year ($18 billion per year in the United States alone). One wildlife economist has estimated that one male lion living to age 7 generates $515,000 in tourist dollars in Kenya; by contrast, if killed for its skin, the lion would bring only about $1,000. Similarly, each of

Figure 17-5 Many species of wildlife, like this bird of paradise in a New Guinea tropical forest, are a source of beauty and pleasure. This species is vulnerable to extinction because it can't mate unless a throng of males assembles and "strut their stuff" in front of females. When numbers decline this is not possible.

Kenya's 20,000 elephants brings in about $20,000 per year in tourist income. Over a lifetime of 60 years a Kenyan elephant is worth close to $1 million in tourist revenue. However, care must be taken to ensure that ecotourists neither damage nor disturb wildlife and ecosystems, nor disrupt local cultures.

- *Scientific importance.* Each species has scientific value because it can help scientists understand how life has evolved and functions—and how it will continue to evolve—on this planet.

- *Ecological importance.* Wild species supply us (and other species) with food from the soil and the sea, recycle nutrients essential to agriculture, and help

Q: What percentage of the world's commercial energy is used by the United States?

Figure 17-6 Some species that have become extinct largely because of human activities, mostly habitat destruction and overhunting.

Passenger pigeon | Great auk | Dodo | Bushy seaside sparrow | Aepyomis (Madagascar)

maintain soil fertility. They also produce oxygen and other gases in the atmosphere, moderate Earth's climate, help regulate water supplies, and store solar energy. Moreover, they detoxify poisonous substances, break down organic wastes, control potential crop pests and disease carriers, and make up a vast gene pool from which we and other species can draw.

■ *Ethics.* To some people, each wild species has an inherent right to exist—or to struggle to exist. According to this view, it is morally wrong for us to hasten the extinction of any species. Some go further and assert that each individual organism—not just each species—has a right to survive without human interference, just as each human being has the right to survive. Some people distinguish between the survival rights of plants and those of animals, mostly for what they consider practical reasons. Poet Alan Watts, for example, once said that he was a vegetarian "because cows scream louder than carrots." Other people distinguish among various types of animals. For instance, they think little about killing a fly, mosquito, cockroach, or sewer rat. Unless they are strict vegetarians, they also think little about having others kill domesticated animals in slaughterhouses to provide them with meat, leather, and other products. These same people, however, might deplore the killing of wild animals such as deer, squirrels, or rabbits.

17-2 HOW SPECIES BECOME DEPLETED AND EXTINCT

The Rise and Fall of Species Extinction is a natural process (Section 5-6). As the planet's surface and climate have changed over the 4.6 billion years of its existence, species have disappeared and new ones have evolved to take their places (Figure 5-36).

This rise and fall of species has not been smooth. Evidence indicates that there have been several periods when mass extinctions reduced Earth's biodiversity—and other periods, called *radiations*, when the diversity of life has increased and spread (Figure 5-40).

The Current Extinction Crisis: Fatal Subtraction Imagine you are driving on an interstate highway at a high speed, and that all the while your passengers are busy dismantling parts of your car and throwing them out the window. How long will it be before they remove enough parts to cause a crash?

This urgent question, applied to the earth, is one that we as a species should be asking ourselves. As we tinker with the only home for us and other species, we are rapidly removing parts of Earth's natural biodiversity, upon which we and other species depend in ways we know little about. We are not heeding Aldo Leopold's warning: "To keep every cog and wheel is the first precaution of intelligent tinkering."

Past mass extinctions took place slowly enough to allow new forms of life to arise as adaptations to an ever-changing world. This process started to change about 40,000 years ago when the latest version of our species came on the scene. Since agriculture began about 10,000 years ago, estimated extinction rates have soared as human settlements have expanded worldwide (Figure 17-6).

It's difficult to document extinctions, for most go unrecorded. But biologists estimate that during 1993 at least 4,000 and as many as 36,000 species became extinct (an average of 11 to 100 species per day); the figure could reach 50,000 species per year by 2000. These scientists warn that if deforestation (especially of tropical forests), desertification, and destruction of wetlands and coral reefs continue at their current rates, within the next few decades we could easily lose at least a quarter of the earth's species. This massive bleeding of life from the planet will rival some of the great natural mass extinctions of the past.

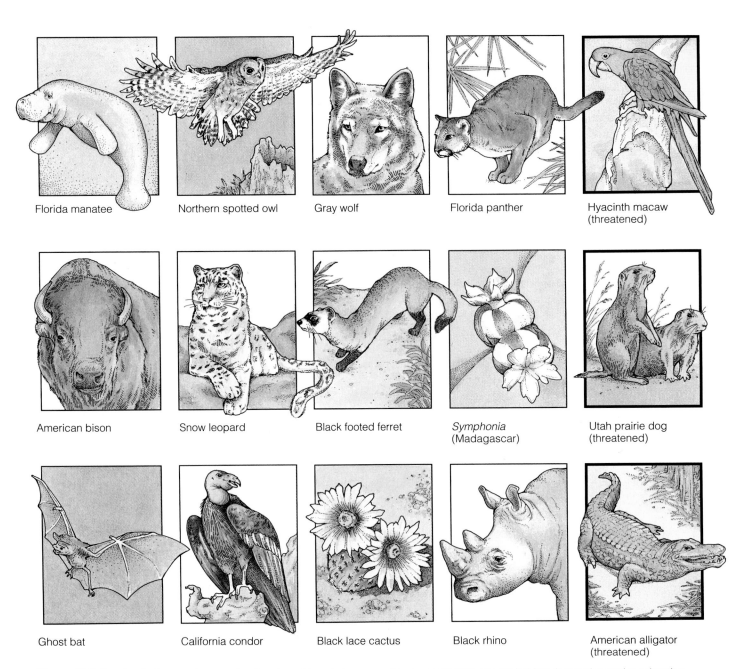

Figure 17-7 Species that are endangered or threatened largely because of human activities, mostly habitat destruction and overhunting.

Mass extinctions occurred long before we arrived, but there are two important differences between the present mass extinction and those of the past:

- *The present extinction crisis is the first to be caused by a single species—our own.* By using 40% of Earth's terrestrial net primary productivity (Figure 4-28), we are crowding out other terrestrial species. What will happen to wildlife and the services they provide, for us and for one another, if our population doubles in the next 40 years and we try to use as much as 80% of the planet's terrestrial net primary productivity?

- *The current wildlife holocaust is taking place in only a few decades, rather than over thousands to millions of*

years. Such rapid extinction cannot be balanced by speciation because it takes 2,000–100,000 generations for new species to evolve. Some analysts, however, claim there is no extinction crisis (Spotlight, p. 456).

Endangered and Threatened Species   Species heading toward extinction are classified as either *endangered* or *threatened* (Figure 17-7). An **endangered species** has so few individual survivors that the species could soon become extinct over all or most of its natural range. Examples are the California condor in the United States (only 5 in the wild—Figure 17-2), the whooping crane in North America (250 left—Figure 17-7), the giant panda in central China (1,000 left—Fig-

454

Q: What are the three most productive types of ecosystems?

Grizzly bear
(threatened)

Arabian oryx

Bald eagle

Kirtland's warbler

African elephant

Mojave desert tortoise
(threatened)

Swallowtail butterfly

Humpback chub

Golden lion tamarin

Siberian tiger

Peregrine falcon
(threatened)

Giant panda

Knowlton cactus

Whooping crane

Blue whale

ure 4-37), the snow leopard in central Asia (2,500 left—Figure 17-7), the black rhinoceros in Africa (2,400 left—Figure 4-41), and the rare swallowtail butterfly (Figure 17-7).

A **threatened species** is still abundant in its natural range but is declining in numbers and likely to become endangered. Examples are the bald eagle (Figure 17-8) the grizzly bear (Figure 17-7), and the American alligator (Figure 17-7 and Connections, p. 84).

Some species have characteristics that make them more vulnerable than others to premature extinction (Table 17-1). One trait that affects the survival of species under different environmental conditions is their reproductive strategy. Species that take years to mature and have few offspring are especially vulnerable. Generally,

in times of stress r-strategist species have an advantage over many K-strategist species (Section 5-4).

Each animal species has a critical population density and size, below which survival may be doubtful because males and females have a hard time finding each other. Once a population falls below its critical size, it continues to decline, even if the species is protected, because its death rate exceeds its birth rate. The remaining small population can easily be wiped out by fire, flood, landslide, disease, or some other single catastrophe. Some species, such as bats, are vulnerable to extinction for a combination of reasons (Connections, p. 459).

The populations of many wild species, while not in danger of extinction, have diminished locally or

A: Estuaries, swamps and marshes, and tropical rain forests

Is There Really an Extinction Crisis?

E. O. Wilson and a number of other prominent biologists have warned that we are involved in a catastrophic loss of species. They also expect this situation to get worse as more people consume more resources and destroy or degrade more forests and other ecosystems that are habitats for wildlife.

Some social scientists and a few biologists, however, question the idea that there is an extinction crisis. They point to several problems in estimating species loss. First, we don't know how many species there are, with estimates ranging from 5 million to 100 million. Second, it is difficult to observe species extinction, especially for species we know little, if anything, about. We know that loss and fragmentation of habitat can lead to extinction for some species, especially highly specialized ones often found in increasingly threatened habitats such as tropical forests and coral reefs. However,

it is difficult to come up with a numerical relationship between habitat loss and species loss.

The annual loss of tropical forest habitat is estimated at about 1.8% per year. E. O. Wilson and several tropical biologists who have counted species in patches of tropical forest before and after destruction or degradation estimate that this 1.8% loss in habitat results in a roughly 0.5% loss of species. However, biomes vary in the number of species they contain (species diversity), and the ratio of habitat loss to species loss varies in different biomes.

Do such estimates add up to an extinction crisis? Let us assume, as Wilson and many other biologists do, that a loss of 1 million species represents an extinction crisis. If we assume a global decline in species of 0.5% per year, then we will lose 25,000 species per year if there are 5 million species; 100,000 per year if there are 20 million species; and 500,000 per year if there are 100 million species. If these assumptions are correct, then we will lose 1 mil-

lion species in 40 years if there are 5 million species; in 10 years with 20 million species; and in only 2 years with 100 million species.

Let's assume however, that the estimate of 0.5% species loss per year is too large for the earth as a whole. If it is 0.25% per year, we will lose 1 million species in 80 years with 5 million species; in 20 years with 20 million species; and in 4 years with 100 million species. Even if we halve the estimated species loss again to 0.125% per year, we can lose 1 million species in 8 to 160 years, easily enough to qualify the situation as an extinction crisis.

The point biologists are trying to make is not that their estimates are precise. Instead, they argue that there is ample evidence that we are destroying and degrading wildlife habitats at an exponentially increasing rate and that this certainly leads to a large loss of species—although the number and rate will vary in different parts of the world.

Figure 17-8 Bald eagle with pink salmon at Kachemak Bay, Alaska. During the late 1960s and early 1970s the bald eagle population in the lower 48 states declined because of loss of habitat, illegal hunting, and fragile eggs caused by uptake of pesticides in fish, their primary diet. Now this protected species is making a comeback.

regionally. Such species may be a better indicator of the condition of entire ecosystems than endangered and threatened species. They can serve as early warnings so that we can prevent species extinction rather than responding to emergencies.

Habitat Loss and Fragmentation The underlying causes of extinction and population reduction of wildlife are human population growth (Figure 1-1) and economic systems and policies that fail to value the environment and its vital ecosystem services (Section 7-1)—and thus promote unsustainable exploitation. As our population grows, we clear, occupy, and damage more land to supply food, fuelwood, timber, and other resources.

Affluence, poverty, and the widening income gap between the rich and the poor (Figure 1-6) all contribute to extinctions. Increasing affluence and economic growth lead to greater average resource use per person, which is a prime factor in taking over wildlife habitats (Figure 1-13). In LDCs the combination of rapid population growth (Figures 6-3 and 6-5) and

Q: What is the name of the atmosphere's innermost layer?

Table 17-1 Characteristics of Extinction-Prone Species

Characteristic	Examples
Low reproduction rate	Blue whale, polar bear, California condor, Andean condor, passenger pigeon, giant panda, whooping crane
Specialized feeding habits	Everglades kite (eats apple snail of southern Florida), blue whale (krill in polar upwellings), black-footed ferret (prairie dogs and pocket gophers), giant panda (bamboo), koala (certain eucalyptus leaves)
Feed at high trophic levels	Bengal tiger, bald eagle, Andean condor, timber wolf
Large size	Bengal tiger, lion, elephant, Javan rhinoceros, American bison, giant panda, grizzly bear
Limited or specialized nesting or breeding areas	Kirtland's warbler (6- to 15-year-old jack pine trees), whooping crane (marshes), orangutan (only on Sumatra and Borneo), green sea turtle (lays eggs on only a few beaches), bald eagle (prefers forested shorelines), nightingale wren (only on Barro Colorado Island, Panama)
Found in only one place or region	Woodland caribou, elephant seal, Cooke's kokio, many unique island species
Fixed migratory patterns	Blue whale, Kirtland's warbler, Bachman's warbler, whooping crane
Preys on livestock or people	Timber wolf, some crocodiles
Behavioral patterns	Passenger pigeon and white-crowned pigeon (nest in large colonies), redheaded woodpecker (flies in front of cars), Carolina parakeet (when one bird is shot, rest of flock hovers over body), Key deer (forages for cigarette butts along highways—it's a "nicotine addict")

poverty (Figure 1-6) push the poor to cut forests, grow crops on marginal land, overgraze grasslands, deplete fish species, and kill endangered animals for their valuable furs, tusks, or other parts in order to survive.

Thus, the greatest threat to most wild species is losing their homes as we increasingly occupy or degrade more of the planet (Figure 17-9). Tropical deforestation (Figure 16-11) is the greatest eliminator of species, followed by destruction of coral reefs (Case Study, p. 104) and wetlands (Section 5-2) and plowing of grasslands (Figure 5-12). Estimated wildlife habitat loss is quite high in a number of countries including Vietnam (80%), the Philippines (79%), Madagascar (75%), Burma (71%), and Ethiopia (70%).

In the lower 48 states 98% of the tall-grass prairies (Figure 5-10) have been plowed, half of the wetlands drained (Section 5-2), 90–95% of old-growth forests cut (Figure 16-9), and overall forest cover reduced by 33%. At least 500 native species have been driven to extinction (Figure 17-6) and others to near-extinction (Figures 17-2 and 17-7), mostly because of habitat loss. Furthermore, much of the remaining wildlife habitat is being fragmented into vulnerable patches, or "habitat islands" (Figure 16-6) and polluted at an alarming rate.

Island species are especially vulnerable to extinction. The endangered *Symphonia* (Figure 17-7) clings to life on the large island of Madagascar, a megadiversity country (Figure 16-26) where 90% of the original vegetation has been destroyed and hundreds of unique species are threatened with extinction (Figure

17-11). Hawaii is fast becoming the global extinction "capital" because of increasing population and development and the introduction of almost 4,000 alien plant and animal species since 1778. Although the islands make up only 0.2% of U.S. land area, they account for 75% of the nation's known bird and plant extinctions. Generally, the smaller an island is and the farther it is from a source of potential colonizing species, the fewer species it supports.

However, any habitat surrounded by a different one is, in effect, an island for the species that live there. Human settlements, roads (Figure 16-5), and clear-cut areas (Figure 16-7) break habitats into patches (Figure 16-6), or "habitat islands" that may be too small to support the minimum breeding populations of some species. Most national parks (Figure 16-2) and other protected areas are habitat islands, many of them surrounded by potentially damaging logging, mining, energy extraction, and industrial activities. Freshwater lakes are also examples of habitat islands. They are especially vulnerable when non-native species are introduced accidentally or deliberately (Case Study, p. 288).

Migrating species face a double habitat problem. Nearly half of the 700 U.S. bird species spend two-thirds of the year in the tropical forests of Central or South America, or the Caribbean islands, and they return to North America during the summer to breed. A U.S. Fish and Wildlife Service study showed that between 1978 and 1987 populations of 44 species of insect-eating, migratory songbirds in North America

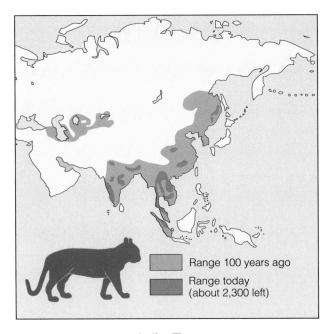

Indian Tiger

Range 100 years ago

Range today
(about 2,300 left)

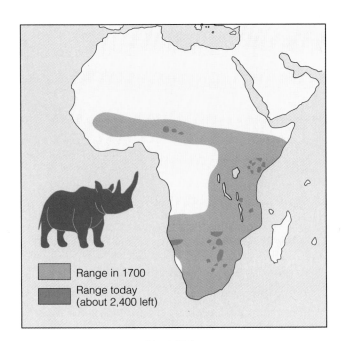

Black Rhino

Range in 1700

Range today
(about 2,400 left)

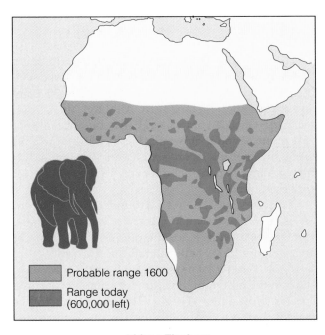

African Elephant

Probable range 1600

Range today
(600,000 left)

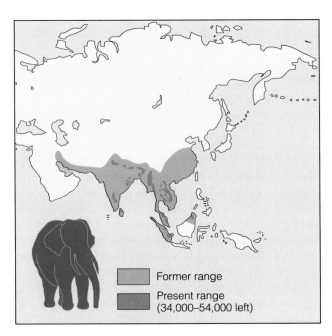

Asian or Indian Elephant

Former range

Present range
(34,000–54,000 left)

Figure 17-9 Reduction in the range of several species, mostly through a combination of habitat loss and hunting. What will happen to these and millions of other species when the global human population doubles in the next few decades? (Data from International Union for the Conservation of Nature and World Wildlife Fund)

declined, with 20 species showing drops of 25–45% (Figure 17-12). The main culprits are logging of tropical forests in their winter habitats (Figure 16-11) and fragmentation of their summer forest habitats in North America. As farms, freeways, and suburbs intrude, they not only remove trees and break forests into patches; they also create more "edge" habitat for opossums, skunks, squirrels, chipmunks, raccoons, and blue jays that feast on the eggs and young of migrant songbirds—as well as habitats for parasitic cowbirds that take over their nests. Three-fourths of the world's 9,000 known bird species are declining in numbers or

Q: When did the first photosynthesizing cells arise on the earth?

Bats: A Bad Rap

Despite their variety (950 species) and worldwide distribution, bats have several traits that expose them to extinction because of human activities. Many bats nest in huge cave colonies, which become vulnerable to destruction when people block the caves' entrances. And once the population of a bat species falls below a certain level, it may not recover because of its slow reproductive rate.

Bats play significant ecological roles and are also of great economic importance. Some bat species help control many crop-damaging insects and other pest species, such as mosquitoes and rodents (Figure 17-10). About 70% of all bat species feed on night-flying insects, making them the primary controls for such insects.

Other species of bats eat pollen; still others eat certain fruits. Bats with this kind of specialized feeding are the chief pollinators for certain trees, shrubs, and other plants; they also spread plants throughout tropical forests by excreting undigested seeds. If these keystone species are eliminated from an area, dependent plants will disappear. U.S. examples of bat-pollinated species are the giant saguaro cactus (Figure 4-11) and agaves (plants of the desert Southwest used in making fiber rope and tequila). In Southeast Asia, a cave-dwelling, nectar-eating bat species is the only known pollinator of durian trees, whose fruit is worth $120 million per year. If you enjoy bananas, cashews, dates, figs, avocados, or mangos, you can thank bats. Research on bats has contributed to the development of birth control and artificial insemination methods, and testing of drugs, studies of disease resistance and aging, vaccine production, and development of navigational aids for the blind.

People mistakenly fear bats as filthy, aggressive, rabies-carrying, bloodsuckers. But most bat species are harmless to people, livestock, and crops. In the United States only 10 people have died of bat-transmitted disease in four decades of record keeping. More Americans die each year from falling coconuts. Only three species of bats (none of them found in the United States) feed on blood, mostly that of cattle or wild animals. These bat species

Figure 17-10 Endangered ghost bat carrying a mouse in tropical northern Australia. This carnivorous night-feeding bat is harmless to people. Bats are considered keystone species in many ecosystems because of their roles in pollinating plants, dispersing seeds, and controlling insect and rodent populations.

can be serious pests to domestic livestock but rarely affect humans.

Because of unwarranted fears of bats and misunderstanding of their vital ecological roles, several species have been driven to extinction, and others, such as the ghost bat (Figure 17-9), are endangered or threatened. We need to see bats as valuable allies, not as enemies.

---

are threatened with extinction, mostly because of habitat loss and fragmentation.

Commercial Hunting and Poaching   Today subsistence hunting (for food) is rare because of the decline in hunting-and-gathering societies (Connections, p. 428). Sport hunting is closely regulated in most countries, and game species are endangered only where protective regulations do not exist or are not enforced.

However, a combination of habitat loss, legal commercial hunting, and illegal commercial hunting (poaching) has driven many species, such as the American bison (Case Study, p. 461) to or over the brink of extinction (Figures 17-6 and 17-7). Bengal tigers are in trouble because a tiger fur sells for $100,000 in Tokyo. A mountain gorilla is worth $150,000 and a chimpanzee $50,000; an ocelot skin, $40,000; an Imperial

Figure 17-11 Endangered ring-tailed lemur in Madagascar. Lemurs are the oldest distant relative of the human species. The 30 species of lemurs on this island are found nowhere else in the wild. Half of them are endangered.

Figure 17-12 Many species of migratory North American songbirds, such as the wood thrush, are suffering serious population declines because of loss of their winter tropical forest habitats in Latin America and the Caribbean islands, and fragmentation of their summer habitats in North America. The wood thrush population, found over most of the eastern United States during the summer, declined 31% between 1978 and 1987. Three-fourths of the world's bird species are declining in population or are threatened with extinction.

Figure 17-13 In Yemen, rhino horns are carved into ornate dagger handles, which sell for $500–$12,000. In China and other parts of Asia, powdered rhino horn is used for medicinal purposes and as an alleged aphrodisiac. All five species of rhinoceros are threatened with extinction because of poachers (who kill them for their horns) and loss of habitat. Efforts to protect these species from extinction are under way, but only about 12,000 rhinos remain in the wild today.

Figure 17-14 Vultures feeding on an elephant carcass in Tanzania. Poachers killed it to cut off its ivory tusks, which are used to make jewelry, piano keys, ornamental carvings, and art objects. The 1990 international ban on the trade of ivory from African elephants may help save this species.

Figure 17-15 Increasing numbers of bull walruses found in the Bering Sea off the coast of Alaska are being killed illegally for their ivory tusks. A pair of tusks is worth $800–$1,500 on the black market. Walruses are easy to kill because they spend the summer on ice floes. Hunters (mostly Inuit) in small boats shoot the slow-moving mammals and use the liver, heart, and other parts for food (legally allowed). Some also sever the head with a chain saw and illegally sell the tusks. With an estimated population of 250,000 in the Pacific, the walrus is not endangered now but could be if hunting for its tusks continues.

Amazon macaw, $30,000; a snow leopard skin, $14,000. Rhinoceros horn sells for as much as $28,600 per kilogram (Figure 17-13). Only about 200 Siberian tigers are left, mostly because these and other tigers are killed for their furs and for their bones, which are ground up and used in Chinese medicines.

For decades poachers have slaughtered elephants for their valuable ivory tusks (Figure 17-14). Legal and illegal trade in ivory—as well as habitat loss (Figure 17-9)—have reduced African elephant numbers from

Q: What is the *first law of human ecology?*

Frontier Expansion and Biological Warfare

CASE STUDY

In 1500, before Europeans settled North America, 60 to 125 million American bison grazed the grassy plains, prairies, and woodlands over much of the continent. A single herd might thunder past for hours. Several Native American tribes depended heavily on bison but did not deplete them, killing just enough for their food, clothing, and shelter. By 1906, however, the once-vast range of the bison had shrunk to a tiny area, and the species was driven nearly to extinction. (Figure 17-16).

As settlers moved west after the Civil War, the sustainable balance between Native Americans and bison was upset. The Sioux, Apaches, Comanches, and other plains tribes traded bison skins to settlers for steel knives and firearms, and they began killing more bison. The most relentless slaughter, however, was carried out by the intruding white settlers.

As railroads spread westward in the late 1860s, railroad companies hired professional bison hunters—like Buffalo Bill Cody—to supply construction crews with meat. Passengers also gunned down bison from train windows for sport, leaving the carcasses to rot. Commercial hunters shot millions of bison for their hides and tongues (considered

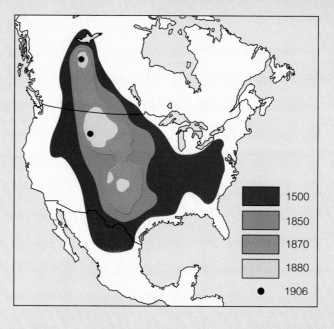

Figure 17-16 The range of the North American bison shrank severely between 1500 and 1906.

■	1500
■	1850
■	1870
□	1880
●	1906

a delicacy), leaving most of the meat to rot. "Bone pickers" collected the bleached bones and shipped them east to be ground up as fertilizer.

Farmers shot bison because they damaged crops, fences, telegraph poles, and sod houses. Ranchers killed them because they competed with cattle and sheep for grass. The U.S. Army killed bison as part of their campaign to subdue the plains tribes by killing off their primary source of food. At least 2.5 million bison perished each year between 1870 and 1875 in this form of biological warfare.

By 1892 only 85 bison were left. They were given refuge in Yellowstone National Park and protected by an 1893 law against the killing of wild animals in national parks.

In 1905, 16 people formed the American Bison Society to protect and rebuild the captive population. Soon thereafter the federal government established the National Bison Range near Missoula, Montana. Today there are an estimated 120,000 bison, about 80% of them on privately owned ranches.

2.5 million in 1970 to about 609,000 today. The 1990 international ban on the trade of ivory from African elephants has caused the bottom to drop out of the price of ivory and may help save this species, but things are not quite that simple (Connections, p. 462). Environmentalists now fear, however, that poachers will start killing Alaska's walruses for their tusks (Figure 17-15). Salespeople in the former Soviet Union are also touting ancient ivory from long-dead mammoths preserved in Siberia's snow.

As more species become endangered, the demand for them on the black market soars, hastening their chances of extinction. Poaching of endangered or

threatened species (many for the Asian market) is increasing in the United States, especially in western national parks and wilderness areas covered by only 22 federal wildlife protection officers.

According to the U.S. Fish and Wildlife Service a poached gyrfalcon sells for $120,000; a bighorn sheep head for $10,000–$60,000; a large saguaro cactus (Figure 4-11) for $5,000–$15,000; a peregrine falcon (Figure 15-5) for $10,000; a polar bear for $6,000; a grizzly bear for $5,000; a mountain goat for $3,500; a bald eagle (Figure 17-8) for $2,500; and a 29-gram (1-ounce) bear gallbladder (used in Asia for medicinal purposes) for $5,000. Most poachers are not caught, and the money

Should All Sales of Elephant Ivory Be Banned?

Most people applaud the international ban on the sale of ivory from elephants as a way to save elephants from extinction. However, increases in elephant populations in areas where their habitat has shrunk have resulted in widespread destruction of vegetation by these animals. This in turn reduces the niches available for other wild species.

To prevent increasing loss of biodiversity in such areas, a certain number of elephants are allowed to be killed each year, both by local people (as a source of meat) and by hunters (as a source of tourist income). In areas where elephant populations were high and needed to be culled, their legally sold ivory was an important source of income for local villagers and small farmers. With the ban on ivory trade this important source of income for the poor has dried up. Some wildlife conservationists argue that people in areas where the elephant populations are not endangered should be allowed to sell ivory from culled elephants in the international marketplace. The ivory would be marked in ways to show that it was obtained legally. Proponents of this approach see this as a way to serve local people and elephants. What do you think?

to be made far outweighs the risk of fines and the much smaller risk of jail.

Predator and Pest Control   People try to exterminate species that compete with us for food and game. For example, U.S. fruit farmers exterminated the Carolina parakeet around 1914 because it fed on fruit crops. The species was easy prey because when one member of a flock was shot, the rest of the birds hovered over its body, making themselves easy targets.

As animal habitats shrink, African farmers kill large numbers of elephants to keep them from trampling and eating food crops (Connections, above). Ranchers, farmers, and hunters in the United States support control of species such as coyotes (Figure 16-24 and Connections, p. 437) and wolves (Figure 16-25 and Spotlight, p. 442) that can prey on livestock and on species prized by game hunters. Since 1929, U.S. ranchers and government agencies have poisoned 99% of North America's prairie dogs (Figure 17-7) because horses and cattle sometimes step into the burrows and

break their legs. This has also nearly wiped out the endangered black-footed ferret (Figure 17-7), which preyed on the prairie dog.

The Market for Pets and Decorative Plants

Worldwide over 5 million live wild birds are captured and sold legally each year, and 2.5 million more are captured and sold illegally. Most are sold in Europe, Japan, and the United States. Over 40 species, mostly parrots, are endangered or threatened because of this wild-bird trade. Collectors of exotic birds may pay $10,000 for a threatened hyacinth macaw (Figure 17-7) smuggled out of Brazil. However, in its lifetime a single macaw might yield $165,000 in tourist income.

About 25 million American households have exotic birds as pets, 85% of them imported. For every wild bird that reaches a pet shop legally or illegally, as many as 10 others die during capture or transport. Keeping pet birds may also be hazardous to human health. A 1992 study suggested that keeping indoor pet birds for more than 10 years doubles a person's chances of getting lung cancer.

Some exotic plants, especially orchids and cacti (such as the black lace cactus, Figure 17-7), are endangered because they are gathered, often illegally, sold to collectors, and used to decorate houses, offices, and landscapes. A collector may pay $5,000 for a single rare orchid. A single rare mature crested saguaro cactus can earn cactus rustlers as much as $15,000. To thwart cactus rustlers, Arizona has put 222 species under state protection, with penalties of up to $1,000 and jail sentences of up to one year. However, only seven people are assigned to enforce this law over the entire state, and the fines are too small to discourage poaching.

Climate Change and Pollution   A potential problem for many species is the possibility of fairly rapid (40–50 years) changes in climate accelerated by deforestation and emissions of heat-trapping gases into the atmosphere (Section 10-2 and Figure 10-9). Wildlife in even the best-protected and best-managed reserves could be depleted in a few decades if such changes in climate take place as a result of projected global warming (Figure 10-10).

Toxic chemicals degrade wildlife habitats, including wildlife refuges, and kill some plants and animals. Slowly degradable pesticides, such as DDT, can be biologically amplified to very high concentrations in food chains and webs (Figure 15-4).

The resulting high concentrations of DDT or other slowly biodegraded, fat-soluble organic chemicals can directly kill the organisms, reduce their ability to reproduce, or make them more vulnerable to diseases,

Q: What continent has the world's highest population growth rate?

parasites, and predators. Such concentrations can also disrupt food webs (Connections, p. 127). During the 1950s and 1960s populations of ospreys, cormorants, brown pelicans, and bald eagles (Figure 17-8) plummeted. These birds feed mostly on fish at the top of aquatic food chains and webs, and thus ingest large quantities of biologically amplified DDT from their prey. Prairie falcons, sparrow hawks, Bermuda petrels, and peregrine falcons (Figure 15-5) also died off when they ate animal prey containing DDT, such as rabbits, ground squirrels, and other crop-damaging small mammals.

Research has shown that the culprit was DDE, a breakdown product of DDT, accumulating in the bodies of the affected birds. This chemical reduces the amount of calcium in the shells of their eggs. The fragile shells break, and the unborn chicks die.

Since the U.S. ban on DDT in 1972, most of these species have made a comeback. In 1980, however, DDT levels were again rising in species such as the peregrine falcon and the osprey. These species may be picking up biologically amplified DDT and other banned pesticides in Latin America, where they winter. In those countries the use of such chemicals is still legal. Illegal use of DDT and other banned pesticides in the United States may also play a role, as well as DDT blown by winds from such countries to the United States.

Introduced Species   Travelers sometimes pick up plants and animals intentionally or accidentally and introduce them to new geographical regions. Many of these introduced species provide food, game, and aesthetic beauty, and they help control pests in their new environments. Some alien species, however, have no natural predators and competitors in their new habitats, which allows them to dominate their new ecosystem and reduce the populations of many native species and are a form of biological pollution (Case Study p. 288). Eventually such aliens can displace or wipe out native species (Table 17-2). According to a 1993 Office of Technology Assessment study, more than 4,500 plant and animal species have been introduced—accidentally or intentionally—into the United States. Annual pest control costs for such species approach $100 billion.

An example of the effects of the accidental introduction of an alien species is the fast-growing water hyacinth (Figure 17-17), which is native to Central and South America. In 1884 a woman took one from a New Orleans exhibition and planted it in her back yard in Florida. The plants spread to waterways and thrived on Florida's nutrient-rich waters. Unchecked by natural enemies, within 10 years the plants (which can double their population every two weeks) rapidly displaced native plants, clogging many ponds, streams,

and canals—first in Florida and later elsewhere in the southeastern United States.

Mechanical harvesters and herbicides have failed to keep the plant in check. Although grazing Florida manatees, or sea cows (Figure 17-18), can control water hyacinths better than mechanical or chemical methods, these gentle and playful herbivores are threatened with extinction. Slashed by powerboat propellers, entangled in fishing gear—even hit on the head by oars—they reproduce too slowly to recover from these assaults and loss of habitat.

Scientists have introduced water hyacinth–eaters to help control its spread. The control agents include a weevil from Argentina, a water snail from Puerto Rico, and the grass carp, a fish from the former Soviet Union. These species can help, but water snails and grass carp also feed on other, desirable aquatic plants.

The good news is that water hyacinths can provide several benefits. They absorb toxic chemicals in sewage treatment lagoons. They can be fermented into a biogas fuel similar to natural gas, added as a mineral and protein supplement to cattle feed, and applied to the soil as fertilizer. They can also be used to clean up polluted ponds and lakes—if their growth can be kept under control.

An example of the impact of deliberately introduced alien species on ecosystems is wild African bees, which were imported to Brazil in 1957 with the false hope that they would increase honey production. Since then these bees have moved northward into Central America and by mid-1993 had reached Texas and Arizona. They are now heading north at 240 kilometers (150 miles) per year, though they will be stopped eventually by cold winters in the central United States. These "killer" bees displace domestic honey bees and reduce the honey supply.

Although they are not the killer bees portrayed in some horror movies, these bees are aggressive and unpredictable, and they had killed thousands of domesticated animals and an estimated 1,000 people in the Western Hemisphere (including one person in the United States) by 1993. Fortunately, most people not allergic to bee stings can run away. Most individuals killed by Africanized honeybees have died because they fell down or became trapped and could not flee.

17-3 SOLUTIONS: PROTECTING WILD SPECIES FROM EXTINCTION

World Conservation Strategy   In 1980 the International Union for Conservation of Nature and Natural Resources (IUCN), the UN Environment Programme, and the World Wildlife Fund developed the

Table 17-2 Damage Caused by Plants and Animals Imported into the United States

Name	Origin	Mode of Transport	Type of Damage
Mammals			
European wild boar	Russia	Intentionally imported (1912), escaped captivity	Destroys habitat by rooting; damages crops
Nutria (cat-sized rodent)	Argentina	Intentionally imported, escaped captivity (1940)	Alters marsh ecology; damages levees and earth dams; destroys crops
Birds			
European starling	Europe	Intentionally released (1890)	Competes with native songbirds; damages crops; transmits swine diseases; causes airport nuisance
House sparrow	England	Intentionally released by Brooklyn Institute (1853)	Damages crops; displaces native songbirds
Fish			
Carp	Germany	Intentionally released (1877)	Displaces native fish; uproots water plants; lowers waterfowl populations
Sea lamprey	North Atlantic Ocean	Entered Great Lakes via Welland Canal (1829)	Wiped out lake trout, lake whitefish, and sturgeon in Great Lakes
Walking catfish	Thailand	Imported into Florida	Destroys bass, bluegill, and other fish
Insects			
Argentine fire ant	Argentina	Probably entered via coffee shipments from Brazil (1918)	Damages crops; destroys native ant species
Camphor scale insect	Japan	Accidentally imported on nursery stock (1920s)	Damaged nearly 200 plant species in Louisiana, Texas, and Alabama
Japanese beetle	Japan	Accidentally imported on irises or azaleas (1911)	Defoliates more than 250 species of trees and other plants, including many of commercial importance
Plants			
Water hyacinth	Central America	Intentionally introduced (1884)	Clogs waterways; shades out other aquatic vegetation
Chestnut blight (fungus)	Asia	Accidentally imported on nursery plants (1900)	Killed nearly all eastern U.S. chestnut trees; disturbed forest ecology
Dutch elm disease (fungus)	Europe	Accidentally imported on infected elm timber used for veneers (1930)	Killed millions of elms; disturbed forest ecology

From *Biological Conservation* by David W. Ehrenfeld. Copyright © 1970 by Holt, Rinehart & Winston, Inc. Modified and reprinted by permission.

World Conservation Strategy, a long-range plan for conserving the world's biological resources. This plan was expanded in 1991. Its primary goals are to:

- Maintain essential ecological processes and life-support systems on which human survival and economic activities depend, mostly by combining wildlife conservation with sustainable development (Figure 16-19)
- Preserve species diversity and genetic diversity

- Ensure that any use of species and ecosystems is sustainable
- Minimize the depletion of nonrenewable resources
- Improve the quality of human life
- Include women and indigenous peoples in the development of conservation plans
- Promote an ethic that includes protection of plants and animals as well as people

Q: How much of the original tall-grass prairies in the United States have been destroyed?

Heather Angel/Biofotos

Figure 17-17 The fast-growing water hyacinth was intentionally introduced into Florida from Latin America in 1884. Since then this plant, which can double its population in only two weeks, has taken over many waterways in Florida and other southeastern states.

Florida Marine Research Institute/Florida Department of Natural Resources

Figure 17-18 The Florida manatee, or sea cow, one of America's most endangered species. Only an estimated 1,000–1,800 of these gentle animals are left in the increasingly polluted bays and streams of southern and central Florida. Most manatees are killed by the propellers of power boats, which outnumber them 500 to 1.

- Monitor the sustainability of development
- Encourage recognition of the harmful environmental effects of armed conflict and economic insecurity
- Encourage rehabilitation of degraded ecosystems upon which humans depend for food and fiber

So far 40 countries have planned or established national conservation programs (the United States is not one of them). If MDCs provide enough support, this conservation strategy offers hope for slowing the loss of biodiversity. However, environmentalists warn that no conservation strategy can protect the planet's biodiversity unless governments act to

- Reduce poverty (Section 7-3)
- Control population growth (Section 6-5)
- Slow projected global warming (Section 10-3) and ozone depletion (Section 10-5)
- Reduce the destruction and degradation of tropical forests (Section 16-4), old-growth forests (Section 16-3), wetlands, and coral reefs (Section 5-2)
- Switch from unsustainable to sustainable forms of agriculture (Sections 14-4 and 15-5)

Methods for Protecting Biodiversity and Managing Wildlife There are three basic approaches to managing wildlife and protecting biodiversity. The *species approach* is based on protecting endangered species by identifying them, giving them legal protection, preserving and managing their critical habitats, propagating them in captivity, and reintroducing them in suitable habitats. The *wildlife management approach* manages game species for sustained yield by using laws to regulate hunting, establishing harvest quotas, developing population management plans, and using international treaties to protect migrating game species such as waterfowl.

The *ecosystem approach*, discussed in Chapter 16, aims to preserve balanced populations of species in their native habitats, establish legally protected wilderness areas and wildlife reserves, and eliminate alien species. Biologists see protecting ecosystems as the best way to preserve ecological, species, and genetic diversity. It is a *prevention* approach to massive loss of biodiversity. By contrast, trying to keep various species whose numbers have sharply declined is an emergency response. It is more costly and less effective than trying to prevent such situations from occurring by leaving large areas of the world's different ecosystems alone.

Using Treaties and Laws to Protect Endangered Species Several international treaties and conventions help protect endangered or threatened wild species. One of the most far-reaching is the 1975 Convention on International Trade in Endangered Species (CITES). This treaty, now signed by 120 countries, lists 675 species that cannot be commercially traded as live specimens or wildlife products because they are endangered or threatened.

Unfortunately, enforcement of this treaty is spotty, convicted violators often pay only small fines, and member countries can exempt themselves from protection of any listed species. Also, much of the $5-billion-per-year illegal trade in wildlife and wildlife products goes on in countries such as Singapore that have not signed the treaty. Other centers of illegal animal trade are Argentina, Indonesia, Spain, Taiwan, and Thailand.

The United States controls imports and exports of endangered wildlife and wildlife products with two important laws. The Lacey Act of 1900 prohibits transporting live or dead wild animals or their parts across state borders without a federal permit. The Endangered Species Act of 1973 (amended in 1982 and 1988) makes it illegal for Americans to import or trade in any product made from an endangered or threatened species unless it is used for an approved scientific purpose or to enhance the survival of the species.

The Endangered Species Act of 1973 is one of the world's toughest environmental laws. It authorizes the National Marine Fisheries Service (NMFS) to identify and list endangered and threatened ocean species; the Fish and Wildlife Service (FWS) identifies and lists all other endangered and threatened species. These species cannot be hunted, killed, collected, or injured in the United States.

Any decision by either agency to list or unlist a species must be based on biology only, not on economic considerations. The act also forbids federal agencies to carry out, fund, or authorize projects that would either jeopardize an endangered or threatened species or destroy or modify its critical habitat—the land, air, and water necessary for its survival.

Between 1970 and 1993 the number of species found only in the United States that have been placed on the official endangered and threatened list increased from 92 to 775 (594 endangered and 181 threatened). Also on the list are 535 species found elsewhere.

Getting listed is only half the battle. Next the FWS or the NMFS is supposed to prepare a plan to help the species recover. However, because of a lack of funds, recovery plans have been developed and approved for only about 55% of the endangered or threatened U.S. species, and half of those plans exist only on paper. Only 5 U.S. species have recovered enough to be unlisted, but 238 of the 775 listed species are stable and recovering. On the losing end, 7 listed domestic species have become extinct.

The annual federal budget for endangered species is less than what beer companies spend on two 30-second TV commercials during the Super Bowl. At this level of funding it will take up to 48 years to evaluate the almost 7,500 species (about 500 of them severely imperiled) now proposed for listing. Wildlife experts estimate that at least 400 of them could vanish while they wait, as did 34 species awaiting listing between 1980 and 1990.

The act requires that all commercial shipments of wildlife and wildlife products enter or leave the coun-

Q: How much money is made each year from illegal trade in wildlife and wildlife products?

Figure 17-19 Confiscated products made from endangered species. Because of a lack of funds and inspectors, probably no more than one-tenth of the illegal wildlife trade in the United States is discovered. The situation is even worse in most other countries.

try through one of nine designated ports. Many illegal shipments slip by, however, because the 60 FWS inspectors can physically examine only about one-fourth of the 90,000 shipments that enter and leave the United States each year (Figure 17-19). Even if caught, many violators are not prosecuted, and convicted violators often pay only a small fine.

Since this act was passed, intense pressure has been exerted by developers, logging and mining companies, and other users of land resources to allow consideration of economic factors, both in evaluating species for listing and in carrying out federally funded projects that threaten the critical habitats of endangered or threatened species. Despite widespread publicity that such species blocked development and resource extraction between 1987 and 1991, only 33 of 118,000 projects evaluated by the FWS were blocked or withdrawn as a result of the Endangered Species Act.

Environmentalists and many members of Congress argue that the Endangered Species Act should be strengthened—not weakened—by:

■ *Emphasizing protection of entire ecosystems to help prevent future declines in species not yet listed as threatened or endangered.*

■ *Carrying out an extensive biodiversity survey of the country's wild species and ecosystems to provide information on what needs to be protected.*

■ *Setting deadlines for development and implementation of recovery plans.*

■ *Giving private landowners tax write-offs or other incentives for assisting in species recovery and helping restore degraded ecosystems.*

■ *Greatly increasing annual funding for endangered and threatened species from the current $55 million to $460 million—the amount needed to get the job done.* This amounts to spending less than $2 per year per U.S. citizen to help protect the entire nation's biological resources.

■ *Allowing citizens to file lawsuits immediately if an endangered species faces serious harm or extinction.*

Such proposals are vigorously opposed by those wishing to have greater access to resources on public and other lands (Spotlight, p. 187).

The federal government has primary responsibility for managing migratory species, endangered species, and wildlife on federal lands. States are responsible for the management of all other wildlife.

Wildlife Refuges   Since 1903, when President Theodore Roosevelt established the first U.S. federal wildlife refuge at Pelican Island, Florida, the National Wildlife Refuge System has grown to 503 refuges (Figure 16-2). About 85% of the area included in these refuges is in Alaska.

Over three-fourths of the refuges are wetlands for protection of migratory waterfowl. Most species on the U.S. endangered list have habitats in the refuge system, and some refuges have been set aside for specific endangered species. These have helped Florida's key deer, the brown pelican (Figure 17-20), and the trumpeter swan to recover. Environmentalists urge the establishment of more refuges for endangered plants.

Congress has not established guidelines (such as multiple use or sustained yield) for management of the National Wildlife Refuge System, as it has for other

A: About $5 billion

Figure 17-20 The first national wildlife refuge was set up off the coast of Florida in 1903 to protect the brown pelican from extinction because of overhunting and loss of habitat. In the 1960s this species was again threatened with extinction when exposure to DDT and other persistent pesticides in the fish it eats caused reproductive losses. Now it has made a comeback.

Figure 17-21 The Arabian oryx barely escaped extinction in 1969 after being overhunted in the deserts of Oman and Jordan (Figure 11-1). Captive breeding programs saved this antelope species from extinction. Some have been reintroduced into protected habitats in the Middle East, with the wild population now about 120.

public lands (Section 16-1). As a result, the Fish and Wildlife Service has allowed many refuges to be used for hunting, fishing, trapping, timber cutting, grazing, farming, oil and gas development (Pro/Con, p. 469), mining, military air exercises, power and air boating, and off-road vehicles. Currently, more than 50% of the refuges are open to hunting, 56% to fishing, and 18% to trapping.

A 1990 report by the General Accounting Office found that activities considered harmful to wildlife occur in nearly 60% of the nation's wildlife refuges. In addition, a 1986 FWS study estimated that one in five federal refuges is contaminated with chemicals from old toxic-waste dump sites (including military bases) and runoff from nearby agricultural land.

Private groups play an important role in conserving wildlife in refuges and other protected areas. For example, since 1951 the Nature Conservancy has preserved over 1 million hectares (2.5 million acres) of forests, marshes, prairies, islands, and other areas of unique ecological or aesthetic significance in the United States.

Gene Banks, Botanical Gardens, and Zoos

Botanists preserve genetic information and endangered plant species by storing their seeds in gene banks—refrigerated, low-humidity environments. Scientists urge that many more such banks be established, especially in LDCs; however, some species can't be preserved in gene banks, and maintaining the banks is very expensive.

The world's 1,500 botanical gardens and arboreta hold about 90,000 plant species. However, these sanc-

tuaries have too little storage capacity and too little funding to preserve most of the world's rare and threatened plants.

Worldwide, 500 zoos house about 540,000 individual animals, many of them from species not threatened or endangered. Zoos and animal research centers are increasingly being used to preserve some individuals of critically endangered animal species, with the long-term goal of reintroducing the species into protected wild habitats.

Two techniques for preserving such species are egg pulling and captive breeding. *Egg pulling* involves collecting eggs laid in the wild by critically endangered bird species and hatching them in zoos or research centers. For *captive breeding* some or all individuals of a critically endangered species still in the wild are captured for breeding in captivity with the hope of being able to reintroduce them into the wild.

Captive breeding programs at zoos in Phoenix, San Diego, and Los Angeles saved the nearly extinct Arabian oryx (Figure 17-21). This large antelope species once lived throughout the Middle East. By the early 1970s, however, it had been hunted to extinction in the wild by people riding in jeeps and helicopters, and wielding rifles and machine guns. Since 1980 small numbers of oryxes bred in captivity have been returned to the wild in protected habitats in the Middle East. Endangered U.S. species now being bred in cap-

Q: What is the world's toughest environmental law?

Should We Develop Oil and Gas in the Arctic National Wildlife Refuge?

The Arctic National Wildlife Refuge on Alaska's North Slope (Figure 17-22), which contains more than one-fifth of all the land in the U.S. wildlife refuge system, has been called the crown jewel of the system. During all or part of the year it is home for more than 160 animal species, including caribou, musk ox, snowy owls, threatened grizzly bears (Figure 17-7), arctic foxes (Figure 5-39), and migratory birds (including as many as 300,000 snow geese, Figure 5-1). It is also home for about 7,000 Inuit (Eskimos), who depend on the caribou for a large part of their diet.

The refuge's coastal plain, its most biologically productive part, is the only stretch of Alaska's arctic coastline not open to oil and gas development. U.S. oil companies hope to change this because they believe that the area *might* contain oil and natural gas deposits. Since 1985 they have been urging Congress to open to drilling some 607,000 hectares (1.5 million acres) along the coastal plain—roughly two-thirds the size of Yellowstone National Park. They argue that such exploration is needed to provide the United States with more oil and natural gas and reduce dependence on oil imports.

Environmentalists oppose this proposal and want Congress to designate the entire coastal plain as wilderness. They cite Interior Department estimates that there is only a 19% chance of finding as much oil in the coastal plain as the United States consumes every three years. Even if the oil does exist, environmentalists do not believe the potential degradation of any portion of this irreplaceable wilderness area would be worth it, especially considering that improvements in energy efficiency would save far more oil at a much lower cost (Section 18-2).

Oil company officials claim they have developed Alaska's Prudhoe Bay oil fields without significant harm to wildlife; they also contend that the area they want to open to oil and gas development is less than 1.5% of the entire coastal plain region—equivalent to an oil field the size of Dulles International Airport in Washington, D.C., within an area approximately the size of South Carolina.

However, the huge 1989 oil spill from the tanker *Exxon Valdez* in Alaska's Prince William Sound cast serious doubt on such claims (Case Study, p. 292). Moreover, a study leaked from the Fish and Wildlife Service in 1988 revealed that oil drilling at Prudhoe Bay has caused much more air and water pollution than was anticipated before drilling began in 1972. According to this study, oil development in the coastal plain could cause the loss of 20–40% of the area's 180,000-head caribou herd, 25–50% of the remaining musk oxen, 50% or more of the wolverines, and 50% of the snow geese that live there part of the year. A 1988 EPA study also found that "violations of state and federal environmental regulations and laws are occurring at an unacceptable rate" in the Prudhoe Bay area where oil fields and facilities have been developed.

Do you think this refuge should be explored and developed for oil and natural gas?

Figure 17-22 Proposed oil-drilling area in Alaska's Arctic National Wildlife Refuge. (Data from U.S. Fish and Wildlife Service)

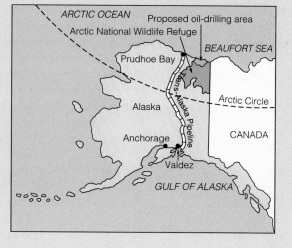

tivity with some returned to the wild include the California condor (Figure 17-2), the peregrine falcon (Figure 15-5), and the black-footed ferret (Figure 17-7). Endangered golden lion tamarins (Figure 17-7) bred at the National Zoo in Washington, D.C., have been released successfully in Brazilian rain forests.

Unfortunately, keeping populations of endangered animal species in zoos and research centers is limited by lack of space and money. The captive population of each species must number 100–500 to avoid extinction through accident, disease, or loss of genetic variability through inbreeding. Moreover, caring for and breeding captive animals is very expensive. Thus the world's zoos now contain only 20 endangered species of animals with populations of 100 or more individuals. It is estimated that today's zoos and research centers have space to preserve healthy and sustainable populations of only 925 of the 2,000 large vertebrate species that could vanish from the planet. It is doubtful that the more than $6 billion needed to care for these animals for 20 years will be available.

For every successful reintroduction, there are many more species that can't go home again. Experience has shown that reintroduction of captive-bred species is not an option for the majority of species threatened with extinction. According to Ben Brock, associate director of the U.S. National Zoo in Washington, D.C., their have been 146 attempts to reintroduce 126 different species (most of them fish). Only 16 of these reintroductions succeeded. Most reintroductions fail because of a lack of suitable habitat and the inability of species bred in captivity to survive in the wild. Thus, many zoos are becoming arks that cannot unload their expensive endangered cargoes.

Because of limited funds and trained personnel, only a few of the world's endangered and threatened species can be saved by treaties, laws, wildlife refuges, gene banks, botanical gardens, and zoos. That means that wildlife experts must decide which species out of thousands of candidates should be saved. Many experts suggest that the limited funds for preserving threatened and endangered wildlife be concentrated on those species that (1) have the best chance for survival, (2) have the most ecological value to an ecosystem, and (3) are potentially useful for agriculture, medicine, or industry.

17-4 WILDLIFE MANAGEMENT

Management Approaches Wildlife management entails (1) manipulating wildlife populations (especially game species) and habitats for their welfare and for human benefit, (2) preserving endangered and threatened wild species, and (3) enforcing wildlife laws.

The first step in wildlife management is to decide which species are to be managed in a particular area—a source of much controversy. Ecologists stress preserving biodiversity. Wildlife conservationists are concerned about endangered species. Bird-watchers want the greatest diversity of bird species. Hunters want large populations of game species for harvest during hunting season. In the United States most wildlife management is devoted to producing surpluses of game animals and game birds.

After goals have been set, the wildlife manager must develop a management plan. Ideally the plan is based on principles of ecological succession (Figures 5-41 and 5-42), wildlife population dynamics (Figures 5-32, 5-33, and 5-34), and an understanding of the cover, food, water, space, and other habitat requirements of each species to be managed. The manager must also consider the number of potential hunters, their success rates, and the regulations available to prevent excessive harvesting.

This information is difficult, expensive, and time-consuming to obtain. In practice it involves much guesswork and trial and error, which is why wildlife management is as much an art as a science. Management plans must also be adapted to political pressures from conflicting groups—and to budget constraints.

In the United States funds for state game management programs come from the sale of hunting and fishing licenses and from federal taxes on hunting and fishing equipment. Two-thirds of the states also have provisions on state income tax returns that allow individuals to contribute money to state wildlife programs. Only 10% of all government wildlife dollars, however, are spent to study or benefit nongame species, which make up nearly 90% of the country's wildlife species.

Manipulation of Vegetation and Water Supplies Wildlife managers can encourage the growth of plant species that are the preferred food and cover for a particular animal species by controlling the ecological succession of vegetation in various areas.

Animal wildlife species can be classified into four types according to the stage of ecological succession at which they are most likely to find sufficient food and other resources: early-successional, mid-successional, late-successional, and wilderness (Figure 17-23). *Early-successional species* find food and cover in weedy pioneer plants. These plants invade an area that has been cleared of vegetation for human activities and then abandoned (Figure 5-42), as well as areas devastated by mining, fires, volcanic lava, or glaciers (Figure 5-41).

Mid-successional species are found around abandoned croplands and partially open areas created by logging of small stands of timber, by controlled burning, and by clearing of vegetation for roads, firebreaks, oil and gas pipelines, and electrical transmission lines. Such openings of the forest canopy promote the growth of vegetation favored by mid-successional mammal and bird species. They also increase the amount of edge habitat, where two communities such as a forest and an open field come together. This transition zone allows animals such as deer to feed on veg-

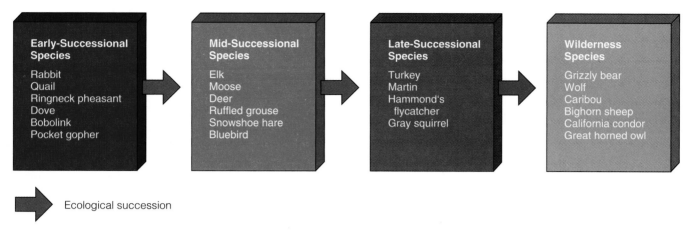

Early-Successional Species

Rabbit
Quail
Ringneck pheasant
Dove
Bobolink
Pocket gopher

Mid-Successional Species

Elk
Moose
Deer
Ruffled grouse
Snowshoe hare
Bluebird

Late-Successional Species

Turkey
Martin
Hammond's
 flycatcher
Gray squirrel

Wilderness Species

Grizzly bear
Wolf
Caribou
Bighorn sheep
California condor
Great horned owl

Ecological succession

Figure 17-23 Preferences of some wildlife species for habitats at different stages of ecological succession.

etation in clearings and quickly escape to cover in the nearby forest.

Late-successional species need old-growth and mature forest habitats to produce the food and cover on which they depend. These animals require the protection of moderate-sized, old-growth forest refuges.

Wilderness species flourish only in fairly undisturbed, mature vegetation communities, such as large areas of old-growth forests, tundra, grasslands, and deserts. They can survive only in large wilderness areas and wildlife refuges.

Various types of habitat improvement can be used to attract a desired species and encourage its population growth, including planting seeds, transplanting certain types of vegetation, building artificial nests, and deliberately setting controlled low-level ground fires to help control vegetation. Wildlife managers often create or improve ponds and lakes in wildlife refuges to provide water, food, and habitat for waterfowl and other wild animals.

Population Management by Sport Hunting

Most MDCs use sport hunting laws to manage populations of game animals, although sport hunting is controversial. Licensed hunters are allowed to hunt only during certain months of the year so as to protect animals during mating season. Hunters can use only certain types of hunting equipment—such as bows and arrows, shotguns, and rifles—for a particular type of game. Limits are set on the size, number, and sex of animals that can be killed, and on the number of hunters allowed in a game refuge.

Close control of sport hunting is difficult. Accurate data on game populations may not exist and may cost too much to get. People in communities near hunting areas, who benefit from money spent by hunters, may push to have hunting quotas raised.

Management of Migratory Waterfowl

In North America migratory waterfowl such as ducks, geese, and swans nest in Canada during the summer. During the fall hunting season they migrate to the United States and Central America along generally fixed routes called **flyways** (Figure 17-24).

Canada, the United States, and Mexico have signed agreements to prevent habitat destruction and overhunting of migratory waterfowl. However, since 1972 the estimated breeding population of North American ducks has dropped 38%, mostly because of prolonged drought in key breeding areas and degradation and destruction of wetland and grassland breeding habitats by farmers. What wetlands remain are used by dense flocks of ducks and geese (Figure 5-1), whose high densities in shrinking wetlands make them more vulnerable to diseases and to predators such as skunks, foxes, coyotes, minks, raccoons, and hunters. Waterfowl in wetlands near croplands are also exposed to pollution from pesticides and other chemicals in irrigation runoff.

Wildlife officials manage waterfowl in several ways. These include regulating hunting, protecting existing habitats, and developing new habitats—as well as by building artificial nesting sites, ponds, and nesting islands. More than 75% of the federal wildlife refuges in the United States are wetlands used by migratory birds. Local and state agencies and private conservation groups such as Ducks Unlimited, the Audubon Society, and the Nature Conservancy have also established waterfowl refuges.

In 1986 the United States and Canada agreed to spend $1.5 billion over a 17-year period, with the goal of almost doubling the continental duck-breeding population. The key elements in this program will be the purchase, improvement, and protection of additional waterfowl habitats in five priority areas.

A: $4 trillion annually

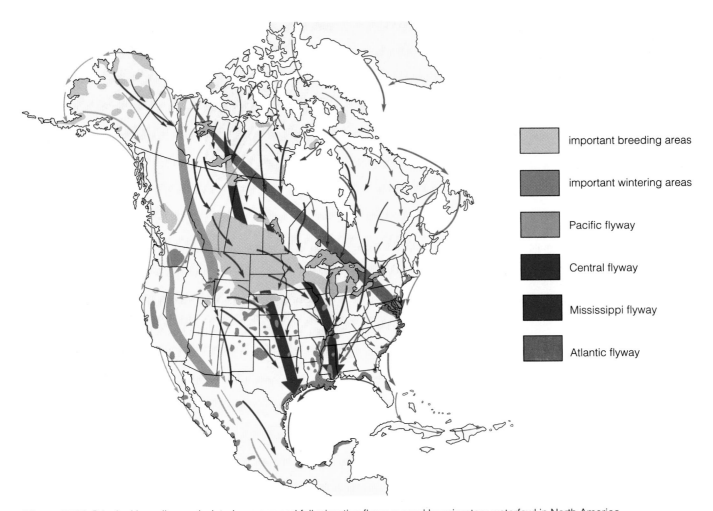

Figure 17-24 Principal breeding and wintering areas and fall migration flyways used by migratory waterfowl in North America.

Legend:
- important breeding areas
- important wintering areas
- Pacific flyway
- Central flyway
- Mississippi flyway
- Atlantic flyway

Since 1934 the Migratory Bird Hunting and Conservation Stamp Act has required waterfowl hunters to buy a duck stamp each season they hunt. Revenue from these sales goes into a fund to buy land and easements for the benefit of waterfowl.

17-5 FISHERY MANAGEMENT AND PROTECTING MARINE BIODIVERSITY

Freshwater Fishery Management The goals of freshwater fish management are to encourage the growth of populations of desirable commercial and sport fish species and to reduce or eliminate populations of less desirable species. A number of techniques are used:

- Regulating the timing and length of fishing seasons

- Setting the size and number of fish that can be taken

- Requiring commercial fishnets to have a large enough mesh to prevent harvesting young fish

- Building reservoirs and farm ponds, and stocking them with game fish

- Fertilizing nutrient-poor lakes and ponds

- Protecting and creating spawning sites

- Protecting habitats from buildup of sediment and other forms of pollution, and removing debris

- Preventing excessive growth of aquatic plants

- Using small dams to control water flow

- Controlling predators, parasites, and diseases by improving habitats, breeding genetically resistant fish varieties, and using antibiotics and disinfectants

Q: What percentage of people in the United States live in urban areas?

- Using hatcheries to restock ponds, lakes, and streams with species such as trout and salmon

Marine Fishery Management   By international law the offshore fishing zone of coastal countries extends to 370 kilometers (200 nautical miles or 230 statute miles) from their shores. Foreign fishing vessels can take certain quotas of fish within such zones, called *exclusive economic zones*, only with government permission. Ocean areas beyond the legal jurisdiction of any country are known as the *high seas*. Any limits on the use of the living and mineral common-property resources in these areas are set by international maritime law and international treaties.

Managers of marine fisheries use several techniques to help prevent commercial extinction and allow depleted stocks to recover. Fishery commissions, councils, and advisory bodies with representatives from countries using a fishery can set annual quotas and establish rules for dividing the allowable catch among the participating countries. These groups may limit fishing seasons and regulate the type of fishing gear that can be used to harvest a particular species; fishing techniques such as dynamiting and poisoning, for example, are outlawed. Fishery commissions may also make it illegal to keep fish below a certain size, usually the average length of the particular fish species when it first reproduces.

In addition, food and game fish species, such as the striped bass along the Pacific and Atlantic coasts of the United States, can be introduced. Finally, artificial reefs can be built from boulders, construction debris, and automobile tires to provide food and cover for commercial and game fish species. About 400 such reefs have been established off U.S. coasts, and Japan has set aside $1 billion to create 2,500 of them.

As voluntary associations, however, fishery commissions don't have any legal authority to compel member states to follow their rules. Nor can they compel all countries fishing in a region to join the commission and submit to its rules. Furthermore, it is very difficult to estimate the sustainable yields of various marine species.

Decline of the Whaling Industry   *Cetaceans* are an order of mammals ranging in size from the 0.9-meter (3-foot) porpoise to the giant 15- to 30-meter (50- to 100-foot) blue whale. They can be divided into two major groups, toothed cetaceans and baleen whales (Figure 17-25).

Toothed cetaceans—such as the porpoise, sperm whale, and killer whale—bite and chew their food. They feed mostly on squid, octopus, and other marine animals. *Baleen whales*, such as the blue, gray, hump-back, and finback, are filter feeders. Instead of teeth, they have several hundred horny plates made of baleen, or whalebone, which hang down from their upper jaw. These plates filter plankton, especially shrimplike krill (Figure 4-22) smaller than your thumb, from seawater. Baleen whales are the most abundant group of cetaceans.

Overharvesting has caused a sharp drop in the population of almost every whale species with commercial value (Figure 17-26). The populations of 8 of the 11 major species of whales once hunted by the whaling industry have been reduced to commercial extinction.

A prime example is the endangered blue whale (Figure 17-25), the world's largest animal. Fully grown, it's more than 30 meters (100 feet) long—longer than three train boxcars—and weighs more than 25 elephants. The adult has a heart the size of a Volkswagen "Beetle" car, and some of its arteries are so big that a child could swim through them.

Blue whales spend about eight months of the year in Antarctic waters. There they find an abundant supply of krill, which they filter daily by the trillions from seawater. During the winter months they migrate to warmer waters, where their young are born.

Once an estimated 200,000 blue whales roamed the Antarctic Ocean. Today the species has been hunted to near biological extinction for its oil, meat, and bone (Figure 17-26). This decline was caused by a combination of prolonged overfishing and certain natural characteristics of blue whales. Their huge size made them easy to spot. They were caught in large numbers because they grouped together in their Antarctic feeding grounds. Also, they take 25 years to mature sexually and have only one offspring every 2–5 years—a reproduction rate that makes it hard for the species to recover once its population falls to a low level.

Blue whales haven't been hunted commercially since 1964 and have been classified as an endangered species since 1975. Despite this protection some marine experts believe that too few blue whales—less than 5,000 and perhaps as few as 1,500—remain for the species to recover. Within a few decades the blue whale could disappear forever.

In 1946 the International Whaling Commission (IWC) was established to regulate the whaling industry. Since 1949 the IWC has set annual quotas to prevent overfishing and commercial extinction. However, these quotas often were based on inadequate scientific information or were ignored by whaling countries. Without any powers of enforcement, the IWC has been unable to stop the decline of most whale species.

In 1970 the United States stopped all commercial whaling and banned all imports of whale products, mostly because of pressure from environmentalists

A: 75%

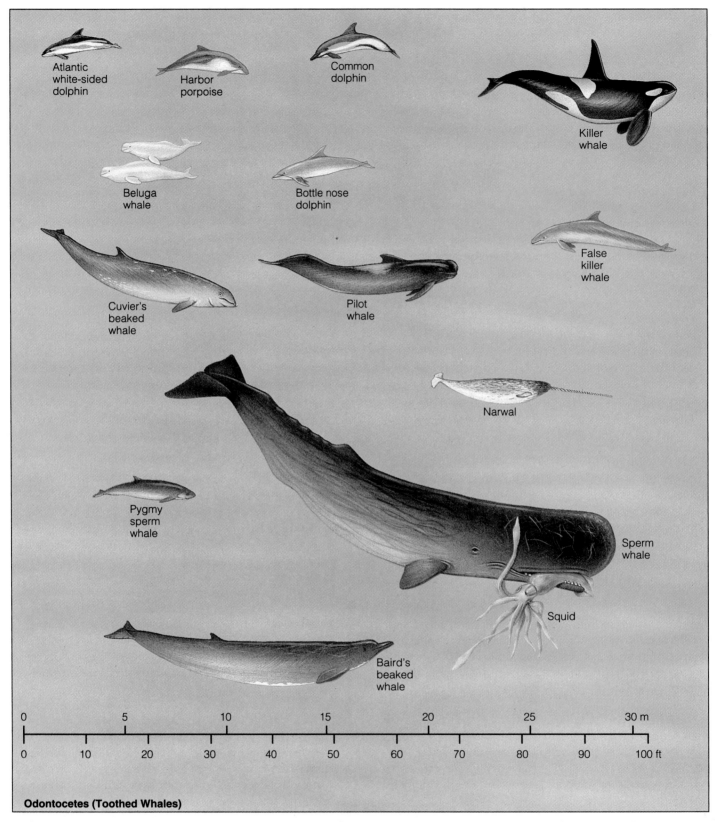

Odontocetes (Toothed Whales)

Figure 17-25 Examples of cetaceans, which can be classified as Odontocetes (toothed whales) and Mysticetes (baleen whales).

Q: Worldwide, how many people are homeless?

Humpback whale

Bowhead whale

Right whale

Minke whale

Blue whale

Feeding on krill

Fin whale

Sei whale

Gray whale

Mysticetes (Baleen Whales)

A: About 150 million (18% of the world's population)

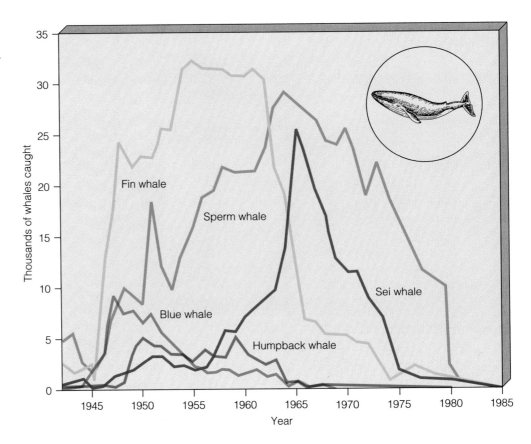

Figure 17-26 Whale harvests, showing the signs of overharvesting. (Data from International Whaling Commission)

and the general public. Under intense pressure from environmentalists and governments of many countries, including the United States, the IWC has imposed a moratorium on commercial whaling since 1986.

Until 1992 the IWC allowed Japan, Norway, and Iceland to kill several hundred minke whales per year for "scientific" purposes. Environmentalists believe that killing whales for such purposes is a sham and that a total ban on whaling should be imposed and vigorously enforced. They also support a French proposal that Antarctic waters be declared as an international marine sanctuary, off limits for harvesting of any whales.

In 1992 Iceland quit the IWC. And in 1993 Norway began commercial whaling of minke whales (Figure 17-25) in defiance of the IWC. Japan has also threatened to resume whaling. Officials of Iceland, Norway, and Japan—and some marine biologists—argue that a small annual harvest of minke whales (with an estimated total of 900,000 worldwide and almost 87,000 off the coast of Norway alone) would have a negligible effect on the population of this nonendangered species. They cite a 1992 analysis by IWC scientists estimating that up to 2,000 minke whales could be harvested annually without endangering the species. All three nations argue that the current moratorium on whaling is governed by emotion, not science.

Most environmentalists disagree. Some argue that whales are intelligent, sensitive mammals that should not be killed. Others fear that opening up the door to any commercial whaling may eventually lead to widespread harvests of minke and other whales by weakening current international disapproval of and economic sanctions against commercial whaling. They cite the earlier failure of the IWC to prevent commercial extinction of many whale species (Figure 17-26).

By law, the United States can put pressure on Iceland and Norway (or on any country violating the IWC ban) by imposing costly economic sanctions on imports from these countries. In 1992, former President George Bush promised such sanctions if whaling resumed. However, by early 1994 the Clinton administration had refused to enact such sanctions and assured Norwegian officials that it would not jeopardize trade relationships with Norway over the whaling issue. A major reason for reluctance to impose sanctions on Norway for illegal whaling is American interest in exploring for oil off the coast of Norway. If Norway can get away with defying the IWC, environmentalists fear that other nations such as Iceland, Japan, Taiwan, Peru, and Chile will also start illegal killing of whales.

Without continuing worldwide pressure from individuals (through political pressure for sanctions,

Q: How many motor vehicles are there in the world?

economic boycotts of all imports from offending nations, not using air and sea travel vessels owned by such countries, and not traveling to offending countries), environmental organizations, and the U.S. government, large-scale commercial whaling may resume in the near future. By 1994 a boycott of Norwegian fish products in the United Kingdom and Germany had cost Norway more than $30 million in canceled contracts.

Protecting Marine Biodiversity   It is difficult to protect marine biodiversity. For one thing, shore-hugging species are adversely affected by coastal development and the accompanying massive inputs of sediment and other wastes from land (Section 5-2). This poses a severe threat to biologically diverse and highly productive coastal ecosystems such as coral reefs (Figures 5-20 and 5-22), marshes (Figure 5-23), and mangrove swamps (Figure 5-24).

Protecting marine biodiversity is also difficult because much of the damage is not visible to people. In addition, the seas are viewed by many as an inexhaustible source of resources and capable of absorbing an almost infinite amount of waste and pollution. Finally, most of the world's ocean area lies outside the legal jurisdiction of countries and thus is an open-access resource subject to overexploitation because of the tragedy of the commons (Connections, p. 16).

Protecting marine biodiversity requires countries to enact and enforce tough regulations to protect coral reefs, mangrove swamps, and other coastal ecosystems from unsustainable use and abuse (Section 11-9). Also, much more effective international agreements are needed to protect biodiversity in the open seas.

What You Can Do to Protect Wildlife Resources and Preserve Biodiversity?

INDIVIDUALS MATTER

- *Improve the habitat on a patch of the earth in your immediate environment (such as a yard or vacant lot), emphasizing the promotion of biological diversity.*

- *Refuse to buy furs, ivory products, reptile-skin goods, tortoiseshell jewelry, rare orchids or cacti, and materials from endangered or threatened animal species.*

- *Leave wild animals in the wild.*

- *Reduce habitat destruction and degradation by recycling paper, cans, plastics, and other household items. Better yet, reuse items and sharply reduce your use of throwaway items (Sections 13-2 and 13-3).*

- *Support efforts to sharply reduce the destruction and degradation of tropical forests (Section 16-4) and old-growth forests (Section 16-3), to slow projected global warming (Section 10-3), and to reduce ozone depletion in the stratosphere (Section 10-5).*

- *Pressure elected officials to pass laws requiring larger fines and longer prison sentences for wildlife poachers—and to provide more funds and personnel for wildlife protection.*

- *Pressure Congress to pass a national biodiversity act and to develop a national conservation program as part of the World Conservation Strategy.*

17-6 SOLUTIONS: INDIVIDUAL ACTION

We are all involved, at least indirectly, in the destruction of wildlife or the degradation of wildlife habitats—any time we buy or drive a car, build a house, consume almost anything, or waste electricity, paper, water, or any other resource.

Modifying our consumption habits is a key goal in protecting wildlife, the environment, and ourselves (Individuals Matter, at right). This also involves supporting efforts to reduce deforestation, projected global warming, ozone depletion, population growth, and poverty—all of which threaten wildlife and ultimately our own species.

During our short time on this planet we have gained immense power over what species—including our own—live or die. We named ourselves the wise (*sapiens*) species. In the next few decades we will learn whether we are indeed a wise species—whether we have the wisdom to learn from and work with nature to protect ourselves and other species.

A greening of the human mind must precede the greening of the Earth. A green mind is one that cares, saves, and shares. These are the qualities essential for conserving biological diversity now and forever.

M. S. SWAMINATHAN

Critical Thinking

1. Discuss your gut-level reaction to this statement: "It doesn't really matter that the passenger pigeon is extinct and that the blue whale, the whooping crane, the California condor, the rhinoceros, and the grizzly bear are endangered mostly because of human activities." Be honest about your reaction and give arguments for your position.

A: About 570 million (460 million cars and 110 million trucks)

2. Make a log of your own consumption of products for a single day. Relate your consumption to the increased destruction and degradation of wildlife and wildlife habitats in the United States, in tropical forests, and in aquatic ecosystems.

3. **a.** Do you accept the ethical position that each *species* has the inherent right to survive without human interference, regardless of whether it serves any useful purpose for humans? Explain.

   b. Do you believe that each *individual* of an animal species has an inherent right to survive? Explain. Would you extend such rights to individual plants and microorganisms? Explain.

4. Should U.S. energy companies be allowed to drill for oil and gas in the Arctic National Wildlife Refuge? Explain.

5. Do you believe that the use of animals to test new drugs, vaccines, and the toxicity of chemicals should be banned? Explain. What are the alternatives? Should animals be used to test cosmetics?

6. Should whaling nations be allowed by the International Whaling Commission to resume a limited annual harvest of minke whales as long as the species' estimated numbers don't significantly decline? Explain. Would you extend this to other commercially valuable whale species if their populations increase significantly? Explain.

*7. Identify examples of habitat destruction or degradation in your local community that have had harmful effects on the populations of various wild plant and animal species. Develop a management plan for the rehabilitation of these habitats and wildlife.

*8. Make a concept map of the key ideas in this chapter using the section heads and subheads and the key terms (shown in boldface type in the chapter). See the inside front cover and Appendix 4 for information on concept maps.

PART V

Energy Resources

A country that runs on energy cannot afford to waste it.
BRUCE HANNON

18 Solutions: Energy Efficiency and Renewable Energy

Houses That Save Energy and Money

Energy experts Hunter and Amory Lovins (Guest Essay, p. 513) have built a large passively heated, superinsulated, partially earth-sheltered home in Old Snowmass, Colorado (Figure 18-1), where winter temperatures can drop to −40°C (−40°F). This structure, which also houses the research center for the Rocky Mountain Institute—an office used by 40 people—gets 99% of its space and water heating and 95% of its daytime lighting from the sun. It uses one-tenth the usual amount of electricity and less than half the usual amount of water for a structure of its size. Total energy savings repaid the cost of its energy-saving features after 10 months and are projected to pay for the entire facility in 40 years.

In energy-efficient houses of the near future, microprocessors will monitor indoor temperatures, sunlight angles, and the people's locations, and they will send heat or cooled air where it is needed. Windows and insulated shutters will automatically open and close to take advantage of solar energy and breezes as well as reduce heat loss from windows at night and on cloudy days. Sensors will turn off lights in unoccupied rooms or dim lights when sunlight is available.

Superinsulating windows, already here, mean that a house can have as many windows as the owner wants in any climate without much heat loss in cold weather or heat gain in hot weather. Thinner insulation material will allow roofs and walls to be insulated far better than in today's best superinsulated houses. Researchers have also developed "smart windows," which change electronically from being clear (which lets sunlight and heat in on cold days) to being reflective (which deflects sunlight when the house gets too warm).

Soon homeowners may be able to get all the electricity they need from rolls of solar cells (that convert sunlight directly into electricity). They can be attached like shingles to a roof or applied to window glass as a coating.

Figure 18-1 The Rocky Mountain Institute in Colorado. This facility is a home and a center for the study of energy efficiency and sustainable use of energy and other resources. It is also an example of energy-efficient passive solar design. (Robert Millman/Rocky Mountain Institute)

If the United States wants to save a lot of oil and money and increase national security, there are two simple ways to do it: stop driving Petropigs and stop living in energy sieves.

AMORY B. LOVINS

These are the questions answered in this chapter:

■ How should we evaluate energy alternatives?

■ What are the benefits and drawbacks of the following energy alternatives?

   a. Improving energy efficiency

   b. Using solar energy to heat buildings and water and to produce electricity

   c. Using flowing water and solar energy stored as heat in water to produce electricity

   d. Using wind to produce electricity

   e. Burning plants and organic waste (biomass) for heating buildings and water, for producing electricity, and for transportation (biofuels)

   f. Extracting heat from Earth's interior (geothermal energy)

   g. Producing hydrogen gas and using it to make electricity, heat buildings and water, and propel vehicles

18-1 EVALUATING ENERGY RESOURCES

The types of energy we use (Figure 18-2) and how we use them are the major factors determining our quality of life and how much we abuse the earth's life-support system. Our current dependence on nonrenewable fossil fuels (Figure 3-5) is the primary cause of air and water pollution, land disruption, and projected global warming and climate change (Section 10-2)—and affordable oil will probably be depleted within 40–80 years and will need to be replaced with other alternatives.

What is our best option for reducing dependence on oil and other fossil fuels? Cut out unnecessary energy waste by improving energy efficiency, as discussed in this chapter. What is our next best energy option? There is disagreement about that.

Some say we should get much more of the energy we need from the sun, wind, flowing water, biomass, heat stored in Earth's interior, and hydrogen gas by making the transition to a new *solar age*. These renewable energy options are evaluated in this chapter.

Others say we should burn more coal and synthetic liquid and gaseous fuels made from coal. Some believe natural gas is the answer, at least as a transition fuel to a new solar age built around improved energy efficiency and renewable energy. Others think nuclear power is the answer. These options are evaluated in the next chapter.

Past experience shows that it usually takes at least 50 years and huge investments to phase in new energy alternatives (Figure 18-3)—with the exception of nuclear power which after almost 50 years still provides only a fairly small amount of our commercial energy. Thus, we have to plan for and begin the shift to a new mix of energy resources now. This involves answering the following questions for each energy alternative:

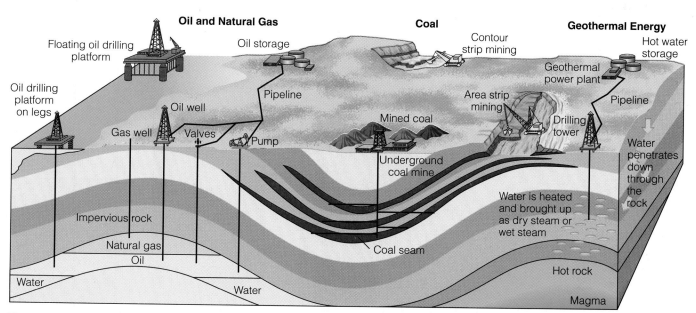

Figure 18-2 Important energy resources from Earth's crust are geothermal energy, coal, oil, and natural gas. Uranium ore (not pictured) is also extracted from the crust and then processed to increase its concentration of uranium-235 (Figure 3-2), which can be used as a fuel in nuclear reactors used to produce electricity.

- How much of the energy source will be available in the near future (the next 15 years), in the intermediate future (the next 30 years), and for the long term (the next 50 years)?

- What is its net energy yield (Spotlight, p. 483)?

- How much will it cost to develop, phase in, and use this energy resource?

- How will extracting, transporting, and using the energy resource affect the environment (Figure 12-13)?

- What will using this energy source do to help sustain the earth for us, for future generations, and for the other species living on this planet?

18-2 IMPROVING ENERGY EFFICIENCY

Doing More with Less   You may be surprised to learn that *84% of all commercial energy used in the United States is wasted* (Figure 18-4). About 41% of this energy is wasted automatically because of the degradation of energy quality imposed by the second law of energy (Section 3-6). However, about 43% is wasted unnecessarily, mostly by using fuel-wasting motor vehicles,

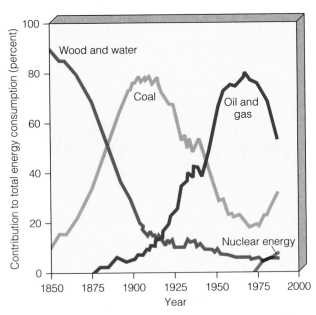

Figure 18-3 Shifts in the use of commercial energy resources in the United States since 1850. Shifts from wood to coal and then from coal to oil and natural gas have each taken about 50 years. Affordable oil is running out, and burning fossil fuels is the primary cause of air pollution and projected warming of the atmosphere. For these reasons most analysts believe we must make a new shift in energy resources over the next 50 years. (Data from U.S. Department of Energy)

Figure 18-4 Flow of commercial energy through the U.S. economy. Note that only 16% of all commercial energy used in the United States ends up performing useful tasks or is converted to petrochemicals. The rest either is automatically and unavoidably wasted because of the second law of energy (41%) or is wasted unnecessarily (43%).

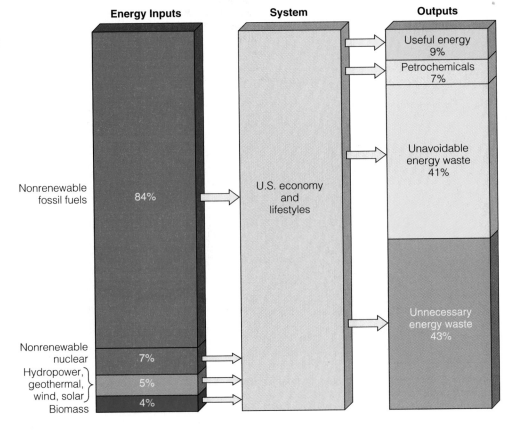

Q: How much of the U.S. working population lives within biking distance of work?

Net Energy: The Only Energy That Really Counts

It takes energy to get energy. For example, oil must be found, pumped up from beneath the ground, transported to a refinery and converted to useful fuels (such as gasoline, diesel fuel, and heating oil), then transported to users, and then burned. All of these steps use energy, and the second law of energy (Section 3-6) tells us that each time we use energy to perform a task some of it is always wasted and is degraded to low-quality energy.

The usable amount of high-quality energy available from a given quantity of an energy resource is its **net energy**—the total useful energy available from the resource over its lifetime minus the amount of energy used (the first law of energy), automatically wasted (the second law of energy), and unnecessarily wasted in finding, process-ing, concentrating, and transporting it to users. For example, if 8 units of fossil fuel energy are needed to supply 10 units of nuclear, solar, or additional fossil fuel energy, the net energy gain is only 2 units of energy. Thus, net energy—not the total energy available—is the only energy that really counts.

We can look at this concept in a different way—as the ratio of useful energy produced to the useful ener-gy used to produce it. In the exam-ple just given, the *net energy ratio* would be 10/8, or approximately 1.2. The higher the ratio, the greater the net energy yield. When the ratio is less than 1, there is a net energy loss over the lifetime of the system.

Figure 18-5 provides estimated net energy ratios for various systems of space heating, high-temperature heat for industrial processes, and transportation. Oil has a relatively high net energy ratio because much of it comes from large, accessible deposits such as those in Saudi Arabia and other parts of the Mid-dle East. When those sources are depleted, the net energy ratio of oil will decline and prices will rise. Then more money and more high-quality fossil fuel will be needed to find, process, and deliver new oil—from widely dispersed small de-posits and deposits buried deep in the earth's crust, or located in re-mote areas like Alaska, the Arctic, and the North Sea.

Conventional nuclear energy has a low net energy ratio because large amounts of energy are required to extract and process uranium ore, to convert it into a usable nuclear fuel, and to build and operate power plants. Energy is also needed to dis-mantle the plants after their 25–30 years of useful life and to store the resulting highly radioactive wastes for thousands of years.

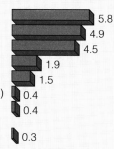

Space Heating

Passive solar — 5.8
Natural gas — 4.9
Oil — 4.5
Active solar — 1.9
Coal gasification — 1.5
Electric resistance heating (coal-fired plant) — 0.4
Electric resistance heating (natural-gas-fired plant) — 0.4
Electric resistance heating (nuclear plant) — 0.3

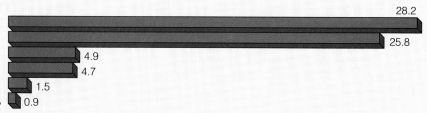

High-Temperature Industrial Heat

Surface-mined coal — 28.2
Underground-mined coal — 25.8
Natural gas — 4.9
Oil — 4.7
Coal gasification — 1.5
Direct solar (highly concentrated by mirrors, heliostats, or other devices) — 0.9

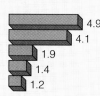

Transportation

Natural gas — 4.9
Gasoline (refined crude oil) — 4.1
Biofuel (ethyl alcohol) — 1.9
Coal liquefaction — 1.4
Oil shale — 1.2

Figure 18-5 Net energy ratios for various energy systems over their estimated lifetimes. (Data from Colorado Energy Research Institute, *Net Energy Analysis*, 1976; and Howard T. Odum and Elisabeth C. Odum, *Energy Basis for Man and Nature*, 3d ed., New York: McGraw-Hill, 1981)

A: More than 50%

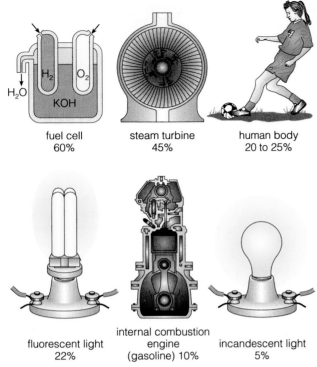

fuel cell
60%

steam turbine
45%

human body
20 to 25%

fluorescent light
22%

internal combustion
engine
(gasoline) 10%

incandescent light
5%

Figure 18-6 Energy efficiency of some common energy conversion devices.

furnaces, and other devices—and by living and working in leaky, poorly insulated buildings. People in the United States unnecessarily waste as much energy as two-thirds of the world's population consumes.

The easiest, fastest, and cheapest way to get more energy with the least environmental impact is to eliminate much of this energy waste. One way to do this is to reduce energy consumption. Examples include walking or biking for short trips, using mass transit, putting on a sweater instead of turning up the thermostat, and turning off unneeded lights.

Another way is to increase the efficiency of the energy conversion devices we use. **Energy efficiency** is the percentage of total energy input that does useful work (is not converted to low-quality, essentially useless heat) in an energy conversion system. The energy conversion devices we use vary considerably in their energy efficiencies (Figure 18-6). We can save energy and money by buying the most energy-efficient home heating systems, water heaters, cars, air conditioners, refrigerators, and other household appliances available, and by supporting research to find even more energy-efficient devices. The energy-efficient models may cost more, but in the long run they usually save money by having a lower **life-cycle cost**: initial cost plus lifetime operating costs.

The net efficiency of the entire energy delivery process for a space heater, water heater, or car is de-

termined by finding the efficiency of each step in the energy conversion process. For example, the sequence of energy-using (and energy-wasting) steps involved in using electricity produced from fossil or nuclear fuels is extraction → transportation → processing → transportation to power plant → electricity generation → transmission → end use.

Figure 18-7 shows the net energy efficiency for heating a well-insulated home **(1)** with electricity produced at a nuclear power plant, transported by wire to the home, and converted to heat (electric resistance heating); and **(2)** passively with an input of direct solar energy through windows facing the sun, with heat stored in rocks or water for slow release. This analysis shows that the process of converting the high-quality energy in nuclear fuel to high-quality heat at several thousand degrees, converting this heat to high-quality electricity, and then using the electricity to provide low-quality heat for warming a house to only about 20°C (68°F), is extremely wasteful of high-quality energy. Burning coal (or any fossil fuel) at a power plant to supply electricity for space heating is also inefficient. By contrast, it is much less wasteful to collect solar energy from the environment, store the resulting heat in stone or water, and—if necessary—raise its temperature slightly to provide space heating or household hot water.

Heating space and water account for 70% of the energy used in a typical U.S. household (space heating 52% and water heating 18%). Physicist and energy expert Amory Lovins (Guest Essay, p. 513) points out that using high-quality electrical energy to provide low-quality heating for living space or household water is like using a chain saw to cut butter or a sledgehammer to kill a fly. As a general rule, he suggests that we not use high-quality energy to do a job that can be done with lower-quality energy (Figure 3-6). The logic of this point is illustrated by looking at the prices for providing heat using various fuels. In 1991, the average price of obtaining 250,000 kilocalories (1 million Btus) for heating either space or water in the United States was $6.05 using natural gas, $7.56 using kerosene, $9.30 using oil, $9.74 using propane, and $24.15 using electricity. As these numbers suggest, if you don't mind throwing away hard-earned dollars, then use electricity to heat your house and bath water. Yet, 22% of U.S. homes are heated by electricity (with 52% being heated by much cheaper natural gas).

Perhaps the three least efficient energy-using devices in widespread use today are **(1)** incandescent light bulbs (which waste 95% of the energy input), **(2)** vehicles with internal combustion engines (which waste 90% of the energy in their fuel), and **(3)** nuclear power plants producing electricity for space heating or water heating (which waste 86% of the energy in their

Q: How much of the air pollution in the United States is produced by motor vehicles?

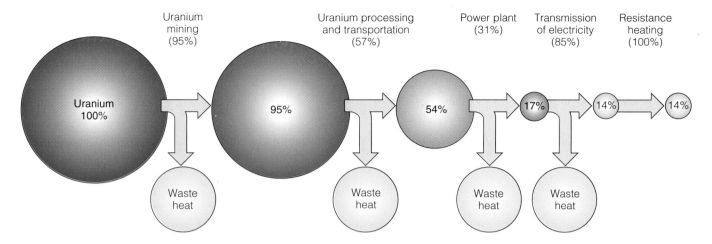

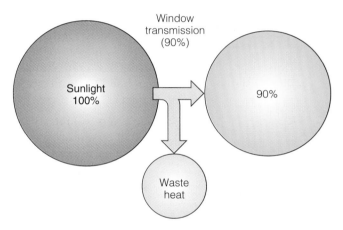

Figure 18-7 Comparison of net energy efficiency for two types of space heating. The cumulative net efficiency is obtained by multiplying the percentage shown inside the circle for each step by the energy efficiency for that step (shown in parentheses). Usually, the greater the number of steps in an energy conversion process, the lower its net energy efficiency. About 86% of the energy used to provide space heating by electricity produced at a nuclear power plant is wasted. By contrast, with passive solar heating, only about 10% of incoming solar energy is wasted.

nuclear fuel; Figure 18-7). These devices were developed when energy was cheap and plentiful. To help sustain ourselves and other species, we will have to replace them or greatly improve their energy efficiency.

Reducing Energy Waste: An Economic and Environmental Offer We Dare Not Refuse

Reducing energy waste is one of the planet's best and most important bargains. Consider the following benefits of reducing energy waste:

- *Making nonrenewable fossil fuels last longer.*

- *Buying time to phase in renewable energy resources.*

- *Decreasing dependence on oil imports* (51% in the United States in 1993).

- *Lessening the need for military intervention in the oil-rich but potentially unstable Middle East.*

- *Reducing environmental damage* because less of each energy resource would provide the same amount of useful energy.

- *Providing the cheapest and quickest way to slow projected global warming* (Solutions p. 246).

- *Saving money, providing more jobs, and promoting more economic growth per unit of energy than other alternatives.* According to energy expert Amory Lovins (Guest Essay, p. 513), if the world really got serious about improving energy efficiency, we could save $1 trillion per year—about 5% of the gross global product. Lovins argues that by using existing energy-efficient technology, the

U.S. economy could be run on one-fourth as much energy as it now uses. A 1993 study by economists estimated that a full-fledged energy-efficiency program could produce 1.3 million jobs in the United States by 2010.

- *Improving competitiveness in the international marketplace.* Currently, the United States spends about 11% of its GNP on energy, whereas Japan spends only 5%, giving Japanese goods a production cost advantage.

Energy analysts and the Office of Technology Assessment estimate that fully implementing *existing* energy efficiency technologies could save the United States about four times as much electricity as all U.S. nuclear power plants now produce, at about one-seventh the cost of just running them, even if it cost nothing to build them. Further, such a program would save more than 40 times as much oil as *might* be under Alaska's Arctic National Wildlife Refuge (Pro/Con, p. 469), at roughly one-tenth the cost of drilling for it.

Why isn't the United States in hot pursuit of improving energy efficiency? There are several reasons. One is the political influence of oil, coal, nuclear power, and automobile companies, with their emphasis on short-term profits regardless of the long-term economic and environmental consequences (Section 7-1). Other reasons include a glut of low-cost fossil fuels, the failure of elected officials to require that external costs of using fossil and nuclear fuels be included in their market prices (Section 7-2), and relatively little federal support since 1980 for improvements in energy efficiency.

Using Waste Heat   Could we save energy by recycling energy? No. The second law of energy tells us that we cannot recycle energy, but we can slow the rate at which waste heat flows into the environment when high-quality energy is degraded. For a house, the best way to do this is to heavily insulate it, eliminate air leaks, and equip it with an air-to-air heat exchanger to prevent buildup of indoor air pollutants. Many homes in the United States are so full of leaks that their heat loss in cold weather and heat gain in hot weather is equivalent to having a large window-size hole in the wall of the house.

In office buildings and stores waste heat from lights, computers, and other machines can be collected and distributed to reduce heating bills during cold weather; during hot weather this heat can be collected and vented outdoors to reduce cooling bills. Waste heat from industrial plants and electrical power plants can be distributed through insulated pipes and used to heat nearby buildings, greenhouses, and fish ponds, as is done in some parts of Europe.

Saving Energy in Industry   Here are some ways to save energy and money in industry:

- *Cogeneration,* *the production of two useful forms of energy (such as steam and electricity) from the same fuel source.* Waste heat from coal-fired and other industrial boilers can be used to produce steam that spins turbines and generates electricity at half the cost of buying it from a utility company. By using the electricity or selling it to the local power company for general use, a plant can save energy and money. Cogeneration, widely used in western Europe for years, could produce more electricity than all U.S. nuclear power plants, do it much more cheaply, and meet projected U.S. electricity needs. Small cogeneration units (about the size of a refrigerator) that run on natural gas or liquefied petroleum gas (LPG) and that can supply a home with all its space heat, hot water, and electricity needs are available and pay for themselves in saved fuel and electricity in four to five years. In Germany, new micro-cogeneration units allow restaurants, apartment buildings, and other buildings to produce their own electric power.

- *Replace energy-wasting electric motors.* About 60–70% of the electricity used in U.S. industry drives electric motors. Most of them are run at full speed with their output "throttled" to match their task—somewhat like driving with the gas pedal to the floor and the brake engaged. According to Amory Lovins, it would be cost-effective to scrap virtually all such motors and replace them with adjustable-speed drives; within a year the costs would be paid back.

- *Switch to high-efficiency lighting.*

- *Use computer-controlled energy management systems to turn off lighting and equipment in nonproduction areas and to make adjustments during periods of low production.*

- *Increase recycling and reuse, and make products that last longer and are easy to repair and recycle (Sections 13-2 and 13-3).*

Despite the potential energy savings, U.S. government support for research and development in industrial energy efficiency was cut nearly 60% between 1981 and 1992.

Saving Energy in Transportation   Today, transportation consumes 63% of all oil used in the United States—up from 50% in 1973, mostly because of increases in population and the average distances traveled by motor vehicles. Americans have 35% of the world's cars and drive about as far each year as the

Q: How much of the oil used in the United States is consumed by motor vehicles?

rest of the world's motorists combined. About one-tenth of the oil consumed in the world carries U.S. motorists to and from work, 75% of them driving alone.

Here are some important ways to save energy (especially oil) and money in transportation—and reduce pollution:

- *Increase the fuel efficiency of motor vehicles.* Between 1973 and 1985 the average fuel efficiency doubled for new American cars and rose 54% for all cars on the road , but it has risen only slightly since then (Figure 18-8). In 1993 U.S. consumers could buy Chevrolet's Geo Metro XFi (built by Suzuki) with a fuel efficiency of 25/22 kpl (58/53 mpg) and the Honda Civic Hatchback HB VX with a fuel efficiency of 24/20 kpl (56/47 mpg). In fact, the technology has existed since the mid-1980s to build even more fuel-efficient cars (Solutions, p. 488). According to the U.S. Office of Technology Assessment (OTA), existing technology could be used to raise the fuel efficiency of the entire U.S. automotive fleet to 15 kilometers per liter (35 miles per gallon) by 2010, to eliminate oil imports, and to save more than $50 billion per year in fuel costs. Buyers of gas-miser cars would get back any extra purchase costs—probably about $500 per car—in fuel savings in about a year.

- *Shift to more energy-efficient ways to move people (Figure 18-9) and freight.* For example, more freight could be shifted from trucks and planes to more energy-efficient trains and ships. General Motors has developed a new truck trailer that converts to a railcar without unloading. This truck-rail combination would use 80% less fuel than trucks alone for hauls of more than 322 kilometers (200 miles).

- *Raise the fuel efficiency of new transport trucks 50% with improved aerodynamic design, turbocharged diesel engines, and radial tires.*

- *Improve the fuel efficiency of planes and ships.* Boeing's new 777 jet will use about half the fuel per passenger seat of a 727. Existing advanced diesel engine technology could improve the fuel efficiency of ships by 30–40% over the next few decades.

Saving Energy in Buildings   In the United States, office buildings account for about one-third of the country's energy use. The 110-story, twin-towered World Trade Center in Manhattan is a monument to energy waste. It uses as much electricity as a city of 100,000 people for about 53,000 employees. Since its windows don't open to take advantage of natural warming and cooling, heating and cooling systems must run around the clock.

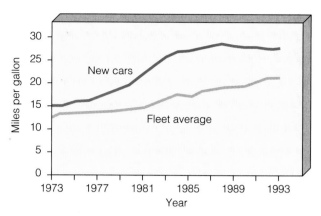

Figure 18-8 Average fuel efficiency of new cars and all cars in the United States between 1973 and 1993. The entire U.S. car fleet averaged 8.5 kilometers per liter (kpl) or 20 miles per gallon (mpg) in 1992, compared to 16 kpl (37 mpg) in Italy—with a gasoline tax of 95¢ per liter ($3.58 per gallon) to encourage fuel efficiency. (Data from U.S. Department of Energy and Environmental Protection Agency)

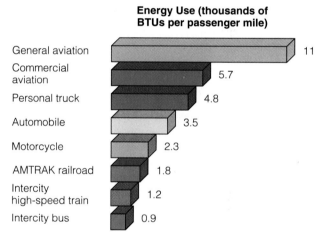

Figure 18-9 Energy efficiency of various types of domestic passenger transportation.

By contrast, Atlanta's 18-story Georgia Power Company building uses 60% less energy than conventional office buildings. The largest surface of the building faces south to capture solar energy. Each floor extends over the one below, which blocks out the higher summer sun to reduce air conditioning costs but allows warming by the low winter sun. Energy-efficient lights focus on desks rather than illuminating entire rooms. Employees working during nonbusiness hours use an adjoining three-story building so the larger structure doesn't have to be heated or cooled just for them. The Georgia Power model and other existing cost-effective technologies could cut energy use by 75% in U.S. buildings by 2010, cut carbon emissions in half, and save more than $130 billion in annual energy bills.

A: 65% (up from 50% in 1973)

Ecocars

SOLUTIONS

Many people believe that fuel-efficient cars will take decades to develop and will be sluggish, small, and unsafe. Wrong! Since 1985 at least 10 companies have had nimble and peppy prototype cars that meet or exceed current safety and pollution standards, with fuel efficiencies of 29–59 kilometers per liter (67–138 miles per gallon). If such cars were mass-produced, their higher costs would be more than offset by their fuel savings.

One such ecocar is Volvo's LCP 2000 prototype, which would be in production today if consumer demand in the United States—the world's largest car market—were there. Its supercharged diesel engine averages 35 kpl (81 mpg) on the highway and 27 kpl (62 mpg) in the city. It can run on various fuels, including diesel, a diesel–gasoline mixture, and vegetable oil—which means the driver could carry a bottle of vegetable oil along as an emergency source of fuel. Using better design and lighter but stronger materials (magnesium, aluminum, and plastics), it exceeds U.S. crash standards. It carries four passengers in comfort, is quiet, resists corrosion better than most of today's cars, and has better-than-average acceleration. This car also meets California's air pollution emission standards—the tightest in the world. It is also designed for easy assembly and for recycling of its materials when it is taken off the road.

Another ecocar is General Motors' sleek, roomy four-passenger Ultralite car (Figure 18-10). This prototype gets 36 kpl (85 mpg) using off-the-shelf technology. What's more, the Ultralite zooms from 0 to 97 kph (0 to 60 mph) in 7.8 seconds, has a 218-kph (135-mph) top speed, and is equipped with four air bags. Its body is made from lightweight carbon-fiber composites that won't dent, scratch, or corrode. The car's three-cylinder, two-stroke engine is much more efficient than conventional four-stroke engines. Furthermore, with 200 fewer parts than a conventional engine, it costs about $400 less to make and saves consumers in maintenance expenses.

We can have roomy, peppy, safe, gas sippers, but only if consumers begin demanding them. In 1992 the market included more than 25 car models with fuel efficiencies of at least 17 kpl (40 mpg), but they made up only 5% of U.S. car sales. There is very little consumer interest in fuel-efficient cars when the inflation-adjusted price of gasoline today is about the same as it was in 1950 and is less than that of bottled water.

Most environmentalists believe that such cars will not be widely used—nor much more efficient ones produced—without significant, government-mandated improvements in fuel efficiency (Figure 18-8) and greatly increased gasoline taxes (coupled with tax relief for the poor and the lower middle class). They argue that both of these actions will eventually save

the country and consumers large amounts of money.

In 1993, the government and the top three U.S. automakers entered into a joint program to build an affordable, safe, and attractive 35-kpl (82-mpg) car by 2003. Many environmentalists call this a sellout by the Clinton administration to automakers. Such cars can already be built, and this program requires only that automakers offer such a car for sale—leading to small energy savings. By contrast, requiring automakers to raise the average fuel efficiency of all new cars to 17 kpl (40 mpg) by 2000—an achievable goal—would save huge amounts of energy and greatly reduce air pollution and emissions of heat-trapping carbon dioxide. Since 1980 automakers have been able to wield enough political power to prevent any increases in government-mandated fuel-efficiency standards (Figure 18-8).

Researchers at the Rocky Mountain Institute (Figure 18-1) project that with proper government support and industry cooperation, *superefficient cars* getting 64–128+ kilometers per liter (150–300+ miles per gallon) could be designed, built, and sold within a decade. Such supercars could gradually eliminate the need for the United States to import oil costing the country $50 billion a year. Worldwide use of supercars would eventually save as much oil as OPEC now produces, permanently depressing oil prices. However, the resulting glut of cheap oil could lead to economic

---

Here are some ways to improve the energy efficiency of buildings:

- *Build more superinsulated houses* (Figure 18-11). Such houses are so heavily insulated and sufficiently airtight that heat from direct sunlight, appliances, and human bodies warms them, with little or no need for a backup heating system. An air-to-air heat exchanger prevents buildup of

indoor air pollution. Although such houses typically cost 5% more to build than conventional houses of the same size, this extra cost is paid back by energy savings within five years and can save a homeowner $50,000–$100,000 over a 40-year period.

- *Improve the energy efficiency of existing houses by adding insulation, plugging leaks, and installing energy-saving windows.* One-third of heated or

Q: What percentage of the Earth's land area is covered by tropical forests?

chaos and political instability in the Middle East, Russia, Mexico, and other areas whose economies depend on large-scale oil production. The auto industry would have to undergo a radical restructuring. However, if supercars become as popular as many analysts expect, the industry could end up bigger and more profitable than today. Supercars would reduce air pollution by being 100 to 1,000 times cleaner than conventional cars. And if half the U.S. fleet were 64-kpl (150-mpg) supercars, the nation's carbon dioxide emissions would drop by about 6%—thus slowing possible global warming.

Two highly promising energy sources for ecocars are electricity and hydrogen. A strong stimulus for producing such cars is the California law requiring that 10% of all new vehicles sold in the state by 2003 have zero emissions. Electric cars could help reduce dependence on oil, especially for urban commuting and short trips. All major U.S. car companies have prototype electric cars and minivans, some to be available by 1995. They are extremely quiet, need little maintenance, can accelerate rapidly with adequate power supplies, and produce no air pollution, except indirectly from the generation of electricity needed to recharge their batteries. However, the plants that generate the electricity are not located in urban areas, as cars are. Overall, electric cars could cut urban air pollution levels by up to 50% and carbon dioxide emissions

General Motors

Figure 18-10 Ultralite concept car built by General Motors. It is fast and much safer than existing cars. When driven at 81 kph (50 mph), it gets 43 kpl (100 mpg).

by 25% per kilometer traveled because power plants burning fossil fuels are about twice as efficient as gasoline engines in urban traffic. If solar cells could be used for recharging, this environmental impact would be eliminated.

On the negative side, current electric cars cost $5,000–$10,000 more than conventional cars, but the costs should come down with mass production. Also, the batteries in current electric cars have to be replaced about every 48,000 kilometers (30,000 miles) at a cost of about $2,000. This requirement, plus the electricity costs for daily recharging, means operating costs double those for gasoline-powered cars. If longer-lasting batteries that hold a higher charge density and last at least 160,000 kilometers (100,000 miles) can be developed, operating costs would be reduced and performance would increase. In 1993 a small company, Energy Conversion

Devices, Inc., announced a new nickel metal hydride car battery that could more than double the range of electric cars and possibly not need replacing over a vehicle's expected life.

Another even more environmentally appealing possibility is to produce electricity to power vehicles using fuel cells powered by solar-produced hydrogen, as discussed in Section 18-7. Fuel cells are compact, make no noise, are safe, and require little maintenance. They are also very efficient, converting 50% or more of the fuel energy to power, compared to about 10% efficiency for gasoline-powered vehicles. More important, fuel-cell cars running on hydrogen come close to being true "zero emission" vehicles because they emit only water vapor and trace amounts of nitrogen oxides, easily controlled with existing technology (Section 18-7).

cooled air in U.S. homes and buildings escapes through closed windows, holes and cracks. This energy waste costs consumers at least $13 billion annually and is equal to the energy in all the oil flowing through the Alaskan pipeline every year. During hot weather these windows also let heat in, increasing the use of air conditioning. A double-pane window has an insulating value of only R-2. (The R value is a measure of resistance to

heat flow). Windows with R values as high as R-8 are now available. They pay for themselves in lower fuel bills within two to four years, and then save money every year thereafter.

■ *Use the most energy-efficient ways to heat houses* (Figure 18-12). The most energy-efficient way to heat space is to build a superinsulated house (Figure 18-11), to use passive solar heating, and to use high-efficiency (85–98% efficient) natural gas furnaces. The most wasteful and expensive way is to use electric resistance heating with the electricity produced by a nuclear or a coal-fired power plant. Some utilities push heat pumps as a more efficient alternative to electric resistance heating and even to natural gas. Heat pumps can save energy and money for space heating in warm climates (where they aren't needed much), but not in cold climates because at low temperatures they switch to wasteful, costly electric resistance heating (which is why some utilities wanting to sell more electricity push them). Also, many heat pumps in their air-conditioning mode are much less efficient than many available stand-alone air conditioning units. Most heat pumps also require expensive repair every few years.

■ *Use the most energy-efficient ways to heat household water.* An efficient method is to use tankless instant water heaters (about the size of bookcase loudspeakers) fired by natural gas or liquefied petroleum gas (LPG). They heat the water instantly as it flows through a small burner chamber, and they provide hot water only when (and as long as) it is needed. Tankless heaters are widely used in many parts of Europe and are slowly beginning to appear in the United States. A well-insulated, conventional natural gas or LPG water heater is also fairly efficient (although

all conventional natural gas and electric resistance heaters keep a large tank of water hot all day and night and can run out after a long shower or two). The least efficient and most expensive way to heat water for washing and bath-

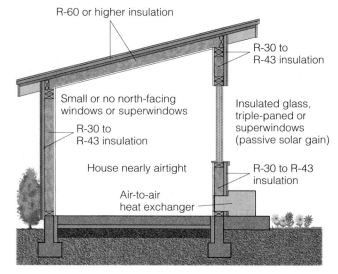

Figure 18-11 Major features of a superinsulated house. Such a house is heavily insulated and nearly airtight. Heat from direct solar gain, appliances, and human bodies warms the house, which requires little or no auxiliary heating. An air-to-air heat exchanger prevents buildup of indoor air pollution.

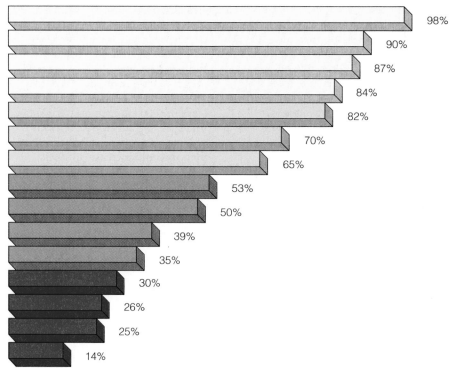

Net Energy Efficiency

Superinsulated house (100% of heat)(R-43)	98%
Passive solar (100% of heat)	90%
Passive solar (50% of heat) plus high-efficiency natural gas furnace (50% of heat)	87%
Natural gas with high-efficiency furnace	84%
Electric resistance heating (electricity from hydroelectric power plant)	82%
Natural gas with typical furnace	70%
Passsive solar (50% of heat) plus high-efficiency wood stove (50% of heat)	65%
Oil furnace	53%
Electric heat pump (electricity from coal-fired power plant)	50%
High-efficiency wood stove	39%
Active solar	35%
Electric heat pump (electricity from nuclear plant)	30%
Typical wood stove	26%
Electric resistance heating (electricity from coal-fired power plant)	25%
Electric resistance heating (electricity from nuclear plant)	14%

Figure 18-12 Net energy efficiencies for various ways to heat an enclosed space such as a house.

Q: How rapidly are the world's remaining tropical forests vanishing?

Light Up Your Life, Help the Earth, and Save Money

SOLUTIONS

One-fourth of the electricity produced in the United States goes to lighting. Since conventional incandescent bulbs are only 5% efficient (Figure 18-6) and last only 750–1,500 hours, they waste enormous amounts of energy and money, and they add to the heat load of houses during hot weather. About half the air conditioning in a typical U.S. office building is used to remove the internal heat gain from inefficient lighting.

Socket-type compact fluorescent light bulbs that use one-fourth as much electricity as conventional bulbs are now available. Although they cost about $10–$20 per bulb, they last 10–20 times longer than conventional incandescent bulbs and save three times their cost over their long life (Figure 18-13). Students in Brown University's environmental studies program showed that the school could save more than $40,000 per year just by replacing the incandescent light bulbs in exit signs with fluorescents.

E-lamp (electronic lamp) bulbs, available in 1995, use high-frequency radio waves to generate electricity. They last even longer than compact fluorescent bulbs because there is no electrode to burn out. Furthermore, they are smaller than compact fluorescents, making them usable in existing light fixtures, and unlike compact fluorescents they can be used with a dimmer switch.

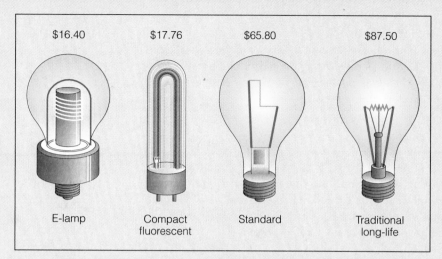

Figure 18-13 Cost of electricity for comparable light bulbs used for 10,000 hours. Because conventional incandescent bulbs are only 5% efficient and last only 750–1,500 hours, they waste enormous amounts of energy and money, and they add to the heat load of houses during hot weather. (Data from Electric Power Research Institute).

Replacing a standard fluorescent bulb with an energy-efficient compact fluorescent or E-lamp bulb saves about $50—or $1,250 for a typical house with 25 light bulbs. Energy-efficient lighting could save U.S. businesses $15–$20 billion per year in electricity bills. By using energy-efficient lighting the Boeing Corporation in Seattle is saving $1 million per year, and Columbia University is saving $2 million per year.

Despite their advantages, compact fluorescents have not caught on in the United States mostly because of a lack of knowledge about their benefits, low availability in stores, too low an intensity for some purposes, high initial cost (despite considerable long-term savings), and a lack of fixtures that will accept their larger size. In contrast, compact fluorescent bulbs provide more than 80% of home lighting in Japan, where a variety of bulb designs and lighting fixtures are widely available in stores.

Tungsten-halogen lights use about 35% less energy than conventional incandescent bulbs, often last longer, and put out an intense, concentrated light. However, they also put out lots of heat, which increases air-conditioning loads. There are also some unsubstantiated studies indicating that long-term exposure to the light they give off may cause some harmful health effects.

ing is to use electricity produced by any type of power plant. A $425 electric hot water heater can cost $5,900 in energy costs over its 13-year life, compared to about $2,900 for a comparable natural gas water heater over the same period.

- *Set higher energy-efficiency standards for buildings.* Building codes could be changed to require that all new houses use 80% less energy than conventional houses of the same size, as has been done in Davis, California (Solutions, p. 150). Because of tough energy-efficiency standards, the average Swedish home consumes about one-third as

much energy as an average American home of the same size.

- *Buy the most energy-efficient appliances and lights* (Solutions, above).* If the most energy-efficient

---

*Each year the American Council for an Energy-Efficient Economy (ACEEE) publishes a list of the most energy-efficient major appliances mass-produced for the U.S. market. To obtain a copy, send $3 to the council at 1001 Connecticut Ave. N.W., Suite 530, Washington, DC 20036. Each year they also publish *A Consumer Guide to Home Energy Savings,* available in bookstores or from the ACEEE.

lights and appliances now available were installed in all U.S. homes over the next 20 years, the savings in energy would equal the estimated energy content of Alaska's entire North Slope oil fields.

- *Give rebates or tax credits for building energy-efficient buildings, for improving the energy efficiency of existing buildings, and for buying high-efficiency appliances and equipment.*

- *Give lower electricity rates for energy-efficient buildings.*

Japan and many western European countries are leading the world in the *energy-efficiency revolution*. A few places in the United States, such as Davis, California (Solutions, p. 150), and Osage, Iowa (p. 40), are leading the way in the United States.

18-3   DIRECT USE OF SOLAR ENERGY FOR HEAT AND ELECTRICITY

The Solar Age: Building a Solar Economy

About 92% of the known reserves and potentially available energy resources in the United States are renewable energy from the sun, wind, flowing water, biomass, and the earth's internal heat. The other 8% of potentially available domestic energy resources are coal (5%), oil (2.5%), and uranium (0.5%). Developing these mostly untapped renewable energy resources could meet 50–80% of projected U.S. energy needs by 2030 or sooner, and it could meet virtually all energy needs if coupled with improvements in energy efficiency.

For example, studies project that all U.S. electricity needs could be provided with solar power plants spread out over 59,000 square kilometers (22,800 square miles)—less than one-third as much land as is now occupied by U.S. military facilities. Alternatively, all of the country's electricity needs could be met by fully exploiting the wind power potential of just three states—Texas, North Dakota, and South Dakota.

Developing the nation's renewable energy resources would save money, create jobs, eliminate the need for oil imports, cause less pollution and environmental damage per unit of energy used, and increase economic, environmental, and military security. According to the Minnesota Department of Energy, each dollar spent on renewable energy generates $2.33–$2.92 of local economic activity, versus only $0.64 for imported oil. In the United States geothermal power plants, wood-fired (biomass) power plants, hydropower plants, wind farms, and solar thermal power plants with natural gas backup can already produce electricity more cheaply than can new nuclear

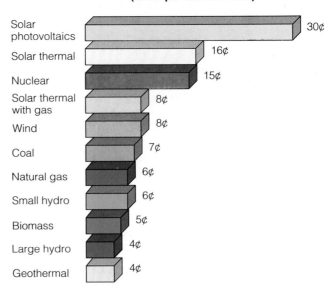

Average Generating Cost (cents per kilowatt-hour)

Solar photovoltaics — 30¢
Solar thermal — 16¢
Nuclear — 15¢
Solar thermal with gas — 8¢
Wind — 8¢
Coal — 7¢
Natural gas — 6¢
Small hydro — 6¢
Biomass — 5¢
Large hydro — 4¢
Geothermal — 4¢

Figure 18-14 Generating costs of electricity per kilowatt-hour by various technologies in 1992. By 2000 costs per kilowatt-hour for wind are expected to fall to 4–5¢, for solar thermal with gas assistance to 6¢, and for solar photovoltaic to 10¢. Costs for other technologies are projected to remain about the same. (Data from U.S. Department of Energy, Council for Renewable Energy Education, and Investor Responsibility Research Center)

power plants, and with far fewer federal subsidies (Figure 18-14).

Heating Houses and Water with Solar Energy

In the United States, solar power helps heat and light more than 100,000 homes. Buildings and water can be heated by solar energy using two methods: passive and active (Figure 18-15). A **passive solar heating system** captures sunlight directly within a structure and converts it into low-temperature heat for space heating (Figures 18-15 and 18-16). Superwindows, greenhouses, and sunspaces face the sun to collect solar energy by direct gain. Thermal mass (heat-storing capacity)—such as walls and floors of concrete, adobe, brick, stone, salt-treated timber, or tile—stores collected solar energy as heat and releases it slowly throughout the day and night. Buildup of moisture and indoor air pollutants is minimized by an air-to-air heat exchanger, which supplies fresh air without much heat loss or gain. A small backup heating system may be used, but is not necessary in many climates.

With available and developing technologies, passive solar designs can provide at least 80% of a building's heating needs and at least 60% of its cooling needs (Solutions, p. 495 and p. 498). Roof-mounted passive solar water heaters can supply all or most of the hot

Q: What country is the largest importer of tropical lumber?

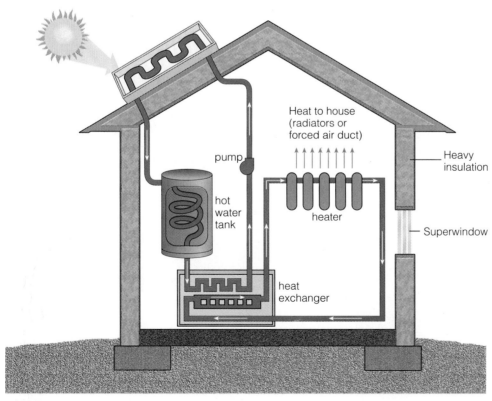

Figure 18-15 Passive and active solar heating for a home.

PASSIVE

Summer sun

Winter sun

Heavy insulation

Superwindow

Superwindow

Stone floor and wall for heat storage

ACTIVE

Heat to house (radiators or forced air duct)

pump

hot water tank

heater

heat exchanger

Heavy insulation

Superwindow

Figure 18-16 Three examples of passive solar design for houses.

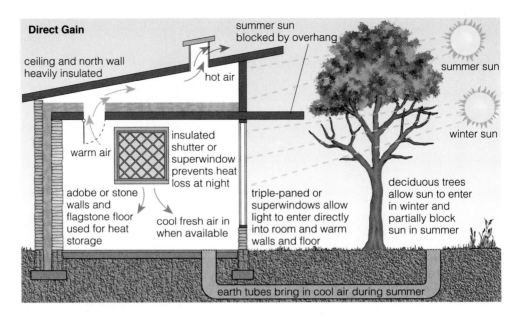

Direct Gain

summer sun blocked by overhang

ceiling and north wall heavily insulated

hot air

summer sun

winter sun

warm air

insulated shutter or superwindow prevents heat loss at night

adobe or stone walls and flagstone floor used for heat storage

triple-paned or superwindows allow light to enter directly into room and warm walls and floor

cool fresh air in when available

deciduous trees allow sun to enter in winter and partially block sun in summer

earth tubes bring in cool air during summer

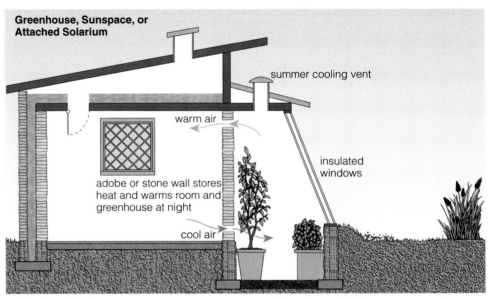

Greenhouse, Sunspace, or Attached Solarium

summer cooling vent

warm air

insulated windows

adobe or stone wall stores heat and warms room and greenhouse at night

cool air

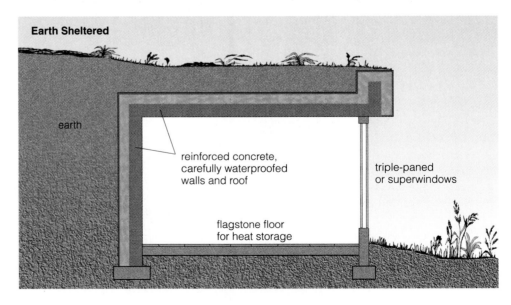

Earth Sheltered

earth

reinforced concrete, carefully waterproofed walls and roof

triple-paned or superwindows

flagstone floor for heat storage

Q: How much of the world's area of tropical forests is managed sustainably?

The Solar Envelope House

Engineer and builder Michael Sykes has designed a solar envelope house that is heated and cooled passively by solar energy and the slow storage and release of energy by massive timbers and the earth beneath the house (Figure 18-18). The front and back sides of this house are double walls of heavy timber impregnated with salt to increase the ability of the wood to store heat. The space between these two walls, plus the basement, forms a convection loop or envelope around the inner shell of the house.

Solar energy enters through windows or a greenhouse on the side of the house facing the sun. It circulates around the loop, is stored in the heavy timber, and is released slowly during the day and at night. In summer, roof vents release heated air from the convection loop throughout the day. At night these roof vents, with the aid of a fan, draw air into the loop, which passively cools the house.

The interior temperature of the house typically stays within 2° of 21°C (70°F) year-round, without any conventional cooling or heating system. In cold or cloudy climates a small wood stove or vented natural

gas heater in the basement can be used as a backup to heat the air in the convection loop.

Sykes sells these houses in kits with all timber precut, which enables quick assembly. Buyers can save money by erecting the inner and outer shells themselves, which requires little experience and few tools. And Michael plants 50 trees for each one used in providing builders with his timber kits.

For his Enertia design he has received both the Department of Energy's Innovation Award and the North Carolina Governor's Energy Achievement Award.

Enertia Building Systems, Rt. 1, Box 67, Wake Forest, NC 27587

Figure 18-17 Solar envelope house that is heated and cooled passively by solar energy and Earth's thermal energy. This patented Enertia design needs no conventional heating or cooling system in most areas. It comes in a precut kit engineered and tailored to the buyer's design goals.

water for a typical house (Figure 18-18). A particularly promising model called the Copper Cricket sells for $2,200.

In hot weather, passive cooling can be provided by blocking the high summer sun with deciduous trees, window overhangs, or awnings (Figure 18-16). For example, one large tree has the cooling power of five average air conditioners running 20 hours per day. Also, windows and fans take advantage of breezes and keep air moving, and a reflective insulating foil sheet suspended in the attic will block heat from radiating down into the house.

Earth tubes can also be used for cooling (Figure 18-16). At a depth of 3–6 meters (10–20 feet), the soil temperature stays at about 5–13°C (41–55°F) all year long in cold northern climates and about 19°C (67°F) in warm southern climates. Several earth tubes—simple plastic (PVC) plumbing pipes with a diameter of 10–15 centimeters (4 to 6 inches)—buried about 0.6 meter (2 feet) apart at this depth can pipe cool and partially dehumidified air into an energy-efficient house at a cost of a few dollars per summer. For a large space, two or three of these geothermal cooling fields running in different directions from the house can be installed.

When heat degrades the cooling effect from one field, homeowners can switch to another. During cold months these geothermal cooling fields are renewed naturally for use during the summer. Initial construction costs (mostly for digging) are high, but operating and maintenance costs are extremely low. People allergic to pollen and molds should add an air purification system, but they would also need to do that with a conventional cooling system.

Solar-powered air conditioners have been developed but thus far are too expensive for residential use. In Reno, Nevada, some buildings are kept cool during summer through the use of large, insulated tanks of water (chilled by cool nighttime air) that keep indoor temperatures comfortable during the day.

On a life-cycle cost basis, good passive solar and superinsulated design is the cheapest way to heat a home or a small building in regions where sunlight is available more than 60% of the time (Figure 18-19). Such a system usually adds 5–10% to the construction cost, but the life-cycle cost of operating such a house is 30–40% lower. Most passive solar systems require that windows and shades be opened or closed to regulate heat flow and distribution, but this can be done by inexpensive microprocessors.

In an **active solar heating system** specially designed collectors absorb solar energy, and a fan or a pump is used to supply part of a building's space-heating or water-heating needs (Figure 18-15). Several connected collectors are usually mounted on a roof with an unobstructed exposure to the sun.

Active solar collectors can also supply hot water (Figure 18-18). In Cyprus, Jordan, and Israel, active solar water heaters supply 25–65% of the hot water for homes. About 12% of houses in Japan and 37% in Australia also use such systems.

With current technology, active solar systems usually cost too much for heating most homes and small buildings, because they use more materials to build, need more maintenance, and eventually deteriorate and must be replaced. In addition, some people consider active solar collectors sitting on rooftops or in yards to be ugly. However, retrofitting existing buildings is often easier with an active solar system. Existing and emerging technologies are expected to make solar heating increasingly attractive (p. 480).

In 1994 a Canadian firm began producing and installing a new type of active solar collector consisting of a thin black aluminum panel perforated with thousands of tiny holes. The panel is mounted a few centimeters from a building's outer walls. A fan draws outside air through the perforations in the panel to the air pocket where it is heated before entering the building's ventilation system. Attached to the sunny side of a building, the panels can preheat ventilation air at efficiencies up to 75%. By comparison, ordinary glass-covered collec-

Passive Solar Water Heating System

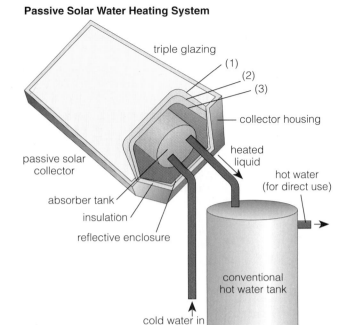

Active Solar Water Heating System

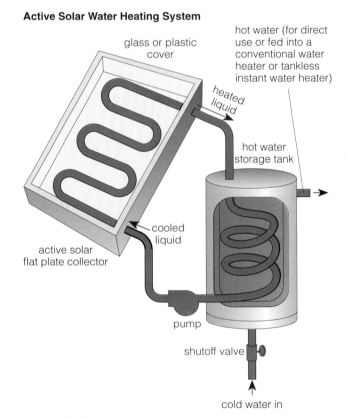

Figure 18-18 Passive and active solar water heaters.

tors (Figure 18-15) are only 35–40% efficient, mostly because glass reflects some of the incoming solar energy. The cost of these new panels is low enough to make them practical for large commercial buildings in colder climates with adequate sunlight.

Q: What percentage of the people in LDCs rely on biomass as their primary fuel for heating and cooking?

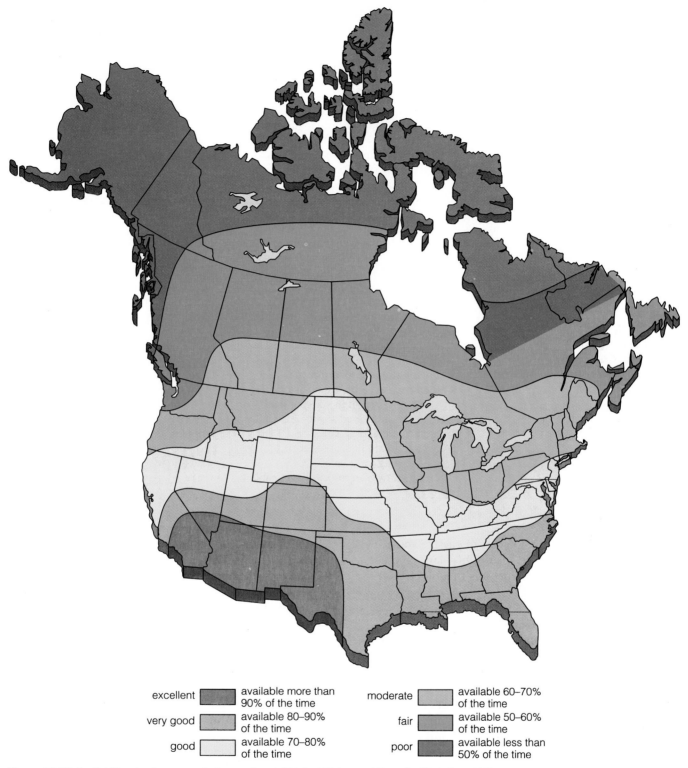

excellent	available more than 90% of the time	
very good	available 80–90% of the time	
good	available 70–80% of the time	
moderate	available 60–70% of the time	
fair	available 50–60% of the time	
poor	available less than 50% of the time	

Figure 18-19 Availability of solar energy in the continental United States and Canada. (Data from U.S. Department of Energy and National Wildlife Federation)

Solar energy for low-temperature heating of buildings, whether collected actively or passively, is free and is naturally available on sunny days; the net energy yield is moderate (active) to high (passive). Both active and passive technology are well developed and can be installed quickly. No heat-trapping carbon dioxide is added to the atmosphere, and environmental impacts from air and water pollution are low. Land disturbance is also minimal because passive systems are built into structures and active solar collectors are

A: Almost 70%

Since the mid-1980s there has been growing interest in building long-lasting, affordable, and easily constructed superinsulated houses with walls made of compacted bales of certain types of straw (available at a low cost almost everywhere). Such houses were widely used in the 1800s in Nebraska and the idea has been revived and improved by several builders and architects.* In recent years, such houses have been built in Canada, Finland, Mexico,

*For information on construction of straw-bale houses contact Out on Bale, 1037 E. Linden St., Tucson, AZ 85719. They publish a quarterly newsletter with an annual subscription price of $21 and provide workshops, videos, and consultations. A video is also available for $29 plus $4 shipping and handling from Carol Escott, P.O. Box 318, Bisbee, AZ 85603, and *Straw-Bay Primer: An Illustrated Guide to Building with Straw Bales* is available for $10 postpaid from Inhabitation Services, P.O. Box 58, Gila, MN 88038.

and Russia, and in fifteen states in the United States, and are spreading.

Most modern straw-bale houses are post-and-beam structures, with a conventional, heavily insulated roof resting on rigid horizontal beams supported by vertical posts and resting on a concrete slab or continuous footings. The walls are easily and quickly constructed by stacking the bales of straw on the slab or footings—a process that can be done by the owner and an untrained volunteer crew of friends in a day or two (much like the old-time barn raisings). Metal bars or wood dowels are usually driven down through the bales to provide rigidity, framing is built for windows and doors, chases are used for plumbing and electrical wires, and the interior and exterior walls are then covered with adobe or plaster.

Depending on the thickness of the bales, plastered straw-bale walls have an insulating value of R-35 to R-150 and are fire resistant and long-lasting (at least 60 years). Using straw—an *annually* renewable

agricultural residue often considered a waste product to be burned—for the walls reduces the use of wood and deforestation. Using straw walls also greatly reduces energy inputs and produces much less pollution than the production and use of lumber or concrete-block walls. There is little ecological impact on land because the straw comes from land already converted to pasture or cropland.

This method of construction is affordable and practical for people in LDCs, for those in MDCs who have been priced out of the conventional housing market, and anyone wanting to save money in construction and long-term heating and cooling costs. It also leads to social cooperation by encouraging people to get together and help one other build their own houses. The main problem is getting banks and other money lenders to recognize the potential of this type of housing and provide homeowners with construction loans.

---

usually placed on rooftops. Owners of passive and active solar systems also need "solar rights" laws to prevent others from building structures that block their access to sunlight.

Using Solar Energy to Generate High-Temperature Heat and Electricity   Several systems are in operation that draw on solar energy to generate electricity and high-temperature heat (Figure 18-20). In one such system, huge arrays of computer-controlled mirrors, called heliostats, track the sun and focus sunlight on a central heat-collection tower (Figure 18-20c). In another system, sunlight is collected and focused on oil-filled pipes running through the middle of curved solar collectors (p. 479 and Figure 18-20b). This concentrated sunlight can generate temperatures high enough for industrial processes or for producing steam to run turbines and generate electricity. Molten salt stores solar heat to produce electricity at night or on cloudy days.

The most promising approach to intensifying solar energy is *nonimaging optics*, which has been

under development since 1965 by American and Israeli scientists. With this technology, the sun's rays are allowed to scramble instead of being focused on a particular point (Figure 18-20d). Experiments show that a nonimaging parabolic concentrator can intensify sunlight striking Earth 80,000-fold—producing the highest intensity of sunlight anywhere in the solar system, including that on the sun's surface.

Because of their high efficiency and ability to generate extremely high temperatures, nonimaging optical concentrators may make solar energy practical for widespread industrial and commercial use within a decade. Giant arrays of these concentrators at solar power plants may power turbines to generate electricity, to produce hydrogen gas for fuel (Section 18-7), and to convert some hazardous wastes into less harmful substances. Inexpensive solar cookers can also be used to focus and concentrate sunlight and cook food, especially in rural villages in sunny LDCs (Figure 18-20e).

The impact of solar power plants on air and water is low. They can be built in 1–2 years, compared to 5–15 years for coal-fired and nuclear power plants.

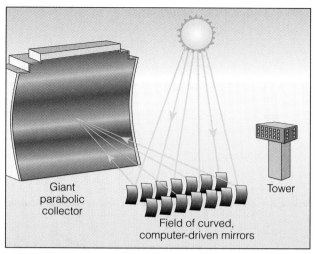

a. Solar Furnace

Giant parabolic collector

Field of curved, computer-driven mirrors

Tower

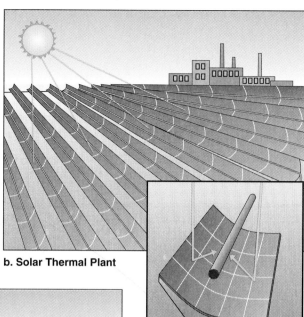

b. Solar Thermal Plant

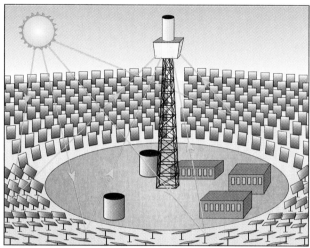

c. Solar Power Tower

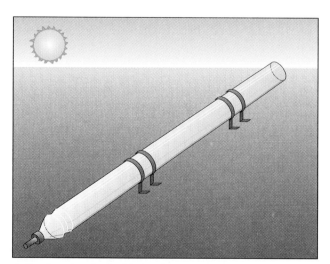

d. Nonimaging Optical Solar Concentrator

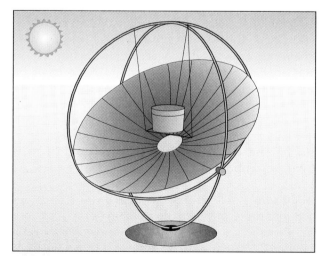

e. Solar Cooker

Figure 18-20 Several ways to collect and concentrate solar energy to produce high-temperature heat and electricity are in use. Today such plants are used mainly to supply reserve power for daytime peak electricity loads, especially in sunny areas with a large demand for air conditioning. On an even economic playing field, such plants could produce electricity more cheaply than a coal-burning or nuclear power plant.

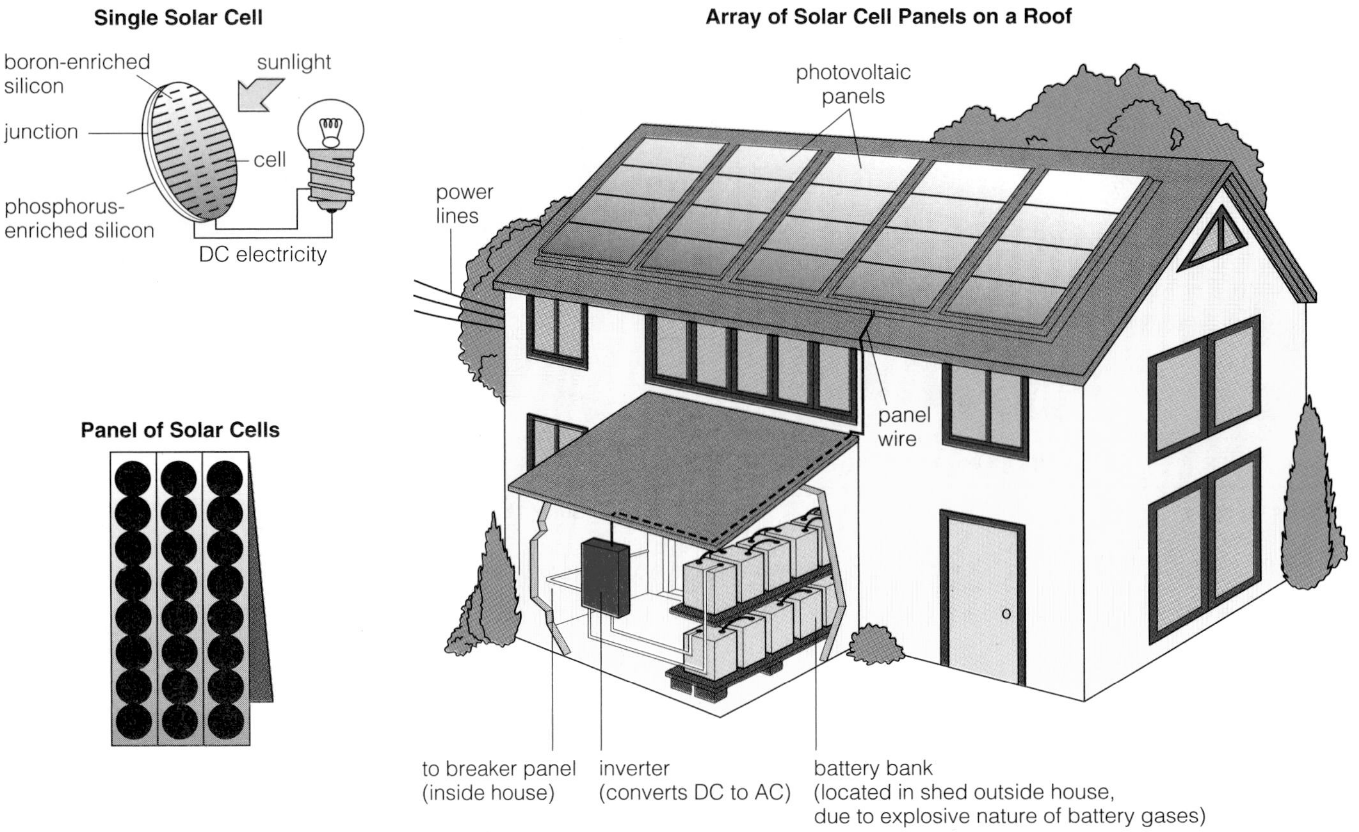

Single Solar Cell

boron-enriched silicon
sunlight
junction
cell
phosphorus-enriched silicon
DC electricity

Panel of Solar Cells

Array of Solar Cell Panels on a Roof

photovoltaic panels
power lines
panel wire

to breaker panel (inside house)
inverter (converts DC to AC)
battery bank (located in shed outside house, due to explosive nature of battery gases)

Figure 18-21 Use of photovoltaic (solar) cells to provide electricity for a house. Small and easily expandable arrays of such cells can provide electricity for urban villages throughout the world without the need for building large power plants and power lines. Massive banks of such cells can also produce electricity at a small power plant. Today at least two dozen U.S. utility companies are using photovoltaic cells in their operations. As the price of such electricity drops, usage will increase dramatically. In 1990 a Florida builder began selling tract houses that get all of their electricity from roof-mounted solar cells. Although the solar-cell systems account for about one-third of the cost of each house, the savings in electric bills will pay this off over a 30-year mortgage period.

Thus builders would save millions in interest on construction loans. Solar thermal power plants produce electricity almost as cheaply as a new nuclear power plant; and, with small turbines burning natural gas as backup, they can produce electricity at almost half the cost of a nuclear power plant (Figure 18-14). Although solar thermal power plants need large collection areas (p. 479), they use one-third less land area than a coal-burning plant (when the land used to extract coal is included) and 95% less land per kilowatt-hour than most hydropower projects.

Producing Electricity from Solar Cells   Solar energy can be converted directly into electrical energy by **photovoltaic cells**, commonly called **solar cells** (Figure 18-21)—an entirely different technology from active solar collectors, which yield heat, not electricity. Because a single solar cell produces a tiny amount of electricity, many cells are wired together in a panel providing 30–100 watts. Several panels, in turn wired together and mounted on a roof or on a rack that tracks the sun, produce electricity for a home or a

building. The DC electricity produced can be stored in batteries and used directly or converted to conventional AC electricity.

Solar cells are reliable and quiet, have no moving parts, and should last 30 years or more if encased in glass or plastic. They can be installed quickly and easily, and maintenance consists of occasional washing to keep dirt from blocking the sun's rays. Small or large solar-cell packages can be built, and they can be easily expanded or moved as needed. Solar cells can be located in deserts and marginal lands, alongside interstate highways, in yards, and on rooftops. Massive banks of such cells can also produce electricity at a small power plant (Figure 18-22).

Solar cells produce no heat-trapping carbon dioxide during use. Air and water pollution during operation is extremely low, air pollution from manufacture is low, and land disturbance is very low for roof-mounted systems. The net energy yield is fairly high and is increasing with new designs. By 2030, electricity produced by solar cells could drop to 4¢ per kilowatt-hour, making it fully cost-competitive. Sanyo, a Japa-

Q: What is the greenhouse effect?

Figure 18-22 This power plant near Sacramento, California, uses photovoltaic cells to produce electricity. The nuclear plant in the background has been closed down. Today at least two dozen U.S. utility companies are using photovoltaic cells in their operations. As the price of such electricity drops, usage will increase dramatically.

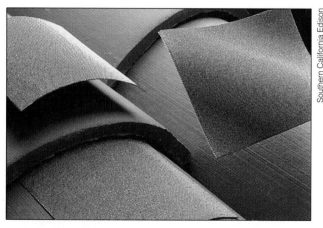

Figure 18-23 New photovoltaic cells that look like metallic sandpaper cost half as much to manufacture as existing solar cells. They are made by sprinkling tiny silicon beads on an aluminum surface. Used as roof panels they should meet a house's electrical needs at a cost competitive with that of conventional electric power.

nese company, has incorporated solar cells into roof shingles. And a German company is selling window panels, suitable for office buildings and homes, that contain a layer of solar cells sandwiched between two panes of conventional glass.

Solar cells are an ideal technology for providing electricity in rural areas in most LDCs. These nations, which cannot afford to build costly nuclear and fossil power plants and electric-line transmission systems, now buy almost two-thirds of the solar cells produced in the United States (with the largest output coming

from an American company that was bought by a German firm several years ago). With financing from the World Bank, India is installing solar-cell systems in 38,000 villages, and Zimbabwe is bringing solar electricity to 2,500 villages.

With an aggressive program starting now, solar cells could supply 17% of the world's electricity by 2010—as much as nuclear power does today—at a lower cost and much lower risk; by 2050, that figure could reach 30% (50% in the United States). That would eliminate the need to build any large-scale power plants and would allow many existing nuclear and coal-fired plants to be phased out.

There are some drawbacks, however. The current costs of solar-cell systems are high (Figure 18-14), but they should become competitive in 5–15 years and are already cost-competitive in some situations. In 1991 Texas Instruments and Southern California Edison developed new low-cost solar-cell roof panels that look like metallic sandpaper (Figure 18-23) and should cut the cost of solar electricity from 30¢ per kilowatt-hour (Figure 18-14) to 12¢ per kilowatt-hour. And in 1994 researchers announced development of more efficient, polycrystalline, thin-film solar cells that should generate electricity at 16¢ per kilowatt-hour, with the cost expected to drop to 12¢ per kilowatt-hour. Moderate levels of water pollution from chemical wastes introduced through the manufacturing process could be a problem, without effective pollution controls. Also, some people find racks of solar cells on rooftops or in yards to be unsightly, but new thin and flexible rolls of cells already developed will eliminate this problem.

A: Warming of the Earth's atmosphere because of the presence of heat-trapping gases

CHAPTER 18   **501**

Despite their enormous potential, the U.S. government cut its research and development budget for solar cells 76% between 1981 and 1990, while the U.S. share of the worldwide solar-cell market fell from 75% to 32%. During that same period, Japan tripled its government expenditures in this area and saw its share of the global solar-cell market grow from 15% to 37%. And in 1989 a U.S. company that was the world's largest manufacturer of solar cells was sold to German interests.

Unless federal and private research efforts on photovoltaic cells are increased sharply, the United States will lose out on a huge global market (at least $5 billion per year by 2010) and may have to import photovoltaic cells from Japan, Germany, Italy, and other countries that have been investing heavily in this promising technology since 1980. If the federal government were to order $500 million worth of solar cells over five years, the U.S. industry could expand. Because of the assured sales and resulting mass-production cost efficiencies, the price of these cells could drop sharply.

There is some good news. In 1993, 68 major utility companies (which serve 40% of U.S. electricity customers) formed a consortium to buy $500 million worth of solar cells between 1994 and 2000. In California the Sacramento Municipal Utility District (SMUD) is putting solar cells on the roofs of 100 homes per year as part of a five-year pilot program. Homeowners pay nothing for the installation but pay a 15% surcharge on their electricity bills to help defray costs.

18-4 PRODUCING ELECTRICITY FROM MOVING WATER AND FROM HEAT STORED IN WATER

Hydroelectric Power In *large-scale hydropower projects*, high dams are built across large rivers to create large reservoirs (Figure 11-10). The stored water then flows through huge pipes at controlled rates, spinning turbines and producing electricity. In *small-scale hydropower projects*, a low dam with no reservoir (or only a small one) is built across a small stream. Because natural water flow generates the electricity, output in small systems can vary with seasonal changes in stream flow.

Falling water can also be used to produce electricity in *pumped-storage hydropower systems*, which are used mainly to supply extra power during times of peak electrical demand. When demand is low, usually at night, pumps using surplus electricity from a conventional power plant pump water uphill from a lake or a reservoir to another reservoir at a higher elevation. When a power company temporarily needs more electricity than its other plants can produce, water in the upper reservoir is released. The water flows through turbines and generates electricity on its downward trip back to the lower reservoir. This is an expensive way to produce electricity, however, and cheaper alternatives are available. Another possibility may be to use solar-powered pumps to raise water to the upper reservoir.

Hydroelectric power, or hydropower, supplies about 18% of the world's electricity and 5% of its total commercial energy. Hydropower supplies Norway with essentially all its electricity, Switzerland with 74%, Austria with 67%, and LDCs with 50%.

Much of the hydropower potential of North America and Europe has been developed. By contrast, Africa has tapped only 5% of its hydropower potential, Latin America 8%, and Asia 9%, although many potential sites are far from where the electricity is needed. Large-scale development of hydropower is planned in many LDCs, including China, India, and Brazil. By the year 2000 China, for example, with one-tenth of the world's hydropower potential, may become the world's largest producer of hydroelectricity.

The United States currently is the world's largest producer of hydroelectricity, which supplies 10% of its electricity and 3–5% of all commercial energy. However, the era of large dams is ending in the United States because construction costs are high, few suitable sites are left, and environmentalists oppose many of the proposed projects because of their harmful effects (Figure 11-10). Any new large supplies of hydroelectric power in the United States will be imported from Canada, which gets more than 70% of its electricity from hydropower and is expanding its capacity under protest from environmentalists and indigenous people (Figure 11-14).

Hydropower has a moderate to high net energy yield and fairly low operating and maintenance costs. Hydroelectric plants rarely need to be shut down, and they emit no heat-trapping carbon dioxide or other air pollutants during operation. They have life spans 2–10 times those of coal and nuclear plants. Large dams also help control flooding and supply a regulated flow of irrigation water to areas below the dam.

However, hydropower also has adverse effects on the environment (Figure 11-10). The reservoirs of large-scale projects flood huge areas, destroy wildlife habitats, uproot people, decrease natural fertilization of prime agricultural land in river valleys below the dam, and decrease fish harvests below the dam. And, small hydroelectric projects can threaten recreational activities and aquatic life, disrupt the flow of wild and scenic rivers, and destroy wetlands.

Producing Electricity from Tides and Waves
Twice a day water that flows into and out of coastal bays and estuaries in high and low tides can be used to spin turbines to produce electricity (Figure 18-24).

Q: What are the principle greenhouse gases?

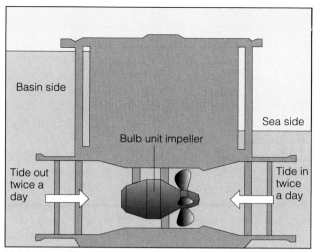

Tidal Power Plant

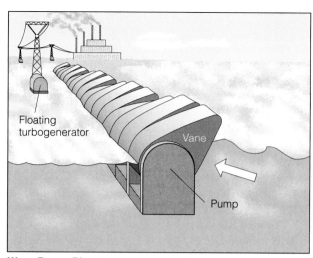

Wave Power Plant

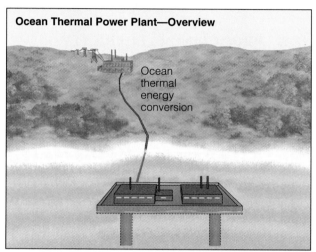

Ocean Thermal Electric Plant

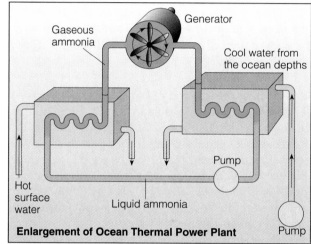

Enlargement of Ocean Thermal Power Plant

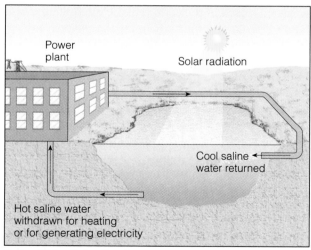

Saline Water Solar Pond

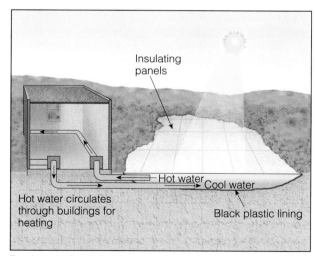

Freshwater Solar Pond

Figure 18-24 Ways to produce electricity from moving water and to tap into solar energy stored in water as heat. None of these are expected to be significant sources of energy in the near future.

A: Water vapor (H_2O), carbon dioxide (CO_2), CFCs, methane (CH_4), nitrous oxide (N_2O)

Currently two large tidal energy facilities are operating: one at La Rance in France, and the other in Canada's Bay of Fundy.

The benefits of tidal power include a free energy source (the moon's gravitational attraction), low operating costs, and a moderate net useful energy yield. Also, no carbon dioxide is added to the atmosphere, air pollution is low, and little land is disturbed. However, most analysts expect tidal power to make only a tiny contribution to world electricity supplies. There are few suitable sites, construction costs are high, and the output of electricity varies daily with tidal flows, which means there must be a backup system.

The kinetic energy in ocean waves, created primarily by wind, is another potential source of electricity (Figure 18-24). Most analysts expect wave power to make little contribution to world electricity production, except in a few coastal areas with the right conditions. Construction costs are moderate to high, and the net useful energy yield is moderate. Also, equipment could be damaged or destroyed by saltwater corrosion and severe storms.

Producing Electricity from Heat Stored in Tropical Oceans   Japan and the United States have been evaluating the use of the large temperature differences (between the cold, deep waters and the sun-warmed surface waters) of tropical oceans for producing electricity. If economically feasible, this would be done in *ocean thermal energy conversion* (OTEC) plants anchored to the bottom of tropical oceans in suitable sites (Figure 18-24).

The energy source for OTEC is limitless at suitable sites; no costly energy storage and backup system is needed; and the floating power plant requires no land area. Nutrients brought up when water is pumped from the ocean bottom might nourish schools of fish and shellfish. However, most energy analysts believe that the large-scale extraction of energy from ocean thermal gradients may never compete economically with other energy alternatives. Despite 50 years of work, the technology is still in the research and development stage.

Solar Ponds   *Saline solar ponds*—usually located near inland saline seas or lakes in areas with ample sunlight—can be used to produce electricity (Figure 18-24). Heat accumulated during the daytime in the bottom layer can be used to produce steam that spins turbines, generating electricity. An experimental saline solar-pond power plant on the Israeli side of the Dead Sea has been operating successfully for several years.

Freshwater solar ponds can be used for water and space heating (Figure 18-24). A shallow hole is dug and lined with concrete. A number of large, black plastic bags, each filled with several centimeters of water, are placed in the hole and then covered with fiberglass insulation panels. The panels let sunlight in and keep most of the heat stored in the water during the daytime from being lost to the atmosphere. When the water in the bags has reached its peak temperature in the afternoon, a computer turns on pumps to transfer hot water from the bags to large, insulated tanks for distribution.

Both saline and freshwater solar ponds require no energy storage and backup systems, emit no air pollution, and have a moderate net useful energy yield. Freshwater solar ponds can be built in almost any sunny area and have moderate construction and operating costs. With adequate research and development support, proponents believe that solar ponds could supply 3–4% of U.S. electricity needs within 10 years.

18-5 PRODUCING ELECTRICITY FROM WIND

Since 1980 wind turbines have become more reliable and less expensive. In 1993, there were nearly 20,000 wind turbines worldwide, most grouped in clusters called *wind farms* (Figure 18-25), producing 2,700 megawatts of electricity. Most are in California (17,000 machines, which produce enough electricity to meet the residential needs of a city as large as San Francisco) and Denmark (which gets 2–3% of its electricity from wind turbines). The island of Hawaii gets about 8% of its electricity from wind, and the use of wind power is spreading to the state's other islands, such as Oahu. If wind farms were built on favorable sites in North Dakota, that state alone could supply 20–36% of the electricity used in the continental United States. Add South Dakota, and the figure jumps to as much as 60%.

Wind power is a virtually unlimited source of energy at favorable sites, and large wind farms can be built in six months to a year and then easily expanded as needed. With a moderate to fairly high net energy yield, these systems emit no heat-trapping carbon dioxide or other air pollutants during operation; they need no water for cooling; and manufacturing them produces little water pollution. The land under wind turbines can be used for grazing cattle and other purposes, and the leases to use the land for wind turbines can provide extra income for farmers and ranchers. Wind power (with much lower subsidies) also has a significant cost advantage over nuclear power (Figure 18-14) and should become competitive with power from natural gas and coal-fired plants in many places before 2000.

However, wind power is economical only in areas with steady winds. When the wind dies down, backup

Q: What are the four principal sources of human emissions of greenhouse gases?

Wind Turbine

Electrical generator
Gearbox
Power cable

Figure 18-25 Using wind to produce electricity. Wind turbines can be used individually or in clusters (wind farms).

electricity from a utility company or from an energy storage system becomes necessary. Backup power could also be provided by linking wind farms with a solar-cell or hydropower system or with efficient natural gas turbines. Other drawbacks to wind farms include visual pollution and noise, although these can be overcome with improved design and location in isolated areas.

Large wind farms might also interfere with the flight patterns of migratory birds in certain areas and have killed large birds of prey (especially hawks, falcons, and eagles) that prefer to hunt along the same ridge lines that are ideal for wind turbines. The killing of birds of prey by wind turbines has pitted environmentalists emphasizing wildlife protection against those promoting renewable wind energy. In 1993 the U.S. Fish and Wildlife Service threatened action against wind power companies planning to construct additional units in California. This has led to cancellations of several wind farm projects. Researchers hope to find ways to eliminate or sharply reduce this problem.

Wind power experts project that by the middle of the next century wind power could supply more than 10% of the world's electricity and 10–25% of the electricity used in the United States. Danish companies, with tax incentives and low-interest loans from their government, have taken over the lion's share of the global market for manufacturing wind turbines.

European governments are presently spending 10 times more for wind energy research and development than the U.S. government in efforts to reduce coal use (and cut acid deposition), decrease reliance on nuclear energy, and launch an industry that can expand worldwide. In the United States most tax incentives for wind power have been removed, and the federal budget for research and development of wind power was cut by 90% between 1981 and 1990. With 25 wind-turbine manufacturers, the European Community plans to produce almost twice as much electricity from the wind by 2000 as the United States.

18-6 PRODUCING ENERGY FROM BIOMASS

Biomass As a Versatile Fuel *Biomass* is organic matter produced directly (in producer organisms) from solar energy through photosynthesis and indirectly by consumers feeding on producers. It includes wood, agricultural wastes, and some components of garbage. Some of this plant matter can be burned as solid fuel or converted into more convenient gaseous or liquid *biofuels* (Figure 18-26). Biomass—mostly from the burning of wood and manure to heat buildings and cook food—supplies about 15% of the world's energy (4–5% in Canada and the United States), and about half of the energy used in LDCs.

Various types of biomass fuels can be used for heating space and water, producing electricity, and propelling vehicles. Biomass is a renewable energy resource so long as trees and plants are not harvested faster than they grow back—a requirement that is not being met in many places (Section 16-4). Also, no net increase in atmospheric levels of heat-trapping carbon

A: Burning fossil fuels (57%), use of CFCs (17%), agriculture (15%), deforestation (8–30%) CHAPTER 18 **505**

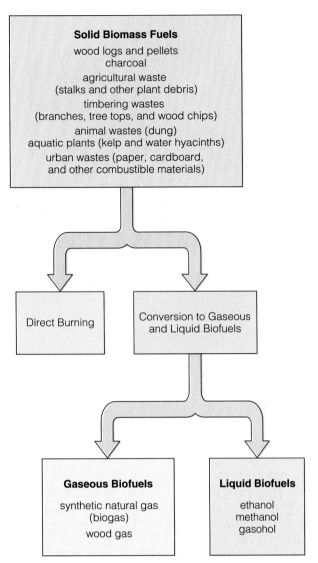

Solid Biomass Fuels
wood logs and pellets
charcoal
agricultural waste
(stalks and other plant debris)
timbering wastes
(branches, tree tops, and wood chips)
animal wastes (dung)
aquatic plants (kelp and water hyacinths)
urban wastes (paper, cardboard,
and other combustible materials)

Direct Burning

Conversion to Gaseous
and Liquid Biofuels

Gaseous Biofuels
synthetic natural gas
(biogas)
wood gas

Liquid Biofuels
ethanol
methanol
gasohol

Figure 18-26 Principal types of biomass fuel.

dioxide occurs so long as the rate of removal and burning of trees and plants—and the rate of loss of belowground organic matter—do not exceed the rate of replenishment. Burning biomass fuels adds much less sulfur dioxide and nitrogen oxides to the atmosphere per unit of energy produced than does the uncontrolled burning of coal, and thus it requires fewer pollution controls.

However, it takes a lot of land to grow biomass fuel—about 10 times as much land as solar cells need to provide the same amount of electricity. Without effective land-use controls and replanting, widespread removal of trees and plants can deplete soil nutrients and cause excessive soil erosion, water pollution, flooding, and loss of wildlife habitat. Biomass resources also have a high moisture content (15–95%). The added weight of the moisture makes collecting

and hauling wood and other plant material fairly expensive and reduces the net energy yield.

Biomass Plantations   One way to produce biomass fuel is to plant large numbers of fast-growing trees (especially cottonwoods, poplars, sycamores, and leucaenas), shrubs, and water hyacinths in *biomass plantations*. After harvest, these "BTU bushes" can be burned directly, converted into burnable gas, or fermented into fuel alcohol. Such plantations can be located on semiarid land not needed to grow crops (although lack of water can limit productivity) and can be planted to reduce soil erosion and help restore degraded lands.

However, this industrialized approach to biomass production usually requires large areas of land (as well as heavy use of pesticides and fertilizers, which can pollute drinking water supplies and harm wildlife); and in some areas the plantations might compete with food crops for prime farmland. Conversion of large forested areas into single-species biomass plantations also reduces biodiversity.

Burning Wood and Wood Wastes   Almost 70% of the people living in LDCs heat their dwellings and cook their food by burning wood or charcoal. However, at least 1.1 billion people in LDCs cannot find, or are too poor to buy, enough fuelwood to meet their needs, and that number may increase to 2.5 billion by 2000 (Figure 16-20).

Sweden leads the world in using wood as an energy source, mostly for district heating plants. In the United States small wood-burning power plants located near sources of their fuel can produce electricity a little more cheaply than burning coal. For example, wood-fired power plants provide 23% of the electricity used in heavily forested Maine and about 4% of the electricity used in New England.

The forest-products industry (mostly paper companies and lumber mills) consumes almost two-thirds of the fuelwood used in the United States. Homes and small businesses burn the rest. Even though only about 4% of U.S. homes burn wood as their main heating fuel (down from 7.5% in 1984), the amount of firewood used in the United States each year would be enough to build a wall 30 meters (100 feet) high from New York City to San Francisco.

Wood has a moderate to high net energy yield when collected and burned directly and efficiently near its source. However, in urban areas where wood must be hauled from long distances, it can cost homeowners more per unit of energy produced than oil or electricity. Harvesting wood can cause accidents (mostly from chain saws), and burning wood in poorly maintained or operated wood stoves can cause house fires.

Q: What country is the largest emitter of greenhouse gases?

Wood stoves can also pollute the air, especially with particulate matter. According to the EPA, wood burning causes as many as 820 cancer deaths a year in the United States. Since 1990 the EPA has required all new wood stoves sold in the United States to emit at least 70% less particulate matter than earlier models.

Fireplaces can also be used for heating, but they usually result in a net loss of energy from a house. The draft of heat and gases rising up the fireplace chimney pushes out warm air and pulls in cold air from cracks and crevices throughout a house. Fireplace inserts with glass doors and blowers help, but they still waste energy compared with an efficient wood-burning stove. Energy loss can be reduced by closing the room off and cracking a window so that the fireplace won't draw much heated air from other rooms. A better solution is to run a small pipe from outside into the front of the fireplace so it gets the air it needs during combustion.

Burning Agricultural and Urban Wastes

In agricultural areas, crop residues and animal manure (Figure 16-21) can be collected and burned or converted into biofuels. Since 1985, for example, Hawaii has been burning *bagasse*, the residue left after sugarcane harvesting and processing, to supply almost 10% of its electricity (58% on the island of Kauai and 33% on the island of Hawaii). Brazil gets 10% of its electricity from burning bagasse and plans to use this crop residue to produce 35% of its electricity by 2000. The ash from biomass power plants can sometimes be used as fertilizer.

This approach makes sense when residues are burned in small power plants located near areas where the residues are produced. Otherwise, it takes too much energy to collect, dry, and transport the residues to power plants. Some ecologists argue that it makes more sense to use animal manure as a fertilizer and to use crop residues to feed livestock, retard soil erosion, and fertilize the soil.

An increasing number of cities in Japan, western Europe, and the United States have built incinerators that burn trash and use the energy released to produce electricity or to heat nearby buildings (Section 13-4). For example, New England obtains about 2% of its electricity from burning trash at 20 plants. However, this approach has been limited by opposition from citizens concerned about emissions of toxic gases and disposal of toxic ash (Section 13-5). Some analysts argue that more energy is saved by composting or recycling paper and other organic wastes than by burning them (Section 13-3).

Converting Biomass into Gaseous and Liquid Fuels
Plants, organic wastes, sewage, pulp and paper mill sludge, and other forms of solid biomass can be converted by bacteria and various chemical processes into gaseous and liquid biofuels (Figure 18-26). Examples include *biogas* (a mixture of 60% methane—the principal component of natural gas—and 40% carbon dioxide), *liquid ethanol* (ethyl, or grain, alcohol), and *liquid methanol* (methyl, or wood, alcohol).

In China, anaerobic bacteria in more than 6 million *biogas digesters* (500,000 of them improved models built in the 1980s) convert organic plant and animal wastes into methane fuel for heating and cooking. After the biogas has been separated, the solid residue is used as fertilizer on food crops or, if contaminated, on trees. When they work, biogas digesters are very efficient. However, they are slow and unpredictable—a problem that could be corrected with development of more reliable models.

Methane gas produced by anaerobic decomposition of organic matter in landfills can be collected by pipes inserted into the ground, separated from other gases, and burned as a fuel (Figure 13-10). Burning this gas instead of allowing it to escape into the atmosphere helps slow projected global warming because methane causes roughly 25 times as much atmospheric global warming per molecule as carbon dioxide (Figure 10-9c).

Methane gas can also be produced by anaerobic digestion of manure from animal feedlots (Figure 14-9) and sludge from sewage treatment plants. In California a plant burning cattle manure from nearby feedlots supplies electricity for as many as 20,000 homes.

Some analysts believe that liquid ethanol and methanol could replace gasoline and diesel fuel when oil becomes too scarce and expensive. Ethanol can be made from sugar and grain crops (sugarcane, sugar beets, sorghum, and corn) by fermentation and distillation. For example, since 1987 ethanol made by fermentation of surplus sugarcane has supplied about 25% of the automotive fuel in Brazil, helped Brazil cut its oil imports, and created an estimated 575,000 full-time jobs. Currently, about 40% of Brazil's cars run on 96% ethanol. However, the government has spent $8 billion to subsidize the country's ethanol industry. Between 1989 and 1991, ethanol's share of the country's transportation fuel fell from 28% to 25%, mostly because of low gasoline prices, a faltering economy, and rampant inflation.

Gasoline mixed with 10–23% pure ethanol makes *gasohol*, which burns in conventional gasoline engines and is sold as super unleaded or ethanol-enriched gasoline. Gasohol now accounts for about 8% of gasoline sales in the United States—25–35% in Illinois, Iowa, Kentucky, and Nebraska. The ethanol used in gasohol is made mostly by fermenting corn. Excluding federal taxes it costs about $1.60 per gallon to produce, compared to about 50¢ per gallon for gasoline; but

A: Unites States (18.4%)

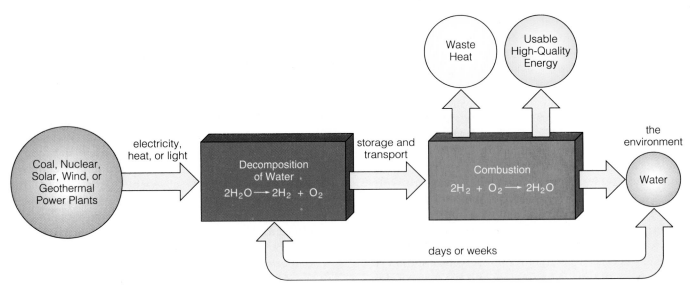

Figure 18-27 The hydrogen energy cycle. The production of hydrogen gas requires electricity, heat, or solar energy to decompose water, thus leading to a negative net energy yield. However, hydrogen is a clean-burning fuel that can be used to replace oil and other fossil fuels and nuclear energy. Using solar energy to produce hydrogen from water could also eliminate most air pollution and greatly reduce the threat of global warming.

new, energy-efficient distilleries could make this fuel competitive with other forms of unleaded gasoline without federal tax breaks. Table 9-3 gives the advantages and disadvantages of using ethanol, methanol, and several other fuels as alternatives to gasoline.

18-7 THE SOLAR-HYDROGEN REVOLUTION

Good-Bye Oil and Smog, Hello Hydrogen

When oil is gone or what's left costs too much to use, what will fuel our vehicles, our industry, and our buildings? Some scientists say the fuel of the future is hydrogen gas (H_2) (Table 9-3).

There is very little hydrogen gas around, but we can get it from something we have plenty of: water, when split by electricity into gaseous hydrogen and oxygen (Figure 18-27). If we can make the transition to an energy-efficient solar-hydrogen age, we could say good-bye to smog, oil spills, acid deposition, and nuclear energy, and perhaps to the threat of global warming. The reason is simple: When hydrogen burns in air it reacts with oxygen gas to produce water vapor (Figure 18-27)—not a bad thing to have coming out of our tailpipes, chimneys, and smokestacks.

What's the Catch?   Using hydrogen as an energy source sounds too good to be true, doesn't it? You're right. We must solve several problems to make hydrogen one of our primary energy resources, and we're making rapid progress in doing this.

One problem is that it takes energy to get this marvelous fuel. The electricity needed to split water might come from coal-burning and nuclear power plants, but this subjects us to the harmful environmental effects associated with using these fuels, and it costs more than the hydrogen fuel is worth.

Most proponents of hydrogen as an energy source believe that the energy to produce hydrogen from water must come from the sun. This means we have to develop solar energy concentrators (p. 479 and Figure 18-20) or solar-cell technology (Figure 18-21) to the point at which the energy they produce can decompose water at an affordable cost.

If we can learn how to use sunlight to decompose water cheaply enough, we will set in motion a *solar-hydrogen revolution* over the next 50 years that will change the world as much as—if not more than—the Agricultural and Industrial Revolutions did. Scientists—especially in Japan, Germany, and the United States—are hard at work trying to bring about this revolution.

Hydrogen gas is much easier to store than electricity. It can be stored in a pressurized tank or in metal powders (including abundant and cheap sponge iron used as an ingredient to make steel) that absorb gaseous hydrogen and release it when heated for use as fuel in a car. If metal powders turn out to be economical, instead of pumping gas you might drive up to a fuel station, pull out a metal rack, replace it with a new one charged with metallic hydrogen, and zoom away. Unlike gasoline, solid metallic hydrogen compounds will not explode or burn if a vehicle's tank is

Q: How much must global CO_2 emissions be cut by 2030 to slow projected global warming to an acceptable rate?

ruptured. The problem is that it's difficult to store enough hydrogen gas in a car as a compressed gas or in a metal powder for it to run very far—a problem similar to the one current electric cars face. Scientists and engineers are seeking solutions to this problem.

We don't need to invent hydrogen-powered vehicles. Mercedes, BMW, and Mazda are already testing prototypes on the roads, and Mercedes hydrogen-powered buses could be on the streets of Hamburg, Germany, by 1997. And in Japan, in 1992 Mazda unveiled a prototype car that runs on hydrogen released slowly from metal hydrides heated by the car's radiator coolant. By 2000, hydrogen cars could be cost-competitive with gasoline cars in the United States, if current, artificially low gasoline prices rise to about 53¢ per liter ($2 per gallon)—half the current price in some European countries. Sometime in the 1990s a German firm also plans to market solar-hydrogen systems that would meet all the heating, cooling, cooking, refrigeration, and electrical needs of a home, as well as providing hydrogen fuel for one or more cars.

Another possibility is to power a car with a *fuel cell* in which hydrogen and oxygen gas combine to produce electrical current. Fuel cells have high energy efficiencies of up to 60% (Figure 18-6)—several times the efficiency of conventional gasoline-powered engines and electric cars. Hydrogen-powered fuel cells may also be the best way to meet the heating and electricity needs of homes. Fuel cells are currently expensive, but this could change with more research and mass production. A fleet of 20 Mercedes-Benz hydrogen-powered test cars have logged over half a million miles powered by electricity from fuel cells. And Energy Partners, a U.S. company based in Florida, has developed a prototype fuel-cell car. Several prototype buses powered by fuel cells are being tested in Canada and in the United States. By 1995 the Southern California Gas Company plans to have 10 small fuel-cell power plants on line, and an even bigger fuel-cell power plant is operating in Japan. Other fuel-cell power plants are under way in Buffalo, Pittsburgh, and Atlanta.

Hydrogen could be phased in gradually by mixing it with natural gas as reserves of this fuel are gradually depleted. Recently researchers have developed a new automobile fuel—dubbed hythane—that consists of roughly 30% hydrogen and 70% natural gas. This clean-burning fuel should cost about the same as gasoline per kilometer traveled. Furthermore, ordinary internal combustion engines can run on hythane with only slight modification. Hythane's major disadvantage so far is that it provides only about half the horsepower of gasoline.

The sunny U.S. Southwest could supply much of the country with electricity produced by solar power (Figures 18-20 and 18-21) or with hydrogen produced by solar power. The gas could then be carried by pipeline to areas where it is needed. Small industrial hydrogen pipelines have been used for decades in several countries.

What's Holding Up the Solar-Hydrogen Revolution? Designing the technology for hydrogen-fueled cars, factories, home furnaces, and appliances is the easy part. The biggest problems are economic and political: figuring out how to replace an economy based largely on oil with entirely new production and distribution facilities and jobs. Overcoming these obstacles involves (1) convincing investors and energy companies with strong vested interests in fossil and nuclear fuels to risk lots of capital on hydrogen; and (2) convincing governments to put up some of the money for developing hydrogen energy, as they have done for decades for fossil fuels and nuclear energy; (3) phasing in full-cost pricing (Section 7-2) so that the much higher total costs of fossil fuels and nuclear energy are reflected in the marketplace.

In the United States, large-scale government funding of hydrogen research is generally opposed by powerful U.S. oil companies, electric utilities, and automobile manufacturers. A solar-hydrogen revolution represents a serious threat to their short-term economic well-being.

By contrast, the Japanese and German governments have been spending 7–8 times more on hydrogen research and development than the United States. Germany and Saudi Arabia have each built a large solar-hydrogen plant, and Germany and Russia have entered into an agreement for the joint development of hydrogen propulsion technology for commercial aircraft. Hydrogen has about 2.5 times the energy by weight of gasoline, making it an especially attractive aviation fuel. Without greatly increased government and private research and development, Americans will be buying solar-hydrogen equipment and fuel cells from Germany and Japan and lose out on a huge global market and source of domestic jobs.

18-8 GEOTHERMAL ENERGY

Tapping Earth's Internal Heat Heat contained in underground rocks and fluids is an important source of energy. Over millions of years this **geothermal energy** from the earth's mantle (Figure 4-2) is transferred to underground concentrations of *dry steam* (steam with no water droplets), *wet steam* (a mixture of steam and water droplets), and *hot water* trapped in fractured or porous rock at various places in the earth's crust (Figure 18-2).

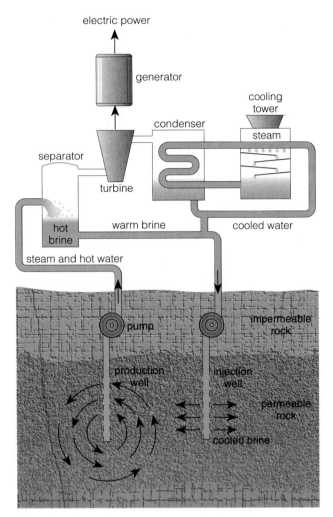

electric power

generator

cooling tower

condenser

steam

separator

turbine

warm brine

cooled water

hot brine

steam and hot water

pump

impermeable rock

production well

injection well

permeable rock

cooled brine

Figure 18-28 Tapping the earth's heat or geothermal energy in the form of wet steam to produce electricity.

If geothermal sites are close to the surface, wells can be drilled to extract the dry steam, wet steam (Figure 18-28), or hot water. This thermal energy can be used for space heating and to produce electricity or high-temperature heat for industrial processes.

Geothermal reservoirs can be depleted if heat is removed faster than it is renewed by natural processes. Thus geothermal resources are nonrenewable on a human time scale, but the potential supply is so vast that it is often classified as a potentially renewable energy resource.

Currently, about 20 countries are extracting energy from geothermal sites and supplying enough heat to meet the needs of over 2 million homes in a cold climate and enough electricity for over 1.5 million homes. The United States accounts for 44% of the geothermal electricity generated worldwide, with most of the favorable sites in the West, especially in California and the Rocky Mountain states. Iceland, Japan, New Zealand, the Philippines, El Salvador, Bolivia, Costa Rica, Kenya, Ethiopia, India, Thailand, and

Indonesia are among the countries with the greatest potential for tapping geothermal energy.

Types of Geothermal Resources   *Dry-steam reservoirs* are the preferred geothermal resource, but they are also the rarest. A large dry-steam well near Larderello, Italy, has been an important source of power for Italy's electric railroads since 1904. Two other large sites are the Matsukawa field in Japan and the Geysers steam field about 145 kilometers (90 miles) northwest of San Francisco. Currently 28 plants economically tap energy from the Geysers field, supplying more than 6% of northern California's electricity.

Wet-steam reservoirs are more common than dry-steam reservoirs but are more difficult and more expensive to convert to electricity. The world's largest wet-steam power plant is in Wairaki, New Zealand; others operate in Mexico, Japan, El Salvador, Nicaragua, and parts of the former Soviet Union.

Hot-water reservoirs are more common than dry-steam or wet-steam reservoirs. Almost all the homes, buildings, and food-producing greenhouses in Reykjavik, Iceland (population 85,000), are heated by hot water drawn from geothermal wells under the city. In Paris, France, the equivalent of 200,000 dwellings are heated the same way. At 180 locations in the United States, mostly in the West, hot-water reservoirs have been used for years to heat homes and farm buildings and to dry crops.

There are also three virtually undepletable sources of geothermal energy: *molten rock* (magma), found near the earth's surface (in volcanoes, for example); *hot dry-rock zones*, where molten rock that has penetrated the earth's crust heats subsurface rock to high temperatures; and low- to moderate-temperature *warm-rock reservoir deposits*, which could be used to preheat water and run heat pumps for space heating and air conditioning. According to the National Academy of Sciences the energy potentially recoverable from such reservoirs would meet U.S. energy needs at current consumption levels for 600–700 years, but the cost may be high.

Pros and Cons   The biggest advantages of geothermal energy include a vast and sometimes renewable supply of energy, for areas near reservoirs; moderate net useful energy yields for large and easily accessible reservoir sites; far less carbon dioxide emission per unit of energy than from fossil fuels; and competitive cost of producing electricity (Figure 18-14).

A serious limitation of geothermal energy is the scarcity of easily accessible reservoir sites. Geothermal reservoirs must also be carefully managed, or they can be depleted within a few decades. Furthermore, geothermal development in some areas can destroy or degrade forests or other ecosystems. In Hawaii, for example, environmentalists are fighting the construction

Q: What is the role of the ozone (O_3) gas in the stratosphere?

Working with Nature: A Personal Progress Report

SPOTLIGHT

I am writing this book deep in the midst of beautiful woods in Eco-Lair, a structure that Peggy, my spouse, and I designed to work with nature. This ongoing experiment is a low-tech, low-cost example of sustainable-Earth design and living.

In 1980 we purchased a 1954 school bus from a nearby school district for $200 and sold the tires for the same price that we paid for the bus. We built an insulated foundation, rented a crane for two hours to lift the gutted bus onto it, placed heavy insulation around the bus, and added a wooden outside frame. The interior is paneled with wood, some of it from the few trees we carefully selected for removal (Figure 18-29). Most visitors don't know the core of the structure is a school bus unless we tell them.

We attached a solar room—a passive solar collector with double-paned conventional sliding glass windows (for ventilation)—to the entire south side of the bus structure (Figure 18-29). Thick concrete floors and filled concrete blocks (on the lower half of the interior wall facing the sun) absorb and slowly release solar energy collected during the day.

The solar room provides 60% of our heat during the cold months. The rest of the heat is provided by water heated by solar collectors and when necessary heated further by a tankless instant water heater fueled by LPG.

Our compact fluorescent light bulbs (Figure 18-13) last an average of ten years and use about 75% less electricity than conventional bulbs. For the time being we are buying electricity from the power company, but we hope to get our electricity from roof-mounted photovoltaic panels (Figures 18-21) within a few years. Electricity bills run around $30 per month (including an $18 minimum charge), compared with $100 or more for conventional structures of the same size.

In moderate weather windows capture breezes. During the summer additional cooling is provided by earth tubes (Figure 18-16). A small variable-speed fan draws outside air at 35°C (95°F) slowly through the buried tubes (which are surrounded by earth at about 16°C or 60°F). Air enters the bus at about 22°C (72°F). This natural air conditioning costs about $1 per summer for running the fan.

Several large oak trees and other deciduous trees in front of the solar room give us additional passive

cooling during summer; in winter, they drop their leaves to let the sun in (Figure 18-16). A used conventional central air conditioning unit (purchased for $200) is our backup. It can be turned on for short periods when excessive pollen or heat and humidity overwhelm our immune systems and the earth tubes.

Eco-Lair is surrounded by natural vegetation, including flowers and low-level ground cover adapted to the climate of the area. This means there is no grass to cut and no lawn mower to push, repair, feed with gasoline, and listen to. Peggy added plants that repel various insects, so we have few insect pest problems. The surrounding trees and other vegetation also provide habitats for various species of insect-eating birds. When ants, mice, and other creatures find their way inside, we use natural alternatives to repel and control them (Section 15-1).

We save water with water-saving faucets, a water-saving shower head, and a low-flush toilet. We have also experimented with a waterless composting toilet that converts waste and garbage scraps into an odorless powder that can be used as a soil conditioner.

We compost kitchen wastes and recycle them to the soil. Paper, tin

(continued)

---

of a large geothermal project in the only lowland tropical rain forest left in the United States. And in some areas the extraction of hot water from hot-water reservoirs causes land to sink. In Wairaki, New Zealand, for instance, land under hot-water geothermal reservoirs tapped for energy is subsiding at a rate of about 0.4 meter (1.3 feet) per year.

Without pollution control, geothermal energy production causes moderate-to-high air and water pollution. Noise, odor, and local climate changes can also be problems. With proper controls, however, most experts consider the environmental effects of geothermal energy to be less, or no greater, than those of fossil-fuel and nuclear power plants.

Making the Transition to a Sustainable Energy Future   There are leaders in making the transition

from the age of oil to the age of energy efficiency and renewable energy. Brazil and Norway get more than half their energy from hydropower, wood, and alcohol fuel. Israel, Japan, the Philippines, and Sweden plan to rely on renewable sources for most of their energy. California has become the world's showcase for solar and wind power. And communities such as Davis, California (Solutions, p. 150), and Osage, Iowa (p. 40), and individuals are taking energy matters into their own hands (Spotlight, above). What are you, your local community, and your state doing to save energy and make more use of renewable energy?

In the long run, humanity has no choice but to rely on renewable energy. No matter how abundant they seem today, eventually coal and uranium will run out.

DANIEL DEUDNEY AND CHRISTOPHER FLAVIN

A: It filters out high-energy ultraviolet (UV) radiation from the sun

Figure 18-29 Eco-Lair is where I work. It is a low-tech, low-cost ongoing experiment in saving energy and money. A south-facing room (left photo) collects solar energy passively and distributes it to a well-insulated, reused 1954 school bus. The photo on the right shows the interior of the refurbished school bus. The large cabinet on the left folds down and serves as a double bed. The bus windows on the right can be opened as needed to allow heat collected in the attached solar room to flow into the bus space.

cans, and nonrefillable glass bottles go to a local recycling center, along with most of the small amount of plastics we use. We try never to use energy-intensive aluminum cans and plastic bags, choosing refillable glass bottles and reusable cloth bags instead. Extra furniture, clothes, and other items we have accumulated or salvaged over the years are stored in three other old school buses and recycled to family, friends, and people in need. For most household chemicals we use more Earth-friendly substitutes (Table 13-2). We use recycled toilet paper (made completely from post-consumer paper waste) and recycled paper for my office.

Because I work at home, I do little driving. Our primary car is a four-door, automatic-transmission version of a GEO Metro that gets 18 kpl (42 mpg). (Peggy has back problems that prevent her from using the even more energy-efficient manual-shift model.) It is peppy and has more interior room than most midsize cars; its only drawback is that it doesn't have air bags (available for the driver on 1995 models). Our backup transportation is a much less fuel-efficient four-

wheel drive vehicle needed mostly when bad weather transforms the dirt road leading to Eco-Lair into a muddy or icy mess.

We wish cars like the Volvo LCP 2000 or GM's Ultralite (Figure 18-10) were available. Eventually, we hope to buy a vehicle that runs on hydrogen gas or on electricity produced by solar photovoltaic cells.

We get most of our food from the grocery store, but recently Peggy has started growing most of our vegetables using raised-bed organic gardening. For health and environmental reasons we have greatly reduced our meat consumption. Ethically, we feel we should be strict vegetarians, but to date we have been unwilling to go quite that far.

We feel a part of the piece of land we live on and love. We feel that its trees, flowers, deer, squirrels, owls, hummingbirds, songbirds, and other wildlife are a part of us, and we a part of them. As temporary caretakers of this small portion of the biosphere, we feel obligated to pass it on to future generations with its ecological integrity preserved.

Our latest project involves wetland restoration. We built a spillway over a creek to allow water to

collect and form a 0.8-hectare (2-acre) wetland that we have stocked with fish to provide a source of protein (aquaculture). The wetland is also an emergency source of water in case of a fire. It is wonderful to see various plants and animals filling the niches in this newly formed aquatic habitat.

Most of our political activities involve thinking globally but acting locally. We also give money to many environmental and conservation organizations working at local, state, national, and global levels.

Working with nature gives us great joy and a sense of purpose. It also saves us money.

Our attempt to work with nature is in a rural area, but people in cities can also have high-quality lifestyles that conserve resources and protect the environment (see Further Readings). Regardless of where you live, you can tread more lightly on the earth and save money by reducing your waste of energy and matter resources, growing some of your own food (in window box planters, for example), and trying to get as much energy as possible (for heating your living space and bathing water) from the sun.

Q: Worldwide, how many people die each year from skin cancer?

Technology Is the Answer (But What Was the Question?)

Amory B. Lovins

GUEST ESSAY

Physicist and energy consultant Amory B. Lovins is one of the world's most recognized and articulate experts on energy strategy. In 1989 he received the Delphi Prize for environmental work; in 1990 The Wall Street Journal *named him as one of the 39 people most likely to change the course of business in the 1990s. He is research director at Rocky Mountain Institute, a nonprofit resource policy center, which he and his wife, Hunter, founded in Old Snowmass, Colorado, in 1982. He has served as a consultant to over 200 utilities, private industries, and international organizations, and to many national, state, and local governments. He is active in energy affairs in more than 35 countries and has published several hundred papers and a dozen books, including* Soft Energy Paths *and the nontechnical version of that work with L. Hunter Lovins,* Energy Unbound: Your Invitation to Energy Abundance *(see Further Readings).*

The answers you get depend on the questions you ask. But sometimes it seems so important to resolve a crisis that we forget to ask what problem we're trying to solve.

It is fashionable to suppose that we're running out of energy and that the solution is obviously to get lots more of it. But asking how to get more energy begs the question of how much we need. That depends not on how much we used in the past but on what we want to do in the future and how much energy it will take to do those things.

How much energy it takes to make steel, run a sewing machine, or keep ourselves comfortable in our house depends on how cleverly we use energy, and the more it costs, the smarter we seem to get. It is now cheaper, for example, to double the efficiency of most industrial electric motor drive systems than to fuel existing power plants to make electricity. (*Just this one saving can more than replace the entire U.S. nuclear power program.*) We know how to make lights five times as efficient as those presently in use [Figure 18-13] and how to make household appliances that give us the same work as now, using one-fifth as much energy (saving money in the process).

Ten automakers have made good-sized, peppy, safe prototype cars averaging 29–59 kilometers per liter (62–138 miles per gallon) [Solutions, p. 488]. We know today how to make new buildings and many old ones so heat-tight (but still well ventilated) that they need essentially no energy to maintain comfort year-round, even in severe climates. In fact, I live in one [Figure 18-1].

These energy-saving measures are uniformly cheaper than going out and getting more energy. Detailed studies in more than a dozen countries have shown that supplying energy services in the cheapest way—by wringing more work from the energy we already have—would let us increase our standard of living while using several times less total energy (and electricity) than we do now.

Those savings cost less than finding new domestic oil or operating existing power plants.

However, the old view of the energy problem included a worse mistake than forgetting to ask how much energy we needed: It sought more energy, in any form, from any source, at any price—as if all kinds of energy were alike. This is like saying, "All kinds of food are alike; we're running short of potatoes and turnips and cheese; but that's okay, we can substitute sirloin steak and oysters Rockefeller."

Some of us have to be more discriminating than that. Just as there are different kinds of food, so there are many different forms of energy, whose different prices and qualities suit them to different uses [Figure 3-6]. There is, after all, no demand for energy as such; nobody wants raw kilowatt-hours or barrels of sticky black goo. People instead want energy services: comfort, light, mobility, hot showers, cold beverages, and the ability to bake bread and make cement. We ought therefore to start at that end of the energy problem and ask, "What tasks do we want energy for, and what amount, type, and source of energy will do each task most cheaply?"

Electricity is a particularly high-quality, expensive form of energy. An average kilowatt-hour delivered in the United States in 1992 was priced at about 7¢, equivalent to buying the heat content of oil costing $116 per barrel—about six times the average world price in 1992. The average cost of electricity from nuclear plants (including fuel and operating expenses) beginning operation in 1988 was 13.5¢ per kilowatt-hour, equivalent on a heat basis to buying oil at about $216 per barrel.

Such costly energy might be worthwhile if it were used only for the premium tasks that require it, such as lights, motors, electronics, and smelters. But those special uses, only 8% of all delivered U.S. energy needs, are already met twice over by today's power stations. Two-fifths of our electricity is already spilling over into uneconomic, low-grade uses such as water heating, space heating, and air conditioning; yet no matter how efficiently we use electricity (even with heat pumps), we can never get our money's worth on these applications.

Thus, *supplying more electricity is irrelevant to the energy problem we have.* Even though electricity accounts for almost all of the federal energy research and development budget and for at least half of national energy investment, it is the wrong kind of energy to meet our needs economically. Arguing about what kind of new power station to build—coal, nuclear, solar—is like shopping for the best buy in antique Chippendale chairs to burn in your stove or brandy to put in your car's gas tank. *It is the wrong question.*

Indeed, *any kind of new power station is so uneconomical that if you have just built one, you will save the country money by writing it off and never operating it.* Why? Because its

(continued)

additional electricity can be used only for low-temperature heating and cooling (the premium "electricity-specific" uses being already filled up) and is the most expensive way of supplying those services. Saving electricity is much cheaper than making it.

The real question is, What is the cheapest way to do low-temperature heating and cooling? The answer is weather-stripping, insulation, heat exchangers, greenhouses, superwindows (which have as much insulating value as the outside wall of a typical house), window shades and overhangs, trees, and so on. These measures generally cost about half a penny per kilowatt-hour; the running costs *alone* for a new nuclear plant will be nearly 4¢ per kilowatt-hour, so it's cheaper not to run it. In fact, under the crazy U.S. tax laws, the extra saving from not having to pay the plant's future subsidies is probably so big that by shutting the plant down society can also recover the capital cost of having built it!

If we want more electricity, we should get it from the cheapest sources first. In approximate order of increasing price, these include:

- Converting to efficient lighting equipment [Figure 18-13]. This would save the United States electricity equal to the output of 120 large power plants plus $30 billion a year in fuel and maintenance costs.

- Using more efficient motors to save half the energy used by motor systems. This would save electricity equal to the output of another 150 large power plants and repay the cost in about a year.

- Eliminating pure waste of electricity, such as lighting empty offices at headache level. Each kilowatt-hour saved can be resold without having to generate it anew.

- Displacing with good architecture, and with passive and some active solar techniques, the electricity now

used for water heating and space heating and cooling. Some U.S. utilities now give low- or zero-interest weatherization loans, which you need not start repaying for ten years or until you sell your house—because it saves the utility millions of dollars to have available the electricity you don't use instead of building new power plants. Most utilities also offer rebates for buying efficient appliances.

- Making appliances, smelters, and the like cost-effectively efficient.

Just these five measures can quadruple U.S. electrical efficiency, making it possible to run today's economy, with no changes in lifestyles, using no power plants, whether old or new and whether fueled with oil, gas, coal, or uranium. We would need only the present hydroelectric capacity, readily available small-scale hydroelectric projects, and a modest amount of wind power. If we still wanted more electricity, the next cheapest sources would include:

- Industrial cogeneration, combined heat-and-power plants, low-temperature heat engines run by industrial waste heat or by solar ponds, filling empty turbine bays and upgrading equipment in existing big dams, modern wind machines or small-scale hydroelectric turbines in good sites, steam-injected natural gas turbines, and perhaps recent developments in solar cells with waste heat recovery.

It is only after we had clearly exhausted all these cheaper opportunities that we would even consider

- Building a new central power station of any kind—the slowest and costliest known way to get more electricity (or to save oil).

To emphasize the importance of starting with energy end uses rather than energy sources, consider a sad little

Critical Thinking

1. What are the five most important things an individual can do to save energy in the home and in transportation? Which, if any, of these do you do? Which, if any, do you plan to do?

2. Should the United States institute a crash program to develop solar photovoltaic cells and solar-produced hydrogen fuel? Explain.

3. Explain why you agree or disagree with the ideas that the United States can get most of the electricity it needs by (a) developing solar power plants; (b) using direct solar energy to produce electricity in solar cells; (c) building new, large hydroelectric plants;

(d) building ocean thermal electric power plants; (e) building wind farms; (f) building power plants fueled by wood, crop wastes, trash, and other biomass resources; (g) tapping geothermal deposits; and (h) recycling some or most of the energy we use.

*4. Make an energy-use study of your school or dorm, and use the findings to develop an energy-efficiency improvement program. Present your plan to school officials.

*5. Make a concept map of the key ideas in this chapter using the section heads and subheads and the key terms (shown in boldface type in the chapter). See the inside front cover and Appendix 4 for information on concept maps.

Q: How much of the stratospheric ozone over the Antarctica is destroyed from September to December each year?

story from France, involving a "spaghetti chart" (or energy flowchart)—a device energy planners often use to show how energy flows from primary sources via conversion processes to final forms and uses. In the mid-1970s energy conservation planners in the French government started, wisely, on the right-hand side of the spaghetti chart. They found that their biggest need for energy was to heat buildings; and that even with good heat pumps, electricity would be the most costly way to do this. So they had a fight with their nationalized utility; they won, and electric heating was supposed to be discouraged or even phased out because it was so wasteful of money and fuel.

Meanwhile, down the street, the energy supply planners (who were far more numerous and influential in the French government) were starting on the left-hand side of the spaghetti chart. They said: "Look at all that nasty imported oil coming into our country! We must replace that oil. Oil is energy.... We need some other source of energy. Voilà! Reactors can give us energy; we'll build nuclear reactors all over the country." But they paid little attention to who would use that extra energy and no attention to relative prices.

Thus the two sides of the French energy establishment went on with their respective solutions to two different, indeed contradictory, French energy problems: *more energy of any kind* versus *the right kind to do each task in the most inexpensive way.* It was only in 1979 that these conflicting perceptions collided. The supply side planners suddenly realized that the only thing they would be able to *sell* all that nuclear electricity for would be electric heating, which they had just agreed not to do.

Every industrial country is in this embarrassing position (especially if we include as "heating" air conditioning, which just means heating the outdoors instead of the indoors). Which end of the spaghetti chart we start on, or *what we think the energy problem is*, is not an academic abstraction: It *determines what we buy.* It is the most fundamental source of disagreement about energy policy.

People starting on the left side of the spaghetti chart think the problem boils down to whether to build coal or nuclear power stations (or both). People starting on the right realize that *no* kind of new power station can be an economic way to meet the needs for using electricity to provide low- and high-temperature heat and for the vehicular liquid fuels that are 92% of our energy problem.

So if we want to provide our energy services at a price we can afford, let's get straight what question our technologies are supposed to answer. Before we argue about the meatballs, let's untangle the strands of spaghetti, see where they're supposed to lead, and find out what we really need the energy *for!*

Critical Thinking

1. The author argues that building more nuclear, coal, or other electrical power plants to supply electricity for the United States is unnecessary and wasteful. Summarize the reasons for this conclusion, and give your reasons for agreeing or disagreeing with this viewpoint.

2. Do you agree or disagree that increasing the supply of energy, instead of concentrating on improving energy efficiency, is the wrong answer to U.S. energy problems? Explain.

19 Nonrenewable Energy Resources

Bitter Lessons from Chernobyl

Chernobyl is a chilling word recognized around the globe as the site of the worst nuclear disaster ever. On April 25, 1986, a series of explosions in a nuclear power plant in Ukraine (then in the Soviet Union) blew the massive roof off the reactor building and flung radioactive debris and dust high into the atmosphere to encircle the planet (Figure 19-1). Here are some consequences of this disaster caused by human error and poor design:

- Thirty-one people died shortly after the accident from massive radiation exposure, and 259 people were hospitalized with acute radiation sickness. By 1994 the accident had caused the premature deaths of at least 8,000 people—mostly members of the Chernobyl cleanup crew.

- Within a few days, 135,000 people were evacuated; 125,000 more were evacuated in 1991; and 2.2 million more people may need to be moved.

- Over half a million people were exposed to dangerous radioactivity, and some may suffer from cancers, thyroid tumors, and eye cataracts.

- Four million people, mostly in Ukraine, Belarus, and northern Europe may suffer health effects.

- Government officials say that the total cost of the accident will reach at least $358 billion.

People evacuated from the region around Chernobyl had to leave their possessions behind and say good-bye, with little or no notice, to lush green wheat fields and blossoming apple trees, to land their families had farmed for generations, to cows and goats that would be shot because the grass they ate was radioactive, and to their radioactive dogs and cats. They will not be able to return.

We will never know the true extent of the harmful health effects of this accident. The names of the thousands of military reservists who helped clean up were never officially recorded. And the computer disk listing the names of evacuees from the contaminated region was stolen in 1990.

The primary lesson of Chernobyl is that *a major nuclear accident anywhere is a nuclear accident everywhere.*

Figure 19-1 Heat-sensing satellite view of the Chernobyl nuclear power plant in Ukraine. The red area near the center shows intense heat from the reactor that exploded in 1986. (© CNES 1991/Spot Image Corporation)

We are an interdependent world and if we ever needed a lesson in that, we got it in the oil crisis of the 1970s.

ROBERT S. MCNAMARA

In this chapter we will seek answers to the following questions:

- How can the following energy alternatives help us?
  a. oil
  b. natural gas
  c. coal
  d. conventional nuclear fission
  e. breeder nuclear fission
  f. nuclear fusion
- What are the best energy options?

19-1   OIL

Conventional Crude Oil   **Petroleum**, or **crude oil**, is a gooey liquid consisting mostly of hydrocarbon compounds, with small amounts of oxygen, sulfur, and nitrogen compounds. Crude oil and natural gas are often trapped together deep within Earth's crust (Figure 18-2), dispersed in pores and cracks in rock formations.

Primary oil recovery involves drilling a well and pumping out the oil that flows by gravity into the bottom of the well. After the flowing oil has been removed, water can be injected into nearby wells to force some of the remaining heavy oil to the surface, a process known as **secondary oil recovery**.

For each barrel of crude oil removed by primary and secondary recovery, two barrels of *heavy oil* are usually left. As oil prices rise, it may become economical to remove about 10% of this heavy oil by **enhanced**, or **tertiary**, **oil recovery**. Steam or carbon dioxide gas can be used to force some of the heavy oil into the well cavity for pumping to the surface. However, enhanced oil recovery is expensive: It takes the energy in one-third of a barrel of oil to retrieve each barrel of heavy oil.

Most crude oil travels by pipeline to a refinery, where it is heated and distilled to separate it into gasoline, heating oil, diesel oil, asphalt, and other components (Figure 19-2). Some of the resulting products, called **petrochemicals**, are used as raw materials in industrial chemicals, fertilizers, pesticides, plastics, synthetic fibers, paints, medicines, and many other products (Figure 18-4).

How Long Will Supplies Last?   Oil is the lifeblood of the global economy (Figure 3-5). Oil *reserves*

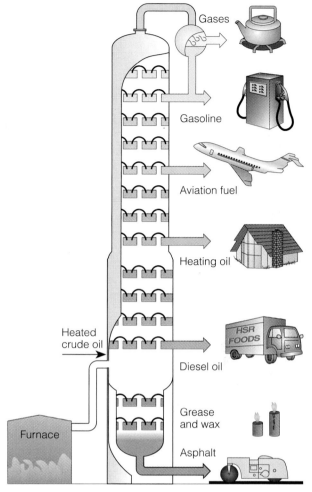

Figure 19-2 Refining of crude oil. Components are removed at various levels, depending on their boiling points, in a giant distillation column.

are identified deposits from which oil can be extracted profitably at current prices with current technology (Figure 12-15). The 13 countries that make up the Organization of Petroleum Exporting Countries (OPEC)* have 67% of these reserves. Saudi Arabia, with 25%, has the largest known crude oil reserves. Geologists believe that the politically volatile Middle East also contains most of the world's undiscovered oil. Therefore, OPEC is expected to have long-term control over world oil supplies and prices.

With only 4% of the world's oil reserves, the United States uses nearly 30% of the oil extracted worldwide each year, 63% of that for transportation. The rest is used by industry (24%), residences and commercial

*OPEC was formed in 1960 so that LDCs with much of the world's known and projected oil supplies could get a higher price for this resource. Today its 13 members are Algeria, Ecuador, Gabon, Indonesia, Iran, Iraq, Kuwait, Libya, Nigeria, Qatar, Saudi Arabia, United Arab Emirates, and Venezuela.

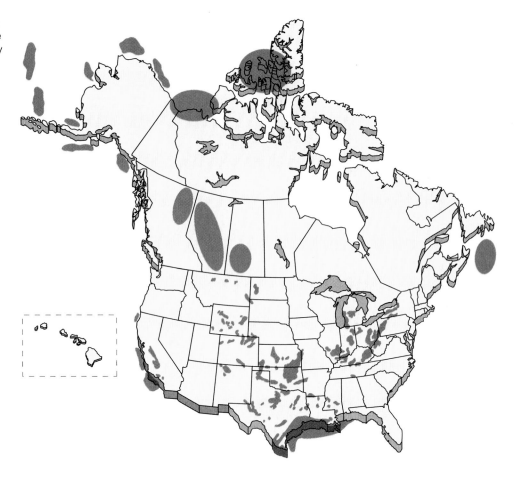

Figure 19-3 Locations of the largest crude oil and natural gas fields in the United States and Canada. Relatively little new oil and natural gas are expected to be found in the United States. (Data from Council on Environmental Quality)

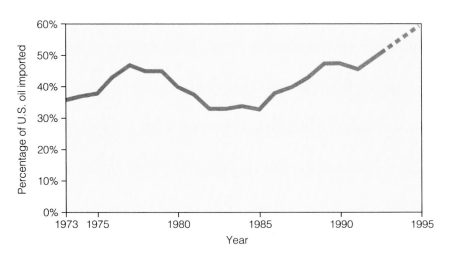

Figure 19-4 Percentage of U.S. oil imported between 1973 and 1992, with projections to 1995. (Data from U.S. Department of Energy and Spears and Associates, Tulsa, Oklahoma)

buildings (8%), and electric utilities (5%). Figure 19-3 shows the locations of the largest crude oil fields in the United States and Canada. Despite an upsurge in exploration and test drilling, U.S. oil extraction has declined since 1985. Moreover, over half of the country's biggest oil fields are 80% depleted, and the net useful energy yield (Figure 18-5) for most domestic new oil is low and falling.

As a result, the United States imported 51% of the oil it used in 1993 (up from 36% in 1973, Figure 19-4), and by 2010 it could be importing 70%. This dependence, and the likelihood of much higher oil prices within 10–20 years, could drain the United States (and other major oil-importing nations) of vast amounts of money, leading to severe inflation and widespread economic recession, perhaps even a major depression.

Q: Worldwide, how fast is soil eroding on farmland?

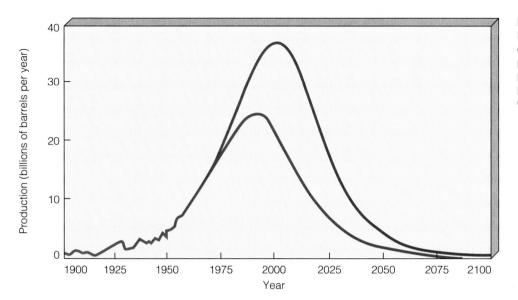

Figure 19-5 The end of the petroleum age is in sight. These two curves show that the world's known petroleum reserves will be 80% depleted between 2025 and 2035, depending on how fast this oil is used. (Data from M. King Hubbert)

Moreover, sabotage of the vulnerable Trans-Alaska oil pipeline—which the Department of Defense says is impossible to protect—could disrupt the entire U.S. economy.

When Iraq invaded Kuwait during the summer of 1990, the United States and other MDCs went to war, mostly to protect their access to oil in the Middle East, especially in Saudi Arabia, which has the world's largest oil reserves. As Christopher Flavin and Nicholas Lenssen put it, "Not only is the world addicted to cheap oil, but the largest liquor store is in a very dangerous neighborhood."

Affordable supplies of oil may be depleted within 35–80 years, depending on how rapidly it is used. At the current rate of consumption, known world oil reserves will last for 42 years. Undiscovered oil that is thought to exist might last another 20–40 years (Figure 19-5). At today's consumption rate, U.S. reserves will be depleted by 2018, and by 2010 if usage rises by 2% per year.

Some analysts argue that rising oil prices will stimulate exploration and that the earth's crust may contain 100 times more oil than is generally thought. Such oil, if it exists, lies 10 kilometers (6 miles) or more below the surface—about twice the depth of today's deepest wells. Most geologists, however, do not believe that this oil exists.

Other analysts argue that such optimistic projections about future oil supplies ignore the consequences of exponentially increasing consumption of oil. Consider the following, assuming we continue to use crude oil at the current rate:

- Saudi Arabia, with the largest known crude oil reserves, could supply all the world's oil needs for only 10 years.

- The estimated reserves under Alaska's North Slope—the largest ever found in North America—would meet world demand for only 6 months or U.S. demand for 3 years.

- All estimated, undiscovered, recoverable deposits of oil in the United States could meet world demand for only 1.7 years and U.S. demand for 10 years.

In short, just to continue using oil at the current rate and not run out, we must discover the equivalent of a new Saudi Arabian supply *every 10 years*. Moreover, if the economies of LDCs grow, the rate of oil consumption will probably exceed current rates and deplete supplies even faster.

Pros and Cons of Conventional Oil Oil is still cheap (when adjusted for inflation it costs about the same as it did in 1950, Figure 19-6). It is easily transported within and between countries, and when extracted from easily accessible deposits it has a high net energy yield (Figure 18-5). Oil's low price has encouraged MDCs and LDCs alike to become heavily dependent on—indeed, addicted to—this important resource. Low prices have also encouraged waste of oil and discouraged the switch to other sources of energy and improvements in energy efficiency.

Oil's fatal flaw, as mentioned above, is that affordable supplies may be depleted within 35–80 years, depending on how rapidly it is used (Figure 19-5). Another drawback of oil (and all fossil fuels) is that burning it releases heat-trapping carbon dioxide, which could alter global climate (Section 10-2), as well as other air pollutants that damage people, crops, trees, fish, and other species (Section 9-4). Oil spills

A: 20–200 times the natural rate of soil renewal

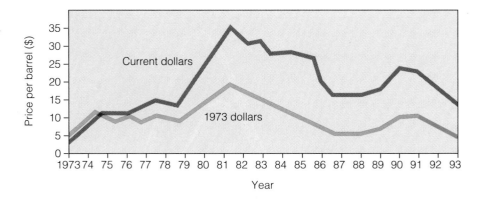

Figure 19-6 Average world crude oil prices, 1973–92. When price is adjusted for inflation, oil has been cheap since 1950. (Data from Department of Energy and Department of Commerce)

(Section 11-8) and leakage of toxic drilling muds pollute water, and the brine solution injected into oil wells can contaminate groundwater (Section 11-7).

Indeed, if all the harmful environmental effects of extracting, processing, and using oil (Figure 12-13) were included in its market price—and if current government subsidies keeping its price artificially low were removed—oil would be too expensive to use and would be replaced by a variety of less harmful and cheaper renewable energy resources.

Heavy Oil from Oil Shale   Oil shale is a fine-grained rock (Figure 19-7) that contains a solid, waxy mixture of hydrocarbon compounds called **kerogen**. After being removed by surface or subsurface mining, the shale is crushed and heated underground to vaporize the kerogen (Figure 19-8). The kerogen vapor is condensed, forming heavy, slow-flowing, dark-brown **shale oil**. Before shale oil can be sent by pipeline to a refinery, it must be heated to increase its flow rate and processed to remove sulfur, nitrogen, and other impurities.

The shale oil potentially recoverable from U.S. deposits—mostly on federal lands in Colorado, Utah, and Wyoming—could probably meet the country's crude oil demand for 41 years at current use levels. Canada, China, and parts of the former Soviet Union also have big oil-shale deposits.

There are problems with shale oil. It has a lower net useful energy yield than that of conventional oil because it takes the energy from almost half a barrel of conventional oil to extract, process, and upgrade one barrel of shale oil (Figure 18-5). Processing the oil requires lots of water, which is scarce in the semiarid locales of the richest deposits. Aboveground mining of shale oil tears up the land, leaving mountains of shale rock (which expands somewhat like popcorn when heated). In addition, salts, cancer-causing substances, and toxic metal compounds can be leached from the processed shale rock into nearby water supplies. Some of these problems can be reduced by extracting shale oil underground (Figure 19-8). However,

Figure 19-7 Oil shale and the shale oil extracted from it. Big U.S. oil shale projects have been canceled because of excessive cost.

this method is too expensive with present technology and produces more sulfur dioxide air pollution than does surface processing.

Heavy Oil from Tar Sand   Tar sand (or oil sand) is a mixture of clay, sand, water, and **bitumen**, a gooey, black, high-sulfur heavy oil. Tar sand is usually removed by surface mining and heated with pressurized steam until the bitumen fluid softens and floats to the top. The bitumen is purified and chemically upgraded into a synthetic crude oil suitable for refining (Figure 19-9).

The world's largest known deposits of tar sands—the Athabasca Tar Sands—lie in northern Alberta, Canada. Economically recoverable deposits there can supply all of Canada's projected oil needs for about 33 years at today's consumption rate, but they would last the world only about two years.

Q: On how much of the world's cropland is soil eroding faster than it forms?

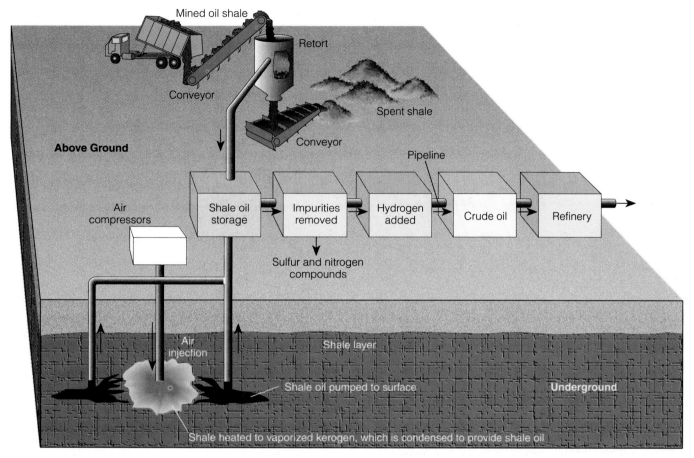

Figure 19-8 Aboveground and underground (*in situ*) methods for producing synthetic crude oil from oil shale.

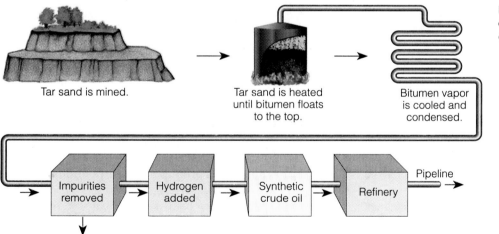

Tar sand is mined.

Tar sand is heated until bitumen floats to the top.

Bitumen vapor is cooled and condensed.

Impurities removed

Hydrogen added

Synthetic crude oil

Refinery

Pipeline

Figure 19-9 Generalized summary of how synthetic crude oil is produced from tar sand.

Producing synthetic crude oil from tar sands has several disadvantages. The net energy yield is low because it takes a lot of energy to extract and process one barrel of bitumen and upgrade it to synthetic crude oil. Also, large quantities of water are needed for processing, and upgrading bitumen to synthetic crude oil releases large quantities of air pollutants. The plants also create huge waste disposal ponds.

19-2 NATURAL GAS

Types, Distribution, and Uses In its underground gaseous state **natural gas** is a mixture of 50–90% by volume of methane (CH_4) and smaller amounts of heavier gaseous hydrocarbons such as propane (C_3H_8) and butane (C_4H_{10}). *Conventional natural gas* lies above most reservoirs of crude oil (Figure 18-2). Some

Figure 19-10 Large quantities of energy are wasted if natural gas found with oil is burned off, as in this Saudi Arabian oil field. This is done because collecting and using the natural gas cost more than it can be sold for in the oil-rich Middle East. Burning this high-quality fuel adds carbon dioxide and other pollutants to the atmosphere, but burning causes less projected global warming than allowing the methane to escape into the atmosphere.

of this is burned off and wasted when the target is oil (Figure 19-10).

Unconventional natural gas is found by itself in other underground sources, including coal seams, Devonian shale rock, deep underground deposits of tight sands, and deep pressurized zones that contain natural gas dissolved in hot water. It is not yet economically feasible to get natural gas from unconventional sources, but the extraction technology is being developed rapidly.

When a natural gas field is tapped, propane and butane gases are liquefied and removed as **liquefied petroleum gas (LPG)**. LPG is stored in pressurized tanks for use mostly in rural areas not served by nat-

ural gas pipelines. The rest of the gas (mostly methane) is dried to remove water vapor, cleaned of hydrogen sulfide and other impurities, and pumped into pressurized pipelines for distribution. At a very low temperature of −184°C (−300°F), natural gas can be converted to **liquefied natural gas (LNG)**. This highly flammable liquid can then be shipped to other countries in refrigerated tanker ships.

The countries making up the former Soviet Union, with 40% of the world's natural gas reserves, are the world's largest extractors of natural gas. Other countries with large proven natural gas reserves are Iran (14%), the United States (5%), Qatar (4%), Algeria (4%), Saudi Arabia (3%), and Nigeria (3%). Geologists expect to find more natural gas, especially in unexplored LDCs.

Most U.S. natural gas reserves are located in the same areas as crude oil (Figure 19-3). About 95% of the natural gas used in the United States is domestic; the other 5% is imported by pipeline from Canada.

About 82% of the natural gas consumed in the United States is used to heat residential and commercial buildings and for drying and other purposes in industry. The rest is used to produce electricity (15%) and to power vehicles (3%) (Table 9-3). Natural gas, burned in new energy-efficient turbines, is the fuel of choice for most new power plants.

Pros and Cons of Natural Gas   Natural gas has a number of advantages over other nonrenewable energy sources. So far it has been cheaper than oil. Known reserves and undiscovered, economically recoverable deposits of conventional natural gas in the United States are projected to last 60 years, and world supplies at least 80 years, at current consumption rates. It is estimated that conventional supplies of natural gas, as well as unconventional supplies available at higher prices, will last about 200 years at the current consumption rate, and 80 years if usage rates rise 2% per year. Natural gas can be transported easily over land by pipeline, has a high net energy yield, and burns hotter and produces less air pollution than any other fossil fuel. Burning natural gas produces just over half as much heat-trapping carbon dioxide per unit of energy produced as coal, and two-thirds that of oil (Figure 19-11). And extracting natural gas damages the environment much less than extracting either coal or uranium ore.

New natural-gas-burning turbines, working like jet engines, can be used to produce electricity much more efficiently (50% efficient) than burning coal (33% efficient). They are very reliable and are easy and quick to install in modular units. Natural gas can also be burned cleanly and efficiently in cogenerators to produce high-temperature heat and electricity.

Q: How much of the world's irrigated cropland has reduced yields because of soil salinization?

One problem with natural gas is that it must be converted to liquid form (LNG) before it can be shipped by tanker from one country to another overseas. This is expensive and dangerous—huge explosions could kill many people and cause much damage in urban areas near LNG loading and unloading facilities. Conversion of natural gas into LNG also reduces the net useful energy yield by one-fourth.

Another problem is leaks of natural gas into the atmosphere from natural gas pipelines, storage tanks, and distribution facilities. Methane, the major component of natural gas, is a greenhouse gas that is more potent per molecule than is carbon dioxide in causing global warming (Figure 10-9c). Improved construction and maintenance, however, could greatly reduce such leaks.

Because of its advantages over oil and coal, some analysts see natural gas as the best fuel to help us make the transition to improved energy efficiency and greatly increased use of renewable energy over the next 50 years. Also, hydrogen gas produced from water by solar-generated electricity (Figures 18-20 and 18-22) could be mixed gradually with natural gas to help smooth the shift to a solar-hydrogen economy (Section 18-7).

19-3 COAL

Types, Distribution, and Uses Coal is a solid formed in several stages as plant remains are subjected to intense heat and pressure over many millions of years. It is a complex mixture of organic compounds, varying amounts of water, and small amounts of nitrogen and sulfur.

Three types of increasingly harder coal are formed over the eons: lignite, bituminous coal, and anthracite (Figure 12-7). Peat, formed in the first stage, is not a coal. It is burned in some places but has a low heat content. Low-sulfur coal (lignite and anthracite) produces less sulfur dioxide when burned than does high-sulfur coal (bituminous). Anthracite is the most desirable type of coal because of its high heat content and low sulfur content.

About 68% of the world's proven coal reserves and 85% of the estimated undiscovered coal deposits are located in the United States, the former Soviet Union, and China (which gets 76% of its commercial energy from coal). Most U.S. coal fields occur in 17 states (Figure 19-12).

Although oil overtook it early this century (Figure 18-3), coal still provides 27% of the world's commercial energy. And since 1950 its use has more than doubled. Coal is used to generate 39% of the world's electricity and make 75% of its steel. Coal supplies 56% of

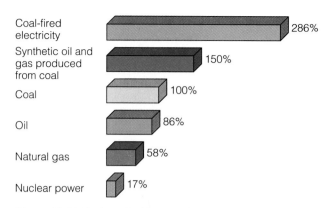

Figure 19-11 Carbon dioxide emissions per unit of energy produced by various fuels, expressed as percentages of emissions produced by coal. Coal contains 80% more carbon per unit of energy than natural gas, and 30% more than oil.

the electricity generated in the United States; the rest is produced by nuclear energy (20%), natural gas (11%), hydropower (9%), oil (3%), and geothermal and solar energy (1%).

Pros and Cons of Solid Coal Coal is the most abundant conventional fossil fuel in the world and in the United States. Identified world reserves of coal should last about 220 years at current usage rates, and 65 years if usage rises 2% per year. The world's unidentified coal reserves are projected to last about 900 years at the current consumption rate, and 149 years if the usage rate increases 2% per year. Identified U.S. coal reserves should last the United States about 300 years at the current consumption rate; unidentified U.S. coal resources could extend those supplies for perhaps 100 years, at a much higher average cost. Coal also has a high net energy yield (Figure 18-5).

Coal also has a number of drawbacks. Coal mining is dangerous because of accidents and black lung disease, a form of emphysema caused by prolonged breathing of coal dust and other particulate matter. Underground mining causes land to sink when a mine shaft collapses during or after mining. Most surface-mined coal is removed by area strip mining or contour strip mining, depending on the terrain (Figures 12-10, 12-11, and 18-2), and these processes cause severe land disturbance and soil erosion (Section 12-3). Restoration of surface-mined land (if required) is expensive, and in arid and semiarid areas the land cannot be fully restored (Figure 12-12). Surface and subsurface coal mining also can severely pollute nearby streams and groundwater from acids and toxic metal compounds (Figure 12-14). Furthermore, once coal is mined it is expensive to move from one place to another; and it cannot be used in solid form as a fuel for cars and trucks.

A: 25%—expected to increase to at least 50% by 2000

Figure 19-12 Major coal fields in the United States and Canada. (Data from Council on Environmental Quality)

Type of Coal
- anthracite
- bituminous
- subbituminous
- lignite

Because coal produces more carbon dioxide per unit of energy than do other fossil fuels (Figure 19-11), burning more coal accelerates projected global warming (Section 10-2). Without expensive air pollution control devices, burning coal produces more air pollution per unit of energy than any other fossil fuel. Each year air pollutants from coal burning kill thousands of people (with estimates ranging from 5,000 to 200,000) and cause at least 50,000 cases of respiratory disease

and several billion dollars in property damage in the United States alone. However, new ways, such as *fluidized-bed combustion* (Figure 19-13) have been developed to burn coal more cleanly and efficiently. In the United States commercial fluidized-bed combustion boilers are expected to begin replacing conventional coal boilers in the mid-1990s.

If all of coal's harmful environmental costs (Figure 12-13) were included in its market price (Section 7-2)

Q: What percentage of the commerial energy used in the United States is consumed by the agricultural system?

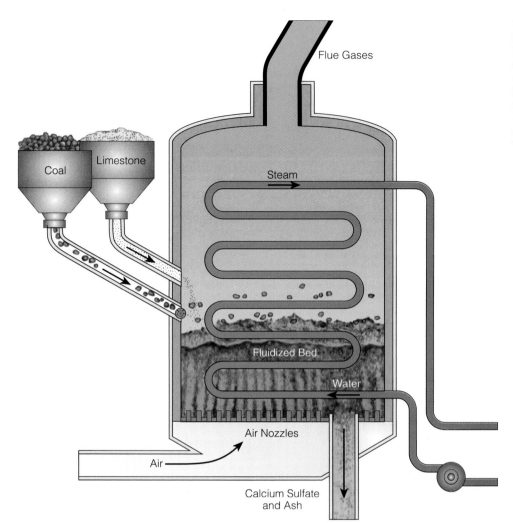

Figure 19-13 Fluidized-bed combustion of coal. A stream of hot air is blown into a boiler to suspend a mixture of powdered coal and crushed limestone. This removes most of the sulfur dioxide, sharply reduces emissions of nitrogen oxides, and burns the coal more efficiently and cheaply than conventional combustion methods.

and if current government subsidies keeping its price artificially low were removed, coal would become so expensive that it would likely be replaced by cheaper and less environmentally harmful renewable energy resources.

Converting Solid Coal into Gaseous and Liquid Fuels Coal can also be converted into gaseous or liquid **synfuels**. **Coal gasification** (Figure 19-14) is the conversion of solid coal into **synthetic natural gas (SNG)**. **Coal liquefaction** is the conversion of solid coal into a liquid fuel such as methanol or synthetic gasoline.

Synfuels can be transported by pipeline. They produce much less air pollution than solid coal. Synfuels can be burned to produce high-temperature heat and electricity, heat houses and water, and propel vehicles.

However, coal gasification has a low net useful energy yield (Figure 18-5), as does coal liquefaction. A synfuel plant costs much more to build and run than an equivalent coal-fired power plant fully equipped

with air pollution control devices. In addition, the widespread use of synfuels would accelerate the depletion of world coal supplies because 30–40% of the energy content of coal is lost in the conversion process. It would also lead to greater land disruption from surface mining because producing a unit of energy from synfuels uses more coal than does burning solid coal. Producing synfuels requires huge amounts of water. Burning synfuels releases larger amounts of carbon dioxide per unit of energy than does burning coal (Figure 19-11). Because of these problems, most analysts expect synfuels to play only a minor role as an energy resource in the next 30–50 years.

19-4 NUCLEAR ENERGY

Nuclear Power: A Fading Dream How did we get into nuclear power plants? The answer is a mixture of technological ingenuity and guilt. After the

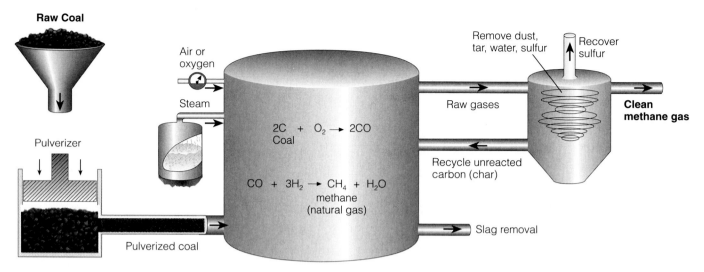

Figure 19-14 Coal gasification. Generalized view of one method for converting solid coal into synthetic natural gas (methane).

United States dropped atomic bombs on Hiroshima and Nagasaki, killing over 200,000 people and ending World War II, the scientists who developed the bomb and the elected officials responsible for its use were determined to show that the peaceful uses of atomic energy could outweigh the immense harm it had done. One part of this "Atoms for Peace" program was to use nuclear power to produce electricity.

Initially, U.S. utility companies were skeptical. However, they began developing nuclear power plants in the late 1950s because the Atomic Energy Commission (which had the conflicting roles of promoting and regulating nuclear power) promised them that nuclear power would produce electricity at a much lower cost than coal and other alternatives, and because the government paid about a quarter of the cost of constructing the first group of commercial reactors. Also, after insurance companies refused to cover more than a small part of the possible damages from a nuclear power-plant accident, Congress passed the Price–Anderson Act, which protected the nuclear industry and utilities from significant liability to the general public in case of accidents. It was an offer utility company officials could not resist.

In the 1950s, researchers predicted that by the end of the century 1,800 nuclear power plants would supply 21% of the world's commercial energy and 25% of that used in the United States. By 1993, after over 40 years of development and enormous government subsidies, 424 commercial nuclear reactors in 25 countries were producing only 17% of the world's electricity and less than 5% of its commercial energy.

In western Europe, plans to build more new nuclear power plants have come to a halt, except in France where the government builds standardized plants and has discouraged public criticism. In France, 56 reactors supply 76% of the country's electricity (a larger share than anywhere else in the world) and 5 new plants are under construction. However, France has more electricity than it needs (Guest Essay, p. 513) and the heavily subsidized government agency that builds and operates France's nuclear plants has lost money for 20 years, accumulating a $35-billion debt, making its financial future is uncertain.

In the United States almost $500 billion (including almost $100 billion in government subsidies) was spent on nuclear power between 1950 and 1994. Yet, no new nuclear power plants have been ordered in the United States since 1978, and all of the 119 plants ordered since 1973 have been canceled. In 1993, the 109 licensed commercial nuclear power plants in the United States generated about 22% of the country's electricity, a percentage that is expected to fall over the next two decades when many of the current reactors reach the end of their useful life.

What happened to nuclear power? The answer is billion-dollar construction cost overruns, high operating costs, frequent malfunctions, false assurances and cover-ups by government and industry officials, inflated estimates of electricity use, poor management, Chernobyl (p. 516), and public concerns about safety, costs, and how to dispose of radioactive waste. For these reasons many people believe that it is unethical, uneconomical, and unnecessary to use nuclear power to generate electricity. According to a 1992 opinion poll, 65% of the U.S. public opposes building any more nuclear power plants. However, proponents of this source of energy strongly disagree. To evaluate nuclear power we need to know how a nuclear power plant works.

Q: How much of the water withdrawn in the United States is used for irrigation?

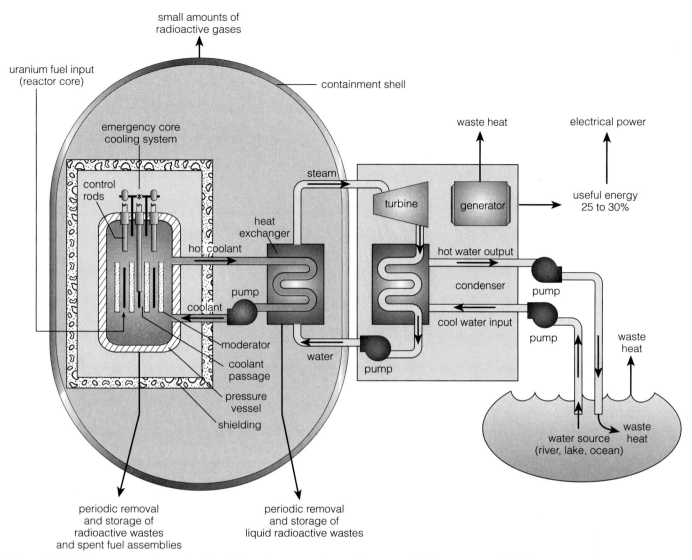

Figure 19-15 Light-water-moderated-and-cooled nuclear power plant with a pressurized water reactor. The ill-fated Chernobyl nuclear reactor in Ukraine (a part of the former Soviet Union) was a carbon-moderated reactor with less extensive safety features than most reactors in the rest of the world.

How Does a Nuclear Fission Reactor Work?

When the nuclei of atoms such as uranium-235 and plutonium-239 are split by neutrons in a nuclear fission chain reaction, energy is released and converted mostly into high-temperature heat (Figure 3-9). The rate at which this happens can be controlled in the nuclear fission reactor of a nuclear power plant, and the heat generated can be used to spin a turbine and produce electricity.

Light-water reactors (LWRs) like the one shown in Figure 19-15 generate about 85% of the world's electricity (100% in the United States) produced by nuclear power plants. These are the key parts of an LWR:

- *Core.* This typically contains 35,000–40,000 long, thin fuel rods bundled in 180 fuel assemblies of

around 200 rods each. Each fuel rod is packed with pellets of uranium oxide fuel the size of a pencil eraser. The uranium-235 in each fuel rod produces energy equal to that in three railroad carloads of coal over its lifetime of about three to four years.

- *Control rods.* These are made of materials such as cadmium or boron that absorb neutrons. They are moved in and out of the reactor core to regulate the rate of fission and thus the amount of power the reactor produces.

- *Moderator.* This is a material, such as liquid water (75% of the world's reactors), solid graphite (20% of reactors, including Chernobyl, p. 516), or heavy water (D_2O, 5% of reactors), used to slow down

Figure 19-16 The nuclear fuel cycle.

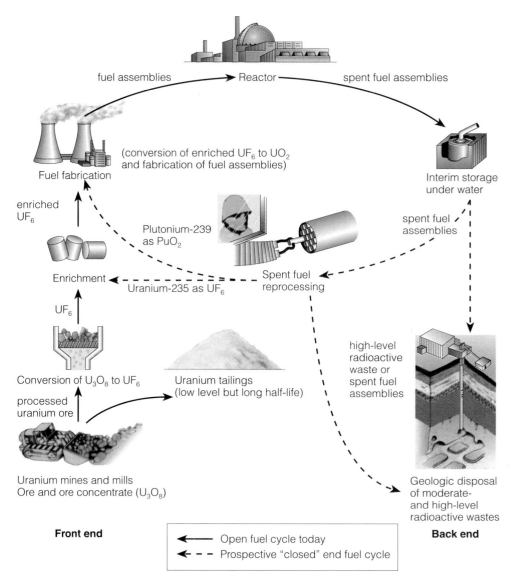

fuel assemblies → Reactor — spent fuel assemblies

(conversion of enriched UF$_6$ to UO$_2$ and fabrication of fuel assemblies)

Fuel fabrication

Interim storage under water

enriched UF$_6$

Plutonium-239 as PuO$_2$

spent fuel assemblies

Enrichment

Uranium-235 as UF$_6$

Spent fuel reprocessing

UF$_6$

high-level radioactive waste or spent fuel assemblies

Conversion of U$_3$O$_8$ to UF$_6$

Uranium tailings (low level but long half-life)

processed uranium ore

Uranium mines and mills
Ore and ore concentrate (U$_3$O$_8$)

Geologic disposal of moderate- and high-level radioactive wastes

Front end

Back end

◄——— Open fuel cycle today
◄ - - - Prospective "closed" end fuel cycle

the neutrons emitted by the fission process so that the chain reaction can be kept going (Figure 3-9). Graphite-moderated reactors can also be used to produce fissionable plutonium-239 for use in nuclear weapons.

- *Coolant.* This material, usually water, is circulated through the reactor's core to remove heat to keep fuel rods and other materials from melting and to produce steam for generating electricity. Most water-moderated and some graphite-moderated reactors use water as a coolant; the few gas-cooled reactors use an unreactive gas such as helium or argon for cooling.

After three to four years in a reactor the concentration of fissionable uranium-235 in a fuel rod becomes too low to keep the chain reaction going, or the rod becomes damaged from exposure to ionizing radiation. Each year about one-third of the spent fuel as-

semblies in a reactor are removed and stored in large, concrete-lined pools of water at the plant site.

After they have cooled for several years and lost some of their radioactivity, the spent fuel rods can be sealed in shielded, supposedly crash-proof casks. They can be transported by truck or train to storage pools away from the reactor, to a nuclear waste repository or dump, or to a fuel-reprocessing plant (Figure 19-16). At a reprocessing plant remaining uranium-235 and plutonium-239 produced as a by-product of the fission process are removed and sent to a fuel fabrication plant. Such plants also handle and ship weapons-grade plutonium-239.

If the fuel is processed to remove plutonium and other radioactive isotopes with long half-lives (Table 3-1), the remaining radioactive waste must be safely stored for at least 10,000 years. Otherwise, the rods must be stored safely for at least 240,000 years—several times as long as our species has been around.

Q: How much of the groundwater withdrawn in the United States is not replenished?

Commercial fuel-reprocessing plants have been built in France, Great Britain, Japan, and Germany, but all have had severe operating and economic problems. The United States has delayed development of commercial fuel-reprocessing plants because of policy concerns (terrorist diversion of bomb-grade materials), technical difficulties, high construction and operating costs, and already adequate domestic supplies of uranium.

Nuclear power has some important advantages, among them the following:

- *Nuclear plants don't emit air pollutants (as do coal-fired plants) so long as they are operating properly.*

- *The entire fuel cycle needed to mine uranium ore, convert it to nuclear fuel, run nuclear plants, and deal with nuclear wastes adds about one-sixth as much heat-trapping carbon dioxide per unit of electricity as does using coal, thus making it more attractive than fossil fuels for reducing the threat of global warming.*

- *Water pollution and disruption of land are low to moderate if the entire nuclear fuel cycle operates normally.*

- *Multiple safety systems with backups greatly decrease the likelihood of a catastrophic accident releasing deadly radioactive material into the environment.*

How Safe Are Nuclear Power Plants?   To greatly reduce the chances of a meltdown or other serious reactor accident, commercial reactors in the United States (and most countries) have many safety features:

- Thick walls and concrete-and-steel shields surrounding the reactor vessel

- A system for automatically inserting control rods into the core to stop fission under emergency conditions

- A steel-reinforced concrete containment building to keep radioactive gases and materials from reaching the atmosphere after an accident

- Large filter systems and chemical sprayers inside the containment building to remove radioactive dust from the air and further reduce chances of radioactivity reaching the environment

- Systems to condense steam released from a ruptured reactor vessel and prevent pressure from rising beyond the holding power of containment building walls

- An emergency core-cooling system to flood the core automatically with huge amounts of water within one minute to prevent meltdown of the reactor core

- Two separate power lines servicing the plant, and several diesel generators to supply backup power for the huge pumps in the emergency core-cooling system

- An automatic backup system to replace each major part of the safety system in the event of a failure

However, a partial or complete meltdown or explosion is possible, as the poorly designed and operated Chernobyl nuclear power plant (p. 516 and Figure 19-17) and the serious accident at the Three Mile Island nuclear power plant have taught us. On March 29, 1979, one of the two reactors at the Three Mile Island (TMI) nuclear plant near Harrisburg, Pennsylvania, lost its coolant water because of a series of mechanical failures and human operator errors not anticipated in safety studies, and the reactor's core became partially uncovered. At least 70% of the core was damaged, and about 50% of it melted and fell to the bottom of the reactor. Unknown amounts of radioactive materials escaped into the atmosphere, 50,000 people were evacuated, and another 50,000 fled the area on their own. Investigators discovered that if a valve had stayed stuck open for another 30–60 minutes, there would have been a complete meltdown.

No one can be shown to have died immediately because of the accident, but data published on the radiation released during the accident are contradictory and incomplete. Partial cleanup of the damaged TMI reactor has cost $1.2 billion so far, more than the $700 million construction cost.

Opponents of nuclear power point to the Three Mile Island accident as a reason to be concerned about the safety of nuclear power. Supporters of nuclear power, however, consider the incident as ample evidence that the built-in safety systems work, although they could be (and since have been) improved.

Scientists in Germany and Sweden project that worldwide there is a 70% chance of another serious core-damaging accident within the next 5.4 years. Especially risky are 15 nuclear reactors in the former Soviet Union with a Chernobyl-type design, which are still operating, along with 10 other older plants with a different design. None of these plants were built using safety systems that meet Western standards. There is a growing international consensus (including the Russian Academy of Sciences) that these and 10 other poorly designed and poorly operated nuclear plants in eastern Europe should be shut down. The World Bank estimates that shutting down these reactors and building new plants fired by Eastern Europe's ample supplies of natural gas would take only three years and cost $18 billion, compared to $24 billion for upgrading these reactors.

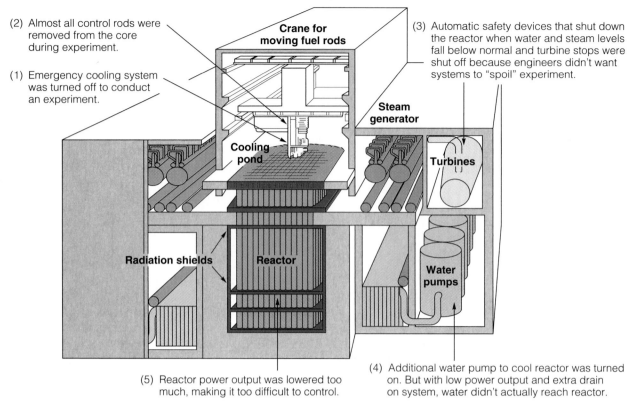

(2) Almost all control rods were removed from the core during experiment.

(1) Emergency cooling system was turned off to conduct an experiment.

(3) Automatic safety devices that shut down the reactor when water and steam levels fall below normal and turbine stops were shut off because engineers didn't want systems to "spoil" experiment.

Crane for moving fuel rods

Steam generator

Cooling pond

Turbines

Radiation shields

Reactor

Water pumps

(5) Reactor power output was lowered too much, making it too difficult to control.

(4) Additional water pump to cool reactor was turned on. But with low power output and extra drain on system, water didn't actually reach reactor.

Figure 19-17 Major events leading to the Chernobyl nuclear power plant accident on April 26, 1986, in the then Soviet Union. The accident happened because engineers turned off most of the reactor's automatic safety and warning systems (to keep them from interfering with an unauthorized safety experiment) and because of insufficient safety design of the reactor.

A 1982 study by the Sandia National Laboratory estimated that a possible, but highly unlikely, *worst-case accident* in a reactor near a large U.S. city might cause 50,000–100,000 immediate deaths, 10,000–40,000 subsequent deaths from cancer, and $100–$150 billion in damages. Most citizens and businesses suffering injuries or property damage from a major nuclear accident would get little if any financial reimbursement because combined government–nuclear industry insurance covers only 7% of the estimated damage from such a worst-case accident. Nuclear critics also contend that plans for dealing with major nuclear accidents in the United States are inadequate.

There is widespread lack of confidence in the ability of the Nuclear Regulatory Commission (NRC) and the Department of Energy to enforce nuclear safety in commercial (NRC) and military (DOE) nuclear facilities. Congressional hearings in 1987 uncovered evidence that high-level NRC staff members destroyed documents and obstructed investigations of criminal wrongdoing by utilities; suggested ways utilities could evade commission regulations; and provided utilities and their contractors with advance notice of surprise inspections.

Since 1986, government studies and once-secret documents have revealed that for decades most of the nuclear weapons production facilities supervised by the Department of Energy (DOE) have been operated with gross disregard for the safety of their workers and people in nearby areas. Since 1957, these facilities have released huge quantities of radioactive particles into the air and dumped tons of radioactive waste and toxic substances into flowing creeks and leaking pits without telling local residents. In 1993, investigators found many bomb plants still violating safety rules.

The DOE released records in 1993 showing that in the 1940s and 1950s the government deliberately released large quantities of radiation into the atmosphere in at least a dozen secret bomb tests. The government also tested the effects of radiation on thousands of weapons plant workers and civilians—including pregnant women—usually without their knowledge or consent. This information was released at the insistence of Energy Secretary Hazel O'Leary—despite strong opposition from DOE officials wishing to preserve the secrecy of government tests. Senator John Glenn of Ohio summed up the situation: "We are poisoning our own people in the name of national security."

Q: What is the key to reducing water waste?

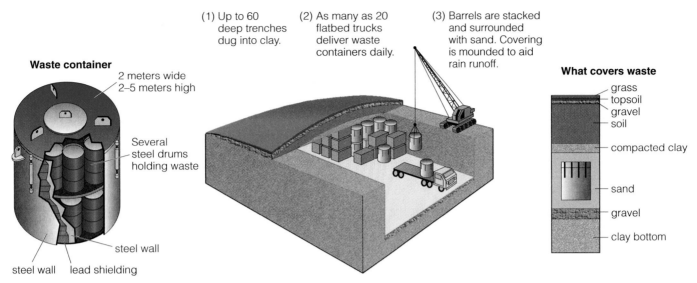

Waste container

2 meters wide
2–5 meters high

(1) Up to 60 deep trenches dug into clay.

(2) As many as 20 flatbed trucks deliver waste containers daily.

(3) Barrels are stacked and surrounded with sand. Covering is mounded to aid rain runoff.

Several steel drums holding waste

steel wall

steel wall   lead shielding

What covers waste

grass
topsoil
gravel
soil

compacted clay

sand

gravel

clay bottom

Figure 19-18 Proposed design of low-level radioactive waste landfill. Such landfills are strongly opposed by local citizens and environmental groups. Since all landfills eventually leak, some environmentalists believe that low-level radioactive waste should be stored in carefully designed aboveground buildings (not shown here). All such buildings would be located at nuclear power plant sites, which produce more than half (as high as 80% in some states) of these wastes (and 80% of the radioactivity) and have the expertise and equipment to manage them. (Data from Atomic Industrial Forum)

Radioactive Wastes   Each part of the nuclear fuel cycle (Figure 19-16) produces low-level and high-level solid, liquid, and gaseous radioactive wastes with various half-lives (Table 3-1). The low-level wastes produced must be stored safely for several decades, and the high-level wastes for thousands of years. Those classified as *low-level radioactive wastes* give off small amounts of ionizing radiation, usually for a long time. From the 1940s to 1970 most low-level radioactive waste produced in the United States (and most other countries) was dumped into the ocean in steel drums; Great Britain and Pakistan still dispose of their low-level wastes in this way. Since 1970 low-level radioactive wastes from U.S. military activities have been buried in government-run landfills.

Today low-level radioactive waste materials from commercial nuclear power plants, hospitals, universities, industries, and other producers are put in steel drums and shipped to regional landfills run by federal and state governments (Figure 19-18). Attempts to build new regional dumps for low-level radioactive waste have met with fierce public opposition. According to the EPA all landfills eventually leak. The best-designed landfills may not leak for several decades, about the time the companies running them are no longer liable for leaks and problems. Then any problems would be passed on to taxpayers and future generations.

High-level radioactive wastes give off large amounts of ionizing radiation for a short time and small amounts

for a long time. Most high-level radioactive wastes are spent fuel rods from commercial nuclear power plants and an assortment of wastes from nuclear weapons plants. Currently most spent fuel rods in the United States are being stored at reactor sites in specially designed cooling ponds, but some plants are running out of storage space.

After 39 years of research and debate, scientists still don't agree on a safe method of storing these wastes (Case Study, p. 532). Regardless of the storage method, most citizens strongly oppose the location of a low-level or high-level nuclear waste disposal facility anywhere near them.

In 1985 the U.S. Department of Energy announced plans to build the first repository for underground storage of high-level radioactive wastes from commercial nuclear reactors (Figure 19-19) on federal land in the Yucca Mountain desert region, 160 kilometers (100 miles) northwest of Las Vegas, Nevada. The facility—which is expected to cost at least $26 billion—was scheduled to open by 2003, but in 1990 it was postponed to at least 2010. There is a strong possibility that the site may never open. A young, active volcano is only 11 kilometers (7 miles) away, and, according to the DOE's own data, there are 32 active earthquake faults on the site itself. Yucca Mountain's many rock fractures also suggest that water flowing through the site could leach out radioactive wastes and carry them 5 kilometers (3.1 miles) or more from the site in 400–500 years.

What Do We Do with Nuclear Waste?

Some scientists believe that the long-term safe storage or disposal of high-level radioactive wastes is technically possible. Others disagree, pointing out that it is impossible to show that any method will work for the 10,000–240,000 years of fail-safe storage needed for such wastes. Here are some of the proposed methods and their possible drawbacks:

- *Bury it deep underground.* This favored strategy is under study by all countries producing nuclear waste. Spent fuel rods would be reprocessed to remove very long-lived radioactive isotopes. The remainder would be fused with glass or a ceramic material and sealed in metal canisters for burial in a deep underground salt, granite, or other stable geological formation that is earthquake-resistant and waterproof (Figure 19-19). However, according to a 1990 report by the National Academy of Sciences, "Use of geological information—to pretend to be able to make very accurate predictions of long-term site behavior—is scientifically unsound."

- *Shoot it into space or into the sun.* Costs would be very high, and a launch accident, such as the explosion of the space shuttle *Challenger*, could disperse high-level radioactive wastes over large areas of the earth's surface. This strategy has been abandoned, for now.

- *Bury it under the Antarctic ice sheet or the Greenland ice cap.* The long-term stability of the ice sheets is not known. They could be destabilized by heat from the wastes, and retrieval of the wastes would be difficult or impossible if the method failed.

This strategy has also been abandoned, for now.

- *Dump it into descending subduction zones in the deep ocean* (Figures 12-2 and 12-4b). Again, our geological knowledge is incomplete; wastes might eventually be spewed out somewhere else by volcanic activity. Also, waste containers might leak and contaminate the ocean before being carried downward, and retrieval would be impossible if the method did not work. This method is under active study by a consortium of 10 countries.

- *Change it into harmless, or less harmful, isotopes.* Currently there is no way to do this. Even if a method were developed, costs would probably be extremely high; and resulting toxic materials and low-level (but very long-lived) radioactive wastes would still have to be disposed of safely.

Meanwhile, the DOE has also asked Congress for permission to build an aboveground interim storage facility for high-level radioactive wastes—a sort of halfway house for wastes awaiting permanent disposal. Some critics fear the temporary facility could become the permanent site.

Critics of nuclear power are appalled that the nuclear industry has been allowed to build hundreds of nuclear reactors and weapons facilities without finding a scientifically and politically acceptable way to deal with radioactive wastes. Proponents of nuclear power contend that the problem of what to do with nuclear wastes is solvable and that if stored properly these wastes pose little threat to human health. Even if the problem is technically solvable, it may be politically unacceptable because most citizens oppose both storage of any radioactive wastes near their communities and the transport of such wastes through their communities on the way to storage sites.

In 1992 the EPA estimated there may be as many as 45,000 sites in the United States contaminated with radioactive materials, 20,000 of these sites belonging to the DOE and the Department of Defense. The General Accounting Office and the DOE estimate that it will cost

taxpayers at least $270 billion over 30 years to get these facilities cleaned up. Some doubt that the necessary cleanup funds will be provided by Congress. The closer the DOE looks at this horrendous nuclear waste problem, the worse it gets. In 1993, Thomas Grumbly, DOE assistant secretary for environmental restoration and waste management, predicted that the total cleanup cost could run $400 to $900 billion.

This situation in the United States pales in comparison to the legacy of nuclear waste and contamination in the former Soviet Union. Mayak, a plutonium production facility in southern Russia, has spewed out 2½ times as much radiation as Chernobyl into the atmosphere and into the nearby Techa River and into Lake Karachay. Today the lake is so radioactive that standing on its shores for about an hour would be fatal. Around the shores of Novaya Zemyla off the coast of Russia, 18 defunct nuclear-powered submarines have been dumped (at least three of them loaded with nuclear fuel), along with as many as 17,000 containers of radioactive waste. If released by corrosion, radioactivity from these submarines and containers threatens Russian citizens in the area as well as many along the northern shores of Finland, Sweden, and Norway.

Q: What percentage of U.S. toilets leak?

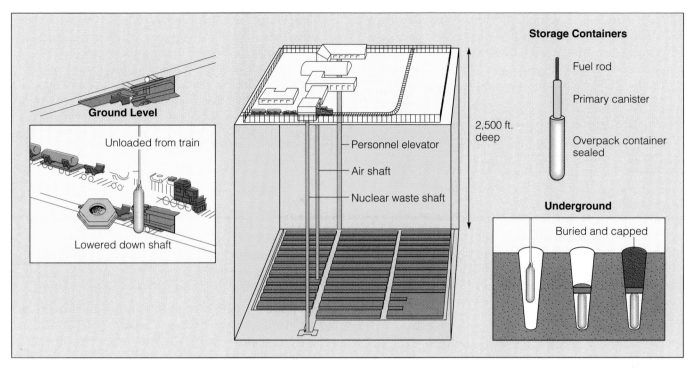

Figure 19-19 Proposed general design for deep-underground permanent storage of high-level radioactive wastes from commercial nuclear power plants in the United States. (Source: U.S. Department of Energy)

In 1993, Greenpeace caught Russians still dumping massive amounts of radioactive material from naval nuclear submarines into the Sea of Japan. A month later members of the London Convention Agreement on Nuclear Waste Dumping at Sea made the ban permanent. However, Russia, Great Britain, France, China, and Belgium opposed the ban.

What Can We Do with Worn-Out Nuclear Plants?   The useful operating life of today's nuclear power plants is supposed to be 40 years, but many plants are wearing out and becoming dangerous faster than anticipated. After many years of bombardment by neutrons released by nuclear fission (Figure 3-9), the walls of a reactor's pressure vessel become brittle and thus more likely to crack, thereby exposing the highly radioactive reactor core (Figure 19-15). Such embrittlement was the primary cause of the early closure of Massachusetts' Yankee Row plant in 1991. And in 1993, the NRC revealed that 18 more U.S. reactors may be at risk because of embrittlement. Moreover, decades of pressure and temperature changes gradually weaken tubes in the reactor's steam generator, which can crack, releasing contaminated water. Finally, corrosion of pipes and valves throughout the system can cause them to crack. Because so many of its parts become radioactive, a nuclear plant cannot be abandoned or demolished by a wrecking ball the way a worn-out coal-fired power plant can.

Decommissioning nuclear power plants and nuclear weapons plants is the last step in the nuclear fuel cycle. Three methods have been proposed: **(1)** *immediate dismantling,* **(2)** *mothballing* for 30–100 years—by putting up a barrier and setting up a 24-hour security system—and then dismantling the plant, and **(3)** *entombment* by covering the reactor with reinforced concrete and putting up a barrier to keep out intruders for several thousand years. Each method involves shutting down the plant, removing the spent fuel from the reactor core, draining all liquids, flushing all pipes, and sending all radioactive materials to an approved waste storage site yet to be built (Figure 19-19).

Worldwide more than 25 commercial reactors (10 in the United States) have been retired and await decommissioning. Another 228 large commercial reactors (20 in the United States) are scheduled for retirement between 2000 and 2012. By 2030 all U.S. reactors will have to be retired based on the life of their operating licenses. Many reactors may be retired early because of mysterious cracks or because they can't be run profitably. There is pressure for the Nuclear Regulatory Commission to extend the operating licenses of such reactors for another 20 years. However, even if this can be done safely (which some nuclear engineers consider doubtful), repairing and maintaining the plants past their original expiration would be so expensive that most utility companies would probably find it cheaper to shut the plants down.

Utility company officials currently estimate that dismantling a typical large reactor should cost about $170 million, and mothballing $225 million, but most analysts place the cost at around $1 billion per reactor.

A: 25%. A leaky toilet can waste 83,000 liters (22,000) gallons of water a year

So far U.S. utilities have only $3.5 billion set aside for decommissioning more than 100 reactors. The balance of the cost could be passed along to ratepayers and taxpayers, adding to the already high cost of electricity produced by nuclear fission (Figure 18-14).

Connections: Reactors and the Spread of Nuclear Weapons   Since 1958 the United States has been giving away and selling to other countries various forms of nuclear technology, mostly in the form of nuclear power plants and research reactors. Today at least 14 other countries sell nuclear technology in the international marketplace.

For decades the U.S. government denied that the information, components, and materials used in the nuclear fuel cycle could be used to make nuclear weapons. In 1981, however, a Los Alamos National Laboratory report admitted, "There is no technical demarcation between the military and civilian reactor and there never was one."

We live in a world with enough nuclear weapons to kill everyone on Earth 60 times over—25 times if current nuclear arms reduction agreements are carried out. By the end of this century 60 countries—one of every three in the world—will have either nuclear weapons or the knowledge and capability to build them, with the fuel and knowledge coming mostly from research and commercial nuclear reactors. The best ways to slow down the spread of bomb-grade material are to abandon civilian reprocessing of power-plant fuel, develop substitutes for highly enriched uranium in research reactors, and tighten international safeguards.

Ironically, the planned dismantlement of thousands of nuclear warheads in the United States and the former Soviet Union can increase this threat by leaving huge amounts of bomb-grade plutonium to be safeguarded or reprocessed as power-plant fuel (which is expensive and produces large amounts of radioactive waste). Because the nuclear genie is out of the bottle, there is no truly effective solution to the serious problem of nuclear weapons proliferation.

Can We Afford Nuclear Power?   Experience has shown that nuclear power is a very expensive way to boil water to produce electricity, even when it is heavily subsidized and enjoys partial protection from free-market competition with other energy sources. Despite massive subsidies, the most modern nuclear power plants built in the United States produce electricity at an average of about 13¢ per kilowatt-hour—the equivalent of burning oil costing well over $100 per barrel (compared to its current price of $16–20) to produce electricity. All methods of producing electricity in the United States (except solar photovoltaic and solar thermal plants) have average costs below those

of new nuclear power plants (Figure 18-14). By 2000, even these methods (with relatively few subsidies) are expected to be cheaper than nuclear power.

Banks and other lending institutions have become leery of financing new U.S. nuclear power plants. The Three Mile Island accident showed that utility companies could lose $1 billion worth of equipment in an hour and at least $1 billion more in cleanup costs, even without any established harmful effects on public health. Abandoned reactor projects have cost U.S. utility investors over $100 billion since the middle 1970s. *Forbes* magazine has called the failure of the U.S. nuclear power program "the largest managerial disaster in U.S. business history." In fact, no U.S. utility company is planning to build any new nuclear power plants because they are no longer a cost-effective or wise investment even with massive government subsidies. The companies and their financial backers are leery of spending the $4 billion to $7 billion it costs to build a nuclear power plant when cheaper (Figure 18-14), less risky, and more reliable alternatives are available.

Despite massive economic and public relations setbacks, the U.S. nuclear power industry hopes for a comeback. Since the Three Mile Island accident, the U.S. nuclear industry and utility companies have financed a $21-million-a-year public relations campaign by the U.S. Council for Energy Awareness. The campaign's goals are to improve the industry's image, resell nuclear power to the American public, and minimize the importance of solar energy, energy efficiency, geothermal energy, wind, and hydropower as alternatives. Its primary goal is to somehow convince a utility to order a new reactor to break the 20-year lack of orders.

Most of the council's ads use the argument that the United States needs more nuclear power to reduce dependence on imported oil and improve national security. Sounds good, doesn't it? But since 1979 only about 5% (3% in 1991) of the electricity in the United States has been produced by burning oil, and 95% of that is residual oil that can't be used for other purposes. Even if all electricity in the United States came from nuclear power—which would require about 500 nuclear plants—this would reduce U.S. oil consumption by less than 5%. The nuclear industry also does not point out that 73% of the uranium used for nuclear fuel in 1992 in the United States was imported.

The U.S. nuclear industry also hopes to persuade the federal government and utility companies to build hundreds of new "second-generation" plants using standardized designs with passive "fail-safe" features, which they claim are safer and can be built more quickly (in three to five years). However, according to *Nucleonics Week*, an important nuclear industry publication, "experts are flatly unconvinced that safety has been achieved—or even substantially increased—by the new designs."

Q: How many species of the world's plants are edible?

Furthermore, none of the new designs solves the problem of what to do with nuclear waste and worn-out nuclear plants, or the problem of using nuclear technology and fuel to build nuclear weapons. Indeed, these problems would become more serious if the number of nuclear plants increased from a few hundred to several thousand. Contrary to President Clinton's campaign promises, his proposed budget for energy research and development provides several hundred million for nuclear power research, including $22 million for a new generation of experimental reactors. According to the Congressional Research Service, nuclear energy soaked up 65% of all federal energy research-and-development funds between 1948 and 1992, and it is still not cost effective.

Breeder Nuclear Fission   Some nuclear power proponents urge the development and widespread use of **breeder nuclear fission reactors**, which generate more nuclear fuel than they consume by converting nonfissionable uranium-238 into fissionable plutonium-239. Because breeders would use over 99% of the uranium in ore deposits, the world's known uranium reserves would last at least for 1,000 years, and perhaps several thousand years.

However, if the safety system of a breeder reactor should fail, the reactor could lose some of its liquid sodium coolant (which ignites when exposed to air and reacts explosively if it comes into contact with water). This could cause a runaway fission chain reaction and perhaps a small nuclear explosion powerful enough to blast open the containment building and release a cloud of highly radioactive gases and particulate matter. Leaks of flammable liquid sodium also can cause fires, as has happened with all experimental breeder reactors built so far.

Since 1966, small experimental breeder reactors have been built in the United Kingdom, the former Soviet Union, Germany, Japan, and France. In December 1986 France opened a $3-billion commercial-size breeder reactor. It cost three times the original estimate to build, and the little electricity it has produced was twice as expensive as that generated by France's conventional fission reactors. In 1989, after 76 accidents, the plant was shut down. Repairs may be so expensive that the reactor will never be put back into operation.

Tentative plans to build full-size commercial breeders in Germany, the former Soviet Union, the United Kingdom, and Japan have been abandoned because of the French experience and an excess of electric generating capacity and conventional uranium fuel. Also, the experimental breeders already built produce plutonium fuel much too slowly. If this problem is not solved, it will take 100–200 years for breeders to begin producing enough plutonium to fuel a significant number of other breeder reactors.

Nuclear Fusion   Scientists hope someday to use controlled nuclear fusion to provide an almost limitless source of high-temperature heat and electricity. For 49 years research has focused on the D-T nuclear fusion reaction in which two isotopes of hydrogen—deuterium (D) and tritium (T)—fuse at about 100 million degrees (Figure 3-10). Another possibility is the D-D fusion reaction (Figure 3-10), in which the nuclei of two deuterium atoms fuse together at even higher temperatures. If developed it would run on virtually unlimited heavy water (D_2O) fuel, obtained from seawater at a cost of about 10¢ per gallon.

Despite almost 50 years of research and huge expenditures of mostly government funds, controlled nuclear fusion is still at the laboratory stage. Deuterium and tritium atoms have been forced together using electromagnetic reactors (the size of 12 locomotives), 120-trillion-watt laser beams, and bombardment with high-speed particles. So far, none of these approaches has produced more energy than it uses. In 1993 press reports of a test run at an experimental fusion reactor at the Princeton Plasma Physics Laboratory gave the impression that a major breakthrough had occurred in nuclear fusion. However, the $1.6-billion machine produced energy from fusion for a mere 4 seconds and generated only one-eighth as much power as it consumed. And in 1989 two chemists claimed to have achieved deuterium-deuterium (D-D) nuclear fusion (Figure 3-10) at room temperature using a simple apparatus, but subsequent experiments could not substantiate their claims.

If researchers eventually can get more energy out of nuclear fusion than they put in, the next step would be to build a small fusion reactor and then scale it up to commercial size—one of the most difficult engineering problems ever undertaken. Also, the estimated cost of a commercial fusion reactor is several times that of a comparable conventional fission reactor.

If things go right, a commercial nuclear fusion power plant might be built by 2030. Even if everything goes right, however, energy experts don't expect nuclear fusion to be a significant energy source until 2100, if then. Meanwhile, we can produce more electricity than we need using several other quicker, cheaper, and safer methods.

19-5   SOLUTIONS: A SUSTAINABLE-EARTH ENERGY STRATEGY

Overall Evaluation of Energy Alternatives

Table 19-1 summarizes the major advantages and disadvantages of the energy alternatives discussed in this chapter, with emphasis on their potential in the United States. Energy experts argue over these and other projections, and new data and innovations may affect the

Table 19-1 Evaluation of Energy Alternatives for the United States (shading indicates favorable conditions)

Energy Resources	Estimated Availability Short Term (1995–2005)	Estimated Availability Intermediate Term (2005–2015)	Estimated Availability Long Term (2015–2045)	Estimated Net Useful Energy of Entire System	Projected Cost of Entire System	Actual or Potential Overall Environmental Impact of Entire System
NONRENEWABLE RESOURCES						
Fossil fuels						
Petroleum	High (with imports)	Moderate (with imports)	Low	High but decreasing	High for new domestic supplies	Moderate
Natural gas	High (with imports)	Moderate (with imports)	Moderate (with imports)	High but decreasing	High for new domestic supplies	Low
Coal	High	High	High	High but decreasing	Moderate but increasing	Very high
Oil shale	Low	Low to moderate	Low to moderate	Low to moderate	Very high	High
Tar sands	Low	Fair? (imports only)	Poor to fair (imports only)	Low	Very high	Moderate to high
Synthetic natural gas (SNG) from coal	Low	Low to moderate	Low to moderate	Low to moderate	High	High (increases use of coal)
Synthetic oil and alcohols from coal	Low	Moderate	High	Low to moderate	High	High (increases use of coal)
Nuclear energy						
Conventional fission (uranium)	Low to moderate	Low to moderate	Low to moderate	Low to moderate	Very high	Very high
Breeder fission (uranium and thorium)	None	None to low (if developed)	Moderate	Unknown, but probably moderate	Very high	Very high
Fusion (deuterium and tritium)	None	None	None to low (if developed)	Unknown, but may be high	Very high	Unknown (probably moderate to high)
Geothermal energy (some are renewable)	Low	Low	Moderate	Moderate	Moderate to high	Moderate to high
RENEWABLE RESOURCES						
Improving energy efficiency	High	High	High	Very high	Low	Decreases impact of other sources
Hydroelectric						
New large-scale dams and plants	Low	Low	Very low	Moderate to high	Moderate to very high	Low to moderate

Q: How many plants feed most of the world's people?

Energy Resources	Estimated Availability			Estimated Net Useful Energy of Entire System	Projected Cost of Entire System	Actual or Potential Overall Environmental Impact of Entire System
	Short Term (1995–2005)	Intermediate Term (2005–2015)	Long Term (2015–2045)			

RENEWABLE RESOURCES (continued)

Hydroelectric (continued)

Energy Resources	Short Term (1995–2005)	Intermediate Term (2005–2015)	Long Term (2015–2045)	Estimated Net Useful Energy of Entire System	Projected Cost of Entire System	Actual or Potential Overall Environmental Impact of Entire System
Reopening abandoned small-scale plants	Moderate	Moderate	Low	Moderate	Moderate	Low to moderate
Tidal energy	Very low	Very low	Very low	Moderate	High	Low to moderate
Ocean thermal gradients	None	Low	Low to moderate (if developed)	Unknown (probably low to moderate)	High	Unknown (probably moderate to high)
Solar energy						
Low-temperature heating (for homes and water)	High	High	High	Moderate to high	Moderate	Low
High-temperature heating	Low	Moderate	Moderate to high	Moderate	High initially, but probably declining fairly rapidly	Low to moderate
Photovoltaic production of electricity	Low to moderate	Moderate	High	Fairly high	High initially but declining fairly rapidly	Low
Wind energy	Low	Moderate	Moderate to high	Fairly high	Moderate	Low
Geothermal energy (low heat flow)	Very low	Very low	Low to moderate	Low to moderate	Moderate to high	Moderate to high
Biomass (burning of wood and agricultural wastes)	Moderate	Moderate	Moderate to high	Moderate	Moderate	Moderate to high
Biomass (urban wastes for incineration)	Low	Moderate	Moderate	Low to fairly high	High	Moderate to high
Biofuels (alcohols and biogas from organic wastes)	Low to moderate	Moderate	Moderate to high	Low to fairly high	Moderate to high	Moderate to high
Hydrogen gas (from coal or water)	Very low	Low to moderate	Moderate to high	Variable but probably low	Variable	Variable, but low if produced with solar energy

A: 30 (with most provided by wheat, rice, corn, and potato)

How to Save Energy and Money

- *Don't use electricity to heat space or water* (unless it is provided by affordable solar cells or hydrogen).

- *Insulate new or existing houses heavily, caulk and weatherstrip to reduce air infiltration and heat loss, and use energy-efficient windows.* Add an air-to-air heat exchanger to minimize indoor air pollution.

- *Obtain as much heat and cooling as possible from natural sources*—especially sun, wind, geothermal energy, and trees (windbreaks and natural shading).

- *Buy the most energy-efficient homes, lights, cars, and appliances available, and evaluate them only in terms of lifetime cost.*

- *Consider walking or riding a bicycle for short trips and buses or trains for long trips.*

- *During cold weather dress more warmly indoors, humidify air, and use fans to distribute heat so that the thermostat setting can be lowered (saves 3% for each °F decrease).*

- *Turn down the thermostat on water heaters to 43°–49° (110°–120°F) and insulate hot water pipes.*

- *Lower the cooling load on an air conditioner by increasing the thermostat setting (saves 3–5% for each °F increase), installing energy-efficient lighting, using floor and ceiling fans, and whole-house window or attic fans to bring in outside air (especially at night when temperatures are cooler).*

- *Turn off lights and appliances when not in use.*

status of certain alternatives, but the table does provide a useful framework for making decisions based on currently available information. Many scientists and energy experts who have evaluated the energy alternatives available to us have come to the following general conclusions:

- *The best short-term, intermediate, and long-term alternatives are a combination of improved energy efficiency and greatly increased use of locally available renewable energy resources.*

- *Future energy alternatives will probably have low-to-moderate net useful energy yields and moderate-to-high development costs.*

- *Because there is not enough financial capital to develop all energy alternatives, projects must be chosen carefully.*

- *We cannot and should not depend mostly on a single nonrenewable energy resource such as oil, coal, natural gas, or nuclear power.*

Economics Cost is the biggest factor determining which commercial energy resources are widely used by consumers. Governments throughout the world use three basic economic and political strategies to stimulate or dampen the short-term and long-term use of a particular energy resource:

1. *Not controlling the price.* Use of a resource depends on free-market competition (assuming all other alternatives also compete in the same way). This is not politically feasible, however, because of well-entrenched government intervention into the marketplace in the form of subsidies, taxes, and regulations. Also, the free-market approach, with its emphasis on short-term gain, inhibits long-term development of new energy resources, which can rarely compete in their developmental stages without government support.

2. *Keeping prices artificially low.* This encourages use and development of a resource. In the United States (and most other countries) the energy marketplace is greatly distorted by huge government subsidies and tax breaks (such as depletion write-offs for fossil fuels) that make the prices of fossil fuels and nuclear power artificially low. At the same time, improving energy efficiency and solar alternatives receive much lower subsidies and tax breaks, creating an uneven economic playing field. In the United States, for example, nuclear power and fossil fuels receive 200 times as much in government subsidies per unit of energy as do energy efficiency and renewable energy resources. Despite the promise of energy efficiency and renewable energy, the DOE's research and development budget for fossil fuels and nuclear power in 1993 and 1994 was still about three times that for energy efficiency and renewable energy. And the controversial North American Free Trade Agreement mandates continued subsidies for fossil fuels and nuclear power, while excluding mention of subsidies for energy efficiency and renewable energy resources. Because keeping energy prices low is popular with consumers, this practice often helps politicians in power. However, artificially low prices encourage waste and rapid depletion of an energy resource (such as oil) and discourage the development of those energy alternatives not getting at least the same level of subsidies and price control. And once an energy industry gets government subsidies, it usually has enough clout to maintain that support long after it becomes unproductive.

Q: What percentage of the U.S. population live on the country's farms?

3. *Keeping prices artificially high.* Governments can raise the price of an energy resource by withdrawing existing tax breaks and other subsidies, or by adding taxes on its use. This provides government revenues, encourages improvements in energy efficiency, reduces dependence on imported energy, and decreases use of an energy resource (like oil) that has a limited future supply. Raising energy prices can stimulate employment because building solar collectors, adding insulation, and most forms of improving energy efficiency are labor-intensive activities. Increasing taxes on energy use, however, can dampen economic growth and put a heavy economic burden on the poor and lower middle class, unless some of the energy tax revenues are used to help offset their increased energy costs.

Environmentalists urge citizens to exert pressure from the bottom up on elected officials—to develop a national energy policy based on much greater improvements in energy efficiency and a more rapid transition to a mix of renewable energy resources. Each of us has a role to play in helping make such a transition (Individuals Matter, p. 538).

Nuclear fission energy is safe only if a number of critical devices work as they should, if a number of people in key positions follow all their instructions, if there is no sabotage, no hijacking of the transport, if no reactor fuel processing plant or repository anywhere in the world is situated in a region of riots or guerrilla activity, and no revolution or war— even a "conventional" one—takes place in these regions. No acts of God can be permitted.

HANNES ALFVEN (Nobel Laureate, Physics)

Critical Thinking

1. Explain why you agree or disagree with the ideas that the United States can get
   a. all of the oil it needs by extracting and processing heavy oil left in known oil wells
   b. all of the oil it needs by extracting and processing heavy oil from oil shale deposits
   c. all of the oil it needs by extracting heavy oil from tar sands
   d. all the natural gas it needs from unconventional sources

2. a. Should air pollution emission standards for *all* new and existing coal-burning plants be tightened significantly? Explain.
   b. Do you favor a U.S. energy strategy based on greatly increased use of coal-burning plants to produce electricity? Explain. What are the alternatives?

3. Explain why you agree or disagree with each of the following proposals made by the nuclear power industry:

   a. The licensing time for new nuclear power plants in the United States should be halved (from an average of 12 years) so they can be built at lower cost and can compete more effectively with coal-burning and other energy-producing facilities or technologies.
   b. A large number of new, better-designed nuclear fission power plants should be built in the United States to reduce dependence on imported oil and slow down projected global warming.
   c. Federal subsidies for commercial nuclear power (already totaling $1 trillion) should be continued so it does not have to compete in the open marketplace with other energy alternatives receiving no, or smaller, federal subsidies.
   d. A comprehensive program for developing the nuclear breeder fission reactor should be developed and funded largely by the federal government to conserve uranium resources and keep the United States from being dependent on other countries for uranium supplies.

4. Explain why you agree or disagree with the following proposals by various energy analysts:
   a. Federal subsidies for all energy alternatives should be eliminated so that all energy choices can compete in a true free-market system.
   b. All government tax breaks and other subsidies for conventional fuels (oil, natural gas, coal), synthetic natural gas and oil, and nuclear power should be removed—and replaced with subsidies and tax breaks for improving energy efficiency and developing solar, wind, geothermal, and biomass energy alternatives.
   c. Development of solar and wind energy should be left to private enterprise with little or no help from the federal government, but nuclear energy and fossil fuels should continue to receive large federal subsidies.
   d. To solve present and future U.S. energy problems, all we need to do is find and develop more domestic supplies of oil, natural gas, and coal— and increase dependence on nuclear power.
   e. A heavy federal tax should be placed on gasoline and imported oil used in the United States.
   f. Between 2000 and 2020 the United States should phase out all nuclear power plants.

*5. Does your campus have a nuclear reactor? Has it had any safety problems? Explain.

*6. How is the electricity in your community produced? How has the cost of that electricity changed since 1970? Do your community and your school have a plan for improving energy efficiency? If so, what is this plan, and how much money has it saved during the past 10 years? If there is no plan, develop one and present it to the appropriate officials.

*7. Make a concept map of the key ideas in this chapter using the section heads and subheads and the key terms (shown in boldface type in the chapter). See the inside front cover and Appendix 4 for information on concept maps.

A: 1.9% (only 650,000 Americans are full-time farmers)

Living Sustainably

When there is no dream, the people perish.
PROVERBS 29:18

BECOMING EARTH CITIZENS

In "Individuals Matter" boxes throughout this book I have listed specific actions you can take. Here are the ones I would put at the top of the list:

- *Respect all life.*
- *Learn all you can about how nature works, human nature, and about the connections between everything.*
- *Listen to and learn from nature and from children.*
- *Evaluate the beneficial and harmful consequences of your lifestyle and profession on the earth today and in the future.*
- *Lead by example.* People are most influenced by what we do, not by what we say.
- *Tread more lightly on the earth by living more simply.* This also saves you money.
- *Do the little things based on thinking globally and acting locally.* Each small measure you take is important, sensitizes you to Earth-sustaining acts, and leads to more such acts.
- *Think and act nationally and globally.* This means targeting the big polluters (industries, industrialized agriculture, governments) and big problems (ozone depletion, biodiversity loss, possible climate change) through political action, economic boycotts, and selective consumption.
- *Get to know, care about, and defend a piece of Earth.* As poet–philosopher Gary Snyder puts it, "Find our place on the planet, dig in, and take responsibility from there."
- *Work with others to help sustain and heal the earth, beginning in your neighborhood and community.*
- *Don't try to do everything.* Focus your energy on the few things that you feel most strongly about and that you can do something about.

- *Don't use guilt and fear to motivate other people, and don't allow other people to do this to you.* We need to nurture, reassure, understand, and love one another—rather than threaten.
- *Have fun and take time to enjoy life.* Don't get so intense and serious that you can't laugh every day and enjoy wild nature, beauty, friendship, and love.

BRINGING ABOUT AN EARTH-WISDOM REVOLUTION

We should design our lives and our societies on the basis of what we know so far about how nature sustains itself. Every country needs a *thousand points of sustainability* that are spotlighted, nurtured, and transplanted so they can grow into *millions of points of sustainability* all intertwined in a global network. These beacons of hope and change can show us how to live sustainably on this wondrous planet.

It is not too late. There is time to deal with the complex, interacting problems we face and to make an orderly rather than catastrophic transition to a sustainable Earth society, if enough of us really care. It's not up to "them," it's up to "us." Don't wait.

Make a difference by caring. Care about the air, water, soil. Care about wild plants, wild animals, wild places. Care about people—young, old, handicapped, black, white, brown—in this generation and generations to come. Let this caring be your guide for doing.

Envision the world as made up of all kinds of matter cycles and energy flows. See these life-sustaining processes as a beautiful and diverse web of interrelationships—a kaleidoscope of patterns, rhythms, and connections whose very complexity and multitude of possibilities remind us that cooperation, sharing, honesty, humility, and love should be the guidelines for our behavior toward one another and the earth.

PUBLICATIONS

The following publications can help you keep well informed and up to date on environmental and resource problems. Subscription prices, which tend to change, are not given.

Alternate Sources of Energy Alternate Sources of Energy, Inc., 107 South Central Ave., Milaca, MN 56353

Ambio: A Journal of the Human Environment Pergamon Press, Fairview Park, Elmsford, NY 10523

American Biology Teacher Journal of the National Association of Biology Teachers, 11250 Roger Bacon Drive, Room 319, Reston, VA 22090

American Forests American Forestry Association, 1516 P St. NW, Washington, DC 20005

American Journal of Alternative Agriculture 9200 Edmonston Rd., Suite 117, Greenbelt, MD 20770

American Rivers American Rivers, 801 Pennsylvania Ave. SE, Suite 303, Washington, DC 20003

Amicus Journal Natural Resources Defense Council, 40 W. 20th St., New York, NY 10011

Annual Review of Energy Department of Energy, Forrestal Building, 1000 Independence Ave. SW, Washington, DC 20585

Audubon National Audubon Society, 950 Third Ave., New York, NY 10022

Biologue National Wood Energy Association, 1730 North Lynn St., Suite 610, Arlington, VA 22209

BioScience American Institute of Biological Sciences, 730 11th St. NW, Washington, DC 20001

Buzzworm: The Environmental Journal P.O. Box 6853, Syracuse, NY 13217-7930

CDIAC Communications Carbon Dioxide Information Analysis Center, Oak Ridge National Laboratory, P.O. Box 2008, Oak Ridge, TN 37831

The CoEvolution Quarterly P.O. Box 428, Sausalito, CA 94965

Conservation Biology Blackwell Scientific Publications, Inc., 52 Beacon St., Boston, MA 02108

Demographic Yearbook Department of International Economic and Social Affairs, Statistical Office, United Nations Publishing Service, United Nations, New York, NY 10017

Earth Kalmbach Publishing Co., 21027 Crossroads Circle, Waukesha, WI 53187

Earth Island Journal Earth Island Institute, 300 Broadway, Suite 28, San Francisco, CA 94133

Earth Journal: Environmental Almanac and Resource Directory Buzzworm, Inc., 2305 Canyon Blvd., Suite 206, Boulder, CO 80302 (Published annually)

Earth Work Student Conservation Association, Dept. 1PR, P.O. Box 550, Charlestown, NH 03603

The Ecologist MIT Press Journals, 55 Hayward St., Cambridge, MA 02142

Ecology Ecological Society of America, Dr. Duncan T. Patten, Center for Environmental Studies, Arizona State University, Tempe, AZ 85281

E Magazine P.O. Box 5098, Westport, CT 06881

Endangered Species Update School of Natural Resources, University of Michigan, Ann Arbor, MI 481090

Environment Heldref Publications, 4000 Albemarle St. NW, Washington, DC 20016

Environment Abstracts Bowker A & I Publishing, 245 W. 17th St., New York, NY 10011. In most libraries

Environmental Action 6930 Carroll Park, 6th Floor, Takoma Park, MD 20912

Environmental Almanac Annual compilation by the World Resources Institute. Published by Houghton Mifflin (Boston)

Environmental Engineering News School of Civil Engineering, Purdue University, West Lafayette, IN 47907

Environmental Ethics Department of Philosophy, University of North Texas, Denton, TX 76203

Environmental Opportunities (Jobs) Sanford Berry, P.O. Box 969, Stowe, VT 05672

The Environmental Professional Editorial Office, Department of Geography, University of Iowa, Iowa City, IA 52242

EPA Journal Environmental Protection Agency. Order from Government Printing Office, Washington, DC 20402

Everyone's Backyard Citizens' Clearinghouse for Hazardous Waste, P.O. Box 926, Arlington, VA 22216

Family Planning Perspectives Planned Parenthood-World Population, 666 Fifth Ave., New York, NY 10019

The Futurist World Future Society, P.O. Box 19285, Twentieth Street Station, Washington, DC 20036

Garbage: The Practical Journal for the Environment P.O. Box 56520, Boulder, CO 80321-6520

Greenpeace Magazine Greenpeace USA, 1436 U St. NW, Washington, DC 20009

In Context Box 11470, Bainbridge Island, WA 98110

International Environmental Affairs University Press of New England, 171/2 Lebanon St., Hanover, NH 03755

International Wildlife National Wildlife Federation, 8925 Leesburg Pike, Vienna, VA 22184

Issues in Science and Technology National Academy of Sciences, 2101 Constitution Ave. NW, Washington, DC 20077-5576

Journal of Environmental Education Heldref Publications, 4000 Albemarle St. NW, Suite 504, Washington, DC 20016

Journal of Environmental Health National Environmental Health Association, 720 S. Colorado Blvd., Suite 970, Denver, CO 80222

Journal of Forestry Society of American Foresters, 5400 Grosvenor Lane, Bethesda, MD 20814

Journal of Pesticide Reform P.O. Box 1393, Eugene, OR 97440

Journal of Range Management Society for Range Management, 1839 York St., Denver, CO 80206

Journal of Soil and Water Conservation Soil and Water Conservation Society, 7515 Northeast Ankeny Rd., Ankeny, IA 50021

Journal of Sustainable Agriculture Food Products Press, 10 Alice Street, Binghampton, NY 13904

Journal of the Water Pollution Control Federation 601 Wythe St., Alexandria, VA 22314

Journal of Wildlife Management Wildlife Society, 5410 Grosvenor Lane, Bethesda, MD 20814

National Geographic National Geographic Society, P.O. Box 2895, Washington, DC 20077-9960

National Parks National Parks and Conservation Association, 1015 31st St. NW, Washington, DC 20007

National Wildlife National Wildlife Federation, 1400 16th St. NW, Washington, DC 20036

Natural Resources Journal University of New Mexico School of Law, 1117 Stanford NE, Albuquerque, NM 87131

Nature 711 National Press Building, Washington, DC 20045

The New Farm Rodale Press, 33 Minor St., Emmaus, PA 18049

The New Internationalist P.O. Box 1143, Lewiston, NY 14092-9984

New Scientist 128 Long Acre, London, WC 2, England

Newsline Natural Resources Defense Council, 122 E. 42nd St., New York, NY 10168

Not Man Apart Friends of the Earth, 530 Seventh St. SE, Washington, DC 20003

Nucleau Union of Concerned Scientists, 26 Church St., Cambridge, MA 02238

Organic Gardening & Farming Magazine Rodale Press, Inc., 33 E. Minor St., Emmaus, PA 18049

Our Planet United Nations Environment Programme (U.S.), 2 United Nations Plaza, Room 803, New York, NY 10017

Planetary Citizen Stillpoint Publishing, Meetinghouse Road, Box 640, Walpole, NH 03608

Pollution Abstracts Cambridge Scientific Abstracts, 7200 Wisconsin Ave., Bethesda, MD 20814. Found in many libraries

Popline The Population Institute, 107 Second St. NE, Suite 207, Washington, DC 20002

Population and Vital Statistics Report United Nations Environment Programme, New York North American Office, Publication Sales Section, United Nations, New York, NY 10017

Population Bulletin Population Reference Bureau, 1875 Connecticut Ave. NW, Suite 520, Washington, DC 20009

Rachel's Hazardous Waste News Environmental Research Foundation, P.O. Box 4878, Annapolis, MD 21403

Rainforest News P.O. Box 140681, Coral Gables, FL 33115

Real World: The Voice of Ecopolitics 91 Nuns Moor Road, Newcastle upon Tyne, NE4 9BA, United Kingdom

Renewable Energy News Solar Vision, Inc., 7 Church Hill, Harrisville, NH 03450

Renewable Resources 5430 Grosvenor Lane, Bethesda, MD 20814

Rocky Mountain Institute Newsletter 1739 Snowmass Creek Rd., Snowmass, CO 81654

Science American Association for the Advancement of Science, 1333 H St. NW, Washington, DC 20005

Science News Science Service, Inc., 1719 N St. NW, Washington, DC 20036

Scientific American 415 Madison Ave., New York, NY 10017

Sierra 730 Polk St., San Francisco, CA 94108

Solar Age Solar Vision, Inc., 7 Church Hill, Harrisville, NH 03450

State of the World Worldwatch Institute, 1776 Massachusetts Ave. NW, Washington, DC 20036 (Published annually)

Statistical Yearbook Department of International Economic and Social Affairs, Statistical Office, United Nations Publishing Service, United Nations, New York, NY 10017

Technology Review Room E219-430, Massachusetts Institute of Technology, Cambridge, MA 02139

Toxic Times National Toxics Campaign, 37 Temple Place, 4th Floor, Boston, MA 02111

Transition Laurence G. Wolf, ed., Department of Geography, University of Cincinnati, Cincinnati, OH 45221

The Trumpeter Journal of Ecosophy P.O. Box 5883 St. B, Victoria, B.C., V8R 6S8, Canada

Vegetarian Journal P.O. Box 1463, Baltimore, MD 21203

Vital Signs: The Trends That Are Shaping Our Future Worldwatch Institute, 1776 Massachusetts Ave. NW, Washington, DC 20036 (Published annually)

Waste Not Work on Waste USA, 82 Judson, Canton, NY 13617

Wild Earth Cenozoic Society, Inc., P.O. Box 455, Richmond, VT 05477

Wilderness The Wilderness Society, 1400 I St. NW, 10th Floor, Washington, DC 20005

Wildlife Conservation New York Zoological Society, 185th St. and Southern Boulevard, Bronx, NY 10460

World Rainforest Report Rainforest Action Network, 300 Broadway, Suite 298, San Francisco, CA 94133

World Resources World Resources Institute, 1735 New York Ave. NW, Washington, DC 20006 (Published every two years)

World Watch Worldwatch Institute, 1776 Massachusetts Ave. NW, Washington, DC 20036

Worldwatch Papers Worldwatch Institute, 1776 Massachusetts Ave. NW, Washington, DC 20036

Yearbook of World Energy Statistics Department of International Economic and Social Affairs, Statistical Office, United Nations Publishing Service, United Nations, New York, NY 10017

ENVIRONMENTAL AND RESOURCE ORGANIZATIONS

For a more detailed list of national, state, and local organizations, see *Conservation Directory* (published annually by the National Wildlife Federation, 1400 16th St. NW, Washington, DC 20036); *Your Resource Guide to Environmental Organizations* (Irvine, CA: Smiling Dolphin Press, 1991); *National Environmental Organizations* (published annually by US Environmental Directories, Inc., P.O. Box 65156, St. Paul, MN 55165); and *World Directory of Environmental Organizations* (published by the California Institute of Public Affairs, P.O. Box 10, Claremont, CA 91711).

African Wildlife Foundation 1717 Massachusetts Ave. NW, Washington, DC 20036

Alan Guttmacher Institute 111 5th Ave., New York, NY 10003

Alliance for Chesapeake Bay 6600 York Rd., Baltimore, MD 21212

Alliance for Environmental Education P.O. Box 368, The Plains, VA 22171

Alliance to Save Energy 1725 K St. NW, Suite 914, Washington, DC 20006-1401

American Cetacean Society P.O. Box 2639, San Pedro, CA 90731-0943

American Council for an Energy Efficient Economy 1001 Connecticut Ave. NW, Suite 535, Washington, DC 20013

American Forestry Association 1516 P St. NW, Washington, DC 20005

American Geographical Society 156 Fifth Ave., Suite 600, New York, NY 10010

American Humane Society 9725 E. Hampden Ave., Denver, CO 80231

American Institute of Biological Sciences, Inc. 730 11th St. NW, Washington, DC 20001

American Rivers 801 Pennsylvania Ave. SE, Suite 303, Washington, DC 20003

American Society for the Prevention of Cruelty to Animals (ASPCA) 441 E. 92nd Street, New York, NY 10128

American Solar Energy Society 2400 Central Ave., Suite G, Boulder, CO 80301

American Water Resources Association 5410 Grosvenor Lane, Suite 220, Bethesda, MD 20814

American Wilderness Alliance 7500 E. Arapahoe Rd., Suite 114, Englewood, CO 80112

American Wildlife Association 1717 Massachusetts Ave. NW, Washington, DC 20036

American Wind Energy Association 777 North Capitol St. NE, Suite 805, Washington, DC 20002

Appropriate Technology International 1331 H St. NW, Washington, DC 20005

Association of Forest Service Employees for Environmental Ethics P.O. Box 11615, Eugene, OR 97440-9958

Bat Conservation International P.O. Box 162603, Austin, TX 78716

Bio-Integral Resource Center P.O. Box 8267, Berkeley, CA 94707

Bioregional Project (North American Bioregional Congress) Turtle Island Office, 1333 Overhulse Rd. NE, Olympia, WA 98502

Carbon Dioxide Information Analysis Center Ms-6335, Building 105, Oak Ridge National Laboratory, P.O. Box 2008, Oak Ridge, TN 37831

Carrying Capacity 1325 G St. NW, Washington, DC 20005

Center for Conservation Biology Department of Biological Sciences, Stanford University, Stanford, CA 94305

Center for Energy and Environmental Studies The Engineering Quadrangle, Princeton University, Princeton, NJ 08544

Center for Marine Conservation 1725 DeSales St. NW, Suite 500, Washington, DC 20036

Center for Plant Conservation P.O. Box 299, St. Louis, MO 63166

Center for Science in the Public Interest 1501 16th St. NW, Washington, DC 20036

Chesapeake Bay Foundation 162 Prince George St., Annapolis, MD 21401

Chipko P.O. Silyara via Ghansale, Tehri-Garwhal, Uttar Pradesh, 249155 India

Citizens' Clearinghouse for Hazardous Waste P.O. Box 6806, Falls Church, VA 22040

Clean Water Action 1320 18th St. NW, Washington, DC 20036

Climate Institute 316 Pennsylvania Ave. SE, Washington, DC 20003

Conservation Foundation 1250 24th St. NW, Suite 500, Washington, DC 20037

Council for Economic Priorities 30 Irving Pl., New York, NY 10003

Cousteau Society 930 W. 21st St., Norfolk, VA 23517

Critical Mass Energy Project 215 Pennsylvania Ave. SE, Washington, DC 20003

Cultural Survival 11 Divinity Ave., Cambridge, MA 02138

Defenders of Wildlife 1244 19th St. NW, Washington, DC 20036

Ducks Unlimited One Waterfowl Way, Long Grove, IL 60047

Earth First! 305 N. Sixth St., Madison, WI 53704

Earth Island Institute 300 Broadway, Suite 28, San Francisco, CA 94133

EarthSave Foundation P.O. Box 949, Felton, CA 95018

Earthwatch 680 Mt. Auburn St., Box 403N, Watertown, MA 02272

Elmwood Institute P.O. Box 5805, Berkeley, CA 94705

Energy Conservation Coalition 1525 New Hampshire Ave. NW, Washington, DC 20036

Environmental Action, Inc. 6930 Carroll Park, 6th Floor, Takoma Park, MD 20912

Environmental Defense Fund, Inc. 257 Park Ave. South, New York, NY 10010

Environmental Law Institute 1616 P St. NW, Suite 200, Washington, DC 20036

Environmental Policy Institute 218 D St. SE, Washington, DC 20003

Florida Solar Energy Center 300 State Road #401, Cape Canaveral, FL 32920

Food First (Institute for Food and Development Policy) 145 Ninth St., San Francisco, CA 94103

Friends of Animals 11 W. 60th St., New York, NY 10023

Friends of the Earth 218 D St. SE, Washington, DC 20003

Friends of the Trees P.O. Box 1466, Chelan, WA 98816

Fund for Animals 200 W. 57th St., New York, NY 10019

Global Greenhouse Network 1130 17th St. NW, Suite 530, Washington, DC 20036

Global Tomorrow Coalition 1325 G St. NW, Suite 915, Washington, DC 20005

Greenhouse Crisis Foundation 1130 17th St. NW, Suite 630, Washington, DC 20036

Greenpeace, Canada 427 Bloor St., West Toronto, Ontario M5S 1X7

Greenpeace, USA, Inc. 1436 U St. NW, Washington, DC 20009

Green Seal 1733 Connecticut Ave, NW, Washington, DC 20009

Humane Society of the United States, Inc. 2100 L St. NW, Washington, DC 20037

INFORM 381 Park Ave. South, New York, NY 10016

Institute for Alternative Agriculture 9200 Edmonston Rd., Suite 117, Greenbelt, MD 20770

Institute for Earth Education P.O. Box 288, Warrenville, IL 60555

Institute for Local Self-Reliance 2425 18th St. NW, Washington, DC 20009

International Alliance for Sustainable Agriculture 1701 University Ave. SE, Newman Center, Room 202, Minneapolis, MN 55414

International Planned Parenthood Federation 105 Madison Ave., 7th Floor, New York, NY 10016

International Union for the Conservation of Nature and Natural Resources (IUCN) 1400 16th St. NW, Washington, DC 20036

Izaak Walton League of America 1401 Wilson Blvd., Level B, Arlington, VA 22209

Land Institute 2440 E. Well Water Road, Salina, KS 67401

League of Conservation Voters 1707 L St. NW, Suite 550, Washington, DC 20036

League of Women Voters of the U.S. 1730 M St. NW, Washington, DC 20036

National Association of Biology Teachers 11250 Roger Bacon Drive, Room 19, Reston, VA 22090

National Audubon Society 950 Third Ave., New York, NY 10022

National Center for Urban Environmental Studies 516 North Charles St., Suite 501, Baltimore, MD 21201

National Clean Air Coalition 801 Pennsylvania Ave. SE, Washington, DC 20003

National Coalition Against the Misuse of Pesticides 701 E St. SE, Suite 200, Washington, DC 20001

National Coalition for Marine Conservation P.O. Box 23298, Savannah, GA 31403

National Environmental Health Association 720 S. Colorado Blvd., South Tower, Suite 970, Denver, CO 80222

National Geographic Society 17th and M Sts. NW, Washington, DC 20036

National Parks and Conservation Association 1776 Massachusetts Ave. NW, Suite 200, Washington, DC 20036

National Recreation and Park Association 2775 S. Quincy St., Suite 300, Arlington, VA 22206

National Recycling Coalition 1101 30th St. NW, Suite 304, Washington, DC 20007

National Science Teachers Association 1742 Connecticut Ave. NW, Washington, DC 20009

National Solid Waste Management Association 1730 Rhode Island Ave. NW, Suite 100, Washington, DC 20036

National Toxics Campaign 37 Temple Place, 4th Floor, Boston, MA 02111

National Wildlife Federation 1400 16th St. NW, Washington, DC 20036

National Wood Energy Association 1730 N. Lynn St., Suite 610, Arlington, VA 22209

Natural Resources Defense Council 40 W. 20th St., New York, NY 10011, and 1350 New York Ave. NW, Suite 300, Washington, DC 20005

Nature Conservancy 1814 N. Lynn St., Arlington, VA 22209

New Alchemy Institute 237 Hatchville Rd., East Falmouth, MA 02536

North American Association for Environmental Education P.O. Box 400, Troy, OH 45373

Nuclear Information and Resource Service 1424 16th St. NW, Suite 601, Washington, DC 20036

The Oceanic Society 218 D St. SE, Washington, DC 20003

Permaculture Association P.O. Box 202, Orange, MA 01364

Permaculture Institute of North America 4649 Sunnyside Ave. N, Seattle, WA 98103

Pesticide Action Network 965 Mission St., No. 514, San Francisco, CA 94103

Physicians for Social Responsibility 639 Massachusetts Ave., Cambridge, MA 02139

Planetary Citizens 325 Ninth St., San Francisco, CA 94103

Planet/Drum Foundation P.O. Box 31251, San Francisco, CA 94131

Planned Parenthood Federation of America 810 Seventh Ave., New York, NY 10019

Population Council 1 Dag Hammarskold Plaza, New York, NY 10017

Population Crisis Committee 1120 19th St. NW, Suite 530, Washington, DC 20036-3605

Population-Environment Balance 1325 G St. NW, Washington, DC 20005

Population Institute 110 Maryland Ave. NE, Suite 207, Washington, DC 20036

Population Reference Bureau 1875 Connecticut Ave. NW, Suite 520, Washington, DC 20009-5728

Public Citizen 215 Pennsylvania Ave. SE, Washington, DC 20003

Rainforest Action Network 3450 Sansome St., Suite 700, San Francisco, CA 94111

Rainforest Alliance 270 Lafayette St., Suite 512, New York, NY 10012

Renewable Natural Resources Foundation 5430 Grosvenor Lane, Bethesda, MD 20814

Renew America 1400 16th St. NW, Suite 710, Washington, DC 20036

Resources for the Future 1616 P St. NW, Washington, DC 20036

Rocky Mountain Institute 1739 Snowmass Creek Rd., Snowmass, CO 81654

Rodale Institute 222 Main St., Emmaus, PA 18098

Save America's Forests 4 Library Court SE, Washington, DC 20003

Scientists' Institute for Public Information 355 Lexington Ave., New York, NY 10017

Sea Shepherd Conservation Society 1314 2nd St., Santa Monica, CA 90401

Sierra Club 730 Polk St., San Francisco, CA 94109, and 408 C St. NE, Washington, DC 20002

Sierra Club Legal Defense Fund 180 Montgomery St., San Francisco, CA 94104

Smithsonian Institution 1000 Jefferson Dr. SW, Washington, DC 20560

Social Investment Forum C.E.R.E.S. Project, 711 Atlantic Ave., Boston, MA 02111

Society of American Foresters 5400 Grosvenor Lane, Bethesda, MD 20814

Society for Conservation Biology Department of Wildlife Ecology, University of Wisconsin, Madison, WI 53706

Soil and Water Conservation Society 7515 N.E. Ankeny Rd., Ankeny, IA 50021

Student Conservation Association, Inc. P.O. Box 550, Charlestown, NH 03603

Student Environmental Action Coalition (SEAC) 217 A Carolina Union, University of North Carolina, Chapel Hill, NC 27599

Survival International 2121 Decatur Place NW, Washington, DC 20008

Treepeople 12601 Mulholland Dr., Beverly Hills, CA 90210

Union of Concerned Scientists 26 Church St., Cambridge, MA 02238

United Nations Population Fund 220 East 42nd St., New York, NY 10017

U.S. Public Interest Research Group 215 Pennsylvania Ave. SE, Washington, DC 20003

Water Pollution Control Federation 601 Wythe St., Alexandria, VA 22314

The Wilderness Society 900 17th St. NW, Washington, DC 20006

Wildlife Conservation International (WCI) New York Zoological Society, 185th St. and Southern Blvd., Bronx, NY 10460

Wildlife Society 5410 Grosvenor Lane, Bethesda, MD 20814

Work on Waste 82 Judson St., Canton, NY 13617

World Future Society 4916 St. Elmo Ave., Bethesda, MD 20814

World Resources Institute 1709 New York Ave. NW, 7th Floor, Washington, DC 20006

Worldwatch Institute 1776 Massachusetts Ave. NW, Washington, DC 20036

World Wildlife Fund 1250 24th St. NW, Suite 500, Washington, DC 20037

Zero Population Growth 1400 16th St. NW, 3rd Floor, Washington, DC 20036

ADDRESSES OF FEDERAL AND INTERNATIONAL AGENCIES

Agency for International Development State Building, 320 21st St. NW, Washington, DC 20523

Bureau of Land Management U.S. Department of Interior, 18th and C Sts., Room 3619, Washington, DC 20240

Bureau of Mines 2401 E St. NW, Washington, DC 20241

Bureau of Reclamation Washington, DC 20240

Congressional Research Service 101 Independence Ave. SW, Washington, DC 20540

Conservation and Renewable Energy Inquiry and Referral Service P.O. Box 8900, Silver Spring, MD 20907, 800-523-2929.

Consumer Product Safety Commission Washington, DC 20207

Department of Agriculture 14th St. and Jefferson Dr. SW, Washington, DC 20250

Department of Commerce 14th St. between Constitution Ave. and E St. NW, Washington, DC 20230

Department of Energy Forrestal Building, 1000 Independence Ave. SW, Washington, DC 20585

Department of Health and Human Services 200 Independence Ave. SW, Washington, DC 20585

Department of Housing and Urban Development 451 Seventh St. SW, Washington, DC 20410

Department of the Interior 18th and C Sts. NW, Washington, DC 20240

Department of Transportation 400 Seventh St. SW, Washington, DC 20590

Environmental Protection Agency 401 M St. SW, Washington, DC 20460

Federal Energy Regulatory Commission 825 N. Capitol St. NE, Washington, DC 20426

Fish and Wildlife Service Department of the Interior, 18th and C Sts. NW, Washington, DC 20240

Food and Agriculture Organization (FAO) of the United Nations 101 22nd St. NW, Suite 300, Washington, DC 20437

Food and Drug Administration Department of Health and Human Services, 5600 Fishers Lane, Rockville, MD 20852

Forest Service P.O. Box 96090, Washington, DC 20013

Government Printing Office Washington, DC 20402

Inter-American Development Bank 1300 New York Ave. NW, Washington, DC 20577

International Whaling Commission The Red House, 135 Station Rd., Histon, Cambridge CB4 4NP England 02203 3971

Marine Mammal Commission 1625 I St. NW, Washington, DC 20006

National Academy of Sciences 2101 Constitution Ave., NW Washington, DC 20418

National Aeronautics and Space Administration 400 Maryland Ave. SW, Washington, DC 20546

National Cancer Institute 9000 Rockville Pike, Bethesda, MD 20892

National Center for Appropriate Technology 3040 Continental Dr., Butte, MT 59701

National Center for Atmospheric Research P.O. Box 3000, Boulder, CO 80307

National Marine Fisheries Service U.S. Dept. of Commerce, NOAA, 1335 East-West Highway, Silver Spring, MD 20910

National Oceanic and Atmospheric Administration Rockville, MD 20852

National Park Service Department of the Interior, P.O. Box 37127, Washington, DC 20013

National Renewable Energy Laboratory 1617 Cole Blvd., Golden, CO 80401

National Science Foundation 1800 G St. NW, Washington, DC 20550

National Solar Heating and Cooling Information Center P.O. Box 1607, Rockville, MD 20850

National Technical Information Service U.S. Department of Commerce, 5285 Port Royal Rd., Springfield, VA 22161

Nuclear Regulatory Commission 1717 H St. NW, Washington, DC 20555

Occupational Safety and Health Administration Department of Labor, 200 Constitution Ave. NW, Washington, DC 20210

Office of Ocean and Coastal Resource Management 1825 Connecticut Ave., Suite 700, Washington, DC 20235

Office of Surface Mining Reclamation and Enforcement 1951 Constitution Ave. NW, Washington, DC 20240

Office of Technology Assessment U.S. Congress, 600 Pennsylvania Ave. SW, Washington, DC 20510

Organization for Economic Cooperation and Development (U.S. Office) 2001 L St. NW, Suite 700, Washington, DC 20036

Soil Conservation Service P.O. Box 2890, Washington, DC 20013

United Nations 1 United Nations Plaza, New York, NY 10017

United Nations Environment Programme Regional North American Office, United Nations Room DC2-0803, New York, NY 10017, and 1889 F St. NW, Washington, DC 20006

U.S. Geological Survey 12201 Sunrise Valley Dr., Reston, VA 22092.

World Bank 1818 H St. NW, Washington, DC 20433

Units of Measurement

LENGTH

Metric

1 kilometer (km) = 1,000 meters (m)
1 meter (m) = 100 centimeters (cm)
1 meter (m) = 1,000 millimeters (mm)
1 centimeter (cm) = 0.01 meter (m)
1 millimeter (mm) = 0.001 meter (m)

English

1 foot (ft) = 12 inches (in)
1 yard (yd) = 3 feet (ft)
1 mile (mi) = 5,280 feet (ft)
1 nautical mile = 1.15 miles

Metric-English

1 kilometer (km) = 0.621 mile (mi)
1 meter (m) = 39.4 inches (in)
1 inch (in) = 2.54 centimeters (cm)
1 foot (ft) = 0.305 meter (m)
1 yard (yd) = 0.914 meter (m)
1 nautical mile = 1.85 kilometers (km)

AREA

Metric

1 square kilometer (km^2) = 1,000,000 square meters (m^2)
1 square meter (m^2) = 1,000,000 square millimeters (mm^2)
1 hectare (ha) = 10,000 square meters (m^2)
1 hectare (ha) = 0.01 square kilometer (km^2)

English

1 square foot (ft^2) = 144 square inches (in^2)
1 square yard (yd^2) = 9 square feet (ft^2)
1 square mile (mi^2) = 27,880,000 square feet (ft^2)
1 acre (ac) = 43,560 square feet (ft^2)

Metric-English

1 hectare (ha) = 2.471 acres (ac)
1 square kilometer (km^2) = 0.386 square mile (mi^2)
1 square meter (m^2) = 1.196 square yards (yd^2)
1 square meter (m^2) = 10.76 square feet (ft^2)
1 square centimeter (cm^2) = 0.155 square inch (in^2)

VOLUME

Metric

1 cubic kilometer (km^3) = 1,000,000,000 cubic meters (m^3)
1 cubic meter (m^3) = 1,000,000 cubic centimeters (cm^3)
1 liter (L) = 1,000 milliliters (mL) = 1,000 cubic centimeters (cm^3)
1 milliliter (mL) = 0.001 liter (L)
1 milliliter (mL) = 1 cubic centimeter (cm^3)

English

1 gallon (gal) = 4 quarts (qt)
1 quart (qt) = 2 pints (pt)

Metric-English

1 liter (L) = 0.265 gallon (gal)
1 liter (L) = 1.06 quarts (qt)
1 liter (L) = 0.0353 cubic foot (ft^3)
1 cubic meter (m^3) = 35.3 cubic feet (ft^3)
1 cubic meter (m^3) = 1.30 cubic yard (yd^3)
1 cubic kilometer (km^3) = 0.24 cubic mile (mi^3)
1 barrel (bbl) = 159 liters (L)
1 barrel (bbl) = 42 U.S. gallons (gal)

MASS

Metric

1 kilogram (kg) = 1,000 grams (g)
1 gram (g) = 1,000 milligrams (mg)
1 gram (g) = 1,000,000 micrograms (µg)
1 milligram (mg) = 0.001 gram (g)
1 microgram (µg) = 0.000001 gram (g)
1 metric ton (mt) = 1,000 kilograms (kg)

English

1 ton (t) = 2,000 pounds (lb)
1 pound (lb) = 16 ounces (oz)

Metric-English

1 metric ton (mt) = 2,200 pounds (lb) = 1.1 tons (t)
1 kilogram (kg) = 2.20 pounds (lb)
1 pound (lb) = 454 grams (g)
1 gram (g) = 0.035 ounce (oz)

ENERGY AND POWER

Metric

1 kilojoule (kJ) = 1,000 joules (J)
1 kilocalorie (kcal) = 1,000 calories (cal)
1 calorie (cal) = 4.184 joules (J)

Metric-English

1 kilojoule (kJ) = 0.949 British thermal unit (Btu)
1 kilojoule (kJ) = 0.000278 kilowatt-hour (kW-h)
1 kilocalorie (kcal) = 3.97 British thermal units (Btu)
1 kilocalorie (kcal) = 0.00116 kilowatt-hour (kW-h)
1 kilowatt-hour (kW-h) = 860 kilocalories (kcal)
1 kilowatt-hour (kW-h) = 3,400 British thermal units (Btu)
1 quad (Q) = 1,050,000,000,000,000 kilojoules (kJ)
1 quad (Q) = 2,930,000,000,000 kilowatt-hours (kW-h)

TEMPERATURE CONVERSIONS

Fahrenheit (°F) to Celsius (°C): $°C = \dfrac{(°F - 32.0)}{1.80}$

Celsius (°C) to Fahrenheit (°F): $°F = (°C \times 1.80) + 32.0$

Major U.S. Resource Conservation and Environmental Legislation

GENERAL

National Environmental Policy Act of 1969 (NEPA)
International Environmental Protection Act of 1983

ENERGY

Energy Policy and Conservation Act of 1975
National Energy Act of 1978, 1980
National Appliance Energy Conservation Act of 1987
Energy Policy Act of 1992

WATER QUALITY

Water Quality Act of 1965
Water Resources Planning Act of 1965
Federal Water Pollution Control Acts of 1965, 1972
Ocean Dumping Act of 1972
Ocean Dumping Ban Act of 1988
Safe Drinking Water Act of 1974, 1984
Water Resources Development Act of 1986
Clean Water Act of 1977, 1987

AIR QUALITY

Clean Air Act of 1963, 1965, 1970, 1977, 1990
Pollution Prevention Act of 1990

NOISE CONTROL

Noise Control Act of 1965
Quiet Communities Act of 1978

RESOURCES AND SOLID WASTE MANAGEMENT

Solid Waste Disposal Act of 1965
Resource Recovery Act of 1970
Resource Conservation and Recovery Act of 1976
Marine Plastic Pollution Research and Control Act of 1987

TOXIC SUBSTANCES

Hazardous Materials Transportation Act of 1975
Toxic Substances Control Act of 1976
Resource Conservation and Recovery Act of 1976
Comprehensive Environmental Response, Compensation, and Liability (Superfund) Act of 1980, 1986
Nuclear Waste Policy Act of 1982

PESTICIDES

Federal Insecticide, Fungicide, and Rodenticide Control Act of 1972, 1988

WILDLIFE CONSERVATION

Lacey Act of 1900
Migratory Bird Treaty Act of 1918
Migratory Bird Conservation Act of 1929
Migratory Bird Hunting Stamp Act of 1934
Pittman-Robertson Act of 1937
Anadromous Fish Conservation Act of 1965
Fur Seal Act of 1966
National Wildlife Refuge System Act of 1966, 1976, 1978
Species Conservation Act of 1966, 1969
Marine Mammal Protection Act of 1972
Marine Protection, Research, and Sanctuaries Act of 1972
Endangered Species Act of 1973, 1982, 1985, 1988
Fishery Conservation and Management Act of 1976, 1978, 1982
Whale Conservation and Protection Study Act of 1976
Fish and Wildlife Improvement Act of 1978
Fish and Wildlife Conservation Act of 1980 (Nongame Act)

LAND USE AND CONSERVATION

Taylor Grazing Act of 1934
Wilderness Act of 1964
Multiple Use Sustained Yield Act of 1968
Wild and Scenic Rivers Act of 1968
National Trails System Act of 1968
National Coastal Zone Management Act of 1972, 1980
Forest Reserves Management Act of 1974, 1976
Forest and Rangeland Renewable Resources Act of 1974, 1978
Federal Land Policy and Management Act of 1976
National Forest Management Act of 1976
Soil and Water Conservation Act of 1977
Surface Mining Control and Reclamation Act of 1977
Antarctic Conservation Act of 1978
Endangered American Wilderness Act of 1978
Alaskan National Interests Lands Conservation Act of 1980
Coastal Barrier Resources Act of 1982
Food Security Act of 1985

APPENDIX 4

Mapping Concepts and Connections*

This textbook emphasizes the connections between environmental principles, problems, and solutions. One way to organize the material in various chapters and to understand such connections is to map the concepts in a particular chapter or portion of a chapter. Inside the front cover you will find a general concept map showing how the parts of this entire textbook are related. At the end of each chapter you will find a question asking you to make a chapter concept map. This appendix provides you with information about how to create such maps. Here are the basic steps in making a concept map:

1. List the key general and specific concepts by indenting specific concepts under more general concepts. Basically, make an outline of the material using section heads, subsection heads, and key terms.

2. Arrange the clusters of concepts with the more general concepts at the top of a page and the more specific concepts at the bottom of the page.

3. Circle key concepts, draw lines connecting them, and write labels on the lines that describe linkages between the concepts.

4. Draw in crosslinking connections (lines that relate various parts of the map) and write labels on the line describing these connections.

5. Study the map to see how you might improve it. You might add clustering concepts and connections of your own. You might eliminate detail to make the big picture stand out. You might add more details to clarify certain concepts. You might add or alter broad crosslinking connections on the map as you think and learn more about the material.

Let's go through these steps using the concept of living systems discussed in Sections 5-1 and 5-2 of Chapter 5 as an example.

1. List the appropriate general and specific concepts. Note that Section 5-1 is about

land systems (called biomes) and Section 5-2 is about aquatic systems. Use the list of section titles, subtitles, words in bold and italics, key drawings, and other terms you think are particularly important. An outline of these two sections in which specific concepts are indented under more general ones would look something like Figure 1.

2. Arrange clusters of the concepts in Figure 1 with the more general ones at the top and more specific ones at the bottom. Concepts of the same general level should be at about the same height on the map. Figure 2 shows what this might look like. You might be able to do this on a single sheet of paper turned sideways or you might find it easier to use large sheets of paper. Use a pencil to draw the map so that you can make changes. Note that I have simplified the diagram by eliminating some of the concepts in Figure 1 such as latitude and altitude, various types of temperate grasslands, and tropical scrub forests.

3. Circle key concepts, draw lines connecting them, and write labels on the lines that describe linkages between the concepts. You might decide to add or delete some of the concepts to make the map more logical, clear, and helpful to you. Figure 3 shows one way to do this. Note that in this figure I put the subcategories of deserts, grasslands, and forests together to improve clarity.

4. Draw in crosslinks that relate various parts of the map. Figure 4 illustrates one possibility. At this point you are trying to understand how the various concepts are connected to and interact with one another. Note that many of the human systems concepts interact positively or negatively with ecosystems and natural resource concepts (in terms of human value judgment). Note that Sections 5-1 and 5-2 focus on natural systems. As you learn more about how human systems interact with natural systems in this chapter and in other chapters, you should be able to add additional crosslinkages.

5. Study your map and try to improve it. You might choose to eliminate excessive

detail or add additional concepts and linkages to improve clarity. You might also find it useful to rearrange the position of concepts vertically or horizontally to eliminate crowding or to reduce confusion over too many connecting links. Figure 5 is a revised version of Figure 4. Note that I have added the general concepts of "fresh water" and "salt water" and divided aquatic systems into two subcategories based on salinity (dissolved salt content) instead of four. To improve clarity, I also added the concepts of "moving" and "still" under fresh water because these characteristics greatly affect the types of organisms living in such systems. Finally, I connected these concepts with some of those in Chapter 4 by including details about photosynthetic and chemosynthetic productivity since these activities form the basis of the food web relationships in terrestrial and aquatic systems.

Remember the following things about concept maps:

- They are particularly useful for showing the big picture and for clarifying particularly difficult concepts. They can also be used in studying for tests to help you sort out what is important and how concepts are related.

- They take time to make. Sometimes they are easy and seem almost to write themselves. Other times, they can be difficult and frustrating to draw.

- There is no single "best" map. Rarely do concept maps developed by different individuals look alike.

- They are works in progress that should change and improve over time as you get better at making them and develop a deeper understanding of how things are connected.

- Comparing your maps with those developed by others is a good way to learn how to improve your map-making skills and to come up with more connections.

- Making them is challenging and fun.

---

*This Appendix was developed by **Jane Heinze-Fry** with assistance from G. Tyler Miller, Jr.

LAND SYSTEMS (Biomes)
climate
  precipitation
  temperature
  latitude
  altitude
deserts
  tropical
  temperate
  cold
  semidesert
grasslands
  tropical
    savanna
  temperate
    tall-grass
    short-grass
    pampas
    veld
    steppes
  polar (arctic tundra)
    permafrost

forests
  tropical
    rain
    deciduous
    scrub
  temperate deciduous
  evergreen coniferous

WATER SYSTEMS
ocean
  coastal
    coral reefs
    estuaries
    wetlands
    barrier islands
    beaches
      rocky shore
      barrier beach

open sea
  euphotic zone
  bathyal zone
  abyssal zone
freshwater lakes
  eutrophic
  oligotrophic
  mesotrophic
  littoral zone
  limnetic zone
  profundal zone
  benthic zone
  thermal stratification
    thermoclines
    turnover
freshwater streams
  surface water
  runoff
  watershed
inland wetlands
  year-round
  seasonal

Figure 1 List of concepts for Living Systems map.

LIVING SYSTEMS

Land Systems (Biomes)
  climate
    precipitation
    temperature
    ~~latitude~~
    ~~altitude~~

Water Systems

ocean

deserts
  tropical
  temperate
  cold
  ~~semidesert~~

grasslands
  tropical
  ~~savanna~~
  temperate
  ~~tall-grass~~
  ~~short-grass~~
  ~~pampas~~
  ~~veld~~
  ~~steppes~~
  polar (arctic tundra)
  ~~permafrost~~

forests
  tropical
  rain
  deciduous
  ~~scrub~~
  temperate deciduous
  evergreen coniferous

coastal
  coral reefs
  estuaries
  wetlands
  barrier islands
  beaches
    rocky shore
    barrier beach

open sea
  euphotic zone
  bathyal zone
  abyssal zone
freshwater lakes
  eutrophic
  oligotrophic
  mesotrophic
  littoral zone
  limnetic zone
  profundal zone
  benthic zone
  thermal stratification
    thermoclines
    turnover

inland wetlands
  year-round
  seasonal

freshwater streams
  ~~surface water~~
  ~~runoff~~
  ~~watershed~~

Figure 2 Clusters of concepts are arranged in a hierarchy from more general at the top to more specific at the bottom.

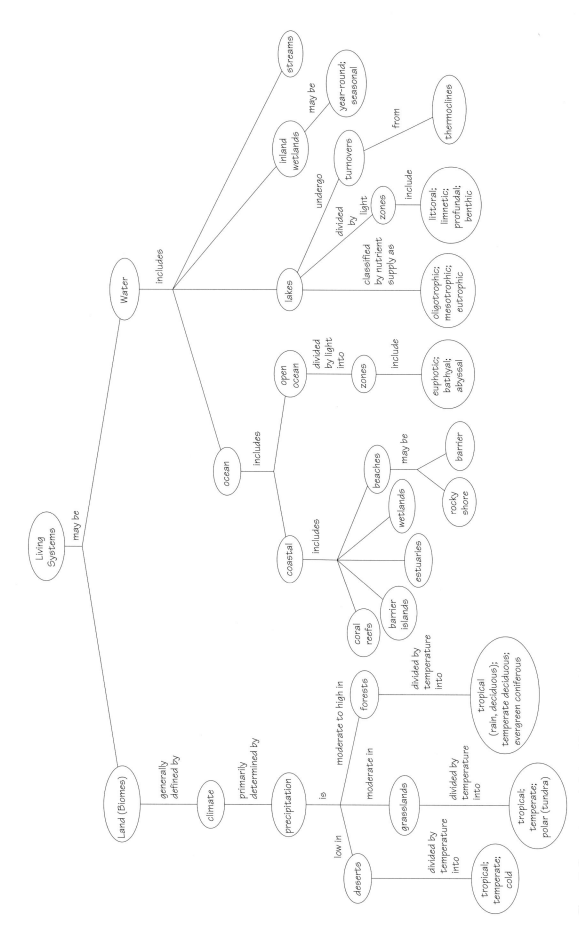

Figure 3 Concepts are circled and linkages are drawn.

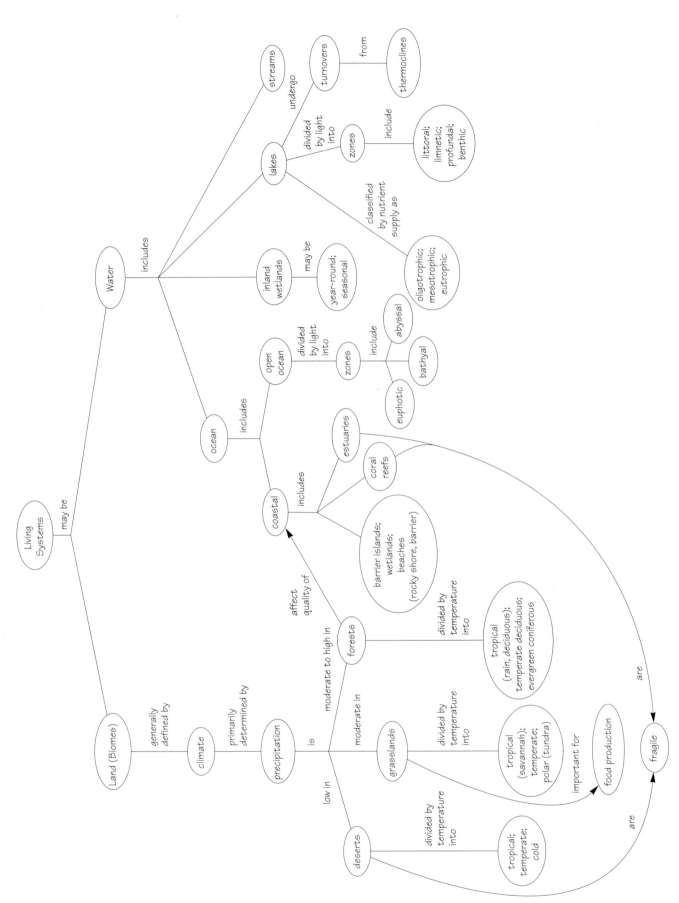

Figure 4 Crosslinkages are drawn and map is improved.

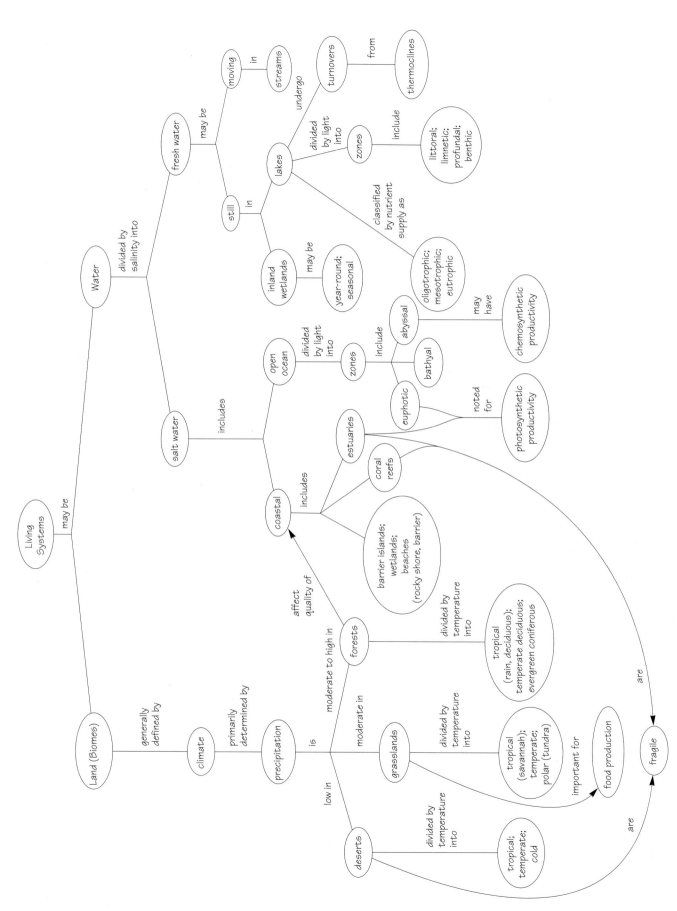

Figure 5 Map is further refined. New hierarchical clusters may emerge.

Further Readings*

GENERAL SOURCES OF ENVIRONMENTAL INFORMATION

Ashworth, William. 1991. *The Encyclopedia of Environmental Studies.* New York: Facts on File.

Beacham, W., ed. 1993. *Beacham's Guide to Environmental Issues and Sources.* 5 vols. Washington, D.C.: Beacham Publishing.

British Petroleum. Annual. *BP Statistical Review of World Energy.* New York: BP America, Inc.

Brown, Lester R., et al. Annual. *State of the World.* New York: Norton.

Brown, Lester R., et al. Annual. *Vital Signs.* New York: Norton.

Brownstone, Frank, Irene Brownstone, and David Brownstone. 1992. *The Green Encyclopedia.* New York: Prentice Hall General Reference.

Buzzworm Magazine Editors. Annual. *Earth Journal: Environmental Almanac and Resource Directory.* Boulder, Colo.: Buzzworm Books.

Kirdon, Michael, and Ronald Segal. 1991. *The New State of the World Atlas.* 4th ed. New York: Simon & Schuster.

Lean, Geoffrey, et al. 1990. *Atlas of the Environment.* Englewood Cliffs, N.J.: Prentice Hall.

Population Reference Bureau. Annual. *World Population Data Sheet.* Washington, D.C.: Population Reference Bureau.

Rittner, Don. 1992. *Ecolinking: Everyone's Guide to Online Environmental Information.* Berkeley, Calif.: Peachpit Press.

United Nations. Annual. *Demographic Yearbook.* New York: United Nations.

United Nations Children's Fund (UNICEF). Annual. *The State of the World's Children.* New York: UNICEF.

United Nations Environment Programme (UNEP). Annual. *State of the Environment.* New York: UNEP.

United Nations Population Fund. Annual. *The State of World Population.* New York: United Nations Population Fund.

U.S. Bureau of the Census. Annual. *Statistical Abstract of the United States.* Washington, D.C.: U.S. Bureau of the Census.

World Health Organization (WHO). Annual. *World Health Statistics.* Geneva, Switzerland: WHO.

World Resources Institute. Annual *The Information Please Environmental Almanac.* Boston: Houghton Mifflin.

World Resources Institute and International Institute for Environment and Development. *World Resources.* New York: Basic Books (Published every two years).

---

*For a more detailed list consult my longer book, *Living in the Environment*, Belmont, Calif.: Wadsworth Publishing.

1 / ENVIRONMENTAL PROBLEMS AND THEIR CAUSES

Asimov, Isaac, and Frederick Pohl. 1991. *Our Angry Earth.* New York: St. Martin's Press.

Bender, David L., and Bruno Leone, eds. 1990. *Environment: Opposing Viewpoints.* Vol. I. San Diego: Greenhaven.

Catton, William R., Jr. 1980. *Overshoot: The Ecological Basis of Revolutionary Change.* Chicago: University of Illinois Press.

Chivian, Eric, et al. 1993. *Critical Condition: Human Health and the Environment.* Cambridge, Mass.: MIT Press.

Commoner, Barry. 1990. *Making Peace with the Planet.* New York: Pantheon.

Ehrlich, Paul R., and Anne H. Ehrlich. 1990. *The Population Explosion.* New York: Doubleday.

Ehrlich, Paul R., and Anne H. Ehrlich. 1991. *Healing the Planet.* Reading, Mass.: Addison-Wesley.

Global Tomorrow Coalition. 1990. *The Global Ecology Handbook: What You Can Do About the Environmental Crisis.* Boston: Beacon Press.

Goldfarb, Theodore D. 1989. *Taking Sides: Clashing Views on Controversial Environmental Issues.* Guilford, Conn.: Dushkin Publishing Group.

Gordon, Anita, and David Suzuki. 1991. *It's a Matter of Survival.* Cambridge, Mass.: Harvard University Press.

Hardin, Garrett. 1968. "The Tragedy of the Commons." *Science,* vol. 162, 1243–48.

Hardin, Garrett. 1993. *Living Within Limits: Ecology, Economics, and Population Taboos.* New York: Oxford University Process.

Harrison, Paul. 1992. *The Third Revolution: Environment, Population, and a Sustainable World.* New York: St. Martin's Press.

Kamieniecki, Sheldon, et al., eds. 1993. *Controversies in Environmental Policy.* Ithaca, N.Y.: State University of New York Press.

Lehr, J., ed. 1992. *Rational Readings on Environmental Concerns.* New York: Van Nostrand Reinhold.

Meadows, Donella. 1991. *The Global Citizen.* Covelo, Calif.: Island Press.

Meadows, Donella H., et al. 1992. *Beyond the Limits: Confronting Global Collapse, Envisioning a Sustainable Future.* Post Mills, Vt.: Chelsea Green.

Myers, Norman, ed. 1993. *Gaia: An Atlas of Planet Management.* Garden City, N.Y.: Anchor Press/Doubleday.

Myers, Norman, and Julian Simon. 1994. *Scarcity or Abundance? A Debate on the Environment.* New York: Norton.

Peterson, D. J. 1993. *Troubled Lands: The Legacy of Soviet Environmental Destruction.* Boulder, Colo.: Westview Press.

Porritt, Jonathan. 1991. *Save the Earth.* Atlanta, Ga.: Turner Publishing.

Ray, Dixie Lee, and Lou Gusso. 1993. *Environmental Overkill: Whatever Happened to Common Sense?* Washington, D.C.: Regnery Gateway.

Repetto, Robert. 1986. *World Enough and Time: Successful Strategies for Resource Management.* New Haven, Conn.: World Resources Institute.

Ruchlis, Hy, and Sandra Oddo. 1990. *Clear Thinking.* New York: Prometheus Books.

Simon, Julian L. 1981. *The Ultimate Resource.* Princeton, N.J.: Princeton University Press.

Simon, Julian L., and Herman Kahn, eds. 1984. *The Resourceful Earth.* Cambridge, Mass.: Basil Blackwell.

Wagner, Travis. 1993. *In Our Backyard: A Guide to Understanding Pollution and Its Effects.* New York: Van Nostrand Reinhold.

2 / CULTURAL CHANGES, WORLDVIEWS, ETHICS, AND SUSTAINABILITY

Berry, Thomas. 1988. *The Dream of the Earth.* San Francisco: Sierra Club.

Berry, Wendell. 1990. *What Are People For?* Berkeley: North Point Press.

Bookchin, Murray. 1990. *Remaking Society: Pathways to a Green Future.* San Francisco: South End Press.

Borrelli, Peter, ed. 1988. *Crossroads: Environmental Priorities for the Future.* Covelo, Calif.: Island Press.

Bronowski, Jacob, Jr. 1974. *The Ascent of Man.* Boston: Little, Brown.

Brown, Lester R., et al. 1991. *Saving the Planet: How to Shape an Environmentally Sustainable Global Economy.* New York: Norton.

Cahn, Robert. 1978. *Footprints on the Planet: A Search for an Environmental Ethic.* New York: Universe Books.

Caldicott, Helen. 1992. *If You Love This Planet: A Plan to Heal the Earth.* New York: Norton.

Carson, Rachel. 1962. *Silent Spring.* Boston: Houghton Mifflin.

Chiras, Daniel D. 1992. *Lessons from Nature: Learning to Live Sustainably on the Earth.* Covelo, Calif.: Island Press.

Clark, M. E. 1989. *Ariadne's Thread: The Search for New Models of Thinking.* New York: St. Martin's Press.

Cohen, Michael J. 1988. *How Nature Works: Regenerating Kinship with Planet Earth.* Walpole, N.H.: Stillpoint.

Desjardins, Joseph R. 1993. *Environmental Ethics.* Belmont, Calif.: Wadsworth.

Devall, Bill, and George Sessions. 1985. *Deep Ecology: Living As If Nature Mattered.* Salt Lake City: Gibbs H. Smith.

Dyson, Freeman. 1992. *From Eros to Gaia.* New York: Pantheon.

Earth Works Group. 1990. *50 Simple Things You Can Do to Save the Earth.* Berkeley: Earthworks Press.

Ehrenfeld, David. 1993. *Beginning Again: People and Nature in the New Millennium.* New York: Oxford University Press.

Fritsch, Albert J. 1980. *Environmental Ethics: Choices for Concerned Citizens.* New York: Anchor Books.

Garbarino, James. 1992. *Toward a Sustainable Society: An Economic, Social and Environmental Agenda for Our Children's Future.* Chicago: Noble.

Gore, Al. 1992. *Earth in the Balance: Ecology and the Human Spirit.* Boston: Houghton Mifflin.

Goudie, Andrew. 1990. *The Human Impact on the Natural Environment.* 3d ed. Cambridge, Mass.: MIT Press.

Hardin, Garrett. 1978. *Exploring New Ethics for Survival.* 2d ed. New York: Viking Press.

Hardin, Garrett. 1993. *Living Within Limits: Ecology, Economics, and Population Taboos.* New York: Oxford University Press.

Hargrove, Eugene C. 1989. *Foundations of Environmental Ethics.* Englewood Cliffs, N.J.: Prentice-Hall.

Harte, John. 1993. *The Green Fuse: An Environmental Odyssey.* Berkeley: University of California Press.

IUCN, UNEP, WWF. 1991. *Caring for the Earth: A Strategy for Sustainable Living.* London: Earthscan.

Johnson, Warren. 1985. *The Future Is Not What It Used to Be: Returning to Traditional Values in an Age of Scarcity.* New York: Dodd, Mead.

Kennedy, Paul. 1993. *Preparing for the Twenty-First Century.* New York: Random House.

Klenig, John. 1991. *Valuing Life.* Princeton, N.J.: Princeton University Press.

Leopold, Aldo. 1949. *A Sand County Almanac.* New York: Oxford University Press.

Levering, Frank, and Wanda Urbanska. 1993. *Simple Living: One Couple's Search for a Better Life.* New York: Penguin.

Livingston, John. 1973. *One Cosmic Instant: Man's Fleeting Supremacy.* Boston: Houghton Mifflin.

Milbrath, Lester W. 1989. *Envisioning a Sustainable Society.* Albany: State University of New York Press.

Mumford, Lewis. 1962. *The Transformations of Man.* New York: Collier.

Naar, Jon. 1990. *Design for a Liveable Planet.* New York: Harper & Row.

Naess, Arne. 1989. *Ecology, Community, and Lifestyle.* New York: Cambridge University Press.

Nash, Roderick. 1988. *The Rights of Nature: A History of Environmental Ethics.* Madison: University of Wisconsin Press.

National Commission on the Environment. 1993. *Choosing a Sustainable Future.* Covelo, Calif.: Island Press.

Norton, Bryan G. 1991. *Toward Unity Among Environmentalists.* New York: Oxford University Press.

Orr, David. 1992. *Ecological Literacy.* Ithaca: State University of New York Press.

Ponting, Clive. 1992. *A Green History of the World. The Environment and the Collapse of Great Civilizations.* New York: St. Martin's Press.

Rampal, Shridath. 1992. *Our Country, The Planet: Forging a Partnership for Survival.* Covelo, Calif.: Island Press.

Rolston, Holmes, III. 1988. *Environmental Ethics: Duties to and Values in the Natural World.* Philadelphia: Temple University Press.

Roszak, Theodore. 1978. *Person/Planet.* Garden City, N.Y.: Doubleday.

Sale, Kirkpatrick. 1990. *Conquest of Paradise.* New York: Alfred A. Knopf.

Schumacher, E. F. 1973. *Small Is Beautiful: Economics As If People Mattered.* New York: Harper & Row.

Shabecoff, Philip. 1993. *A Fierce Green Fire: The American Environmental Movement.* New York: Hill & Wang.

Starke, Linda. 1990. *Signs of Hope: Working Towards Our Common Future.* New York: Oxford University Press.

Stone, Christopher. 1993. *The Gnat Is Older Than Man: Global Environment and Human Agenda.* Princeton, N.J.: Princeton University Press.

Swimme, Brian, and Thomas Berry. 1992. *The Universe Story: From the Primordial Flaring Forth to the Ecozoic Era.* San Francisco, Calif.: HarperCollins.

Taylor, Paul W. 1986. *Respect for Nature: A Theory of Environmental Ethics.* Princeton, N.J.: Princeton University Press.

Thomas, Lewis. 1992. *The Fragile Species.* New York: Scribner's (Macmillan).

Worster, Donald. 1992. *Under Western Skies: Nature and History in the American West.* New York: Oxford University Press.

3 / MATTER AND ENERGY RESOURCES: TYPES AND CONCEPTS

Bauer, Henry H. 1992. *Scientific Literacy and the Myth of the Scientific Method.* Urbana: University of Illinois Press.

Christensen, John W. 1990. *Global Science: Energy, Resources, and Environment.* 3d ed. Dubuque, Iowa: Kendall/Hunt.

Fowler, John M. 1984. *Energy and the Environment.* 2d ed. New York: McGraw-Hill.

Fumento, Michael. 1993. *Science Under Siege: Balancing Technology and the Environment.* New York: William Morrow.

Miller, G. Tyler, Jr., and David G. Lygre. 1991. *Chemistry: A Contemporary Approach.* 3d ed. Belmont, Calif.: Wadsworth.

Odum, Howard T., and Elisabeth C. Odum. 1981. *Energy Basis for Man and Nature.* 3d ed. New York: McGraw-Hill.

Rifkin, Jeremy. 1989. *Entropy: Into the Greenhouse World: A New World View.* New York: Bantam.

Rothman, Milton A. 1992. *The Science Gap: Dispelling Myths and Understanding the Reality of Science.* New York: Prometheus Books.

4 / ECOSYSTEMS AND HOW THEY WORK

Bolin, B., and R. B. Cook. 1983. *The Major Biogeochemical Cycles and Their Interactions.* New York: John Wiley.

Bradbury, Ian. 1991. *The Biosphere.* New York: Belhaven Press.

Brewer, R. 1993. *The Science of Ecology.* 2d ed. Philadelphia, Pa.: Saunders.

Burton, Robert, ed. 1991. *Animal Life.* New York: Oxford University Press.

Colinvaux, Paul A. 1978. *Why Big Fierce Animals Are Rare.* Princeton, N.J.: Princeton University Press.

Ehrlich, Anne H., and Paul R. Ehrlich. 1987. *Earth.* New York: Franklin Watts.

Ehrlich, Paul R. 1986. *The Machinery of Life: The Living World Around Us and How It Works.* New York: Simon & Schuster.

Elsom, Derek. 1992. *Earth: The Making, Shaping, and Working of a Planet.* New York: Macmillan.

Kormondy, Edward J. 1984. *Concepts of Ecology.* 3d ed. Englewood Cliffs, N.J.: Prentice-Hall.

Macmillan Publishing. 1992. *The Way Nature Works.* New York: Macmillan.

Moore, D. M., ed. 1991. *Plant Life.* New York: Oxford University Press.

Odum, Eugene P. 1993. *Ecology and Our Endangered Life-Support Systems.* 2d ed. Sunderland, Mass.: Sinauer.

Rickleffs, Robert E. 1990. *Ecology.* 3d ed. San Francisco: W. H. Freeman.

Smith, Robert L. 1990. *Elements of Ecology.* 4th ed. New York: Harper & Row.

Tudge, Colin. 1991. *Global Ecology.* New York: Oxford University Press.

Watt, Kenneth E. F. 1982. *Understanding the Environment.* Newton, Mass.: Allyn & Bacon.

5 / ECOSYSTEMS: WHAT ARE THE MAJOR TYPES, AND WHAT CAN HAPPEN TO THEM?

Aber, John, and Jerry Melito. 1991. *Terrestrial Ecosystems.* Philadelphia: Saunders.

Akin, Wallace E. 1991. *Global Patterns: Climate, Vegetation, and Soils.* Norman, Okla.: University of Oklahoma Press.

Attenborough, David, et al. 1989. *The Atlas of the Living World.* Boston: Houghton Mifflin.

Berger, John J. 1986. *Restoring the Earth.* New York: Alfred A. Knopf.

Botkin, Daniel. 1990. *Discordant Harmonies: A New Ecology for the Twenty-First Century.* New York: Oxford University Press.

Brown, J. H., and A. C. Gibson. 1983. *Biogeography.* St. Louis: C. V. Mosby.

Cook, L. M. 1991. *Genetic and Ecological Diversity.* New York: Chapman & Hall.

Elder, Danny, and John Pernetta, eds. 1991. *Oceans.* London: Michael Beazley.

Goldsmith, Edward, et al. 1990. *Imperiled Planet: Restoring Our Endangered Ecosystems.* Cambridge, Mass.: MIT Press.

Hardin, Garrett. 1985. "Human Ecology: The Subversive, Conservative Science." *American Zoologist,* vol. 25, 469–76.

Lovelock, James E. 1988. *The Ages of Gaia: A Biography of Our Living Earth.* New York: Norton.

Lovelock, James E. 1991. *Healing Gaia: Practical Medicine for the Planet.* New York: Random House.

Maltby, Edward. 1986. *Waterlogged Wealth.* Washington, D.C.: Earthscan.

Meyer, Christine, and Faith Moosang. 1992. *Living with the Land: Communities Restoring the Earth.* Philadelphia: New Society.

National Academy of Sciences. 1986. *Ecological Knowledge and Environmental Problem-Solving.* Washington, D.C.: National Academy Press.

Nisbet, E. G. 1991. *Living Earth.* San Francisco: HarperCollins.

Pilkey, Orin H., Jr, and William J. Neal, eds. 1987. *Living with the Shore.* Durham, N.C.: Duke University Press.

Pimm, Stuart L. 1992. *The Balance of Nature?* Chicago: University of Chicago Press.

Starr, Cecie, and Ralph Taggart. 1992. *Biology: The Unity and Diversity of Life.* 6th ed. Belmont, Calif.: Wadsworth.

Teal, J., and M. Teal. 1969. *Life and Death of a Salt Marsh.* New York: Ballantine.

Tudge, Colin. 1988. *The Environment of Life.* New York: Oxford University Press.

Wallace, David. 1987. *Life in the Balance.* New York: Harcourt Brace Jovanovich.

Weber, Peter. 1993. *Abandoned Seas: Reversing the Decline of the Oceans.* Washington, D.C.: Worldwatch Institute.

Wilson, E. O. 1992. *The Diversity of Life.* Cambridge, Mass.: Harvard University Press.

Wilson, E. O., ed. 1988. *Biodiversity.* Washington, D.C.: National Academy Press.

6 / THE HUMAN POPULATION: GROWTH, URBANIZATION, AND REGULATION

Barna, George. 1992. *The Invisible Generation: Baby Busters.* New York: Barna Research Group.

Berg, Peter, et al. 1989. *A Green City Program for San Francisco Bay Area Cities and Towns.* San Francisco: Planet/Drum Foundation.

Bouvier, Leon F. 1992. *Peaceful Invasions: Immigration and Changing America.* Lanham, Md.: University Press of America.

Bouvier, Leon F., and Carol J. De Vita. 1991. "The Baby Boom—Entering Midlife." *Population Bulletin,* vol. 6, no. 3, 1–35.

Brown, Lester R., and Jodi Jacobson. 1986. *Our Demographically Divided World.* Washington, D.C.: Worldwatch Institute.

Brown, Lester R., and Jodi Jacobson. 1987. *The Future of Urbanization: Facing the Ecological and Economic Restraints.* Washington, D.C.: Worldwatch Institute.

Cadman, D., and G. Payne, eds. 1990. *The Living City: Towards a Sustainable Future.* London: Routledge.

Davis, Kingsley, et al., eds. 1989. *Population and Resources in a Changing World.* Stanford, Calif.: Morrison Institute for Population and Resource Studies.

Day, Lincoln H. 1992. *The Future of Low-Birthrate Populations.* New York: Routledge.

Donaldson, Peter J., and Amy Ong Tsui. 1990. "The International Family Planning Movement." *Population Bulletin,* vol. 43, no. 3, 1–42.

Ehrlich, Paul R., and Anne H. Ehrlich. 1990. *The Population Explosion.* New York: Doubleday.

Farallones Institute. 1979. *The Integral Urban House: Self-Reliant Living in the City.* San Francisco: Sierra Club Books.

Formos, Werner. 1987. *Gaining People, Losing Ground: A Blueprint for Stabilizing World Population.* Washington, D.C.: Population Institute.

Gordon, Deborah. 1991. *Steering a New Course: Transportation, Energy, and the Environment.* Covelo, Calif.: Island Press.

Grant, Lindsey. 1992. *Elephants in the Volkswagen: Facing the Tough Questions About Our Overcrowded Country.* New York: W. H. Freeman.

Gupte, Pranay. 1984. *The Crowded Earth: People and the Politics of Population.* New York: Norton.

Hardin, Garrett. 1993. *Living Within Limits: Ecology, Economics, and Population Taboos.* New York: Oxford University Press.

Harrison, Paul. 1992. *The Third Revolution: Environment, Population, and a Sustainable World.* New York: I. B. Tauris.

Hart, John. 1992. *Saving Cities, Saving Money: Environmental Strategies That Work.* Washington, D.C.: Resource Renewal Institute.

Hartmann, Betsy. 1987. *Reproductive Rights and Wrongs: The Global Politics of Population Control and Contraceptive Choice.* San Francisco: Harper & Row.

Haub, Carl. 1987. "Understanding Population Projections." *Population Bulletin,* vol. 42, no. 4, 1–41.

Haupt, Arthur, and Thomas T. Kane. 1985. *The Population Handbook: International.* 2d ed. Washington, D.C.: Population Reference Bureau.

Heathcote, Willimas. 1991. *Autogeddon.* London: Jonathan Cape.

Honachefsky, William B. 1992. *Land Planner's Environmental Handbook.* New York: Noyes.

Jacobs, Jane. 1984. *Cities and the Wealth of Nations.* New York: Random House.

Jacobson, Jodi. 1991. *Women's Reproductive Health: The Silent Emergency.* Washington, D.C.: Worldwatch Institute.

Jones, Elsie F., et al. 1986. *Teenage Pregnancy in Industrialized Countries.* New Haven, Conn.: Yale University Press.

Keyfitz, Nathan, and Wilhelm Finger. 1991. *World Population Growth and Aging: Demographic Trends in the Late 20th Century.* Chicago: University of Chicago Press.

Lowe, Marcia D. 1990. *Alternatives to the Automobile: Transport for Living Cities.* Washington, D.C.: Worldwatch Institute.

Lowe, Marcia D. 1991. *Shaping Cities: The Environmental and Human Dimensions.* Washington, D.C.: Worldwatch Institute.

Lowe, Marcia D. 1993. "Rediscovering Rail." In Lester R. Brown, et al. *State of the World 1993.* New York: Norton, 120–38.

Lowe, Marcia D. 1994. "Reinventing Transport." In Lester R. Brown, et al. *State of the World 1994.* New York: Norton, 81–98.

MacKenzie, James J., et al. 1992. *The Going Rate: What It Really Costs to Drive.* Washington, D.C.: World Resources Institute.

Makower, Joel. 1992. *The Green Commuter.* Washington, D.C.: National Press.

Mantrell, Michael L., et al. 1989. *Creating Successful Communities: A Guidebook to Growth Management Strategies.* Covelo, Calif.: Island Press.

McFalls, Joseph A., Jr. 1991. "Population: A Lively Introduction." *Population Bulletin,* vol. 46, no. 2, 1–43.

McHarg, Ian L. 1969. *Design with Nature.* Garden City, N.Y.: Natural History Press.

Morris, David. 1982. *Energy and the Transformation of Urban America.* San Francisco: Sierra Club Books.

Nadis, Stephen, and James J. Mackenzie. 1992. *Car Trouble.* Washington, D.C.: World Resources Institute.

Nafis, Sadik, ed. 1991. *Population Policies and Programmes: Lessons Learned from Two Decades of Experience.* New York: United Nations Population Fund.

Olshansky, S. Jay, et al. 1993. "The Aging of the Human Species." *Scientific American,* April, 46–52.

Population Reference Bureau. 1990. *World Population: Fundamentals of Growth.* Washington, D.C.: Population Reference Bureau.

Population Reference Bureau. Annual. *World Population Data Sheet.* Washington, D.C.: Population Reference Bureau.

Register, Richard. 1992. *Ecocities.* Berkeley, Calif.: North Atlantic Books.

Renner, Michael. 1988. *Rethinking the Role of the Automobile.* Washington, D.C.: Worldwatch Institute.

Ryn, Sin van der, and Peter Calthorpe. 1986. *Sustainable Communities: A New Design Synthesis for Cities, Suburbs, and Towns.* San Francisco: Sierra Club.

Simon, Julian L. 1989. *Population Matters: People, Resources, Environment, and Immigration.* New Brunswick, N.J.: Transaction.

Thornton, Richard D. 1991. "Why the U.S. Needs a MAGLEV System." *Technology Review,* April, 31–42.

Tien, H. Yuan, et al. 1992. "China's Demographic Dilemmas." *Population Bulletin,* vol. 47, no. 1, 1–44.

Todd, John, and Nancy Jack Todd. 1993. *From Ecocities to Living Machines: Precepts for Sustainable Technologies.* Berkeley, Calif.: North Atlantic Books.

Todd, John, and George Tukel. 1990. *Reinhabiting Cities and Towns: Designing for Sustainability.* San Francisco: Planet/Drum Foundation.

United Nations. 1991. *Consequences of Rapid Population Growth in Developing Countries.* New York: United Nations.

United Nations. 1992. *Long-Range World Population Projections: Two Centuries of Population Growth, 1950–2150.* New York: United Nations.

United Nations Population Division. 1991. *World Urbanization Prospects.* New York: United Nations.

Walter, Bob, et al., eds. 1992. *Sustainable Cities: Concepts and Strategies for Eco-City Development.* Los Angeles, Calif.: Eco-Home Media.

Weber, Susan, ed. 1988. *USA by Numbers: A Statistical Portrait of the United States.* Washington, D.C.: Zero Population Growth.

Weeks, John R. 1992. *Population: An Introduction to Concepts and Issues.* 5th ed. Belmont, Calif.: Wadsworth.

Yang, Linda. 1990. *The City Gardener's Handbook.* New York: Random House.

Zero Population Growth. 1990. *Planning the Ideal Family: The Small Family Option.* Washington, D.C.: Zero Population Growth.

Zuckerman, Wolfgang. 1991. *End of the Road: The World Car Crisis and How We Can Solve It.* Post Mills, Vt.: Chelsea Green.

7 / ENVIRONMENTAL ECONOMICS AND POLITICS

Andersen, Terry, and Donald R. Leal. 1991. *Free-Market Environmentalism.* Boulder, Colo.: Westview Press.

Anderson, Victor. 1991. *Alternative Economic Indicators.* New York: Routledge.

Arnold, Ron, and Alan Gottlieb. 1993. *Trashing the Economy: How Runaway Environmentalism Is Wrecking America.* Bellevue, Wash.: Free Enterprise Press.

Atkinson, Adrian. 1991. *Principles of Political Ecology.* London: Belhaven Press.

Basta, Nicholas. 1991. *The Environmental Career Guide: Job Opportunities with the Earth in Mind.* New York: Wiley.

Bennett, Steven J. 1991. *Ecopreneuring: The Green Guide to Small Business Opportunities from the Environmental Revolution.* New York: John Wiley.

Berle, Gustav. 1991. *The Green Entrepreneur.* New York: McGraw-Hill.

Bowden, Elbert V. 1990. *Principles of Economics: Theory, Problems, Policies.* 5th ed. Cincinnati: South-Western.

Brown, Lester R., et al. 1991. *Saving the Planet: How to Shape an Environmentally Sustainable Global Economy.* New York: Norton.

Brundtland, G. H., et al. 1987. *Our Common Future: World Commission on Environment and Development.* New York: Oxford University Press.

Cairncross, Frances. 1992. *Costing the Earth.* Boston, Mass.: Harvard Business School.

Caldwell, Lynton K. 1990a. *Between Two Worlds: Science, the Environmental Movement, and Policy Choice.* New York: Cambridge University Press.

Caldwell, Lynton K. 1990b. *International Environmental Policy.* 2d ed. Durham, N.C.: Duke University Press.

Campbell, Monica E., and William M. Glenn. 1982. *Profit from Pollution Prevention.* Willowdale, Ontario: Firefly Books.

Cohn, Susan. 1992. *Green at Work: Finding a Business Career that Works for the Environment.* Covelo, Calif.: Island Press.

Constanza, Robert, ed. 1992. *Ecological Economics: The Science and Management of Sustainability.* New York: Columbia University Press.

Court, T. de la. 1990. *Beyond Bruntland: Green Development in the 1990s.* London: Zed Books.

Dahlberg, Kenneth A., et al. 1985. *Environment and the Global Arena.* Durham, N.C.: Duke University Press.

Daly, Herman E. 1991. *Steady-State Economics.* 2d ed. Covelo, Calif.: Island Press.

Daly, Herman E., and John B. Cobb, Jr. 1989. *For the Common Good: Redirecting the Economy Toward Community, the Environment, and a Sustainable Future.* Boston: Beacon Press.

Daly, Herman E., and Kenneth W. Townsend, eds. 1993. *Valuing the Earth: Economics, Ecology, Ethics.* Cambridge, Mass.: MIT Press.

Davis, John. 1991. *Greening Business: Managing for Sustainable Development.* New York: Basil Blackwell.

Day, David. 1990. *The Environmental Wars: Reports from the Front Line.* New York: St. Martin's Press.

Douthwaite, Richard. 1992. *The Growth Illusion.* Devon, UK: Green Books.

Durning, Alan T. 1989. *Poverty and the Environment: Reversing the Downward Spiral.* Washington, D.C.: Worldwatch Institute.

Durning, Alan T. 1992. *How Much Is Enough? The Consumer Society and the Earth.* New York: Norton.

Elkington, John, et al. 1990. *The Green Consumer.* New York: Penguin Books.

Elkins, Paul, and Jakob von Uexhull. 1992. *Grassroots Movements for Global Change.* New York: Routledge.

Environmental Careers Organization. 1993. *The New Complete Guide to Environmental Careers.* Covelo, Calif.: Island Press.

Flavin, Christopher, and John E. Young. 1993. "Shaping the Next Industrial Revolution." In Lester R. Brown, et al. *State of the World 1993.* New York: Norton, 180–99.

Foreman, Dave. 1991. *Confessions of an Eco-Warrior.* New York: Harmony Books.

French, Hilary F. 1992. *After the Earth Summit: The Future of Environmental Governance.* Washington, D.C.: Worldwatch Institute.

French, Hilary F. 1993. "Reconciling Trade and Environment." In Lester R. Brown, et al. *State of the World 1993.* New York: Norton, 158–79.

Garbarino, James. 1992. *Toward a Sustainable Society: An Economic, Social and Environmental Agenda for Our Children's Future.* Chicago: Noble.

George, Susan. 1992. *The Debt Boomerang: How Third World Debt Harms Us All.* Boulder, Colo.: Westview Press.

Georgescu-Roegen, Nicholas. 1971. *The Entropy Law and the Economic Process.* Cambridge, Mass.: Harvard University Press.

Ghandi, M. K. 1961. *Non-Violent Resistance.* New York: Schocken.

Gottlieb, Robert. 1993. *Forcing the Spring: The Transformation of the Environmental Movement.* Covelo, Calif.: Island Press.

Greenpeace. 1993. *The Greenpeace Guide to Anti-Environmental Organizations.* Berkeley, Calif.: Odionan Press.

Greider, William. 1992. *Who Will Tell the People? The Betrayal of American Democracy.* New York: Simon & Schuster.

Hall, Bob. 1990. *Environmental Politics: Lessons from the Grassroots.* Durham, N.C.: Institute for Southern Studies.

Hawken, Paul. 1993. *The Ecology of Commerce.* New York: Harper-Collins.

Henderson, Hazel. 1988. *The Politics of the Solar Age.* Chicago: Knowledge Systems.

Henderson, Hazel. 1991. *Paradigms in Progress: Life Beyond Economics.* Chicago: Knowledge Systems.

Henning, Daniel H., and William R. Mangun. 1989. *Managing the Environmental Crisis.* Durham, N.C.: Duke University Press.

Hirschorn, Joel S., and Kirsten U. Oldenberg. 1990. *Prosperity Without Pollution: The Prevention Strategy for Industry and Consumers.* New York: Van Nostrand Reinhold.

Institute for Local Self-Reliance. 1990. *Proven Profits from Pollution Prevention.* Washington, D.C.: Institute for Local Self-Reliance.

Irvine, Sandy, and A. Ponton. 1988. *A Green Manifesto.* London: Optima.

Jacobs, Michael. 1991. *The Green Economy: Environment, Sustainable Development, and Politics.* New York: Pluto Press.

Johnson, R. J. 1990. *Environmental Problems: Nature, Economy, and State.* New York: Belhaven Press.

Kazis, Richard, and Richard L. Grossman. 1991. *Fear at Work: Job Blackmail, Labor, and the Environment.* Philadelphia: New Society.

Keene, Ann T. 1993. *Observers and Protectors of Nature.* New York: Oxford University Press.

Lamay, Craig L., and Everette E. Dennis, eds. 1991. *Media and the Environment.* Covelo, Calif.: Island Press.

Landy, Marc K., et al. 1990. *The Environmental Protection Agency: Asking the Wrong Questions.* New York: Oxford University Press.

Lewis, Martin W. 1992. *Green Delusions: An Environmentalist Critique of Radical Environmentalism.* Durham, N.C.: Duke University Press.

List, Peter C. 1993. *Radical Environmentalism: Philosophy and Tactics.* Belmont, Calif.: Wadsworth.

Makower, Joel. 1993. *The E-Factor: The Bottom-Line Approach to Environmentally Responsible Business.* New York: Times Books.

Mathews, Jessica Tuchman, ed. 1989. "Redefining Security." *Foreign Affairs,* Spring, 162–77.

McCoy, Michael, and Patrick McCully. 1993. *The Road from Rio.* East Haven, Conn.: Inbook.

Mikesell, Raymond F., and Lawrence F. Williams. 1992. *International Banks and the Environment.* San Francisco: Sierra Club Books.

Myers, Norman. 1993. *Ultimate Security: The Environmental Basis of Political Stability.* New York: Norton.

Myers, Norman. 1994. "What Ails the Globe?" *International Wildlife,* Mar./Apr., 34–41.

Office of Technology Assessment. 1992. *Trade and Environment: Conflicts and Opportunities.* Washington, D.C.: Office of Technology Assessment.

Ophuls, William, and A. Stephen Boyan, Jr. 1992. *Ecology and the Politics of Scarcity Revisited: The Unravelling of the American Dream.* San Francisco: W. H. Freeman.

Paehilke, Robert C. 1989. *Environmentalism and the Future of Progressive Politics.* New Haven, Conn.: Yale University Press.

Paepke, C. Owen. 1993. *The Evolution of Progress: The End of Economic Growth and the Beginning of the Human Transformation.* New York: Random House.

Pearce, David, et al. 1991. *Blueprint 2: Greening the World Economy.* East Haven, Conn.: Earthscan.

Pearce, Fred. 1991. *Green Warriors: The People and Politics Behind the Environmental Revolution.* Cornwall, England: WEC Books.

Peet, John. 1992. *Energy and the Ecological Economics of Sustainability.* Covelo, Calif.: Island Press.

Ramphal, Shridath. 1992. *Our Country, the Planet: Forging a Partnership for Survival.* Washington, D.C.: Island Press.

Renner, Michael. 1989. *National Security: The Economic and Environmental Dimensions.* Washington, D.C.: Worldwatch Institute.

Renner, Michael. 1991. *Jobs in a Sustainable Economy.* Washington, D.C.: Worldwatch Institute.

Repetto, Robert. 1992. "Accounting for Environmental Assets." *Scientific American,* June, 94–100.

Repetto, Robert, et al. 1991. *Transforming Technology: An Agenda for Environmentally Sustainable Growth in the Twenty-First Century.* Washington, D.C.: World Resources Institute.

Repetto, Robert, et al. 1992. *Green Fees: How a Tax Shift Can Work for the Environment and the Economy.* Washington, D.C.: World Resources Institute.

Rifkin, Jeremy. 1991. *Biosphere Politics: A New Consciousness for a New Century.* New York: Crown.

Robertson, James. 1990. *Future Wealth: A New Economics for the 21st Century.* New York: Bootstrap Press.

Roddick, Anita. 1991. *Body and Soul: Profits With Principles.* New York: Crown.

Rogers, Adam. 1993. *The Earth Summit, A Planetary Reckoning.* Los Angeles, Calif.: Global View Press.

Rosenbaum, Walter A. 1990. *Environment, Politics, and Policy.* 2d ed. Washington, D.C.: Congressional Quarterly.

Sachs, Wolfgang, ed. 1993. *Global Ecology: A New Arena of Political Conflict.* Atlantic Highlands, N. J.: Zed Books.

Sanjor, William. 1992. *Why the EPA Is Like It Is and What Can Be Done About It.* Washington, D.C.: Environmental Research Foundation.

Schmidheiny, Stephan. 1992. *Changing Course: A Global Business Perspective on Development and the Environment.* Cambridge, Mass.: MIT Press.

Schumacher, E. F. 1973. *Small Is Beautiful: Economics As If the Earth Mattered.* New York: Harper & Row.

Shabecoff, Philip. 1993. *A Fierce Green Fire: The American Environmental Movement.* New York: Hill & Wang.

Shiva, Vandana. 1989. *Staying Alive: Women, Ecology, and Development.* London: Zed.

Smart, Bruce, ed. 1992. *Beyond Compliance: A New Industry View of the Environment.* Washington, D.C.: World Resources Institute.

Stead, W. Edward, and John Garner Stead. 1992. *Management for a Small Planet.* New York: Sage.

Tietenberg, Tom. 1992. *Environmental and Resource Economics.* 3d ed. Glenview, Ill.: Scott, Foresman.

Warner, David J. 1992. *Environmental Careers: A Practical Guide to Opportunities in the 1990s.* Boca Raton, Fla.: Lewis.

Watson, Paul. 1993. *Earthforce: An Earth Warrior's Guide to Strategy.* La Cañada, Calif.: Chaco Press.

Weinstein, Mirriam. 1993. *1993 Making a Difference College Guide: Education for a Better World.* San Anselmo, Calif.: Sage Press.

World Resources Institute. 1993. *A New Generation of Environmental Leadership: Action for the Environment and the Economy.* Washington, D.C.: World Resources Institute.

8 / RISK, TOXICOLOGY, AND HUMAN HEALTH

Ames, Bruce N., et al. 1987. "Ranking Possible Carcinogenic Hazards." *Science,* vol. 236, 271–79.

Aral, Sevgi O., and King K. Holmes. 1991. "Sexually Transmitted Diseases in the AIDS Era." *Scientific American,* vol. 264, no. 2, 62–69.

Bernarde, Melvin A. 1989. *Our Precarious Habitat: Fifteen Years Later.* New York: John Wiley.

Clarke, Lee. 1989. *Acceptable Risk? Making Decisions in a Toxic Environment.* Berkeley and Los Angeles: University of California Press.

Cohen, Mark N. 1989. *Health and the Rise of Civilization.* New Haven, Conn.: Yale University Press.

Crone, Hugh D. 1986. *Chemicals and Society.* New York: Cambridge University Press.

Douglas, Mary, and Aaron Wildavsky. 1982. *Risk and Culture.* Berkeley: University of California Press.

Efron, Edith. 1984. *The Apocalyptics: Cancer and the Big Lie.* New York: Simon & Schuster.

Environmental Protection Agency. 1987. *Unfinished Business: A Comparative Assessment of Environmental Problems.* Washington, D.C.: Environmental Protection Agency.

Environmental Protection Agency. 1990. *Reducing Risk: Setting Priorities and Strategies for Environmental Protection.* Washington, D.C.: Environmental Protection Agency.

Environmental Protection Agency. 1991. *The Environmental Challenge of the 1990s.* Washington, D.C.: Environmental Protection Agency.

Environmental Protection Agency. 1992. *Respiratory Health Effects of Passive Smoking: Lung Cancer and Other Disorders.* Washington, D.C.: Environmental Protection Agency.

Fischoff, Baruch, et al. 1984. *Acceptable Risk: Science and Determination of Safety.* New York: Cambridge University Press.

Foster, Kenneth R., et al., eds. 1993. *Phantom Risk: Scientific Inference and Law.* Cambridge, Mass.: MIT Press.

Fox, Michael W. 1992. *Superpigs and Wondercorn: The Brave New World of Biotechnology and Where It All May Lead.* New York: Lyons & Burford.

Freudenburg, William R. 1988. "Perceived Risk, Real Risk: Social Science and the Art of Probabilistic Risk Assessment." *Science,* vol. 242, 44–49.

Freudenthal, Ralph I., and Susan L. Freudenthal. 1989. *What You Need to Know to Live with Chemicals.* Greens Farms, Conn.: Hill & Garnett.

Hall, Ross Hume. 1990. *Health and the Global Environment.* Cambridge, Mass.: Basil Blackwell.

Harris, John. 1992. *Wonderwoman and Superman: The Ethics of Human Biotechnology.* New York: Oxford University Press.

Harris, Stephen L. 1990. *Agents of Chaos: Earthquakes, Volcanoes, and Other Natural Disasters.* Missoula, Mont.: Mountain Press.

Harte, John, et al. 1992. *Toxics A to Z: A Guide to Everyday Pollution Hazards.* Berkeley: University of California Press.

Hunter, Linda Mason. 1989. *The Healthy House: An Attic-to-Basement Guide to Toxin-Free Living.* Emmaus, Pa.: Rodale Press.

Imperato, P. J., and Greg Mitchell. 1985. *Acceptable Risks.* New York: Viking Press.

Kamarin, M. A. 1988. *Toxicology: A Primer on Toxicology Principles and Applications.* Boca Raton, Fla.: Lewis.

Lappé, Marc. 1991. *Chemical Deception: The Toxic Threat to Health and Environment.* San Francisco: Sierra Club.

Lewis, H. W. 1990. *Technological Risk.* New York: Norton.

Merrell, Paul, and Carol Van Strum. 1990. "Negligible Risk or Premeditated Murder?" *Journal of Pesticide Reform,* vol. 10, Spring, 20–22.

Misch, Ann. 1994. "Assessing Environmental Health Risks." In Lester R. Brown, et al. *State of the World 1994.* New York: Norton, 117–136.

Moeller, Dade W. 1992. *Environmental Health.* Cambridge, Mass.: Harvard University Press.

National Academy of Sciences. 1986. *Environment Tobacco Smoke: Measuring Exposures and Assessing Health Effects.* Washington, D.C.: National Academy Press.

National Academy of Sciences. 1989. *Diet and Health: Implications for Reducing Chronic Disease Risk.* Washington, D.C.: National Academy Press.

National Academy of Sciences. 1991a. *Malaria: Obstacles and Opportunities.* Washington, D.C.: National Academy Press.

National Academy of Sciences. 1991b. *A Safer Future: Reducing the Impacts of Natural Disasters.* Washington, D.C.: National Academy Press.

National Academy of Sciences. 1992a. *Eat for Life: The Food and Nutrition Board's Guide to Reducing Your Risk of Chronic Disease.* Washington, D.C.: National Academy Press.

National Academy of Sciences. 1992b. *The Social Impact of AIDS.* Washington, D.C.: National Academy Press.

National Academy of Sciences. 1993. *Science and Judgement of Risk Assessment.* Washington, D.C.: National Academy Press.

Nelkin, M. M., and M. S. Brown. 1984. *Workers at Risk: Voices from the Workplace.* Chicago: University of Chicago Press.

Olson, Steve. 1989. *Shaping the Future: Biology and Human Values.* Washington, D.C.: National Academy Press.

Ottoboni, M. Alice. 1991. *The Dose Makes the Poison: A Plain-Language Guide to Toxicology.* 2d ed. New York: Van Nostrand Reinhold.

Piller, Charles. 1991. *The Fail-Safe Society.* New York: Basic Books.

Rifkin, Jeremy. 1983. *Algeny.* New York: Viking/Penguin.

Rifkin, Jeremy. 1985. *Declaration of a Heretic.* Boston: Routledge & Kegan Paul.

Rodricks, Joseph V. 1992. *Calculated Risks: The Toxicity and Human Health Risks of Chemicals in the Environment.* New York: Cambridge University Press.

Russell, Dick, et al. 1992. *Inconclusive by Design: Waste, Fraud, and Abuse in Federal Environmental Health Research.* Boston: Environmental Health Network.

Shrader-Frechette, K. S. 1991. *Risk and Rationality.* Berkeley: University of California Press.

Suzuki, David, and Peter Knudtson. 1989. *Genethics: The Clash Between the New Genetics and Human Values.* Cambridge, Mass.: Harvard University Press.

U.S. Department of Health and Human Services. 1986. *The Health Consequences of Involuntary Smoking: A Report of the Surgeon General.* Rockville, Md.: U.S. Department of Health and Human Services.

U.S. Department of Health and Human Services. 1988. *The Surgeon General's Report on Nutrition and Health.* Washington, D.C.: Government Printing Office.

U.S. Department of Health and Human Services. Annual. *The Health Consequences of Smoking.* Washington, D.C.: Government Printing Office.

Whelan, Elizabeth M. 1993. *Toxic Terror: The Truth Behind the Cancer Scares.* 2d ed. Buffalo, N.Y.: Prometheus Books.

Wilson, Richard, and E. A. C. Crouch. 1987. "Risk Assessment and Comparisons: An Introduction." *Science,* vol. 236, 267–70.

9 / AIR

Borman, F. H. 1985. "Air Pollution and Forests: An Ecosystem Perspective." *BioScience,* vol. 35, no. 7, 434–41.

Bridgman, Howard. 1991. *Global Air Pollution: Problems for the 1990s.* New York: Belhaven Press.

Brookins, Douglas G. 1990. *The Indoor Radon Problem.* Irvington, N.Y.: Columbia University Press.

Bryner, Gary. 1992. *Blue Skies, Green Politics: The Clean Air Act of 1990.* Washington, D.C.: Congressional Quarterly Press.

Coffel, Steve, and Karyn Feiden. 1991. *Indoor Pollution.* New York: Random House.

Cole, Leonard D. 1993. *Element of Risk: The Politics of Radon.* Washington, D.C.: AAAS Press.

Elsom, Derek. 1987. *Atmospheric Pollution: Causes, Effects, and Control Policies.* Cambridge, Mass.: Basil Blackwell.

Environmental Protection Agency. 1988. *The Inside Story: A Guide to Indoor Air Quality.* Washington, D.C.: Environmental Protection Agency.

EPA Journal, vol. 17, no 1, 1991. Entire issue devoted to 1990 Clean Air Act.

French, Hilary F. 1990. *Clearing the Air: A Global Agenda.* Washington, D.C.: Worldwatch Institute.

Geller, H., et al. 1986. *Acid Rain and Energy Conservation.* Washington, D.C.: American Council for an Energy-Efficient America.

Hunter, Linda Mason. 1989. *The Healthy House: An Attic-to-Basement Guide to Toxin-Free Living.* Emmaus, Pa.: Rodale Press.

MacKenzie, James J., and Mohamed T. El-Ashry. 1990. *Air Pollution's Toll on Forests and Crops.* Washington, D.C.: World Resources Institute.

Mello, Robert A. 1987. *Last Stand of the Red Spruce.* Covelo, Calif.: Island Press.

Mohnen, Volker A. 1988. "The Challenge of Acid Rain." *Scientific American,* vol. 259, no. 2, 30–38.

Mossman, B. T., et al. 1990. "Asbestos: Scientific Developments and Implications for Public Policy." *Science,* vol. 251, 247–300.

National Academy of Sciences. 1988. *Air Pollution, the Automobile, and Human Health.* Washington, D.C.: National Academy Press.

Nero, Anthony V. 1992. "A National Strategy for Indoor Radon." *Issues in Science and Technology,* Fall, 33–40.

Office of Technology Assessment. 1989. *Catching Our Breath: Next Steps for Reducing Urban Ozone.* Washington, D.C.: Government Printing Office.

Pawlick, Thomas. 1986. *A Killing Rain: The Global Threat of Acid Precipitation.* San Francisco: Sierra Club Books.

Regens, James L., and Robert W. Rycroft. 1988. *The Acid Rain Controversy.* Pittsburgh: University of Pittsburgh Press.

Smith, William H. 1991. "Air Pollution and Forest Damage." *Chemistry and Engineering News,* November 11, 30–43.

10 / CLIMATE, GLOBAL WARMING, AND OZONE LOSS

American Forestry Association. 1993. *Forests and Global Warming.* vol. 1. Washington, D.C.: American Forestry Association.

Ausubel, Jesse H. 1991. "A Second Look at the Impacts of Climate Change." *American Scientist,* vol. 79, 210–21.

Balling, Robert C., Jr. 1992. *The Heated Debate: Greenhouse Predictions Versus Climate Reality.* San Francisco: Pacific Research Institute for Public Policy.

Barry, R. C., and J. Chorley. 1987. *Atmosphere, Weather, and Climate.* 5th ed. London: Methuen.

Bates, Albert K. 1990. *Climate in Crisis: The Greenhouse Effect and What We Can Do.* Summertown, Tenn.: The Book Publishing Company.

Benedick, Richard Eliot. 1991. *Ozone Diplomacy: New Directions in Safeguarding the Planet.* Cambridge, Mass.: Harvard University Press.

Bernard, Harold W., Jr. 1993. *Global Warming Unchecked: Signs to Watch For.* Bloomington: University of Indiana Press.

Bloch, Ben, and Harold Lyons. 1993. *Apocalypse Not: Science, Economics, and Environmentalism.* Washington, D.C.: Cato Institute.

Broome, John. 1992. *Counting the Cost of Global Warming.* Isle of Harris, UK: White Horse Press.

Cogan, Douglas. 1992. *The Greenhouse Gambit: Business and Investment Responses to Climate Change.* Washington, D.C.: Investor Responsibility Research Center.

Council for Agricultural Science and Technology. 1992. *Preparing U.S. Agriculture for Global Climate Change.* Ames, Iowa: Council for Agricultural Science and Technology.

Dotto, Lydia. 1990. *Thinking the Unthinkable: Civilization and Rapid Climate Change.* Waterloo, Ontario: Wilfrid Lanier University Press.

Edgerton, Lynne T. 1990. *The Rising Tide: Global Warming and World Sea Levels.* Covelo, Calif.: Island Press.

Environmental Protection Agency. 1988. *The Potential Effects of Global Climate Change on the United States.* Washington, D.C.: Environmental Protection Agency.

Environmental Protection Agency. 1989. *Policy Options for Stabilizing Global Climate.* Washington, D.C.: Environmental Protection Agency.

Fisher, David E. 1990. *Fire and Ice: The Greenhouse Effect, Ozone Depletion, and Nuclear Winter.* New York: Harper & Row.

Fishman, Albert, and Robert Kalish. 1990. *Global Alert: The Ozone Pollution Crisis.* New York: Plenum.

Flavin, Christopher. 1989. *Slowing Global Warming: A Worldwide Strategy.* Washington, D.C.: Worldwatch Institute.

Gates, David M. 1993. *Climate Change and Its Biological Consequences.* Sunderland, Mass.: Sinauer.

Graedel, T. C., and Paul J. Cutzen. 1992. *Atmospheric Change.* New York: W. H. Freeman.

Greenhouse Crisis Foundation. 1990. *The Greenhouse Crisis: 101 Ways to Save the Earth.* Washington, D.C.: Greenhouse Crisis Foundation.

Hileman, Bette. 1992. "Web of Interactions Makes It Difficult to Untangle Global Warming Data." *Chemistry & Engineering News,* April 27, 7–19.

Intergovernmental Panel on Climate Change (IPCC). 1990. *Climate Change: The IPCC Assessment.* New York: Cambridge University Press.

Intergovernmental Panel on Climate Change (IPCC). 1992. *The Supplementary Report to the IPCC Scientific Assessment.* New York: Cambridge University Press.

Jones, Philip D., and Tom M. L. Wiglet. 1990. "Global Warming Trends." *Scientific American,* August, 84–91.

Karplus, Walter J. 1992. *The Heavens Are Falling: The Scientific Prediction of Catastrophes in Our Time.* New York: Plenum.

Lovins, Amory B., et al. 1989. *Least-Cost Energy: Solving the CO_2 Problem.* 2d ed. Snowmass, Colo.: Rocky Mountain Institute.

Lyman, Francesca, et al. 1990. *The Greenhouse Trap: What We're Doing to the Atmosphere and How We Can Slow Global Warming.* Washington, D.C.: World Resources Institute.

McKibben, Bill. 1989. *The End of Nature.* New York: Random House.

Mintzer, Irving, and William R. Moomaw. 1991. *Escaping the Heat Trap: Probing the Prospects for a Stable Environment.* Washington, D.C.: World Resources Institute.

Mintzer, Irving, et al. 1990. *Protecting the Ozone Shield: Strategies for Phasing Out CFCs During the 1990s.* Washington, D.C.: World Resources Institute.

National Academy of Sciences. 1989. *Ozone Depletion, Greenhouse Gases, and Climate Change.* Washington, D.C.: National Academy Press.

National Academy of Sciences. 1990a. *Confronting Climate Change.* Washington, D.C.: National Academy Press.

National Academy of Sciences. 1990b. *Sea Level Change.* Washington, D.C.: National Academy Press.

National Academy of Sciences. 1992. *Global Environmental Change.* Washington, D.C.: National Academy Press.

National Audubon Society. 1990. *CO_2 Diet for a Greenhouse Planet: A Citizen's Guide to Slowing Global Warming.* New York: National Audubon Society.

Office of Technology Assessment. 1991. *Changing by Degrees: Steps to Reduce Greenhouse Gases.* Washington, D.C.: Government Printing Office.

Oppenheimer, Michael, and Robert H. Boyle. 1990. *Dead Heat: The Race Against the Greenhouse Effect.* New York: Basic Books.

Peters, Robert L., and Thomas E. Lovejoy, eds. 1992. *Global Warming and Biological Diversity.* New Haven, Conn.: Yale University Press.

Ray, Dixie Lee, and Lou Guzzo. 1993. *Environmental Overkill: Whatever Happened to Common Sense?* Washington, D.C.: Regnery Gateway.

Revkin, Andrew. 1992. *Global Warming: Understanding the Forecast.* New York: Abbeville Press.

Roan, Sharon L. 1989. *Ozone Crisis: The 15-Year Evolution of a Sudden Global Emergency.* New York: John Wiley.

Rowland, F. Sherwood. 1989. "Chlorofluorocarbons and the Depletion of Stratospheric Ozone." *American Scientist,* vol. 77, 36–45.

Rubin, G., et al. 1992. "Realistic Mitigation Options for Global Warming." *Science,* vol. 257, 148–49, 26–66.

Schneider, Stephen H. 1987. "Climate Modeling." *Scientific American,* vol. 256, no. 5, 72–80.

Schneider, Stephen. 1989. *Global Warming: Are We Entering the Greenhouse Century?* New York: Random House.

Shea, Cynthia Pollack. 1988. *Protecting Life on Earth: Steps to Save the Ozone Layer.* Washington, D.C.: Worldwatch Institute.

Weiner, Jonathan. 1990. *The Next One Hundred Years: Shaping the Fate of Our Living Earth.* New York: Bantam.

Zurer, Pamela S. 1993 "Ozone Depletion's Recurring Surprises Challenge Atmospheric Scientists." *Chemistry & Engineering News,* May 24, 8–18.

11 / WATER

Allaby, Michael. 1992. *Water: Its Global Nature.* New York: Facts on File.

Anderson, T. L. 1986. *Water Rights: Scarce Resource Allocation, Bureaucracy, and the Environment.* San Francisco: Pacific Institute for Public Policy.

Ashworth, William. 1982. *Nor Any Drop to Drink.* New York: Summit Books.

Ashworth, William. 1986. *The Late, Great Lakes: An Environmental History.* New York: Alfred A. Knopf.

Boon, P. J., et al., eds. 1992. *River Conservation and Management.* New York: John Wiley.

Borgese, Elisabeth Mann. 1986. *The Future of the Oceans.* New York: Harvest House.

Bullock, David K. 1989. *The Wasted Ocean.* New York: Lyons & Burford.

Center for Marine Conservation. 1989. *The Exxon Valdez Oil Spill: A Management Analysis.* Washington, D.C.: Center for Marine Conservation.

Clarke, Robin. 1993. *Water: The International Crisis.* Cambridge, Mass.: MIT Press.

Colborn, Theodora E., et al. 1989. *Great Lakes, Great Legacy?* Washington, D.C.: Conservation Foundation.

Costner, Pat, and Glenna Booth. 1986. *We All Live Downstream: A Guide to Waste Treatment That Stops Water Pollution.* Berkeley, Calif.: Bookpeople.

Davidson, Art. 1990. *In the Wake of the Exxon Valdez.* San Francisco: Sierra Club.

Edmonson, W. T. 1991. *The Uses of Ecology: Lake Washington and Beyond.* Seattle: University of Washington Press.

El-Ashry, Mohamed, and Diana C. Gibbons, eds. 1988. *Water and Arid Lands of the Western United States.* Washington, D.C.: World Resources Institute.

Ellefson, Connie, et al. 1992. *Xeriscape Gardening: Water Conservation for the American Landscape.* New York: Macmillan.

Environmental and Energy Study Institute. 1993. *New Policy Directions to Sustain the Nation's Water Resources.* Washington, D.C.: Environmental and Energy Study Institute.

Environmental Protection Agency. 1990. *Citizen's Guide to Ground-Water Protection.* Washington, D.C.: Environmental Protection Agency.

Falkenmark, Malin, and Carl Widstand. 1992. "Population and Water Resources: A Delicate Balance." *Population Bulletin,* vol. 47, no. 3, 1–36.

Feldman, David Lewis. 1991. *Water Resources Management: In Search of an Environmental Ethic.* Baltimore, Md.: Johns Hopkins University Press.

Gabler, Raymond. 1988. *Is Your Water Safe to Drink?* New York: Consumer Reports Books.

Gleick, Peter H. 1993. *Water in Crisis: A Guide to the World's Freshwater Resources.* Berkeley, Calif.: Pacific Institute for Studies in Development, Environment, and Security.

Goldsmith, Edward, and Nicholas Hidyard, eds. 1986. *The Social and Environmental Effects of Large Dams.* (3 vols.). New York: John Wiley.

Gottlieb, Robert. 1988. *A Life of Its Own: The Politics and Power of Water.* New York: Harcourt Brace Jovanovich.

Gray, N. F. 1992. *Biology of Wastewater Treatment.* New York: Oxford University Press.

Hansen, Nancy R., et al. 1988. *Controlling Nonpoint-Source Water Pollution.* New York: National Audubon Society and The Conservation Society.

Horton, Tom, and William Eichbaum. 1991. *Turning the Tide: Saving the Chesapeake Bay.* Covelo, Calif.: Island Press.

Hundley, Norris, Jr. 1992. *The Great Thirst: Californians and Water, 1770s–1990s.* Berkeley: University of California Press.

Ingram, Colin. 1991. *The Drinking Water Book.* Berkeley, Calif.: Ten Speed Press.

Irwin, Frances H. 1989. "Integrated Pollution Control." *International Environmental Affairs,* vol. 1, no. 4, 255–74.

Ives, J. D., and B. Messeric. 1989. *The Himalayan Dilemma: Reconciling Development and Conservation.* London: Routledge.

Jefferies, Michael, and Derek Mills. 1991. *Freshwater Ecology: Principles and Applications.* New York: Belhaven Press.

Jorgensen, Eric P., ed. 1989. *The Poisoned Well: New Strategies for Groundwater Protection.* Covelo, Calif.: Island Press.

Keeble, John. 1991. *Out of the Channel: The Exxon Valdez Oil Spill in Prince William Sound.* New York: HarperCollins.

King, Jonathan. 1985. *Troubled Water: The Poisoning of America's Drinking Water.* Emmaus, Pa.: Rodale Press.

Knopman, Debra S., and Richard A. Smith. 1993. "20 Years of the Clean Water Act." *Environment,* vol. 35, no. 1, 17–20, 34–41.

Kotlyakov, V. M. 1991. "The Aral Sea Basin: A Critical Environmental Zone." *Environment,* vol. 33, no. 1, 4–9, 36–39.

Kourik, Robert. 1988. *Gray Water Use in the Landscape.* Santa Rosa, Calif.: Edible Publications.

Kourik, Robert. 1992. *Drip Irrigation.* Santa Rosa, Calif.: Edible Publications.

Marquardt, Sandra, et al. 1989. *Bottled Water: Sparkling Hype at a Premium Price.* Washington, D.C.: Environmental Policy Institute.

McCutcheon, Sean. 1992. *Electric Rivers: The Story of the James River Project.* New York: Paul & Co.

Mitchell, Bruce, ed. 1990. *Integrated Water Management.* New York: Belhaven Press.

Munasinghe, Mohan. 1992. *Water Supply and Environmental Management.* Boulder, Colo.: Westview Press.

Naiman, Robert J., ed. 1992. *Watershed Management: Balancing Sustainability and Environmental Change.* New York: Springer-Verlag.

National Academy of Sciences. 1986. *Drinking Water and Health.* Washington, D.C.: National Academy Press.

National Academy of Sciences. 1992. *Water Transfers in the West: Efficiency, Equity, and the Environment.* Washington, D.C.: National Academy Press.

Natural Resources Defense Council. 1989. *Ebb Tide for Pollution: Actions for Cleaning Up Coastal Waters.* Washington, D.C.: Natural Resources Defense Council.

Office of Technology Assessment. 1989. *Coping with Oiled Environments*. Washington, D.C.: Government Printing Office.

Patrick, Ruth, E. Ford, and J. Quarles, eds. 1987. *Groundwater Contamination in the United States*. Philadelphia: University of Pennsylvania Press.

Patrick, Ruth, et al. 1992. *Surface Water Quality: Have the Laws Been Successful?* Princeton: Princeton University Press.

Pearce, Fred. 1992. *The Dammed: Rivers, Dams, and the Coming World Water Crisis*. London: Bodley Head.

Postel, Sandra. 1992. *The Last Oasis: Facing Water Scarcity*. New York: Norton.

Reisner, Marc. 1986. *Cadillac Desert: The American West and Its Disappearing Water*. New York: Viking Press.

Reisner, Marc, and Sara Bates. 1990. *Overtapped Oasis: Reform or Revolution for Western Water*. Covelo, Calif.: Island Press.

Rocky Mountain Institute. 1990. *Catalog of Water-Efficient Technologies for the Urban/Residential Sector*. Old Snowmass, Colo.: Rocky Mountain Institute.

Sierra Club Defense Fund. 1989. *The Poisoned Well: New Strategies for Groundwater Protection*. Covelo, Calif.: Island Press.

Simon, Anne W. 1985. *Neptune's Revenge: The Ocean of Tomorrow*. New York: Franklin Watts.

Starr, Joyce R. 1991. "Water Wars." *Foreign Policy*, Spring, 12–45.

Waggoner, Paul E., ed. 1990. *Climate Change and U.S. Water Resources*. New York: John Wiley.

Wilkinson, Charles F. 1992. *Crossing the Next Meridian: Land, Water, and the Future of the West*. Covelo, Calif.: Island Press.

Worster, Donald. 1985. *Rivers of Empire: Water, Aridity, and the Growth of the American West*. New York: Pantheon.

Zwingle, Erla. 1993. "Ogallala Aquifer: Wellspring of the High Plains." *National Geographic*, March, 80–107.

12 / MINERALS AND SOIL

Brady, Nyle C. 1989. *The Nature and Properties of Soils*. 10th ed. New York: Macmillan.

Brown, Bruce, and Lane Morgan. 1990. *The Miracle Planet*. Edison, N.J.: W. H. Smith.

Brown, Lester R., and Edward C. Wolf. 1984. *Soil Erosion: Quiet Crisis in the World Economy*. Washington, D.C.: Worldwatch Institute.

Cameron, Eugene N. 1986. *At the Crossroads: The Mineral Problems of the United States*. New York: John Wiley.

Campbell, Stu. 1990. *Let It Rot: The Gardener's Guide to Composting*. Pownal, Vt.: Storey Communications.

Donahue, Roy, et al. 1990. *Soils and Their Management*. 5th ed. Petaluma, Calif.: Inter Print.

Dorr, Ann. 1984. *Minerals—Foundations of Society*. Montgomery County, Md.: League of Women Voters of Montgomery County Maryland.

Dregnue, Harold E. 1985. "Aridity and Land Degradation." *Environment*, vol. 27, no. 8, 33–39.

Ernst, W. G. 1990. *The Dynamic Planet*. New York: Columbia University Press.

Gordon, Robert B., et al. 1988. *World Mineral Exploration: Trends and Issues*. Washington, D.C.: Resources for the Future.

Grainger, Alan. 1990. *The Threatening Desert: Controlling Desertification*. East Haven, Conn.: Earthscan.

Hamblin, W. Kenneth. 1989. *The Earth's Dynamic Systems: A Textbook in Physical Geology*. 5th ed. New York: Macmillan.

Harben, Peter. 1992. "Strategic Minerals." *Earth*, July, 36–43.

Hillel, Daniel. 1992. *Out of the Earth: Civilization and the Life of the Soil*. New York: Free Press.

Keller, Edward A. 1988. *Environmental Geology*. 5th ed. Columbus, Ohio: Charles E. Merrill.

Little, Charles E. 1987. *Green Fields Forever: The Conservation Tillage Revolution in America*. Covelo, Calif.: Island Press.

Mainguet, Monique. 1991. *Desertification: Natural Background and Human Mismanagement*. New York: Springer-Verlag.

McLaren, Digby J., and Brian J. Skinner, eds. 1987. *Resources and World Development*. New York: John Wiley.

Mollison, Bill. 1990. *Permaculture*. Covelo, Calif.: Island Press.

National Academy of Sciences. 1986. *Soil Conservation*. (2 vols.). Washington, D.C.: National Academy Press.

Paddock, Joe, et al. 1987. *Soil and Survival: Land Stewardship and the Future of American Agriculture*. San Francisco: Sierra Club Books.

Pimentel, David, ed. 1993. *World Soil Erosion and Conservation*. New York: Cambridge University Press.

Tapp, B. A., and J. R. Watkins. 1989. *Energy and Mineral Resource Systems: An Introduction*. New York: Cambridge University Press.

Wild, Alan. 1993. *Soils and the Environment: An Introduction*. New York: Cambridge University Press.

Wilson, G. F., et al. 1986. *The Soul of the Soil: A Guide to Ecological Soil Management*. 2d ed. Quebec, Canada: Gaia Services.

Young, John E. 1992. *Mining the Earth*. Washington, D.C.: Worldwatch Institute.

Youngquist, Walter. 1990. *Mineral Resources and the Destinies of Nations*. Portland, Ore.: National Book.

13 / WASTES: REDUCTION AND PREVENTION

Agency for Toxic Substances and Disease Registry. 1988. *The Nature and Extent of Lead Poisoning in Children in the United States*. Atlanta: U.S. Department of Health and Human Services.

Allen, Robert. 1992. *Waste Not, Want Not*. London: Earthscan.

Blumberg, Louis, and Robert Grottleib. 1988. *War on Waste—Can America Win Its Battle with Garbage?* Covelo, Calif.: Island Press.

Bryant, Bunyan, and Paul Mohai, eds. 1992. *Race and the Incidence of Environmental Hazards: A Time for Discourse*. Boulder, Colo.: Westview Press.

Bullard, Robert D. 1990. *Dumping in Dixie: Race, Class, and Environmental Quality*. Boulder, Colo.: Westview Press.

Bullard, Robert D., ed. 1993. *Confronting Environmental Racism: Voices from the Grassroots*. Boston, Mass.: South End Press.

Carless, Jennifer. 1992. *Taking Out the Trash: A No-Nonsense Guide to Recycling*. Covelo, Calif.: Island Press.

Centers for Disease Control and Prevention. 1991. *Preventing Lead Poisoning in Young Children*. Atlanta, Ga.: Centers for Disease Control and Prevention.

Chepesiuk, Ron. 1991. "From Ash to Cash: The International Trade in Toxic Waste." *E Magazine*, July/August, 31–63.

Clarke, Marjorie J., et al. 1991. *Burning Garbage in the U.S.: Practice vs. State of the Art*. New York: Inform.

Clean Sites. 1991. *What Works? Alternative Strategies for Superfund Cleanups*. Alexandria, Va.: Clean Sites.

Cohen, Gary, and John O'Connor. 1990. *Fighting Toxics: A Manual for Protecting Family, Community, and Workplace*. Covelo, Calif.: Island Press.

Commoner, Barry. 1990. *Making Peace with the Planet*. New York: Pantheon.

Connett, Paul H. 1992. "The Disposable Society." In F. H. Bormann and Stephen R. Kellert, eds. *Ecology, Economics, Ethics*. New Haven, Conn.: Yale University Press, 99–122.

Costner, Pat, and Joe Thornton. 1989. *Sham Recyclers, Part 1: Hazardous Waste Incineration in Cement and Aggregate Kilns*. Washington, D.C.: Greenpeace.

Dadd, Debra Lynn. 1990. *Nontoxic, Natural & Earthwise*. Los Angeles: Jeremy Tarcher.

Denison, Richard A., and John Ruston. 1990. *Recycling and Incineration: Evaluating the Choices*. Covelo, Calif.: Island Press.

Dorfman, Mark H., et al. 1992. *Environmental Dividends: Cutting More Chemical Wastes*. New York: INFORM.

Durning, Alan Thein. 1992. *How Much Is Enough? The Consumer Society and the Future of the Earth*. New York: Norton.

Earth Works Group. 1990. *The Recycler's Handbook: Simple Things You Can Do*. Berkeley: Earth Works Press.

Environment Canada. 1990. *Reduction and Reuse: The First 2Rs of Waste Management*. Ottawa: Environment Canada.

Environmental Defense Fund. 1985. *To Burn or Not to Burn?* New York: Environmental Defense Fund.

Environmental Protection Agency. 1987. *The Hazardous Waste System*. Washington, D.C.: Environmental Protection Agency.

Environmental Protection Agency. 1990. *Methods to Manage and Control Plastic Wastes*. Washington, D.C.: Government Printing Office.

Environmental Protection Agency. 1992a. *Characterization of Municipal Solid Waste in the United States*. Washington, D.C.: Government Printing Office.

Environmental Protection Agency. 1992b. *The Consumer's Handbook for Reducing Solid Waste.* Washington, D.C.: Environmental Protection Agency.

Gibbs, Lois. 1982. *The Love Canal: My Story.* Albany, N.Y.: State University of New York Press.

Gibbs, Lois, and Will Collette. 1987. *Solid Waste Action Project Guidebook.* Arlington, Va.: Citizen's Clearinghouse for Hazardous Waste.

Goldman, Benjamin A. 1991. *The Truth About Where You Live: An Atlas for Action on Toxins and Mortality.* New York: Random House.

Gordon, Ben, and Peter Montague. 1989. *Zero Discharge: A Citizen's Toxic Waste Manual.* Washington, D.C.: Greenpeace.

Gourlay, K. A. 1992. *World of Waste: Dilemmas of Industrial Development.* Atlantic Heights, N.J.: Zed Books.

Harte, John, et al. 1991. *Toxics A to Z: A Guide to Everyday Pollution Hazards.* Berkeley: University of California Press.

Hershkowitz, Allen, and Eugene Salermi. 1987. *Garbage Management in Japan: Leading the Way.* New York: INFORM.

Hilz, Christoph. 1993. *The International Toxic Waste Trade.* New York: Van Nostrand Reinhold.

Hofrichter, Richard, ed. 1993. *Toxic Struggles: The Theory and Practice of Environmental Justice.* Philadelphia, Penn.: New Society Publishers.

Institute for Local Self-Reliance. 1990. *Proven Profits from Pollution Prevention.* Washington, D.C.: Institute for Local Self-Reliance.

Kenworthy, Lauren, and Eric Schaeffer. 1990. *A Citizen's Guide to Promoting Toxic Waste Reduction.* New York: INFORM.

Kharbanda, O. P., and E. A. Stallworthy. 1990. *Waste Management: Toward a Sustainable Society.* New York: Auburn House.

Lappé, Marc. 1991. *Chemical Deception: The Toxic Threat to Health and Environment.* San Francisco: Sierra Club.

League of Women Voters. 1993a. *The Garbage Primer: A Handbook for Citizens.* New York: Lyons and Burford.

League of Women Voters. 1993b. *A Plastic Waste Primer.* New York: Lyons and Burford.

Lester, Stephen, and Brian Lipsett. 1989. *Track Record of the Hazardous Waste Incineration Industry.* Arlington, Va.: Citizen's Clearinghouse for Hazardous Waste.

Lipsett, B., and D. Farrell. 1990. *Solid Waste Incineration Status Report.* Arlington, Va.: Citizen's Clearinghouse for Hazardous Waste.

Mazmanian, Daniel, and David Morrell. 1992. *Beyond Superfailure: America's Toxics Policy for the 1990s.* Boulder, Colo.: Westview Press.

McLenighan, Valjean. 1991. *Sustainable Manufacturing: Saving Jobs, Saving the Environment.* Chicago: Center for Neighborhood Technology.

Minnesota Mining and Manufacturing. 1988. *Low- or Non-Pollution Technology Through Pollution Prevention.* St. Paul, Minn.: 3M Company.

Moyers, Bill. 1990. *Global Dumping Ground: The International Traffic in Hazardous Waste.* Cabin John, Md.: Seven Locks Press.

National Academy of Sciences. 1991. *Environmental Epidemiology Vol. 1: Public Health and Hazardous Wastes.* Washington, D.C.: National Academy Press.

National Toxics Campaign Fund. 1991. *The U.S. Military's Toxic Legacy: America's Worst Environmental Enemy.* Boston, Mass.: National Toxics Campaign Fund.

Newsday. 1989. *Rush to Burn: Solving America's Garbage Crisis?* Covelo, Calif.: Island Press.

Office of Technology Assessment. 1986. *Serious Reduction of Hazardous Waste.* Washington, D.C.: Government Printing Office.

Office of Technology Assessment. 1987. *From Pollution to Prevention: A Progress Report on Waste Reduction.* Washington, D.C.: Government Printing Office.

Office of Technology Assessment. 1992. *Green Products By Design.* Washington, D.C.: Government Printing Office.

Ortbal, John. 1991. *Buy Recycled! Your Practical Guide to the Environmentally Responsible Office.* Chicago: Services Marketing Group.

Piasecki, Bruce, and Gary Davis. 1987. *America's Future in Toxic Waste Management: Lessons from Europe.* New York: Quorum Books.

Platt, Brenda, et al. 1991. *Beyond 40 Percent: Record-Setting Recycling and Composting Programs.* Covelo, Calif.: Island Press.

Pollack, Cynthia. 1987. *Mining Urban Wastes: The Potential for Recycling.* Washington, D.C.: Worldwatch Institute.

Pollack, Stephanie. 1989. "Solving the Lead Dilemma." *Technology Review,* October, 22–31.

Poore, Patricia. 1993. "Is Garbage an Environmental Problem?" *Garbage,* Nov./Dec., 40–45.

Portney, Kent E. 1992. *Siting Hazardous Waste Treatment Facilities: The NIMBY Syndrome.* New York: Auburn House.

Postel, Sandra. 1987. *Defusing the Toxics Threat: Controlling Pesticides and Industrial Waste.* Washington, D.C.: Worldwatch Institute.

Rathje, William, and Cullen Murphy. 1992. *Rubbish! The Archaeology of Garbage: What Our Garbage Tells Us About Ourselves.* San Francisco: HarperCollins.

Regenstein, Lewis G. 1993. *Cleaning Up America the Poisoned.* Washington, D.C.: Acropolis Books.

Reynolds, Michael. 1990–91. *Earthship.* Vols. I and II. Taos, N.M.: Solar Survival Architecture.

Setterberg, Fred, and Lonny Shavelson. 1993. *Toxic Nation: The Fight to Save Our Communities from Chemical Contamination.* New York: Wiley.

Shulman, Seth. 1992. *The Threat at Home: Confronting the Toxic Legacy of the U.S. Military.* Boston: Beacon Press.

Theodore, Louis, and Young C. McGuinn. 1992. *Pollution Prevention.* New York: Van Nostrand Reinhold.

Water Pollution Control Federation. 1989. *Household Hazardous Waste: What You Should and Shouldn't Do.* Alexandria, Va.: Water Pollution Control Federation.

Whelan, Elisabeth M. 1993. *Toxic Terror.* 2d ed. Ottawa, Ill.: Jameson Books.

Wolf, Nancy, and Ellen Feldman. 1990. *Plastics: America's Packaging Dilemma.* Covelo, Calif.: Island Press.

Young, John E. 1991. *Discarding the Throwaway Society.* Washington, D.C.: Worldwatch Institute.

Young, John E. 1992. "Aluminum's Real Tab." *World Watch,* Mar./Apr., 26–34.

14 / FOOD RESOURCES

Aliteri, Miguel A., and Susanna B. Hecht. 1990. *Agroecology and Small Farm Development.* New York: CRC Press.

Ausubel, Kenny. 1994. *Seeds of Change.* New York: Harper-Collins.

Bartholomew, Mel. 1987. *Square Foot Gardening.* Emmaus, Pa.: Rodale Press.

Berry, Wendell. 1990. *Nature as Measure.* Berkeley, Calif.: North Point Press.

Berstein, Henry, et al., eds. 1990. *The Food Question: Profits Versus People.* East Haven, Conn.: Earthscan.

Bingaarts, John. 1994. "Can the Growing Human Population Feed Itself?" *Scientific American,* Mar., 36–42.

Brown, Lester R. 1994. "Facing Food Scarcity." In Lester Brown, et al. *State of the World 1994.* New York: Norton, 177–197.

Carroll, C. Ronald, et al. 1990. *Agroecology.* New York: McGraw-Hill.

Coleman, Elliot. 1992. *The New Organic Grower's Four-Season Harvest.* Post Mills, Vt.: Chelsea Green.

Crosson, Pierre R., and Norman J. Rosenberg. 1989. "Strategies for Agriculture." *Scientific American,* September, 128–35.

Dover, Michael J., and Lee M. Talbot. 1988. "Feeding the Earth: An Agroecological Solution." *Technology Review,* Feb./Mar., 27–35.

Doyle, Jack. 1985. *Altered Harvest: Agriculture, Genetics, and the Fate of the World's Food Supply.* New York: Viking Press.

Dunning, Alan B., and Holly W. Brough. 1991. *Taking Stock: Animal Farming and the Environment.* Washington, D.C.: Worldwatch Institute.

The Ecologist, vol. 21, no. 2, 1991. Entire issue devoted to world hunger.

Edwards, Clive A., et al., eds. 1990. *Sustainable Agricultural Systems.* Ankeny, Iowa: Soil and Water Conservation Society.

Faeth, Paul, et al. 1993. *Agricultural Policy and Sustainability: Case Studies from India, Chile, the Philippines, and the United States.* Washington, D.C.: World Resources Institute.

Foster, Phillips. 1992. *The World Food Problem.* Boulder, Colo.: Rienner.

Fowler, Cary, and Pat Mooney. 1990. *Shattering: Food, Politics, and the Loss of Genetic Diversity.* Tucson: University of Arizona Press.

Fukuoka, Masanobu. 1985. *The Natural Way of Farming: The Theory and Practice of Green Philosophy.* New York: Japan Publications.

Goering, Peter, et al. 1993. *From the Ground Up: Rethinking Industrial Agriculture.* Atlantic Highlands, N.J.: Zed Books.

Gordon, R. Conway, and Edward R. Barbier. 1990. *After the Green Revolution: Sustainable Agriculture for Development.* East Haven, Conn.: Earthscan.

Granatstein, David. 1988. *Reshaping the Bottom Line: On-Farm Strategies for a Sustainable Agriculture.* Stillwater, Minn.: Land Stewardship Project.

Hendry, Peter. 1988. "Food and Population: Beyond Five Billion." *Population Bulletin,* Apr., 1–55.

Hunger Project. 1985. *Ending Hunger: An Idea Whose Time Has Come.* New York: Praeger.

Jackson, Wes. 1980. *New Roots for Agriculture.* San Francisco: Friends of the Earth.

Jacobson, Michael, et al. 1991. *Safe Food: Eating Wisely in a Risky World.* Washington, D.C.: Planet Earth Press.

Jeavons, John. 1982. *How to Grow More Vegetables.* Berkeley, Calif.: Ten Speed Press.

Landau, Matthew. 1992. *Introduction to Aquaculture.* New York: Wiley.

Lappé, Francis M., et al. 1988. *Betraying the National Interest.* San Francisco: Food First.

League of Women Voters, 1991. *U.S. Farm Policy: Who Benefits? Who Pays? Who Decides?* Washington, D.C.: League of Women Voters.

McGoodwin, Russell. 1990. *Crisis in the World's Fisheries: People, Problems, and Politics.* Stanford, Calif.: Stanford University Press.

Mollison, Bill. 1990. *Permaculture: A Practical Guide for a Sustainable Future.* Covelo, Calif.: Island Press.

National Academy of Sciences. 1989. *Alternative Agriculture.* Washington, D.C.: National Academy Press.

National Academy of Sciences. 1992. *Marine Aquaculture: Opportunities for Growth.* Washington, D.C.: National Academy Press.

Pimentel, David, and Carl W. Hall. 1989 *Food and Natural Resources.* Orlando, Fla.: Academic Press.

Rifkin, Jeremy. 1992. *Beyond Beef: The Rise and Fall of the Cattle Culture.* New York: Dutton.

Ritchie, Mark. 1990. "GATT, Agriculture, and the Environment." *The Ecologist,* vol. 20, no. 6, 214–20.

Robbins, John. 1987. *Diet for a New America.* Walpole, N.H.: Stillpoint Publishing.

Rodale Press. 1984. *The Encyclopedia of Organic Gardening.* Emmaus, Pa.: Rodale Press.

Schell, Orville. 1984. *Modern Meat: Antibiotics, Hormones, and the Pharmaceutical Farm.* New York: Random House.

Shiva, Vandana. 1991. *The Violence of the Green Revolution.* Atlantic Highlands, N.J.: Zed Books.

Soil and Water Conservation Society. 1990. *Sustainable Agricultural Systems.* Ankeny, Iowa: Soil and Water Conservation Society.

Soule, Judith D., and Jon Piper. 1992. *Farming in Nature's Image: An Ecological Approach to Agriculture.* Covelo, Calif.: Island Press.

Tarrant, J. R., ed. 1991. *Farming and Food.* New York: Oxford University Press.

Todd, Nancy J., and John Todd. 1984. *Bioshelters, Ocean Arks, City Farming: Ecology as a Basis for Design.* San Francisco: Sierra Club Books.

Tudge, Colin. 1988. *Food Crops for the Future: Development of Plant Resources.* Oxford, England: Blackwell.

Whelan, Elizabeth M., and Frederick J. Stare. 1983. *The 100% Natural, Purely Organic, Cholesterol-Free, Megavitamin, Low-Carbohydrate Nutrition Hoax.* New York: Atheneum.

Wolf, Edward C. 1986. *Beyond the Green Revolution: New Approaches for Third World Agriculture.* Washington, D.C.: Worldwatch Institute.

Yang, Linda. 1990. *The City Gardener's Handbook: From Balcony to Backyard.* New York: Random House.

15 / PROTECTING FOOD RESOURCES: PESTICIDES AND PEST CONTROL

Bogard, William. 1989. *The Bhopal Tragedy: Language, Logic, and Politics in the Production of a Hazard.* Boulder, Colo.: Westview Press.

Bormann, F. Herbert, et al. 1993. *Redesigning the American Lawn.* New Haven, Conn.: Yale University Press.

Bosso, Christopher. 1987. *Pesticides and Politics.* Pittsburgh: University of Pittsburgh Press.

Briggs, Shirley, and the Rachel Carson Council. 1992. *Basic Guide to Pesticides: Their Characteristics and Hazards.* Washington, D.C.: Taylor & Francis.

Carson, Rachel. 1962. *Silent Spring.* Boston: Houghton Mifflin.

Dinham, Barbara. 1993. *The Pesticide Hazard: A Global Health and Environmental Audit.* New York: Zed Books.

Dover, Michael J. 1985. *A Better Mousetrap: Improving Pest Management for Agriculture.* Washington, D.C.: World Resources Institute.

Flint, Mary Louise. 1990. *Pests of the Garden & a Small Farm: A Grower's Guide to Using Less Pesticide.* Oakland, Calif.: ANR Publications.

Friends of the Earth. 1990. *How to Get Your Lawn and Garden Off Drugs.* Ottawa, Ontario: Friends of the Earth.

Garland, Ann W. 1989. *For Our Kids' Sake: How to Protect Your Child Against Pesticides.* San Francisco: Sierra Club Books.

Gips, Terry. 1987. *Breaking the Pesticide Habit.* Minneapolis: IASA.

Heylin, Michael, ed. 1991. "Pesticides: Costs Versus Benefits." *Chemistry & Engineering News,* 7 January, 27–56.

Horn, D. J. 1988. *Ecological Approach to Pest Management.* New York: Guilford Press.

Hynes, Patricia. 1989. *The Recurring Silent Spring.* New York: Pergamon Press.

Kourik, Robert. 1990. "Combatting Household Pests Without Chemical Warfare." *Garbage,* March/April, 22–29.

League of Women Voters. 1989. *America's Growing Dilemma: Pesticides in Food and Water.* Washington, D.C.: League of Women Voters.

Marco, G. J., et al. 1987. *Silent Spring Revisited.* Washington, D.C.: American Chemical Society.

Marquardt, Sandra. 1989. *Exporting Banned Pesticides: Fueling the Circle of Poison.* Washington, D.C.: Greenpeace.

National Academy of Sciences. 1986. *Pesticide Resistance: Strategies and Tactics for Management.* Washington, D.C.: National Academy Press.

National Academy of Sciences. 1993. *Pesticides in the Diet of Infants and Children.* Washington, D.C.: National Academy Press.

Pimentel, David, et al. 1991. "Environmental and Economic Effects of Reducing Pesticide Use." *BioScience,* vol. 41, no. 6, 402–9.

Pimentel, David, et al. 1992. "Environmental and Economic Cost of Pesticide Use." *BioScience,* vol. 42, no. 10, 750–60.

Preston-Mafham, Rod, and Ken Preston-Mafham. 1984. *Spiders of the World.* New York: Facts on File.

Schultz, Warren. 1989. *The Chemical-Free Lawn.* Emmaus, Pa.: Rodale Press.

Van den Bosch, Robert. 1978. *The Pesticide Conspiracy.* Garden City, N.Y.: Doubleday.

Wiles, Richard, and Christopher Campbell. 1993. *Pesticides in Children's Food.* Washington, D.C.: Environmental Working Group.

Yepsen, Roger B., Jr. 1987. *The Encyclopedia of Natural Insect and Pest Control.* Emmaus, Pa.: Rodale Press.

16 / BIODIVERSITY: SUSTAINING ECOSYSTEMS

Allin, Craig W. 1982. *The Politics of Wilderness Preservation.* Westport, Conn.: Greenwood.

Anderson, Patrick. 1989. "The Myth of Sustainable Logging: The Case for a Ban on Tropical Timber Imports." *The Ecologist,* vol. 19, no. 5, 166–68.

Barber, Charles V., et al. 1993. *Breaking the Deadlock: Obstacles to Forest Reform in Indonesia and the United States.* Washington, D.C.: World Resources Institute.

Barber, Chip. 1991. *Cutting Our Losses: Policy Reform to Sustain Tropical Forest Resources.* Washington, D.C.: World Resources Institute.

Booth, Douglas E. 1993. *Valuing Nature: The Decline and Preservation of Old Growth Forests.* Lantham, Md.: University Press of America.

Browder, John O. 1992. "The Limits of Extractivism." *BioScience,* vol. 42, no. 3, 174–82.

Burger, Julina. 1990. *The Gaia Atlas of First Peoples.* New York: Anchor Books.

Caufield, Catherine. 1985. *In the Rainforest.* New York: Alfred A. Knopf.

Chagnon, Napoleon A. 1992. *Yanomano: The Last Days of Eden.* New York: Harcourt Brace Jovanovich.

Chase, Alston. 1986. *Playing God in Yellowstone: The Destruction of America's First National Park.* New York: Atlantic Monthly Press.

Collins, Mark, ed. 1990. *The Last Rainforests: A World Conservation Atlas.* Emmaus, Pa.: Rodale Press.

Dietrich, William. 1992. *The Final Forest.* New York: Simon & Schuster.

DiSilvestro, Roger L. 1993. *Reclaiming the Last Wild Places: A New Agenda for Biodiversity.* New York: Wiley.

Durning, Alan Thein. 1992. *Guardians of the Land: Indigenous Peoples and the Health of the Earth.* Washington, D.C.: Worldwatch Institute.

Durning, Alan Thein. 1993. *Saving the Forests: What Will It Take?* Washington, D.C.: Worldwatch Institute.

Dysart, Benjamin, III, and Marion Clawson. 1988. *Managing Public Lands in the Public Interest.* New York: Praeger.

Earth Island Institute and Sierra Club Books. 1993. *Clearcut: The Tragedy of Industrial Forestry.* San Francisco, Calif.: Sierra Club Books.

Eckholm, Erik, et al. 1984. *Fuelwood: The Energy Crisis That Won't Go Away.* Washington, D.C.: Earthscan.

Ervin, Keith. 1989. *Fragile Majesty: The Battle for North America's Last Great Forest.* Seattle: The Mountaineers.

Foreman, Dave, and Howie Wolke. 1989. *The Big Outside.* Tucson: Nedd Ludd Books.

Friends of the Earth. 1992. *The Rainforest Harvest: Sustainable Strategies for Saving Tropical Forests.* Washington, D.C.: Friends of the Earth.

Fritz, Edward. 1983. *Sterile Forest: The Case Against Clearcutting.* Austin, Texas: Eakin Press.

Frome, Michael. 1974. *The Battle for the Wilderness.* New York: Praeger.

Frome, Michael. 1983. *The Forest Service.* Boulder, Colo.: Westview Press.

Frome, Michael. 1992. *Regreening the National Parks.* Tucson: University of Arizona Press.

Goodland, Robert, ed. 1990. *Race to Save the Tropics: Ecology and Economics for a Sustainable Future.* Covelo, Calif.: Island Press.

Gradwohl, Judith, and Russell Greenberg. 1988. *Saving the Tropical Forests.* Covelo, Calif.: Island Press.

Greenpeace. 1990. *The Greenpeace Guide to Paper.* Washington, D.C.: Greenpeace.

Head, Suzanne, and Robert Heinzman. 1990. *Lessons of the Rainforest.* San Francisco: Sierra Club Books.

Hendee, John, et al. 1991. *Principles of Wilderness Management.* Golden, Colo.: Fulcrum Publishing.

Hess, Karl, Jr. 1992. *Visions Upon the Land: Man and Nature on the Western Range.* Covelo, Calif.: Island Press.

Hunter, Malcolm L., Jr. 1990. *Wildlife, Forests, and Forestry: Principles of Managing Forests for Biodiversity.* Englewood Cliffs, N.J.: Prentice-Hall.

International Union for Conservation of Nature and Natural Resources (IUCN). 1992. *Tropical Deforestation and the Extinction of Species.* Gland, Switzerland: IUCN.

Jacobs, Lynn. 1992. *Waste of the West: Public Lands Ranching.* Tucson, Ariz.: Lynn Jacobs.

Johnson, Nels, and Brice Cabarle. 1993. *Looking Ahead: Sustainable Natural Forest Management in the Humid Tropics.* Washington, D.C.: World Resources Institute.

Kelly, David, and Gary Braasch. 1988. *Secrets of the Old Growth Forest.* Salt Lake City: Peregrine Smith.

Lansky, Mitch. 1992. *Beyond the Beauty Strip: Saving What's Left of Our Forests.* Gardiner, Maine: Tilbury House.

Leopold, Aldo. 1949. *A Sand County Almanac.* New York: Oxford University Press.

Mahony, Rhona. 1992. "Debt-for-Nature Swaps: Who Really Benefits?" *The Ecologist,* vol. 22, no. 3, 97–103.

Manning, Richard. 1993. *Last Stand.* New York: Penguin Books.

Margolis, Marc. 1992. *The Last New World: The Conquest of the Amazon Frontier.* New York: Norton.

Maser, Chris. 1989. *Forest Primeval.* San Francisco: Sierra Club.

Maybury-Lewis, David. 1992. *Millennium: Tribal Wisdom and the Modern World.* New York: Viking Press.

Miller, Kenton, and Laura Tangley. 1991. *Trees of Life: Saving Tropical Forests and Their Biological Wealth.* Washington, D.C.: World Resources Institute.

Myers, Norman. 1984. *The Primary Source: Tropical Forests and Our Future.* New York: Norton.

Naar, Jon, and Alex J. Naar. 1993. *This Land Is Your Land: A Guide to North America's Endangered Ecosystems.* New York: Harper.

Nash, Roderick. 1982. *Wilderness and the American Mind.* 3d ed. New Haven, Conn.: Yale University Press.

National Parks and Conservation Association. 1988. *Blueprint for National Parks.* (9 vols.). Washington, D.C.: National Parks and Conservation Association.

National Park Service. 1992. *National Parks for the 21st Century.* Washington, D.C.: National Park Service.

Newman, Arnold. 1990. *The Tropical Rainforest: A World Survey of Our Most Valuable Endangered Habitats.* New York: Facts on File.

Norse, Elliot A. 1990. *Ancient Forests of the Pacific Northwest.* Covelo, Calif.: Island Press.

Oelschlager, Max. 1991. *The Idea of Wilderness from Prehistory to the Age of Ecology.* New Haven, Conn.: Yale University Press.

Office of Technology Assessment. 1992a. *Combined Summaries: Technologies to Sustain Tropical Forest Resources and Biological Diversity.* Washington, D.C.: Office of Technology Assessment.

Office of Technology Assessment. 1992b. *Forest Service Planning: Accommodating Uses, Producing Outputs, Protecting Ecosystems.* Washington, D.C.: Office of Technology Assessment.

Onis, Juan de. 1992. *The Green Cathedral: Sustainable Development of Amazonia.* New York: Oxford University Press.

O'Toole, Randal. 1987. *Reforming the Forest Service.* Covelo, Calif.: Island Press.

Panayotou, Theodore, and Peter S. Ashton. 1992. *Not By Timber Alone: Economics and Ecology for Sustaining Tropical Forests.* Covelo, Calif.: Island Press.

Patterson, Alan. 1990. "Debt for Nature Swaps and the Need for Alternatives." *Environment,* vol. 32, no. 10, 5–13, 31–32.

Perlin, John. 1989. *A Forest Journey: The Role of Wood in the Development of Civilization.* New York: Norton.

Pimentel, David, et al. 1992. "Conserving Biological Diversity in Agricultural/Forestry Systems." *BioScience,* vol. 42, no. 5, 354–62.

Postel, Sandra, and Lori Heise. 1988. *Reforesting the Earth.* Washington, D.C.: Worldwatch Institute.

Primack, Richard B. 1993. *Essentials of Conservation Biology.* Sunderland, Mass.: Sinauer.

Rainforest Action Network. 1991. *Wood User's Guide.* San Francisco, Calif.: Rainforest Action Network.

Reiss, Bob. 1992. *The Road to Extrema.* New York: Summit Books.

Repetto, Robert. 1990. "Deforestation in the Tropics." *Scientific American,* vol. 262, no. 4, 36–42.

Revkin, Andrew. 1990. *The Burning Season: The Murder of Chico Mendes and the Fight for the Amazon.* Boston: Houghton Mifflin.

Robinson, Gordon. 1987. *The Forest and the Trees: A Guide to Excellent Forestry.* Covelo, Calif.: Island Press.

Romme, William H., and Don G. Despain. 1989. "The Yellowstone Fires." *Scientific American,* vol. 261, no. 5, 37–46.

Runte, Alfred. 1987. *National Parks: The American Experience.* 2d ed. Lincoln: University of Nebraska Press.

Runte, Alfred. 1990. *Yosemite: The Embattled Wilderness.* Lincoln: University of Nebraska Press.

Runte, Alfred. 1991. *Public Lands, Public Heritage: The National Forest Idea.* Niwot, Colo.: Roberts Rinehart.

Rush, James. 1991. *The Last Tree: Reclaiming the Environment in Tropical Asia.* Boulder, Colo.: Westview Press.

Ryan, John C. 1992. *Life Support: Conserving Biological Diversity.* Washington, D.C.: Worldwatch Institute.

Shanks, Bernard. 1984. *This Land Is Your Land.* San Francisco: Sierra Club.

Shiva, Vandana. 1993. *Monocultures of the Mind: Perspectives on Biodiversity and Biotechnology.* Atlantic Highlands, N.J.: Zed Books.

Society of American Foresters. 1981. *Choices in Silviculture for American Forests.* Washington, D.C.: Society of American Foresters.

Stoddard, Charles H., and Glenn M. Stoddard. 1987. *Essentials of Forestry Practice.* 4th ed. New York: John Wiley.

Teitel, Martin. 1992. *Rain Forest in Your Kitchen: The Hidden Connection Between Extinction and Your Supermarket.* Covelo, Calif.: Island Press.

Thompson, Claudia. 1992. *Recycled Papers: The Essential Guide.* Cambridge, Mass.: MIT Press.

Tobin, Richard J. 1990. *The Expendable Future: U.S. Politics and the Protection of Biodiversity.* Durham, N.C.: Duke University Press.

Tree People. 1990. *The Simple Act of Planting a Tree.* Los Angeles: Jeremy Tarcher.

United Nations Development Programme. 1992. *Global Biodiversity Strategy.* Washington, D.C.: World Resources Institute.

Valentine, John E., ed. 1990. *Grazing Management.* San Diego: Academic Press.

Wellner, Pamela, and Eugene Dickey. 1991. *The Wood Users Guide.* San Francisco: Rainforest Action Network.

Wilcove, David S. 1988. *National Forests: Policies for the Future.* Vols. 1 & 2. Washington, D.C.: Wilderness Society.

Wilderness Society. 1988. *Ancient Forests: A Threatened Heritage.* Washington, D.C.: The Wilderness Society.

Wilderness Society. 1991. *Keeping It Wild: A Citizen Guide to Wilderness Management.* Washington, D.C.: Wilderness Society.

Wilson, E. O. 1992. *The Diversity of Life.* Cambridge, Mass.: Harvard University Press.

Wilson, E. O., ed. 1988. *Biodiversity.* Cambridge, Mass.: Harvard University Press.

World Resources Institute (WRI), The World Conservation Union (IUCN), and Wuerthner, George. 1989. *Yellowstone and the Fires of Change.* Salt Lake City: Dream Garden Press.

Zaslowsky, Dyan, and The Wilderness Society. 1986. *These American Lands.* New York: Henry Holt.

Zuckerman, Seth. 1991. *Saving Our Ancient Forests.* Federalsburg, Md.: Living Planet Press.

17 / BIODIVERSITY: SUSTAINING WILD SPECIES

Ackerman, Diane. 1991. *The Moon by Whalelight.* New York: Random House.

Barker, Rocky. 1993. *Saving All the Parts: Reconciling Economics and the Endangered Species Act.* Covelo, Calif.: Island Press.

Bonner, Raymond. 1993. *At the Hand of Man: Peril and Hope for Africa's Wildlife.* New York: Knopf.

Boo, Elizabeth. 1990. *Ecotourism: The Potential and the Pitfalls.* Vols. 1, 2. Washington, D.C.: World Wildlife Fund.

Cox, George W. 1993. *Conservation Ecology.* Dubuque, Iowa: Wm. C. Brown.

Credlund, Arthur G. 1983. *Whales and Whaling.* New York: Seven Hills Books.

DeBlieu, Jan. 1991. *Meant to Be Wild: The Struggle to Save Endangered Species Through Captive Breeding.* New York: Fulcrum.

DiSilvestro, Roger L. 1990. *Fight for Survival.* New York: John Wiley.

DiSilvestro, Roger L. 1992. *Rebirth of Nature.* New York: John Wiley.

Dunlap, Thomas R. 1988. *Saving America's Wildlife.* Princeton, N.J.: Princeton University Press.

Durrell, Lee. 1986. *State of the Ark: An Atlas of Conservation in Action.* Garden City, N.Y.: Doubleday.

Ehrlich, Paul, and Anne Ehrlich. 1981. *Extinction.* New York: Random House.

Eldredge, Niles. 1991. *The Miner's Canary.* New York: Prentice Hall.

Elliot, David K. 1986. *Dynamics of Extinction.* New York: John Wiley.

Gilbert, Frederick F., and Donald G. Dodds. 1992. *The Philosophy and Practice of Wildlife Management.* 2d ed. Malabar, Fla.: Robert E. Krieger.

Gray, Gary G. 1993. *Wildlife and People.* Champaign, Ill.: University of Illinois Press.

Grumbine, R. Edward. 1992. *Ghost Bears: Exploring the Biodiversity Crisis.* Covelo, Calif.: Island Press.

Hargrove, Eugene C. 1992. *The Animal Rights/Environmental Ethics Debate.* Ithaca: State University of New York.

Jasper, James M., and Dorothy Nelkin. 1992. *The Animal Rights Crusade: The Growth of a Moral Protest.* New York: Macmillan.

Kaufman, Les, and Kenneth Mallay. 1993. *The Last Extinction.* 2d ed. Cambridge, Mass.: MIT Press.

Klenig, John. 1992. *Valuing Life.* Princeton, N.J.: Princeton University Press.

Kohm, Kathryn A., ed. 1990. *Balancing on the Brink of Extinction: The Endangered Species Act and Lessons for the Future.* Covelo, Calif.: Island Press.

Koopowitz, Harold, and Hilary Kaye. 1983. *Plant Extinctions: A Global Crisis.* Washington, D.C.: Stone Wall Press.

Lackey, R. T., and L. A. Nielson. 1980. *Fisheries Management.* New York: John Wiley.

Leopold, Aldo. 1933. *Game Management.* New York: Charles Scribner's.

Livingston, John A. 1981. *The Fallacy of Wildlife Conservation.* Toronto: McClelland and Stewart.

Luoma, Jon. 1987. *A Crowded Ark: The Role of Zoos in Wildlife Conservation.* Boston: Houghton Mifflin.

Mathiessen, Peter. 1992. *Shadows of Africa.* New York: Frank Abrams.

McNeely, Jeffery A., et al. 1989. *Conserving the World's Biological Resources: A Primer on Principles and Practice for Development Action.* Washington, D.C.: World Resources Institute.

Miller, Kenton R. 1993. *Balancing the Scales: Managing Biodiversity at the Bioregional Level.* Washington, D.C.: World Resources Institute.

Myers, Norman. 1983. *A Wealth of Wild Species: Storehouse for Human Welfare.* Boulder, Colo.: Westview Press.

Nash, Roderick F. 1988. *The Rights of Nature: A History of Environmental Ethics.* Madison: University of Wisconsin Press.

National Wildlife Federation. 1987. *The Arctic National Wildlife Refuge Coastal Plain: A Perspective for the Future.* Washington, D.C.: National Wildlife Federation.

Office of Technology Assessment. 1989. *Oil Production in the Arctic National Wildlife Refuge.* Washington, D.C.: Government Printing Office.

Oldfield, Margery L., and Janis B. Alcorn, eds. 1992. *Biodiversity: Culture, Conservation, and Ecodevelopment.* Boulder, Colo.: Westview Press.

Owens, Delia, and Mark Owens. 1992. *The Eye of the Elephant: An Epic Adventure in the African Wilderness.* Boston, Mass.: Houghton Mifflin.

Passmore, John. 1974. *Man's Responsibility for Nature.* New York: Charles Scribner's.

Peterson, Melvin N. A., ed. 1992. *Diversity of Oceanic Life.* Boulder, Colo.: Westview Press.

Primack, Richard B. 1993. *Essentials of Conservation Biology.* Sunderland, Mass.: Sinauer.

Reagan, Tom. 1983. *The Case for Animal Rights.* Berkeley: University of California Press.

Reid, Walter V. C., and Kenton R. Miller. 1989. *Keeping Options Alive: The Scientific Basis for Conserving Biodiversity.* Washington, D.C.: World Resources Institute.

Reisner, Marc. 1991. *Game Wars: The Undercover Pursuit of Wildlife Poachers.* New York: Viking Press.

Roe, Frank G. 1970. *The North American Buffalo.* Toronto: University of Toronto Press.

Shaw, J. H. 1985. *Introduction to Wildlife Management.* New York: McGraw-Hill.

Soulé, Michael E. 1986. *Conservation Biology: Science of Scarcity and Diversity.* Sunderland, Mass.: Sinauer.

Tudge, Colin. 1988. *The Environment of Life.* New York: Oxford University Press.

Tudge, Colin. 1992. *Last Animals at the Zoo: How Mass Extinction Can be Stopped.* Covelo, Calif.: Island Press.

Tuttle, Merlin D. 1988. *America's Neighborhood Bats: Understanding and Learning to Live in Harmony with Them.* Austin: University of Texas Press.

Wallace, David Rains. 1987. *Life in the Balance.* New York: Harcourt Brace Jovanovich.

Western, David, and Mary Pearl, eds. 1989. *Conservation for the Twenty-First Century.* New York: Oxford University Press.

Whelan, Tensie, ed. 1991. *Nature Tourism: Managing for the Environment.* Covelo, Calif.: Island Press.

Wilson, E. O. 1992. *The Diversity of Life.* Cambridge, Mass.: Harvard University Press.

Wolf, Edward C. 1987. *On the Brink of Extinction: Conserving the Diversity of Life.* Washington, D.C.: Worldwatch Institute.

World Resources Institute, et al. 1992. *Global Biodiversity Strategy: Guidelines for Action to Save, Study, and Use Earth's Biotic Wealth Sustainably and Equitably.* Washington, D.C.: World Resources Institute.

World Wildlife Fund. 1992. *The Official World Wildlife Fund Guide to Endangered Species of North America.* Washington, D.C.: Beacham.

18 / SOLUTIONS: ENERGY EFFICIENCY AND RENEWABLE ENERGY

Also see readings for Chapter 3.

American Council for an Energy Efficient Economy. 1991. *Energy Efficiency and Environment: Forging the Link.* Washington, D.C.: American Council for an Energy Efficient Economy.

American Council for an Energy Efficient Economy. Annual. *The Most Energy-Efficient Appliances.* Washington, D.C.: American Council for an Energy Efficient Economy.

American Physical Society. 1975. *Efficient Use of Energy.* New York: American Institute of Physics.

Anderson, Bruce. 1990. *Solar Building Architecture.* Cambridge, Mass.: MIT Press.

Blackburn, John O. 1987. *The Renewable Energy Alternative: How the United States and the World Can Prosper Without Nuclear Energy or Coal.* Durham, N.C.: Duke University Press.

Brower, Michael. 1992. *Cool Energy: The Renewable Solution to Global Warming.* Cambridge, Mass.: Union of Concerned Scientists.

Carless, Jennifer. 1993. *Renewable Energy: A Concise Guide to Green Alternatives.* New York: Walker.

Colorado Energy Research Institute. 1976. *Net Energy Analysis: An Energy Balance Study of Fossil Fuel Resources.* Golden, Colo.: Colorado Energy Research Institute.

Davidson, Joel. 1987. *The New Solar Electric Home.* Ann Arbor, Mich.: Aatec Publications.

Echeverria, John, et al. 1989. *Rivers at Risk: The Concerned Citizen's Guide to Hydropower.* Covelo, Calif.: Island Press.

Flavin, Christopher, and Alan B. Durning. 1988. *Building on Success: The Age of Energy Efficiency.* Washington, D.C.: Worldwatch Institute.

Flavin, Christopher, and Rock Piltz. 1990. *Sustainable Energy.* Washington, D.C.: Renew America.

Gever, John, et al. 1991. *Beyond Oil: The Threat to Food and Fuel in Coming Decades.* Boulder: University of Colorado Press.

Gipe, Paul. 1993. *Wind Power for Home and Business: Renewable Energy for the 1990s and Beyond.* Post Mills, Vt.: Chelsea Green.

Goldenberg, Jose, et al. 1988. *Energy for a Sustainable World.* New York: John Wiley.

Hollander, Jack M., ed. 1992. *The New Energy-Environment Connection.* Covelo, Calif.: Island Press.

Johansson, Thomas B., et al, eds. 1992. *Renewable Energy: Sources for Fuels and Electricity.* Covelo, Calif.: Island Press.

Kozloff, Keith, and Roger C. Dower. 1993. *Energy Policies for the 1990s and Beyond.* Washington, D.C.: World Resources Institute.

Lovins, Amory B. 1977. *Soft Energy Paths.* Cambridge, Mass.: Ballinger.

Lovins, Amory B. 1989. *Energy, People, and Industrialization.* Old Snowmass, Colo.: Rocky Mountain Institute.

Lovins, Amory B., and L. Hunter Lovins. 1986. *Energy Unbound: Your Invitation to Energy Abundance.* San Francisco: Sierra Club Books.

Mackenzie, James L. 1993. *Electric and Hydrogen Vehicles: Transportation Technologies for the Twenty-first Century.* Washington, D.C.: World Resources Institute.

McKeown, Walter. 1991. *Death of the Oil Age and the Birth of Hydrogen America.* San Francisco: Wild Bamboo Press.

National Academy of Sciences. 1988. *Geothermal Energy Technology.* Washington, D.C.: National Academy Press.

National Academy of Sciences. 1992. *Automotive Fuel Efficiency: How Far Can We Go?* Washington, D.C.: National Academy Press.

Office of Technology Assessment. 1990. *Replacing Gasoline: Alternative Fuels for Light-Duty Vehicles.* Washington, D.C.: Government Printing Office.

Office of Technology Assessment. 1991. *Improving Automobile Fuel Economy: New Standards, New Approaches.* Washington, D.C.: Government Printing Office.

Office of Technology Assessment. 1992a. *Building Energy Efficiency.* Washington, D.C.: Government Printing Office.

Office of Technology Assessment. 1992b. *Fueling Development: Energy Technologies for Developing Countries.* Washington, D.C.: Government Printing Office.

Ogden, Joan M., and Robert H. Williams. 1989. *Solar Hydrogen: Moving Beyond Fossil Fuels.* Washington, D.C.: World Resources Institute.

Oppenheimer, Michael, and Robert H. Boyle. 1990. *Dead Heat: The Race Against the Greenhouse Effect.* New York: Basic Books.

Pimentel, David, et al. 1984. "Environmental and Social Costs of Biomass Energy." *BioScience,* Feb., 89–93.

Potts, Michael. 1993. *The Independent Home: Living Well with Power from the Sun, Wind, and Water.* Post Mills, Vt.: Chelsea Green.

Real Goods Trading Corporation. Annual. *Alternative Energy Sourcebook.* Ukiah, Calif.: Real Goods Trading Corporation.

Rocky Mountain Institute. 1991. *Practical Home Energy Savings.* Snowmass, Colo.: Rocky Mountain Institute.

Røstvik, Harald N. 1992. *The Sunshine Revolution.* Oslo, Norway: SUN-LAB.

Scientific American. 1990. *Energy for Planet Earth.* Entire Sept. issue.

Starr, Chauncey, et al. 1992. "Energy Sources: A Realistic Outlook." *Science,* vol. 256, 981–86.

Vale, Brenda, and Robert Vale. 1991. *Green Architecture.* Boston: Little, Brown.

Wade, Herb. 1983. *Building Underground: The Design and Construction Handbook for Earth-Sheltered Houses.* Emmaus, Pa.: Rodale Press.

Wells, Malcolm. 1991. *How to Build an Underground House.* Brewster, Mass.: Malcolm Wells.

Zweibel, Ken. 1990. *Harnessing Solar Power: The Challenge of Photovoltaics.* New York: Plenum.

19 / NONRENEWABLE ENERGY RESOURCES

See also the readings for Chapter 3.

Ahearne, John F. 1993. "The Future of Nuclear Power." *American Scientist,* vol. 81, 24–35.

Campbell, John L. 1988. *Collapse of an Industry: Nuclear Power and the Contradictions of U.S. Policy.* Ithaca, N.Y.: Cornell University Press.

Chernousenko, Vladimir M. 1991. *Chernobyl: Insight from the Inside.* New York: Springer-Verlag.

Clark, Wilson, and Jake Page. 1983. *Energy, Vulnerability, and War.* New York: Norton.

Cohen, Bernard L. 1990. *The Nuclear Energy Option: An Alternative for the 90s.* New York: Plenum.

Flavin, Christopher. 1987. *Reassessing Nuclear Power: The Fallout from Chernobyl.* Washington, D.C.: Worldwatch Institute.

Flavin, Christopher. 1992. "Building a Bridge to Sustainable Energy." In Lester Brown, et al. *State of the World 1992.* New York: Norton, 27–55.

Flavin, Christopher, and Nicholas Lenssen. 1994. "Reshaping the Power Industry." In Lester R. Brown, et al. *State of the World 1994.* New York: Norton, 61–80.

Ford, Daniel F. 1986. *Meltdown.* New York: Simon & Schuster.

Fund for Renewable Energy and the Environment. 1987. *The Oil Rollercoaster.* Washington, D.C.: Fund for Renewable Energy and the Environment.

Gershey, Edward L., et al. 1993. *Low-Level Radioactive Waste: From Cradle to Grave.* New York: Van Nostrand Reinhold.

Gever, John, et al. 1991. *Beyond Oil: The Threat to Food and Fuel in Coming Decades.* Boulder: University of Colorado Press.

Golay, Michael W., and Neil E. Todreas. 1990. "Advanced Light-Water Reactors." *Scientific American,* April, 82–89.

Gould, Jay M., and Benjamin A. Goldman. 1991. *Deadly Deceit: Low-Level Radiation, High-Level Cover-up.* New York: Four Walls Eight Windows.

Heede, H. Richard, et al. 1985. *The Hidden Costs of Energy.* Washington, D.C.: Center for Renewable Resources.

Hubbard, Harold H. 1991. "The Real Costs of Energy." *Scientific American,* vol. 265, no. 4, 36–41.

Hughes, Barry B., et al. 1985. *Energy in the Global Arena: Actors, Values, Policies, and Futures.* Durham, N.C.: Duke University Press.

Huizenga, John R. 1992. *Cold Fusion: The Scientific Fiasco of the Century.* Rochester, N.Y.: University of Rochester Press.

Lasper, James M. 1990. *Nuclear Politics: Energy and the State in the United States, Sweden, and France.* Princeton, N.J.: Princeton University Press.

League of Women Voters Education Fund. 1985. *The Nuclear Waste Primer.* Washington D.C.: League of Women Voters.

Lenssen, Nicholas. 1991. *Nuclear Waste: The Problem That Won't Go Away.* Washington, D.C.: Worldwatch Institute.

Lovins, Amory B. 1986. "The Origins of the Nuclear Power Fiasco." *Energy Policy Studies,* vol. 3, 7–34.

Lovins, Amory B., and L. Hunter Lovins. 1982. *Brittle Power: Energy Strategy for National Security.* Andover, Mass.: Brick House.

May, John. 1990. *The Greenpeace Book of the Nuclear Age.* New York: Pantheon.

Medvedev, Grigori. 1990. *The Truth About Chernobyl.* New York: Basic Books.

Morone, Joseph G., and Edward J. Woodhouse. 1989. *The Demise of Nuclear Energy?* New Haven, Conn.: Yale University Press.

Murray, Raymond L. 1989. *Understanding Radioactive Waste.* 3d ed. Columbus, Ohio: Batelle Press.

National Academy of Sciences. 1990. *Energy: Production, Consumption, and Consequences.* Washington, D.C.: National Academy Press.

National Academy of Sciences. 1991a. *Nuclear Power: Technical and Institutional Options for the Future.* Washington, D.C.: National Academy Press.

National Academy of Sciences. 1991b. *Undiscovered Oil and Gas Resources.* Washington, D.C.: National Academy Press.

Office of Technology Assessment. 1991a. *Complex Cleanup: The Environmental Legacy of Nuclear Weapons Production.* Washington, D.C.: Government Printing Office.

Oppenheimer, Ernest J. 1990. *Natural Gas, the Best Energy Choice.* New York: Pen & Podium.

Patterson, Walter C. 1984. *The Plutonium Business and the Spread of the Bomb.* San Francisco: Sierra Club.

President's Commission on the Accident at Three Mile Island. 1979. *Report of the President's Commission on the Accident at Three Mile Island.* Washington, D.C.: Government Printing Office.

Read, Piers Paul. 1993. *Ablaze: The Story of Chernobyl.* New York: Random House.

Renner, Michael. 1994. "Cleaning Up After the Arms Race." In Lester R. Brown, et al. *State of the World 1994.* New York: Norton, 137–156.

Rhodes, Richard. 1993. *Nuclear Renewal: Common Sense About Energy.* New York: Viking Penguin.

Rosenbaum, Walter A. 1987. *Energy, Politics, and Public Policy.* 2d ed. Washington, D.C.: Congressional Quarterly.

Schobert, Harold H. 1987. *Coal: The Energy Source of the Past and Future.* Washington, D.C.: American Chemical Society.

Shea, Cynthia Pollack. 1989. "Decommissioning Nuclear Plants: Breaking Up Is Hard to Do." *World Watch,* July/August, 10–16.

Sierra Club. 1991. *Kick the Oil Habit: Choosing a Safe Energy Future for America.* San Francisco: Sierra Club.

Squillace, Mark. 1990. *Strip Mining Handbook.* Washington, D.C.: Friends of the Earth.

Strohmeyer, John K. 1993. *Extreme Conditions: Big Oil and the Transformation of Alaska.* New York: Simon and Schuster.

Union of Concerned Scientists. 1990. *Safety Second: The NRC and America's Nuclear Power Plants.* Bloomington, Ind.: Indiana University Press.

Watson, Robert K. 1988. *Fact Sheet on Oil and Conservation Resources.* New York: Natural Resources Defense Council.

Winteringham, F. P. W. 1991. *Energy Use and the Environment.* Boca Raton, Fla.: Lewis Publishers.

Yeargin, Daniel. 1990. *The Prize: The Epic Quest for Oil, Money, and Power.* New York: Simon & Schuster.

Glossary

abiotic Nonliving. Compare *biotic*.

acclimation Adjustment to slowly changing new conditions. Compare *threshold effect*.

acid deposition The falling of acids and acid-forming compounds from the atmosphere to Earth's surface. Acid deposition is commonly known as *acid rain*, a term that refers only to wet deposition of droplets of acids and acid-forming compounds.

acid rain See *acid deposition*.

acid solution Any water solution that has more hydrogen ions (H^+) than hydroxide ions (OH^-); any water solution with a pH less than 7. Compare *basic solution*, *neutral solution*.

active solar heating system System that uses solar collectors to capture energy from the sun and store it as heat for space heating and heating water. A liquid or air pumped through the collectors transfers the captured heat to a storage system such as an insulated water tank or rock bed. Pumps or fans then distribute the stored heat or hot water throughout a dwelling as needed. Compare *passive solar heating system*.

adaptation Any genetically controlled structural, physiological, or behavioral characteristic that enhances members of a population's chances of surviving and reproducing in its environment. It usually results from a beneficial mutation. See *biological evolution*, *differential reproduction*, *mutation*, *natural selection*.

adaptive radiation Period of time (usually millions of years) during which numerous new species evolve to fill vacant and new ecological niches in changed environments, usually after a mass extinction.

advanced sewage treatment Specialized chemical and physical processes that reduce the amount of specific pollutants left in wastewater after primary and secondary sewage treatment. This type of treatment is usually expensive. See also *primary sewage treatment*, *secondary sewage treatment*.

aerobic respiration Complex process that occurs in the cells of most living organisms, in which nutrient organic molecules such as glucose ($C_6H_{12}O_6$) combine with oxygen (O_2) and produce carbon dioxide (CO_2), water (H_2O), and energy. Compare *photosynthesis*.

age structure Percentage of the population (or the number of people of each sex) at each age level in a population.

Agricultural Revolution Gradual shift from small, mobile hunting-and-gathering bands to settled agricultural communities, where people survived by learning how to breed and raise wild animals and to cultivate wild plants near where they lived. It began 10,000–12,000 years ago. Compare *Industrial Revolution*.

agroforestry Planting trees and crops together.

air pollution One or more chemicals in high enough concentrations in the air to harm humans, other animals, vegetation, or materials. Excess heat or noise can also be considered as forms of air pollution. Such chemicals or physical conditions are called air pollutants. See *primary pollutant*, *secondary pollutant*.

alien species See *immigrant species*.

alley cropping Planting of crops in strips with rows of trees or shrubs on each side.

alpha particle Positively charged matter, consisting of two neutrons and two protons, that is emitted as a form of radioactivity from the nuclei of some radioisotopes. See also *beta particle*, *gamma rays*.

altitude Height above sea level. Compare *latitude*.

ancient forest See *old-growth forest*.

animal manure Dung and urine of animals that can be used as a form of organic fertilizer. Compare *green manure*.

animals Eukaryotic, multicelled organisms such as sponges, jellyfishes, arthropods (insects, shrimp, lobsters), mollusks (snails, clams, oysters, octopuses), fish, amphibians (frogs, toads, salamanders), reptiles (turtles, lizards, alligators, crocodiles, snakes), birds, mammals (kangaroos, bats, cats, rabbits, elephants, whales, porpoises, monkeys, apes, humans). See *carnivores*, *herbivores*, *omnivores*.

annual rate of natural population change Annual rate at which the size of a population changes. It is usually expressed as a percentage.

aquaculture Growing and harvesting of fish and shellfish for human use in freshwater ponds, irrigation ditches, and lakes, or in cages or fenced-in areas of coastal lagoons and estuaries. See *fish farming*, *fish ranching*.

aquatic Pertaining to water. Compare *terrestrial*.

aquifer Porous, water-saturated layers of sand, gravel, or bed rock that can yield an economically significant amount of water. See *confined aquifer*, *unconfined aquifer*.

arable land Land that can be cultivated to grow crops.

arid Dry. A desert or other area with an arid climate has little precipitation.

asthenosphere Portion of the mantle that is capable of solid flow. See *crust*, *lithosphere*, *mantle*.

atmosphere The whole mass of air surrounding the earth. See *stratosphere*, *troposphere*.

atomic number Number of protons in the nucleus of an atom. Compare *mass number*.

atoms Minute units made of subatomic particles that are the basic building blocks of all chemical elements and thus all matter; the smallest unit of an element that can exist and still have the unique characteristics of that element. Compare *ion*, *molecule*.

autotroph See *producer*.

bacteria Prokaryotic, one-celled organisms. Some transmit diseases. Most act as decomposers and get the nutrients they need by breaking down complex organic compounds in the tissues of living or dead organisms into simpler inorganic nutrient compounds.

barrier islands Long, thin, low, offshore islands of sediment that generally run parallel to the shore along some coasts.

basic solution Water solution with more hydroxide ions (OH^-) than hydrogen ions (H^+); water solution with a pH greater than 7. Compare *acid solution*, *neutral solution*.

beta particle Swiftly moving electron emitted by the nucleus of a radioactive isotope. See also *alpha particle*, *gamma rays*.

bioaccumulation The retention or accumulation of nonbiodegradable or slowly biodegradable chemicals in the body, often in a particular part of the body. Compare *biological amplification*.

biodegradable pollutant Material that can be broken down into simpler substances (elements and compounds) by bacteria or other decomposers. Paper and most organic wastes such as animal manure are biodegradable but can take decades to biodegrade in modern landfills. Compare *degradable pollutant, nondegradable pollutant, slowly degradable pollutant*.

biodiversity See *biological diversity*.

biofuel Gas or liquid fuel (such as ethyl alcohol) made from plant material (biomass).

biogeochemical cycle Natural processes that recycle nutrients in various chemical forms from the nonliving environment to living organisms, and then back to the nonliving environment. Examples are the carbon, oxygen, nitrogen, phosphorus, sulfur, and hydrologic cycles.

biological amplification Increase in concentration of DDT, PCBs, and other slowly degradable, fat-soluble chemicals in organisms at successively higher trophic levels of a food chain or web. See also *bioaccumulation*.

biological community See *community*.

biological diversity Variety of different species (*species diversity*), genetic variability among individuals within each species (*genetic diversity*), and variety of ecosystems (*ecological diversity*).

biological evolution Change in the genetic makeup of a population of a species in successive generations. Note that populations—not individuals—evolve. See also *adaptation, differential reproduction, natural selection, theory of evolution*.

biological oxygen demand (BOD) Amount of dissolved oxygen needed by aerobic decomposers to break down the organic materials in a given volume of water at a certain temperature over a specified time period.

biological pest control Control of pest populations by natural predators, parasites, or disease-causing bacteria and viruses (pathogens).

biomass Organic matter produced by plants and other photosynthetic producers; total dry weight of all living organisms that can be supported at each trophic level in a food chain or web; dry weight of all organic matter in plants and animals in an ecosystem; plant materials and animal wastes used as fuel.

biome Terrestrial regions inhabited by certain types of life, especially vegetation. Examples are various types of deserts, grasslands, and forests.

bioregion A unique life-place with its own soils, landforms, watersheds, climates, native plants and animals, and many other distinct natural characteristics.

biosphere Zone of Earth where life is found. It consists of parts of the atmosphere (the troposphere), hydrosphere (mostly surface water and groundwater), and lithosphere (mostly soil and surface rocks and sediments on the bottoms of oceans and other bodies of water) where life is found. It is also called the ecosphere.

biotic Living organisms make up the biotic parts of ecosystems. Compare *abiotic*.

biotic potential Maximum rate at which the population of a given species can increase when there are no limits of any sort on its rate of growth. See *environmental resistance*.

birth rate See *crude birth rate*.

bitumen Gooey, black, high-sulfur, heavy oil extracted from tar sand and then upgraded to synthetic fuel oil. See *tar sand*.

breeder nuclear fission reactor Nuclear fission reactor that produces more nuclear fuel than it consumes, by converting nonfissionable uranium-238 into fissionable plutonium-239.

calorie Unit of energy; amount of energy needed to raise the temperature of 1 gram of water 1°C. See also *kilocalorie*.

cancer Group of more than 120 different diseases—one for each type of cell in the human body. Each type of cancer produces a tumor in which cells multiply uncontrollably and invade surrounding tissue.

capital goods See *manufactured capital*.

capitalism See *pure market economic system*.

carbon cycle Cyclic movement of carbon in different chemical forms from the environment to organisms, and then back to the environment.

carcinogen Chemicals, ionizing radiation, and viruses that cause or promote the growth of cancer. See *cancer, mutagen, teratogen*.

carnivore Animal that feeds on other animals. Compare *herbivore, omnivore*.

carrying capacity (K) Maximum population of a particular species that a given habitat can support over a given period of time. See *consumption overpopulation, people overpopulation*.

cell Smallest living unit of an organism. Each cell is encased in an outer membrane or wall and contains genetic material (DNA) and other parts to perform its life function. Organisms such as bacteria consist of only one cell, but most of the organisms we are familiar with contain many cells. See *eukaryotic cell, prokaryotic cell*.

centrally planned economy See *pure command economic system*.

CFCs See *chlorofluorocarbons*.

chain reaction Multiple nuclear fissions, taking place within a certain mass of a fissionable isotope, that release an enormous amount of energy in a short time.

chemical One of the millions of different elements and compounds found naturally and synthesized by humans. See *compound, element*.

chemical change Interaction between chemicals in which there is a change in the chemical composition of the elements or compounds involved. Compare *physical change*.

chemical evolution Period of about 1 billion years prior to biological evolution. It involves the formation of the earth and its early crust and atmosphere, evolution of the biological molecules necessary for life, and evolution of systems of chemical reactions needed to produce the first living cells. Compare *biological evolution*.

chemical formula Shorthand way to show the number of atoms (or ions) in the basic structural unit of a compound. Examples are H_2O, $NaCl$, and $C_6H_{12}O_6$.

chemical reaction See *chemical change*.

chemosynthesis Process in which certain organisms (mostly specialized bacteria) extract inorganic compounds from their environment and convert them into organic nutrient compounds without the presence of sunlight. Compare *photosynthesis*.

chlorinated hydrocarbon Organic compound made up of atoms of carbon, hydrogen, and chlorine. Examples are DDT and PCBs.

chlorofluorocarbons (CFCs) Organic compounds made up of atoms of carbon, chlorine, and fluorine. An example is Freon-12 (CCl_2F_2), used as a refrigerant in refrigerators and air conditioners and in making plastics such as Styrofoam. Gaseous CFCs can deplete the ozone layer when they slowly rise into the stratosphere and their chlorine atoms react with ozone molecules.

chromosome A grouping of various genes and associated proteins in plant and animal cells that carry certain types of genetic information. See *genes*.

clear-cutting Method of timber harvesting in which all trees in a forested area are removed in a single cutting. Compare *selective cutting, seed-tree cutting, shelterwood cutting, strip logging*.

climate General pattern of atmospheric or weather conditions, seasonal variations, and weather extremes in a region over a long period—at least 30 years; average weather of an area. Compare *weather*.

climax community See *mature community*.

coal Solid, combustible mixture of organic compounds with 30% to 98% carbon by weight, mixed with varying amounts of water and small amounts of sulfur and nitrogen compounds. It is formed in several stages as the remains of plants are subjected to heat and pressure over millions of years.

coal gasification Conversion of solid coal to synthetic natural gas (SNG).

coal liquefaction Conversion of solid coal to a liquid hydrocarbon fuel such as synthetic gasoline or methanol.

coastal wetland Land along a coastline, extending inland from an estuary that is covered with salt water all or part of the year. Examples are marshes, bays, lagoons, tidal flats, and mangrove swamps. Compare *inland wetland*.

coastal zone Relatively warm, nutrient-rich, shallow part of the ocean that extends from the high-tide mark on land to the edge of a shelflike extension of continental land masses known as the continental shelf. Compare *open sea*.

coevolution Evolution when two or more species interact and exert selective pressures on each other that can lead each species to undergo various adaptations. See *evolution, natural selection*.

cogeneration Production of two useful forms of energy, such as high-temperature heat or steam and electricity, from the same fuel source.

commensalism An interaction between organisms of different species in which one type of organism benefits, while the other type is neither helped nor harmed to any great degree. Compare *mutualism*.

commercial extinction Depletion of the population of a wild species used as a resource to a level where it is no longer profitable to harvest the species.

commercial inorganic fertilizer Commercially prepared mixtures of plant nutrients such as nitrates, phosphates, and potassium applied to the soil to restore fertility and increase crop yields. Compare *organic fertilizer*.

common-property resource Resource that people are normally free to use; each user can deplete or degrade the available supply. Most are potentially renewable and are owned by no one. Examples are clean air, fish in parts of the ocean not under the control of a coastal country, migratory birds, gases of the lower atmosphere, and the ozone content of the upper atmosphere. See *tragedy of the commons*. Compare *private-property resource, public land resources*.

community Populations of all species living and interacting in an area at a particular time.

community development See *ecological succession*.

competition Two or more individual organisms of a single species (*intraspecific competition*) or two or more individuals of different species (*interspecific competition*) attempting to use the same scarce resources in the same ecosystem.

competitive exclusion principle No two species can occupy exactly the same fundamental niche indefinitely in a habitat where there is not enough of a particular resource to meet the needs of both species. See *ecological niche, fundamental niche, realized niche*.

compost Partially decomposed organic plant and animal matter that can be used as a soil conditioner or fertilizer.

compound Combination of atoms, or oppositely charged ions, of two or more different elements held together by attractive forces called chemical bonds. Compare *element*.

concentration Amount of a chemical in a particular volume or weight of air, water, soil, or other medium.

confined aquifer Aquifer between two layers of relatively impermeable Earth materials, such as clay or shale. Compare *unconfined aquifer*.

coniferous trees Cone-bearing trees, mostly evergreens, that have needle-shaped or scalelike leaves. They produce wood known commercially as softwood. Compare *deciduous plants*.

conservation-tillage farming Crop cultivation in which the soil is disturbed little (minimum-tillage farming) or not at all (no-till farming) to reduce soil erosion, lower labor costs, and save energy. Compare *conventional-tillage farming*.

constancy Ability of a living system, such as a population, to maintain a certain size. Compare *inertia*, *resilience*. See *homeostasis*.

consumer Organism that cannot synthesize the organic nutrients it needs and gets its organic nutrients by feeding on the tissues of producers or of other consumers; generally divided into *primary consumers* (herbivores), *secondary consumers* (carnivores), *tertiary (higher-level) consumers*, *omnivores*, and *detritivores* (decomposers and detritus feeders). In economics, one who uses economic goods.

consumption overpopulation Situation in which people in the world or in a geographic region use resources at such a high rate and without sufficient pollution prevention and control that significant pollution, resource depletion, and environmental degradation occur. Compare *people overpopulation*.

continental shelf Submerged part of a continent.

contour farming Plowing and planting across the changing slope of land, rather than in straight lines, to help retain water and reduce soil erosion.

contraceptive Physical, chemical, or biological method used to prevent pregnancy.

controlled burning Deliberately set, carefully controlled surface fires to reduce flammable litter and decrease the chances of damaging crown fires. See *crown fire*, *ground fire*, *surface fire*.

conventional-tillage farming Making a planting surface by plowing land, breaking up the exposed soil, and then smoothing the surface. Compare *conservation-tillage farming*.

convergent plate boundary Area where Earth's lithospheric plates are pushed together. See *subduction zone*. Compare *divergent plate boundary*, *transform fault*.

coral reef Formation produced by massive colonies containing billions of tiny coral animals, called polyps, which secrete a stony substance (calcium carbonate) around themselves for protection. When the corals die, their empty outer skeletons form layers that cause the reef to grow. They are found in the coastal zones of warm tropical and subtropical oceans.

core Inner zone of the earth. It consists of a solid inner core and a liquid outer core. Compare *crust*, *mantle*.

cost-benefit analysis Estimates and comparison of short-term and long-term costs (losses) and benefits (gains) from an economic decision. If the estimated benefits exceed the estimated costs, the decision to buy an economic good or provide a public good is considered worthwhile.

critical mass Amount of fissionable nuclei needed to sustain a nuclear fission chain reaction.

critical mineral A mineral necessary to the economy of a country.

crop rotation Planting a field, or an area of a field, with different crops from year to year to reduce depletion of soil nutrients. A plant such as corn, tobacco, or cotton, which removes large amounts of nitrogen from the soil, is planted one year. The next year a legume such as soybeans, which add nitrogen to the soil, is planted.

crown fire Extremely hot forest fire that burns ground vegetation and tree tops. Compare *controlled burning*, *ground fire*, *surface fire*.

crude birth rate Annual number of live births per 1,000 persons in the population of a geographical area at the midpoint of a given year. Compare *crude death rate*.

crude death rate Annual number of deaths per 1,000 persons in the population of a geographical area at the midpoint of a given year. Compare *crude birth rate*.

crude oil Gooey liquid consisting mostly of hydrocarbon compounds and small amounts of compounds containing oxygen, sulfur, and nitrogen. Extracted from underground accumulations, it is sent to oil refineries, where it is converted to heating oil, diesel fuel, gasoline, tar, and other materials.

crust Solid outer zone of the earth. It consists of oceanic crust and continental crust. Compare *core*, *mantle*.

cultural eutrophication Overnourishment of aquatic ecosystems with plant nutrients (mostly nitrates and phosphates) because of human activities such as agriculture, urbanization, and discharges from industrial plants and sewage treatment plants. See *eutrophication*.

cyanobacteria Single-celled, prokaryotic, microscopic organisms. Before being reclassified as monera, they were called blue-green algae.

DDT Dichlorodiphenyltrichloroethane, a chlorinated hydrocarbon that has been widely used as a pesticide, but is now banned in some countries.

death rate See *crude death rate*.

debt-for-nature swap Agreement in which a certain amount of foreign debt is canceled in exchange for local currency investments that will improve natural resource management or protect certain areas from harmful development in the debtor country.

deciduous plants Trees, such as oaks and maples, and other plants that survive during dry seasons or cold seasons by shedding their leaves. Compare *coniferous trees*, *succulent plants*.

decomposer Organism that digests parts of dead organisms and cast-off fragments and wastes of living organisms by breaking down the complex organic molecules in those materials into simpler inorganic compounds and then absorbing the soluble nutrients. Most of these chemicals are returned to the soil and water for reuse by producers. Decomposers consist of various bacteria and fungi. Compare *consumer*, *detritivore*, *producer*.

deforestation Removal of trees from a forested area without adequate replanting.

degradable pollutant Potentially polluting chemical that is broken down completely or reduced to acceptable levels by natural physical, chemical, and biological processes. Compare *biodegradable pollutant*, *nondegradable pollutant*, *slowly degradable pollutant*.

degree of urbanization Percentage of the population in the world, or a country, living in areas with a population of more than 2,500 people (higher in some countries). Compare *urban growth*.

democracy Government "by the people" through their elected officials and appointed representatives. In a *constitutional democracy*, a constitution provides the basis of governmental authority and puts restraints on governmental power through free elections and freely expressed public opinion.

demographic transition Hypothesis that countries, as they become industrialized, have declines in death rates followed by declines in birth rates.

demography Study of characteristics and changes in the size and structure of the human population in the world or other geographical area.

depletion time How long it takes to use a certain fraction—usually 80%—of the known or estimated supply of a nonrenewable resource at an assumed rate of use. Finding and extracting the remaining 20% usually costs more than it is worth.

desalination Purification of salt water or brackish (slightly salty) water by removing dissolved salts.

desert Biome where evaporation exceeds precipitation and the average amount of precipitation is less than 25 centimeters (10 inches) a year. Such areas have little vegetation or have widely spaced, mostly low vegetation. Compare *forest*, *grassland*.

desertification Conversion of rangeland, rain-fed cropland, or irrigated cropland to desertlike land, with a drop in agricultural productivity of 10% or more. It is usually caused by a combination of overgrazing, soil erosion, prolonged drought, and climate change.

desirability quotient A number expressing the results of risk-benefit analysis by dividing the estimate of the benefits to society of using a particular product or technology by its estimated risks. See *risk-benefit analysis*. Compare *cost-benefit analysis*.

detritivore Consumer organism that feeds on detritus, parts of dead organisms and cast-off fragments and wastes of living organisms. The two principal types are detritus feeders and decomposers.

detritus Parts of dead organisms and cast-off fragments and wastes of living organisms.

detritus feeder Organism that extracts nutrients from fragments of dead organisms and their cast-off parts and organic wastes. Examples are earthworms, termites, and crabs. Compare *decomposer*.

deuterium (D; hydrogen-2) Isotope of the element hydrogen, with a nucleus containing one proton and one neutron, and a mass number of 2. Compare *tritium*.

dieback Sharp reduction in the population of a species when its numbers exceed the carrying capacity of its habitat. See *carrying capacity*, *consumption overpopulation*, *overshoot*, *people overpopulation*.

differential reproduction Ability of individuals with adaptive genetic traits to produce more living offspring than individuals without such traits. See also *natural selection*.

dioxins Family of 75 different chlorinated hydrocarbon compounds formed as by-products in chemical reactions involving chlorine and hydrocarbons, usually at high temperatures.

discount rate How much economic value a resource will have in the future compared with its present value.

dissolved oxygen (DO) content   Amount of oxygen gas (O_2) dissolved in a given volume of water at a particular temperature and pressure, often expressed as a concentration in parts of oxygen per million parts of water.

divergent plate boundary   Area where Earth's lithospheric plates move apart in opposite directions. Compare *convergent plate boundary, transform fault.*

DNA (deoxyribonucleic acid)   Large molecules in the cells of organisms; carries genetic information in living organisms.

dose   The amount of a potentially harmful substance an individual ingests, inhales, or absorbs through the skin. Compare *response.* See *dose—response curve, lethal dose, median lethal dose.*

dose–response curve   Plot of data showing effects of various doses of a toxic agent on a group of test organisms. See *dose, lethal dose, median lethal dose, response.*

doubling time   The time it takes (usually in years) for the quantity of something growing exponentially to double. It can be calculated by dividing the annual percentage growth rate into 70. See *rule of 70.*

drainage basin   See *watershed.*

dredge spoils   Materials scraped from the bottoms of harbors and streams to maintain shipping channels. They are often contaminated with high levels of toxic substances that have settled out of the water. See *dredging.*

dredging   Type of surface mining in which chain buckets and draglines scrape up sand, gravel, and other surface deposits covered with water. It is also used to remove sediment from streams and harbors to maintain shipping channels. See *dredge spoils.*

drift-net fishing   Catching fish in huge nets that drift in the water.

drought   Condition in which an area does not get enough water because of lower than normal precipitation, higher than normal temperatures that increase evaporation, or both.

dust dome   Dome of heated air that surrounds an urban area and traps and keeps pollutants, especially suspended particulate matter. See also *urban heat island.*

dust plume   Elongation of a dust dome by winds, which can spread a city's pollutants hundreds of kilometers downwind.

Earth capital   Earth's natural resources and processes that sustain us and other species.

earthquake   Shaking of the ground resulting from the fracturing and displacement of rock, producing a fault, or from subsequent movement along the fault

Earth-wisdom worldview   See *sustainable-Earth worldview.*

ecological diversity   The variety of forests, deserts, grasslands, oceans, streams, lakes, and other biological communities interacting with one another and with their nonliving environment. See *biological diversity.* Compare *genetic diversity, species diversity.*

ecological land-use planning   Method for deciding how land should be used; development of an integrated model that considers geological, ecological, health, and social variables.

ecological niche   Total way of life or role of a species in an ecosystem. It includes all physical, chemical, and biological conditions a species needs to live and reproduce in an ecosystem. See *fundamental niche, realized niche.*

ecological population density   Number of individuals of a population per unit of habitat area. Compare *population density.*

ecological succession   Process in which communities of plant and animal species in a particular area are replaced over time by a series of different and often more complex communities. See *primary succession, secondary succession.*

ecology   Study of the interactions of living organisms with one another and with their nonliving environment of matter and energy; study of the structure and functions of nature.

economic decision   Deciding what goods and services to produce, how to produce them, how much to produce, and how to distribute them to people.

economic depletion   Exhaustion of 80% of the estimated supply of a nonrenewable resource. Finding, extracting, and processing the remaining 20% usually costs more than it is worth; may also apply to the depletion of a potentially renewable resource, such as a species of fish or trees.

economic growth   Increase in the real value of all final goods and services produced by an economy; an increase in real GNP.

economic needs   Types and amounts of certain economic goods—food, clothing, water, oxygen, shelter—that each of us must have to survive and to stay healthy. Compare *economic wants.* See *poverty.*

economic resources   Natural resources, capital goods, and labor used in an economy to produce material goods and services. See *Earth capital, human capital, manufactured capital.*

economics   Study of how individuals and groups make decisions about what to do with economic resources to meet their needs and wants.

economic system   Method that a group of people uses to choose *what* goods and services to produce, *how* to produce them, *how much* to produce, and *how to distribute* them to people. See *mixed economic system, pure command economic system, pure market economic system.*

economic wants   Economic goods that go beyond our basic economic needs. These wants are influenced by the customs and conventions of the society we live in and by our level of affluence. Compare *economic needs.*

economy   System of production, distribution, and consumption of economic goods.

ecosphere   Earth's collection of living organisms interacting with one another and their nonliving environment (energy and matter) throughout the world; all of Earth's ecosystems. Also called the *biosphere.*

ecosystem   Community of different species interacting with one another and with the chemical and physical factors making up its nonliving environment.

efficiency   Measure of how much output of energy or of a product is produced by a certain input of energy, materials, or labor. See *energy efficiency.*

electromagnetic radiation   Forms of kinetic energy traveling as electromagnetic waves. Examples are radio waves, TV waves, microwaves, infrared radiation, visible light, ultraviolet radiation, X rays, and gamma rays. Compare *ionizing radiation, nonionizing radiation.*

electron (e)   Tiny particle moving around outside the nucleus of an atom. Each electron has one unit of negative charge (⁻) and almost no mass.

element   Chemical, such as hydrogen (H), iron (Fe), sodium (Na), carbon (C), nitrogen (N), or oxygen (O), whose distinctly different atoms serve as the basic building blocks of all matter. There are 92 naturally occurring elements. Another 15 have been made in laboratories. Two or more elements combine to form compounds that make up most of the world's matter. Compare *compound.*

emigration   Migration of people out of one country or area to take up permanent residence in another country or area. Compare *immigration.*

endangered species   Wild species with so few individual survivors that the species could soon become extinct in all or most of its natural range. Compare *threatened species.*

energy   Capacity to do work by performing mechanical, physical, chemical, or electrical tasks or to cause a heat transfer between two objects at different temperatures.

energy conservation   Reduction or elimination of unnecessary energy use and waste. See *energy efficiency.*

energy efficiency   Percentage of the total energy input that does useful work and is not converted into low-quality, usually useless, heat in an energy conversion system or process. See *energy quality, net energy.*

energy quality   Ability of a form of energy to do useful work. High-temperature heat and the chemical energy in fossil fuels and nuclear fuels are concentrated high-quality energy. Low-quality energy such as low-temperature heat is dispersed or diluted and cannot do much useful work. See *high-quality energy, low-quality energy.*

enhanced oil recovery   Removal of some of the heavy oil left in an oil well after primary and secondary recovery. Compare *primary oil recovery, secondary oil recovery.*

environment   All external conditions and factors, living and nonliving (chemicals and energy), that affect an organism or other specified system during its lifetime.

environmental degradation   Depletion or destruction of a potentially renewable resource such as soil, grassland, forest, or wildlife by using it at a faster rate than it is naturally replenished. If such use continues, the resource can become nonrenewable on a human time scale or nonexistent (extinct). See also *sustainable yield.*

environmental resistance   All the limiting factors jointly acting to limit the growth of a population. See *biotic potential, limiting factor.*

Environmental Revolution   See *Sustainable-Earth Revolution.*

environmental science   Study of how we and other species interact with each other and with the nonliving environment of matter and energy. It is a holistic science that uses and integrates knowledge from physics, chemistry, biology (especially ecology), geology, geography, resource technology and engineering, resource conservation and management, demography (the study of population dynamics), economics, politics, and ethics.

EPA Environmental Protection Agency; responsible for managing federal efforts in the United States to control air and water pollution, radiation and pesticide hazards, environmental research, hazardous waste, and solid waste disposal.

epidemiology Study of the patterns of disease or other harmful effects from toxic exposure within defined groups of people to find out why some people get sick and some do not.

epiphytes Plants that use their roots to attach themselves to branches high in trees, especially in tropical forests.

erosion Process or group of processes by which earth materials, loose or consolidated, are dissolved, loosened, and worn away, removed from one place and deposited in another. See *weathering.*

estuary Partially enclosed coastal area at the mouth of a river where its fresh water, carrying fertile silt and runoff from the land, mixes with salty seawater.

ethics What we believe to be right or wrong behavior.

eukaryotic cell Cell containing a *nucleus*, a region of genetic material surrounded by a membrane. Membranes also enclose several of the other internal parts found in a eukaryotic cell. Compare *prokaryotic cell.*

eutrophication Physical, chemical, and biological changes that take place after a lake, an estuary, or a slow-flowing stream receives inputs of plant nutrients—mostly nitrates and phosphates—from natural erosion and runoff from the surrounding land basin. See *cultural eutrophication.*

eutrophic lake Lake with a large or excessive supply of plant nutrients—mostly nitrates and phosphates. Compare *mesotrophic lake, oligotrophic lake.*

evaporation Physical change in which a liquid changes into a vapor or gas.

even-aged management Method of forest management in which trees, sometimes of a single species in a given stand, are maintained at about the same age and size and are harvested all at once so a new stand may grow. Compare *uneven-aged management.*

even-aged stand Forest area where all trees are about the same age. Usually, such stands contain trees of only one or two species. See *even-aged management, tree farm.* Compare *uneven-aged management, uneven-aged stand.*

evergreen plants Plants that keep some of their leaves or needles throughout the year. Examples are ferns and cone-bearing trees (conifers) such as firs, spruces, pines, redwoods, and sequoias. Compare *deciduous plants, succulent plants.*

evolution See *biological evolution.*

exhaustible resources See *nonrenewable resources.*

exploitation competition Situation in which two competing species have equal access to a specific resource but differ in how quickly or efficiently they exploit it. See *interference competition, interspecific competition.* Compare *intraspecific competition.*

exponential growth Growth in which some quantity, such as population size or economic output, increases by a fixed percentage of the whole in a given time period; when the increase in quantity over time is plotted, this type of growth yields a curve shaped like the letter *J.* Compare *linear growth.*

external benefit Beneficial social effect of producing and using an economic good that is not included in the market price of the good. Compare *external cost, full cost, true cost.*

external cost Harmful social effect of producing and using an economic good that is not included in the market price of the good. Compare *external benefit, full cost, true cost.*

externalities Social benefits ("goods") and social costs ("bads") not included in the market price of an economic good. See *external benefit, external cost.* Compare *full cost, internal cost.*

extinction Complete disappearance of a species from the earth. This happens when a species cannot adapt and successfully reproduce under new environmental conditions or when it evolves into one or more new species. Compare *speciation.* See also *endangered species, threatened species.*

family planning Providing information, clinical services, and contraceptives to help individuals or couples choose the number and spacing of children they want to have.

famine Widespread malnutrition and starvation in a particular area because of a shortage of food, usually caused by drought, war, flood, earthquake, or other catastrophic event that disrupts food production and distribution.

feedlot Confined outdoor or indoor space used to raise hundreds to thousands of domesticated livestock. Compare *rangeland.*

fertilizer Substance that adds inorganic or organic plant nutrients to soil and improves its ability to grow crops, trees, or other vegetation. See *commercial inorganic fertilizer, organic fertilizer.*

first law of energy See *first law of thermodynamics.*

first law of human ecology We can never do merely one thing. Any intrusion into nature has numerous effects, many of which are unpredictable.

first law of thermodynamics (energy) In any physical or chemical change, no detectable amount of energy is created or destroyed, but in these processes energy can be changed from one form to another; you can't get more energy out of something than you put in; in terms of energy quantity, you can't get something for nothing (there is no free lunch). This law does not apply to nuclear changes, in which energy can be produced from small amounts of matter. See also *second law of thermodynamics.*

fishery Concentrations of particular aquatic species suitable for commercial harvesting in a given ocean area or inland body of water.

fish farming Form of aquaculture in which fish are cultivated in a controlled pond or other environment and harvested when they reach the desired size. See also *fish ranching.*

fish ranching Form of aquaculture in which members of a fish species such as salmon are held in captivity for the first few years of their lives, released, and then harvested as adults when they return from the ocean to their freshwater birthplace to spawn. See also *fish farming.*

fissionable isotope Isotope that can split apart when hit by a neutron at the right speed and thus undergo nuclear fission. Examples are uranium-235 and plutonium-239.

floodplain Flat valley floor next to a stream channel. For legal purposes, the term is often applied to any low area that has the potential for flooding, including certain coastal areas.

flyway Generally fixed route along which waterfowl migrate from one area to another at certain seasons of the year.

food additive A natural or synthetic chemical deliberately added to processed foods to retard spoilage, to provide missing amino acids and vitamins, or to enhance flavor, color, and texture.

food chain Series of organisms, each eating or decomposing the preceding one. Compare *food web.*

food web Complex network of many interconnected food chains and feeding relationships. Compare *food chain.*

forage Vegetation eaten by animals, especially grazing and browsing animals.

forest Biome with enough average annual precipitation (at least 76 centimeters, or 30 inches) to support growth of various species of trees and smaller forms of vegetation. Compare *desert, grassland.*

fossil fuel Products of partial or complete decomposition of plants and animals that occur as crude oil, coal, natural gas, or heavy oils as a result of exposure to heat and pressure in Earth's crust over millions of years. See *coal, crude oil, natural gas.*

fossils Skeletons, bones, shells, body parts, leaves, seeds, or impressions of such items that provide recognizable evidence of organisms that lived long ago.

Freons See *chlorofluorocarbons.*

full cost Cost of a good when its internal costs and its estimated short- and long-term external costs are included in its market price. Compare *external cost, internal cost.*

fundamental niche The full potential range of the physical, chemical, and biological factors a species can use, if there is no competition from other species. See *ecological niche.* Compare *realized niche.*

fungi Eukaryotic, mostly multicelled organisms such as mushrooms, molds, and yeasts. As decomposers, they get the nutrients they need by secreting enzymes that break down the organic matter in the tissue of other living or dead organisms. Then they absorb the resulting nutrients.

fungicide Chemical that kills fungi.

Gaia hypothesis Proposal that Earth is alive and can be considered a system that operates and changes by feedback of information between its living and nonliving components.

game species Type of wild animal that people hunt or fish for, for sport and recreation and sometimes for food.

gamma rays A form of ionizing, electromagnetic radiation with a high energy content emitted by some radioisotopes. They readily penetrate body tissues.

gasohol Vehicle fuel consisting of a mixture of gasoline and ethyl or methyl alcohol—typically 10% to 23% ethanol or methanol by volume.

gene mutation See *mutation.*

gene pool The sum total of all genes found in the individuals of the population of a particular species.

generalist species Species with a broad ecological niche. They can live in many different places, eat a variety of foods, and tolerate a wide range of environmental conditions. Examples are flies, cockroaches, mice, rats, and human beings. Compare *specialist species.*

genes Segments of DNA molecules found in chromosomes that impart certain inheritable traits in organisms.

genetic adaptation Changes in the genetic makeup of organisms of a species that allow the species to reproduce and gain a competitive advantage under changed environmental conditions. See *differential reproduction*, *evolution*, *mutation*, *natural selection*.

genetic diversity Variability in the genetic makeup among individuals within a single species. See *biodiversity*. Compare *ecological diversity*, *species diversity*.

geothermal energy Heat transferred from the earth's underground concentrations of dry steam (steam with no water droplets), wet steam (a mixture of steam and water droplets), or hot water trapped in fractured or porous rock.

GNP See *gross national product*.

GNP per capita Annual gross national product (GNP) of a country divided by its total population. See *gross national product*, *real GNP per capita*.

grassland Biome found in regions where moderate annual average precipitation (25 to 76 centimeters, or 10 to 30 inches) is enough to support the growth of grass and small plants, but not enough to support large stands of trees. Compare *desert*, *forest*.

greenhouse effect A natural effect that traps heat in the atmosphere (troposphere) near Earth's surface. Some of the heat flowing back toward space from Earth's surface is absorbed by water vapor, carbon dioxide, ozone, and several other gases in the lower atmosphere (troposphere) and then radiated back toward the earth's surface. If the atmospheric concentrations of these greenhouse gases rise, and are not removed by other natural processes, the average temperature of the lower atmosphere will gradually increase.

greenhouse gases Gases in Earth's lower atmosphere (troposphere) that cause the greenhouse effect. Examples are carbon dioxide, chlorofluorocarbons, ozone, methane, water vapor, and nitrous oxide.

green manure Freshly cut or still-growing green vegetation that is plowed into the soil to increase the organic matter and humus available to support crop growth. Compare *animal manure*.

green revolution Popular term for introduction of scientifically bred or selected varieties of grain (rice, wheat, maize) that, with high enough inputs of fertilizer and water, can greatly increase crop yields.

gross national product (GNP) Total market value in current dollars of all goods and services produced by an economy for final use usually during a year. Compare *GNP per capita*, *real NEW per capita*, *real GNP*, *real GNP per capita*.

gross primary productivity The rate at which an ecosystem's producers capture and store a given amount of chemical energy as biomass in a given length of time. Compare *net primary productivity*.

ground fire Fire that burns decayed leaves or peat deep below the ground surface. Compare *crown fire*, *surface fire*.

groundwater Water that sinks into the soil and is stored in slowly flowing and slowly renewed underground reservoirs called aquifers; underground water in the zone of saturation, below the water table. See *confined aquifer*, *unconfined aquifer*. Compare *runoff*, *surface water*.

growth rate (r) Increase in the size of a population per unit of time (such as a year).

gully reclamation Restoring land suffering from gully erosion by seeding gullies with quick-growing plants, building small dams to collect silt and gradually fill in the channels, and building channels to divert water away from the gully.

habitat Place or type of place where an organism or a population of organisms lives. Compare *ecological niche*.

half-life Time needed for one-half of the nuclei in a radioisotope to emit its radiation. Each radioisotope has a characteristic half-life, which may range from a few millionths of a second to several billion years.

hazard Something that can cause injury, disease, economic loss, or environmental damage.

hazardous chemical Chemical that can cause harm because it is flammable or explosive, or that can irritate or damage the skin or lungs (such as strong acidic or alkaline substances) or cause allergic reactions of the immune system (allergens). See *toxic chemical*.

hazardous waste Any solid, liquid, or containerized gas that can catch fire easily, is corrosive to skin tissue or metals, is unstable and can explode or release toxic fumes, or has harmful concentrations of one or more toxic materials that can leach out. See also *toxic waste*.

heat island See *urban heat island*.

herbicide Chemical that kills a plant or inhibits its growth.

herbivore Plant-eating organism. Examples are deer, sheep, grasshoppers, and zooplankton. Compare *carnivore*, *omnivore*.

heterotroph See *consumer*.

high-quality energy Energy that is organized or concentrated and has great ability to perform useful work. Examples are high-temperature heat and the energy in electricity, coal, oil, gasoline, sunlight, and nuclei of uranium-235. Compare *low-quality energy*.

high-quality matter Matter that is organized, concentrated, and contains a high concentration of a useful resource. Compare *low-quality matter*.

high-waste society See *throwaway society*.

homeostasis Maintenance of favorable internal conditions in a system despite changes in external conditions. See *constancy*, *inertia*, *resilience*.

host Plant or animal upon which a parasite feeds.

human capital Physical and mental talents of people used to produce, distribute, and sell an economic good. Compare *Earth capital*, *manufactured capital*.

humification Process in which organic matter in the upper soil layers is reduced to finely divided pieces of humus or partially decomposed organic matter.

humus Slightly soluble residue of undigested or partially decomposed organic material in topsoil. This material helps retain water and water-soluble nutrients, which can be taken up by plant roots. See *humification*.

hunter-gatherers People who get their food by gathering edible wild plants and other materials and by hunting wild animals and fish.

hydrocarbon Organic compound of hydrogen and carbon atoms.

hydroelectric power plant Structure in which the energy of falling or flowing water spins a turbine generator to produce electricity.

hydrologic cycle Biogeochemical cycle that collects, purifies, and distributes the earth's fixed supply of water from the environment to living organisms, and then back to the environment.

hydropower Electrical energy produced by falling or flowing water. See *hydroelectric power plant*.

hydrosphere Earth's liquid water (oceans, lakes and other bodies of surface water, and underground water), Earth's frozen water (polar ice caps, floating ice caps, and ice in soil known as permafrost), and small amounts of water vapor in the atmosphere.

identified resources Deposits of a particular mineral-bearing material of which the location, quantity, and quality are known or have been estimated from direct geological evidence and measurements. Compare *undiscovered resources*.

igneous rock Rock formed when molten rock material (magma) wells up from Earth's interior, cools, and solidifies into rock masses. Compare *metamorphic rock*, *sedimentary rock*. See *rock cycle*.

immature community Community at an early stage of ecological succession. It usually has a low number of species and ecological niches and cannot capture and use energy and cycle critical nutrients as efficiently as more complex, mature ecosystems. Compare *mature community*.

immigrant species Species that migrate into an ecosystem or that are deliberately or accidentally introduced into an ecosystem by humans. Some of these species are beneficial, while others can take over and eliminate many native species. Compare *indicator species*, *keystone species*, *native species*.

immigration Migration of people into a country or area to take up permanent residence. Compare *emigration*.

indicator species Species that serve as early warnings that a community or an ecosystem is being degraded. Compare *immigrant species*, *keystone species*, *native species*.

industrialized agriculture Using large inputs of energy from fossil fuels (especially oil and natural gas), water, fertilizer, and pesticides to produce large quantities of crops and livestock for domestic and foreign sale. Compare *subsistence farming*.

Industrial Revolution Use of new sources of energy from fossil fuels and later from nuclear fuels, and use of new technologies, to grow food and manufacture products.

industrial smog Type of air pollution consisting mostly of a mixture of sulfur dioxide, suspended droplets of sulfuric acid formed from some of the sulfur dioxide, and a variety of suspended solid particles. Compare *photochemical smog*.

inertia Ability of a living system to resist being disturbed or altered. Compare *constancy*, *resilience*.

infant mortality rate Number of babies out of every 1,000 born each year that die before their first birthday.

infiltration Downward movement of water through soil.

inland wetland Land away from the coast, such as a swamp, marsh, or bog, that is covered all or part of the year with fresh water. Compare *coastal wetland*.

inorganic fertilizer See *commercial inorganic fertilizer*.

input pollution control See *pollution prevention.*

insecticide Chemical that kills insects.

integrated pest management (IPM) Combined use of biological, chemical, and cultivation methods in proper sequence and timing to keep the size of a pest population below the size that causes economically unacceptable loss of a crop or livestock animal.

intercropping Growing two or more different crops at the same time on a plot. For example, a carbohydrate-rich grain that depletes soil nitrogen and a protein-rich legume that adds nitrogen to the soil may be intercropped. Compare *monoculture, polyculture, polyvarietal cultivation.*

interference competition Situation in which one species limits access of another species to a resource, regardless of whether the resource is abundant or scarce. See *exploitation competition, interspecific competition.* Compare *intraspecific competition.*

intermediate goods See *manufactured capital.*

internal cost Direct cost paid by the producer and the buyer of an economic good. Compare *external cost.*

interplanting Simultaneously growing a variety of crops on the same plot. See *agroforestry, intercropping, polyculture, polyvarietal cultivation.*

interspecific competition Members of two or more species trying to use the same limited resources in an ecosystem. See *competition, competitive exclusion principle, exploitation competition, interference competition, intraspecific competition.*

intraspecific competition Two or more individual organisms of a single species trying to use the same limited resources in an ecosystem. See *competition, interspecific competition.*

inversion See *thermal inversion.*

invertebrates Animals that have no backbones. Compare *vertebrates.*

ion Atom or group of atoms with one or more positive (+) or negative (−) electrical charges. Compare *atom, molecule.*

ionizing radiation Fast-moving alpha or beta particles or high-energy radiation (gamma rays) emitted by radioisotopes. They have enough energy to dislodge one or more electrons from atoms they hit, forming charged ions in tissue that can react with and damage living tissue.

isotopes Two or more forms of a chemical element that have the same number of protons but different mass numbers due to different numbers of neutrons in their nuclei.

J-shaped curve Curve with a shape similar to that of the letter *J;* represents exponential growth.

kerogen Solid, waxy mixture of hydrocarbons found in oil shale rock. When the rock is heated to high temperatures, the kerogen is vaporized. The vapor is condensed, purified, and then sent to a refinery to produce gasoline, heating oil, and other products. See also *oil shale, shale oil.*

keystone species Species that play roles affecting many other organisms in an ecosystem. Compare *immigrant species, indicator species, native species.*

kilocalorie (kcal) Unit of energy equal to 1,000 calories. See *calorie.*

kilowatt (kw) Unit of electrical power equal to 1,000 watts. See *watt.*

kinetic energy Energy that matter has because of its motion and mass. Compare *potential energy.*

K-strategists Species that produce a few, often fairly large offspring but invest a great deal of time and energy to ensure that most of those offspring will reach reproductive age. Compare *r-strategists.*

kwashiorkor Type of malnutrition that occurs in infants and very young children when they are weaned from mother's milk to a starchy diet low in protein. See *marasmus.*

labor See *human capital.*

lake Large natural body of standing fresh water formed when water from precipitation, land runoff, or groundwater flow fills a depression in the earth created by glaciation, earth movement, volcanic activity, or a giant meteorite. See *eutrophic lake, mesotrophic lake, oligotrophic lake.*

landfill See *sanitary landfill.*

land-use planning Process for deciding the best present and future use of each parcel of land in an area.

latitude Distance from the equator. Compare *altitude.*

law of conservation of energy See *first law of thermodynamics.*

law of conservation of matter In any physical or chemical change, matter is neither created nor destroyed, but merely changed from one form to another; in physical and chemical changes, existing atoms are rearranged into either different spatial patterns (physical changes) or different combinations (chemical changes).

law of pollution prevention If you don't put something into the environment, it isn't there.

law of tolerance The existence, abundance, and distribution of a species in an ecosystem are determined by whether the levels of one or more physical or chemical factors fall within the range tolerated by the species. See *threshold effect.*

LD$_{50}$ See *median lethal dose.*

LDC See *less developed country.*

leaching Process in which various chemicals in upper layers of soil are dissolved and carried to lower layers and, in some cases, to groundwater.

less developed country (LDC) Country that has low to moderate industrialization and low to moderate GNP per person. Most are located in Africa, Asia, and Latin America. Compare *more developed country.*

lethal dose Amount of a toxic material per unit of body weight of the test animals that kills all of the test population in a certain time. See *median lethal dose.*

life-cycle cost Initial cost plus lifetime operating costs of an economic good.

life expectancy Average number of years a newborn infant can be expected to live.

limiting factor Single factor that limits the growth, abundance, or distribution of the population of a species in an ecosystem. See *limiting factor principle.*

limiting factor principle Too much or too little of any abiotic factor can limit or prevent growth of a population of a species in an ecosystem, even if all other factors are at or near the optimum range of tolerance for the species.

linear growth Growth in which a quantity increases by some fixed amount during each unit of time. Compare *exponential growth.*

liquefied natural gas (LNG) Natural gas converted to liquid form by cooling to a very low temperature.

liquefied petroleum gas (LPG) Mixture of liquefied propane and butane gas removed from natural gas.

lithosphere Outer shell of the Earth, composed of the crust and the rigid, outermost part of the mantle outside of the asthenosphere; material found in Earth's plates. See *crust, mantle.*

loams Soils containing a mixture of clay, sand, silt, and humus. Good for growing most crops.

low-quality energy Energy that is disorganized or dispersed and has little ability to do useful work. An example is low-temperature heat. Compare *high-quality energy.*

low-quality matter Matter that is disorganized, dilute, or dispersed, or contains a low concentration of a useful resource. Compare *high-quality matter.*

LPG See *liquefied petroleum gas.*

magma Molten rock below the earth's surface.

malnutrition Faulty nutrition. Caused by a diet that does not supply an individual with enough proteins, essential fats, vitamins, minerals, and other nutrients needed for good health. Compare *overnutrition, undernutrition.*

mantle Zone of the earth's interior between its core and its crust. Compare *core, crust.* See *lithosphere.*

manufactured capital Manufactured items made from Earth capital and used to produce and distribute economic goods and services bought by consumers. These include tools, machinery, equipment, factory buildings, and transportation and distribution facilities. Compare *Earth capital, human capital.*

manure See *animal manure, green manure.*

marasmus Nutritional-deficiency disease caused by a diet that does not have enough calories and protein to maintain good health. See *kwashiorkor, malnutrition.*

market equilibrium State in which sellers and buyers of an economic good agree on the quantity to be produced and the price to be paid.

mass The amount of material in an object.

mass extinction A catastrophic, widespread—often global—event in which major groups of species are wiped out over a relatively short time compared to normal (background) extinctions.

mass number Sum of the number of neutrons and the number of protons in the nucleus of an atom. It gives the approximate mass of that atom. Compare *atomic number.*

mass transit Buses, trains, trolleys, and other forms of transportation that carry large numbers of people.

matter Anything that has mass (the amount of material in an object) and takes up space. On Earth, where gravity is present, we weigh an object to determine its mass.

matter quality Measure of how useful a matter resource is based on its availability and concentration. See *high-quality matter, low-quality matter.*

matter-recycling society Society that emphasizes recycling the maximum amount of all resources that can be recycled. The goal is to allow economic growth to continue without depleting matter resources and without producing excessive pollution and environmental degradation. Compare *sustainable-Earth society, throwaway society.*

mature community Fairly stable, self-sustaining community in an advanced stage of ecological succession; usually has a diverse array of species and ecological niches; captures and uses energy and cycles critical chemicals more efficiently than simpler, immature communities. Compare *immature community*.

maximum sustainable yield See *sustainable yield*.

MDC See *more developed country*.

median lethal dose (LD$_{50}$) Amount of a toxic material per unit of body weight of test animals that kills half the test population in a certain time. Compare *lethal dose*.

meltdown The melting of the core of a nuclear reactor.

mesosphere Third layer of the atmosphere; found above the stratosphere. Compare *stratosphere, thermosphere, troposphere*.

mesotrophic lake Lake with a moderate supply of plant nutrients. Compare *eutrophic lake, oligotrophic lake*.

metabolic reserve Lower half of rangeland grass plants; can grow back as long as it is not consumed by herbivores.

metamorphic rock Rock produced when a preexisting rock is subjected to high temperatures (which may cause it to melt partially), high pressures, chemically active fluids, or a combination of these agents. Compare *igneous rock, sedimentary rock*. See *rock cycle*.

metastasis Spread of malignant (cancerous) cells from a cancer to other parts of the body.

mineral Any naturally occurring inorganic substance found in the earth's crust as a crystalline solid. See *mineral resource*.

mineralization Process taking place in soil in which decomposers turn organic materials into inorganic ones.

mineral resource Concentration of naturally occurring solid, liquid, or gaseous material, in or on Earth's crust, in such form and amount that its extraction and conversion into useful materials or items is currently or potentially profitable. Mineral resources are classified as metallic (such as iron and tin ores) or nonmetallic (such as fossil fuels, sand, and salt).

minimum-tillage farming See *conservation-tillage farming*.

mixed economic system Economic system that falls somewhere between pure market and pure command economic systems. Virtually all of the world's economic systems fall into this category, with some closer to a pure market system and some closer to a pure command system. Compare *pure command economic system, pure market economic system*.

mixture Combination of one or more elements and compounds.

model See *scientific model*.

molecule Combination of two or more atoms of the same chemical element (such as O$_2$) or different chemical elements (such as H$_2$O) held together by chemical bonds.

monera See *bacteria, cyanobacteria*.

monoculture Cultivation of a single crop, usually on a large area of land. Compare *polyculture*.

more developed country (MDC) Country that is highly industrialized and has a high GNP per person. Compare *less developed country*.

multiple use Principle of managing public land, such as a national forest, so it is used for a variety of purposes, such as timbering, mining, recreation, grazing, wildlife preservation, and soil and water conservation. See also *sustainable yield*.

municipal solid waste Solid materials discarded by homes and businesses in or near urban areas. See *solid waste*.

mutagen Chemical, or form of ionizing radiation, that causes inheritable changes in the DNA molecules in the genes found in chromosomes (mutations). See *carcinogen, mutation, teratogen*.

mutation A random change in DNA molecules making up genes that can yield changes in anatomy, physiology, or behavior in offspring. See *mutagen*.

mutualism Type of species interaction in which both participating species generally benefit. Compare *commensalism*.

native species Species that normally live and thrive in a particular ecosystem. Compare *immigrant species, indicator species, keystone species*.

natural gas Underground deposits of gases consisting of 50% to 90% by weight methane gas (CH$_4$) and small amounts of heavier gaseous hydrocarbon compounds such as propane (C$_3$H$_8$) and butane (C$_4$H$_{10}$).

natural ionizing radiation Ionizing radiation in the environment from natural sources.

natural radioactive decay Nuclear change in which unstable nuclei of atoms spontaneously shoot out particles (usually alpha or beta particles), energy (gamma rays), or both at a fixed rate.

natural recharge Natural replenishment of an aquifer by precipitation, which percolates downward through soil and rock. See *recharge area*.

natural resource capital See *Earth capital*.

natural resources Area of the earth's solid surface; nutrients and minerals in the soil and deeper layers of the earth's crust; water; wild and domesticated plants and animals; air; and other resources produced by the earth's natural processes. Compare *human capital, manufactured capital*. See *Earth capital*.

natural selection Process by which a particular beneficial gene or set of beneficial genes is reproduced more than others in a population, through adaptation and differential reproduction. It is the major mechanism leading to biological evolution. See *adaptation, biological evolution, differential reproduction, mutation*.

nematocide Chemical that kills nematodes (roundworms).

net economic welfare (NEW) Measure of annual change in quality of life in a country. It is obtained by subtracting the value of all final products and services that decrease the quality of life from a country's GNP. See *NEW per capita*.

net energy Total amount of useful energy available from an energy resource or energy system over its lifetime minus the amount of energy used (the first energy law), automatically wasted (the second energy law), and unnecessarily wasted in finding, processing, concentrating, and transporting it to users.

net primary productivity Rate at which all the plants in an ecosystem produce net useful chemical energy; equal to the difference between the rate at which the plants in an ecosystem produce useful chemical energy (primary productivity) and the rate at which they use some of that energy through cellular respiration. Compare *primary productivity*.

net useful energy See *net energy*.

neutral solution Water solution containing an equal number of hydrogen ions (H$^+$) and hydroxide ions (OH$^-$); water solution with a pH of 7. Compare *acid solution, basic solution*.

neutron (n) Elementary particle in the nuclei of all atoms (except hydrogen-1). It has a relative mass of 1 and no electrical charge.

NEW See *net economic welfare*.

NEW per capita Annual net economic welfare (NEW) of a country divided by its total population. See *net economic welfare, real NEW per capita*.

niche See *ecological niche*.

nitrogen cycle Cyclic movement of nitrogen in different chemical forms from the environment to organisms, and then back to the environment.

nitrogen fixation Conversion of atmospheric nitrogen gas into forms useful to plants, by lightning, bacteria, and cyanobacteria; it is part of the nitrogen cycle.

noise pollution Any unwanted, disturbing, or harmful sound that impairs or interferes with hearing, causes stress, hampers concentration and work efficiency, or causes accidents.

nondegradable pollutant Material that is not broken down by natural processes. Examples are the toxic elements lead and mercury. Compare *biodegradable pollutant, degradable pollutant, slowly degradable pollutant*.

nonionizing radiation Forms of radiant energy such as radio waves, microwaves, infrared light, and ordinary light that do not have enough energy to cause ionization of atoms in living tissue. Compare *ionizing radiation*.

nonpersistent pollutant See *degradable pollutant*.

nonpoint source Large or dispersed land areas such as crop fields, streets, and lawns that discharge pollutants into the environment over a large area. Compare *point source*.

nonrenewable resource Resource that exists in a fixed amount (stock) in various places in the earth's crust and has the potential for renewal only by geological, physical, and chemical processes taking place over hundreds of millions to billions of years. Examples are copper, aluminum, coal, and oil. We classify these resources as exhaustible because we are extracting and using them at a much faster rate than the geological time scale on which they were formed. Compare *potentially renewable resource*.

nontransmissible disease A disease that is not caused by living organisms and that does not spread from one person to another. Examples are most cancers, diabetes, cardiovascular disease, and malnutrition. Compare *transmissible disease*.

no-till farming See *conservation-tillage farming*.

nuclear change Process in which nuclei of certain isotopes spontaneously change, or are forced to change, into one or more different isotopes. The three principal types of nuclear change are natural radioactivity, nuclear fission, and nuclear fusion. Compare *chemical change*.

nuclear energy Energy released when atomic nuclei undergo a nuclear reaction such as the spontaneous emission of radioactivity, nuclear fission, or nuclear fusion.

nuclear fission Nuclear change in which the nuclei of certain isotopes with large mass numbers (such as uranium-235 and plutonium-239) are split apart into lighter nuclei when struck by a neutron. This process releases more neutrons and a large amount of energy. Compare *nuclear fusion.*

nuclear fusion Nuclear change in which two nuclei of isotopes of elements with a low mass number (such as hydrogen-2 and hydrogen-3) are forced together at extremely high temperatures until they fuse to form a heavier nucleus (such as helium-4). This process releases a large amount of energy. Compare *nuclear fission.*

nucleus Extremely tiny center of an atom, making up most of the atom's mass. It contains one or more positively charged protons and one or more neutrons with no electrical charge (except for a hydrogen-1 atom, which has one proton and no neutrons in its nucleus).

nutrient Any food or element an organism must take in to live, grow, or reproduce.

nutrient cycle See *biogeochemical cycle.*

oil See *crude oil.*

oil shale Fine-grained rock containing varying amounts of kerogen, a solid, waxy mixture of hydrocarbon compounds. Heating the rock to high temperatures converts the kerogen into a vapor that can be condensed to form a slow-flowing heavy oil called shale oil. See *kerogen, shale oil.*

old-growth forest Virgin and old, second-growth forests containing trees that are often hundreds, sometimes thousands, of years old. Examples include forests of Douglas fir, western hemlock, giant sequoia, and coastal redwoods in the western United States. Compare *second-growth forest, tree farm.*

oligotrophic lake Lake with a low supply of plant nutrients. Compare *eutrophic lake, mesotrophic lake.*

omnivore Animal organism that can use both plants and other animals as food sources. Examples are pigs, rats, cockroaches, and people. Compare *carnivore, herbivore.*

open-pit mining Removal of ores of metals such as iron and copper by digging them out of the earth's surface and leaving a pit.

open sea The part of an ocean that is beyond the continental shelf. Compare *coastal zone.*

optimum yield Amount of fish (or other potentially renewable resource) that can be economically harvested on a sustained basis; it is usually less than the sustainable yield. See *sustainable yield.*

ore Part of a metal-yielding material that can be economically and legally extracted at a given time. An ore typically contains two parts: the ore mineral, which contains the desired metal, and waste mineral material (gangue).

organic farming Producing crops and livestock naturally by using organic fertilizer (manure, legumes, compost) and natural pest control (bugs that eat harmful bugs, plants that repel bugs, and environmental controls such as crop rotation) instead of using commercial inorganic fertilizers and synthetic pesticides and herbicides.

organic fertilizer Organic material such as animal manure, green manure, and compost, applied to cropland as a source of plant nutrients. Compare *commercial inorganic fertilizer.*

organism Any form of life.

other resources Identified and unidentified resources not classified as reserves.

output pollution control See *pollution cleanup.*

overburden Layer of soil and rock overlying a mineral deposit; removed during surface mining.

overconsumption Situation where some people consume much more than they need at the expense of those who cannot meet their basic needs and at the expense of Earth's present and future life-support systems.

overfishing Harvesting so many fish of a species, especially immature fish, that there is not enough breeding stock left to replenish the species, so it is not profitable to harvest them.

overgrazing Destruction of vegetation when too many grazing animals feed too long and exceed the carrying capacity of a rangeland area.

overnutrition Diet so high in calories, saturated (animal) fats, salt, sugar, and processed foods, and so low in vegetables and fruits that the consumer runs high risks of diabetes, hypertension, heart disease, and other health hazards. Compare *malnutrition, undernutrition.*

overpopulation State in which there are more people than can live on Earth or in a geographic region in comfort, happiness, and in health and still leave the planet or region a fit place for future generations. It is a result of growing numbers of people, growing affluence (resource consumption), or both. See *carrying capacity, consumption overpopulation, dieback, overshoot, people overpopulation.*

overshoot Condition in which population size of a species temporarily exceeds the carrying capacity of its habitat. This leads to a sharp reduction in its population. See *carrying capacity, consumption overpopulation, dieback, people overpopulation.*

oxygen cycle Cyclic movement of oxygen in different chemical forms from the environment to organisms, and then back to the environment.

oxygen-demanding wastes Organic materials that are usually biodegraded by aerobic (oxygen-consuming) bacteria, if there is enough dissolved oxygen in the water. See also *biological oxygen demand.*

ozone layer Layer of gaseous ozone (O_3) in the stratosphere that protects life on Earth by filtering out harmful ultraviolet radiation from the sun.

PANs Peroxyacyl nitrates. Group of chemicals found in photochemical smog.

parasite Consumer organism that lives on or in and feeds on a living plant or animal, known as the host, over an extended period of time. The parasite draws nourishment from and gradually weakens its host. This may or may not kill the host.

particulate matter Solid particles or liquid droplets suspended or carried in the air.

parts per billion (ppb) Number of parts of a chemical found in one billion parts of a particular gas, liquid, or solid.

parts per million (ppm) Number of parts of a chemical found in one million parts of a particular gas, liquid, or solid.

passive solar heating system System that captures sunlight directly within a structure and converts it into low-temperature heat for space heating or for heating water for domestic use without the use of mechanical devices. Compare *active solar heating system.*

pathogen Organism that produces disease.

PCBs See *polychlorinated biphenyls.*

people overpopulation Situation in which there are more people in the world or a geographic region than available supplies of food, water, and other vital resources can support. It can also occur where the rate of population growth so exceeds the rate of economic growth, or the distribution of wealth is so inequitable, that a number of people are too poor to grow or buy enough food, fuel, and other important resources. Compare *consumption overpopulation.*

per capita GNP See *GNP per capita.*

per capita NEW See *NEW per capita.*

per capita real NEW See *real NEW per capita.*

permafrost Permanently frozen underground layers of soil in tundra.

permeability The degree to which underground rock and soil pores are interconnected with each other, and thus a measure of the degree to which water can flow freely from one pore to another. Compare *porosity.*

persistent pollutant See *slowly degradable pollutant.*

pest Unwanted organism that directly or indirectly interferes with human activities.

pesticide Any chemical designed to kill or inhibit the growth of an organism that people consider to be undesirable. See *fungicide, herbicide, insecticide.*

pesticide treadmill Situation in which the cost of using pesticides increases while their effectiveness decreases, mostly because the pest species develop genetic resistance to the pesticides.

petrochemicals Chemicals obtained by refining (distilling) crude oil. They are used as raw materials in the manufacture of most industrial chemicals, fertilizers, pesticides, plastics, synthetic fibers, paints, medicines, and many other products.

petroleum See *crude oil.*

pH Numeric value that indicates the relative acidity or alkalinity of a substance on a scale of 0 to 14, with the neutral point at 7. Acid solutions have pH values lower than 7, and basic or alkaline solutions have pH values greater than 7.

phosphorus cycle Cyclic movement of phosphorus in different chemical forms from the environment to organisms, and then back to the environment.

photochemical smog Complex mixture of air pollutants produced in the lower atmosphere by the reaction of hydrocarbons and nitrogen oxides under the influence of sunlight. Especially harmful components include ozone, peroxyacyl nitrates (PANs), and various aldehydes. Compare *industrial smog.*

photosynthesis Complex process that takes place in cells of green plants. Radiant energy from the sun is used to combine carbon dioxide (CO_2) and water (H_2O) to produce oxygen (O_2) and carbohydrates (such as glucose, $C_6H_{12}O_6$), and other nutrient molecules. Compare *aerobic respiration, chemosynthesis.*

photovoltaic cell (solar cell) Device in which radiant (solar) energy is converted directly into electrical energy.

physical change Process that alters one or more physical properties of an element or a compound without altering its chemical composition. Examples are changing the size and shape of a sample of matter (crushing ice and cutting aluminum foil) and changing a sample of matter from one physical state to another (boiling and freezing water). Compare *chemical change*.

phytoplankton Small, drifting plants, mostly algae and bacteria, found in aquatic ecosystems. Compare *plankton, zooplankton*.

pioneer community First integrated set of plants, animals, and decomposers found in an area undergoing primary ecological succession. See *immature community, mature community*.

pioneer species First hardy species, often microbes, mosses, and lichens, that begin colonizing a site as the first stage of ecological succession. See *ecological succession, pioneer community*.

planetary management worldview Belief that Earth is a place of unlimited resources that we can manage for our use. Any type of resource conservation that hampers short-term economic growth is unnecessary because if we pollute or deplete resources in one area, we will find substitutes, control the pollution through technology, and, (if necessary) get resources from the moon and asteroids in the "new frontier" of space. See *spaceship-Earth worldview*. Compare *sustainable-Earth worldview*.

plankton Small plant organisms (phytoplankton) and animal organisms (zooplankton) that float in aquatic ecosystems.

plantation agriculture Growing specialized crops such as bananas, coffee, and cacao in tropical LDCs, primarily for sale to MDCs.

plants (plantae) Eukaryotic, mostly multicelled organisms such as algae (red, blue, and green), mosses, ferns, flowers, cacti, grasses, beans, wheat, rice, and trees. These organisms use photosynthesis to produce organic nutrients for themselves and for other organisms feeding on them. Water and other inorganic nutrients are obtained from the soil for terrestrial plants and from the water for aquatic plants.

plates Various-sized areas of Earth's lithosphere that move slowly around with the mantle's flowing asthenosphere. Most earthquakes and volcanoes occur around the boundaries of these plates. See *asthenosphere, lithosphere, plate tectonics*.

plate tectonics Theory of geophysical processes that explains the movements of lithospheric plates and the processes that occur at their boundaries. See *lithosphere, plates*.

poaching Illegal commercial hunting or fishing.

point source A single identifiable source that discharges pollutants into the environment. Examples are the smokestack of a power plant or an industrial plant, the drainpipe of a meat-packing plant, the chimney of a house, and the exhaust pipe of an automobile. Compare *nonpoint source*.

politics Process through which individuals and groups try to influence or control the policies and actions of governments that affect the local, state, national, and international communities.

pollution An undesirable change in the physical, chemical, or biological characteristics of air, water, soil, or food that can adversely affect the health, survival, or activities of humans or other living organisms.

pollution cleanup Device or process that removes or reduces the level of a pollutant after it has been produced or has entered the environment. Examples are automobile emission-control devices and sewage treatment plants. Compare *pollution prevention*.

pollution prevention Device or process that prevents a potential pollutant from forming or from entering the environment or that sharply reduces the amounts entering the environment. Compare *pollution cleanup*.

polychlorinated biphenyls (PCBs) Group of 209 different toxic, oily, synthetic chlorinated hydrocarbon compounds that can be biologically amplified in food chains and webs.

polyculture Complex form of intercropping in which a large number of different plants maturing at different times are planted together. See also *intercropping*. Compare *monoculture, polyvarietal cultivation*.

polyvarietal cultivation Planting a plot of land with several varieties of the same crop. Compare *intercropping, monoculture, polyculture*.

population Group of individual organisms of the same species living within a particular area.

population crash Large number of deaths over a fairly short time, brought about when the number of individuals in a population is too large to be supported by available environmental resources.

population density Number of organisms in a particular population found in a specified area. Compare *ecological population density*.

population dispersion General pattern in which the members of a population are arranged throughout its habitat.

population distribution Variation of population density over a particular geographical area. For example, a country has a high population density in its urban areas and a much lower population density in rural areas.

population dynamics Major abiotic and biotic factors that tend to increase or decrease the population size and age and sex composition of a species.

population size Number of individuals making up a population's gene pool.

porosity The pores (cracks and spaces) in rocks or soil, or the percentage of the rock's or soil's volume not occupied by the rock or soil itself. Compare *permeability*.

potential energy Energy stored in an object because of its position or the position of its parts. Compare *kinetic energy*.

potentially renewable resource Resource that theoretically can last indefinitely without reducing the available supply because it is replaced more rapidly through natural processes than are nonrenewable resources or because it is essentially inexhaustible (solar energy). Examples are trees in forests, grasses in grasslands, wild animals, fresh surface water in lakes and streams, most groundwater, fresh air, and fertile soil. If such a resource is used faster than it is replenished, it can be depleted and converted into a nonrenewable resource. Compare *nonrenewable resource*. See also *environmental degradation*.

poverty Inability to meet basic needs for food, clothing, and shelter.

ppb See *parts per billion*.

ppm See *parts per million*.

precipitation Water in the form of rain, sleet, hail, and snow that falls from the atmosphere onto the land and bodies of water.

predation Situation in which an organism of one species (the predator) captures and feeds on parts or all of an organism of another species (the prey).

predator Organism that captures and feeds on parts or all of an organism of another species (the prey).

predator–prey relationship Interaction between two organisms of different species in which one organism, called the predator, captures and feeds on parts or all of another organism, called the prey.

prey Organism that is captured and serves as a source of food for an organism of another species (the predator).

primary consumer Organism that feeds directly on all or parts of plants (*herbivore*) or other producers. Compare *detritivore, omnivore, secondary consumer*.

primary oil recovery Pumping out the crude oil that flows by gravity or under gas pressure into the bottom of an oil well. Compare *enhanced oil recovery, secondary oil recovery*.

primary pollutant Chemical that has been added directly to the air by natural events or human activities and occurs in a harmful concentration. Compare *secondary pollutant*.

primary productivity See *gross primary productivity*. Compare *net primary productivity*.

primary sewage treatment Mechanical treatment of sewage in which large solids are filtered out by screens and suspended solids settle out as sludge in a sedimentation tank. Compare *advanced sewage treatment, secondary sewage treatment*.

primary succession Sequential development of communities in a bare area that has never been occupied by a community of organisms. Compare *secondary succession*.

prime reproductive age Years between ages 20 and 29, during which most women have most of their children. Compare *reproductive age*.

principle of connectedness Everything is connected to and intermingled with everything else; we are all in it together.

principle of multiple use See *multiple use*.

prior appropriation Legal principle by which the first user of water from a stream establishes a legal right to continued use of the amount originally withdrawn. Compare *riparian rights*.

private-property resource Resource owned by an individual or a group of individuals other than the government. Compare *common-property resource, public-land resource*.

probability A mathematical statement about how likely it is that something will happen.

producer Organism that uses solar energy (green plant) or chemical energy (some bacteria) to manufacture the organic compounds it needs as nutrients from simple inorganic compounds obtained from its environment. Compare *consumer, decomposer*.

prokaryotic cell Cell that doesn't have a distinct nucleus. Other internal parts are also not enclosed by membranes. Compare *eukaryotic cell*.

protists (protista) Eukaryotic, mostly single-celled organisms such as diatoms, amoebas, some algae (golden brown and yellow-green), protozoans, and slime molds. Some protists produce their own organic nutrients through photosynthesis. Others are decomposers, and some feed on bacteria, other protists, or cells of multicellular organisms.

proton (p) Positively charged particle in the nuclei of all atoms. Each proton has a relative mass of 1 and a single positive charge.

public land resources Land that is owned jointly by all citizens but is managed for them by an agency of the local, state, or federal government. Examples are state and national parks, forests, wildlife refuges, and wilderness areas. Compare *common-property resource, private-property resource*.

pure capitalism See *pure market economic system*.

pure command economic system System in which all economic decisions are made by the government or some other central authority. Compare *mixed economic system, pure market economic system*.

pure market economic system System in which all economic decisions are made in the market, where buyers and sellers of economic goods freely interact, with no government or other interference. Compare *mixed economic system, pure command economic system*.

pyramid of biomass Diagram representing the biomass, or total dry weight of all living organisms, that can be supported at each trophic level in a food chain or food web. See *pyramid of energy flow, pyramid of numbers*.

pyramid of energy flow Diagram representing the flow of energy through each trophic level in a food chain or food web. With each energy transfer, only a small part (typically 10%) of the usable energy entering one trophic level is transferred to the organisms at the next trophic level. Compare *pyramid of biomass, pyramid of numbers*.

pyramid of numbers Diagram representing the number of organisms of a particular type that can be supported at each trophic level from a given input of solar energy at the producer trophic level in a food chain or food web. Compare *pyramid of biomass, pyramid of energy flow*.

radiation Fast-moving particles (particulate radiation) or waves of energy (electromagnetic radiation).

radioactive decay Change of a radioisotope to a different isotope by the emission of radioactivity.

radioactive isotope See *radioisotope*.

radioactive waste Radioactive waste products of nuclear power plants, research, medicine, weapons production, or other processes involving nuclear reactions. See *radioactivity*.

radioactivity Nuclear change in which unstable nuclei of atoms spontaneously shoot out "chunks" of mass, energy, or both, at a fixed rate. The three principal types of radioactivity are gamma rays and fast-moving alpha particles and beta particles.

radioisotope Isotope of an atom that spontaneously emits one or more types of radioactivity (alpha particles, beta particles, gamma rays).

range condition Estimate of how close a particular area of rangeland is to its potential for producing vegetation that can be consumed by grazing or browsing animals.

rangeland Land that supplies forage or vegetation (grasses, grasslike plants, and shrubs) for grazing and browsing animals and that is not intensively managed. Compare *feedlot*.

range of tolerance Range of chemical and physical conditions that must be maintained for populations of a particular species to stay alive and grow, develop, and function normally. See *law of tolerance*.

real GNP gross national product adjusted for inflation. Compare *GNP per capita, gross national product, real GNP per capita*.

real GNP per capita Per capita GNP adjusted for inflation. See *GNP per capita*.

real NEW per capita Per capita NEW adjusted for inflation. See *net economic welfare, NEW per capita*.

realized niche Parts of the fundamental niche of a species that are actually used by that species. See *ecological niche, fundamental niche*.

recharge area Any area of land allowing water to pass through it and into an aquifer. See *aquifer, natural recharge*.

recycling Collecting and reprocessing a resource so it can be made into new products. An example is collecting aluminum cans, melting them down, and using the aluminum to make new cans or other aluminum products. Compare *reuse*.

reforestation Renewal of trees and other types of vegetation on land where trees have been removed; can be done naturally by seeds from nearby trees or artificially by planting seeds or seedlings.

relative humidity Measure (as a percentage) of the amount of water vapor in a certain mass of air compared with the maximum amount it could hold at that temperature.

renewable resource See *potentially renewable resource*.

replacement-level fertility Number of children a couple must have to replace themselves. The average for a country or the world is usually slightly higher than 2 children per couple (2.1 in the United States and 2.5 in some LDCs) because some children die before reaching their reproductive years. See also *total fertility rate*.

reproduction Production of offspring by one or more parents.

reproductive age Ages 15 to 44, when most women have all their children. Compare *prime reproductive age*.

reproductive isolation Long-term geographic separation of members of a particular sexually reproducing species.

reserves Resources that have been identified and from which a usable mineral can be extracted profitably at present prices with current mining technology. See *identified resources, undiscovered resources*.

resilience Ability of a living system to restore itself to original condition after being exposed to an outside disturbance that is not too drastic. See *constancy, inertia*.

resource Anything obtained from the living and nonliving environment to meet human needs and wants. It can also be applied to other species.

resource partitioning Process of dividing up resources in an ecosystem so species with similar requirements (overlapping ecological niches) use the same scarce resources at different times, in different ways, or in different places. See *ecological niche, fundamental niche, realized niche*.

resource recovery Salvaging usable metals, paper, and glass from solid waste and selling them to manufacturing industries for recycling or reuse.

respiration See *aerobic respiration*.

response The amount of health damage caused by exposure to a certain dose of a harmful substance or form of radiation. See *dose, dose–response curve, lethal dose, median lethal dose*.

reuse To use a product over and over again in the same form. An example is collecting, washing, and refilling glass beverage bottles. Compare *recycling*.

riparian rights System of water law that gives anyone whose land adjoins a flowing stream the right to use water from the stream as long as some is left for downstream users. Compare *prior appropriation*.

riparian zones Thin strips and patches of vegetation that surround streams. They are very important habitats and resources for wildlife.

risk The probability that something undesirable will happen from deliberate or accidental exposure to a hazard. See *risk assessment, risk-benefit analysis, risk management*.

risk analysis Identifying hazards, evaluating the nature and severity of risks (*risk assessment*), using this and other information to determine options and make decisions about reducing or eliminating risks (*risk management*), and communicating information about risks to decision makers and the public (*risk communication*).

risk assessment Process of gathering data and making assumptions to estimate short- and long-term harmful effects on human health or the environment from exposure to hazards associated with the use of a particular product or technology. See *risk, risk-benefit analysis*.

risk-benefit analysis Estimate of the short- and long-term risks and benefits of using a particular product or technology. See *desirability quotient, risk*.

risk communication Communicating information about risks to decision makers and the public. See *risk, risk analysis, risk-benefit analysis*.

risk management Using risk assessment and other information to determine options and make decisions about reducing or eliminating risks. See *risk, risk analysis, risk-benefit analysis, risk communication*.

rock Any material that makes up a large, natural, continuous part of Earth's crust. See *mineral*.

rock cycle Largest and slowest of the earth's cycles, consisting of geologic, physical, and chemical processes that form and modify rocks and soil in the earth's crust over millions of years.

rodenticide Chemical that kills rodents.

r-strategists Species that reproduce early in their life span and that produce large numbers of usually small and short-lived offspring in a short period of time. Compare *K-strategists*.

rule of 70 Doubling time = 70/percentage growth rate. See *doubling time, exponential growth*.

runoff Fresh water from precipitation and melting ice that flows on the earth's surface into nearby streams, lakes, wetlands, and reservoirs. See *surface runoff, surface water*. Compare *groundwater*.

rural area Geographical area in the United States with a population of less than 2,500 people per unit of area. The number of people used in this definition may vary in different countries. Compare *urban area*.

salinity Amount of various salts dissolved in a given volume of water.

salinization Accumulation of salts in soil that can eventually make the soil unable to support plant growth.

saltwater intrusion Movement of salt water into freshwater aquifers in coastal and inland areas as groundwater is withdrawn faster than it is recharged by precipitation.

sanitary landfill Waste disposal site on land in which waste is spread in thin layers, compacted, and covered with a fresh layer of clay or plastic foam each day.

scavenger Organism that feeds on dead organisms that either were killed by other organisms or died naturally. Examples are vultures, flies, and crows. Compare *detritivore*.

science Attempts to discover order in nature and then use that knowledge to make predictions about what will happen in nature. See *scientific data*, *scientific hypothesis*, *scientific law*, *scientific methods*, *scientific theory*.

scientific data Facts obtained by making observations and measurements. Compare *scientific hypothesis*, *scientific law*, *scientific model*, *scientific theory*.

scientific hypothesis An educated guess that attempts to explain a scientific law or certain scientific observations. Compare *scientific data*, *scientific law*, *scientific model*, *scientific theory*.

scientific law Summary of what scientists find happening in nature over and over in the same way. See *first law of thermodynamics*, *second law of thermodynamics*, *law of conservation of matter*. Compare *scientific data*, *scientific hypothesis*, *scientific model*, *scientific theory*.

scientific methods The ways scientists gather data and formulate and test scientific laws and theories. See *scientific data*, *scientific hypothesis*, *scientific law*, *scientific model*, *scientific theory*.

scientific model A simulation of complex processes and systems. Many are mathematical models that are run and tested using computers.

scientific theory A well-tested and widely accepted scientific hypothesis. Compare *scientific data*, *scientific hypothesis*, *scientific model*, *scientific law*.

secondary consumer Organism that feeds only on primary consumers. Most secondary consumers are animals, but some are plants. Compare *detritivore*, *omnivore*, *primary consumer*.

secondary oil recovery Injection of water into an oil well after primary oil recovery to force out some of the remaining, usually thicker, crude oil. Compare *enhanced oil recovery*, *primary oil recovery*.

secondary pollutant Harmful chemical formed in the atmosphere when a primary air pollutant reacts with normal air components or with other air pollutants. Compare *primary pollutant*.

secondary sewage treatment Second step in most waste treatment systems, in which aerobic bacteria break down up to 90% of degradable, oxygen-demanding organic wastes in wastewater. This is usually done by bringing sewage and bacteria together in trickling filters or in the activated sludge process. Compare *advanced sewage treatment*, *primary sewage treatment*.

secondary succession Sequential development of communities in an area in which natural vegetation has been removed or destroyed but the soil is not destroyed. Compare *primary succession*.

second-growth forest Stands of trees resulting from secondary ecological succession. Compare *ancient forest*, *old-growth forest*, *tree farm*.

second law of energy See *second law of thermodynamics*.

second law of thermodynamics In any conversion of heat energy to useful work, some of the initial energy input is always degraded to a lower-quality, more dispersed, less useful energy—usually low-temperature heat that flows into the environment; you can't break even in terms of energy quality. See *first law of thermodynamics*.

sedimentary rock Rock that forms from the accumulated products of erosion and in some cases from the compacted shells, skeletons, and other remains of dead organisms. Compare *igneous rock*, *metamorphic rock*. See *rock cycle*.

seed-tree cutting Removal of nearly all trees on a site in one cutting, with a few seed-producing trees left uniformly distributed to regenerate the forest. Compare *clear-cutting*, *selective cutting*, *shelterwood cutting*, *strip logging*.

selective cutting Cutting of intermediate-aged, mature, or diseased trees in an uneven-aged forest stand, either singly or in small groups. This encourages the growth of younger trees and maintains an uneven-aged stand. Compare *clear-cutting*, *seed-tree cutting*, *shelterwood cutting*, *strip logging*.

septic tank Underground tank for treatment of wastewater from a home in rural and suburban areas. Bacteria in the tank decompose organic wastes and the sludge settles to the bottom of the tank. The effluent flows out of the tank into the ground through a field of drain pipes.

sewage sludge Gooey mixture of toxic chemicals, infectious agents, and settled solids removed from wastewater at sewage treatment plants.

shale oil Slow-flowing, dark brown, heavy oil obtained when kerogen in oil shale is vaporized at high temperatures and then condensed. Shale oil can be refined to yield gasoline, heating oil, and other petroleum products. See *kerogen*, *oil shale*.

shelterbelt See *windbreak*.

shelterwood cutting Removal of mature, marketable trees in an area in a series of partial cuttings to allow regeneration of a new stand under the partial shade of older trees, which are later removed. Typically, this is done by making two or three cuts over a decade. Compare *clear-cutting*, *seed-tree cutting*, *selective cutting*, *strip logging*.

shifting cultivation Clearing a plot of ground in a forest, especially in tropical areas, and planting crops on it for a few years (typically two to five years) until the soil is depleted of nutrients or until the plot has been invaded by a dense growth of vegetation from the surrounding forest. Then a new plot is cleared and the process is repeated. The abandoned plot cannot successfully grow crops for 10 to 30 years. See also *slash-and-burn cultivation*.

slash-and-burn cultivation Cutting down trees and other vegetation in a patch of forest, leaving the cut vegetation on the ground to dry, and then burning it. The ashes that are left add nutrients to the nutrient-poor soils found in most tropical forest areas. Crops are planted between tree stumps. Plots must be abandoned after a few years (typically two to five years) because of loss of soil fertility or invasion of vegetation from the surrounding forest. See also *shifting cultivation*.

slowly degradable pollutant Material that is slowly broken down into simpler chemicals or reduced to acceptable levels by natural physical, chemical, and biological processes. Compare *biodegradable pollutant*, *degradable pollutant*, *nondegradable pollutant*.

sludge Gooey mixture of toxic chemicals, infectious agents, and settled solids, removed from wastewater at a sewage treatment plant.

smelting Process in which a desired metal is separated from the other elements in an ore mineral.

smog Originally a combination of smoke and fog, but now used to describe other mixtures of pollutants in the atmosphere. See *industrial smog*, *photochemical smog*.

soil Complex mixture of inorganic minerals (clay, silt, pebbles, and sand), decaying organic matter, water, air, and living organisms.

soil conservation Methods used to reduce soil erosion, to prevent depletion of soil nutrients, and to restore nutrients already lost by erosion, leaching, and excessive crop harvesting.

soil erosion Movement of soil components, especially topsoil, from one place to another, usually by exposure to wind, flowing water, or both. This natural process can be greatly accelerated by human activities that remove vegetation from soil.

soil horizons Horizontal zones that make up a particular mature soil. Each horizon has a distinct texture and composition that varies with different types of soils.

soil permeability Rate at which water and air move from upper to lower soil layers.

soil porosity See *porosity*.

soil profile Cross-sectional view of the horizons in a soil.

soil structure How the particles that make up a soil are organized and clumped together. See also *soil permeability*, *soil texture*.

soil texture Relative amounts of the different types and sizes of mineral particles in a sample of soil.

soil water Underground water that partially fills pores between soil particles and rocks within the upper soil and rock layers of the earth's crust, above the water table. Compare *groundwater*.

solar capital Solar energy from the sun reaching Earth. Compare *Earth capital*.

solar cell See *photovoltaic cell*.

solar collector Device for collecting radiant energy from the sun and converting it into heat. See *active solar heating system*, *passive solar heating system*.

solar energy Direct radiant energy from the sun and a number of *indirect* forms of energy produced by the direct input. Principal indirect forms of solar energy include wind, falling and flowing water (hydropower), and biomass (solar energy converted into chemical energy stored in the chemical bonds of organic compounds in trees and other plants).

solar pond Fairly small body of fresh water or salt water from which stored solar energy can be extracted, because of temperature difference between the hot surface layer exposed to the sun during daylight and the cooler layer beneath it.

solid waste Any unwanted or discarded material that is not a liquid or a gas. See *municipal solid waste*.

spaceship-Earth worldview Earth is viewed as a spaceship—a machine that we can understand, control, and change at will by using advanced technology. See *planetary management worldview*. Compare *sustainable-Earth worldview*.

specialist species Species with a narrow ecological niche. They may be able to live in only one type of habitat, tolerate only a narrow range of climatic and other environmental conditions, or use only one or a few types of food. Compare *generalist species*.

speciation Formation of two species from one species as a result of divergent natural selection in response to changes in environmental conditions; usually takes thousands of years. Compare *extinction*.

species Group of organisms that resemble one another in appearance, behavior, chemical makeup and processes, and genetic structure. Organisms that reproduce sexually are classified as members of the same species only if they can actually or potentially interbreed with one another and produce fertile offspring.

species diversity Number of different species and their relative abundances in a given area. See *biological diversity*. Compare *ecological diversity*, *genetic diversity*.

spoils Unwanted rock and other waste materials produced when a material is removed from the earth's surface or subsurface by mining, dredging, quarrying, or excavation.

S-shaped curve Leveling off of an exponential, J-shaped curve when a rapidly growing population exceeds the carrying capacity of its environment and ceases to grow in numbers. See also *overshoot*, *population crash*.

stability Ability of a living system to withstand or recover from externally imposed changes or stresses. See *constancy*, *inertia*, *resilience*.

stewardship View that because of our superior intellect and power or because of our religious beliefs we have an ethical responsibility to manage and care for domesticated plants and animals as well as for the rest of nature. Compare *sustainable-Earth worldview*, *planetary management worldview*.

stocking rate Number of a particular kind of animal grazing on a given area of grassland.

strategic mineral A fuel or nonfuel mineral vital to the industry and defense of a country. Ideally, supplies are stockpiled to cushion against supply interruptions and sharp price rises.

stratosphere Second layer of the atmosphere, extending from about 17 to 48 kilometers (11 to 30 miles) above the earth's surface. It contains small amounts of gaseous ozone (O_3), which filters out about 99% of the incoming harmful ultraviolet (UV) radiation emitted by the sun. Compare *troposphere*.

stream Flowing body of surface water. Examples are creeks and rivers.

strip cropping Planting regular crops and close-growing plants, such as hay or nitrogen-fixing legumes, in alternating rows or bands to help reduce depletion of soil nutrients.

strip logging A variation of clear-cutting in which a strip of trees is clear-cut along the contour of the land, with the corridor narrow enough to allow natural regeneration within a few years. After regeneration another strip is cut above the first, and so on. Compare *clear-cutting*, *seed-tree cutting*, *selective cutting*, *shelterwood cutting*.

strip mining Form of surface mining in which bulldozers, power shovels, or stripping wheels remove large chunks of the earth's surface in strips. See *surface mining*. Compare *subsurface mining*.

subatomic particles Extremely small particles—electrons, protons, and neutrons—that make up the internal structure of atoms.

subduction zone Area in which oceanic lithosphere is carried downward (subducted) under the island arc or continent at a convergent plate boundary. A trench ordinarily forms at the boundary between the two converging plates. See *convergent plate boundary*.

subsidence Slow or rapid sinking down of part of Earth's crust that is not slope-related.

subsistence farming Supplementing solar energy with energy from human labor and draft animals to produce enough food to feed oneself and family members; in good years there may be enough food left over to sell or put aside for hard times. Compare *industrialized agriculture*.

subsurface mining Extraction of a metal ore or fuel resource such as coal from a deep underground deposit. Compare *surface mining*.

succession See *ecological succession*.

succulent plants Plants, such as desert cacti, that survive in dry climates by having no leaves, thus reducing the loss of scarce water. They store water and use sunlight to produce the food they need in the thick fleshy tissue of their green stems and branches. Compare *deciduous plants*, *evergreen plants*.

sulfur cycle Cyclic movement of sulfur in different chemical forms from the environment to organisms, and then back to the environment.

superinsulated house House that is heavily insulated and extremely airtight. Typically, active or passive solar collectors are used to heat water, and an air-to-air heat exchanger is used to prevent buildup of excessive moisture and indoor air pollutants.

surface fire Forest fire that burns only undergrowth and leaf litter on the forest floor. Compare *crown fire*, *ground fire*.

surface mining Removing soil, subsoil, and other strata, and then extracting a mineral deposit found fairly close to the earth's surface. Compare *subsurface mining*.

surface runoff Water flowing off the land into bodies of surface water.

surface water Precipitation that does not infiltrate the ground or return to the atmosphere by evaporation or transpiration. See *runoff*. Compare *groundwater*.

survivorship curve Graph showing the number of survivors in different age groups for a particular species.

sustainable agriculture See *sustainable-Earth agricultural system*.

sustainable development See *sustainable economic development*.

sustainable-Earth agricultural system Method of growing crops and raising livestock based on organic fertilizers, soil conservation, water conservation, biological control of pests, and minimal use of nonrenewable fossil-fuel energy.

sustainable-Earth economy Economic system in which the number of people and the quantity of goods are maintained at some constant level. This level is ecologically sustainable over time and meets at least the basic needs of all members of the population.

Sustainable-Earth Revolution Cultural change involving halting population growth and altering lifestyles, political and economic systems, and the way we treat the environment so that we can help preserve the Earth for ourselves and other species and help heal some of the wounds we have inflicted on the Earth. See *sustainable-Earth society*.

sustainable-Earth society Society based on working with nature by recycling and reusing discarded matter; by pollution prevention; by conserving matter and energy resources through reducing unnecessary waste and use; by not degrading renewable resources, by building things that are easy to recycle, reuse, and repair; by not allowing population size to exceed the carrying capacity of the environment; and by preserving biodiversity. See *sustainable-Earth worldview*. Compare *matter-recycling society*, *planetary management worldview*, *spaceship Earth worldview*, *throwaway society*.

sustainable-Earth worldview Belief that Earth is a place with finite room and resources, so continuing population growth, production, and consumption inevitably put severe stress on natural processes that renew and maintain the resource base of air, water, and soil. To prevent environmental overload, environmental degradation, and resource depletion, people should work with nature by controlling population growth, reducing unnecessary use and waste of matter and energy resources, and not causing the premature extinction of any other species. Compare *spaceship-Earth worldview*, *planetary management worldview*.

sustainable economic development Forms of economic growth and activities that do not deplete or degrade natural resources upon which present and future economic growth and life depend.

sustainable living Taking no more potentially renewable resources from the natural world than can be replenished naturally and not overloading the capacity of the environment to cleanse and renew itself by natural processes.

sustainable society A society that manages its economic development and population growth in ways that do no irreparable environmental harm. It satisfies the needs of its people without depleting Earth capital and thus jeopardizing the prospects of future generations of people or other species.

sustainable yield (sustained yield) Highest rate at which a potentially renewable resource can be used without reducing its available supply throughout the world or in a particular area. See also *environmental degradation*.

symbiotic relationship Species interaction in which two kinds of organisms live together in an intimate association, with members of one or both species benefiting from the association. See *commensalism*, *mutualism*.

synfuels Synthetic gaseous and liquid fuels produced from solid coal or sources other than natural gas or crude oil.

synthetic natural gas (SNG) Gaseous fuel containing mostly methane produced from solid coal.

tailings Rock and other waste materials removed as impurities when waste mineral material is separated from the metal in an ore.

tar sand Deposit of a mixture of clay, sand, water, and varying amounts of a tarlike heavy oil known as bitumen. Bitumen can be extracted from tar sand by heating. It is then purified and upgraded to synthetic crude oil. See *bitumen*.

technology Creation of new products and processes that are supposed to improve our survival, comfort, and quality of life. Compare *science*.

temperature inversion See *thermal inversion*.

teratogen Chemical, ionizing agent, or virus that causes birth defects. See *carcinogen, mutagen*.

terracing Planting crops on a long, steep slope that has been converted into a series of broad, nearly level terraces with short vertical drops from one to another that run along the contour of the land to retain water and reduce soil erosion.

terrestrial Pertaining to land. Compare *aquatic*.

tertiary (higher-level) consumers Animals that feed on animal-eating animals. They feed at high trophic levels in food chains and webs. Examples are hawks, lions, bass, and sharks. Compare *detritivore, primary consumer, secondary consumer*.

tertiary oil recovery See *enhanced oil recovery*.

tertiary sewage treatment See *advanced sewage treatment*.

theory of evolution Widely-accepted idea that all life forms developed from earlier forms of life. Although this theory conflicts with the creation stories of most religions, it is the way biologists explain how life has changed over the past 3.6–3.8 billion years and why it is so diverse today.

thermal inversion Layer of dense, cool air trapped under a layer of less dense warm air. This prevents upward-flowing air currents from developing. In a prolonged inversion, air pollution in the trapped layer may build up to harmful levels.

thermocline Zone of gradual temperature decrease between warm surface water and colder deep water in a lake, reservoir, or ocean.

thermosphere Fourth layer of the atmosphere; found above the mesosphere. Compare *mesosphere, stratosphere, troposphere*.

threatened species Wild species that is still abundant in its natural range but is likely to become endangered because of a decline in numbers. Compare *endangered species*.

threshold effect The harmful or fatal effect of a small change in environmental conditions that exceeds the limit of tolerance of an organism or population of a species. See *law of tolerance*.

throwaway society The situation in most advanced industrialized countries, in which ever-increasing economic growth is sustained by maximizing the rate at which matter and energy resources are used, with little emphasis on pollution prevention, recycling, reuse, reduction of unnecessary waste, and other forms of resource conservation. Compare *matter-recycling society, sustainable-Earth society*.

total fertility rate (TFR) Estimate of the average number of children that will be born alive to a woman during her lifetime if she passes through all her childbearing years (ages 15–44) conforming to age-specific fertility rates of a given year. In simpler terms, it is an estimate of the average number of children a woman will have during her childbearing years.

totally planned economy See *pure command economic system*.

toxic chemical Chemical that is fatal to humans in low doses, or fatal to over 50% of test animals at stated concentrations. Most are neurotoxins, which attack nerve cells. See *carcinogen, hazardous chemical, mutagen, teratogen*.

toxic waste Form of hazardous waste that causes death or serious injury (such as burns, respiratory diseases, cancers, or genetic mutations). See *hazardous waste*.

traditional intensive agriculture Producing enough food for a farm family's survival and perhaps a surplus that can be sold. This type of agriculture requires higher inputs of labor, fertilizer, and water than traditional subsistence agriculture. See *traditional subsistence agriculture*.

traditional subsistence agriculture Production of enough crops or livestock for a farm family's survival and, in good years, a surplus to sell or put aside for hard times. Compare *traditional intensive agriculture*.

tragedy of the commons Depletion or degradation of a resource to which people have free and unmanaged access. An example is the depletion of commercially desirable species of fish in the open ocean beyond areas controlled by coastal countries. See *common-property resource*.

transform fault Area where Earth's lithospheric plates move in opposite but parallel directions along a fracture (fault) in the lithosphere. Compare *convergent plate boundary, divergent plate boundary*.

transmissible disease A disease that is caused by living organisms such as bacteria, viruses, and parasitic worms and that can spread from one person to another by air, water, food, body fluids, or in some cases by insects or other organisms. Compare *nontransmissible disease*.

transpiration Process in which water is absorbed by the root systems of plants, moves up through the plants, passes through pores (stomata) in their leaves or other parts, and then evaporates into the atmosphere as water vapor.

tree farm Site planted with one or only a few tree species in an even-aged stand. When the stand matures, it is usually harvested by clear-cutting and replanted. Normally used to grow rapidly growing tree species for fuelwood, timber, or pulpwood. See *even-aged management*. Compare *old-growth forest, second-growth forest, uneven-aged management, uneven-aged stand*.

tritium (T; hydrogen-3) Isotope of hydrogen with a nucleus containing one proton and two neutrons, thus having a mass number of 3. Compare *deuterium*.

trophic level All organisms that are the same number of energy transfers away from the original source of energy (for example, sunlight) that enters an ecosystem. For example, all producers belong to the first trophic level, and all herbivores belong to the second trophic level in a food chain or a food web.

troposphere Innermost layer of the atmosphere. It contains about 75% of the mass of Earth's air and extends about 17 kilometers (11 miles) above sea level. Compare *stratosphere*.

true cost See *full cost*.

unconfined aquifer Collection of groundwater above a layer of Earth material (usually rock or clay) through which water flows very slowly (low permeability). Compare *confined aquifer*.

undernutrition Consuming insufficient food to meet one's minimum daily energy requirement, for a long enough time to cause harmful effects. Compare *malnutrition, overnutrition*.

undiscovered resources Potential supplies of a particular mineral resource, believed to exist because of geologic knowledge and theory, though specific locations, quality, and amounts are unknown. Compare *identified resources, reserves*.

uneven-aged management Method of forest management in which trees of different species in a given stand are maintained at many ages and sizes to permit continuous natural regeneration. Compare *even-aged management*.

uneven-aged stand Stand of trees in which there are considerable differences in the ages of individual trees. Usually, such stands have a variety of tree species. See *uneven-aged management*. Compare *even-aged stand, tree farm*.

upwelling Movement of nutrient-rich bottom water to the ocean's surface. This can occur far from shore but usually occurs along certain steep coastal areas where the surface layer of ocean water is pushed away from shore and replaced by cold, nutrient-rich bottom water.

urban area Geographic area with a population of 2,500 or more people. The number of people used in this definition may vary, with some countries setting the minimum number of people at 10,000 to 50,000.

urban growth Rate of growth of an urban population. Compare *degree of urbanization*.

urban heat island Buildup of heat in the atmosphere above an urban area. This heat is produced by the large concentration of cars, buildings, factories, and other heat-producing activities. See also *dust dome*.

urbanization See *degree of urbanization*.

vertebrates Animals with backbones. Compare *invertebrates*.

volcano Emission of magma through a central vent or long fissure in the earth's surface, releasing ejecta, liquid lava, and gases into the environment.

water consumption Water that is not returned to the surface water or groundwater from which it came, mostly because of evaporation and transpiration. As a result, this water is not available for use again in the area from which it came. See *water withdrawal*.

water cycle See *hydrologic cycle*.

waterlogging Saturation of soil with irrigation water or excessive precipitation, so the water table rises close to the surface.

water pollution Any physical or chemical change in surface water or groundwater that can harm living organisms or make water unfit for certain uses.

watershed Land area that delivers the water, sediment, and dissolved substances via small streams to a major stream (river).

water table Upper surface of the zone of saturation, in which all available pores in the soil and rock in the earth's crust are filled with water.

water withdrawal Removing water from a groundwater or surface water source and transporting it to a place of use. Compare *water consumption*.

watt Unit of power, or rate at which electrical work is done. See *kilowatt*.

weather Short-term changes in the temperature, barometric pressure, humidity, precipitation, sunshine (solar radiation), cloud cover, wind direction and speed, and other conditions in the troposphere at a given place and time. Compare *climate*.

weathering Physical and chemical processes in which solid rock exposed at Earth's surface is changed to separate solid particles and dissolved material, which can then be moved to another place as sediment. See *erosion*.

wetland Land that is covered all or part of the year with salt water or fresh water, excluding streams, lakes, and the open ocean. See *coastal wetland*, *inland wetland*.

wilderness Area where the earth and its community of life have not been seriously disturbed by humans and where humans are only temporary visitors.

wildlife All free, undomesticated species. Sometimes the term is used to describe only free, undomesticated species of animals.

wildlife management Manipulation of populations of wild species (especially game species) and their habitats for human benefit, the welfare of other species, and the preservation of threatened and endangered wildlife species.

wildlife resources Species of wildlife that have actual or potential economic value to people.

windbreak Row of trees or hedges planted to partially block wind flow and reduce soil erosion on cultivated land.

wind farm Cluster of small to medium-sized wind turbines in a windy area to capture wind energy and convert it into electrical energy.

work What happens when a force is used to move a sample of matter over some distance or to raise its temperature. Energy is defined as the capacity to do such work.

worldview How individuals think the world works and what they think their role in the world should be. See *planetary management worldview*, *spaceship-Earth worldview*, *sustainable-Earth worldview*.

zero population growth (ZPG) State in which the birth rate (plus immigration) equals the death rate (plus emigration), so that the population of a geographical area is no longer increasing.

zone of saturation Area where all available pores in soil and rock in the earth's crust are filled by water. See *water table*.

zoning Regulating how various parcels of land can be used.

zooplankton Animal plankton. Small floating herbivores that feed on plant plankton (phytoplankton). Compare *phytoplankton*.

Index